INSTRUMENTATION AND CONTROL

INSTRUMENTATION AND CONTROL

FUNDAMENTALS AND APPLICATIONS

Edited by
CHESTER L. NACHTIGAL
Kistler-Morse Corporation
Redmond, Washington

WILEY SERIES IN MECHANICAL ENGINEERING PRACTICE

CONSULTING EDITOR

Marvin D. Martin
President, Marvin D. Martin, Inc., Consulting Engineers, Tucson, Arizona

A WILEY-INTERSCIENCE PUBLICATION
John Wiley & Sons, Inc.
NEW YORK / CHICHESTER / BRISBANE / TORONTO / SINGAPORE

Library of Congress Cataloging in Publication Data:

Main entry under title:

Nachtigal, Chester L.
    Instrumentation and control : fundamentals and applications / Chester
L. Nachtigal.
        p.   cm. — (Wiley series in mechanical engineering practice)
    "Wiley-Interscience publication."
    Includes bibliographical references.
    ISBN 0-471-88045-0
    1. Engineering instruments.   2. Automatic control.   I. Title.
II. Title: Instrumentation and control.   III. Series.
TA165.N22   1990
681'.2—dc20   90-30083
                                                                  CIP

Printed in the United States of America

10 9 8 7 6 5 4 3 2 1

CONTENTS

SERIES PREFACE

The Wiley Series in Mechanical Engineering Practice is written for the practicing engineer. Students and academicians may find it useful, but its primary thrust is for the working engineer who needs a convenient and comprehensive reference on hand.

Two kinds of information are contained in the several volumes:

1. Numerical information such as strengths of materials, thermodynamic properties of fluids, standard pipe sizes, thread systems, and so on.
2. Descriptive and mathematical information typical of the state-of-the-art of the many facets and specialties encompassed by the broad term "mechanical engineering."

The profession has expanded to cover such a broad range of engineering activities that no one can be knowledgeable in more than a fraction of the whole field. Yet, in day-to-day work, practicing engineers frequently have to use, or at least interface with, specialty areas outside their normal sphere of competence. This book was written to provide readers with the state-of-the-art information and standard practices in these areas.

The task of covering a vast amount of material has dictated the decision to split the series into five separate volumes:

Design and Manufacturing
Power and Energy Systems
Instrumentation and Control
Fluids and Fluid Machinery
Mechanics, Materials, and Structures

Each volume is designed to stand alone but the five complement each other in providing the broad coverage mentioned above. Within each volume chapter and section headings are designed to help the user find the material being sought.

A serious attempt was made to provide state-of-the-art material at the time of writing. Since many of the areas are in a state of rapid change, there will be some obsolescence by the time printing is complete. It is planned to revise and update at reasonable intervals so that users may purchase newer editions to keep their references up to date.

The many editors and contributors who have made this series possible join me in the hope that the several volumes will turn out to be really useful tools for the practicing engineer.

MARVIN D. MARTIN

Tucson, Arizona
January 1985

PREFACE

This reference book is intended primarily for the engineer who has had perhaps one introductory course in automatic and feedback control. At the same time, it is recognized that many university engineering curricula have for many years offered this subject matter only as an elective in the fourth year. This is especially true for non-electrical engineers. Therefore, this book is also intended for the practicing engineer who has never had formal training in feedback control. For a first exposure to automatic control, Chapter 3, "Dynamic Systems Analysis," is recommended as the entry point, followed by Chapters 14 and 15. The practicing engineer who knows and understands the fundamentals of classical control may go directly to Chapters 16 and 17 for the needed material. Chapter 16 presents a variety of actuators and power modulators that provide the muscle for a feedback control system designed and built for positioning very large or very small masses. Chapter 17 details the design procedures needed to configure the intelligence of the control system.

An alternative route to the objective of analyzing and configuring a feedback control system is through the application of the so-called methods of modern control. This approach is closely linked with the use of a digital computer for simulation and analysis of the proposed automatic control system. Chapters 19 and 20 are the repository of this information.

This reference book contains yet another body of knowledge that is often overlooked or taken for granted in control system design. The problem of selecting or designing an instrumentation system is certainly an important one from the viewpoint of the control engineer—if a physical variable (such as position, velocity, or force) cannot be accurately measured, it most certainly cannot be accurately controlled. But apart from its incorporation in a feedback control system, the field of measurement and instrumentation is a science in its own right. The economics of industrial processes dictate the accuracy to which a process is monitored by measurement devices. Pressure, temperature, force, torque, size, position, velocity, and acceleration are but a few of the variables that a process engineer is called upon to instrument. Understanding vendor specifications and selecting an appropriate transducer consistent with the needs of the application both require a clear understanding of the language of measurement and instrumentation. The last section of Chapter 1 presents definitions and terminology in the field of instrumentation. Chapter 4 adds to this terminology in the course of developing the static characterization of instruments.

Because the physical variables to be monitored in most industrial processes exhibit nonstationary behavior, a significant issue for the instrument engineer is to be able to interpret dynamic response specifications for transducers and other components in the signal processing chain of events. These specifications must then be matched with the requirements of the system. Just as Chapter 3 serves the dynamic analysis needs for automatic control systems, it also serves these same needs for instrumentation systems.

Another chapter that serves the needs of both instrumentation and automatic control is Chapter 5, "Input and Output Characteristics." The concept of knowing the output characteristics of a component that supplies an input to another component of the system is extremely important for both automatic control systems and instrumentation systems. For example, the requirement for optimum power transfer from the modulator to the actuator of a control system is that the output and input impedances of the two components must be closely matched. On the other hand, the requirement for optimum signal transfer from one component to another in an instrumentation system is that the respective output and input characteristics must be as widely mismatched as possible. This reference book presents the basis for the crucial understanding of these two different sets of output-input conditions for both instrumentation and automatic control systems.

The microcomputer, perhaps the most significant technological development of the past decade, has had and is continuing to have enormous impact on the fields of instrumentation and control. Automated data acquisition systems, display devices, digital filtering, remote data transmission, microcomputer-based signal conditioners, and intelligent transducers are but a few examples of instrumentation components that have benefited by the advent of the microprocessor or microcomputer. Today it is inconceivable to purchase a controller for a feedback control system that is not micro-computer-based. Sections 6.7 and 6.8 introduce the microprocessor as a component and Chapters 12 and 13 present some of its features and benefits for instrumentation systems. The effects of using a discrete calculation device such as a microcomputer in a control system are presented in Chapters 15 and 18.

This book contains three groups of subject matter: general topics, Part I; measurement and instrumentation, Part II; and control, Part III. The material in Part I is useful for applying the material of both Parts II and III.

The editor thanks all of the contributing authors and the Editorial Board for helping to make this book a reality. A special note is due Frank Cerra, John Wiley Scientific Editor, for his patience and endurance throughout my association with him on this project.

CHESTER L. NACHTIGAL

Redmond, Washington
February 1990

CONTRIBUTORS

Airpax Corporation *Engineering Staff, Ft. Lauderdale, Florida*

Adam C. Bell *Department of Mechanical Engineering, Technical University of Nova Scotia, Halifax, Nova Scotia*

A. G. Bolton *South Australian Institute of Technology*

Albert E. Brendel *Sensor Developments, Inc., Lake Orion, Michigan*

William Brenner *MTS Systems Corporation, Floral Park, New York*

John W. G. Budd *Schaevitz Engineering, Pennsauken, New Jersey*

Sujeet Chand *Rockwell International Corporation, Thousand Oaks, California*

James H. Christensen *Allen-Bradley Company, Inc., Highland Heights, Ohio*

Stanley J. Domanski *Nashua, NH*

Richard A. Downs *Department of Mechanical Engineering, University of Washington, Seattle, Washington*

Dale K. DuVall *Scantek, Incorporated, Irving, Texas*

Kenneth C. Fischer *Allen-Bradley Company, Inc., Milwaukee, Wisconsin*

Joseph L. Garbini *Department of Mechanical Engineering, University of Washington, Seattle, Washington*

R. Eugene Goodson *Johnson Controls, Inc., Ann Arbor, Michigan*

Jerry Lee Hall *Department of Mechanical Engineering, Iowa State University, Ames, Iowa*

Syed Hamid *Halliburton Services, Duncan, Oklahoma*

David A. Hullender *Department of Mechanical Engineering, University of Texas at Arlington, Arlington, Texas*

Howard Jaslow *Digital Signal Corporation, Bohemia, New York*

Suhada Jayasuriya *Department of Mechanical Engineering, Texas A&M University, College Station, Texas*

Lynn A. Kauffman *(Formerly with Motorola, Inc., Austin, Texas)*

William J. Kerwin *Department of Electrical and Computer Engineering, University of Arizona, Tucson, Arizona*

Robert W. Lally *Orchard Park, New York*

Jackie Marsh *SEMATEC, Austin, Texas*

Georg F. Mauer *Department of Mechanical and Civil Engineering, University of Nevada at Las Vegas, Las Vegas, Nevada*

Michael D. McEvoy *Nematron Corp., Ann Arbor, Michigan*

Mark A. McQuilken *Motorola, Inc., Austin, Texas*

Daniel T. Miklovic *Weyerhaeuser Company, Tacoma, Washington*

Philip C. Milliman *Weyerhaeuser Company, Tacoma, Washington*

Robert J. Moffat *Department of Mechanical Engineering, Stanford University, Stanford, California*

Chester L. Nachtigal *Kistler-Morse Corporation, Redmond, Washington*

Mahmood Naim *Union Carbide Corporation, Indianapolis, Indiana*

T. Peter Neal *Moog, Inc., East Aurora, New York*

Henry W. Ott *Henry Ott Consultants, Livingston, New Jersey*

Robert J. Pfeifer *Combustion Engineering, Inc., Columbus, Ohio*

George A. Quinn *Farrand Controls, Valhalla, New York*

Karl N. Reid *Oklahoma State University, Stillwater, Oklahoma*

Frank S. Ruhle   *Farrand Controls, Valhalla, New York*

William R. Seitz   *Control System Technology, Inc., Clawson, Michigan*

J. Barrie Sellers   *Geokon, Inc., West Lebanon, New Hampshire*

Bernard H. Shapiro   *Dynisco, Sharon, Massachusetts*

Krishnaswamy Srinivasan   *Department of Mechanical Engineering, The Ohio State University, Columbus, Ohio*

Odo J. Struger   *Allen-Bradley Company, Inc., Highland Heights, Ohio*

Jeffrey F. Tonn   *Heart Technology, Inc., Bellevue, Washington*

Patrick L. Walter   *Sandia National Laboratories, Albuquerque, New Mexico*

Anthony D. Wang   *Burr-Brown Corporation, Tucson, Arizona*

David N. Wormley   *Department of Mechanical Engineering, Massachusetts Institute of Technology, Cambridge, Massachusetts*

CHAPTER 1

INTRODUCTION TO THE HANDBOOK

CHESTER L. NACHTIGAL

Kistler-Morse Corporation
Redmond, Washington

1.1 OVERVIEW

The twentieth century will soon close on an era of technological development and change of staggering proportions. The single development that has touched nearly everyone in the western world is the digital computer and especially the microcomputer. It is hard to imagine what future innovations may impact us as strongly as the digital computer has in the last decade. But the certainty is that we will continue to see and experience dramatic changes in how we communicate, how our machines work for us, and how we work and earn our livelihoods, all because of the existence of the computer.

Instrumentation and control, the two main topics of this handbook, are at the forefront of embracing the digital computer and spurring the evolutionary changes enabled by this technology. More and more instrumentation systems are becoming available with some form of embedded digital signal processing and/or data transmission. A digital computer was first used in closed-loop control systems as a supervisory control system.[1] That is, the computer changed the set points to the analog controllers and did not directly send signals to the process actuators. Shortly after these first experiments in digital process control, direct digital control (DDC)[1] came into existence and, as the cost of digital computing on a per-loop basis decreased, more and more loop-dedicated computers came into use. A natural by-product of the use of dedicated computers is the increased sophistication of the control tasks performed by the computer. One example of this increased sophistication is mentioned in Section 14.1, the design of aircraft that are inherently unstable without the aid of computerized closed-loop control. The smart transducer, introduced in Section 9.8, is a very recent example of a digital computer embedded in an instrument system to enhance its capabilities. Artificial intelligence,[2] or AI, is a relatively new technology that could not possibly have come into existence without the digital computer.

This handbook appears at a time when the application of digital computers in instrumentation and closed-loop control is a rapidly moving target and therefore difficult to characterize as a mature topic. An argument could be made for allowing the field to settle down a bit before capturing it in print. On the other hand, now that the microprocessor is firmly established, it is appropriate that a reference such as this be available to help users take advantage of recent developments. Additionally, many textbooks on automatic control continue to be authored. Yet they are devoted predominantly to the theory and not the design of working control systems. Instrumentation, on the other hand, continues in scarcity so far as hardcover books are concerned. Topics such as impedance matching for designing control systems actuators and impedance mismatching for designing signal processing and transmission networks are rarely treated in the hardcover book realm. These are but a few of the deficiencies that are addressed in this reference handbook.

Although this book is part of a series for mechanical engineers, much of the material is useful to a wider audience. It is assumed that the reader has had some exposure to the systems language of transforming linear time-domain differential equations into linear algebraic equations via the Laplace transform. This transform theory is presented in this book but perhaps not in as much detail as the first-time reader would like.

1.1.1 Objectives of this Handbook

The first objective of this reference book is to update the material available for selecting and/or designing both instrumentation and control systems. Because most feedback control systems require one or more measurement transducers, the instrumentation material is appropriately included in this book. It stands on its own as a separate discipline and also as a necessary adjunct to the feedback control material. Systems language and modeling of dynamic systems complete the array of tools needed for specifying and designing a complete control system. Thus, the second objective of this book is to present the entire spectrum of control systems design in one reference—modeling, systems language (Laplace transforms and block diagrams), instrumentation theory and practice, and feedback control theory and practice.

A third major objective of this handbook is to present the advantages and importance of the digital computer for both instrumentation and control systems. Transmitting and acquiring data, storing calibration parameters, controlling and automating data acquisition processes, and creating intelligent transducers and transmitters are but a few of the many ways a digital computer may be applied in the measurement area. *Control* in this book most often refers to feedback or closed-loop control. But machine control as in directing robots and machine tools, where the digital computer is a necessity, is also covered.

1.1.2 Systems Approach

The first step in applying a systems approach to characterize the behavior of a device, component, or machine is to identify the input(s) and output(s) of interest. The focus here is to model the causality between one or more inputs and one or more outputs. Even more specifically, much of the systems presentation in this book centers on a single input-output pair. The advantage of this seemingly narrow focus is that the causality between such a single input-output pair can be readily portrayed in a visual form using block diagrams. Other visual forms are also available, such as bond graphs and signal flow graphs. This book uses only the block diagram visual form. Lest the reader be overly concerned about the narrowness of a single input-output approach, Chapters 19 and 20 on state-space description of dynamic and feedback control systems deal with the multi-input, multi-output problem.

After the input-output pair of interest has been identified, the next step is to develop a set of algebraic and ordinary differential equations relating the input and output. The focus is to describe the macroscopic behavior of the system, not the details of the intermediate behavior. (Chapter 3 and its references present this process in greater detail.) The next steps are to linearize these algebraic and differential equations and to transform them from the time domain to the Laplace transform domain. In this form they can be visually represented by a serial array of blocks as shown in Fig. 1.1(a). The transformed equations may also be combined mathematically by eliminating intermediate variables until an overall transfer function is obtained as shown in Fig. 1.1(b). The block diagram of Fig. 1.1(a) is most useful for displaying the intermediate interactions of the system. An overall block diagram format as shown in Fig. 1.1(b) is most useful for subsequent mathematical operations on the input-output relationship, such as frequency response analysis.

One of the attractions of the block diagram format for systems analysis is that the interaction between an incoming variable and an individual block is simply multiplication. This holds true for not only a variable and a single block but for a series of blocks as well. At any point in the string of events in Fig. 1.1(a), the variable at the output of a block is simply the multiplicative product of the system's input and each of the preceding blocks. Lest this sound too good to be true, the reader is reminded that substantial simplifying assumptions need to be made to get to this point, including the assumption that the behavior of the system can be adequately characterized by a linear or linearized set of equations.

The serial block diagram of Fig. 1.1(a) is especially suited to a measurement system. One can readily visualize the flow of events for a physical variable (the input variable) that needs to be measured and displayed or captured (the output variable) for later study. The successive blocks may

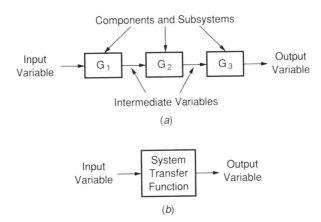

(a)

(b)

Fig. 1.1(a) A typical block diagram using individual transfer functions; **(b)** Block diagram with an overall transfer function.

be "sensor/transducer," "signal conditioner," and "recorder" as shown in Fig. 1.2. Note that this figure shows a two-input block diagram where the second input is a disturbance. The effect of either of the two inputs taken one at a time is simply the input in question multiplied by the intervening blocks between it and the output. Because of the linearity assumption, the total output is the sum of the effects of the individual inputs determined separately.

A rather different form of the block diagram is required to characterize a feedback (also called *closed-loop*) control system. This type of system is self-correcting in that a measurement of the output variable is fed back and compared with the command input variable as shown in Fig. 1.3. Because of the loop structure of the system, it is possible for a momentary disturbance within the loop to recirculate and cause an increasing output which persists even though the original disturbance has disappeared. Since all real systems have unknowable and unpredictable disturbances imposed on them, a sustained or growing response is unacceptable. This type of response must be eliminated either by redesign or by addition of components. In any case, the block diagram is a valuable systems tool which enables the designer to predict behaviors and responses before committing a design to hardware.

The block diagrams of Figs. 1.1*a*, 1.2, and 1.3 are good vehicles for differentiating *control* from *closed-loop control*. Some systems have a control structure (not feedback control) that resembles that of Fig. 1.1*a*. Clearly if a disturbance enters the serial chain of blocks as in Fig. 1.2, the input command has no knowledge of the disturbance and cannot make a correction to compensate for this input. The system designer, given an open-loop control structure, can at best be aware of the most likely disturbance entry points and can build into the system design some means for attenuating the disturbance inputs, such as a filter at their inputs. The closed-loop structure of Fig. 1.3, however, is self-correcting by comparing the actual output with the desired output. (The command input is also the desired output.) Examples where both open- and closed-loop control are used include machine tool and robot control. A master computer derives the path of the robot or the tool point and sends commands in an open-loop sense to a local control computer or programmable logic controller for execution. If disturbances enter the system, transducers mounted as near to the working end point as possible feed back real-time information about actual position, velocity, force, and so on. The closed-loop control system corrects the disturbed output variable to the best of its capability by generating an error signal that is the difference between the command input and a measure of the fed back actual output variable.

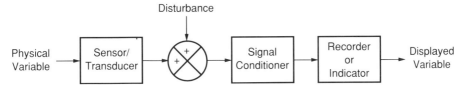

Fig. 1.2 Instrumentation system block diagram structure with disturbance input.

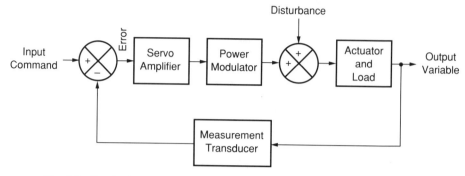

Fig. 1.3 Feedback control system block diagram structure with disturbance input.

1.1.3 Impact of the Digital Computer on Instrumentation Systems

Data acquisition systems were among the first beneficiaries of digital computers in the instrumentation field. As mass memory dropped in price and in size, larger blocks of data could be acquired and stored for subsequent inspection and analysis. As computational speed grew faster, more dynamic systems qualified for digital data acquisition. This approach to data gathering simplified the setup procedure and enabled much more variety and creativity in experimentation with physical systems. The early data acquisition systems were minicomputer based but this approach quickly became obsolete with the advent of the microcomputer An interesting side commentary: the mainframe digital computer had very little impact on instrumentation systems.

A second area where the digital computer has made significant inroads in the instrumentation field is in microprocessor-based signal conditioners. Ease in calibration, storage of calibration parameters, semiautomatic zero calibration, as well as parameter communication with and data transmission to central management information system computers are but a few tasks that are now possible and are much simpler as a result of incorporating a microprocessor into the signal conditioner.

The most recent development in instrumentation systems as a result of the digital computer is the introduction of a microprocessor into the transducer itself. Two separate endeavors have evolved as a result of this approach: the creation of a digital transducer from a fundamentally analog building block, and the implementation of a smart sensor which has greatly improved performance characteristics as well as digital data transmission capability. These topics are introduced in Sections 9.7 and 9.8.

1.1.4 Impact of the Digital Computer on Control Systems

The mainframe computer was first used in feedback control systems as a supervisory computer to change the set points of the process analog controllers[1] in the late 1950s and early 1960s.[1,3] From these early machines to the present distributed hierarchical systems with microcomputers at the process controller level, there are five traceable generations of digital computers.[1,4] (See Fig. 1.4 for an example of hierarchical control.) The early drum-memory computers such as the RW 300 and the GE 312 quickly gave way to core-memory machines of essentially the same electronic make-up except for the higher speed memory. These second-generation computers in turn were made obsolete by centralized direct digital control computers such as the IBM 1800, the CDC 1700, and the FOX1A. Then along came the fourth generation of digital computers, the minicomputer. Some early examples in this class were the Westinghouse P50 and the Digital Equipment Corporation PDP8 and PDP11. Certainly one of the major arguments favoring the use of minicomputers was the move away from centralized computer control, except for the higher level hierarchical tasks such as downloading set points to the process control computers in the network and collecting data for management information. The fifth generation of computers, the microprocessor, microcomputer, or computer-on-a-chip, has certainly enabled the concept of a single-loop digital controller. Now practically every feedback loop can afford a dedicated digital computer in the form of a microprocessor.

Industry has long been divided into two rather different classes: the continuous process industries such as petrochemical, paper, glass, and power, and the discrete parts manufacturing industries such as metal parts manufacturing, automobile and machinery production, and clothing manufacturing.[5] Formerly, control equipment manufacturers, suppliers, and vendors were aligned with either the discrete parts manufacturing industry or the continuous process industry, but not both. The advent of the microprocessor has blurred the two distinct lines by enabling the programmable logic controller

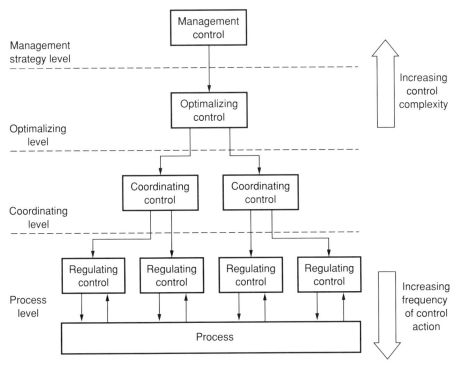

Fig. 1.4 Multilevel hierarchical control structure.

(serving the discrete parts industry) to take on more of the features of the electronic process controller (serving the process control industry) and vice versa.[5]

The concept of distributed computer-based control under the supervision of and reporting to a hierarchical arrangement of digital computers is now completely feasible, and the dream of plant-wide control and automation can be realized. Development of the microcomputer as well as significant advances in mini- and main-frame computers have joined forces to make this dream a possibility.

In the decade of the 1980s the theory of adaptive or self-tuning control that had been developed over the past two decades finally became a reality. Numerous microprocessor-based self-tuning controllers made their commercial debut. This technology established itself rather quickly and it has proven itself economically. Given the rapid development of digital computer technology and usage in the recent past, it is difficult to extrapolate into the future, near term or otherwise.

However difficult such prediction may be, one technology that will definitely have a major impact on systems and control in the decade of the 1990s is Artificial Intelligence. Despite its newness, AI has already established itself in a number of applications.[2] Industrial product offerings are available that provide advice for more optimal total system operation, such as adjustments in process, tool wear, fixture alignment, and problem diagnosis. *Just in time* training that offers aid in diagnosing an equipment fault and then delivers on-the-spot training to repair the fault reduces the laborious process of trouble-shooting and service manual consultation. Expert Systems, a subset of AI, has made some commercial inroads in the area of computer numerical control (CNC). One example is a productivity-enhancing tool that speeds up programming of parts and minimizes the chance of making bad parts by guiding the machine-tool operator through set-up and machine operating variable choices. Expert Systems, although it has been mostly static and off-line to date, will emphasize real-time applications to a greater extent.[2]

1.1.5 Organization of this Reference Book

This book contains three groups of subject matter: general topics, instrumentation, and control. Sections 1.2, 1.3, and 1.4 describe the contents of these three groups of chapters. The Table of Contents shows the chapter titles within each of these subject groups.

1.2   GENERAL TOPICS

The first part of Chapter 1 is an overview of the approach taken in this handbook, particularly of the impact of the digital computer on the areas of instrumentation and control. The last section of Chapter 1 presents some definitions and terminology for both instrumentation and control systems.

Chapter 2 is a delightful introduction to feedback control through historical examples that shaped the development of this field as we know it today. Feedback control cannot work without some form of measurement. Each example of control presented in this chapter has associated measurement, but the *transducer* is not obvious because many of these measurement devices do not change the domain of the measured signal as do most modern measurement devices. For example, James Watts' steam engine has a flyball governor to control its speed. This speed-measuring transducer simply converts a rotary speed to a linear translation of a slide valve—all in the mechanical domain. Examples of the computational and theoretical development of systems and control are also presented. The last part of the chapter, Section 2.3, outlines the vocabulary and definitions of systems engineering and control.

Modeling of dynamic systems and the specifics of the systems approach form the basis for Chapter 3. Mechanical, fluid, and electrical systems are all treated under a unified theory applicable to both instrumentation and control systems. The Laplace transform is presented and is used as the vehicle for developing input/output transfer functions.

Chapter 4 deals with static analysis of instruments. Because transducers and measurement systems are designed to operate at response rates considerably greater than those of the systems under their scrutiny, the problem of characterizing an instrument's static performance rightfully deserves considerable attention. Statistics and best-fit calibration parameters are treated in this chapter.

A pivotal concept in both instrumentation and control is that of input and output characteristics and impedances. Chapter 5 is specifically devoted to introducing this concept. It is particularly important in designing, selecting, and integrating the hardware components of instrumentation and control systems. This subject matter receives only a cursory treatment in most textbooks on instrumentation and control.

Electronic devices, including operational amplifier theory and practice, as well as analog-to-digital and digital-to-analog signal converters are discussed in Sections 6.1 to 6.4. Section 6.5 presents the logic devices needed for interfacing digital systems with analog systems. Sections 6.6 and 6.7 delve into the architecture and the utilization of microprocessors and microcomputers.

Chapter 7, the closing chapter of the general topics portion, deals with grounding and cabling techniques for the purpose of noise reduction in signal transmission. Ground loops, proper grounding, and shielding are some example topics from this chapter.

1.3   INSTRUMENTATION

The chapters under this heading have the same order as the flow of events in a real measurement system, as shown in Fig. 1.5. Bridge-type transducer circuits, sensors, transducers, signal conditioners, and measurement systems are the subject matter of Chapters 8, 9, 10, and 11. Digital and smart transducers are introduced in Sections 9.7 and 9.8. This material is the starting point in designing an instrumentation system. The intent in these four chapters is to present the fundamentals of transducer design and to discuss the most essential transducers used in mechanical engineering. These include displacement, profile imaging, velocity, acceleration, force, torque, pressure, temperature, and flow transducers. Examples from instrument manufacturers' literature are presented here.

Chapter 12 takes up the issues of analog and digital filtering and digital signal transmission. The most common method of analog signal transmission, the current loop, is presented in Section 10.6 in conjunction with pressure measurement systems.

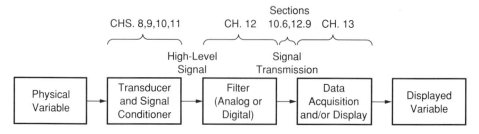

Fig. 1.5   A block diagram showing the components of a measurement system.

The final element in a measurement system is a data acquisition system and/or a display device, taken up in Chapter 13, which concludes this section.

1.4   CONTROL

Chapters 14, 15, 16, and 17 constitute a compact "booklet" on the theory, analysis, modeling, and design of closed-loop control systems using classical methodology. The first two of these chapters present the theory and analysis starting with Laplace transforms and block diagram manipulation. Given that microprocessors are an important control system building block, z transforms deal with the unavoidable problem of time delays introduced by digital sampling.

Chapter 16 focuses on modeling electromechanical and electrohydraulic servoactuators, the muscle element of a control system. Other actuator forms are briefly mentioned as well.

Chapter 17 is an in-depth design-oriented presentation of controllers for closed-loop control systems. Performance indices, frequency domain compensation, derivative feedback, and PID controllers are discussed.

A different type of controller is required for controlling the path of a machine such as a robot or a machine tool. In a hierarchical sense, a machine controller of the type presented in Chapter 18 is the master to a closed-loop controller. It calculates the desired displacements and velocities and sends these signals as setpoints to the closed-loop controller for execution. Programmable logic controllers, numerical controllers, and robot controllers are all discussed.

Chapters 19 and 20 present the modern version of feedback control (still called *modern control* even though this material has been available for at least 25 years). This material is presented as a separate entity because many control engineers still active in industry have not had the benefit of assimilating this approach. Many recent automatic control textbooks weave the two topics together in such a way that the reader has no choice but to wade through both the classical and the modern approach to studying feedback control. In this handbook the reader is given a choice: Chapters 14 through 17 present the classical view of control while Chapters 19 and 20 present the modern or state-space analysis.

1.5   INSTRUMENTATION AND CONTROL DEFINITIONS, TERMINOLOGY, AND CONCEPTS

1.5.1   Static versus Dynamic Characteristics

The terms "static" and "dynamic" are frequently used synonymously with "stationary" and "nonstationary" inputs and outputs. Instrumentation and control require different definitions for these two terms. Note that the definitions given below are based on the system under study and not its input or output variables.

A system is said to be *static* if its input/output relationship is independent of the rate of change of the input. All physical systems eventually violate this definition as the input rate increases. Therefore *static* is usually accompanied by a qualifying statement that specifies the range for which the system is static, such as a band of frequencies extending from zero to some limiting value. An example is a mechanical spring operating at such a slow input rate that the force/displacement relationship is invariant. At higher rates of the input, the mass of the spring becomes a significant factor and the spring no longer behaves as a static device.

A system is called a *dynamic* system if its input/output relationship depends on the rate of change of the input. Systems in this category have energy storage and must be described by one or more differential equations. Section 2.3.2 offers further explanation of the term *dynamic*. The response time of a dynamic system is characterized by its time constants and/or natural frequencies. Instrumentation systems are unavoidably dynamic systems, but they are designed to have time constants smaller than and natural frequencies larger than those of the system being measured. For example, a feedback control system requires at least one measurement transducer. Once the response speed of the control system has been specified, the response time of the transducer can be designed or selected so that it is substantially faster than that of the measured system.

Figure 1.6 illustrates some of the individual components that make up a complete study of the statics and dynamics of a system.

1.5.2   Accuracy and Resolution

A measurement systems's accuracy is invariably reported by providing information about its maximum error. For example, "This transducer's accuracy is 0.1% of full scale" really means that its error is no greater than 0.1% of its maximum output, called its *full-scale* response. Section 4.1.2 describes

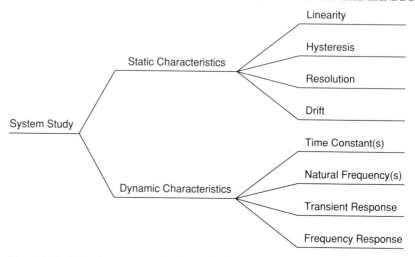

Fig. 1.6   Individual components in the study of the statics and dynamics of a system.

a number of error sources that contribute to the overall error and aptly names the totality of the system's *error band*. Included here are precision, nonlinearity (frequently referred to as linearity), hysteresis, and repeatability, all of which are assumed to be independent of changes in the environment.

Two additional sources of error, often referred to as *zero drift* or "shift" and *sensitivity drift* or "shift," result from changes in the environment, changes in electrical conditions such as a slight reduction in the AC supply voltage, and aging of components. In fact, the most significant of these modifying inputs is a change in the temperature of the environment. This results in a change in the offset of the transducer, called temperature zero shift, and is expressed in units of change in the output signal per degree of temperature change. Similarly, the effect of temperature change of the environment on the sensitivity is called temperature sensitivity shift and is expressed in percent change of the instrument's sensitivity per degree of temperature change.

Resolution and threshold are also defined in Section 4.1.2. The former is important in establishing requirements for the instrumentation components needed in a measurement system. The latter is not generally a significant factor in instrument performance but invariably is encountered in control system components, especially in the power transfer portion of the system.

1.5.3   Calibration

Calibrating an instrument consists of presenting a standard or known value of the input variable to a transducer for the purpose of establishing the change in the output variable per unit change of the input variable. Once this relationship is known, it is used to infer the value of an unknown input variable during the actual measurement process. If the measurement transducer can be characterized by a linear input/output relationship, the slope of this line is called its *sensitivity* and carries with it the units of the output signal, such as voltage or current, per unit of the input variable, such as pressure or temperature. Some transducers require an excitation input, in which case the output becomes a function of not only the input measured variable but also the magnitude of the excitation. The sensitivity of this type of transducer is expressed in units of output signal per unit of input variable and per unit of excitation. Or it may be expressed in units of output signal per unit of excitation at *rated input*, where rated input is the maximum or full-scale value of the input. The input variable is often called the *measurand*. Chapter 4 presents the mathematics of this calibration process, including the case where nonlinear input/output relationships prevail.

A term often incorrectly used in the context of calibrating an instrument is *gain*, a quantity that is also the ratio of an output to an input variable. The use of gain in instrument terminology is limited to a dimensionless ratio, such as volts per volt. However, gain used in the context of feedback control is often applied to the ratio of two different types of variables, such as forward loop gain (see Chapters 14 and 15). When it is used to characterize the loop gain, it again refers to the ratio of like quantities and is therefore dimensionless.

Scale factor, a term largely restricted to the instrument world, is similar to sensitivity in that it is a ratio of two different variables. Its usage is restricted to a portion of the signal conditioner, an electronic device that receives the output signal from the measurement transducer and produces a

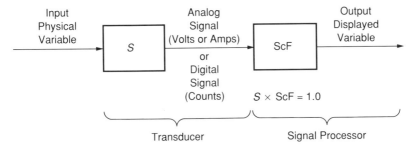

Fig. 1.7 Sensitivity and scale factor.

high-level output signal which may be accurately recorded or displayed by an indicating meter (see Chapters 8, 9, and 10 for more details). The block diagram of a basic measurement system is as shown in Figs. 1.2 and 1.5. Using the notion of a sensitivity (S) to characterize the transducer and scale factor (ScF) to describe the remainder of the system, the block diagram of Fig. 1.7 may be constructed.

In nearly all measurement systems, the displayed variable ideally has the same value as the input physical variable. It therefore follows that the product $S \times \text{ScF} = 1.0$, a dimensionless quantity. If the sensitivity S is known or can be experimentally determined, the scale factor ScF of the signal conditioner can be calculated and set or keyed into the signal conditioner/signal processor: $\text{ScF} = 1/S$. This is a powerful and useful concept in the event that it is difficult or impossible to calibrate the instrument system by presenting a known value of the input physical variable to the transducer.

As a minor comment, microprocessors are more often being incorporated into the signal conditioner. Consequently, the concept of keying in a value for the scale factor becomes more natural and obvious. This is true because an alpha/numeric display of the original and new values of the scale factor lessens the likelihood of entering an incorrect value. Furthermore, the scale factor of a digital signal processor is most likely to be stated in units of the measured physical variable per count value of the input signal from the transducer (see Fig. 1.7). At the present time very few transducers produce an output signal in the digital domain. Hence the input stage of the digitally-based signal processor must include an analog-to-digital converter, and it is quite natural to think of this conversion element as a part of the transducer itself. Thus the sensitivity S of Fig. 1.7 may be thought of as having units of counts per unit of the measured variable. In the future more and more transducers will be available with an on-board A/D converter and the sensitivity of the transducer stage of the instrumentation system will truly have a digital output signal. Then the role of an analog voltage or a current will be greatly diminished. See Section 9.7 for an introduction to digital transducers.

1.5.4 Full-Scale, Range, and Span

These three terms are most frequently used in the context of a measurement system but may also be used in a control system. Full scale, when referring to the physical variable, means the maximum value that it can take on. When referring to the output of the transducer or the readout device, it means the value of the signal or indicated value when the input to the measurement system is at its maximum value, such as the weight of the contents of a vessel when it is completely full. Range means the entire array of values that a measurand can take on. Similarly, range may also refer to the entire array of transducer output values or signal processor output values. In its broadest sense, span and range are synonymous. However, some instrument manufacturers use the word "span" when referring to the control that sets the value of the signal conditioner scale factor, as in "coarse span adjust" and "fine span adjust."

1.5.5 DC and AC

The use of "DC" has several meanings in instrumentation and control terminology. "DC" means literally *direct current*, but it seldom refers to current. One usage of "DC" refers to a stationary or constant value, such as "the DC component of a signal." Another meaning of "DC" refers to the fact that the output of a device varies proportionally with the input variation as opposed to an output consisting of a sinusoid at constant frequency where the amplitude of the sine wave is instantaneously related to the input signal. AC servos are examples of the latter. "AC" means literally *alternating current*, but here too "AC" hardly ever refers to an alternating current. It may mean a signal of both constant frequency and constant amplitude, or it may mean a signal with constant frequency but variable amplitude as in the case just cited.

REFERENCES

1. T. J. Williams, "Two Decades of Change—A Review of the 20 Year History of Digital Control," *Control Engineering*, Vol. 24, No. 9, September 1977.
2. F. J. Bartos, "AI Now More Realistically at Work in Industry," *Control Engineering*, Vol. 36, No. 7, July 1989.
3. P. Oldenburger, "Automatic Control—A State-of-the-Art Report," *Mechanical Engineering*, April 1965.
4. Personal conversation with Prof. T. J. Williams, Director of the Purdue Laboratory for Applied Industrial Control (PLAIC), Purdue University, West Lafayette IN, October 1989.
5. T. J. Williams, "Recent Developments in the Application of Plant-Wide Computer Control," *Computers in Industry*, Vol. 8, Nos. 2 and 3, March 1987.

CHAPTER 2

SYSTEMS ENGINEERING CONCEPTS

R. EUGENE GOODSON

Johnson Controls, Inc.
Ann Arbor, Michigan

WILLIAM R. SEITZ

Control System Technology, Inc.
Clawson, Michigan

2.1 INTRODUCTION

The concept of a *system* is central to engineering, mathematics, science, and indeed to many fields. The concept is imprecise, however, since a complete and all-encompassing system to one observer may be only a subsystem to another. In fact, one of the central themes in engineering and science—and indeed of civilization—has been the expanding boundaries of the interrelated systems that compose our universe. The broadening of our concept of the universe has been profound. For example, we were sure not too long ago that the earth was at the center of the universe (Galileo was imprisoned for theorizing that the earth might not be at the center of the solar system). Today, it is accepted that the earth is but one planet (although quite striking when viewed from space) among many in one solar system, among billions in one galaxy, among the innumerable galaxies we can "see." It seems that new phenomena and systems are being discovered at an increasing rate.

Probably the oldest concept of a system is life itself. Every living thing has a cycle of birth, reproduction, and death. Darwin's modification of the life concept showed that all life, which had seemed independent, was in fact interrelated and deeply dependent—so much so that each life system evolved in a direction influenced by the combined and interrelated effects of the myriad individual systems.

The concept of systems is at the heart of the ability of engineers to conceptualize and construct increasingly more complex buildings, transport machines, energy and raw material conversion plants, factories, agricultural implements, computers, medical aids, recreational equipment, and products on which our modern civilization relies for its everyday functioning. It should not be surprising that more powerful analysis tools have been developed along with the ability to construct more complex systems; the discipline of *systems engineering* has evolved along with the expanding concept of a "system." The purpose of Chapter 2 is to discuss briefly those generic tools and concepts in use today that allow systems engineers to construct, modify, and control a broad variety of systems.

A historical development of the concepts and tools of systems engineering is given below. This historical record illustrates the evolution of concepts, measuring instruments, mathematical analysis tools, and the technology that allows increasingly more precise and capable systems to be constructed. The people in this field were often interesting characters as well as innovators. The systems discussed in this historical record illustrate both the diversity of the phenomena to which systems engineering can be applied and the expanding conceptual context, which allowed increasingly more difficult jobs to be successfully attacked. The historical record also strongly suggests that nature is infinitely rich and that no state of knowledge can be reached where the concepts and tools are sufficient for the future.

Probably the most famous example of a systems engineering achievement is the Apollo project. While it is true that the moon landing project was blessed with many resources (both talented people and money), completion was mandatory prior to full understanding of the magnitude of the undertaking. Further, system failure was entirely possible and would have been disastrous.

All the previous tools of systems engineering were used in the Apollo project, and new tools and concepts were even developed. One analytical triumph was the decision to land on the moon with staged propulsion units rather than relying on one main transport unit to make the total round trip. The actual mission proceeded in stages. The total system was first placed in earth orbit by the Saturn rocket. Then everything was jettisoned, except those modules needed for the remaining mission. The next stage was a lunar orbit, followed by the trip of only the lunar module to the moon's surface. A portion of the module was used to go back into lunar orbit, leaving on the moon those propulsion and other units needed to land on and explore the surface. Finally, only that portion of the unit necessary to return to earth orbit and reenter the atmosphere made the trip from the moon to earth. Thus, only a fraction of the original spaceship returned to earth.

The decision to implement this strategy was made after extensive simulation and analysis of alternatives. The concept of mathematical models for systems, which allow alternatives to be simulated and their efficiency and performance evaluated, is a fundamental tool of systems engineering. Models have been in use since the early stages of the field. Such a mathematical model was used in the Apollo project to support the decision of staging the moon landing and to evaluate the risks. The energy use and therefore the size of the original rocket necessary to complete the lunar mission were minimized by the model actually employed.

A famous engineering disaster, the Tacoma Narrows suspension bridge failure, is a graphic teacher of the necessity to analyze more than obvious phenomena. Most of us have seen the movie of the destruction of this bridge by winds in the Narrows. The static analysis tools of bridge construction showed that this bridge was safe and that the design was more than adequate for any conceivable static load. A dynamic analysis was not performed, however. Such an analysis, using tools then available, might have shown that the bridge was unstable under sufficiently high and steady winds. This phenomenon is best illustrated by the flag that waves back and forth in a steady wind or a power line cable that oscillates in the wind. It is now universally known that every system has vibrational characteristics that can be excited by many such stimuli and that no systems engineering job is complete until such stimuli are hypothesized and the resulting system response analyzed.

Prior to mathematical models of systems, experimental models that simulated the behavior of the completed system were the rule. Experiments still play a vital role in systems engineering, but they are now integrated with mathematical models to enhance both. Clearly the European cathedrals of the 11th through 14th centuries evolved from experience and observation of actual structures. Today, the use of scale models in major construction projects such as chemical plants and bridges, and the concept of separating complex systems into simpler subsystems are important tools of systems engineering. The latter concept is used in major engineering projects such as construction of power and chemical plants, dams, and major transport systems. Joining subsystems together correctly to predict the performance of the whole is then required.

Two areas where systems engineering tools are beginning to be applied are in behavioral and environmental systems. In behavioral systems such as economics, political science, and business, the concepts and tools have been applied with relatively little success to date. More success has been demonstrated in environmental systems, where the effect of a man-made system on the local ecology is analyzed and where systems engineering tools have been applied to the systematic preparation of environmental impact statements.

Systems engineering concepts are embodied most successfully in the field of automatic control, which has been at the forefront of the analytical approach to systems for over 50 years. Much of the discussion below will focus on this area of systems engineering.

2.2 HISTORICAL SYSTEMS CONTROL HIGHLIGHTS

The history of control systems engineering reaches back to Greek antiquity. It covers areas from timekeeping to transportation, from prime movers to prime numbers, and from pure mathematics to business and profit. The impact of automatic controls on our daily lives is ubiquitous. Our homes are replete with automatic controls. They govern ovens, furnaces, water heaters, air conditioners, toilet bowls, humidifiers, watches, audio and video equipment, garage door openers, water softeners, refrigerators, washers, and dryers. In our automobiles, engine temperature, air/fuel ratio, spark advance, cruise control, and automatic transmission operation were all designed and built incorporating automatic controls. Our factories and processing plants could not function without automatic control. The air traffic and train control systems maintain efficient operations and the safety of our transport systems.

Why should we study the history of automatic control? What is so interesting about this field? First, automatic control and systems engineering are concepts not anchored to a single physical science or phenomenon. Systems engineering concepts and tools have been applied to many different problems quite successfully, and the problems are often central to the discipline involved. Second, in the well-recorded history of science, many great advancements have been relatively unheralded because they were implementations of rather simple concepts. As any scientist or engineer will attest, the implementation phase of any project brings the most advancement. This is certainly true of systems engineering and automatic control, a field where rather simple concepts have had profound implications.

Finally, in the history of systems engineering we have included the development of the computer. Systems engineers were among the first to develop and make use of computers, both analog and digital, in predicting the behavior of complex systems. Many control synthesis methods are best performed by computers, and some methods even demand computers. And of course, the implementation of modern control systems invariably involves some sort of computational element to perform the comparator and compensation functions. It would be hard to imagine practicing systems engineering without computers.

A short survey of the history of any technical specialty must, by necessity, be limited to the highlights, and even some of them must be left out. Not all important contributors to control system engineering are discussed in the following stories. If your own favorite hero is missing from our list, we apologize. If you meet someone new here, or if you gain a different perspective on one that you already were acquainted with, then our effort will have been worthwhile. A few of the giants of the past have piqued our interest, and their stories are related here.

2.2.1 Ctesibios (250 B.C.): The Water Clock Regulator

The ancient city of Alexandria, built on the northern coast of Egypt by Alexander the Great in 355 B.C., quickly became a center of scholars and "engineers." Ctesibios began his adult life there as a barber but soon became fascinated with the mechanical arts. It is said that he built a movable mirror for his barber shop that made use of a counterweight and a set of pulleys. He fitted this weight into a tube concealed in the wall but found that the fit was close enough to trap air in the bottom of the tube, hindering the movement of the weight. When he bored a hole in the bottom of the tube, the air was allowed to move freely, and the mirror could be positioned more easily. This simple success led Ctesibios to a change in career, and he joined the School of Alexandria to study hydraulics, mechanics, and mathematics.

Ctesibios holds the honor of inventing what has been called the first feedback control system, a water pressure regulator for a clepsydra, or water clock. Ctesibios' work is not well documented since his own writings were destroyed when Julius Caesar sacked the city and burned down the museum in 47 B.C. Fortunately, the descriptions written by the Roman historian, Vitruvius, of Ctesibios' mechanisms have survived and have been interpreted by Otto Mayer in his book *The Origins of Feedback Control*. Mayer speculates that Ctesibios' clepsydra made use of a float-operated valve to regulate the water pressure which, through a calibrated orifice, supplied the constant flow rate that was the heart of accurate timekeeping in that age.

The water clocks that had been in use for hundreds of years before Ctesibios' time were not very accurate and were unsophisticated in their manner of displaying the time. Ctesibios added an elaborate set of dials and gearing to the float mechanism to make reading the time of day easy. He applied the principles of hydraulics developed by his predecessor, Archimedes, to devise the flow rate control. The concept of the float regulator, still in use today in carburetors and toilet tanks, appears to be Ctesibios' most important original contribution to the art of feedback control systems.

2.2.2 Cornelis Drebbel (1624): Temperature Control

Cornelis Drebbel, a mechanic and chemist in Holland in 1624, needed a constant temperature incubator for hatching chicken eggs. He devised an oven to produce the heat and a thermostat to provide the necessary regulation of the oven's temperature. Drebbel's concept is described in vague terms in an English patent (number 75) issued in 1634, a year after his death, when patent law was new and did not require the detailed teaching it does today. A better description is found in a manuscript written by Drebbel's grandson, Augustus Kuffler. A U-tube filled with alcohol in the bottom and mercury in the top served as the temperature sensor, with the level of the mercury rising or falling with the thermal expansion of the alcohol below it. A metal rod floating in the mercury transferred this motion to a damper, which controlled the airflow through a flue leading into the firebox. If the incubator temperature was too high, the alcohol/mercury sensor would react by moving the damper to close off the airflow and choke off the fire. The reverse action raised the fire's heat output if the temperature fell, thus regulating it to a setpoint.

2.2.3 Denis Papin (1707): Pressure Regulation

As the use of steam for power generation became feasible, the need to regulate steam pressure became essential for safety. Steam engine explosions were common, and a prior invention, intended for quite a different application—cooking—came to the rescue. Denis Papin had first devised a pressure cooker in 1681. His cooker, which he called a "digester," used a weight and a relief valve to sense and control the pressure in the vessel. By changing the lever arm on the weight, any vessel pressure could be maintained. His invention earned him membership in the Royal Society, and to show his appreciation, Papin cooked a meal in his digester for the other members of the Society.

Papin later became interested in steam-powered engines. Around 1700, he was commissioned by the Duke of Hesse in Germany to build a steam-powered water pump for the fountains in the Duke's gardens in Kassel. Papin used his pressure cooker idea in the steam boiler and a reciprocating piston with check valves for the pump. He was plagued by the water pipes continually breaking, and did not successfully perfect his ideas until 1707, in England. His safety valve became standard equipment on all steam engines, including that of Thomas Newcomen, the predecessor of James Watt.

2.2.4 Edmund Lee (1745): Windpower Control

Windmill power predated steam power by a considerable margin, the first windmill dating back to 600 A.D. in Persia. But is was over a thousand years later, in 1745, that automatic control was first applied to the problem of keeping the sails aligned with the wind as it changed direction. Edmund Lee added a small windmill, known alternatively as a "fantail" or "rosette," to the main windmill and geared it in such a way as to rotate the entire top tower of the mill. When the main sails were out of line with the wind, the fantail would catch the wind, rotate the tower, and bring the main sails around into the wind. Once there, the fantail would no longer catch the wind, and rotation of the tower would cease. This device can be considered as more than a simple servo; rather, it was a peak-seeking controller that optimized the extraction of the wind's available power. Lee's invention gained wide popularity in his home country of England as well as in Germany. Lee also invented a control device for the main sails to limit their speed of rotation to a safe level during particularly high winds. This device did not work well at first but was perfected in 1807 by William Cubitt.

2.2.5 James Watt (1789): Steam Engine Speed Regulation

James Watt began his technical career as a mathematical instrument maker for the University of Glasgow. One of his first assignments was to repair a steam engine designed by Thomas Newcomen. On analyzing the device, he hit upon a new concept for this prime mover that would greatly improve its efficiency. His improvements were perfected, and in 1769, he was granted an important patent entitled, "A New Method of Lessening the Consumption of Steam and Fuel in Fire Engines." In order to exploit his discoveries, he sought financial backing from an iron merchant, Mathew Boulton. Obtaining an extension to the patent, the two men formed the Boulton & Watt Company in 1775, the first major producer of prime movers. Their first products had one-third the fuel consumption of the Newcomen machines and were an instant hit. Boulton & Watt made a fortune by leasing its engines for a fee equal to the customer's savings in fuel costs.

After several improvements to the power transmission mechanism, the first steam engine with a rotary output motion was installed by Boulton & Watt at the Albion Mill in London in 1786. Andrew Meikle had built the mill machinery, including a centrifugal governor to maintain the spacing between the millstones, based on a design by Thomas Mead. Mead's device, patented in 1787, held the millstones apart when not rotating and brought them together by centrifugal action as the speed of rotation increased. Boulton visited the Albion Mill in 1788 and wrote to Watt, back at the factory in Birmingham, that he had seen the millstone regulator and described its operation.

Watt immediately saw the possibility of applying the invention to the problem of speed regulation of his steam engine and set about designing his famous flyweight governor. This time, for some unknown reason, Watt did not seek a patent and instead relied on secrecy to protect his ideas. Perhaps he was afraid that a patent application would draw attention to the fact that he was using Mead's discovery and he would be required to pay Mead a royalty.

In any event, when Watt's original steam engine patent expired in 1800, the centrifugal governor began appearing on everyone else's steam engines as well. Even though Watt's centrifugal governor was not wholly original, its occurrence at that time during the industrial revolution did have a great impact on the field of feedback control. New applications of the feedback principle started appearing with increasing frequency, and the use of mathematics as an aid to design was just around the corner.

2.2.6   Charles Babbage (1822): The Analytical Engine

Before describing the Analytical Engine, which was never built because of technical problems of the day, it is first necessary to mention its ancestors. The first mechanization of arithmetic was accomplished by Wilhelm Schickard in 1623. His machine was able to add, subtract, multiply, and divide and was apparently unknown to future generations until a letter from Schickard to Kepler, the astronomer, describing it was discovered in 1957. For this reason, Blaise Pascal has often been incorrectly named as the inventor of the first mechanical calculator. His device, called a "Pascaline" and first built in 1642, could add and subtract. It is said that his father's computational labors as a tax collector inspired Pascal to this invention.

The mathematician Gottfried Leibniz was also ignorant of Schickard's machine but did know of Pascal's when he invented the Leibniz wheel, a machine that could multiply and divide by repeated addition or subtraction. Leibniz refers to Pascal's "calculating box" in his writing and is noted for his foreshadowing quote; "it is unworthy of excellent men to lose hours like slaves in the labor of calculations which could safely be relegated to anyone else if machines were used."

In 1812, Charles Babbage began thinking about mechanical computers. Working manually on calculations for an astronomical table as a young man, he is quoted as saying that such work should be done by "a soul-less steam engine." It was not until 1822 that he began building such a device. The Difference Engine was a special-purpose adding machine that could be used to calculate polynomials, based on the principle of repeated differences. Construction was hampered by the precision required for the parts, being somewhat beyond that obtainable in English machine shops at the time. Babbage was compelled to spend a large amount of his private wealth (he had inherited a fortune from his father, a banker) in order to devise new tools for making the intricate mechanisms.

The British Government was finally persuaded to fund part of the development costs, based on the Navy's interest in computing nautical tables, but the device was never completed by Babbage (although a six-decimal-place version had been built, Babbage felt that 20 decimal places were needed to make the thing worthwhile). A Swedish engineer, Per Georg Scheutz, read Babbage's description of the Difference Engine and with his son, Edvard, began working on a simplified version. Their first machine was completed in 1843, and further refinements were made until 1854, when the Scheutzes visited Babbage on their way to display the finished machine at the Great Exhibition in Paris.

All of this leads to Babbage's more visionary, but ultimately frustrating, work on what he called the Analytical Engine. Babbage was eccentric, sometimes going off in new directions before finishing the tasks at hand. In 1833, while the Difference Engine was still incomplete, Babbage conceived a grander device, which would consume the remainder of his life. This development was chronicled by Ada Byron, daughter of Lord Byron the poet. Ada, who would have a programming language funded by the U.S. Department of Defense named for her in 1979, was herself a mathematician.

Babbage envisioned a new machine with four main parts: the store, in which a thousand 50-digit numbers would be stored; the mill, for carrying out the calculations; a transfer device to shuttle the data back and forth between the store and the mill; and finally a mechanism for getting the data out of the machine. For this last task, according to Ada, Babbage had chosen the method of punched cards, similar to those used by Joseph Jacquard to automate his weaving looms in France. He also planned to use the punched cards to program the operations of the engine, which included the ability to perform conditional branching to an alternate set of cards.

The mechanical complexity of the Analytical Engine was its downfall. In a simplified form, over 50,000 moving parts would have been required to carry out the operations. The drawings alone numbered in the thousands. Babbage tried in vain to complete the machine. His son, Henry, constructed models of some of the parts. The famous mechanical master, Joseph Whitworth, was employed to produce the precision parts, but even his great skills were not up to the enormous task. Finally, the government withdrew its financial support, having spent some 17,000 pound with no useful results. Babbage and Ada then attempted to devise a mathematical method of betting on horse races to raise funds to continue their work, but to no avail. After almost 40 years of work on the Analytical Engine, Babbage died in 1871. The technical breakthroughs required to achieve his goal would have to wait until the next century.

2.2.7   Macfarlane Gray (1867): Automatic Helmsman

Steam power soon became the engine of choice for rail and water transportation because it did not depend on the weather. The giant steamship, Great Eastern, was guided along its ocean journeys by the first automatic helmsman, invented in 1867 by Macfarlane Gray. Gray's device used the ship's steam power to actuate the huge rudder. A mechanical differential mechanism compared the pilot's input command with the rudder position; the output of this mechanism opened a steam valve to bring them into agreement.

The success of Gray's invention led others to emulate it, and Joseph Farcot in France worked on similar ideas. In 1868, Farcot used a very sensitive centrifugal governor as a speed sensor for another application but found it not powerful enough to actuate a large steam valve directly. He hit upon the idea of using an intermediate steam-operated cylinder to amplify the power and titled his patent application, "*Servomoteur, ou Moteur Asservi,*" adding the much-used word *servomotor* to our technical vocabulary.

2.2.8 James Maxwell (1868): Early Control Theory

Most of the control system work up to this point had been inventive rather than analytical in nature, as was then common in the other sciences. The great intellects up to the 18th century were usually involved in pure science, not to be bothered with the mundane things that tinkerers dealt with. Things began to change, however, in the 1800s as science started dealing with the real world and tackling some of the difficulties encountered in control system design, most notably in the area of stability.

In 1840 Sir George Airy, the astronomer, encountered a problem with his telescope's speed-regulating mechanism. In order to track the stars, the telescope must move at a fixed rate, but Airy's speed control exhibited an instability that he dubbed "hunting." He diverted his scholarly attention to this phenomenon and discovered that the amplitude of the hunting could be mathematically related to the natural frequency of the regulator mechanism and to the speed of the telescope. Whether this led designers of speed controls to better designs is not recorded, but it serves as the first record of an analytical approach to a practical control system problem.

James Clerk Maxwell was probably more successful in influencing contemporary control engineers when he published a paper entitled, "On Governors," in 1868, almost a hundred years after Mead's invention of the flyball millstone governor and Watt's subsequent adaptation of it to speed regulation. Maxwell had started his career as a mathematician and later turned to physics. Among his many contributions to science and engineering are works in astronomy, stress in materials, kinetic gas theory, and his famous equations of electromagnetic radiation. In his analysis of the governor, Maxwell formulated and solved the differential equations describing the flyball mechanism and explained the conditions that would lead to instability, or hunting. He further proposed the application of anticipation (rate feedback) as a means of achieving smoother performance of the device and the use of friction as a simple means to damp out oscillations. It is not recorded whether Maxwell's theories were any better received by engineers than Airy's. But it is certain that a fundamental concept, one that would eventually serve as the basis for much of control engineering, had been observed and publicized. It showed that mathematics could be a crucial aid to the inventors and developers of feedback devices. Soon after Maxwell's work, mathematicians such as E. J. Routh at Cambridge (1877), A. M. Liapunov in Russia (1892), and A. Hurwitz in Germany (1895) began analyzing linear and nonlinear systems and establishing important stability criteria that are still in use today by control systems people.

2.2.9 Lord Kelvin (1876): The Mechanical Analog Computer

The study of dynamic phenomena and their mathematical equations was still esoteric, far above the heads or patience of practical engineers. Many results of elegant analyses would lie dormant until almost a century later, when differential equations would become required training for engineers. The solution of these equations was often nearly impossible by hand calculation, so their lack of use is understandable.

One area of application requiring detailed solution of dynamic equations was tidal motion for marine navigation. As discussed above, Charles Babbage had interested the British Navy in the application of his machines to those calculations but had failed because of the lack of efficient mechanisms for his calculator concept.

James Thomson had experimented about the same time as Babbage with mechanization of mathematical integration, the procedure used for solving differential equations. He followed J. H. Hermann, who had invented the planimeter in 1814 in Germany. Maxwell had also invented such an instrument in 1855, and Thomson was inspired by this device and discussions with its inventor. In the 1860s, Thomson designed a ball-disc-roller mechanism for mechanical integration but dropped the idea because he saw no practical use for the device.

In 1876, James Thomson's younger brother, William, persuaded James to resurrect the mechanical integration device because he thought it could be used to solve the equations necessary to predict tidal motion. William Thomson, who became Lord Kelvin, was a mathematician and physicist who would make fundamental contributions to thermodynamics and electricity. William Thomson was familiar with Babbage's computing machines but thought them impractical. He wrote that his goal was to "substitute brass for brain in the great mechanical labor of calculating."

The two brothers then implemented that scheme and produced the first mechanical analog computer. Kelvin determined that the machine, limited to the single problem of computing tides, was ten times faster than a skilled arithmetician. Due to the limitations on the output torque of the integrators, only eight terms of the infinite Fourier series could be computed. Even so, the accuracy was approximately 2 percent. The brothers then tried to make a more flexible machine that could solve a variety of differential equations of higher order but ran into the same type of difficulty that had stymied Babbage—the current technology was not sufficient to implement the theory. It would be another 50 years before these concepts would be perfected by Vannevar Bush.

2.2.10   Elmer Sperry and Sons (1911): The Autopilot

Elmer Sperry, Sr. was a tinkerer and inventor from his boyhood in Cortland, New York. An interest in electricity led him, at age 15, to volunteer his help to Cornell University in building a large dynamo in 1876. In 1883 he moved to Chicago to begin a business, the first of many that he would start in his lifetime, ambitiously naming it The Sperry Electric Light, Motor and Car Brake Company. After 25 years of accumulating some 200 patents and founding and selling six more companies, Sperry became interested in the gyroscope, up until that time a scientific toy with no serious applications.

In 1908 Sperry, speaking before the Society of Naval Architects, predicted that the gyroscope would someday become a useful tool in industry and transportation. Thinking in terms of the navigation of ships, Sperry corresponded with the U.S. Navy about the possibilities of a gryocompass and stabilizer to reduce the rolling motion of ships. After founding the Sperry Gyroscope Company, Sperry completed the first gyrocompass in 1911 and successfully tested it at sea on the USS Delaware.

In 1913 Sperry began fitting gyroscopes into warships as well as merchant vessels. He showed that the rolling of a 26,000-ton ship could be virtually eliminated by a 180-ton gyro. Soon, however, he hit upon the idea of replacing this massive device with a more elegant solution, the use of a small gyro to sense vertical and active fins to provide the reaction torques.

Sperry's three sons—Edward, Lawrence, and Elmer, Jr.—would soon follow in their father's path, joining him in the development of many of his ideas. Lawrence became deeply interested in 1911 in designed and flying his own plane. His father had been working since 1909 on adapting gyroscopes to the problem of stabilizing the shaky aircraft of that era and had become acquainted with Glenn Curtiss.

Curtiss was building military planes in Hammondsport, New York, and shared the elder Sperry's view that the gyroscope could be used to maintain automatic level flight by a connection to the control surfaces. Lawrence went to San Diego late in 1913 to work with Curtiss on the idea, and the gyrostabilizer was ready for testing by the summer of 1914. A Curtiss type F flying boat was equipped with the device and demonstrated to the French War Department, which had offered a prize of 50,000 francs for a stable airplane.

In addition to winning this prize, Lawrence was awarded the *Concours de la Sécurité en Aeroplane*. For his part, Elmer, Sr. won the First Prize in 1914 from the Aero Club of France as well as the Collier Trophy. Nine years later Lawrence died when the aircraft he was flying went down in the English Channel.

In 1915 Edward Sperry graduated from Cornell University and joined his father's company to develop ship stabilizers. His younger brother, Elmer, Jr., followed suit two years later and was instrumental in the design of automatic ship steering systems, many of which are still in operation. Elmer, Jr. also played a key role in the development of flight instruments based on the gyroscope. His contributions to Jimmy Doolittle's blind flying experiments won him a medal from the Franklin Institute in 1936.

2.2.11   Vannevar Bush (1927): The Analog Computer

At the Massachusetts Institute of Technology (MIT), Vannevar Bush began a program in 1927 to build a general-purpose mechanical analog computer for solving the differential equations of importance in electrical network analysis. The machines produced by Bush and his colleagues over the next 15 years at MIT were modern versions of the earlier ideas of Lord Kelvin. Using electric motors and an ingenious torque amplification device invented by C. W. Niemann from Bethlehem Steel, Bush's team first built a model similar to a large erector set. Interconnected gears, shafts, motors, Thomson integrators, mechanized function tables, and a mechanical multiplier formed this machine, which could solve up to a sixth-order nonlinear differential equation. By 1942, the newest version of these machines, called differential analyzers, used punched paper tape to implement the interconnections among the electrical elements. The paper tape reduced the time required to program the machine so that a new problem could be set up in five minutes rather than the two days typically required to assemble shaft connectors. This machine played a major role in the computation of artillery tables during World War II.

Digital computing was emerging around this same time in several locations. At the International Business Machines Company, Wallace J. Ekert was forming and promoting ideas that would revolutionize the direction of his company. George Stibnitz at Bell Laboratories developed an electromechanical machine called the Relay Interpolator. Professor Howard Aiken at Harvard led an effort to design a relay-based machine that started in 1939 and accelerated during the war years. Called Mark I, it was completed in 1944 and was used for ballistics calculations. Weighing five tons, the machine had a memory of seventy-two 23-digit words and could multiply two numbers in 4.5 seconds. In Germany, Konrad Zuse began developing a relay computer in 1936, but perhaps the most interesting story of automatic computations was taking place in England, where code breakers were busy trying to read the Germans' Enigma enciphered radio communications.

2.2.12 Alan Turing (1936): Modern Digital Computers

Alan Turing, a mathematician from King's College in Britain, won a scholarship to study at Princeton in 1936. While there, he became interested in the algebraic specialty called Group Theory and published a paper entitled "On Computable Numbers." In this paper, Turing envisioned mechanization of computing by a sequential process on a machine that came to be known as a Universal Turing Machine.

Although Turing built a relay multiplier as a basic element of the machine, his idea was then dismissed due to the emergence of war in Europe. Meanwhile the Polish Secret Service, also preparing for the war, had started employing mathematicians to break the secret of the Germans' moving rotor cipher machine, known as the Enigma. The invasion of Poland cut the work short, but the Poles sent their key people and machinery to France and later to England to continue working. The British quickly picked up the Poles' ideas and built a more advanced version of their machine, called a Bombe. The Bombe could mechanically run through the many possible cipher combinations that could be produced by an Enigma, but it was mechanical and not very fast in light of the large number of possibilities. The task was simplified by Turing, who worked out mathematical shortcuts, greatly reducing the number of operations and allowing quicker entry into German communications.

In 1942, the Germans added another rotor to their Enigma machines, putting the Bombes out of business overnight. An improved vacuum tube version of the Bombe that made it more computer-like had been under development, and its completion was now hastened. The machine, called Colossus, produced the desired results from 1943 until the end of the war. After the war, Turing became involved in the development of Britain's first general-purpose digital computer, the ACE, the most advanced machine of its time.

2.2.13 Black, Nyquist, and Bode (1930s): Theoretical Foundations

The advent of the telephone and radio led to contributions to control theory by researchers in the new technology of electronics. Amplifiers used in communications made use of the feedback principle. Positive, or regenerative, feedback was first applied by Edwin Armstrong in 1912 to greatly increase the gain of a vacuum tube amplifier. The regenerative radio receiver was one of the most popular of the time, and Lee DeForest also claimed authorship. DeForest eventually won a long legal battle that wound up in the Supreme Court.

In 1923, H.S. Black of Bell Laboratories applied the principle of negative feedback to problems of long-distance telephone repeater amplifiers. Feedback allowed the production of constant gain devices despite wide variation in the values of the individual electronic components. Harry Nyquist and J. Bode, also from Bell, published works in 1932 and 1945, respectively, on stability analysis of feedback systems using frequency response techniques. These methods, now considered classical control technology, predict the total closed-loop performance of a system by considering the frequency response characteristics of the individual system components, whether they be amplifiers, servomotors, airplanes, or whatever. These powerful techniques have been used for many years on a wide variety of single-input, single-output problems and have been extended to multivariable applications.

2.2.14 The Modern Era of Theoretical Developments

Wartime pressures accelerated developments of control systems of all types, many in the area of gunfire control. During this time, servomechanism theory was applied to human tracking tasks in the United States by workers at Columbia and Iowa State. In England, A. Tustin was a pioneer in the modeling of the human operator's dynamic response characteristics as they applied to gunnery. Data smoothing and prediction became an area of research, involving people such as Norbert Wiener and Claude Shannon.

After the war, Wiener collected his thoughts in his famous book *Cybernetics: Control and Communications in the Animal and the Machine*. The word cybernetics, which has come to be synonymous with controls, comes from the Greek word for steersman. In this work, Wiener developed the theory of optimum nonlinear filtering and also emphasized the interdisciplinary nature of his theories, which have application to computers, neuromuscular response, politics, and economics. Interestingly, the German biologist Ludwig von Bertalanffy published the work that established him as the father of General Systems Theory at about the same time.

One of Wiener's students, Claude Shannon, published *The Mathematical Theory of Communication*, his now famous development of information theory, in 1948. His theory of the statistical nature of information and its quantization into a digital form has broad application to circuits, controls, computers, communications, biology, psychology, and semantics. Shannon's co-worker at Bell Labs, John Tukey, coined the word *bit* (for binary digit) to represent the smallest quantum of information in a message.

The war also saw the beginning of a new area of research in optimal control. Rufus Oldenburger at the Woodward Governor Company designed a constant-speed propeller control that had an optimum transient response. The optimum in this case was defined as the simultaneous minimization of transient overshoot and settling time. In the systems that Oldenburger treated, the resulting optimal control is bang-bang and makes use of his so-called "absquare" feature in the switching function to obtain stability around the zero error condition.

In 1949, Charles Draper and Y. T. Li of MIT first published their work on self-optimizing controls. Going beyond the usual servo function, this type of control measured a higher level of objective function and adjusted the servo setpoints in order to search out the optimum overall performance of a system. This type of control has also been termed *adaptive*, and many published articles have argued the exact semantic classifications far beyond the point of usefulness.

Draper and Li implemented their scheme on a two-dimensional problem: the adjustment of spark advance and air-fuel ratio of internal combustion engines to achieve best fuel economy. They showed, by graphical analysis and experiment, that their optimizing control could find the peak of a surface (in their example, output horsepower plotted versus the two control inputs) by dithering the controls and measuring the effect on the objective variable, thus obtaining gradient data that was used to climb the performance hill automatically.

The year 1957 saw the publication of Richard Bellman's *Dynamic Programming*, which describes his method of solving multistage, or dynamic, optimization problems. The theory leads to an iterative search while working backward from the final state of the system. At about the same time, the Russian mathematician L. S. Pontryagin published his work on the Maximum Principle, an analytical approach to the optimization of the performance of general dynamic systems. This theory leads to an open-loop solution.

In 1960, Rudolph Kalman from Columbia University published his work on optimal control, dealing with linear, multivariable systems. The minimization of a quadratic cost function yields a linear feedback control law. In 1959, H. P. Whitaker at MIT described work done there on Model-Reference Adaptive Control, sparking a long series of additional research in this area. Applications of optimal and adaptive controls were limited initially to high-cost military and aerospace systems due to the complexity involved in implementing the algorithms. The continually lowering cost of computing with microelectronics, however, has begun to create a wider application base for these theories.

2.2.15   Historical Perspective

The development of systems and control engineering has followed a path similar to many other areas of technology. The first work was done in ancient times. Contributions by individuals dominated in the early part of the industrial revolution, when scholars began contributing a mathematical framework to the field. The space age inspired rapid and all-encompassing advances, especially in theoretical groundings. Now the computer age is enabling the implementation of many of the concepts heretofore developed only in theory.

The people who contributed to systems engineering and automatic control are a mixture of inventors, businessmen, mathematicians, tinkerers, scientists, and engineers. Because the field is rich in concepts, techniques, and devices, each has been able to contribute something of lasting value. The contributors of the future will be of the same diversity, except that they will come more from the social and behavioral sciences, the life sciences, and the business community.

Inclusion of these disciplines implies more integration of people in the concepts and framework of systems engineering and automatic control. This integration is best explained by reviewing in broad terms the role of people in industrial endeavor. From ancient times to rather recent times, people were divided into organizers and workers. The role of the worker has oscillated from craftsman to drone to automaton to someone who had to have monetary incentive alone to work efficiently.

The most exciting concept being implemented in various modern societies is that each person in an organization has responsibilities and potential for creative action and that the opportunity to innovate and contribute is a fundamental incentive for improving the output and environment of the group. This concept has been implemented successfully in many Japanese companies as well as in the United States. It means that control systems involving people can and probably should be adaptive. Groups where each member has responsibilities as part of a system will tend to seek a goal and perform better than those rigidly structured systems where each action is prescribed in advance. This is a fundamental characteristic of most biological systems able to adapt their responses to the environment. They adapt in short time frames individually and evolve over longer ones.

The basic concepts of systems engineering, however, are able to encompass systems where people are an unpredictable part of the response and the adaptation is inherent. One new concept that has not been explored enough is that the system itself seeks its own goal as well. All of these concepts can be explored, simulated, discussed, and tried in practice. The only sure concept is that there are richer fields and more complex systems yet to be tackled and there always will be.

2.3   SYSTEMS ENGINEERING AND CONTROL DEFINITIONS

2.3.1   Overview

Specialists in automatic control systems have their own vocabulary—as do professionals in every field. The words in this language are rich in meaning, having evolved over more than 50 years. There is some ambiguity in the descriptive language of automatic control. This is not unusual since no language is absolutely clear, including mathematics and logic, as was shown by Kurt Gödel.

The descriptors for this field will be categorized into three general sets. The first set of terms defines concepts fundamental to the understanding of control and systems engineering. The second set of terms denotes elements of a control system. These describe characterics that are an essential part of every real process under control. The third set of terms describes those tools that allow the performance of a control system to be understood and changed prior to its actual construction.

The relationship among these terms is shown in Fig. 2.1. The circular aspect of this figure indicates the iterative nature of control system design. The basic concepts of automatic control revolve around

1. The desire to bring some process under control
2. An objective for its behavior

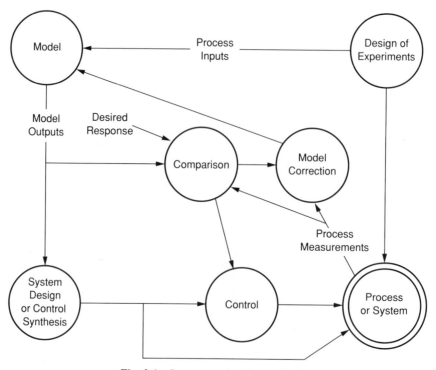

Fig. 2.1   Systems engineering synthesis.

3. Some means to change the system action
4. A way to measure the system state
5. An analysis capability or behavioral history to predict the response of the system
6. The synthesis of a control to compensate for undesirable response and bring the output to the desired value.

The basic elements of control systems are illustrated in Fig. 2.2. They are generic in that every control system will have at least these elements. More complex systems with multiple loops and interacting controls are combinations of the basic elements defined below. The elements derive from the goal of synthesizing a feedback loop to control a physical process to a desired output.

The tools of automatic control are based on the concept of a mathematical model representing the behavior of a process to be controlled. The dynamic behavior of the process is important, so differential equations are common tools of control engineers. Analysis of the process by manipulation of its mathematical description is at the heart of control system and systems engineering. Today the process and controlled system and almost certainly simulated by computers.

2.3.2 Basic Concepts

Block Diagrams. A schematic depicting the major elements in a control system and how they are connected. Figure 2.2 is an example of a block diagram. A block diagram illustrates power and information flow as well as interconnections. In their strictest form, the mathematical relationships between inputs and outputs are given in the blocks of the diagram.

Compensation. The filter or device (and the accompanying design procedure) that compensates for one or more poor response areas in the frequency domain of a process to be controlled.

Controller. The name given to that element of the control system that operates on the comparison of the desired and actual output of the system and provides inputs to the process actuator to bring the output closer to the desired response. The controller incorporates the *intelligence* of the control system. Synthesis in a control system is primarily concerned with defining the controller.

Dynamic System. A process that requires differential equations for its model with time as an independent variable. A dynamic system has energy storage which is out of phase with its motion. In order to synthesize a control, then, information on the time relationship of the energy storages and the motions must be understood. This knowledge is embodied in the solution of time-dependent differential equations.

Disturbance. An influence on a process or control that is generally not predictable except statistically. A control system compensates for such disturbances by either measuring them directly or detecting their influence on the output of the system and countering their effects through the controller and actuator.

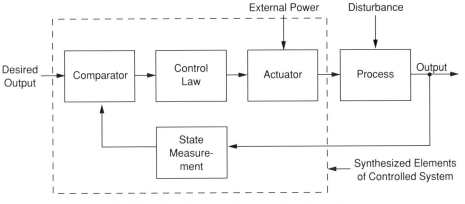

Fig. 2.2 Control system elements and configuration.

Error.   The difference between the desired value of an output of a controlled system and the actual value. The resulting signal is most often used as an input to the controller, whose output is used to correct the error.

Feedback.   Measuring the process output and transmitting back to the originating input location the results of a controller and acutator and the disturbances changing the behavior. This is often described as "closing the loop." An open-loop system is one where there is no feedback. Negative feedback is the common type; positive feedback is used occasionally for increasing the gain. The feedback signal is basic to automatic control, and the accuracy of the required measurement transduction is the limiting factor in the accuracy of an automatic control system.

Filter.   The electric, fluidic, or mechanical circuit that selectively reduces or amplifies certain frequency ranges. It is used to reduce noise, compensate for poor response of a process, or smooth the response of a system. It is usually part of the controller or feedback loop.

Gain.   The multiplier of the error in a controller to produce the required unit of actuation per unit change in the error. High gain, then, denotes a quick-responding system.

Inputs.   The variables that are the causal factors in changing a system's outputs. Inputs vary with time in the same frequency range as the system response.

Linear and Nonlinear Systems.   Systems described by linear and nonlinear differential equations, respectively. Linear response is characterized by a proportional relationship between inputs and outputs. There is a whole body of control theory developed for linear systems, which allows a methodical and complete synthesizing procedure for control system development. Nonlinear systems are all of those systems that are not linear. Natural physical processes are almost all nonlinear. Fortunately, their response in the neighborhood of an equilibrium state can usually be described adequately by a linear system of equations.

Measurement.   The determination of the value of variables in a control system. This is generally accomplished by transducers that transform physical variables to signals that can be monitored or read. Measurement of the output and other variables in a controlled system is central to its accurate performance. Measurements may be either continuous or discrete in time.

Model.   The representation of a process or system by a set of mathematical equations or a smaller-scale physical model. This depiction predicts the process or system performance in its more important characteristics. It is used to simulate the process or system so that a controller or other means of varying the performance of the system can be synthesized with less cost and risk than by experimenting iteratively with the actual process. A model needs to be close to but not necessarily equivalent to an actual process. There are always many more variables than can be reasonable modeled, and even those that can be modeled are accurately described over a limited range and by simple interactions with other variables.

On-Off Control.   A controller that turns the actuator full on or full off (or full positive or negative). It is the common control system in heating, ventilating, and air-conditioning systems, where the heating or cooling source is either on or off. The controller continuously monitors the error signal and determines when the actuation should be off or on. On-off systems are sometimes called *bang-bang* control systems.

Outputs.   The variables defined as those of primary interest to be measured and controlled. These variables are defined as part of the development of the model. Experiments with the actual system will indicate which of the many variables are important and essential in characterizing the system's response.

PID Control.   Proportional plus integral plus derivative control. This term denotes a control common to the process industry. In fact, hardware implementing PID control has been available for many years. Measurement signals and set points are inputs to the PID control. The output to the actuator is the sum of a proportional constant times the error, an integral (or reset) constant times the integration of the error signal, and a rate constant times the derivative of the error. Proportional control is the most used controller. The error will not become zero in a proportional controller; if it were zero, there would not be any signal to the actuator to hold the process output to the desired value. Integral control, however, can yield output to the actuator with zero error. Integral control adds another energy storage element, however, and will tend to destabilize the overall system. Derivative control embodies a prediction of the output variation and tends to stabilize the system unless there are high-frequency noise sources.

Process Control.   Automatic control as applied in the process industries. These are the chemical, metal, glass, paper, and other basic processing industries. They are characterized by the conversion of raw materials to basic products that become the material for industries that produce items more closely related to consumer products. The control problems and solutions in these industries have some intrinsic commonality. Flows, temperatures, chemical reactions, pressure, and chemical and material composition are common characterizing variables. Massive equipment and large capital investments are common. Automatic control is essential to lowering risk and producing consistently high quality. As a result, automatic control in these industries has a long history. Computer control was first widely used in these industries.

Servomechanism.   A power-amplifying element of a control system derived from the French term meaning servant. It is a general concept describing the class of automatic controls that embody an intermediate power amplifier. The controller and comparator are often incorporated in servomechanism devices.

Set Point.   A process industry term denoting the desired output value for a controlled system. The set point temperature in a chemical reactor is the value the controller and actuator would attempt to maintain in the reactor.

Simulation.   The procedure of representing an actual process and hypothesized control system by mathematical or scale physical models and iteratively solving the equations or running the models to characterize the behavior of the actual system from the model. Simulation implies digital or analog computers that embody the solution of the differential equations representing the process to be controlled and its control elements.

Stability.   A system characteristic such that for small disturbances or inputs, the output will remain close to its initial value. Unstable systems will not remain in equilibrium and will destroy themselves or move to a new state that is unacceptable. Much of control theory and simulation is concerned with stability since feedback itself can be a destabilizing factor.

Statistical Process Control.   A theory and procedure to bring discrete manufacturing processes under control. The procedure relies on a mixture of human and machine interaction with data taken from the controlled process output. The statistical part of the control is the sampling of a relatively small number of products of the process and inferring the behavior of the entire process by averaging sample results. Feedback is performed by an operator relying on the control charts developed from the statistical data. *Process capability* defines the ability of a process to produce products within specifications. *In control* defines a process that is capable and is being controlled so that the output meets the requirements. This theory was developed and codified in the United States by Deming but was first implemented successfully in Japan.

Supervisory Control.   Control of the set points in a multivariable, interacting control system. It implies change in the set points to optimize the system configuration, assuming that the dynamic response is handled by single controllers.

Synthesis.   The steps leading to development of a controlled system primarily in choosing the measurements and defining the controller. Modeling and simulation are one step in synthesis. The term was first used in the electronics industry, where a control, or filtering, circuit was synthesized to accomplish some desired objective.

Transfer Functions.   The key mathematical elements in a block diagram denoting the transformation of input to output for that block. Transfer functions are usually Laplace or Fourier transforms of linear differential equations relating inputs and outputs.

2.3.3   Advanced Concepts

Adaptive Control.   A controller that can change its character to adapt to functional changes in the process being controlled. It was first developed for prime mover controls, where the response characteristics of the engine varied significantly with load and speed. Changes in the controller were needed to compensate for varying engine characteristics. The difficult part of implementing the concept has always been identifying the change in the process while the engine is in operations. This is accomplished by adding an additional small input to the system and measuring the correspondingly small response of the system to infer the change.

Computer Control.   A control system where a digital computer is the primary means of control and communication. Computer control was first used widely in the process industries, where there are many processes to be controlled, hundreds of measurements, and the need for centralized setpoint control and measurement communication. It is now widely used in most modern control systems.

Cybernetics. A term first coined by Norbert Wiener to denote intelligent control systems, where human reasoning might be embodied in the system.

Discrete Systems. Systems where sampling theory is used to model the process and synthesize the control. It is often used to describe computer control systems where analog signals must be discretized to work with a computer and the resulting actuation commands converted into continuous variables by digital-to-analog conversion.

Distributed Systems. Systems where partial differential equations are necessary to describe the behavior and response, most frequently where spatially independent variables are important in addition to the time variable. Examples are acoustics and traveling waves, heat transfer and chemical reacting systems, and geological systems. Partial differential equations are often approximated by ordinary differential ones. This is entirely appropriate since all systems are distributed in nature.

Identification. Denotes the many methods and procedures for determining the differential equations or models for systems from measurements of inputs and outputs. It also denotes the actual on-line procedure of determining the equations or models. Much effort has been expended to develop identification tools since the knowledge of a system model is essential to the development of a control system.

Interacting Controls. A control system where there are two or more outputs and corresponding inputs that are interrelated. The controller then must compute the inputs to the actuators that account for this interaction and be able to control each process output separately.

Large-Scale Systems. Systems with a very large number of interrelated variables and great complexity. The telephone system and the economy are examples. Theories for simulating and controlling such systems are incomplete.

Man-Machine Systems. Systems where a person is an integral part of the control system, either as a controller or measurement transducer. Examples include airplanes, where the pilot is in the control loop as the primary controller; statistical process control, where the person takes measurements, computes the controls, and implements the actuation change; and most economic systems, where decisions by experts based on economic trends actuate change.

Multivariate Systems. Systems with two or more inputs and outputs that are interrelated.

Optimal Control. Denotes that the controller has been synthesized to minimize (or maximize) some performance measure, usually a time integration of a function of the error signals. This concept has been at the heart of the advanced work in the control and systems engineering field for almost 40 years.

Performance Criteria. A function whose value describes the overall optimality of a controlled system. It is usually a time integration of a linear combination of squares of the differences between the desired values for system state variables and their actual values.

Robustness. The ability of a controlled system to operate after the loss of one or more measurement or actuator.

State Estimation. The determination of the *state* of a process from differential equations relating the inputs and outputs and measurements of portions of the inputs and outputs. Kalman was the first to develop the theory to its full potential. It has always been an important concept in stochastic control and filtering.

State Variables. The variables equivalent to the initial conditions for differential equations. This variable set, if known at any point in time, is sufficient (with the describing equations) to predict the future behavior of the system. State variables can be used as feedback variables to optimize linear systems. State estimation is the technique whereby state variables are estimated from those variables that can actually be measured.

Stochastic Control. Control derived from a statistical model of the process. It is also the body of theory associated with the synthesis of control systems using statistical and stochastic techniques.

2.3.4 Control System Elements

Actuator. The element that implements the controller signal. It receives its input from the controller, and its output goes to the process to be controlled. Its power supply will determine the system capability to actuate the process.

Comparator. The element in a control system that compares the desired output with the measured output. It is often a subtracting element with a gain.

Control Computer. The computer, usually digital, that is the controller of a process. In the process industries, the control computer is the central controller, which receives all measured outputs, computes all control outputs, and monitors the system.

Instrumentation. The set of transducers, amplifiers, wiring, and configuration for the measurements in a system. It implies the analysis of errors, noise effects, signal levels, accuracy, and a knowledge of the various transduction elements and available hardware for measurements.

Interfaces. Impedance matching devices that are placed in the signal paths of control and other systems. For example, the analog-to-digital interface for a computer conditions the analog signals and discretizes them for digital processing.

Power Supply. Device or devices that supply controlled power to the actuator in a control system. It is most often electrical power for a motor, pneumatic pressure for a valve, or hydraulic pressure for a motor or linear actuator. Rocket engine gases are more exotic examples.

Process. The physical device or system whose output is to be controlled. One of the strengths of systems engineering is that vastly different processes can be approached with a standard set of concepts and tools.

Transducers. Devices that transform one physical variable into another. Examples include thermocouples for temperature, moving coils and quartz crystals for pressure and motion, magnetic fields for motion, and heated wires for flow.

2.3.5 Control System Tools

Analog and Digital Computers. Analog computers are interconnected electronic components able to simulate the solution of differential equations. They were the primary means of simulating control systems for over 25 years. Digital computers have been used in the systems field since their inception, primarily for system simulation. Digital computers were always used when there were many interrelated variables with complex algebraic relationships. In recent years, they have become predominant for all system simulation because of their increased speed and capability, falling prices for hardware, and expanded software.

Bode Diagrams. Frequency domain approximation techniques whereby straight lines in a log-log plot replace actual frequency plots. These approximations are quickly drawn from the system transfer function. The term *Bode analysis* implies the whole body of technique and theory associated with stability and synthesis on a frequency domain plot of amplitude and phase versus frequency.

Decision Theory. The whole body of theory associated with alternative decision paths and the risks associated with them. This theory is often used in economic systems, where costs and benefits are the decision criteria.

Design of Experiments. The technique to structure an experimental approach to identify models of multivariate processes. The theory is often associated with statistical process control. It embodies a sequential procedure to identify variables and relate them with the minimum amount of data.

Differential Equations. The primary mathematical tool for systems analysis. Linear, ordinary differential equations with time as the independent variable have been the main focus of theoretical development in the field.

Frequency Domain. The analysis domain for most electrical circuit control systems and many mechanical systems. The term is derived from the Fourier transform of linear differential equations. It implies the whole body of frequency domain analysis techniques developed in electronics and electric circuit theory.

Game Theory. A formal approach to competing systems analysis assuming each is intelligent and independent, but receiving information from the behavior of the other.

Identification Theory. That body of knowledge which aids in the determination of system or process models from measurements of system inputs and outputs. The theories encompass parameter determination from frequency domain data to differential equation synthesis.

Information Theory. That body of knowledge associated with signal transmission along noisy channels. Sampling theorems are part of this body of knowledge. Examples include the fact that the minimum sampling period is half the period of the highest frequency of interest in a signal. Much of the theory has been developed for signal radio and telephone transmission.

Laplace Transforms. The theory of mapping differential equations to the Laplace domain so that algebraic techniques may be used for manipulation. This has been the mainstay of control system analysis tools for 50 years.

Multivariate Analysis. Techniques to design and synthesize controllers for systems with multiple interacting variables. Matrices and vectors are a basic part of such tools.

Numerical Methods. Techniques to convert continuous variables to discrete values, which can be manipulated by digital computers. The solutions of differential equations by numerical means has been a major area of concentration in numerical methods.

Nyquist Criterion. A stability criterion, developed by Nyquist of Bell Telephone Laboratories in the 1930s, relying on Laplace transforms and complex variable theory. It has been widely used in automatic control throughout its development.

Optimization Theory. The body of theory used to synthesize a controller that minimizes a performance criterion. The calculus of variations, Pontryagin's Maximum Principle, Bellman's dynamic programming, and the Kalman/Wiener linear optimization theory are examples of techniques in wide use at one time or another.

Queuing Theory. The theory associated with staging processes where the optimum configurations and waiting times are determined. Its is used in manufacturing and service industries to control inventories and work in process as well.

Project Control. The body of theory associated with project scheduling, costing, manning, and supervision. It includes Pert analysis, whereby the critical path is determined from a formal listing of the individual tasks in a project and their dependence on other tasks.

Risk Assessment. A formal approach to evaluating risk among alternative decisions. Probabilistic estimates of the outcome of individual decisions are used to construct an overall assessment.

Stochastic Theory. Theory that combines probability theory with variations over time. Thus, differential equations with solutions that represent expected rather than actual values are central to stochastic theory.

Z-Transforms. The transforms analogous to the Laplace transform for discrete systems. They are useful for analyzing digital computer control systems. They have been largely replaced by matrix discrete analysis techniques.

BIBLIOGRAPHY

Cannon, R. H., *Dynamics of Physical Systems*, McGraw-Hill, New York, 1967.

Gödel, K., *On Formally Undecidable Propositions*, Basic Books, New York, 1962.

Goldstine, H. H., *The Computer from Pascal to Von Neumann*, Princeton University Press, Princeton, 1972.

Hodges, A., *Alan Turing: The Enigma*, Simon and Schuster, New York, 1983.

Karnopp, D., and Rosenberg, R., *System Dynamics: A Unified Approach*, Wiley, New York, 1975.

Mayr, O., *The Origins of Feedback Control*, The Colonial Press, Cambridge, Massachusetts, 1970.

Nassau County Historical Journal, Vol. XXI, Fall, 1960.

Potter, J. H., *Handbook of the Engineering Sciences*, Vols. 1 and 2, Van Nostrand, Princeton, 1967.

Strandh, S., *A History of the Machine*, A and W Publishers, New York, 1979.

CHAPTER 3

DYNAMIC SYSTEMS ANALYSIS

DAVID N. WORMLEY

Department of Mechanical Engineering
Massachusetts Institute of Technology
Cambridge, Massachusetts

3.1 INTRODUCTION

Dynamic systems analysis is central to the design of instrumentation and control systems that interact with physical systems. An understanding of the time histories, the characteristic times, and the amplitude, phase, and frequency content of critical system variables is key to the correct specifications of instruments to record performance and to the design of control systems to achieve desired performance specifications. The elements of a dynamic system analysis are described in the following sections and include

- Formulation of a dynamic system model
- Reduction of the model to a mathematical description
- Determination of the model's critical features in terms of time or frequency domain characteristics

3.2 PHYSICAL SYSTEM DESCRIPTIONS

3.2.1 General Concepts

Definitions

A general system is sketched in Fig. 3.1 with a set of input variables, output variables, and state variables. The selection of these variables is one of the most critical elements of system analysis. The definitions of a system and these variables are

  System—a collection of matter identified by a real or imaginary closed surface or boundary

  Input variables—the set of variables prescribed *independently* by the environment to the system

  Output variables—the set of system variables of interest

  State variables—the minimum set of variables required to fully define the system state

An illustration of these concepts is shown in Fig. 3.2, where a heating coil is used to increase the temperature of a cylindrical metal mass in a furnace. The example illustrates a number of

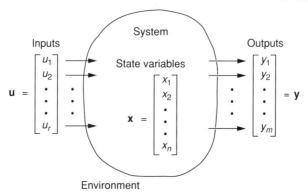

Fig. 3.1 General system.

independent system input variables, several possible output variables and one possible set of system state variables, assuming the coil has electrical energy storage and the metal mass and case have thermal energy storage.

Distributed and Lumped Element Models

The variables in dynamic system analyses are in general a function of time and spatial location, that is, an output variable $y = y(z, t)$, where z designates a spatial location and t is time. When detailed models are formulated in which variables are functions of both space and time, the model is designated a *distributed parameter system model* and is usually described by sets of partial

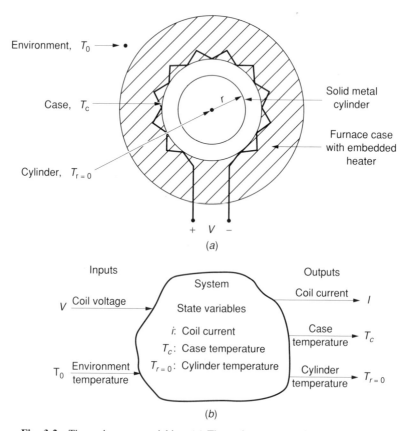

Fig. 3.2 Thermal system variables. (*a*) Thermal system. (*b*) System variables.

differential equations. When the approximation is made that variables at discrete points may be used to approximate spatially distributed variables, then a *lumped parameter system model* is adopted in which, for a defined lump in space, the variable represents the value of the quantity of interest for the entire lump. Ordinary differential equations with time as the independent variable are used to describe dynamic lumped parameter models. In Fig. 3.2, for example, if the metal mass temperature as a function of time and location from the mass surface, $T_r = T_r(z, t)$ is desired, a distributed parameter system model is required. If, however, only temperatures at discrete locations are desired, then the variables are functions of only time, and a lumped parameter system model may be employed. If the mass is considered as a set of concentric cylinders, each of which has a relatively uniform temperature, then a simple, lumped model may be formulated, and the system variables for each lump are functions only of time. Many lumps may be used to approximate a distributed system model, with guidelines cited in the literature for the selection of the appropriate number of lumps to represent heat exchangers and fluid processes as well as mechanical structural elements such as beams.[1-3]

3.2.2 Lumped Parameter Element Models

Lumped Elements

Formal methodologies have been developed for the formulation of lumped parameter system models.[4-6] Sets of lumped elements for mechanical translational, mechanical rotational, electrical, fluid, and thermal systems have been defined which encompass

- Active energy sources
- Passive energy storage elements
- Passive energy dissipation elements

The variables required to describe sets of lumped elements in each media are summarized in terms of *through* and *across* variables and *integrated through* and *across* variables in Table 3.1. Nonlinear constitutive equations describing lumped energy storage and dissipation elements, as well as the elemental equations describing ideal, linear lumped forms of the elements are summarized in Table 3.2. In across and through source elements, the across and through variables are respectively specified as independent functions of time.

Common nonlinear dissipation elements are illustrated in Fig. 3.3 while formulas for computation of selected energy storage elements are summarized in Table 3.3.

In addition to basic elements describing energy supply, storage, and dissipation within a single medium, transduction elements exist for energy transfer between media. Common devices in which energy is transferred include motors (electrical-mechanical), generators (mechanical-electrical), pumps (mechanical-fluid), and hydraulic rams. The equations for ideal, linear, lossless energy transfer between two media are summarized in Fig. 3.4. Two ideal forms of transduction elements are defined—(1) transformers in which the transduction is across variable to across variable and through variable to through variable and (2) gyrators in which the transformation is across variable to through variable and through variable to across variable.

System Equations

Lumped system models are described by a set of ordinary, nonlinear differential equations. The equations are derived from

TABLE 3.1 LUMPED ELEMENT THROUGH AND ACROSS VARIABLES

System	Across Variables	Integrated Across	Through Variables	Integrated Through
Mechanical translational	v : velocity	x : displacement	F : force	p : momentum
Mechanical rotational	ω: angular velocity	θ : angular displacement	T : torque	h : angular momentum
Electrical	V: voltage	Λ: flux linkage	I : current	q : charge
Fluid	P: pressure	Γ : pressure-momentum	q : volume flow rate	V : volume
Thermal	T: temperature	—	Q: heat flow rate	H: heat
General	V: across variable	X: integrated across	F: through variable	h : integrated through

TABLE 3.2   PURE AND IDEAL ELEMENTS

Element	Symbol	Constitutive Equation	Ideal Elemental Equation	Ideal Energy: E or Power: $\mathcal{P}$
Energy Storage				
Electrical Inductance		$\lambda = f(i)$	$V = L\dfrac{di}{dt}$	$E = \dfrac{1}{2}Li^2$
Translational Spring		$x = f(F)$	$V = \dfrac{1}{K}\dfrac{dF}{dt}$	$E = \dfrac{1}{2}\dfrac{F^2}{K}$
Rotational Spring		$\theta = f(T)$	$\omega = \dfrac{1}{K}\dfrac{dT}{dt}$	$E = \dfrac{1}{2}\dfrac{T^2}{K}$
Fluid Inertia		$\Gamma = f(g)$	$P = I\dfrac{dq}{dt}$	$E = \dfrac{1}{2}Iq^2$
Energy Storage				
Electrical Capacitance		$q = f(v)$	$i = C\dfrac{dV}{dt}$	$E = \dfrac{1}{2}CV^2$
Translational Mass		$p = f(v)$	$F = m\dfrac{dv}{dt}$	$E = \dfrac{1}{2}mv^2$
Rotational Mass		$h = f(\omega)$	$T = I\dfrac{d\omega}{dt}$	$E = \dfrac{1}{2}I\omega^2$
Fluid Capacitance		$v = f(P)$	$q = C\dfrac{dP}{dt}$	$E = \dfrac{1}{2}CP^2$
Thermal Capacitance		$H = f(T)$	$Q = C\dfrac{dT}{dt}$	$E = CT$

TABLE 3.2 *(Continued)*

Element	Symbol	Constitutive Equation	Ideal Elemental Equation	Ideal Energy: E or Power: $\mathcal{P}$
	Dissipators			
Electrical Resistance	$\begin{array}{c}+\;\;\;V\;\;\;-\\ \text{—}\!\!\bigwedge\!\!\bigwedge\!\!\text{—}\\ i\end{array}$	$i = f(v)$	$i = \dfrac{v}{R}$	$\mathcal{P} = \dfrac{v^2}{R}$
Translational Damper	$\begin{array}{c}+\;\;\;v\;\;\;-\\ \text{—}\boxed{\text{ }}\text{—}\\ \overrightarrow{F}\end{array}$	$F = f(v)$	$F = bv$	$\mathcal{P} = bv^2$
Rotational Damper	$\begin{array}{c}+\;\;\;\omega\;\;\;-\\ \text{—}\boxed{\text{ }}\text{—}\\ \overrightarrow{T}\end{array}$	$T = f(\omega)$	$T = b\omega$	$\mathcal{P} = b\omega^2$
Fluid Resistance	$\begin{array}{c}+\;\;\;P\;\;\;-\\ \text{—}\!\!\bigwedge\!\!\bigwedge\!\!\text{—}\\ \overrightarrow{q}\end{array}$	$q = f(P)$	$q = \dfrac{P}{R}$	$\mathcal{P} = \dfrac{P^2}{R}$
Thermal Resistance	$\begin{array}{c}+\;\;\;T\;\;\;-\\ \text{—}\!\!\bigwedge\!\!\bigwedge\!\!\text{—}\\ \overrightarrow{Q}\end{array}$	$Q = f(T)$	$Q = \dfrac{T}{R}$	$\mathcal{P} = \dfrac{T}{R}$
	Sources			
Through Variable Source	$\text{—}\!\!\bigcirc\!\!\overrightarrow{F}\!\!\bigcirc\!\!\text{—}$	$F = f(t)$	$F = f(t)$	$\mathcal{P} = F \cdot v$
Across Variable Source	$\text{—}\!\!\bigcirc\!\!\overset{+}{\,v\,}\!\!\bigcirc\!\!\text{—}$	$v = f(t)$	$v = f(t)$	$\mathcal{P} = F \cdot v$

1. The set of equations describing each element
2. A set of equations describing the interconnection of the elements. These equations represent
   a. *Generalized continuity:* the summation of through variables at any junction of elements is zero.
   b. *Generalized compatibility:* the summation of across variable potentials around any closed loop of elements is zero.

For system models consisting of the basic elements, a set of n first-order, ordinary differential equations may be derived by combining the elemental equations with continuity and compatibility equations, where n represents the number of *independent* energy storage elements in the system and is the number of initial conditions which must be specified in the model. Models may contain energy storage elements which are dependent because of the system configuration. For example, if two capacitors are connected directly in parallel, one is independent, the other is dependent, and only one voltage may be selected as a state variable for the pair.

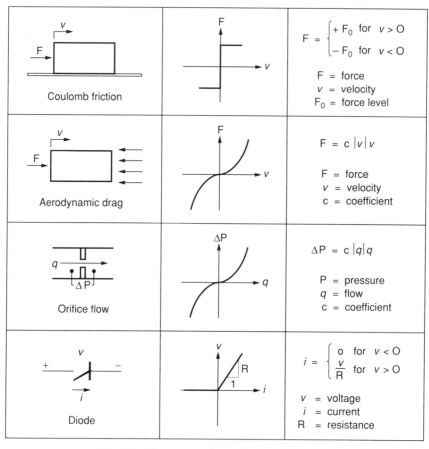

Fig. 3.3 Common nonlinear dissipation elements.

TABLE 3.3 ENERGY STORAGE ELEMENT PARAMETERS

Thermal Capacitance	Fluid Capacitance	Fluid Capacitance
Thermal Mass	Liquid-filled tank	Gas-filled tank
$C = C_P m$ m = Mass C_P = Specific Heat	$C = A/\rho g$ A = Tank Area ρ = Fluid Density g = Acceleration of Gravity	$C = V/B$ V = Tank Volume B = Bulk Modulus of Fluid
Spring Stiffness	**Rotary Inertia**	**Fluid Inertance**
Beam	Flywheel	Fluid flowing in pipe
$K = 3EI/l^3$ E = Elastic Modulus I = Moment of Inertia l = Beam Length	$I = mr^2/2$ m = Mass r = Radius	$I = \rho l/A$ A = Pipe Area l = Pipe Length ρ = Fluid Density

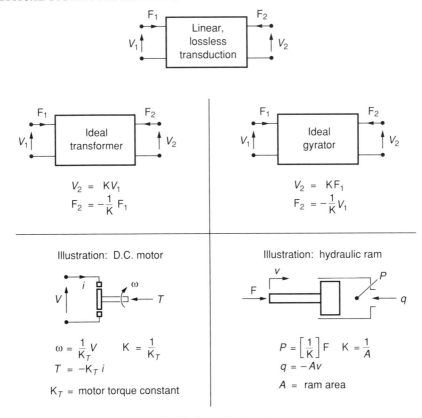

Fig. 3.4 Ideal transduction elements.

An example of the derivation of state equations for a system in which the pressure in a port with trapped air is measured to infer the pressure in a liquid stream is illustrated in Fig. 3.5. A linear fluid capacitance represents the trapped air compressibility, and a linear resistance and inertance represent the connecting tube. Two state equations are required to describe the system, which has two independent energy storage elements.

3.2.3 General Form of State Equations

The state equations describing a system model may be written for the general nonlinear case as n first-order nonlinear equations.

$$\dot{\mathbf{x}} = \mathbf{f}(\mathbf{x}, \mathbf{u}) \tag{3.1}$$

where $\mathbf{x}$ = is the n-dimensional state vector
   $\dot{\mathbf{x}}$ = is the time derivative of the state vector
   $\mathbf{u}$ = is the input vector containing inputs that are specified functions of time t
   $\mathbf{f}$ = is a set of n nonlinear functions

At an initial time t_0, the system initial state $\mathbf{x}_0$ must be specified.

$$\mathbf{x}_0 = \mathbf{x}(t_0) \tag{3.2}$$

With the specification of the input functions $\mathbf{u}(t)$, the solution to the n state equations may be determined as a function of time by numerical integration as described in Section 3.6.

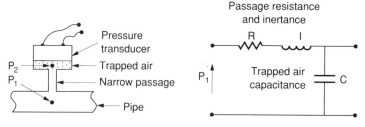

Fig. 3.5 State equation derivation.

(a) Elemental equations
$$\begin{cases} P_1 = P_s(t) \;:\; \text{Source} \\ P_I = I\dfrac{dq_I}{dt} \;:\; \text{Inertance} \end{cases}$$

Continuity: $q_r = q_L \qquad q_L = q_c$

(b) Elemental equations
$$\begin{cases} P_r = Rq_r \;:\; \text{Resistance} \\ q_c = C\dfrac{dP_c}{dt} \;:\; \text{Capacitance} \end{cases}$$

Compatability: $P_1 = P_r + P_I + P_c$

State equations
$$\begin{cases} \dfrac{dP_c}{dt} = \dfrac{1}{C}[q_I] \\ \dfrac{dq_I}{dt} = \dfrac{1}{I}[P_s - Rq_I - P_c] \end{cases}$$

Standard Form $\qquad \dot{\mathbf{x}} = \mathbf{Ax} + \mathbf{Bu}$

$$\begin{bmatrix} \dot{p}_c \\ \dot{q}_r \end{bmatrix} = \begin{bmatrix} 0 & 1/C \\ -1/I & -R/I \end{bmatrix} \cdot \begin{bmatrix} p_c \\ q_I \end{bmatrix} + \begin{bmatrix} 0 \\ 1/I \end{bmatrix} [p_s]$$

3.3   LINEARITY AND SUPERPOSITION

3.3.1   Linearization of Nonlinear Equations

The general form of the system equations is nonlinear. However, because analysis methods for linear systems are so advanced, the consideration of a linearized form of the full set of nonlinear equations is desired. Many techniques[7] exist for linearizing a nonlinear equation set with a number of common methods derived from Taylor series expansion of the state variable excursions $\Delta\mathbf{x}$ and input function excursions $\Delta\mathbf{u}$ from an equilibrium operating point $(\mathbf{x}_e, \mathbf{u}_e)$. Formally for

$$\mathbf{x} = \mathbf{x}_e + \Delta\mathbf{x} \tag{3.3}$$

$$\mathbf{u} = \mathbf{u}_e + \Delta\mathbf{u} \tag{3.4}$$

substitution into Eq. 3.1 yields

$$\dot{\mathbf{x}}_e + \Delta\dot{\mathbf{x}} = \mathbf{f}(\mathbf{x}_e, \mathbf{u}_e) + \left.\frac{\Delta\mathbf{f}}{\Delta\mathbf{x}}\right|_e \Delta\mathbf{x} + \left.\frac{\Delta\mathbf{f}}{\Delta\mathbf{u}}\right|_e \Delta\mathbf{u} + \text{ higher order terms} \tag{3.5}$$

Introducing the matrices **A** and **B** which are defined as

$$A = \begin{bmatrix} \dfrac{\Delta f_1}{\Delta x_1}\Big|_e & \cdots & \dfrac{\Delta f_1}{\Delta x_n}\Big|_e \\ \vdots & & \vdots \\ \dfrac{\Delta f_n}{\Delta x_1}\Big|_e & \cdots & \dfrac{\Delta f_n}{\Delta x_n}\Big|_e \end{bmatrix} \tag{3.6}$$

$$B = \begin{bmatrix} \dfrac{\Delta f_1}{\Delta u_1}\Big|_e & \cdots & \dfrac{\Delta f_1}{\Delta u_r}\Big|_e \\ \vdots & & \vdots \\ \dfrac{\Delta f_n}{\Delta u_1}\Big|_e & \cdots & \dfrac{\Delta f_n}{\Delta u_r}\Big|_e \end{bmatrix} \tag{3.7}$$

where it is assumed that the nonlinear functions are continuous so that each of the incremental quantities can be computed. The linear incremental equations are

$$\Delta \dot{x} = A \Delta x + B \Delta u \tag{3.8}$$

for incremental motion from the equilibrium

$$\dot{x}_e = f(x_e, u_e) \tag{3.9}$$

For systems which have been linearized or have been represented with all linear elements, the system state equations are expressed in the standard form

$$\dot{x} = Ax + Bu \tag{3.10}$$

where the incremental notation is omitted for convenience. The system outputs of interest may be expressed in terms of linear combinations of state variables and input variables as

$$y = Cx + Du \tag{3.11}$$

where C is a matrix relating the n state variables and D is a matrix relating the r input variables to the output vector y containing outputs $y_1, y_2, y_3 \ldots$.

In the most general case, the matrices A, B, C, and D may be time varying, in which case the system is designated a *linear time-varying system*. For a wide class of systems the matrices A, B, C, and D have constant parameter values, and the system is a *linear time-invariant* (LTI) system, where the linearized system parameters are constant.

For cases in which the relationship of a single output y to a single input u is of interest, the general Eqs. 3.10 and 3.11 may be reduced and combined algebraically or numerically (see Section 3.5) to form a single nth-order equation of the form

$$\frac{d^n y}{dt^n} + a_{n-1}\frac{d^{n-1}y}{dt^{n-1}} + \cdots + a_0 y = b_m \frac{d^m u}{dt^m} + b_{m-1}\frac{d^{m-1}u}{dt^{m-1}} + \cdots + b_0 u \tag{3.12}$$

where the number of input terms depends specifically upon the input selected and where the output contains derivatives of order n. The terms on the left-hand side of Eq. 3.12 represent the nth-order system characteristic equation.

3.3.2 Principle of Superposition

Linear systems such as described by Eqs. 3.10 and 3.11 satisfy the principle of superposition which states that the response of a linear system to any combination of inputs is identical to the response obtained by summing together the responses to each individual input. Formally, if y_i is the solution of the system equations for input u_i where all initial conditions are zero, then the total solution y_T to a general input u_T

$$u_T = \sum_{i=1}^{l} L_i[u_i] \tag{3.13}$$

where u_T is composed of l inputs and where L_i is a linear operator such as a constant, differentiation, or integration, may be written using the principle of superposition as

$$y_T = \sum_{i=1}^{l} L_i[y_i] \qquad (3.14)$$

where y_i is the solution for input u_i. For linear systems, once the solution for a single input is determined, the solution for any other input related by a linear operation may be determined by performing the operation on the output. For example, if the system response to a unit step input u_1 is determined as y_1, the system response to a step of value A is Ay_1, and the response to a unit ramp which is the integral of a unit step is $\int y_1 \, dt$, where all initial conditions are zero.

3.4 SYSTEM VARIABLE REPRESENTATION AND TRANSFORMATION

3.4.1 Input Variable Characterization

The input variables which excite systems may be classified for engineering analysis as

1. *Deterministic* in which the variable is a specified function of time, i.e., $u(t) = f(t)$, where $f(t)$ is prescribed. Within this general classification two subsets are common:
   a. *Periodic variables* in which the variable is periodic, repeating each time period T.
   b. *Transient variables* in which a nonperiodic time function is specified.
2. *Random* in which a set of time functions, an ensemble of time functions, may represent the variable. Within this classification two categories are of interest.
   a. *Stationary random processes* in which the characteristics of the process averages, such as the mean value and variance, computed at any point in time for a set of sample functions are time invariant. The correlation of a sample with itself or another function at some time from the sampled time is independent of the absolute sample time t and depends only on the time displacement. *Stationary, ergodic processes* have the additional property that a statistical average of any single time sample function such as the mean, variance or autocorrelation is identical to the statistical value across the set of ensemble samples.
   b. *Nonstationary random processes* in which statistical averages across a set of sample functions are a function of time t.

Most engineering analysis techniques have been developed for random variables which are generated by stationary, ergodic processes.[1] For example, the techniques used in spectrum analysis to characterize experimental data are based upon the assumption that the data are derived from is a stationary, ergodic process.

3.4.2 Singularity Functions

The singularity functions displayed in Fig. 3.6 are one class of transient functions commonly used to evaluate system characteristics. These functions are related by differentiation and integration and include ramp, step, and impulse functions where the unit impulse is a convenient mathematical function of infinite height, zero time duration, and unit area.

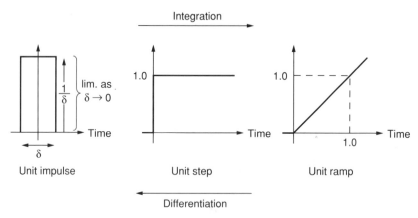

Fig. 3.6 Time domain singularity functions.

3.4.3 Fourier Series Representation of Periodic Functions

Sine-Cosine Fourier Series

Functions which are periodic over the time period T may be represented in terms of a Fourier series:

$$f(t) = a_0 + \sum_{n=1}^{\infty} a_n \cos n\omega_0 t + \sum_{n=1}^{\infty} b_n \sin n\omega_0 t \tag{3.15}$$

where: $\omega_0 = 2\pi/T$ = frequency in radians/s, and where the coefficients in the series are defined as

$$a_0 = \frac{1}{T} \int_0^T f(t) \, dt \tag{3.16}$$

which represents the time average value of the signal, and for $n = 1, 2, 3, \ldots,$

$$a_n = \frac{2}{T} \int_0^T f(t) \cos n\omega_0 t \, dt \tag{3.17}$$

$$b_n = \frac{2}{T} \int_0^T f(t) \sin n\omega_0 t \, dt \tag{3.18}$$

with a_n and b_n, respectively, representing the cosine and sine harmonic coefficients of the series. The practical use of the Fourier series results from the ability to represent many periodic signals with only a few harmonic components. An illustration of the representation of a square wave with a Fourier series is shown in Fig. 3.7 for one and three terms in the series.

Exponential Fourier Series

The sine-cosine Fourier series may be written in compact exponential form which relates to the Fourier transform:

$$f(t) = \sum_{n=-\infty}^{\infty} A_n e^{jn\omega_0 t} \tag{3.19}$$

where $j = \sqrt{-1}$ and the coefficients A_n are complex numbers defined for $n = 0, \pm 1, \pm 2, \ldots,$

$$A_n = \frac{1}{T} \int_0^T f(t) \, e^{-j\omega t} \, dt \tag{3.20}$$

with Euler's relationship as

$$e^{jn\omega_0 t} = \cos n\omega_0 t + j \sin n\omega_0 t \tag{3.21}$$

The complex amplitude contains both magnitude and phase information

$$A_n = |A_n| e^{-j\theta_n} \tag{3.22}$$

where θ_n = the phase angle of the nth harmonic and where the magnitude and phase are related to the sine-cosine series terms. For real time functions $f(t)$, the magnitude and phase are

$$|A_n| = |A_{-n}| = \frac{1}{2} \sqrt{a_n^2 + b_n^2} \tag{3.23}$$

$$\theta_n = -\theta_{-n} = \tan^{-1}\left(\frac{b_n}{a_n}\right) \tag{3.24}$$

3.4.4 Fourier Transforms

Transform methods facilitate the representation of variables and system descriptions in convenient forms for system analysis. Functions of time $f(t)$ which satisfy

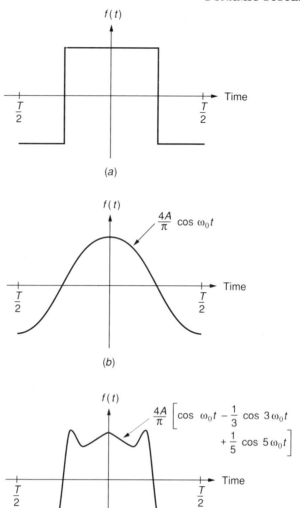

Fig. 3.7 (a) Fourier series for a square wave; (b) one-term series; (c) three-term series.

$$\int_{-\infty}^{\infty} |f(t)| \, dt < \infty \qquad (3.25)$$

have a Fourier transform. The condition of Eq. 3.25 requires that as $t \rightarrow \infty$, the time function $f(t) \rightarrow 0$ for the Fourier transform to exist. The Fourier transform is

$$F(\omega) = \frac{1}{2\pi} \int_{-\infty}^{\infty} f(t) \, e^{-j\omega t} \, dt \qquad (3.26)$$

and the inverse Fourier transform is

$$f(t) = \int_{-\infty}^{\infty} F(\omega) \, e^{j\omega t} \, d\omega \qquad (3.27)$$

Alternate forms of the Fourier transform pair exist in which the $1/2\pi$ term in Eq. 3.26 is removed from the forward transform and instead associated with the inverse transform of Eq. 3.27 or in which $1/\sqrt{2\pi}$ is associated with each of the transform pairs. For the convention adopted in Eq. 3.26 and 3.27, the Fourier transform in terms of frequency $f = \omega/2\pi$ is

$$F(f) = 2\pi F(\omega) \tag{3.28}$$

where f is frequency in Hertz. The Fourier transform of a time function is in general a complex number containing real and imaginary parts or equivalently magnitude and phase information.

3.4.5 Laplace Transforms

The Laplace transform converts a time variable into a variable in the complex domain. Letting the Laplace operator s be defined as

$$s = \sigma + j\omega \tag{3.29}$$

where σ is the real part of s and ω is the imaginary part of s, the Laplace transform $F(s)$ of a variable $f(t)$ is

$$F(s) = \int_0^\infty f(t) e^{-st} dt \tag{3.30}$$

and the inverse transform is

$$f(t) = \frac{1}{2\pi j} \int_{\sigma - j\infty}^{\sigma + j\infty} F(s) e^{st} ds \tag{3.31}$$

where it is required that the following condition be satisfied for existence of the transform:

$$\int_0^\infty |f(t)| e^{-\sigma_0 t} dt < \infty \tag{3.32}$$

for some positive value of σ_0. Table 3.4 summarizes pairs of Laplace transforms. An important transformation is the representation of the derivatives and integrals of variables. Formally these transformations require the evaluation of initial conditions; when initial conditions are zero, however, the following correspondence exists between the time and Laplace domains:

$$\frac{d}{dt}[\quad] \leftrightarrow s[\quad] \tag{3.33}$$

$$\int [\quad] dt \leftrightarrow \frac{1}{s}[\quad] \tag{3.34}$$

with the result that in general with zero initial conditions

$$\frac{d^n}{dt^n}[\quad] = s^n[\quad] \tag{3.35}$$

and a differential equation may be directly transformed into a form where s is effectively the differential operator.

3.4.6 Random Variable Characterization

The characterization of random variables which are stationary and ergodic is described in terms of their statistical properties. For a stationary, ergodic process, a single time sample has statistical properties identical to the statistical properties at any time t computed by averaging across all the sample functions in an ensemble. Basic properties may be related directly to the probability density of a random process. The probability that a time function $y_0(t)$ lies between specified limits at a given time t is

$$\text{PROB } [y < y_0 < y + dy] = p(y) \, dy \tag{3.36}$$

TABLE 3.4 LAPLACE TRANSFORMS

Time Domain: $f(t)$		Laplace Domain: $F(s)$
$u_I(t)$:	unit impulse	1
$u_s(t)$:	unit step	$\dfrac{1}{s}$
$u_r(t)$:	unit ramp	$\dfrac{1}{s^2}$
e^{-at}:	exponential	$\dfrac{1}{s + a}$
$\sin at$:	sine	$\dfrac{a}{s^2 + a^2}$
$\cos at$:	cosine	$\dfrac{s}{s^2 + a^2}$
$\dfrac{d^n}{dt^n}[f(t)]$:	differentiation	$sF(s) - s^{n-1}f(0) - S^{n-2}\dfrac{df(0)}{dt} \cdots \dfrac{d^{n-1}f(0)}{dt^{n-1}}$
$\int f(t)dt$:	integration	$\dfrac{F(s)}{s} + \dfrac{\displaystyle\int_{-\infty}^{0} f(t)dt}{s}$

where $p(y)$ is the probability density of the variable y. For an ensemble of time functions or a single time function, the probability density for a set of samples $p(y)dy$ is the fraction of the total samples which lie between y and $y + dy$. Two common probability density functions are shown in Fig. 3.8, the uniform density and the Gaussian density which is given by

$$p(y) = \frac{1}{\sqrt{2\pi}\,\sigma}\, e^{-(y-m)^2/2\sigma^2} \tag{3.37}$$

where m = mean of process and σ = standard deviation.

The statistical properties of a random time function are expressed as the statistical expectation of the variable $E[\ ]$:

Mean Value—$E[y]$—expectation of the mean value of y

$$E[y] = m = \lim_{T \to \infty}\left[\frac{1}{T}\int_0^T y(t)\,dt\right] = \int_{-\infty}^{\infty} yp(y)\,dy \tag{3.38}$$

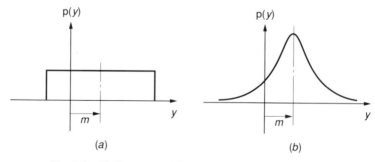

(a)   (b)

Fig. 3.8 Uniform (*a*) and Gaussian (*b*) probability densities.

Mean Square Value—$E[y^2]$—expectation of value of y^2

$$E[y^2] = \lim_{T \to \infty} \left[\frac{1}{T} \int_0^T y^2(t)\, dt \right] = \int_{-\infty}^{\infty} y^2 p(y)\, dy \tag{3.39}$$

Variance—$E[(y - m)^2]$—expectation of mean square of the deviation of y from its mean value, which may be written

$$\sigma^2 = E[(y - m)^2] = \lim_{T \to \infty} \left[\frac{1}{T} \int_0^T (y(t) - m)^2\, dt \right] \tag{3.40}$$

or

$$\sigma^2 = E[y^2] - m^2 \tag{3.41}$$

and for a function which has zero mean, the variance is equal to the mean square of the signal. The square root of the variance is the standard deviation.

Autocorrelation and Cross-Correlation. The correlation of the amplitude of a random variable with itself displaced by a time interval τ or with another random variable is central to system analyses and may be defined for the two variables $x(t)$ and $y(t)$ as

$$R_{xy}(\tau) = \lim_{T \to \infty} \left[\frac{1}{T} \int_0^T x(t)\, y(t + \tau)\, d\tau \right] \tag{3.42}$$

where $R_{xy}(\tau)$ is the cross-correlation between $x(t)$ and $y(t + \tau)$. If $x(t)$ is selected as $y(t)$, then the autocorrelation function is

$$R_{yy}(\tau) = \lim_{T \to \infty} \left[\frac{1}{T} \int_0^T y(t)\, y(t + \tau)\, d\tau \right] \tag{3.43}$$

For the cross-correlation function

$$R_{xy}(\tau) = R_{yx}(-\tau) \tag{3.44}$$

while the autocorrelation function is an even function with

$$R_{yy}(\tau) = R_{yy}(-\tau) \tag{3.45}$$

In addition at $\tau = 0$, the autocorrelation function is equal to the mean square of $y(t)$

$$R_{yy}(0) = \sigma_y^2 + m_y^2 \tag{3.46}$$

and as $\tau \to \infty$, the autocorrelation function approaches the mean value squared

$$R_{yy}(\infty) = m_y^2 \tag{3.47}$$

Covariance. The covariance C_{xy} of x and y is defined as

$$C_{xy} = \lim_{T \to \infty} \left[\frac{1}{T} \int_0^T (x(t) - m_x)(y(t + \tau) - m_y)\, d\tau \right] \tag{3.48}$$

and may be written as

$$C_{xy}(\tau) = R_{xy}(\tau) - m_x m_y \tag{3.49}$$

and for the case $x(t) = y(t)$

$$C_{yy}(\tau) = R_{yy}(\tau) - m_y^2 \tag{3.50}$$

Spectral Density

The spectral density of a random time function provides information concerning the mean square signal amplitude as a function of frequency. The development of spectrum analyzers[8,9] has greatly facilitated the analysis of random signals. Formally, spectral densities may be defined in terms of the Fourier transform of the cross- or autocorrelation functions.

Cross-Spectral Density (S_{xy}). To satisfy the condition of Eq. 3.25 for existence of the Fourier transform, the spectral density for zero mean, stationary, ergodic processes is defined as:

$$S_{xy}(\omega) = \frac{1}{2\pi} \int_{-\infty}^{\infty} R_{xy}(\tau) e^{-j\omega\tau} \, d\tau \tag{3.51}$$

The cross-spectral density is in general a complex number that contains magnitude and phase information relating $x(t)$ and $y(t + \tau)$ and has the property

$$S_{xy}(-\omega) = S_{yx}(\omega) = S_{yx}^*(-\omega) \tag{3.52}$$

where S^* is the complex conjugate of S.

Auto-Spectral Density (S_{yy}). The auto-spectral density or (power) spectral density of a variable is

$$S_{yy} = \frac{1}{2\pi} \int_{-\infty}^{\infty} R_{yy}(\tau) e^{-j\omega\tau} \, d\tau \tag{3.53}$$

where the spectral density is an even function which is real and contains only magnitude information:

$$S_{yy}(\omega) = S_{yy}(-\omega) \tag{3.54}$$

The integral of S_{yy} over all frequencies yields the mean square value of the variable y or the variance

$$\sigma^2 = \int_{-\infty}^{\infty} S_{yy}(\omega) \, d\omega = R_{yy}(0) \tag{3.55}$$

Spectral Density Functions. The spectral densities of derivatives of a stationary, ergodic random function are related. If the spectral density of y is $S_{yy}(\omega)$, then the spectral density of $\dot{y} = dy/dt$ is

$$S_{\dot{y}\dot{y}} = \omega^2 S_{yy} \tag{3.56}$$

and the spectral density of $\ddot{y} = d^2y/dt^2$ is

$$S_{\ddot{y}\ddot{y}} = \omega^4 S_{yy} \tag{3.57}$$

White Noise. A special case of spectral density is a signal which has no correlation except at $\tau = 0$; this signal, white noise, has a spectral density which is constant over all frequencies:

$$S_{yy}(\omega) = \text{CONSTANT} = S_0 \tag{3.58}$$

The autocorrelation function of white noise is an impulse:

$$R_{yy}(\tau) = \text{UNIT Impulse of area } 2\pi S_0 \tag{3.59}$$

Engineering Form of Spectral Density. The spectral density is formally defined by the Fourier transform over the interval $-\infty$ to $+\infty$ in terms of rad/s. This form is commonly referred to as the double-sided spectral density. For engineering use, the single-sided spectral density over the interval of 0 to ∞ is common and is usually expressed in terms of frequency f in hertz. The single-sided spectral density in hertz may be written as

$$G_{yy}(f) = 2(2\pi) S_{yy}(\omega) \tag{3.60}$$

and the single-sided cross-spectral density may be written

$$G_{xy}(f) = 2(2\pi) S_{xy} \tag{3.61}$$

The engineering forms $G_{xx}(f)$ and $G_{xy}(f)$ are usually computed by spectrum analyzers. It is noted that the mean square amplitude of a signal between frequencies f_1 and f_2 is

$$E(y^2)|_{f_2-f_1} = \int_{f_1}^{f_2} G_{yy}(f)\, df \tag{3.62}$$

and the spectral density is the mean square amplitude per unit of frequency at a given frequency.

Computation of Spectral Densities. The computation of cross- and auto-spectral densities for time functions is performed using fast Fourier transform (FFT) techniques. These techniques[8,9] have been directly incorporated in spectrum analyzers and in computer codes available for mainframe and personal computers such as the programs cited in Section 3.5.3.

3.5   TRANSFER FUNCTION REPRESENTATION OF LINEAR SYSTEMS

3.5.1   Definition

Single Input-Single Output System

A transfer function relating the input to the output of a single input–single output linear time-invariant system may be derived directly by Laplace transforming the nth-order differential equation describing the system, Eq. 3.12, to form the transfer function $G(s)$:

$$G(s) = \frac{s^m + b_{m-1}s^{m-1} + \cdots + b_0}{s^n + a_{n-1}s^{n-1} + \cdots + a_0} \tag{3.63}$$

where all initial conditions have been set to zero in the transformation. The transfer function denominator contains the system characteristic equation, and the numerator contains the terms operating on the input function. The value of m is typically less than n.

Multiple Input-Output Systems

The transfer functions for multiple input-output, linear time-invariant systems are obtained directly from the state equations. Formally the state equations (3.10) may be Laplace transformed, using the Laplace transform operator s, to obtain

$$s\mathbf{x} = \mathbf{A}\mathbf{x} + \mathbf{B}\mathbf{u} \tag{3.64}$$

then $\mathbf{x}$ may be extracted using matrix algebra and introducing the unity matrix $\mathbf{I}$ as:

$$\mathbf{x} = [s\mathbf{I} - \mathbf{A}]^{-1}\mathbf{B}\mathbf{u} \tag{3.65}$$

and finally the relationship between the system outputs and inputs may be written using Eq. 3.11 with $\mathbf{D} = 0$; that is, no direct transfers of the input to the outputs are considered

$$\mathbf{y} = \mathbf{G}(s)\mathbf{u} \tag{3.66}$$

where

$$\mathbf{G}(s) = \mathbf{C}[s\mathbf{I} - \mathbf{A}]^{-1}\mathbf{B} \tag{3.67}$$

and $\mathbf{G}(s)$ is the system transfer function matrix relating each input to each output. The form of $\mathbf{G}(s)$ is

$$\mathbf{G}(s) = \begin{bmatrix} G_{11}(s) & \cdots & G_{1r}(s) \\ \vdots & & \vdots \\ G_{q1}(s) & \cdots & G_{qr}(s) \end{bmatrix} \tag{3.68}$$

in which each $G_{ij}(s)$ is an individual transfer function relating the input u_j to the output y_i. If matrix relationships are employed, $\mathbf{G}(s)$ may also be expressed as:

$$\mathbf{G}(s) = \frac{\mathbf{C}\ \mathrm{ADJ}[s\mathbf{I} - \mathbf{A}]\mathbf{B}}{\mathrm{DET}[s\mathbf{I} - \mathbf{A}]} \tag{3.69}$$

where ADJ[ ] is the adjoint which is a matrix and DET[ ] is the determinant and is a polynominal of order n in s. Noting Eq. 3.69, it may be shown that every transfer function $G(s)$ contained in $G(s)$ is of the form:

$$G(s) = b_m \frac{s^m + b_{m-1}s^{m-1} + \cdots + b_0}{s^n + a_{n-1}s^{n-1} + \cdots + a_0} \tag{3.70}$$

3.5.2   System Transfer Function Synthesis from Operational Block Diagrams

System models may be formulated by combining individual transfer functions of system components. When system components are represented by linear, time-invariant models, operational block diagram algebra provides a convenient method of deriving a complete system model. For transfer functions of linear time invariant systems, the combination laws are summarized in Fig. 3.9 and illustrate that (1) a combination of parallel systems

$$y = G_1(s)u + G_2(s)u \tag{3.71}$$

is equivalent to

$$y = G(s)u \tag{3.72}$$

with $G(s) = G_1(s) + G_2(s)$ and (2) a combination of cascaded systems

$$y = [G_2(s)][G_1(s)u] \tag{3.73}$$

is equivalent to

$$y = G(s)u \tag{3.74}$$

where $G(s) = G_2(s) G_1(s)$.
   For linear time-invariant systems the combination of transfer functions is commutative and

$$G_1(s) G_2(s) = G_2(s) G_1(s) \tag{3.75}$$

Also illustrated in Fig. 3.9 is the combination of transfer functions for closed-loop systems. Block diagram algebra allows the output to be written

$$y = G_1(s) [u - G_2(s)y] \tag{3.76}$$

which yields the closed-loop relationship

$$\frac{y}{u} = \frac{G_1}{1 + G_1G_2} \tag{3.77}$$

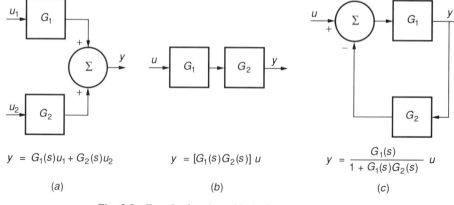

$$y = G_1(s)u_1 + G_2(s)u_2 \qquad\qquad y = [G_1(s)G_2(s)]\, u \qquad\qquad y = \frac{G_1(s)}{1 + G_1(s)G_2(s)}\, u$$

(a)   (b)   (c)

Fig. 3.9   Transfer function—block diagram representation.

When system models are synthesized from individual element transfer functions, it is assumed that as elements are combined their individual transfer functions are unaltered. For example, when the measurement of an instrument is recorded on a chart recorder, it is assumed that the instrument transfer function is not altered by the presence of a recorder. Transfer function models of elements and subsystems are valid representations of the element or subsystem only when they are unaltered when combined with other subsystem models.

3.5.3 Transfer Function Properties

For a linear time-invariant system, every transfer function relating an input and output pair has an identical denominator, corresponding to the left side of the differential equation (3.12), which is the system characteristic equation

$$s^n + a_{n-1}s^{n-1} + \cdots + a_0 = 0 \tag{3.78}$$

while each transfer function for a system may have a different numerator corresponding to the right side of Eq. 3.12 that determines the influence of the input function on the output.

The system characteristic equation is independent of which input-output pair is selected and depends only on the elements of the system A matrix and is the determinant of $(s\mathbf{I} - \mathbf{A})$, the characteristic equation of the system $\mathbf{A}$ matrix.

The system characteristic equation may be factored and written as the product of n terms:

$$(s - r_1)(s - r_2) \cdots (s - r_n) = 0 \tag{3.79}$$

where $r_1, r_2, \ldots, r_n$ are the roots of the characteristic equation, and the eigenvalues of the system $\mathbf{A}$ matrix and are referred to as the system poles $p_1, p_2, \ldots, p_n$. Since all of the coefficients in the characteristic equation are real, the roots are either real or appear as complex conjugate pairs.

The numerator of the transfer function may also be factored, and the resulting roots are designated zeros of the transfer function, $z_1, z_2, \ldots, z_m$, where the number of zeros depends on the specific input-output pair and in general is different for different outputs within the same system. A transfer function may be written in terms of its factors as

$$G(s) = \frac{K(s - z_1)(s - z_2) \cdots (s - z_m)}{(s - p_1)(s - p_2) \cdots (s - p_n)} \tag{3.80}$$

where K is a constant. The number of zeros depends on the specific input, while the number of poles is n. Computer programs[10-14] for personal and mainframe computers are available that accept system descriptions in terms of state equations and generate system transfer functions and roots of the characteristic equation by determining the $\mathbf{A}$ matrix eigenvalues.

3.5.4 Standard Forms

The factors of the system characteristic equation may be written as products of first-order terms of the form

$$\frac{1}{\tau}(\tau s + 1) = (s - r) \tag{3.81}$$

where $\tau = -(1/r)$ and second-order terms containing complex conjugate pairs r_1 and r_2:

$$s^2 + 2\zeta\omega_n s + \omega_n^2 = (s - r_1)(s - r_2) \tag{3.82}$$

also written as

$$s^2 + 2\zeta\omega_n s + \omega_n^2 = \omega_n^2 \left(\frac{s^2}{\omega_n^2} + \frac{2\zeta s}{\omega_n} + 1 \right)$$

where

$$\omega_n = \sqrt{r_1 r_2}$$

$$\zeta = \frac{-(r_1 + r_2)}{2\sqrt{r_1 r_2}}$$

The characteristic equation consists of products of first- and second-order standard forms

$$s^n + \cdots + a_0 = K(\tau_1 s + 1)(\tau_2 s + 1) \cdots (s^2 + 2\zeta\omega_n s + \omega_n^2) \cdots \qquad (3.83)$$

where K is a constant. The quantities defined are

$$\tau = \text{first-order time constant}$$
$$\omega_n = \text{second-order undamped natural frequency}$$
$$\zeta = \text{second-order damping ratio}$$

A general nth-order system may be characterized in terms of time constants and sets of pairs of natural frequencies and damping ratios.

The transfer functions and time constants of a number of first-order systems are tabulated in Fig. 3.10, and the transfer functions, natural frequencies, and damping ratios for a number of second-order systems are shown in Fig. 3.11.

3.6   SYSTEM TRANSIENT RESPONSE

3.6.1   Solution Techniques

General Techniques

The time histories of system outputs for a specified set of inputs may be determined for the general nonlinear (Eq. 3.1) or linear (Eq. 3.10) state equations by direct numerical integration, where in the nonlinear case it is assumed a solution exists and in the linear case the existence and uniqueness of the solution are guaranteed. Many numerical integration techniques and standard computer programs have been developed[15-17] to provide numerical solutions to differential equations.

Direct numerical integration techniques compute the values of the system state variables at a time $t + \Delta t$ using information concerning the previous system states at time t and before. Two major types of direct numerical integration techniques have been developed. Explicit methods express state variables at time $t + \Delta t$ solely in terms of values of the states and inputs at times t and before. Common explicit techniques include Euler integration, the central difference method, and Runge-Kutta Methods. Implicit methods employ conditions not only at t and before but also utilize equilibrium conditions at $t + \Delta t$ in their formulation and include methods such as the Newmark beta, Wilson theta, and Houbolt techniques.

A common class of explicit techniques used to solve nonlinear systems may be derived from Taylor series expansion techniques in which the state at $x(t + \Delta t)$ is expressed as

$$x(t + \Delta t) = x(t) + \dot{x}|_t \Delta t + \ddot{x}|_t \frac{\Delta t^2}{2!} + \cdots \qquad (3.84)$$

Euler Integration

When Euler integration is employed, only first order terms in Δt in the Taylor series are retained, and the numerical approximation may be determined by replacing $\dot{x}$ using Eq. 3.1 to form

$$x(t + \Delta t) = x(t) + f(x(t), u(t)) \Delta t \qquad (3.85)$$

Runge-Kutta Integration

Runge-Kutta integration techniques employ higher order approximations and the fourth-order Runge-Kutta technique which includes terms of fourth order in Δt

$$\mathbf{x}(t + \Delta t) = \mathbf{x}(t) + \frac{1}{6}[\mathbf{K}_1 + 2\mathbf{K}_2 + 2\mathbf{K}_3 + \mathbf{K}_4]\Delta t \qquad (3.86)$$

where $\mathbf{K}_1 = \mathbf{f}(\mathbf{x}, t)$

$$\mathbf{K}_2 = \mathbf{f}\left(t + \frac{\Delta t}{2}, \mathbf{x} + \mathbf{K}_1 \frac{\Delta t}{2}\right)$$

$$\mathbf{K}_3 = \mathbf{f}\left(t + \frac{\Delta t}{2}, \mathbf{x} + \mathbf{K}_2 \frac{\Delta t}{2}\right)$$

$$\mathbf{K}_4 = \mathbf{f}(t + \Delta t, \mathbf{x} + \mathbf{K}_3 \Delta t)$$

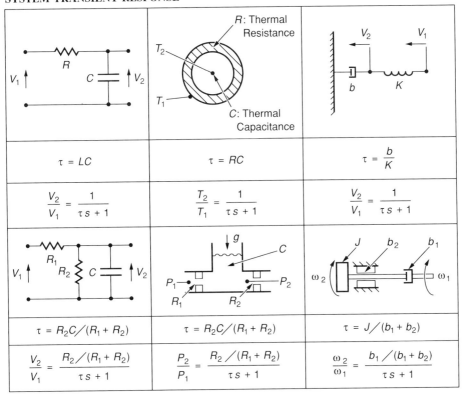

Fig. 3.10 First-order system transfer functions.

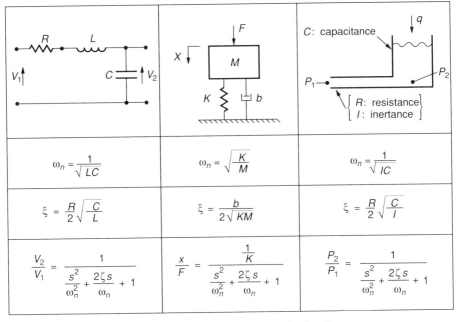

Fig. 3.11 Second-order system transfer functions.

In the numerical integration techniques, the accuracy of the solution depends upon the time interval Δt, the types of system nonlinearities, and the type of numerical technique employed. In the Runge-Kutta method, the selection of the time interval as a fraction (one-fifth to one-tenth) of the smallest characteristic time in the system is common as an initial estimate for the integration time interval.

Linear Systems

For linear time-invariant systems, the general solution to the state equations can be derived as[18]

$$\mathbf{x}(t) = \mathbf{\Phi}(t) \mathbf{x}(0) + \int_0^T \mathbf{\Phi}(t - \tau)\mathbf{B} \mathbf{u}(\tau) \, d\tau \tag{3.87}$$

where $\mathbf{\Phi}$ is the state transition matrix defined as

$$\mathbf{\Phi}(t) = \exp(At) = \mathbf{I} + \mathbf{A}t + \frac{\mathbf{A}^2 t^2}{2!} + \cdots \tag{3.88}$$

and the matrix exponential is defined by the exponential series of Eq. 3.88. The first term on the right side of Eq. 3.87 represents the response to the initial conditions $\mathbf{x}(0)$ while the second term represents the portion of the response due to the input functions $\mathbf{u}$.

The solutions for the case when the inputs are a set of unit impulses or a set of unit steps are for the unit impulses:

$$\mathbf{x}(t) = \mathbf{\Phi}(t) \mathbf{x}(0) + \mathbf{\Phi}(t)\mathbf{B} \tag{3.89}$$

and for the unit steps:

$$\mathbf{x}(t) = \mathbf{\Phi}(t) \mathbf{x}(0) + \mathbf{A}^{-1}[\mathbf{\Phi}(t) - \mathbf{I}]\mathbf{B} \tag{3.90}$$

With the evaluation of the matrix exponential, the solutions to the state equations for these singularity functions may be determined directly by matrix multiplication to determine the solution at any time t.

3.6.2 Solutions to First- and Second-Order Systems

Since a general nth-order linear system consists of combinations of first- and second-order characteristic equations, it is useful to consider the response of systems described by first- and second-order transfer functions in detail. For a system described by the transfer function

$$\frac{y}{u} = \frac{K}{\tau s + 1} \tag{3.91}$$

the solution for a unit step input with the system initially at rest is

$$y(t) = K[1 - e^{-t/\tau}] \tag{3.92}$$

which is plotted in Fig. 3.12. The system response monotonically increases to its final value, reaching 63 percent in $t/\tau = 1$, 87 percent in $t/\tau = 2$, 95 percent in $t/\tau = 3$, and 98 percent in $t/\tau = 4$. Thus a first-order system requires four time constants to reach 98 percent of its final value. The step responses of the following two first-order transfer functions are also displayed in Fig. 3.12

$$\frac{y}{u} = \frac{Ks}{\tau s + 1} \tag{3.93}$$

with the step responses

$$y(t) = K e^{-t/\tau} \tag{3.94}$$

and

$$\frac{y}{u} = K \frac{\tau_1 s + 1}{\tau_2 s + 1} \tag{3.95}$$

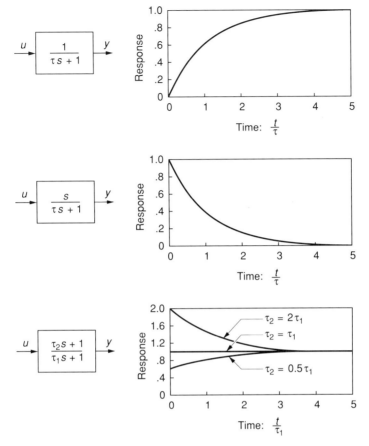

Fig. 3.12 First-order system unit step response.

with the response

$$y(t) = K\,[1 + (\tau_1 - 1)\,e^{-t/\tau_2}]\tag{3.96}$$

which also illustrate the dominant role of the time constant in reaching the steady state response. The three first-order system step responses can be used to determine the responses to an input which is related to the step response by a linear operation such as differentiation or integration or to an arbitrary input approximated by a series of steps using the principle of superposition.

The unit step response of the general second-order system described by the transfer function

$$\frac{y}{u} = \frac{K\omega_n^2}{s^2 + 2\zeta\omega_n s + \omega_n^2} = \frac{K}{(s^2/\omega_n^2) + (2\zeta s/\omega_n) + 1}\tag{3.97}$$

can be derived for the case of zero initial conditions as

$$y = K\left[1 + \left(\frac{r_2}{r_1 - r_2}\right)e^{r_1 t} + \left(\frac{r_1}{r_2 - r_1}\right)e^{r_2 t}\right]\tag{3.98}$$

where

$$r_1,\, r_2 = \frac{-\zeta\omega_n \pm \sqrt{\zeta^2 - 1}}{2}\tag{3.99}$$

Four cases need to be considered:

1. $\zeta > 1$, r_1 and r_2 are real and negative: the system is considered *overdamped*.
2. $\zeta = 1$, r_1 and r_2 are equal, and the solution of (3.98) needs to be augmented: the system is *critically damped*.
3. $0 < \zeta < 1$, r_1 and r_2 are complex conjugates: the system is *underdamped*.
4. $\zeta < 0$, the roots have positive real parts: the system is *unstable*.

The detailed solution of second-order systems is plotted in nondimensional form in Fig. 3.13. For the underdamped case, the solution overshoots the final value, the percent overshoot being determined by the damping ratio. For the system with $0 < \zeta < 1$ the peak overshoot occurs at time t_p:

$$t_p = \frac{\pi}{\omega_n (1 - \zeta^2)^{1/2}} \tag{3.100}$$

and the overshoot has a percent magnitude, M_p, of

$$M_p = 100 \frac{e^{-\pi \zeta}}{(1 - \zeta^2)^{1/2}} \tag{3.101}$$

as $\zeta \to 0$, the time at which the peak overshoot occurs approaches π/ω_n, and the value approaches 100 percent.

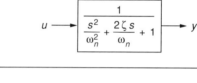

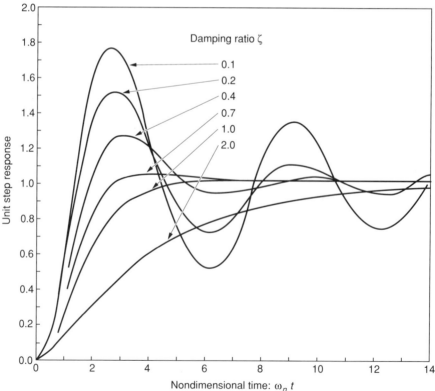

Fig. 3.13 Second-order system unit step responses.

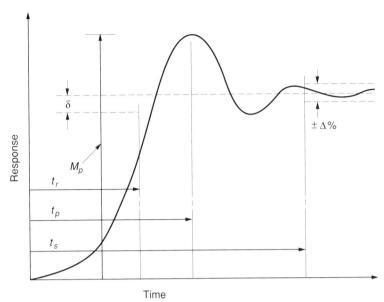

Fig. 3.14   Time-domain unit step response system specifications.

3.6.3   System Specifications in the Time Domain

The step response of a system provides a basis for the specification of system dynamic requirements in the time domain. The primary measures of system performance in response to a unit step input are indicated in Fig. 3.14 where the system response is normalized to provide nominally unity gain and include

- The percent overshoot, P.O. $= (M_p - 1)$ (100%), where M_p is the value of the maximum response
- The peak time, t_p, where t_p is the time at which the peak response occurs
- The rise time, t_r, where t_r is the time required for the response to rise to within a fixed percent of the steady-state response amplitude
- The settling time, t_s, where t_s is the time required for the system to settle within a fixed percent of the input amplitude

The performance measures t_r and t_p essentially indicate the speed of response of the system, while M_p indicates the degree of damping. The settling time t_s provides some measure of both speed of response and damping. For linear first- and second-order systems, the time to settle within 2 percent of the final value is $t_s = 4\tau$ for first-order systems and $t_s = 4/\zeta\omega_n$ for second-order systems.

3.7   FREQUENCY RESPONSE

3.7.1   Analysis Technique

Many control and instrumentation system requirements are specified and evaluated in the frequency domain. Analytically, any periodic function may be represented using a Fourier series as a function of frequency, and stationary random variables may be represented by spectral densities as a function of frequency. Experimentally, the developments of spectrum analyzers and digital analysis and control instruments have facilitated the widespread use of frequency domain analysis techniques.

The frequency response of a linear, time-invariant system may be developed by considering a system subjected to a general sinusoidal input, where the input is the complex exponential

$$u(t) = [Ae^{st}] = A[\cos \omega t + j \sin \omega t] \qquad (3.102)$$

$$s \rightarrow j\omega$$

where A is the real amplitude and where using Euler's relationship and the real, Re[ ], and imaginary, Im[ ], operators, it is noted that $\text{Re}[Ae^{st}]$ for $s \to j\omega = A \cos \omega t$ and $\text{Im}[Ae^{st}]$ for $s \to j\omega = A \sin \omega t$.

The steady state response of the output $y(t)$ to the sinusoidal input $u(t)$ for a system represented by the transfer function

$$\frac{y}{u} = G(s) \tag{3.103}$$

may be obtained as

$$y(t) = A \, |G(j\omega)| e^{j\omega t + \phi} \tag{3.104}$$

where in the system transfer function the Laplace operator has been replaced by the Fourier operator, $s \to j\omega$, and where $|G(j\omega)|$ is the magnitude of $G(j\omega)$ and ϕ is the phase angle of $G(j\omega)$.

For example, if the response to a cosine input is required, the cosine is the real part of $u(t)$ in Eq. 3.102, and thus the real part of $y(t)$ in Eq. 3.104 is extracted to yield

$$y(t) = A \, |G(j\omega)| \cos (\omega t + \phi) \tag{3.105}$$

as the response to a cosine of amplitude A while the response to a sine is extracted by taking the imaginary part of Eq. 3.104

$$y(t) = A \, |G(j\omega)| \sin (\omega t + \phi) \tag{3.106}$$

For a sine or cosine wave input of amplitude A, the steady-state output is an equivalent sine or cosine of the same frequency ω, altered in magnitude by the magnitude of the transfer function and shifted in phase by the angle of the transfer function. Thus plots of a transfer function magnitude and phase contain all of the essential information required to determine a system steady state response to a sine or cosine wave.

The transfer function may be written in terms of real and imaginary parts:

$$G(j\omega) = \frac{G_{NR} + j \, G_{NI}}{G_{DR} + j \, G_{DI}} \tag{3.107}$$

where from Eq. 3.63 with $s \to j\omega$, the real part of the numerator is

$$G_{NR} = b_0 - b_2\omega^2 + b_4\omega^4 - b_6\omega^6 \cdots \tag{3.108}$$

the imaginary part of the numerator is

$$G_{NI} = b_1\omega - b_3\omega^3 + b_5\omega^5 \cdots \tag{3.109}$$

the real part of the denominator is

$$G_{DR} = a_0 - a_2\omega^2 + a_4\omega^4 - a_6\omega^6 \cdots \tag{3.110}$$

the imaginary part of the denominator is

$$G_{DI} = a_1\omega - a_3\omega^3 + a_5\omega^5 \cdots \tag{3.111}$$

and where the series for the numerator is extended through order m and for the denominator through order n.

The transfer function magnitude is

$$|G(j\omega)| = \sqrt{\frac{G_{NR}^2 + G_{NI}^2}{G_{DR}^2 + G_{DI}^2}} \tag{3.112}$$

and the transfer function angle is

$$\phi = \tan^{-1} \left(\frac{G_{NI}}{G_{NR}} \right) - \tan^{-1} \left(\frac{G_{DI}}{G_{DR}} \right) \tag{3.113}$$

Plots of system magnitude and phase are presented in many different forms[19,20] including polar or Nyquist plots in which magnitude as a function of phase with frequency as a parameter are plotted and plots in which magnitude and phase are each plotted versus frequency using either linear, semi-log or log-log scales. A common method of plotting the data is the Bode plot, in which log magnitude and linear phase are plotted versus the log of frequency, where log is of base 10. Often the magnitude of the transfer function is presented in terms of decibels (db) which are defined as

$$|G|_{db} = 20 \log_{10} |G| \qquad (3.114)$$

and plots of magnitude in db and phase in degrees versus log frequency are presented. Alternatively, plots of magnitude versus frequency may be presented on log-log coordinates and phase angle versus frequency on linear-log coordinates. This latter plotting method is adopted in this section.

3.7.2 First- and Second-Order System Frequency Response

The magnitudes and phase angles of a set of first- and second-order transfer functions are summarized in Table 3.5 and plotted in Figs. 3.15 and 3.16 where for first-order terms the nondimensional frequency $\tau\omega$, where τ is a characteristic time, is used and for second-order terms ω/ω_n, where ω_n is the undamped natural frequency, is used.

For the first-order lead and lag terms, a break frequency ω_b may be defined as

$$\omega_b = 1/\tau$$

For frequencies much less than the break frequency, $\omega \ll \omega_b$, the magnitude of the lead and lag is approximately unity with no phase shift:

$$|G(j\omega)| \approx 1.0$$

$$\phi \approx 0.0°$$

At the break frequency, $\omega \approx \omega_b$, the lead has increased in magnitude to 1.414 ($+3$ db) and in phase by 45°, and the lag has decreased in magnitude to 0.707 (-3 db) and in phase to $-45°$:

$$\text{Lead:} \quad |G(j\omega)| = 1.414 \quad \text{and} \quad \phi = 45°$$

$$\text{Lag:} \quad |G(j\omega)| = 0.707 \quad \text{and} \quad \phi = -45°$$

For frequencies significantly above the break frequency, $\omega \gg \omega_b$, the lead and lag asymptotically approach the pure differentiator and pure integrator, respectively:

TABLE 3.5 TRANSFER FUNCTION MAGNITUDES AND PHASE ANGLES

Transfer Function	Magnitude	Phase Angle
Constant: K	K	$0.0°$
Integrator: $\dfrac{1}{s}$	$\dfrac{1}{\omega}$	$-90°$
Differentiator: s	ω	$+90°$
Lead: $\tau s + 1$	$\sqrt{(\tau\omega)^2 + 1}$	$\tan^{-1}(\tau\omega)$
Lag: $\dfrac{1}{\tau s + 1}$	$\dfrac{1}{\sqrt{(\tau\omega)^2 + 1}}$	$-\tan^{-1}(\tau\omega)$
Second-Order: $\dfrac{K}{(s^2/\omega_n^2) + (2\zeta s/\omega_n) + 1}$	$\dfrac{K}{\sqrt{\left[1 - (\omega^2/\omega_n^2)\right]^2 + (2\zeta\omega/\omega_n)^2}}$	$-\tan^{-1}\left[\dfrac{2\zeta\omega/\omega_n}{1 - (\omega/\omega_n)^2}\right]$

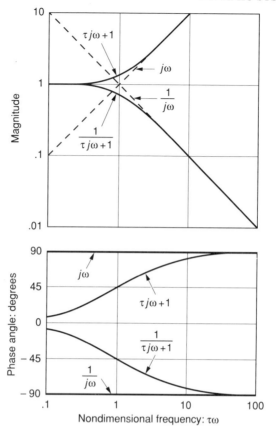

Fig. 3.15 First-order system frequency responses.

Lead: $|G(j\omega)| \approx \omega$ and $\phi \approx 90°$

Lag: $|G(j\omega)| \approx 1/\omega$ and $\phi \approx -90°$

The second-order denominator responds differently. For frequencies much less than the system's natural frequency, $\omega \ll \omega_n$, the transfer function is unity:

$$|G(j\omega)| \approx 1.0 \quad \text{and} \quad \phi \approx 0.0°$$

For frequencies much greater than the system's natural frequency, $\omega \gg \omega_n$, the magnitude decreases as $1/\omega^2$ with a constant phase of $-180°$:

$$|G(j\omega)| \approx \frac{1}{\omega^2} \quad \text{and} \quad \phi \approx -180°$$

For $\omega \approx \omega_n$, the response depends strongly on the damping ratio. A peak in the response occurs at the resonant frequency

$$\frac{\omega_r}{\omega_n} = \sqrt{1 - 2\zeta^2} \tag{3.115}$$

for $0 < \zeta < 0.707$. The frequency at which the peak occurs is at ω_n for zero damping and decreases to zero at 0.707 damping. The magnitude of the peak is

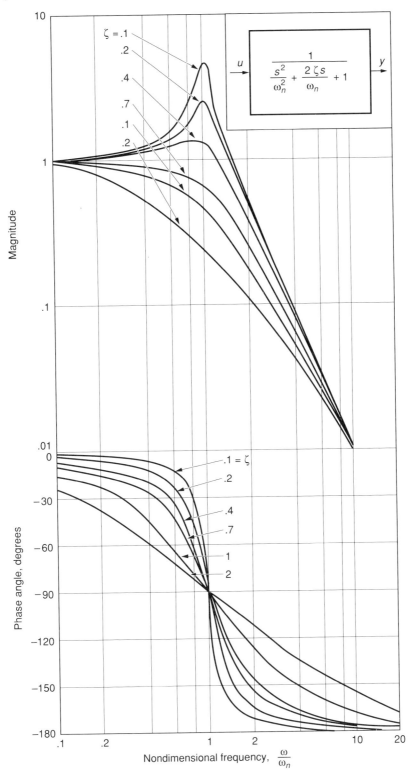

Fig. 3.16   Second-order system frequency responses.

$$\left|G(j\omega)\right|\bigg|_{\omega=\omega_r} = \frac{1}{2\zeta} \frac{1}{\sqrt{1-\zeta^2}} \tag{3.116}$$

and becomes infinite for zero damping. For systems with small values of damping, significant resonant amplification may occur if the system is excited at $\omega = \omega_r$.

3.7.3 Generalized Frequency Response Techniques

A system transfer function can be factored into products of gains, first-order and second-order terms as

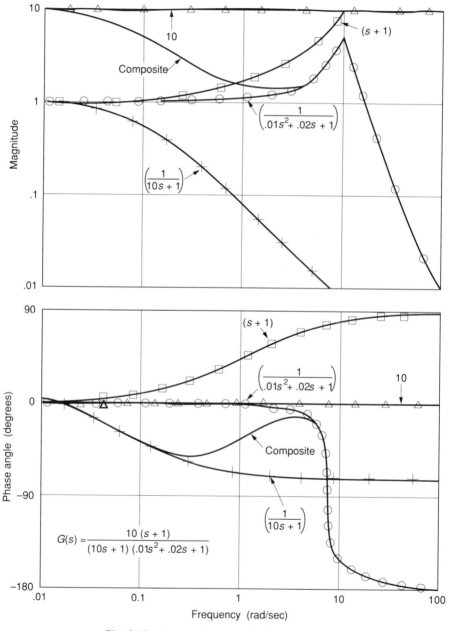

Fig. 3.17 Construction of system frequency response.

$$G(s) = \left[\frac{K}{s^{-p}}\right] \frac{\left[\prod_{i=1} (\tau s + 1)_i\right]\left[\prod_i \left((s^2/\omega_n^2) + (2\zeta s/\omega_n) + 1\right)_i\right]}{\left[\prod_{j=1} (\tau s + 1)_j\right]\left[\prod_j \left((s^2/\omega_n^2) + (2\zeta s/\omega_n) + 1\right)_j\right]} \qquad (3.117)$$

where the first term is a gain with all free s terms collected and the power p representing the number of free differentiators ($p = +$) or integrators ($p = -$) in the system and all other terms are either first- or second-order terms. When these terms are represented in log-log plots, the products of the magnitudes may be combined and the phases added at any given frequency to produce the overall system frequency response. For a given transfer function, the low frequency asymptote is as $\omega \to 0$

$$|G(0)| = K s^p$$

If $p = 0$, then when $\omega \to 0$, the steady state constant gain is approached. If $p \neq 0$, the asymptote approaches a slope of p, which is equal to the net number of differentiators less integrators in the system.

The representation of a system as products of component parts is useful both in constructing frequency responses for systems and in interpreting experimentally measured frequency responses of physical systems. The determination of the complete system frequency response from component parts is illustrated in Fig. 3.17 for the system

$$G(s) = \frac{10(s + 1)}{(10s + 1)\left[(s^2/100) + (0.2s/10) + 1\right]} \qquad (3.118)$$

3.7.4 System Specifications in the Frequency Domain

The performance requirements of systems may be specified in the frequency domain. For instruments and many types of closed-loop control systems, the specifications are given assuming the system has a finite gain at zero frequency. Typical specifications are shown in Fig. 3.18. These include

1. Amplitude gain at $\omega = 0$, $G(0) = K$, the system zero frequency gain.
2. Maximum amplitude with respect to zero frequency gain $G(j\omega_r)/G(0)$, where ω_r is the frequency at which the peak in amplitude occurs. For some systems a requirement is specified that no peaking occur in the amplitude plot. For systems with a dominant second-order pole, the damping ratio of the pole must be 0.707 or greater to avoid peaking.
3. a. The system bandwidth ω_b, the frequency at which the amplitude has decreased to 0.707 (-3 db) of its zero frequency value. The system bandwidth is an indication of the range of frequencies over which it is responsive.
   b. The frequency at which the system phase shift reaches a prescribed value. Often the $-90°$ phase shift point is prescribed.

The specifications relate directly to the time domain specifications with the gain at zero frequency representing the system steady state gain, the peak amplitude at resonance relating directly to overshoot in the time response, and the system bandwidth relating to the speed of response where for a system dominated by a second-order pole, the response time may be approximated as

$$T \approx \pi/\omega_b \qquad (3.119)$$

3.8 SYSTEM RESPONSE TO RANDOM INPUTS

3.8.1 Basic Relationships

The response of a linear time-invariant system to stationary, ergodic inputs may be specified in terms of input spectral densities and the system transfer functions. For a single input $u(t)$, single output $y(t)$ system with transfer function $G(s)$, the output spectral density is related to the input spectral density through the system transfer function as

$$S_{yy}(\omega) = |G(j\omega)|^2 S_{xx}(\omega) \qquad (3.120)$$

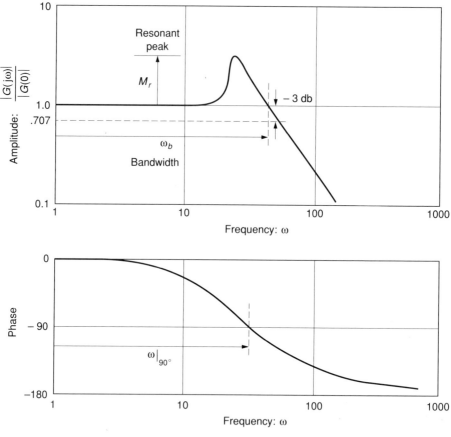

Fig. 3.18 Frequency domain system specifications.

The output spectral density contains magnitude information and is computed from the magnitude squared of the transfer function

$$|G(j\omega)|^2 = G^*(j\omega)\, G(j\omega) \tag{3.121}$$

where $G^*(j\omega)$ is the complex conjugate of $G(j\omega)$. The output cross-spectral density with respect to the input contains both magnitude and phase information and is given by

$$S_{uy}(\omega) = G(j\omega)\, S_u(\omega) \tag{3.122}$$

Spectrum analyzers commonly use this relationship to determine the transfer function of an experimental system as a function of frequency.

$$G(j\omega) = \frac{S_{uy}(\omega)}{S_u(\omega)} \tag{3.123}$$

where to compute the cross-spectral density, simultaneous measurement of the input $u(t)$ and output $y(t)$ are required.

3.8.2 Second-Order System Response

The output spectral density of a second-order system response to a white noise input with spectral density S_0 is shown in Fig. 3.19 as a function of damping ratio. Systems with damping ratios less than 0.4 have significant resonant amplification even when excited by white noise.

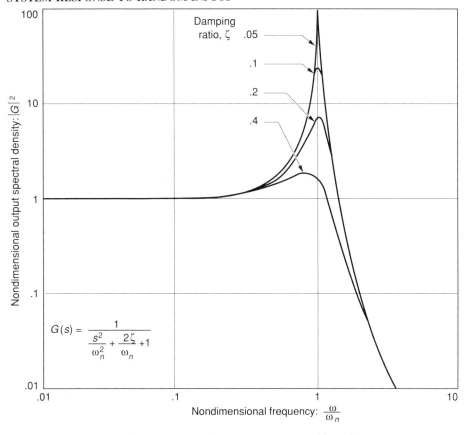

Fig. 3.19   Second-order system response to white noise.

3.8.3   Multiple Input System

The extension of random input response methods to a system with two inputs is shown in Fig. 3.20. For this system the spectral density of the output is

$$S_{yy}(\omega) = G_1^*(j\omega)\, G_1(j\omega)\, S_{u_1}(\omega) + G_1^*(j\omega)\, G_2(j\omega)\, S_{u_1 u_2}$$

$$+ G_2^*(j\omega)\, G_1(j\omega)\, S_{u_2 u_1} + G_2^*(j\omega)\, G_2(j\omega)\, S_{u_2} \tag{3.124}$$

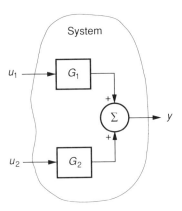

Fig. 3.20   Two-input system.

and the cross-spectral densities are

$$S_{u_1 y}(\omega) = G_1(j\omega) S_{u_1}(\omega) + G_2(j\omega) S_{u_1 u_2}(\omega) \tag{3.125}$$

$$S_{u_2 y}(\omega) = G_2(j\omega) S_{u_2}(\omega) + G_1(j\omega) S_{u_2 u_1}(\omega) \tag{3.126}$$

If the inputs u_1 and u_2 are uncorrelated then

$$S_{u_1 u_2}(\omega) = S_{u_2 u_1}(\omega) = 0$$

and each output cross-spectrum is related only to a single input. In the identification of transfer functions from experimental measurements, Eqs. 3.125 and 3.126 form a basis for determining system transfer functions and show that the transfer function between an output and input may be computed in the presence of a second input using cross-spectral density techniques. Detailed numerical methods have been developed to perform these computations.[21,22] If the input of interest $u(t)$ is uncorrelated with all other inputs (including noise) then the determination of the transfer function reduces to Eq. 3.123, even when many inputs are present.

REFERENCES

1. G. Stephanopoulos, *Chemical Process Control: An Introduction to Theory and Practice*. Prentice-Hall, Englewood Cliffs, NJ, 1984.

2. W. H. Ray, *Advanced Process Control*, McGraw-Hill, New York, 1981.

3. K.-J. Bathe, *Finite Element Procedures in Engineering Analysis*, Prentice-Hall, Englewood Cliffs, NJ, 1982.

4. J. L. Shearer, A. Murphy, and H. H. Richardson, *Introduction to Systems Dynamics*, Addison-Wesley, Reading, MA, 1967.

5. D. Karnopp and R. C. Rosenberg, *System Dynamics: A Unified Approach*, Wiley-Interscience, New York, 1975.

6. R. H. Cannon, *Dynamics of Physical Systems*, McGraw-Hill, New York, 1967.

7. N. Minorsky, *Theory of Nonlinear Control Systems*, McGraw-Hill, New York, 1969.

8. J. S. Bendat and A. G. Piersol, *Engineering Applications of Correlation and Spectral Analysis*, John Wiley & Sons, New York, 1980.

9. D. E. Newland, *Random Vibrations and Spectral Analysis*, Longman, New York, 1975.

10. M. Jamshidi, *Computer Aided Control Systems Engineering*, Pergamon, New York, 1985.

11. Matrix-X, Integrated Systems Inc., Palo Alto, CA.

12. CTRL-C, Systems Control Technology, Palo Alto, CA.

13. CC, Systems Technology, Inc., Hawthorne, CA.

14. Matlab, The MathWorks, Sherborn, MA.

15. K.-J. Bathe, *Finite Element Procedures in Engineering Analysis*, Prentice-Hall, Englewood Cliffs, NJ, 1982.

16. F. D'Sousa and V. K. Garg, *Advanced Dynamics*, Prentice-Hall, New Jersey, 1984.

17. M. A. Forsythe, et al., *Computer Methods of Mathematical Computations*, Prentice-Hall, Englewood Cliffs, NJ, 1977.

18. B. C. Kuo, *Automatic Control Systems*, Prentice-Hall, Englewood Cliffs, NJ, 1982.

19. G. F. Franklin, J. D. Powell, and A. Emani-Noeini, *Feedback Control of Dynamic Systems*, Addison-Wesley, Reading, MA, 1986.

20. R. C. Dorf, *Modern Control Systems*, Addison-Wesley, Reading, MA, 1980.

21. J. S. Bendat and A. G. Piersol, *Engineering Applications of Correlation and Spectral Analysis*, John Wiley & Sons, New York, 1980.

22. D. E. Newland, *Random Vibrations and Spectral Analysis*, Longman, New York, 1975.

CHAPTER 4
INSTRUMENT STATICS

JERRY LEE HALL

Department of Mechanical Engineering
Iowa State University
Ames, Iowa

MAHMOOD NAIM

Union Carbide Corporation
Indianapolis, Indiana

4.1 TERMINOLOGY

4.1.1 Transducer Characteristics

The measurement of any variable is accomplished by an instrumentation system composed of transducers. Each transducer is an energy conversion device and requires energy transformation into the device before the variable of interest can be detected. Each variable of interest must be accompanied by an additional quantity such that the combination of the two provides energy to be transferred into the sensing transducer. Therefore, pressure cannot be measured without an accompanying change in volume, force cannot be measured without an accompanying change in length, and voltage cannot be measured without an accompanying flow of charge. This energy is then converted into another form of energy as required by a signal conditioning transducer and a readout transducer or controller such that the input signal can be presented and appropriately interpreted at the output of the measuring system.

The Instrument Society of America (ISA) defines *transducer* as "a device that provides a usable output in response to a specified measurand." The *measurand* is "a physical quantity, property or condition which is measured." The *output* is "the electrical quantity, produced by a transducer, which is a function of the applied measurand."[1] Transducer characteristics are categorized as follows:

Static characteristics describe performance at room temperature conditions, with very slow changes in measurand. Room conditions are established, in general, to be a temperature of 25 ± 10°C, a relative humidity of 90% or less, and a barometric pressure of 880 to 1080 mbar.

Dynamic characteristics relate the performance of a transducer to variations of the measurand with time.

Environmental characteristics relate the performance of a transducer to its exposure to external conditions such as temperature, vibration, and shock.

Reliability characteristics relate to the life expectancy of a transducer and to any hazards that may be presented by its malfunction to the system in which it is intended to operate.

Theoretical characteristics indicate the ideal behavior of a transducer as defined by a relationship between the output and the input in the form of a table of values, a graph, or a mathematical equation.

Noise characteristics cause modification of a transducer response due to variation in temperature, humidity, electromagnetic surroundings, corrosive atmosphere, aging of transducer elements, and other effects. Those effects having nothing to do with the input signal of interest are called noise.

4.1.2 Definitions

The description of a transducer and its role in a measuring system is based on most of the definitions that follow. Further details of these definitions can be found in other works.[2,3]

Static calibration is the process of measuring the static characteristics of an instrument. This involves applying a range of known values of static input to the instrument and recording the corresponding outputs. The data obtained are presented in a tabular or graphical form.

Range is defined by the upper and lower limits of the measurand values that an instrument can measure. Instruments are designed to provide predictable performance and, often, enhanced linearity over the range specified.

Sensitivity is defined as the change in the output signal relative to the change in the input signal at an operating point. Sensitivity may be constant over the range of the input signal to a transducer, or it can vary. Instruments that have a constant sensitivity are called "linear."

Resolution is defined as the smallest change in the input signal that will yield a readable change in the output of the measuring system at its operating point.

Threshold of an instrument is the minimum input for which there will be an output. Below this minimum input the instrument will read zero.

Zero of an instrument refers to a selected datum. The output of an instrument is adjusted to read zero at a predefined point in the measured range. For example, the output of a Celsius thermometer is zero at the freezing point of water; the output of a pressure gage may be zero at atmospheric pressure.

Zero-drift is the change in output from its set zero value over a specified period of time. Zero-drift occurs due to changes in ambient conditions, changes in electrical conditions, aging of components, or mechanical damage. The error introduced may be significant when a transducer is used for long-term measurement.

Creep is a change in output occurring over a specific time period while the measurand is held constant at a value other than zero, and all environmental conditions are held constant.

Accuracy is the maximum amount of difference between a measured variable and its true value. It is usually expressed as a percentage of full-scale output. In the strictest sense, accuracy is never known because the true value is never really known.

Precision is the difference between a measured variable and the best estimate (as obtained from the measured variable) of the true value of the measured variable. It is a measure of repeatability. Precise measurements have small dispersion but may have poor accuracy if they are not close to the true value.

Linearity describes the maximum deviation of the output of an instrument from a best-fitting straight line through the calibration data. Most instruments are designed so that the output is a linear function of the input. Linearity is based on the type of straight line fitted to the calibration data. For example, "Least-Squares linearity" is referenced to that straight line for which the sum of the squares of the residuals is minimized. The term "residual" refers to the deviations of output readings from their corresponding values on the straight line fitted through the data.

Hysteresis is the maximum difference in output, at any measurand value within the specified range, when the value is approached first with increasing and then with decreasing measurand. Hysteresis is typically caused by a lag in the action of the sensing element of the transducer. Loading the instrument through a cycle of first increasing values, then decreasing values, of the measurand provides a hysteresis loop. Hysteresis is usually expressed in percent of full-scale output.

Error Band is the band of maximum deviation of output values from a specified reference line or curve. A static error band (see Fig. 4.1) is obtained by static calibration. It is determined on the basis of maximum deviations observed over at least two consecutive calibration cycles so as to include repeatability. Error band accounts for deviations that may be due to nonlinearity,

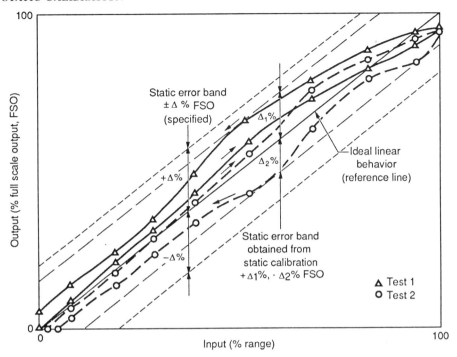

Fig. 4.1 Static error band referred to ideal behavior.

nonrepeatability, hysteresis, zero-shift, sensitivity shift, and so forth. It is a convenient way to specify transducer behavior when individual types of deviations need not be specified nor determined.[4]

4.2 STATIC CALIBRATION

4.2.1 The Calibration Process

Calibration is the process of comparison of the output of a measuring system to the values of a range of known inputs. For example, a pressure gage is calibrated by a device called a "dead weight" tester, where known pressures are applied to the gage and the output of the gage is recorded over its complete range of operation.

The calibration signal should, as closely as possible, be the same as the type of input signal to be measured. A measurement system used for dynamic signals should be calibrated using known dynamic inputs. Most calibrations are performed by means of static or level calibration signals since they are usually easy to produce and maintain accurately. However, a measuring system calibrated with static signals may not read correctly when subjected to the dynamic input signals since the natural dynamic characteristics and the forcing function response characteristics of the measurement system would not be accounted for with a static calibration.

A static calibration should include both increasing and decreasing values of the known input signal and a repetition of the input signal.[5] This allows one to determine hysteresis as well as the repeatability of the measuring system. The sensitivity of the measuring system is obtained from the slope of a suitable line or curve plotted through the calibration points at any level of the input signal. A typical static calibration data plot for a measurement system is shown in Fig. 4.2.

4.2.2 Fitting Equations to Calibration Data

The calibration plot of a specific measurement system may not be linear and may require a choice of a functional form for the curve to be selected for characterization of the calibration data. The functional form may be of the standard polynomial type or may be one of a transcendental function

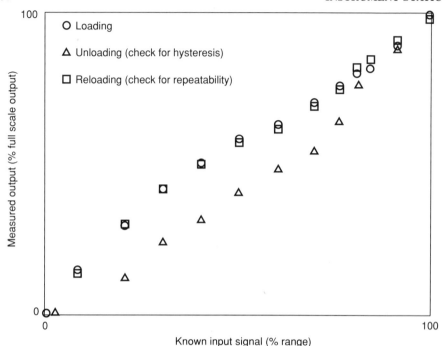

Fig. 4.2 Static calibration data plot.

type. Statistics can be used to fit a desired function to calibration data. A common method is to use the "Least-Squares curve fit technique."[6] The principle used in making this type of curve fit is to minimize the sum of the squares of deviations of the data from the assumed curve. These deviations from the assumed curve may be due to errors in one variable or errors in many variables. If the error is in one variable, the technique is called *linear regression*. If several variables are involved, the technique is called *multiple regression*. Two assumptions that are often used with the Least-Squares technique are that the x variable has relatively little error as compared to the y variable, and the magnitude of the uncertainty in y is not dependent upon the magnitude of the x variable.

An equation that fits the data must be assumed before the Least-Squares procedure can be used. Table 4.1 can be used as an aid in making this determination. First, a plot of the data should be made. Values of y at equal increments of x are then obtained from a smooth curve sketched through the data points. Successive differences such as Δy, $\Delta(\Delta y)$, $\Delta(\Delta(\Delta y))$, and so on, are then formed according to the outline given by Table 4.1. The particular successive difference that has an approximately constant value gives the indication of the appropriate equation form that will adequately fit the data.

To illustrate the Least-Squares technique, assume that an equation of the following polynomial form will fit a given set of data.

$$y = a + bx^1 + cx^2 + \cdots + mx^k \tag{4.1}$$

If the data points are denoted by (x_i, y_i), where i ranges from 1 to n, then the expression for summation of the residuals is

$$\sum_{i=1}^{n} (y_i - y)^2 = R \tag{4.2}$$

The Least-Squares method requires that R be minimized. The parameters used for the minimization are the unknown coefficients $a, b, c, \ldots, m$ in the assumed equation. The following differentiation

$$\frac{\partial R}{\partial a} = \frac{\partial R}{\partial b} = \frac{\partial R}{\partial c} = \cdots = \frac{\partial R}{\partial m} = 0 \tag{4.3}$$

TABLE 4.1 METHOD FOR TESTING THE SUITABILITY OF VARIOUS ASSUMED FORMS OF EQUATIONS TO REPRESENT GIVEN DATA

Case	Assumed Form of Equation	Plot	Obtain these Successive Differences at Constant Values of Δx, $\Delta \log x$, etc.	Criterion of Suitability*
1	$y = a + bx + cx^2 + \cdots + qx^n$	$y = f(x)$	$\Delta y; \Delta^2 y; \Delta^3 y; \ldots; \Delta^n y$	$\Delta^n y$
2	$y = a + \dfrac{b}{x} + \dfrac{c}{x^2} + \cdots + \dfrac{q}{x^n}$	$y = f\left(\dfrac{1}{x}\right)$	$\Delta y; \Delta^2 y; \Delta^3 y; \ldots; \Delta^n y$	$\Delta^n y$
3	$y^2 = a + bx + cx^2 + \cdots + qx^n$	$y^2 = f(x)$	$\Delta y^2; \Delta^2 y^2; \Delta^3 y^2; \ldots; \Delta^n y^2$	$\Delta^n y^2$
4	$\log y = a + bx + cx^2 + \cdots + qx^n$	$\log y = f(x)$	$\Delta(\log y); \Delta^2(\log y); \ldots; \Delta^n(\log y)$	$\Delta^n(\log y)$
5	$y = a + b(\log x) + c(\log x)^2$	$y = f \log x$	$\Delta y; \Delta^2 y$	$\Delta^2 y$
6	$y = ab^x = ae^{b'x}$	$\log y = f(x)$	$\Delta(\log y)$	$\Delta(\log y)$
7	$y = a + bc^x = a + be^{c'x}$	$y = f(x)$	$\Delta y; \log \Delta y; \Delta(\log \Delta y)$	$\Delta(\log \Delta y)$
8	$y = a + bx + cd^x = a + bx + cd^{e'x}$	$y = f(x)$	$\Delta y; \Delta^2 y; \log \Delta^2 y; \Delta(\log \Delta^2 y)$	$\Delta(\log \Delta^2 y)$
9	$y = ax^b$	$\log y = f(\log x)$	$\Delta(\log y)$	$\Delta(\log y)$
10	$y = a + bx^c$	$y = f(\log x)$	$\Delta y; \log \Delta y; \Delta(\log \Delta y)$	$\Delta(\log \Delta y)$
11	$y = axe^{bx}$	$\ln y = f(x)$	$\Delta \ln y; \Delta \ln x$	$(\Delta \ln y - \Delta \ln x)$

*Item that should be approximately constant

yields $k + 1$ equations called "normal equations" to determine the $k + 1$ coefficients in the assumed relation. The coefficients $a, b, c, \ldots, m$ are found by solving the normal equations simultaneously. For example, if $k = 1$, then the polynomial is of first degree (a straight line) and the normal equations become

$$\sum y_i = a(n) + b \sum x_i$$
$$\sum x_i y_i = a \sum x_i + b \sum x_i^2$$

(4.4)

and the coefficients a and b are

$$a = \frac{\sum x^2 \sum y - \sum x \sum xy}{n \sum x^2 - (\sum x)^2}$$
$$b = \frac{n \sum xy - \sum x \sum y}{n \sum x^2 - (\sum x)^2}$$

(4.5)

The resulting curve ($y = a + bx$) is called the regression curve of y on x.

It can be shown that a regression curve fit by the Least-Squares method passes through the centroid $(\bar{x}, \bar{y})$ of the data.[6] If two new variables X and Y are defined as

$$X = x - \bar{x} \quad \text{and} \quad Y = y - \bar{y}$$

(4.6)

then

$$\sum X = 0 = \sum Y$$

(4.7)

Substitution of these new variables in the normal equations for a straight line yields the following result for a and b.

$$a = 0$$
$$b = \frac{\sum XY}{\sum X^2}$$

(4.8)

The regression line becomes

$$Y = bX$$

(4.9)

The technique described above will yield a curve based on an assumed form that will fit a set of data. This curve may not be the best one that could be found, but it will be the best based on the assumed form. Therefore, the "goodness-of-fit" must be determined to check that the fitted curve follows the physical data as closely as possible.

Example 4.1 Choice of functional form. Find a suitable equation to represent the following calibration data.

$$x = [3, 4, 5, 7, 9, 12, 13, 14, 17, 20, 23, 25, 34, 38, 42, 45]$$

$$y = [5.5, 7.75, 10.6, 13.4, 18.5, 23.6, 26.2, 27.8, 30.5, 33.5, 35, 35.4, 41, 42.1, 44.4, 46.2]$$

Solution: A computer program can be written to fit the data to several assumed forms as given in Table 4.2. The data can be plotted and the best-fitting curve selected on the basis of minimum residual error, maximum correlation coefficient, or smallest maximum absolute deviation as shown in Table 4.2.

The analysis shows that the assumed equation $y = a + (b)\log(x)$ represents the best fit through the data as it has the smallest maximum deviation and the highest correlation coefficient. Also note that the equation $y = 1/a + bx$ is not appropriate for this data because it has a negative correlation coefficient.

Example 4.2 Nonlinear regression. Find the regression coefficients a, b, and c if the assumed behavior of the (x, y) data is $y = a + bx + cx^2$

$$x = [2, 3, 4, 5, 6, 7, 8, 9, 10, 11, 12, 13, 14, 15]$$

$$y = [0.26, 0.38, 0.55, 0.70, 1.05, 1.36, 1.75, 2.20, 2.70, 3.20, 3.75, 4.40, 5.00, 6.00]$$

TABLE 4.2 STATISTICAL ANALYSIS FOR EXAMPLE 4.1

Assumed Equation	Regression Coefficient		Residual Error, R	Maximum Deviation	Correlation* Coefficient
	a	b			
$y = bx$	—	1.254	56.767	10.245	0.700
$y = a + bx$	9.956	0.907	20.249	7.178	0.893
$y = ae^{bx}$	10.863	0.040	70.247	18.612	0.581
$y = 1/(a + bx)$	0.098	−0.002	14257.327	341.451	−74.302
$y = a + b/x$	40.615	−133.324	32.275	9.326	0.830
$y = a + b \log x$	−14.188	15.612	1.542	2.791	0.992
$y = ax^b$	3.143	0.752	20.524	8.767	0.892
$y = x/(a + bx)$	0.496	0.005	48.553	14.600	0.744

*Correlation coefficient defined in Section 4.3.7

From Eqs. 4.1, 4.2, and 4.3

$$y = a + bx + cx^2$$

$$xy = ax + bx^2 + cx^3 \qquad (4.10)$$

$$x^2y = ax^2 + bx^3 + cx^4$$

A simultaneous solution of the above equations provides the desired regression coefficients.

$$a = 0.1959$$

$$b = -0.0205$$

$$c = 0.0266 \qquad (4.11)$$

4.3   STATISTICS IN THE MEASUREMENT PROCESS

4.3.1   Unbiased Estimates

Data sets typically have two very important characteristics: *central tendency* (or most representative value) and *dispersion* (or scatter). Other characteristics such as skewness and flatness (or peakness) may also be of importance but will not be considered here.[7]
   A basic problem in every quantitative experiment is that of obtaining an unbiased estimate of the true value of a quantity as well as an unbiased measure of the dispersion or uncertainty in the measured variable. Philosophically, in any measurement process a deterministic event is observed through a "foggy" window. If so, ultimate refinement of the measuring system would result in all values of measurements to be the true value, μ. Because errors occur in all measurements, one can never exactly measure the true value of any quantity. Continued refinement of the methods used in any measurement will yield closer and closer approximations, but there is always a limit beyond which refinements cannot be made. To determine the relation that a measured value has with the true value, we must specify the unbiased estimate $\bar{x}$ of the true value, μ, of a measurement and its uncertainty (or precision) interval, W_x, based on a desired confidence level (or probability of occurrence).
   An *unbiased estimator*[6] exists if the mean of its distribution is the same as the quantity being estimated. Thus, for sample mean, $\bar{x}$, to be an unbiased estimator of population mean, μ, the mean of the distribution of sample means, $\bar{\bar{x}}$, must be equal to the population mean.

4.3.2   Sampling

Unbiased estimates for determining population mean, population variance, and variance of the sample mean depend on the type of sampling procedure used.
   Sampling with replacement (random sampling):

$$\hat{\mu} = \bar{x} \qquad (4.12)$$

where $\bar{x}$ = sample mean and $\hat{\mu}$ is the unbiased estimate of the population mean, μ.

$$\hat{\sigma}^2 = S^2\left(\frac{n}{n-1}\right)$$

where S^2 = sample variance

$$S^2 = \frac{\sum(x_i - \bar{x})^2}{n} \tag{4.13}$$

$$\hat{\sigma}_{\bar{x}}^2 = \frac{\hat{\sigma}^2}{n} \tag{4.14}$$

where $\sigma_{\bar{x}}^2$ is the variance of the mean.

Sampling without replacement (the usual case):

$$\hat{u} = \bar{x} \tag{4.15}$$

$$\hat{\sigma}^2 = S^2\left(\frac{n}{n-1}\right)\left(\frac{N-1}{N}\right) \tag{4.16}$$

where N = population size and n = sample size,

$$\hat{\sigma}_{\bar{x}}^2 = \frac{\hat{\sigma}^2}{n}\left(\frac{N-n}{N-1}\right) \tag{4.17}$$

Note that sampling without replacement from an extremely large population is equivalent to random sampling.

4.3.3 Types of Errors

There are at least three types of errors that one must consider in making measurements. They are systematic (or fixed) errors, illegitimate errors, and random errors.

Systematic errors are of consistent form. They result from conditions or procedures that are correctable. This type of error may generally be eliminated by calibration.

Illegitimate errors are mistakes and should not exist. They may be eliminated by using care in the experiment, proper laboratory procedures, and repetition of the measurement.

Random errors are accidental errors that occur in all measurements. They are characterized by their inconsistent nature, and their origin cannot be determined in the measurement process. These errors are estimated by statistical analysis.

If the illegitimate errors can be eliminated by care and proper laboratory procedures and the systematic errors can be eliminated by calibrating the measurement system, then the random errors remain to be determined by statistical analysis to yield the precision of the measurement.

4.3.4 Propagation of Error or Uncertainty

In many cases the desired quantity cannot be measured directly but must be calculated from the most representative value (e.g., the mean) of two or more measured quantities. It is desirable to know the uncertainty or precision of such calculated quantities.

Precision is specified by quantities called *precision indexes* (denoted by W_x) that are calculated from the random errors of a set of measurements. A $\pm W_x$ should be specified for every measured variable. The confidence limits or probability for obtaining the range $\pm W_x$ is generally specified directly or is implied by the particular type of precision index being used.

The precision index of a calculated quantity depends on the precision indexes of the measured quantities required for the calculations.[6] If the measured quantities are determined independently and if their distribution about a measure of central tendency is approximately symmetrical, the following "propagation-of-error" equation is valid.[8]

$$W_R^2 = \sum\left(\frac{\partial R}{\partial x_i}\right)^2 W_{x_i}^2 \tag{4.18}$$

In this equation, R represents the calculated quantity and $x_1, x_2, \ldots, x_n$ represent the measured independent variables so that mathematically we have $R = f(x_1, x_2, \ldots, x_n)$. The precision index is a measure of dispersion about the central tendency and is denoted by W in Eq. 4.18. The standard deviation is often used for W, however, any precision index will do as long as the same type of precision index is used in each term of the equation.

A simplified form of this propagation-of-error equation results if the function R has the form

$$R = kx_1^a x_2^b x_3^c \ldots x_n^m \qquad (4.19)$$

where the exponents $a, b, \ldots, m$ may be positive or negative, integer or noninteger. The simplified result is

$$\left(\frac{W_R}{R}\right)^2 = a^2 \left(\frac{W_{x_1}}{x_1}\right)^2 + b^2 \left(\frac{W_{x_2}}{x_2}\right)^2 + \cdots + m^2 \left(\frac{W_{x_n}}{x_n}\right)^2 \qquad (4.20)$$

The propagation-of-error equation is also used in planning experiments. If a certain precision is desired on the calculated result, R, the precision of the measured variables can be determined from this equation. Then, the cost of a proposed measurement system can be determined as it is directly related to precision.

Example 4.3 Propagation of uncertainty. Determine the resistivity and its uncertainty for a conducting wire of circular cross section from the measurements of resistance, length, and diameter. Given

$$R = \rho \frac{L}{A} = \rho \frac{4L}{\pi D^2} \quad \text{or} \quad \rho = \frac{\pi D^2 R}{4L} \qquad (4.21)$$

$$R = 0.0959 \pm 0.0001 \text{ ohms}$$

$$L = 250 \pm 2.5 \text{ cm}$$

$$D = 0.100 \pm 0.001 \text{ cm}$$

where

$R =$ wire resistance in ohms

$L =$ wire length in centimeters

$A =$ cross-sectional area in square centimeters $= \dfrac{\pi D^2}{4}$

$\rho =$ wire resistivity in ohm-centimeters

Solution:

$$\rho = \frac{(\pi)(0.100)^2(0.0959)}{4(250)} = 3.01 \times 10^{-6} \text{ ohm-cm}$$

The propagation of variance (or precision index) equation for ρ reduces to the simplified form, that is,

$$\left(\frac{W_\rho}{\rho}\right)^2 = 4\left(\frac{W_D}{D}\right)^2 + \left(\frac{W_R}{R}\right)^2 + \left(\frac{W_L}{L}\right)^2$$

$$= 4\left(\frac{0.001}{0.10}\right)^2 + \left(\frac{0.0001}{0.0959}\right)^2 + \left(\frac{2.5}{250}\right)^2$$

$$= 4.00 \times 10^{-4} + 1.09 \times 10^{-6} + 1.00 \times 10^{-4}$$

$$= 5.01 \times 10^{-4}$$

$$W_\rho = \rho\sqrt{(5.01)10^{-4}} = \pm 6.74 \times 10^{-8}$$

$$\rho = (3.01 \pm 0.07) \times 10^{-6} \text{ ohm-cm}$$

4.3.5   Uncertainty Interval

When several measurements of a variable have been obtained to form a data set (multisample data), the best estimates of the most representative value (mean) and dispersion (standard deviation) are obtained from the formulae in Section 4.3.2. When a single measurement exists (or when the data are taken so that they are equivalent to a single measurement) the standard deviation cannot be determined and the data are said to be "single-sample" data. Under these conditions the only estimate of the true value is the single measurement, and the uncertainty interval must be estimated by the observer.[9] It is recommended that the precision index be estimated as the maximum error and that it corresponds approximately to the 99% confidence level associated with multisample data.

Uncertainty Interval Considering Random Error.   Once the unbiased estimates of mean and variance are determined from the data sample, the uncertainty interval for μ is

$$\mu = \hat{\mu} \pm \hat{W} = \hat{\mu} \pm k(\nu, \gamma)\hat{\sigma} \qquad (4.22)$$

where $\hat{\mu}$ represents the most representative value of μ from the measured data and $\hat{W}$ is the uncertainty interval or precision index associated with the estimate of μ. The magnitude of the precision index or uncertainty interval depends on the confidence level, γ (or probability chosen), the amount of data, n, and the type of probability distribution governing the distribution of measured items.

The uncertainty interval, $\hat{W}$, can be replaced by $k\hat{\sigma}$, where $\hat{\sigma}$ is the standard deviation (measure of dispersion) of the population as estimated from the sample and k is a constant that depends on the probability distribution function, the confidence level, γ, and the amount of data, n. For example, with a Gaussian distribution the 95% confidence limits are $\hat{W} = 1.96\hat{\sigma}$, where $k = 1.96$ is independent of n. For a t-distribution, $k = 2.78$, 2.06, and 1.96 with a sample size of 5, 25, and ∞, respectively, at the 95% level of confidence probability. The t-distribution is the same as the Gaussian distribution as $n \to \infty$.

Uncertainty Interval Considering Random Error with Resolution, Truncation, and Significant Digits.   The uncertainty interval, $\hat{W}$, in Eq. 4.22 assumes a set of measured values with only random error present. Furthermore, the set of measured values is assumed to have unbounded significant digits and to have been obtained with a measuring system having infinite resolution. When finite resolution exists and truncation of digits occurs, the uncertainty interval may be larger than that predicted by consideration of the random error only. The uncertainty interval can never be less than the resolution limits or truncation limits of the measured values.

Resolution and Truncation.   Let $\{s_n\}$ be the theoretically possible set of measurements of unbound significant digits from a measuring system of infinite resolution, and let $\{x_n\}$ be the actual set of measurements expressed to m significant digits from a measuring system of finite resolution. Then the quantity $s_i - x_i = \pm e_i$ is the resolution or truncation deficiency caused by the measurement process. The unbiased estimates of mean and variance are

$$\hat{\mu} = \frac{\sum s_i}{n} = \bar{s} \quad \text{and} \quad \hat{\sigma}^2 = \frac{\sum (s_i - \bar{s})^2}{n - 1} \qquad (4.23)$$

Noting that the set $\{x_n\}$ is available rather than $\{s_n\}$, the required mean and variance are

$$\hat{\mu} = \frac{\sum x_i}{n} \pm \frac{\sum e_i}{n} = \bar{x} \pm \frac{\sum e_i}{n} \quad \text{and} \quad \hat{\sigma}^2 = \frac{\sum (x_i - \bar{x})^2}{n - 1} \qquad (4.24)$$

The truncation or resolution has no effect on the estimate of variance but does affect the estimate of the mean. The truncation error, e_i, is not necessarily distributed randomly and may all be of the same sign. Thus $\bar{x}$ can be biased as much as $\sum e_i/n = \bar{e}$ high or low from the unbiased estimate of the value of μ so that $\hat{\mu} = \bar{x} \pm \bar{e}$.

If e_i is a random variable observed through a "cloudy window" with a measuring system of finite resolution, the value of e_i may be plus or minus but its upper bound is R (the resolution of the measurement). Thus the resolution error is no larger than R and $\hat{\mu} = \bar{x} \pm Rn/n = \bar{x} \pm R$.

If the truncation is never more than that dictated by the resolution limits (R) of the measurement system, the uncertainty in $\bar{x}$ as a measure of the most representative value of μ is never larger than R plus the uncertainty due to the random error. Thus $\hat{\mu} = \bar{x} \pm (\hat{W} + R)$. It should be emphasized that the uncertainty interval can never be less than the resolution bounds of the measurement. The resolution bounds cannot be reduced without changing the measurement system.

Significant Digits. When x_i is observed to m significant digits, the uncertainty (except for random error) is never more than $\pm 5/10^m$ and the bounds on s_i are equal to $x_i \pm 5/10^m$ so that

$$x_i - \frac{5}{10^m} < s_i < x_i + \frac{5}{10^m} \tag{4.25}$$

The relation for $\hat{\mu}$ for m significant digits is then

$$\hat{\mu} = \bar{x} \pm \frac{\sum e_i}{n} = \bar{x} \pm \frac{\sum 5/10^m}{n} = \bar{x} \pm \frac{5}{10^m} \tag{4.26}$$

The estimated value of variance is not affected by the constant magnitude of $5/10^m$.

When the uncertainty due to significant digits is combined with the resolution limits and random error the uncertainty interval on $\hat{\mu}$ becomes

$$\hat{\mu} = \bar{x} \pm \left(\hat{W} + R + \frac{5}{10^m} \right) \tag{4.27}$$

This illustrates that the number of significant digits of a measurement should be carefully chosen in relation to the resolution limits of the measuring system so that $5/10^m$ has about the same magnitude as R. Additional significant digits would imply more accuracy to the measurement than would actually exist based on the resolving ability of the measuring system.

4.3.6 Amount of Data to Take

Exactly what data to take and how much data to take are two important questions to be answered in any experiment. Assuming that the correct variables have been measured, the amount of data to be obtained can be determined by using the relation

$$\mu = \bar{\bar{x}} \pm \left(\hat{W}_{\bar{x}} + R + \frac{5}{10^m} \right) \tag{4.28}$$

where it is presumed that several samples may exist for estimation of μ. This equation can be rewritten using Eqs. 4.13 and 4.14 (assuming random sampling).

$$\mu = \bar{\bar{x}} \pm \left(k(\nu, \gamma)\hat{\sigma}_{\bar{x}} + R + \frac{5}{10^m} \right)$$

$$= \bar{\bar{x}} \pm \left(k(\nu, \gamma)\frac{\hat{\sigma}}{\sqrt{n}} + R + \frac{5}{10^m} \right) \tag{4.29}$$

The value of n to achieve the difference in $\mu - \bar{\bar{x}}$ within a stated percent of μ (i.e., $(\%/100)\mu = \mu - \bar{x}$) can be determined from

$$n = \left[\frac{k(\nu, \gamma)\hat{\sigma}}{(\%/100)\hat{\mu} - R - (5/10^m)} \right]^2 \tag{4.30}$$

This equation can only yield valid values of n once valid estimates of $\hat{\mu}, \hat{\sigma}, k, R$, and m are available. This means that the most correct values of n can only be obtained once the measurement system and data-taking procedure have been specified so that R and m are known. Furthermore, either a preliminary experiment or a portion of the actual experiment should be performed to obtain good estimates of $\hat{\mu}$ and $\hat{\sigma}$. Because k not only depends on the type of distribution the data follows but also on the sample size, n, the solution is iterative. Thus the most valid estimates of the amount of data to take can only be obtained after the experiment has begun. However, the equation can be quite useful for prediction purposes if one wishes to estimate values of $\hat{\mu}, \hat{\sigma}, k, R$, and m. This is especially important in experiments for which the cost of a single run may be relatively high.

Example 4.4 Amount of data to take. The life for a certain type of automotive tire is to be established. The mean and standard deviation of the life estimated for these tires is 84,000 km and $\pm 7,230$ km, respectively, from a sample of nine tires. On the basis of the sample, how many data are required to establish the life of this type of tire to within $\pm 10\%$ with 90% confidence and a resolution of five kilometers?

Solution:

Confidence limits

$$\mu = \hat{\mu} \pm \left(t\hat{\sigma}_{\bar{x}} + R + \frac{5}{10^m} \right)$$

$$\mu - \hat{\mu} = (0.10)\bar{x} = t\hat{\sigma}_{\bar{x}} + R + \frac{5}{10^m} = t\frac{\hat{\sigma}_x}{\sqrt{n}} + R + \frac{5}{10^m}$$

$$\frac{t}{\sqrt{n}} = \frac{(0.10)\bar{x} - R - (5/10^m)}{\hat{\sigma}_x} = \frac{(0.10)(84000) - 5}{230} = 1.16$$

n	ν	$t(\nu, 0.10)*$	$t/\sqrt{n}$
2	1	6.31	3.65
3	2	2.92	1.46
5	4	2.13	0.87

*From a *t*-statistic table.[6]

Conclusion: A sample of 5 tires is sufficient to establish the tire life within $\pm 10\%$ at a 90% level of confidence.

4.3.7 Goodness-of-Fit

Statistical methods can be used to fit a curve to a given data set. In general, the Least-Squares principle is used to minimize the sum of the squares of deviations away from the curve to be fitted. The deviations from an assumed curve, $y = f(x)$, are due to errors in y, in x, or in both y and x. In most cases the errors in the independent variable, x, are much smaller than the dependent variable, y. Therefore, only the errors in y are considered for the Least-Squares curve. The goodness-of-fit of an assumed curve is defined by the correlation coefficient, r, where

$$r = \pm \sqrt{\frac{\sum (y - \bar{y})^2}{\sum (y_i - \bar{y})^2}} = \pm \sqrt{1 - \frac{\sum (y_i - y)^2}{\sum (y_i - \bar{y})^2}}$$

$$= \pm \sqrt{1 - \frac{\hat{\sigma}_{y,x}^2}{\hat{\sigma}_y^2}}$$

(4.31)

where

$$\sum (y_i - \bar{y})^2 = \text{total variation (variation about the mean)}$$

$$\sum (y_i - y)^2 = \text{unexplained variation (variation about regression)}$$

$$\sum (y - \bar{y})^2 = \text{explained variation (variation based on the assumed regression equation)}$$

$\hat{\sigma}_y$ is the estimated population standard deviation of the y variable, and $\hat{\sigma}_{y,x}$ is the standard error of the estimate of y on x.

When the correlation coefficient, r, is zero, the data cannot be explained by the assumed curve. However, when r is close to ± 1, the variation of y with respect to x can be explained by the assumed curve and a good correlation is indicated between the variables x and y.

The probabilistic statement for the goodness-of-fit test is given by

$$P[r_{\text{calc}} > r] = \alpha = 1 - \gamma$$

(4.32)

where r_{calc} is calculated from Eq. 4.31 and the null and alternate hypotheses are:

H_0: No correlation of assumed regression equation with data
H_1: Correlation of regression equation with data

The goodness-of-fit for a straight line is determined by comparing the r_{calc} with the value of r obtained at $n - 2$ degrees of freedom at a selected confidence level, γ, from tables. If r_{calc} is greater than r, the null hypothesis is rejected and a significant fit of the data within the confidence level specified is inferred. However, if r_{calc} is less than r, the null hypothesis cannot be rejected and no correlation of the curve fit with the data is inferred at the chosen confidence level.

Example 4.5 Goodness-of-fit test. The given x-y data were fitted to a curve $y = a + bx$ by the method of Least-Squares linear regression. Determine the goodness-of-fit at a 5% significance level.

$$x = [56, 58, 60, 70, 72, 75, 77, 77, 82, 87, 92, 104, 125]$$

$$y = [51, 60, 60, 52, 70, 65, 49, 60, 63, 61, 64, 84, 75]$$

At $\alpha = 0.05$, the value of r is 0.55. Thus $P[r_{calc} > 0.55] = 0.05$. The Least-Squares regression equation is calculated to be $y = 39.32 + 0.30x$, and the correlation coefficient is calculated as $r_{calc} = 0.61$. Therefore, a satisfactory fit of the regression line to the data is inferred at the 5% significance level.

4.3.8 Probability Density Functions

Consider the measurement of a quantity x. Let,

$x_i = i$th measurement of the quantity

$\bar{x} = $ most representative value of the measured quantity

$d_i = x_i - \bar{x} = $ deviation of the ith measurement from $\bar{x}$

$n = $ total number of measurements

$\Delta x = $ smallest measurable change of x, also known as the "least count"

$m_j = $ number of measurements in x_j size group

A *histogram* and the corresponding *frequency polygon* obtained from the data are shown in Fig. 4.3 where the x_j size group is taken to be inclusive at the lower limit and exclusive at the upper limit. Thus each x_j size group corresponds to the following range of values.

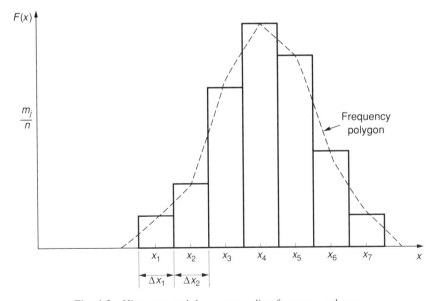

Fig. 4.3 Histogram and the corresponding frequency polygon.

$$x_j - (\Delta x_j/2) \le x < x_j + (\Delta x_j/2) \tag{4.33}$$

The height of any rectangle of the histogram shown is denoted as the relative number m_j/n and is equal to the statistical probability $F(x)$ that a measured value will have the size $x_j \pm (\Delta x_j/2)$. The area of each rectangle can be made equal to the relative number by transforming the ordinate of the histogram in the following way: *area = relative number* $= F(x) = m_j/n = p_j(x)\Delta x$. Thus

$$p_j(x) = \frac{m_j}{n\,\Delta x} \tag{4.34}$$

The shape of the histogram is preserved in this transformation since the ordinate is merely changed by a constant scale factor $(1/\Delta x)$. The resulting diagram is called the *probability density diagram*. The sum of the areas underneath all rectangles is then equal to one (i.e., $\sum p_j(x)\Delta x_j = 1$). In the limit, as the number of data approaches infinity and the least count becomes very small, a smooth curve called the *probability density function* is obtained. For this smooth curve we note that

$$\int_{-\infty}^{+\infty} p(x)\,dx = 1 \tag{4.35}$$

and that the probability of any measurement, x, having values between x_a and x_b is found from

$$P(x_a \le x \le x_b) = \int_{x_a}^{x_b} p(x)\,dx \tag{4.36}$$

In order to integrate this expression, the exact probability density function $p(x)$ is required. Based on the assumptions made, several forms of frequency distribution laws have been obtained. The distribution of a proposed set of measurements is usually unknown in advance. However, the Gaussian (or normal) distribution fits observed data distributions in a large number of cases.

The Gaussian probability density function is given by the expression

$$p(x) = \left(1/\sigma\sqrt{2\pi}\right)e^{-(x-\bar{x})^2/2\sigma^2} \tag{4.37}$$

where σ is the standard deviation. The standard deviation is a measure of dispersion and is defined by the following relation

$$\sigma = \frac{\int_{-\infty}^{+\infty} x^2 p(x)\,dx}{\int_{-\infty}^{+\infty} p(x)\,dx} = \sqrt{\frac{\sum (x_i - \mu)^2}{n}} \tag{4.38}$$

4.3.9 Determination of Confidence Limits on μ

If a set of measurements is given by a random variable, x, then the central limit theorem[10] states that the distribution of means, $\bar{x}$, of the samples of size n is Gaussian (normal) with mean, μ, and variance, $\sigma_{\bar{x}}^2 = \sigma^2/n$, that is, $\bar{x} \sim G(\mu, \sigma^2/n)$. [Also, the random variable $z = (\bar{x} - \mu)/(\sigma/\sqrt{n})$ is Gaussian with a mean of zero and a variance of unity, that is, $z \sim G(0, 1)$]. The random variable, z, is used to determine the confidence limits on μ due to random error of the measurements when σ is known.

The confidence limit is determined from the following probabilistic statement and the Gaussian distribution for a desired confidence level, γ.

$$P\left[-z < \frac{\bar{x} - \mu}{\sigma/\sqrt{n}} < z\right] = \gamma \tag{4.39}$$

It shows a γ probability or confidence that the experimental value of z will be between $\pm z$ obtained from the Gaussian distribution table. For a 95% confidence level, $z = \pm 1.96$ from the Gaussian table[6] and $P[-1.96 < z < 1.96] = \gamma$ where $z = (\bar{x} - \mu)/(\sigma/\sqrt{n})$. Therefore, the expression for the 95% confidence limits on μ is

$$\mu = \bar{x} \pm 1.96\frac{\sigma}{\sqrt{n}} \tag{4.40}$$

In general

$$\mu = \bar{x} \pm k \frac{\sigma}{\sqrt{n}} \tag{4.41}$$

where k is found from the Gaussian table for the specified value of confidence, γ.

If the population variance is not known and must be estimated from a sample, the statistic $(\bar{x} - \mu)/(\sigma/\sqrt{n})$ is not distributed normally but follows the t-distribution. When n is very large the t-distribution is the same as the Gaussian distribution. When n is finite the value of k is the "t" value obtained from the t-distribution table.[6] The probabilistic statement then becomes

$$P\left[-t < \frac{\bar{x} - \mu}{\sigma/\sqrt{n}} < +t\right] = \gamma \tag{4.42}$$

and the inequality yields the expression for the confidence limits on μ.

$$\mu = \bar{x} \pm t \frac{\sigma}{\sqrt{n}} \tag{4.43}$$

If the effects of resolution and significant digits are included, the expression becomes as previously indicated in Eq. 4.29.

$$\mu = \bar{x} \pm \left(t \frac{\sigma}{\sqrt{n}} + R + \frac{5}{10^m}\right) \tag{4.44}$$

4.3.10 Confidence Limits on Regression Lines

The Least-Squares method is used to fit a straight line to data that is either linear or transformed to a linear relation.[6,11] The following method assumes that the uncertainty in the variable x is negligible compared to the uncertainty in the variable y and that the uncertainty in the variable y is independent of the magnitude of the variable x.

Figure 4.4 and the definitions that follow are used to obtain confidence levels relative to regression lines fitted to experimental data.

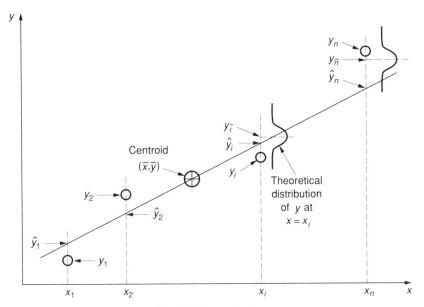

Fig. 4.4 Regression line.

a = intercept of the regression line

b = slope of the regression line = $\sum XY/\sum X^2$

y_i = value of y from data at $x = x_i$

$\hat{y}_i$ = value of y from regression line at $x = x_i$. Note that $y_i = a + bx_i$ for a straight line.

$\overline{y}_i$ = mean estimated value of y_i at $x = x_i$. This is the mean of the distribution of y values at $x = x_i$. If there is only one measurement of y at $x = x_i$ then that value of y (i.e., y_i) is the best estimate of $\overline{y}_i$.

ν = degrees of freedom in fitting regression line to data ($\nu = n - 2$ for a straight line)

$\sum (y_i - \hat{y}_i)^2/\nu = \hat{\sigma}^2_{y,x}$ = unexplained variance (for the regression line) where $\hat{\sigma}_{y,x}$ is the standard error of the estimate

$\sigma^2_{\overline{y},x} = \sigma^2_{y,x}/n$, from the central limit theorem

$\sigma^2_b = \sigma^2_{y,x}/\sum X^2$ = estimate of the variance on slope

Slope-Centroid Approximation. This method assumes that the placement uncertainty of the regression line is due to uncertainties in the centroid $(\overline{x}, \overline{y})$ of the data and the slope, b, of the regression line passing through this centroid. These uncertainties are determined from the following relations and are shown in Fig. 4.5.

$$Centroid:\ \mu_{\overline{y}} = \hat{\mu}_{\overline{y}} \pm t(\nu, \gamma)\hat{\sigma}_{\overline{y},x} = \overline{y} \pm t\frac{\hat{\sigma}_{y,x}}{\sqrt{n}} \tag{4.45}$$

$$Slope:\ \mu_b = \hat{b} \pm t(\nu, \gamma)\hat{\sigma}_b = \hat{b} \pm t\frac{\hat{\sigma}_{y,x}}{\sum X^2} \tag{4.46}$$

Point-by-Point Approximation. This is a better approximation than the slope-centroid technique. It gives confidence limits of the points, y_i, where

$$\mu_{y_i} = \hat{y}_i \pm t(\nu, \gamma)\hat{\sigma}_{y_i} \tag{4.47}$$

and

$$\hat{\sigma}^2_{y_i} = \hat{\sigma}^2_{y,x}\left[\frac{1}{n} + \frac{X_i^2}{\sum X_i^2}\right] \tag{4.48}$$

At the centroid where $X_i = 0$, this relation reduces to the result for $\mu_{\overline{y}}$ given in Eq. 4.45.

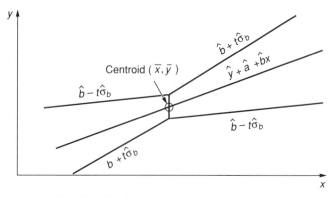

Fig. 4.5 Confidence limits on a regression line.

Line as a Whole. More uncertainty is involved when all points for the line are taken collectively. The "price" paid for the additional uncertainty is given by replacing $t(\nu, \gamma)$ in the confidence interval relation by $\sqrt{2F}$, where $F = F(2, \nu, \gamma)$. Thus the uncertainty interval for placement of the line as a whole confidence limits at $x = x_i$ is found by the formula

$$\mu_{\text{Line}} = \hat{y}_i \pm \sqrt{2F} \hat{\sigma}_{y.x} \sqrt{\frac{1}{n} + \frac{X_i}{\sum X_i^2}} \tag{4.49}$$

Future Estimate of a Single Point at $x = x_i$. This gives the expected confidence limits on a future estimate value of y at $x = x_i$ in relation to a prior regression estimate. This confidence limit can be found from the relation

$$\mu_y = \hat{y}_i \pm t(\nu, \gamma)\hat{\sigma}_{y_i} \tag{4.50}$$

where $\hat{y}_i$ is the best estimate of y and

$$\hat{\sigma}_{y_i}^2 = \hat{\sigma}_{y.x}^2 \left[1 + \frac{1}{n} + \frac{X_i^2}{\sum X_i^2} \right] \tag{4.51}$$

If one uses $\gamma = 0.99$, nearly all (99%) of the observed data points should be within the uncertainty limits calculated by Eq. 4.51.

Example 4.6 Confidence limits on a regression line. Calibration data for a copper-constantan thermocouple are given:

Temperature, $x = [0, 10, 20, 30, 40, 50, 60, 70, 80, 90, 100]$ °C

Voltage, $y = [-0.89, -0.53, -9.15, 0.20, 0.61, 1.03, 1.45, 1.88, 2.31, 2.78, 3.22]$ mV

If the variation of y is expected to be linear over the range of x and the uncertainty in temperature is much less than the uncertainty in voltage, then:

1. Determine the linear relation between y and x.
2. Test the goodness-of-fit at the 95% confidence level.
3. Determine the 90% confidence limits of point on the line by the slope-centroid technique.
4. Determine the 90% confidence limits on the intercept of the regression line.
5. Determine the 90% confidence limits on a future estimated point of temperature at 120°C.
6. Determine the 90% confidence limits on the whole line.
7. How much data would be required to determine the centroid of the voltage values within 1% at the 90% confidence level?

Calculations:

$$\bar{x} = 50.0$$

$$\bar{y} = 1.0827$$

$$y = a + bx$$

$$\sum x_i = 550$$

$$\sum y_i = 11.91$$

$$\sum x_i y_i = 1049.21$$

$$\sum (x_i)^2 = 38,500.0$$

$$\sum (y_i) = 31.6433$$

$$\sum (x_i - \bar{x})^2 = \sum X_i^2 = 11,000.00$$

$$\sum (y_i - \bar{y})^2 = \sum Y_i^2 = 18.7420$$

$$\sum (x_i - \bar{x})(y_i - \bar{y}) = \sum X_i Y_i = 453.7000$$

$$\sum (y_i - \hat{y}_i)^2 = 0.0299$$

1. $b = \sum XY / \sum X^2 = 453.7/11,000 = 0.0412$

   $a = \bar{y} - b\bar{x} = 1.0827 - (0.0412)(50.00) = 1.0827 - 2.0600$

   　　$= -0.9773$

2. $r_{\exp} = \sqrt{\sum (\hat{y} - \bar{y})^2 / \sum (y_i - \bar{y})^2} = \sqrt{1 - \sum (y_i - \hat{y}_i)^2 / \sum (y_i - \bar{y})^2} = \sum XY / \sqrt{\sum X^2 \sum Y^2}$

   　$= 0.998$

   $r_{\text{Table}} = r(\alpha, \nu) = r(0.05, 9) = 0.602$

   $P[r_{\exp} > r_{\text{Table}}] = \alpha$ with H_0 of no correlation, therefore reject H_0 and infer significant correlation

3. $t = t(\gamma, \nu) - t(0.90, 9) = 1.833$　　　(see Ref. 6)

   $\hat{\sigma}_{y.x}^2 = \sum (y_i - \hat{y}_i)^2 / \nu = 0.0299/9 = 0.00333$

   $\hat{\sigma}_{y.x} = \pm 0.0577, \quad \hat{\sigma}_{\bar{y}.x} = \pm 0.0173$

   $\hat{\sigma}_b = \hat{\sigma}_{y.x} / \sqrt{\sum X^2} = \pm 0.000550$

   $\mu_{\bar{y}} = \hat{\mu}_{\bar{y}} \pm t\hat{\sigma}_{\bar{y}.x} = y \pm t\hat{\sigma}\bar{y}, x$

   　　$= 1.0827 \pm (1.833)(0.01737) = 1.0827 \pm 0.0318$

   $\mu_b = \hat{b} \pm t\hat{\sigma}_b = 0.0412 \pm (1.833)(0.00055)$

   　　　$= 0.0412 \pm 0.00101$

4. $\mu_{y_i} = \hat{\mu}_{y_i} \pm t\hat{\sigma}_{y_i} = y_i \pm t\hat{\sigma}_{y.x} \sqrt{1/n + X^2 / \sum X^2}$

   　$= -0.9773 \pm (1.833)(0.0325)$

   　$= -0.9773 \pm 0.0596$

5. $\mu_{y_i} = \hat{\mu}_{y_i} \pm t\hat{\sigma}_{y_i} = \hat{y}_i + t\hat{\sigma}_{y.x} \sqrt{1 + 1/n + X^2 / \sum X^2}$

   　$= (-0.9773 + 4.9500) \pm (1.833)(0.0715)$

   　$= 3.9727 \pm 0.1310$

6. $\mu_{y_i \text{Line}} = \hat{\mu}_{y_i} \pm \sqrt{2F(2, \nu)}\hat{\sigma}_{y_i}$　　For any given point compare with number 4

   　$= \hat{y}_i \pm \sqrt{(2)(3.01)}\hat{\sigma}_{y_i}$

   　$= \hat{y}_i \pm (2.46)\hat{\sigma}_{y_i}$

   　$= -0.9773 \pm 0.0799$　　For point of number 4

7. $\mu_{\bar{y}_i} = \hat{\mu}_y \pm t\hat{\sigma}_{\bar{y}.x} \qquad \sigma_{\bar{y}.x} = \sigma\bar{y}, x / \sqrt{n}$

   　$= \bar{y} \pm t\dfrac{\hat{\sigma}_{y.x}}{\sqrt{n}} \qquad \hat{\sigma}_{y.x} = \pm 0.0577$

   $\mu_{\bar{y}} - \bar{y} = (1\%)(\mu_y) \cong (1\%)(\bar{y}) = 0.010827$

   $\therefore t/\sqrt{n} = (1\%)\bar{y}/\hat{\sigma}_{y.x} = 0.010827/0.05770 = 0.188$

From t table[6]:

ν	$t(0.10, \nu)$	$t/\sqrt{n}$
60	1.671	0.213
90	1.663	0.174
75	1.674	0.188

therefore $n = \nu + 1 = 77$

4.3.11　Inference and Comparison

Events are not only deterministic but probabilistic. Under certain specified conditions some events will always happen. Some other events, however, may or may not happen. Under the same specified conditions, the latter ones depend on chance and, therefore, the probability of occurrence of such events is of concern. For example, it is quite certain that a tossed unbiased die will fall down.

However, it is not at all certain which face will appear on top when the die comes to rest. The probabilistic nature of some events is apparent when questions of the following type are asked. Does Medicine A cure a disease better than Medicine B? What is the ultimate strength of 1020 steel? What total mileage will Brand X tire yield? Which heat treatment process is better for a given part?

Answering such questions involves designing experiments, performing measurements, analyzing the data, and interpreting the results. In this endeavor two common phenomena are observed: (1) repeated measurements of the same attribute differ due to measurement error and resolving capability of the measurement system, and (2) corresponding attributes of identical entities differ due to material differences, manufacturing tolerances, tool wear, and so on.

Conclusions based on experiments are statistical inferences and can only be made with some element of doubt. Experiments are performed to make statistical inferences with minimum doubt. Therefore, experiments are designed specifying the data required, amount of data needed, and the confidence limits desired in drawing conclusions. In this process an instrumentation system is specified, a data-taking procedure is outlined, and a statistical method is used to make conclusions at preselected confidence levels.

In statistical analysis of experimental data, the descriptive and inference tasks are considered. The descriptive task is to present a comprehensible set of observations. The inference task determines the truth of the whole by examination of a sample. The inference task requires sampling, comparison, and a variety of statistical tests to obtain unbiased estimates and confidence limits to make decisions.

Statistical Testing. A statistical hypothesis is an assertion relative to the distribution of a random variable. The test of a statistical hypothesis is a procedure to accept or reject the hypothesis A hypothesis is stated such that the experiment attempts to nullify the hypothesis Therefore, the hypothesis under test is called the *null* hypothesis and symbolized by H_0. All alternatives to the null hypothesis are termed the *alternate* hypothesis and are symbolized by H_1.[12]

If the results of the experiment cannot reject H_0, the experiment cannot detect the differences in measurements at the chosen probability level.

Statistical testing determines if a process or item is better than another with some stated degree of confidence. The concept can be used with a certain statistical distribution to determine the confidence limits.

The following procedure is used in statistical testing.

1. Define H_0 and H_1.
2. Choose the confidence level, γ, of the test.
3. Form an appropriate probabilistic statement.
4. Using the appropriate statistical distribution perform the indicated calculation.
5. Make a decision concerning the hypothesis and/or determine confidence limits.

Two types of error are possible in statistical testing. A Type I error is that of rejecting a true null hypothesis (rejecting truth). A Type II error is that of accepting a false null hypothesis (embracing fiction). The confidence levels and sample size are chosen to minimize the probability of making a Type I or a Type II error. Figure 4.6 illustrates the Type I (α) and Type II (β) errors, where

H_0—Sample with $n = 1$ comes from $N(10,16)$
H_1—Sample with $n = 1$ comes from $N(17,16)$
α = probability of Type I error (concluding the data come from $N(17,16)$ when the data actually come from $N(10,16)$)
β = probability of a Type II error (accepting that the data come from $N(10,16)$ when the data actually come from $N(17,16)$)

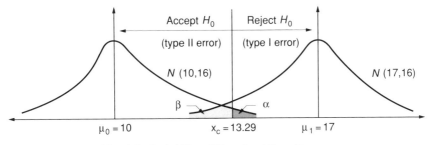

Fig. 4.6 Probability of Type I and Type II errors.

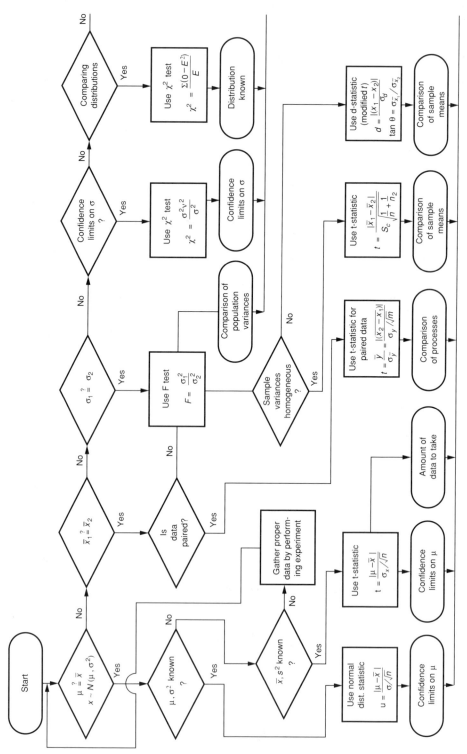

Fig. 4.7 Flow chart for statistical tests.

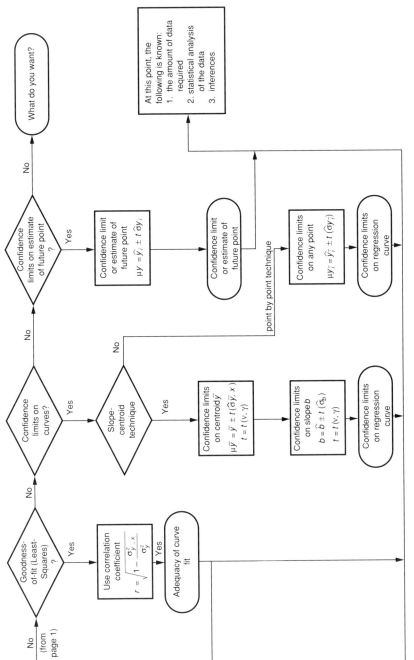

Fig. 4.7 *(Continued)*

When $\alpha = 0.05$ and $n = 1$, the critical value of x is 16.58. If the value of x obtained from the measurement is less than 16.58, accept H_0. If x is larger than 16.58, reject H_0 and infer H_1. For $\beta = 0.458$ there is a large chance for a Type II error. To minimize this error, increase α or the sample size. For example, when $\alpha = 0.05$ and $n = 4$ the critical value of x is 13.29 and $\beta = 0.032$. Thus the chance of a Type II error is significantly decreased by increasing the sample size.

Various statistical tests are summarized in the flow chart shown in Fig. 4.7.

Comparison of Variability.   To test whether two samples are from the same population, their variability or dispersion characteristics are compared using F statistics.[6,12]

If x and y are random variables from two different samples, the parameters $U = \sum (x_i - \bar{x})^2/\sigma_x^2$ and $V = \sum (y_i - \bar{y})^2/\sigma_y^2$ are also random variables and have chi-square distributions (see Fig. 4.8) with $\nu_1 = n_1 - 1$ and $\nu_2 = n_2 - 1$ degrees of freedom, respectively. The random variable, W, formed by the ratio $(U/\nu_1)/(V/\nu_2)$ has an F-distribution with ν_1 and ν_2 degrees of freedom (i.e., $W \sim F(\gamma, \nu_1, \nu_2)$). The quotient W is symmetric; therefore, $1/W$ has an F-distribution with ν_2 and ν_1 degrees of freedom (i.e., $1/W \sim F(\alpha, \nu_2, \nu_1)$). Figure 4.9 shows the F-distribution and its probabilistic statement.

$$P[F_L < W < F_R] = \gamma \tag{4.52}$$

where $W = (U/\nu_1)/(V/\nu_2) = \hat{\sigma}_1^2/\hat{\sigma}_2^2$ with

$$H_0 \text{ as } \sigma_1^2 = \sigma_2^2 \quad \text{and} \quad H_1 \text{ as } \sigma_1^2 \neq \sigma_2^2 \tag{4.53}$$

W is calculated from values of $\hat{\sigma}_1^2$ and $\hat{\sigma}_2^2$ obtained from the samples.

Example 4.7   Testing for homogeneous variances.   Test the hypothesis at the 90% level of confidence that $\sigma_1^2 = \sigma_2^2$ when the samples of $n_1 = 16$ and $n_2 = 12$ yielded $\hat{\sigma}_1^2 = 5.0$ and $\hat{\sigma}_2^2 = 2.5$. Here H_0 is $\sigma_1 = \sigma_2$ and H_1 is $\sigma_1 \neq \sigma_2$

$$P[F_L(\nu_1, \nu_2) < \left(\frac{\hat{\sigma}_1}{\hat{\sigma}_2}\right)^2 < F_R(\nu_1, \nu_2)] = \gamma$$

$$P[F_{0.95}(15, 11) < \left(\frac{\hat{\sigma}_1}{\hat{\sigma}_2}\right)^2 < F_{0.05}(15, 11)] = 0.90$$

$$F_{0.95}(15, 11) = \frac{1}{F_{0.05}(11, 15)} = \frac{1}{2.51} = 0.398$$

$$P[0.398 < \left(\frac{\hat{\sigma}_1}{\hat{\sigma}_2}\right)^2 < 2.72] = 0.90$$

Since $\hat{\sigma}_1^2/\hat{\sigma}_2^2 = 2.0$, H_0 cannot be rejected at the 90% confidence level.

Comparison of Means.   Industrial experimentation often compares two treatments of a set of parts to determine if a part characteristic such as strength, hardness, or lifetime has been improved. If it is assumed that the treatment does not change the variability of items tested (H_0), then the t-distribution determines if the treatment had a significant effect on the part characteristics (H_1). The t-statistic is

$$t = \frac{d - \mu_d}{\sigma_d} \tag{4.54}$$

where $d = \bar{x}_1 - \bar{x}_2$ and $\mu_d = \mu_1 - \mu_2$. From the propagation of variance, σ_d^2 becomes

$$\sigma_d^2 = \sigma_{\bar{x}_1}^2 + \sigma_{\bar{x}_2}^2 = \frac{\sigma_1^2}{n_1} + \frac{\sigma_2^2}{n_2} \tag{4.55}$$

where σ_1^2 and σ_2^2 are each estimates of the population variance σ^2. A better estimate of σ^2 is the combined variance σ^2, and it replaces both σ_1^2 and σ_2^2 in Eq. 4.55. The combined variance is determined by weighting the individual estimates of variance based on their degrees of freedom according to the relation

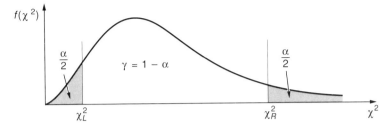

Fig. 4.8 Chi-square distribution.

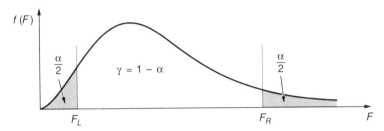

Fig. 4.9 F-distribution.

$$\sigma_c^2 = \frac{\hat{\sigma}_1^2 \nu_1 + \hat{\sigma}_2^2 \nu_2}{\nu_1 + \nu_2} \tag{4.56}$$

Then

$$\sigma_d^2 = \frac{\hat{\sigma}_c^2}{n_1} + \frac{\hat{\sigma}_c^2}{n_2} = \sigma_c^2 \left(\frac{1}{n_1} + \frac{1}{n_2} \right) \tag{4.57}$$

Under the hypothesis, H_0, that $\mu_1 = \mu_2$ (no effect due to treatment) the resulting probabilistic statement is

$$P\left[-t\sigma_c \sqrt{\frac{1}{n_1} + \frac{1}{n_2}} < \bar{x}_1 - \bar{x}_2 < t\sigma_c \sqrt{\frac{1}{n_1} + \frac{1}{n_2}} \right] = \gamma \tag{4.58}$$

If the variances of the items being compared are not equal (homogeneous), a modified t- (or d-) statistic is used,[6,12] where d depends on confidence level, γ, degrees of freedom, ν, and a parameter, θ. The parameter θ depends on the ratio of standard deviations according to

$$\tan \theta = \frac{\hat{\sigma}_1 / \sqrt{n_1}}{\hat{\sigma}_2 / \sqrt{n_2}} \tag{4.59}$$

The procedure for using the d-statistic is the same as described for the t-statistic.

Example 4.8   Testing for homogeneous means.   A transducer manufacturer has the following data:

Sample number	Number of transducers	Mean lifetime (hours)	Variance (hours)
1	15	2530	490
2	11	2850	360

Determine if the variance in the lifetime of the transducer is due to chance at a 10% significance level.

Solution:

$$\hat{\sigma}_1^2 = 490 \left(\frac{15}{14} \right) = 525$$

$$\hat{\sigma}_2^2 = 360 \left(\frac{11}{10} \right) = 396$$

$$\sigma_c^2 = \frac{(14)(525) + (10)(396)}{24} = 471$$

Check variance first (H_0—homogeneous variances, and H_1—nonhomogeneous variances)

$$P[F_L < F_{calc} < F_R] = \gamma$$

$$F_{exp} = \frac{\hat{\sigma}_1^2}{\hat{\sigma}_2^2} = \frac{525}{396} = 1.33$$

$$F_R = F_{0.05}(\nu_1, \nu_2) = F_{0.05}(14.10) = 2.8$$

$$F_L = F_{0.95}(\nu_1, \nu_2) = \frac{1}{F_{0.05}(10, 14)} = \frac{1}{2.60} = 0.385$$

Therefore, accept H_0 (variances are homogeneous).
Check means next (H_0 is $\mu_1 = \mu_2$, and H_1 is $\mu_2 \neq \mu_1$)

$$P[-t < t_{calc} < t] = \gamma$$

$$t = t(\nu_1 + \nu_2, \gamma) = t(24, 0.90) = 1.711$$

Therefore, $P[-1.71 < t_{exp} < 1.71] = 0.90$

$$t_{exp} = \frac{|\bar{x}_1 - \bar{x}_2|}{\sqrt{\dfrac{\sigma_1^2}{n_1} + \dfrac{\sigma_2^2}{n_2}}} = \frac{|\bar{x}_1 - \bar{x}_2|}{\sqrt{\sigma_c^2 \left(\dfrac{1}{n_1} + \dfrac{1}{n_2} \right)}}$$

$$t_{exp} = \frac{|2530 - 2850|}{\sqrt{(471)(0.1576)}}$$

$$= \frac{|-320|}{\sqrt{73.6}} = \frac{320}{8.58}$$

$$= 32.3$$

Therefore, reject H_0 and accept H_1 of $\mu_1 \neq \mu_2$ and infer differences in samples are not due to chance alone but due to some real cause.

Comparing Distributions. A chi-square distribution is also used for testing whether or not an observed phenomenon fits an expected or theoretical behavior.[12] The chi-square statistic is defined as

$$\sum \left[\frac{(O_j - E_j)^2}{E_j} \right]$$

where O_j is the observed frequency of occurrence in the jth class interval and E_j is the expected frequency of occurrence in the jth class interval. The expected class frequency is based on an assumed distribution or a hypothesis, H_0. This statistical test is used to compare the performance of machines or other items. For example, lifetimes of vacuum tubes, locations of hole center lines on rectangular plates, and locations of misses from targets in artillery and bombing missions, etc., follow chi-square distributions.

The probabilistic statements depend on whether a one-sided test or a two-sided test is being performed.[13] The typical probabilistic statements are

$$P[\chi^2_{\exp} > \chi^2(\alpha, \nu)] = \alpha \qquad (4.60)$$

for the one-sided test, and

$$P[\chi^2_L < \chi^2_{\exp} < \chi^2_R] = \gamma \qquad (4.61)$$

for the two-sided test where

$$\chi^2_{\exp} = \frac{\sum(O_j - E_j)^2}{E_j} \qquad (4.62)$$

In using Eq. 4.62 at least five items per class interval are normally used, and a continuity correction must be made if less than three class intervals are used.

Example 4.9   Use of the chi-square distribution. In a calibration laboratory the following record shows the number of failures of a certain type of transducer.

No. of Failures	Observed No. of Transducers	Expected No. of Transducers*
0	364	325
1	376	396
2	218	243
3	89	97
4	33	30
5	13	7
6	4	1
7	3	0
8	2	0
9	1	0
	$\sum = 1103$	

*From an assumed statistical model that the failures are random.

At a 95% confidence level, determine whether the failures are attributable to chance alone or to some real cause. For H_0, assume failures are random and not related to a cause so that the expected distribution of failures follows the Poisson distribution. For H_1, assume failures are not random and are cause related. The Poisson probability is $P_r = e^{-\mu} \mu^r / r!$, where $\mu = \sum x_i P_i$

$$\mu = (0)\left(\frac{364}{1103}\right) + (1)\left(\frac{376}{1103}\right) + (2)\left(\frac{218}{1103}\right) + (3)\left(\frac{89}{1103}\right) + (4)\left(\frac{33}{1103}\right) + (5)\left(\frac{13}{1103}\right)$$

$$+ \frac{24}{1103} + \frac{21}{1103} + \frac{16}{1103} + \frac{9}{1103}$$

$$= \frac{1346}{1103} = 1.22$$

Then $P_0 = 0.295$ and $E_0 = P_0 n = 325$. Similarly $P_1 = 0.360, P_2 = 0.220, P_3 = 0.088 \ldots$ and, correspondingly, $E_1 = 396$, $E_2 = 243$, $E_3 = 97 \ldots$ are tabulated above for the expected number of transducers to have the number of failures listed.

Goodness-of-fit test: H_0 represents random failures, and H_1 represents real cause

$$P[\chi^2_{\exp} > \chi^2_{\text{Table}}] = \alpha = 0.05$$

$$\chi^2_{\text{Table}} = \chi^2(\nu, \alpha) = \chi^2(5, 0.05) = 11.070$$

$$\chi^2_{\exp} = \frac{\sum(0 - E)^2}{E}$$

O	E	$O - E$	$(O - E)^2$	$(O - E)^2/E$
364	325	39	1520	4.6800
376	396	-20	400	1.0100
218	243	-25	625	2.5700
89	97	-8	64	0.6598
33	30	3	9	0.3000
23	8	15	225	28.1000

$$\chi^2_{exp} = 37.320$$

Since there is only a 5% chance of obtaining a calculated chi-square statistic as large as 11.07 under the hypothesis chosen, the null hypothesis is rejected and it is inferred that some real (rather than random) cause exists for the failures.

REFERENCES

1. E. J. Minnar (Ed.), *ISA Transducer Compendium,* Instrument Society of America, Plenum Press, New York, 1963.
2. "Electrical Transducer Nomenclature and Terminology," ANSI Standard MC 6.1-1975 (ISA S37.1). Instrument Society of America, Research Triangle Park, NC, 1975.
3. M. T. Yothers (Ed.), *Standards and Practices for Instrumentation,* 5th ed., Instrument Society of America, PA, 1977.
4. H. N. Norton, *Sensor and Analyzer Handbook,* Prentice-Hall, Englewood Cliffs, NJ, 1982.
5. A. R. Graham, *An Introduction to Engineering Measurements,* Prentice-Hall, Englewood Cliffs, NJ, 1975.
6. J. B. Kennedy and A. M. Neville, *Basic Statistical Methods for Engineers and Scientists,* 2nd ed., Harper and Row, New York, 1976.
7. A. G. Worthing and J. Geffner, *Treatment of Experimental Data,* John Wiley, New York, 1943.
8. C. R. Mischke, *Mathematical Model Building (an introduction to engineering),* 2nd ed., Iowa State University Press, Ames, IA, 1980.
9. C. Lipson and N. J. Sheth, *Statistical Design and Analysis of Engineering Experiments,* McGraw-Hill, New York, 1973.
10. B. Ostle and R. W. Mensing, *Statistics in Research,* 3rd ed., Iowa State University Press, Ames, IA, 1975.
11. W. W. Hines and D. C. Montgomery, *Probability and Statistics in Engineering and Management Science,* 2nd ed., John Wiley, New York, 1972.
12. R. M. Bethea, B. S. Duran, and T. L. Boullion, *Statistical Methods for Engineers and Scientists,* 2nd ed., Marcel Dekker, New York, 1985.
13. R. E. Walpole and R. H. Myers, *Probability and Statistics for Engineers and Scientists,* Macmillan, New York, 1972.

CHAPTER 5
INPUT AND OUTPUT CHARACTERISTICS

ADAM C. BELL

Department of Mechanical Engineering
Technical University of Nova Scotia
Halifax, Nova Scotia

5.1 INTRODUCTION

Everyone is familiar with the interaction of devices connected to form a system, although they may not think of their observations in those terms. Familiar examples include the following:

1. Dimming of the headlights while starting a car.
2. Slowdown of an electric mixer lowered into heavy batter.
3. Freezing a showerer by starting the dishwasher.
4. Speedup of a vacuum cleaner when the hose plugs.
5. Two-minute wait for a fever thermometer to rise.
6. Special connectors required for TV antennas.
7. Speedup of a fan in the window with the wind against it.
8. Shifting of an automatic transmission on a hill.

These effects happen because one part of a system loads another. Most mechanical engineers would guess that weighing an automobile by placing a bathroom-type scale under its wheels one at a time and summing the four measurements will yield a higher result than would be obtained if the scale was flush with the floor. Most electrical engineers understand that loading a potentiometer's wiper with too low a resistance makes its dial nonlinear for voltage division. Instrumentation engineers know that a heavy accelerometer mounted on a thin panel will not measure the true natural frequencies of the panel. Audiophiles are aware that loudspeaker impedances must be matched to amplifier impedance. We have all seen the 75 Ω and 300 Ω markings under the antenna connections on TV sets, and most cable subscribers have seen balun transformers for connecting a coaxial cable to the flat-lead terminals of an older TV.

Every one of these examples involves a desired or undesirable interaction between a source and a receiver of energy. In every case, there are properties of the source part and the load part of the system that determine the efficiency of the interaction. This chapter deals exclusively with interactions between static and dynamic subsystems intended to function together in a task, and with how best to configure and characterize those subsystems.

Chapter 3 considers the analysis of dynamic systems. To create mathematical models of these systems requires that we idealize our view of the physical world. First, the system must be identified and separated from its environment. The environment of a system is the universe outside the free body, control volume, or isolated circuit. The combination of these, which is the system under study and the external sources, provide or remove energy from the system in a known way. Next, in the system itself, we must arrange a restricted set of ideal elements connected in a way that will correctly represent the energy storages and dissipations of the physical system while, at the same time, we need the mathematical handles that explore the system's behavior in its environment. The external environment of the system being modeled must then itself be modeled and connected, and is usually represented by special ideal elements called sources.

We expect, as a result of these sources, that the system under study will not alter the important variables in its environment. The water rushing from a kitchen faucet will not normally alter the atmospheric pressure; our electric circuit will not measurably slow the turbines in the local power plant; the penstock will not draw down the level of the reservoir (in a time frame consistent with a study of penstock dynamics, anyway); the cantilever beam will not distort the wall it is built into; and so on. In this last instance, for example, the wall is a special source of zero displacement and zero rotation no matter what forces and moments are applied.

In this chapter, we consider, instead of the behavior of a single system in a known environment, the interaction between pairs of connected dynamic systems at their interface, often called the driving point. The fundamental currency is, as always, the energy or power exchanged through the interface. In an instrumentation or control system, the objective of the energy exchange might be information transmission, but this is not considered here (we would like information exchanges to take place at the lowest possible energy costs, but the second law of thermodynamics rules out a free transmission).

As always, energy factors into two variables, such as voltage and current in electrical systems, and we are concerned with the behavior of these in the energetic interaction. The major difference in this perspective is that the system supplying energy cannot do so at a fixed value. Neither the source nor the system receiving energy can fix its values for a changing demand without a change in the value of a supply variable. The two subsystems are in an equilibrium with each other and are forced by their connection to have the same value of both of the appropriate energy variables. We concern ourselves with determining and controlling the value of these energy variables at the interface where, obviously, only one is determined by each of the interacting systems.

5.2 FAMILIAR EXAMPLES[*] OF INPUT-OUTPUT INTERACTIONS

5.2.1 Power Exchange

In the real world, pure sources and sinks are difficult to find. They are idealized, convenient constructs or approximations that give our system analyses independent forcing functions. We commonly think of an automobile storage battery as a source of 12.6 V independent of the needed current, and yet we have all observed dimming headlights while starting an engine. Clearly, the voltage of this battery is a function of the current demanded by its load. Similarly, we cannot charge the battery unless our alternator provides more than 12.6 V, and the charging rate depends on the over-voltage supplied. Thus, when the current demanded or supplied to a battery approaches its limits, we must consider that the battery really looks like an ideal 12.6 V source in series with a small resistance. The voltage at the battery terminals is a function of the current demanded and is not independent of the system loading or charging it in the interaction. This small internal resistance is termed the output impedance (or input impedance or driving-point impedance) of the battery.

If we measure the voltage on this battery with a voltmeter, we should draw so little current that the voltage we see is truly the source voltage without any loss in the internal resistance. The power delivered from the battery to the voltmeter is negligible (but not zero) because the current is so small. Alternatively, if we do a short-circuit test of the battery, its terminal voltage should fall to zero while we measure the very large current that results. Again, the power delivered to the ammeter is negligible because, although the current is very large, the voltage is vanishingly small.

At these two extremes the power delivered is essentially zero, so clearly at some intermediate load the power delivered will be a maximum. We will show later that this occurs when the load resistance is equal to the internal resistance of the battery (a point at which batteries are usually not designed to operate). The discussion above illustrates a simple concept: impedances should be matched to maximize power or energy transfer but should be maximally mismatched for making a measurement without loading the system in which the measurement is to be made. We will return to the details of this statement later.

---

[*] Many of the examples in this chapter are drawn from Chapter 6 of a manuscript of unpublished notes "Dynamic Systems and Measurements," C. L. Nachtigal, used in the School of Mechanical Engineering, Purdue University, © 1978.

5.2.2 Energy Exchange

Interactions between systems are not restricted to resistive behavior, nor is the concept of impedance matching restricted to real, as opposed to reactive, impedances. Consider a pair of billiard balls on a frictionless table (to avoid the complexities of spin), and consider that their impact is governed by a coefficient of restitution, ϵ. Before impact, only one ball is moving, but afterwards both may be. The initial and final energies are as follows:

$$\text{Initial Energy} = \frac{1}{2}M_1v_{1i}^2$$

$$\text{Final Energy} = \frac{1}{2}M_1v_{1f}^2 + \frac{1}{2}M_2v_{2f}^2 \tag{5.1}$$

where the subscript 1 refers to the striker and 2 to the struck ball; M is mass; v is velocity; and the subscripts i and f refer to initial and final conditions respectively.

Since no external forces act on this system of two balls during their interaction, the total momentum of the system is conserved:

$$M_1v_{1i} + M_2v_{2i} = M_1v_{1f} + M_2v_{2f}$$

or

$$v_{1i} + \mathbf{m}v_{2i} = v_{1f} + \mathbf{m}v_{2f} \tag{5.2}$$

where $\mathbf{m} = M_2/M_1$. The second equation, required to solve for the final velocities, derives from impulse and momentum considerations for the balls considered one at a time. Since no external forces act on either ball during their interaction except those exerted by the other ball, the impulses,[*] or integrals of the force acting over the time of interaction on the two, are equal.[†] From this it can be shown that the initial and final velocities must be related:

$$\epsilon(v_{1i} - v_{2i}) = (v_{1f} - v_{2f}) \tag{5.3}$$

where $v_{2i} = 0$ in this case, and where the coefficient of restitution, ϵ, is a number between 0 and 1. A 0 corresponds to a plastic impact while a 1 corresponds to a perfectly elastic impact. Equations 5.2 and 5.3 can be solved for the final velocities of the two balls:

$$v_{1f} = \frac{1 - \mathbf{m}\epsilon}{1 + \mathbf{m}}v_{1i} \quad \text{and} \quad v_{2f} = \frac{1 + \epsilon}{1 + \mathbf{m}}v_{1i} \tag{5.4}$$

Now assume that one ball strikes the other squarely[‡] and that the coefficient of restitution, ϵ, is unity (perfectly elastic impact). Consider three cases:

1. The two balls have equal mass, so $\mathbf{m} = 1$, and $\epsilon = 1$. Then the striking ball, M_1, will stop, and the struck ball, M_2, will move away from the impact with exactly the initial velocity of the striking ball. *All the initial energy is transferred.*
2. The struck ball is more massive than the striking ball $\mathbf{m} > 1$, $\epsilon = 1$. Then the striker will rebound along its initial path, and the struck ball will move away with less than the initial velocity of the striker. *The initial energy is shared between the balls.*
3. The striker is the more massive of the two, $\mathbf{m} < 1$, $\epsilon = 1$. Then, the striker, M_1, will follow at reduced velocity behind the struck ball after their impact, and the struck ball will move away faster than the initial velocity of the striker (because it has less mass). Again, *the initial energy is shared between the balls.*

Thus, the initial energy is conserved in all of these transactions. But the energy can be transferred completely from one ball to the other *if, and only if*, the two balls have the same mass.

---

[*] Impulse $= \int_{t=0}^{t}$ Force dt, where Force is the vector sum of all the forces acting over the period of interaction, t.
[†] See *Impact* in virtually any dynamics text.
[‡] Referred to in dynamics as *direct central impact*.

If these balls were made of clay so that the impact was perfectly plastic (no rebound whatsoever), then $\epsilon = 0$ so the striker and struck balls would move off together at the same velocity after impact no matter what the masses of the two balls. They would be effectively stuck together. The final momentum of the pair would equal the initial momentum of the striker because, on a frictionless surface, there are no external forces acting, but energy could not be conserved because of the losses in plastic deformation during the impact. The final velocities for the same three cases are

$$v_f = \frac{1}{1 + \mathbf{m}} v_i \tag{5.5}$$

Since the task at hand, however, is to transfer kinetic (KE) from the first ball to the second, we are interested in maximizing the energy in the second ball after impact with respect to the energy in the first ball before impact.

$$\frac{KE_{(M_2,\text{after})}}{KE_{(M_1,\text{before})}} = \frac{\frac{1}{2}M_2(v_{2f})^2}{\frac{1}{2}M_1(v_{1i})^2} = \frac{M_2\big(1/(1+\mathbf{m})\big)^2(v_{1i})^2}{M_1(v_{1i})^2} = \frac{\mathbf{m}}{(1+\mathbf{m})^2} \tag{5.6}$$

This takes on a maximum value of one-fourth when $\mathbf{m} = 1$, and falls off rapidly as $\mathbf{m}$ departs from one.

Thus, after the impact of two clay balls of equal mass, one-fourth of the initial energy remains in the striker, one-fourth is transferred to the struck ball, and one-half of the initial energy of the striker is lost in the impact. If the struck ball is either larger or smaller than the striker, however, then a greater fraction of the initial energy is dissipated in the impact and a smaller fraction is transferred to the second ball. The reader should reflect on how this influences the severity of automobile accidents between vehicles of different sizes.

5.2.3 A Human Example

Those in good health can try the following experiment. Run up a long flight of stairs one at a time and record the elapsed time. After a rest, try again, but run the stairs two at a time. Still later, try a third time, but run three steps at a time. Most runners will find that their best time is recorded for two steps at a time.

In the first test, the runner's legs are velocity limited: too much work is expended simply moving legs and feet, and the forces required are too low to use the full power of the legs effectively. In the third test, although the runner's legs do not have to move very quickly, they are on the upper edge of their force capabilities for continued three-step jumps; the forces required are too high and the runner could, at lower forces, move his legs much faster. In the intermediate case there is a match between the task and the force-velocity characteristics of the runner's legs.

Bicycle riders assure this match with a variable speed transmission that they adjust so they can crank the pedals at approximately 60 RPM. We will later look at other means of ensuring the match between source capabilities and load requirements when neither of them is changeable, but the answer is always a transformer or gyrator of some type (a gear ratio in this case).

5.3 ENERGY, POWER, IMPEDANCE

5.3.1 Definitions and Analogies

Energy is the fundamental currency in the interactions between elements of a physical system no matter how the elements are defined. In engineering systems, it is convenient to describe these transactions in terms of a complementary pair of variables whose product is the power or flow rate of the energy in the transaction. These product pairs are familiar to all engineers: voltage × current = power, force × displacement = energy, torque × angular velocity = power, pressure × flow = power, and pressure × time rate of change of volume exchanged = power. Some are less familiar: flux linkage × current = energy, charge × voltage = energy, and absolute temperature × entropy flux = thermal power. Henry M. Paynter's[*] *tetrahedron of state* shows how these are related (Fig. 5.1). Typically, one of these factors is *extensive*, a flux or *flow*, such as current, velocity, volume flow rate, or angular velocity. The other is *intensive*, a potential or *effort*,[†] such as voltage, force, pressure, or torque. Thus $\mathcal{P} = extensive \times intensive$ for any of these domains of physical activity.

---

[*] Paynter, H. M. *Analysis and Design of Engineering Systems*, MIT Press, Cambridge, MA, 1960.
[†] This is Paynter's terminology, used with reference to his "Bond Graphs."

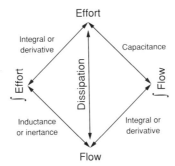

Fig. 5.1 H. M. Paynter's *Tetrahedron of State.*

This factoring is quite independent of the analogies between the factors of power in different domains, for which any arbitrary selection is acceptable. In essence, velocity is not like voltage nor force like current, just as velocity is not like current nor force like voltage. It is convenient, however, before defining impedance and working with it, to choose an analogy so that generalizations can be made across the domains of engineering activity. There are two standard ways to do this: the *Firestone* analogy[*] and the *mobility* analogy. Electrical engineers are most familiar with the Firestone analogy, while mechanical engineers are probably more comfortable with the mobility analogy. The results derived in this chapter are independent of the analogy chosen. To avoid confusion both will be introduced, but only the mobility analogy will be used in this chapter.

The Firestone analogy gives circuitlike properties to mechanical systems: all systems consist of nodes like a circuit and only of lumped elements considered to be two-terminal or four-terminal devices. For masses and tanks of liquid, one of the terminals must be understood to be ground, the inertial reference frame, or atmosphere. Then one of the energy variables is measured *across* the terminals of the element and the other passes *through* the element. In a circuit, voltage is *across* and current passes *through*. For a spring, however, velocity difference is *across* and the force passes *through*. Thus this analogy linked voltage to velocity, angular velocity, and pressure as across variables and linked current to force, torque, and flow rate as *through* variables. Clearly, *across* × *through* = Power.

The mobility analogy, in contrast, considers the complementary power variables to consist of a potential and a flux, an intrinsic and extrinsic variable. The potentials or *efforts* are voltage, force, torque, and pressure, while the fluxes or *flows* are current, velocity, angular velocity, and fluid flow rate.

5.3.2 Impedance and Admittance

Impedance, in the most general sense, is the relationship between the factors of power. Because only the constitutive relationships for the dissipative elements are expressed directly in terms of the power variables, $\Delta V_R = R \cdot i_R$ for example, while the equations for the energy storage elements are expressed in terms of the derivative of one of the power variables[†] with respect to the other, $i_C = C(dV_C/dt)$ for example, these are most conveniently expressed in Laplace Transform terms. Impedances are really self-transfer functions at a point in a system. Since the concept was probably defined first for electrical systems, that definition is most standardized: electrical impedance, $Z_{\text{electrical}}$, is defined as the rate of change of voltage with current:

$$Z_{\text{electrical}} = \frac{d(\text{voltage})}{d(\text{current})} = \frac{d(\text{effort})}{d(\text{flow})} \tag{5.7}$$

By analogy, impedance can be similarly defined for the other engineering domains.

$$Z_{\text{translation}} = \frac{d(\text{force})}{d(\text{velocity})} \tag{5.8}$$

---

[*] Firestone, F. A. "A New Analogy between Mechanical and Electrical Systems," *J. Acoustic Soc. Am.* Vol. 4, 1932–33, pp. 249–267.
[†] See Fig. 5.1 again. Capacitance is a relationship between the integral of the flow and the effort, which is the same as saying that capacitance relates the flow to the derivative of the effort.

$$Z_{\text{rotation}} = \frac{d(\text{torque})}{d(\text{angular velocity})} \tag{5.9}$$

$$Z_{\text{fluid}} = \frac{d(\text{pressure})}{d(\text{flow rate})} \tag{5.10}$$

An impedance table, using these definitions of the fundamental lumped linear elements, is in Table 5.1. Note that these are derived from the Laplace transforms of the constitutive equations for these elements; they are the transfer functions of the elements and are expressed in terms of the Laplace operator, s. The familiar $F = M \cdot a$, for example, becomes, in power variable terms, $F = M(dv/dt)$; it transforms as $F(s) = Msv(s)$, so

$$(Z_{\text{translation}})_{\text{mass}} = \frac{dF_{\text{mass}}}{dv_{\text{mass}}} = Ms \tag{5.11}$$

Because each of these involve the Laplace operator, s, they can be manipulated algebraically to derive combined impedances. The reciprocal of the impedance, the *admittance*, is also useful. Formally, admittance is defined as:

$$\text{admittance:} \quad Y = \frac{1}{Z} = \frac{d(\text{flow})}{d(\text{effort})} \tag{5.12}$$

5.3.3   Combining Impedances and/or Admittances

Elements in series are those for which the *flow* variable is common to both elements and the *efforts* sum. Elements in parallel are those for which the *effort* variable is common to both elements and the *flows* sum. By analogy to electrical resistors, we can deduce that the impedance sum for series elements and the admittance sum for parallel elements form the combined impedance or admittance of the elements. Thus

Impedances in **series:** $Z_1 + Z_2 = Z_{\text{total}}$ (common flow)

$$\frac{1}{Y_1} + \frac{1}{Y_2} = \frac{1}{Y_{\text{total}}} \tag{5.13}$$

Impedances in **parallel:** $Y_1 + Y_2 = Y_{\text{total}}$ (common effort)

$$\frac{1}{Z_1} + \frac{1}{Z_2} = \frac{1}{Z_{\text{total}}} \tag{5.14}$$

When applying these relationships to electrical or fluid elements, there is rarely any confusion about what constitutes series and parallel. In the mobility analogy, however, a pair of springs connected end-to-end are in parallel *because they experience a common force*, regardless of the

TABLE 5.1   IMPEDANCES OF THE LUMPED LINEAR ELEMENTS

Domain	Kinetic Storage	Dissipation	Potential Storage
Translational	Mass: Ms	Damping: b	Spring: $\dfrac{k}{s}$
Rotational	Inertia: Js	Damping: B	Torsion spring: $\dfrac{k_f}{s}$
Electrical	Inductance: Ls	Resistance: R	Capacitance: $\dfrac{1}{Cs}$
Fluid	Inertance: Is	Fluid Resistance: R	Fluid Capacitance: $\dfrac{1}{Cs}$

topological appearance, whereas springs connected side-by-side are in series *because they experience a common velocity difference.*[*] For a pair of springs *end-to-end*, the total admittance is:

$$\frac{s}{k_{total}} = \frac{s}{k_1} + \frac{s}{k_2},$$

so the impedance is:

$$\frac{k_{total}}{s} = \frac{k_1 k_2}{s(k_1 + k_2)}.$$

For the same springs *side-by-side*, the total impedance is:

$$\frac{k_{total}}{s} = \frac{k_1 + k_2}{s}.$$

5.3.4   Computing Impedance or Admittance at an Input or Output

There are basically two ways in which an input or output admittance can be computed. The first, and most direct, is to compute the transfer function between the *effort* and the *flow* at the driving point and take the derivative with respect to the *flow*. For a mechanical rotational system, for example, torque as a function of angular velocity is expressed and differentiated with respect to angular velocity. This method must be used if the system being considered is nonlinear because the derivative must be taken at an operating point. If the system is linear, then the ratio of *flow* or *effort* will suffice; in the rotational system, the impedance is simply the ratio: τ/ω (torque/speed).

The second method takes the impedances of the elements one at a time and combines them. This approach is particularly useful for linear (or linearized) systems. The question then arises of determining the impedance of any sources in the subsystem being considered. *Flow* sources, such as current sources, velocity sources, angular velocity sources, and fluid flow sources, all have a relationship: flow = constant. Their impedance is therefore infinite ($Z_{flow\ source} = \infty$) because any change in *effort* results in zero change in *flow*. *Effort* sources, such as voltage sources, force sources, torque sources, and pressure sources, will provide any flow to maintain the effort required; the change in effort for a change in flow remains zero, so their impedance is $Z_{effort\ source} = 0$. An *effort* source therefore represents a short circuit from an impedance point of view; it connects together two nodes that were separate. A *flow* source represents a null element; since its impedance is infinite, it represents an open circuit. *Flow* sources are simply removed in impedance calculations.

An example will distinguish between these two approaches. Figure 5.2 shows, on the left, a simple circuit disconnected at a driving point from its load. The load is of no consequence in this calculation; we simply require the driving point impedance of the circuit. In the first approach, we derive the voltage at the driving point as a function of the current leaving those terminals and the source, V_s and I_s, to obtain the relationship:

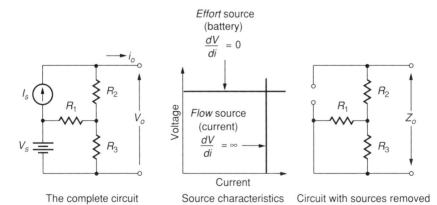

Fig. 5.2   A simple circuit as a source.

---

[*] For many, the appeal of the Firestone analogy is that springs are equivalent to inductors, and there can be no ambiguity about series and parallel connections. End-to-end is series.

$$V_o = \frac{R_3}{R_1 + R_3} V_s + \left\{ \frac{R_1 R_2 + R_2 R_3 + R_1 R_3}{R_1 + R_3} \right\} (I_s - i_o) \qquad (5.15)$$

Clearly, the negative derivative of the voltage (V_o) with respect to the current (i_o) is given by

$$-\frac{dV_o}{di_o} = Z_o = \frac{R_1 R_2 + R_2 R_3 + R_1 R_3}{R_1 + R_3} \qquad (5.16)$$

In the second method, the voltage source (V_s) can be set to zero, a short circuit, without affecting the impedances, and the current source (I_s) can be removed to yield the circuit on the right in Fig. 5.2. Then the impedances need only be combined as R_2 in series with the parallel pair, R_1 and R_3. Thus,

$$Z_o = R_2 + \frac{R_1 R_3}{R_1 + R_3} = \frac{R_1 R_2 + R_2 R_3 + R_1 R_3}{R_1 + R_3} \qquad \text{as before.} \qquad (5.17)$$

5.3.5 Transforming or Gyrating Impedances

Ideal transformers, transducers, and gyrators play an important part in dynamic systems, and an equally important part in obtaining optimal performance from source-load combinations. They share several vital features: all are two-port (or four-terminal) devices; all are energetically conservative; and all are considered lumped elements. Each of the many types requires two equations for its description, always of the same form. Table 5.2 lists many of the more common linear two-ports, and Fig. 5.3 illustrates them.

All of these transducing devices alter the *effort-flow* relationships of elements connected at their far end. Consider Fig. 5.4, which shows a simple model of a front wheel of an automobile suspension. Because the spring and damper are mounted inboard of the wheel, their effectiveness is reduced by the mechanical disadvantage of the suspension arm. What is the impedance of the spring and shock absorber at ( )$_1$ as viewed from the wheel at ( )$_2$?

Since the spring and shock absorber share a common velocity (both ends share nodes or points of common velocity), their impedances add to give the impedance of the pair at their point of attachment, ( )$_1$, to the lever:

$$(Z_1)_{\text{total}} = Z_{\text{spring}} + Z_{\text{damper}} = \left\{ b + \frac{k}{s} \right\} = \frac{bs + k}{s} \qquad (5.18)$$

At the ( )$_2$ end of the lever, the force, from Table 5.2, is $F_2 = (L_1/L_2)F_1$, but from the definition of Z for linear elements, $F_1 = (Z_1)_{\text{total}} v_1$. So we get the following:

$$F_2 = \frac{L_1}{L_2} \left\{ \frac{bs + k}{s} \right\} v_1 \qquad (5.19)$$

The second equation for the lever is $v_1 = (L_1/L_2)v_2$, and substituting this into Eq. 5.19 yields:

$$F_2 = \frac{L_1}{L_2} \left\{ \frac{bs + k}{s} \right\} \left\{ \frac{L_1}{L_2} \right\} v_2 = \left\{ \frac{L_1}{L_2} \right\}^2 \left\{ \frac{bs + k}{s} \right\} v_2 \qquad (5.20)$$

$$(Z_2)_{\text{total}} = \left\{ \frac{L_1}{L_2} \right\}^2 (Z_1)_{\text{total}} \qquad (5.21)$$

Thus, the impedance of the suspension, observed from the wheel end of the lever arm, is multiplied by the *square* of the lever ratio. This general result applies to all transduction elements.

5.3.6 Source Equivalents: Thévenin and Norton

Thévenin's and Norton's theorems were both originally developed for electric circuits. They are duals; that is, while Thévenin uses a series voltage source, Norton uses a parallel current source to construct an equivalent to a real subsystem being considered as a source. Both of these theorems are completely generalizable to any system, provided that the appropriate sources are used. In this chapter, the mobility analogy has been adapted so that Thévenin equivalents are constructed using sources of voltage, force, pressure, or torque, and Norton equivalents are constructed using sources of current, velocity, flow, or angular velocity. In the development below, these two classes of sources will be referred to as *effort* and *flow* sources.

TABLE 5.2 THE IDEAL LINEAR LUMPED TWO-PORTS

Domain to Domain	Name	See Figure	Governing Equations (Left to Right in Figures)
Translation–Translation	Lever L = distance to fulcrum	5.3a	$F_1 = \dfrac{L_2}{L_1} F_2$ $v_1 = \dfrac{L_1}{L_2} v_2$
Rotation–Rotation	Gears N = number of teeth	5.3b	$\tau_1 = \dfrac{N_1}{N_2} \tau_2$ $\omega_1 = \dfrac{N_2}{N_1} \omega_2$
Rotation–Translation	Crank ($\theta \ll 1$) or Rack & Pinion	5.3c 5.3d	$\begin{cases} \tau = RF \\ \omega = \dfrac{1}{R} v \end{cases}$
	Screw (p = length/rad)	5.3e	$\begin{cases} \tau = pF \\ \omega = \dfrac{1}{p} v \end{cases}$
Electrical–Electrical	Transformer	5.3f	$V_1 = \dfrac{N_1}{N_2} V_2$ $i_1 = \dfrac{N_2}{N_1} i_2$
Electrical–Rotation	PM DC Motor	5.3g	$V_a = K_b \omega$ $i_a = \dfrac{1}{K_m} \tau$
Electrical–Translation	Voice Coil	5.3h	$V_a = K_c v$ $i_a = \dfrac{1}{K_c} F$
Fluid–Translation	Piston Area = A	5.3j	$P = \dfrac{1}{A} F$ $Q = Av$
Rotation–Fluid	Fixed-Displacement Pump-Motor Displ/Rad = D	5.3k	$\tau = DP$ $\omega = \dfrac{1}{D} Q$

Thévenin Equivalent

Assume that a subsystem being considered as a source contains within its structure one or more ideal sources. Insofar as can be measured externally at the driving point (the point of connection between any two systems considered to be source and load), any active subsystem, no matter what its load, can be replaced by a new *effort* source added in series with the original subsystem; all the original internal sources are set to zero. The value of the new source *effort* variable is the output that would appear at the driving point if the load were disconnected from the original subsystem. Setting an *effort* source to zero is equivalent to connecting its nodes together; setting a *flow* source to zero is equivalent to removing it from the system.

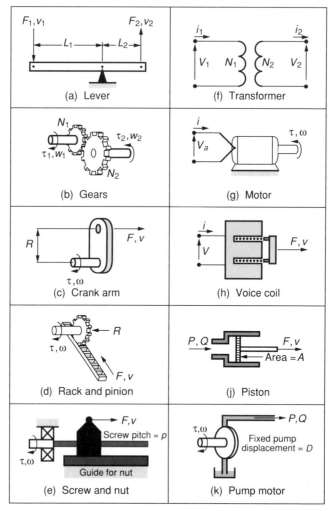

Fig. 5.3 The ideal linear lumped two-ports: *transformers* and *gyrators*.

The output impedance of this Thévenin equivalent is clearly the impedance looking in at the driving point; it is the derivative of the driving point *effort* variable with respect to the driving point *flow* variable with the load disconnected. Thévenin's theorem simplifies this calculation because it tells us that the internal sources can be set to zero before the calculation is made, and the system topology is often substantially simplified by this move. We have already seen an example of this

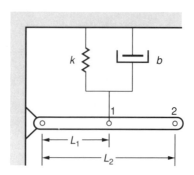

Fig. 5.4 An abstraction of an automobile suspension linkage.

(see Eqs. 5.15 through 5.17 in Section 5.3.4). The Thévenin equivalent of this circuit is simply a new source determined with the current, $i_0 = 0$ (see Fig. 5.2).

$$V_{\text{Thévenin}} = \frac{R_3}{R_1 + R_3} V_s + \left\{ \frac{R_1 R_2 + R_2 R_3 + R_1 R_3}{R_1 + R_3} \right\} I_s \tag{5.22}$$

in series with a resistance determined with sources V_s and I_s zeroed:

$$R_{\text{Thévenin}} = \frac{R_1 R_2 + R_2 R_3 + R_1 R_3}{R_1 + R_3} \tag{5.23}$$

Norton Equivalent

Assume the same subsystem considered above. Insofar as can be measured externally at the driving point, any active subsystem, no matter what its load, can be replaced by a new *flow* source added in parallel with the original subsystem; all the original sources set to zero. The value of the new source *flow* variable is the *flow* that would pass through a short circuit substituted at the driving point for the original load.

Referring again to Eq. 5.15, with V_0 set to zero, we obtain the following:

$$0 = \frac{R_3}{R_1 + R_3} V_s + \left\{ \frac{R_1 R_2 + R_2 R_3 + R_1 R_3}{R_1 + R_3} \right\} (I_s - i_0) \tag{5.24}$$

we see that i_0 for a short circuit would be

$$i_0 = I_s + \left\{ \frac{R_3}{R_1 R_2 + R_2 R_3 + R_1 R_3} \right\} V_s \tag{5.25}$$

and, as for the Thévenin equivalent, the circuit impedance is

$$R_{\text{Norton}} = \frac{R_1 R_2 + R_2 R_3 + R_1 R_3}{R_1 + R_3} \tag{5.26}$$

Note that $R_{\text{Thévenin}} = R_{\text{Norton}}$, *always*.

5.4   THE OPERATING POINT OF STATIC SYSTEMS

A static system is a system without energy storage, a system in which there are only sources, transducers, and dissipation elements. Such systems have no transient response: they respond instantly to their inputs algebraically. The relationships among any of their variables are proportionalities—simple static gains. If the inputs to a stable dynamic system are held constant for long enough, it will become stationary; its variables will not change with time provided only that there is sufficient dissipation in the system to damp out any oscillations. Such a system is not static, it is at steady state. There is no exchange of energy among its energy storage elements, and the dissipative elements completely determine the state of the system.

5.4.1   Exchange of Real Power

If one system is supplying real power to another system in steady-state operation, then for the purposes of a static analysis, energy storage elements can be ignored. Both the source and the load can be considered to be purely resistive. If the source is separated from the load at the point of interest (at least conceptually), then the characteristics of source and load can be measured or computed. The load will be a power absorber: an electrical fluid resistance or a mechanical damper, and its characteristics can be represented as a line in a power plane coordinate system: voltage versus current, force versus velocity, torque versus angular velocity, or pressure versus flow. There is no requirement that this line be straight, and except that the measurement might be more difficult, there is no necessity that this line be single-valued or that it start at the origin. Figure 5.5 shows a selection of common dissipative load characteristics.

Similarly, when the source portion of the system is loaded with a variable dissipation, the line representing its output characteristics can be plotted on the same coordinates as the load. Such characteristics are often given for pumps, servomotors, transistors, hydraulic valves, electrical supplies, and fans. Again, there is no requirement that this line be straight or that it be single-valued, but it very unusual for it to pass through the origin. If the source or the load are not

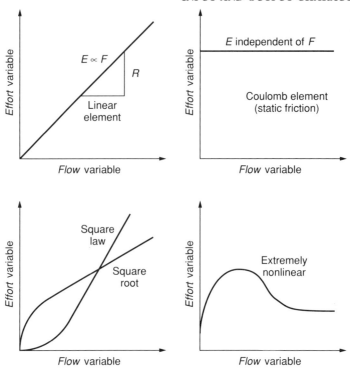

Fig. 5.5   Common dissipative load characteristics.

constant in time or can be controlled, then a family of these characteristics will be required for variations in the parameters of the source or load. Figure 5.6 shows a selection of these sources. With very few exceptions, real sources cannot operate at the maximum values of their power variables simultaneously: a battery cannot deliver its maximum voltage and current at the same time. In more general terms, this means that in spite of local variations, real source characteristics tend to droop from left to right in the power plane; their average slope is negative.

5.4.2   Operating Points in an Exchange of Power or Energy

When resistive source and load characteristics are plotted on the same coordinates, they intersect at least once. The coordinate values at that point, or at those points if there are several, are the values of the power variables at which that combination of source and load must operate if they are connected. This is called an operating point. From a computational point of view, the source *causes* one of the power variables given the other, and the load *causes* the other given the source. They must operate at the same point in the power plane to satisfy continuity (common *flows*) and compatibility (common *efforts*) conditions.

When there are multiple intersections, all are possible operating points, but not all will be stable operating points; for example, any disturbance from equilibrium might result in a transition to another operating point. The condition for a stable intersection is best seen graphically in Fig. 5.7. For a stable intersection, as shown on the left, it is required that a small perturbation of the load, which increases its demand for power, be countered by a shortage of power from the supply side of the system, and that a small perturbation of the load, which decreases its demand for power, be met with an excess from the source. In either case, the load will be driven back to the intersection by the excess or deficit in the source capability. A reversal of these conditions is an unstable operating point because disturbances will be driven further in the direction of their initial departure.

At the unstable intersection in Fig. 5.7 (2, on the right hand side), a slight increase in the *flow* demand of the load will result in an overwhelming increase in the supply *flow* available to drive the load, which will then cause a traversal to point 1 in the figure. Similarly, if the *effort* decreases slightly from operating point 2, then the source will be starved compared to the demand of the load at that *effort* level, so the *effort* will fall until point 3 is reached.

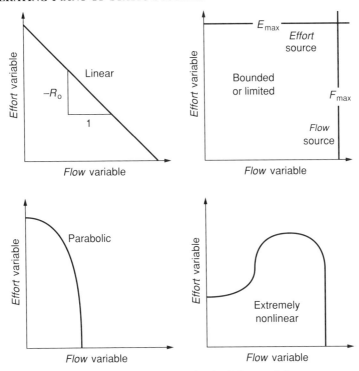

Fig. 5.6 Common dissipative load characteristics.

5.4.3 Input and Output Impedance at the Operating Point

Lines of constant power are hyperbolas in the power plane, with increasing values of the power at increasing distance from the origin. The usual sign conventions imply that sources deliver power in the first and third quadrants while loads absorb power in those quadrants. Conversely, sources absorb power in the second and fourth quadrants and loads return it. The output impedance of a source is defined as minus the slope of the output characteristic. For nonlinear characteristics, the output impedance at any point is defined as minus the slope at that point. For loads, the input impedance is defined as the slope of the load line, but for nonlinear characteristics there are two possibilities of importance: the slope of the line at a point (the incremental or local input impedance) and the

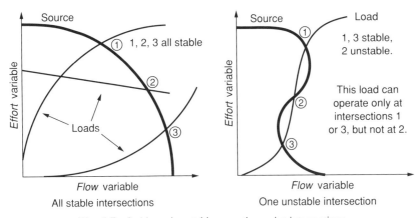

Fig. 5.7 Stable and unstable operating point intersections.

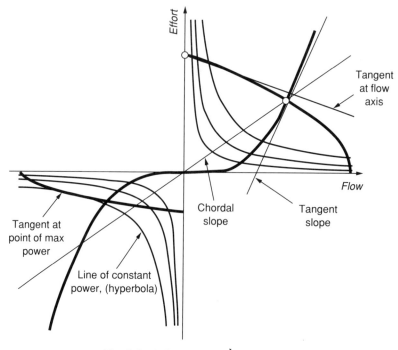

Fig. 5.8   Definitions in the power plane.

slope of the chord to the point from the origin (the chordal impedance). Figure 5.8 summarizes these features of the power plane.

5.4.4   Operating Point and Load for Maximum Transfer of Power

Consider a battery with the voltage-current characteristic shown in Fig. 5.9. The maximum unloaded terminal voltage (open circuit) of the battery is V_{oc} volts and the short circuit current is i_{sc} amps. The equation of the line shown is

$$V_t = V_{oc} - R_o i_b$$

or

$$i_b = i_{sc} - \frac{V_t}{R_o} \qquad (5.27)$$

where V_t is the terminal voltage, and i_b is the battery current.

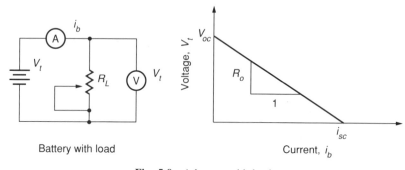

Fig. 5.9   A battery with load.

The output impedance is $Z_o = -dV_t/di_b = R_o = V_{oc}/i_{sc}$ (a pure resistance in this case). If the battery is loaded by a resistor across its terminals (R_L), the terminal voltage must be

$$V_t = R_L i_b \qquad (5.28)$$

Solving Eqs. 5.27 and 5.28 simultaneously for V_t and i_b yields the following operating point coordinates:

$$V_t = \frac{V_{oc}}{1 + R_o/R_L} \qquad i_b = \frac{V_{oc}}{R_o + R_L} \qquad (5.29)$$

The output power at the operating point is (from 5.29):

$$\mathcal{P} = V_t i_b = \frac{(V_{oc})^2 R_L}{(R_o + R_L)^2} \qquad (5.30)$$

Clearly, the output power depends on the load resistance. If R_L is zero or infinite, no power is drawn from the battery. To measure V_{oc}, we would want a voltmeter with an infinite input impedance, and to measure i_{sc}, we would want an ammeter with zero resistance. In practice, we would use a voltmeter with an input resistance very large compared to R_o and an ammeter with a resistance very small compared to R_o.

If our objective is to deliver power, then a best value of R_L is that for which the derivative $d\mathcal{P}/dR_L = 0$. This value is the point at which $R_L = R_o$. Alternatively, the maximum power output of the battery for any load occurs at the current (i_b) that maximizes $V_t i_b$. Equation 5.27 can be restated as

$$V_t = V_{oc}\left\{ 1 - \frac{i_b}{i_{sc}} \right\} \qquad (5.31)$$

so that

$$\mathcal{P} = V_t i_b = V_{oc}\left\{ 1 - \frac{i_b}{i_{sc}} \right\} i_b \qquad (5.32)$$

which is maximized at $\mathcal{P} = V_{oc} i_{sc}/4$ when $i_b = i_{sc}/2$.

For the battery characteristic, the operating point $i_b = i_{sc}/2$ yields $V_t = V_{oc}/2$ (substitution in 5.29). A loading resistor characteristic must pass through this point to draw maximum power from the battery, so that the load resistance must be $R_L = V_{oc}/i_{sc} = R_o$. At this operating point, the equivalent resistor within the battery (representing the internal losses in the battery) is dissipating exactly as much power as is being delivered to the load.

If we want maximum power delivery, impedance should match the load to the source, but if we want to minimize power delivery from a source, then impedance mismatching is the key. Impedance matching assures that the source and load will divide the power equally; all other impedances will result in less power transfer.

5.4.5   An Unstable Energy Exchange: Tension Testing Machine

Although tensile studies of material properties require only a simple test apparatus, it is not simple to interpret the data from such tests. The problem is that the tensile test machine and the specimen can interact[*] in an unstable way. Almost any desired stress-strain curve can be obtained in a given material by a suitable choice of the test machine's elastic compliance compared to the specimen.

A tensile test involves the interaction between two springs, one that represents the specimen, and the other combined with a velocity source that represents the testing machine. Figure 5.10 shows this simple model. While the testing machine is linearly elastic and does not yield, the test specimen is not elastic. It undergoes a large plastic deformation in a typical load-elongation test. Normally in such a test, the specimen is to be elongated at a constant cross-head velocity (v). The test machine, however, is not a velocity source as is commonly supposed; that source is really in series with a spring (K) representing the elastic deformations of the testing machine structure between the source of the motion and the jaws of the machine.

[*] A. C. Bell and S. Ramalingam, "Design and Application of a Tensile Testing Stage for the SEM," *Jour. Eng. Matl. and Tech.* ASME Trans., Vol. 96, July, 1974, pp. 157–162.

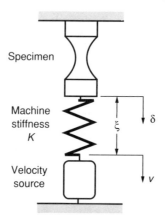

Fig. 5.10 Simple model of a tensile testing machine.

In the course of a test, the specimen undergoes an elongation (δ) comprised of both elastic and plastic displacements. The test machine undergoes only an elastic displacement (ξ) given by F/K, where F is the load applied to the specimen by the machine at a given cross-head displacement (y), which is really the sum of both the specimen (δ) and the machine (ξ) displacements. Figure 5.11 shows the components of this force-displacement situation.

The cross-head velocity of the test machine is made up of two components as well:

$$v = \frac{dy}{dt} = \left\{ \frac{d\delta}{dt} + \frac{d\xi}{dt} \right\} \tag{5.33}$$

The machine displacement (ξ), however, is a function of the force being applied, so it is therefore a function of the elongation of the specimen (δ). Equation 5.33 must be rewritten:

$$v = \left\{ \frac{d\delta}{dt} + \frac{d\xi}{d\delta}\frac{d\delta}{dt} \right\} = \frac{d\delta}{dt}\left\{ 1 + \frac{d\xi}{d\delta} \right\} \tag{5.34}$$

Since $\xi = F/K$, presuming linearity for the machine, we get the following derivative:

$$\frac{d\xi}{d\delta} = \frac{1}{K}\frac{dF}{d\delta} \tag{5.35}$$

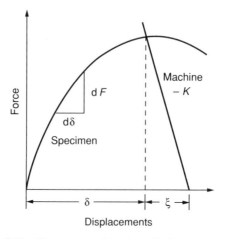

Fig. 5.11 Components of the force-displacement curve.

where the term $dF/d\delta$ is the slope of the force-elongation curve of the specimen. The slope, in other words, is the driving point stiffness at any point along the test curve. If these last two equations are combined, the specimen's elongation rate is found in terms of the cross-head velocity (v), which is normally constant, the machine stiffness (K), and the driving point stiffness of the machine ($dF/d\delta$).

$$\frac{d\delta}{dt} = \frac{v}{\left\{1 + \frac{1}{K}\frac{dF}{d\delta}\right\}} \qquad (5.36)$$

As a test proceeds, however, the specimen starts to neck down, its stiffness begins to decrease, and then it becomes negative. When the driving point stiffness ($dF/d\delta$) passes through zero, the specimen has reached the maximum force. The stiffness thereafter decreases until it equals $-K$ and at that point the specimen elongation rate ($d\delta/dt$) becomes infinite for any cross-head rate (v), and the system is mechanically unstable. This instability point in a tensile test is the point at which the machine load line is tangential to the load-elongation curve of the material being tested. At that point, the specimen breaks and the energy stored in the machine structure dumps into the specimen at this unstable intersection.

Clearly, percent elongation at failure specifications is not very meaningful because it depends on the test machine stiffness. Figure 5.12 shows that for accurate measurements, the stiffness of the test machine must be many times as large as the stiffness of the specimen near failure or the machine will dump its energy into the specimen and force a premature failure.

5.4.6 Fatigue in Bolted Assemblies

The mechanical engineering reader will perhaps recall that preloading a bolt stretches the bolt and compresses the part being bolted, but it is the relative stiffness of the bolt and part that determines what fraction of external loads applied to separate the part from bolt will be felt by the bolt.[*] If the objective is to relieve the bolt of these loads, as it is in the head bolts of an automobile engine, then the designer tries to make the part much stiffer than the bolt; he mismatches the stiffness of the bolt and the part by specifying a hard head gasket. If the stiffness of the bolt and the part were the same, then they would share the external load equally: the bolt tension would increase by half the applied load while the part compression would decrease by half the applied load. If the gasket were very soft compared to the bolts, then the bolts would see virtually all of the applied load.

5.4.7 Operating Point for Nonlinear Characteristics

Let us continue to use the battery as an example of a linear source. It should be obvious the maximum power point for the battery is independent of the load it must drive, but the load characteristic, however nonlinear, must pass through this point for maximum power transfer. Figure 5.13 illustrates this. It is not the *slope* of the load impedance that must match the source impedance, it is the *chordal*

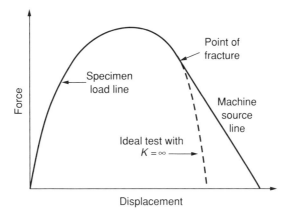

Fig. 5.12   The point of instability in a tensile test.

[*] Refer to any text on Machine Design under the indexed heading "Fatigue in Bolts."

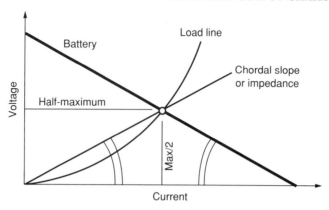

Fig. 5.13 Chordal impedance matching.

slope, the slope of a line from the origin of the power plane to the maximum power point of the source.

An orifice supplied from a constant upstream pressure and loaded at its output is a good example of a nonlinear source. For a short, sharp-edged orifice, the orifice equation[*] applies.

$$Q_o = C_d A_o \sqrt{\frac{2}{\rho}\Delta P} = C_d A_o \sqrt{\frac{2}{\rho}(P_{\text{up}} - P_{\text{down}})} \tag{5.37}$$

where Q_o is the orifice flow, C_d is the discharge coefficient of the orifice, A_o is the orifice area, ρ is the density of the fluid, and $\Delta P = P_{\text{up}} - P_{\text{down}}$ is the pressure drop across the orifice. If the upstream pressure is kept constant, then P_{down} may be considered the output pressure, and Q_o may be considered the output flow from this orifice, a characteristic typical of many hydraulic valves. Where is the maximum power point on this characteristic?

Hydraulic power $\mathcal{P}$ is the product $Q \cdot P$. It is a maximum when it is stationary with respect to either Q or P. If we nondimensionalize Eq. 5.37 with respect to the maximum flow through the orifice when $P_{\text{down}} = 0$, then the orifice equation becomes the following:[†]

$$\mathbf{Q*} = \frac{Q_o}{Q_{\text{max}}} = \sqrt{1 - \frac{P_{\text{down}}}{P_{\text{up}}}} = \sqrt{1 - \mathbf{P*}} \tag{5.38}$$

$$\mathcal{P}\mathbf{*} = \frac{\mathcal{P}}{\mathcal{P}_{\text{max}}} = \mathbf{P*Q*} = \mathbf{P*}\sqrt{1 - \mathbf{P*}} \tag{5.39}$$

$$\frac{d\mathcal{P}\mathbf{*}}{d\mathbf{P*}} = \sqrt{1 - \mathbf{P*}} - \frac{\mathbf{P*}}{2\sqrt{1 - \mathbf{P*}}} = 0 \tag{5.40}$$

for which $\mathbf{P*} = 2/3$, and by substitution into 5.38, the flow at that point is $\mathbf{Q*} = 1/\sqrt{3}$, and the maximum power delivered is $(\sqrt{2}/3)P_{\text{max}}Q_{\text{max}}$. The output admittance of the orifice is the slope of the curve at any operating point:

$$Z_o = -\frac{d\mathbf{Q*}}{d\mathbf{P*}} = \frac{1}{2\sqrt{1 - \mathbf{P*}}} = \left.\frac{\sqrt{3}}{2}\right|_{\mathbf{P*} = 2/3} \tag{5.41}$$

The load for which maximum power will be delivered, whether it has a linear or nonlinear characteristic, must pass through the maximum power point. Its *chordal admittance* must be

---

[*] This is derived from Bernoulli's equation. The discharge coefficient (C_d) corrects for viscous effects not considered and for the existence of a vena contracta or convergence in the flow through the orifice, which makes the area of the jet smaller than the orifice itself. For most oils $C_d \approx 0.62$.
[†] Where bold letters will be used to indicate the nondimensional forms.

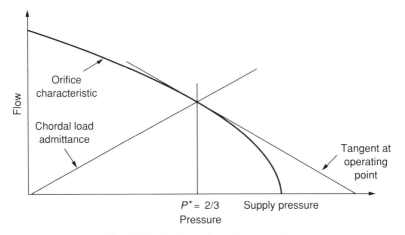

Fig. 5.14 Nonlinear impedance matching.

$$Z_{chordal} = \frac{\mathbf{Q}_{op}^*}{\mathbf{P}_{op}^*} = \frac{1/\sqrt{3}}{2/3} = \frac{\sqrt{3}}{2} \tag{5.42}$$

which is exactly the slope in Eq. 5.41. Figure 5.14 shows this graphically.

Note that for this power plane, the ordinate is *flow* and abscissa is *pressure* so that the slopes shown are admittances rather than impedances. While in other figures the abscissa has always been *effort* and the ordinate *flow*, it is conventional in hydraulic systems to show these figures the other way probably because pressure is usually the independent variable in hydraulic system characteristics: pressure is controlled and flow is measured.

5.4.8 Graphical Determination of Output Impedance for Nonlinear Systems

A three-way hydraulic valve supplies or drains its load through a pair of variable orifices, one connecting the supply to the load and the other connecting the drain or tank to the load. These orifices are operated (typically on a single moving part of the valve) in a push-pull fashion; that is, as one opens, the other closes. If both orifices are partially open together when the valve is centered, the three-way valve is open-centered. If one is wide open just as the other closes, the valve has no overlap.

Suppose it is required to find the *P-Q* characteristics of the load port for flows both into and out of the valve. The system is shown in Fig. 5.15. Thus for the upstream and downstream orifices respectively, Eq. 5.37 becomes

$$Q_u = C_d A_u \sqrt{\frac{2}{\rho}(P_S - P_L)} \quad \text{and} \quad Q_d = C_d A_d \sqrt{\frac{2}{\rho}(P_L - P_T)} \tag{5.43}$$

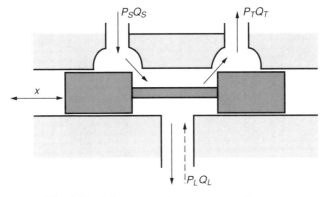

Fig. 5.15 A three-way valve geometry and output.

The load flow (Q_L) is the difference between Q_u and Q_d. Also, the tank pressure (P_T) can be taken as zero. These equations combined become

$$Q_L = C_d A_u \sqrt{\frac{2}{\rho}(P_S - P_L)} - C_d A_d \sqrt{\frac{2}{\rho}P_L} \qquad (5.44)$$

It is much more convenient to work with Eq. 5.44 in dimensionless form. If we assume that the maximum upstream and downstream areas are the same and that they are truly push-pull, then we can nondimensionalize the area, using A_{max}. The discharge coefficient (C_d) is a constant for conventional valve geometries. The supply pressure (P_S) is a convenient term for the pressure nondimensionalization. Flows can be nondimensionalized with respect to a maximum flow that would pass through either orifice at full area A_{max} with P_S acting across it.

$$Q_{max} = C_d A_{max} \sqrt{\frac{2}{\rho}P_S} \qquad (5.45)$$

If we set $P_L/P_S = \mathbf{P}$ and $Q_L/Q_S = \mathbf{Q}$ and express the upstream and downstream orifice sizes as push-pull fractions of $A_{max}, 0.5(1 + x)A_{max}$ upstream and $0.5(1 - x)A_{max}$ downstream where $-1 \leq x \leq 1$ and the valve is centered for $x = 0$, then Eq. 5.44 combined with 5.45 yields:

$$\mathbf{Q} = \mathbf{Q}_u - \mathbf{Q}_d = 0.5(1 + x)\sqrt{1 - \mathbf{P}} - 0.5(1 - x)\sqrt{\mathbf{P}} \qquad (5.46)$$

which is one of those unfortunate equations in which the radical cannot be eliminated by squaring both sides. While Eq. 5.46 can be readily plotted, $\mathbf{Q}$ vs $\mathbf{P}$ with x (the valve stroke) as a parameter, it is instructive to construct it instead from its parts.

The term $0.5(1 + x)\sqrt{1 - \mathbf{P}}$ is the characteristic family for the upstream orifice, and the term $0.5(1 - x)\sqrt{\mathbf{P}}$ is the characteristic family for the downstream orifice. The first of these are parabolas to the left (on their sides because they are roots), starting at ($\mathbf{Q}_u = 0.5(1 + x)$, $\mathbf{P} = 0$) and ending at ($\mathbf{Q}_u = 0, \mathbf{P} = 1$) (there is no flow when the downstream pressure equals the upstream pressure). The second term starts at ($\mathbf{Q}_d = 0, \mathbf{P} = 0$) and rises to the right to the points ($\mathbf{Q}_d = 0.5(1 - x), \mathbf{P} = 1$). All this is shown in Fig. 5.16. If there is no load *flow*, then the curves for the upstream orifice show its output characteristic while those for the downstream orifice represent the only load. The intersections predict the operating pressures for $\mathbf{Q} = 0$ as the valve is stroked, $-1 \leq x \leq 1$.

If there is a load flow ($\mathbf{Q}$), then continuity must be served. This requires that

$$\mathbf{Q} = \mathbf{Q}_u - \mathbf{Q}_d \rightarrow \mathbf{Q}_u = \mathbf{Q} + \mathbf{Q}_d \qquad (5.47)$$

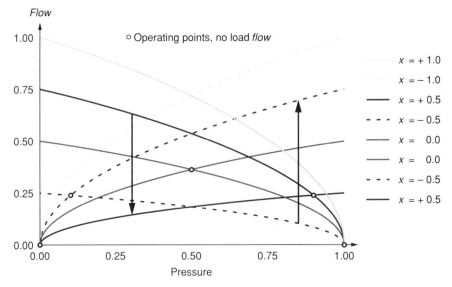

Fig. 5.16 *Flow versus pressure-drop characteristics for the valve orifices.*

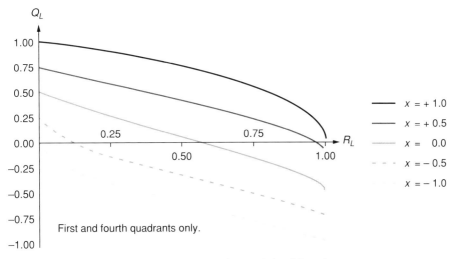

Fig. 5.17 The output characteristic of the valve.

In Fig. 5.16, **Q** is simply a vertical bar between the curves for $\mathbf{Q}_u$ and $\mathbf{Q}_d$, whose length is the load flow. Thus for any load pressure value on the abscissa, the load flow is determined as the vertical distance between the input orifice curve and the output orifice curve. The load flow is positive if the upstream orifice curve is above the downstream orifice curve, but negative otherwise. These data can be picked off and plotted as the **Q-P** characteristic of the valve as shown in Fig. 5.17.

Figure 5.17 illustrates that an operating point on an input-output characteristic forces compatibility and continuity, but as long as those are preserved, a second energy exchange as at the $P_L - Q_L$ port of this valve can still be accommodated. This approach can be extremely useful in systems such as air conditioning ducting where the fan characteristic is known only graphically, and then its output impedance at some point along the ducting must be determined. Exactly the same procedure is used.

5.5 TRANSFORMING THE OPERATING POINT

It is often not among the system designer's options to choose either the output impedance of the power source in the system or the input impedance of the load that must drive. The only recourse at that point is to insert a transformer, transducer, or gyrator in the system if it does not already contain one, or to vary the modulus of the two-port element, if it does. In the old tube-type audio systems, the output impedance of the push-pull power tubes exceeded 1000 Ω while the input impedance of the speakers available then was four, eight, or sixteen Ω. To match the amplifier to the load, each channel had a large transformer for speaker connections with taps having turns ratios of $\sqrt{1000/4}$, $\sqrt{1000/8}$, and $\sqrt{1000/16}$. With this arrangement, maximum power transfer was assured down to the lowest frequencies for which the transformers were designed.

5.5.1 Transducer-Matched Impedances

Suppose a permanent dc magnet servomotor is to drive a screw that, in turn, drives a mass, perhaps a machine-tool table. If the objective is to minimize the move time from stationary start to full stop, then the optimal trajectory for the servo, assuming equal and constant acceleration and deceleration, is well known: maximum acceleration to either half of the move distance or maximum velocity, whichever comes first, followed by maximum deceleration to the finish. These trajectories are illustrated in Fig. 5.18.

The motor operates in two modes: at maximum acceleration (maximum torque), which is set by the maximum short-term current permitted by the coercivity of the motor magnets and the capacity of the commutation, and at maximum speed, if that is reached, set by the maximum voltage available or by whatever the commutation allows. Often, servo designers accomplish these two modes by using an overvoltage (two to three times rating) during acceleration and deceleration, which runs the power amplifier as a current source, and then as maximum speed is attained, switching the amplifier

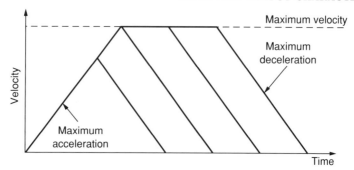

Fig. 5.18   Optimal trajectories for point-to-point move.

voltage limit to the motor rating, which runs the amplifier as a voltage source. This achieves a constant acceleration and a constant top speed.[*]

During the acceleration or deceleration phase of the trajectory, the electrical torque available is accelerating two inertias: the motor itself and the load. Since the motor is acting as an electromagnetic torque source while constant current is supplied, $\tau_{\text{motor}} = K_m i_{\text{armature}}$, the mechanical output impedance viewed at the motor shaft is the rotor inertia between the output shaft and the electromagnetic torque source:

$$Z_o\bigg|_{\substack{\text{motor shaft with} \\ \text{constant armature current}}} = J_{\text{armature}}\, s \tag{5.48}$$

During acceleration and deceleration, the optimal load impedance (Z_{load}) will be equal to the motor inertia, and the available electromagnetic torque will be shared equally by the motor armature and the load.

The load given in this example is primarily massive, so with reflection through the screw, the load inertia is computed as the load inertia times the square of the transducer ratio (p for the screw, see Table 5.2):

$$Z_{\text{load}} = p^2 M_{\text{load}} s \tag{5.49}$$

To the extent possible, the pitch of the screw or the inertia of the motor should be chosen to achieve this match. Failing both of those options, a gear box should be placed between the screw and the motor to accomplish the match required.

5.5.2   Impedance Requirements for Mixed Systems

When a source characteristic is primarily real, or static (so that the source impedance is resistive) and the load is reactive, or dynamic (dominated by energy storage elements), then impedance matching in the strictest sense is impossible, and the concept of passing the load line through the source characteristic at the maximum power point does not make sense. How then does one match a static source to a dynamic load or the reverse?

Figure 5.19 shows an electrohydraulic position servo driving a mass load with negligible damping

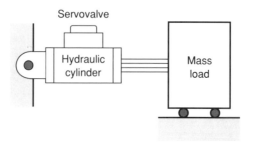

Fig. 5.19   Electrohydraulic position servo.

[*] Actually, designers rarely use maximum acceleration or maximum velocity because such full-scale values leave no overhead for control. A servo running in saturation is an open-loop system.

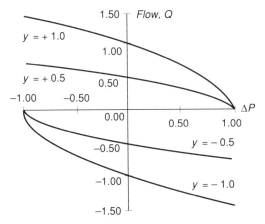

Fig. 5.20 Servovalve characteristic.

losses, and Fig. 5.20 shows the pressure-flow characteristics of the servovalve: a family of parabolas in the power plane used for hydraulic systems. The transducer between this hydraulic power plane and the force-velocity power plane in which the load operates is a piston of area A. There are two equations:

$$P = \frac{1}{A}F \quad \text{and} \quad Q = Av \tag{5.50}$$

With these equations, a load in F-v coordinates can be transformed to a load in P-Q coordinates for superposition on the source characteristics.

The impedance of the mass load, however, is simply $Z_{\text{load}} = (M)(s)$, but this cannot be plotted in an F-v coordinate system because it is the slope of a straight line in energetic coordinates: F vs. dv/dt. To match load to source, we need the force-velocity relationship for the load. A servo designer normally has a frequency response in mind for his application, or can determine one from a linearization of the system. Suppose, in this example, that the valve operating between a supply pressure (P_s) and a maximum load *flow* (Q_{max}) must drive the mass load (M) through the piston of the area (A), with a frequency response flat in position to a frequency of ω (radians/second), at an amplitude of D_{max}. With these conditions, the force-velocity relationship for the load can be found as follows.

If y is the displacement, $\dot{y}$ is the velocity, and $\ddot{y}$ is the acceleration, then the most taxing demand on the servo will be $y = D_{\text{max}}\sin(\omega t)$, where t is time. Then $\dot{y} = D_{\text{max}}\omega\cos(\omega t)$ and $\ddot{y} = -D_{\text{max}}\omega^2\sin(\omega t)$. Given that $F = M\ddot{y}$ is the load equation, however, the load can now be expressed parametrically as a pair of equations:

$$v = \dot{y} = D_{\text{max}}\omega\cos(\omega t) \quad \text{and} \quad F = -MD_{\text{max}}\omega^2\sin(\omega t) \tag{5.51}$$

If these are cross-plotted in the force-velocity plane with time as a parameter, the plot traces out an ellipse as time goes from zero through multiples of $2\pi/\omega$. Transforming these to the P-Q plane requires application of the piston equations (5.50) to yield

$$Q = AD_{\text{max}}\omega\cos(\omega t) \quad \text{and} \quad P = -\frac{MD_{\text{max}}\omega^2}{A}\sin(\omega t) \tag{5.52}$$

If the valve is to drive the mass through the piston around the trajectory, $y = D_{\text{max}}\sin(\omega t)$, then the valve output characteristic for the maximum valve stroke must *entirely enclose* the elliptical load characteristic derived in Eqs. 5.52. Furthermore, if the valve and load are to be perfectly matched, then the valve characteristic and the load ellipse must be *tangent* at the maximum power point for the valve: $(P = (2/3)P_S, Q = (1/\sqrt{3})Q_{\text{max}}$. This is shown in Fig. 5.21.

Note that this requirement means that neither maximum valve output pressure nor maximum valve *flow* will ever be reached while the load executes its maximal sinusoid. If $P_s = -MD_{\text{max}}\omega^2/A$ and $Q_{\text{max}} = AD_{\text{max}}\omega$, the valve sizing would have provided inadequate power for reaching any but those points on the trajectory, which would therefore have followed the valve characteristic instead. The correct matching relationships are those which satisfy the following equations:

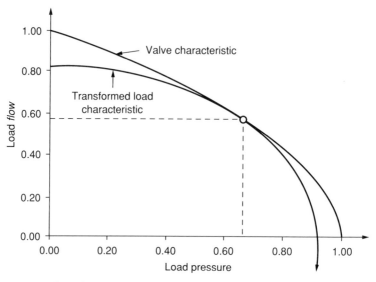

Fig. 5.21 Matching power requirements for a dynamic load.

$$\frac{1}{\sqrt{3}} Q_{max} = AD_{max}\omega\cos(\omega t) \quad \text{and} \quad \frac{2}{3}P_s = \frac{MD_{max}\omega^2}{A}\sin(\omega t) \tag{5.53}$$

at the appropriate time. These equations relate $D_{max}, \omega, M, Q_{max}, P_s,$ and A, so any five of these determine t and the sixth. If the load mass, supply pressure, piston area, peak frequency, and maximum amplitude are all known, for example, Eqs. 5.53 size the valve by determining its maximum required *flow*. If the valve has been selected, these equations will size the piston.

In the example above, the load was purely massive. Most real loads are dissipative as well. The procedure outlined above, however, need only be modified by adding the damping term to the force equation:

$$F = M\ddot{y} + B\dot{y} = -MD_{max}\omega^2\sin(\omega t) + BD_{max}\omega\cos(\omega t) \tag{5.54}$$

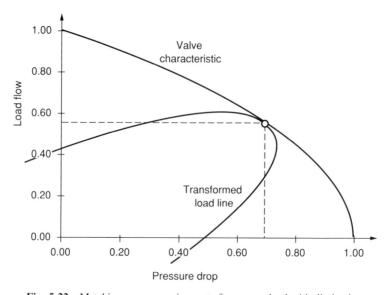

Fig. 5.22 Matching power requirements for a mass load with dissipation.

where B is the damping coefficient. This has the effect of tipping the axis of the elliptical load line up to the right, but the remainder of the development follows as before with the added term. Figure 5.22 shows the result. If the load mass were to be negligible, then the elliptical transformed load line in Fig. 5.22 would collapse to a line passing through the maximum power point of the valve characteristic, as discussed previously.

5.6 MEASUREMENT SYSTEMS

Measurement extracts information about the state of a measured system, usually in the form of one of the factors in the measurement domain whose product is power or energy flux, such as voltage or current, pressure or flow, etc. A measurement usually extracts some energy from the system being measured, if only in a transient sense; each link in the chain from measurement to display involves a further exchange of energy or power. If we measure voltage in a circuit, for example, we must draw some current, at least instantaneously, so that the power is dissipated by effective resistances, or energy is stored capacitively, inductively, or both.

At the measurement interface, we wish to disturb the measured system as little as possible by extracting as little energy or power as can be managed. It is usually our objective to pass power or energy along the chain of elements that form the measurement system, and there is a best combination of the energy variables to attain the optimum transfer. This chapter has dealt with these issues: the maximization of energy transfer within the system where we want it, and the minimization of energy theft at the measurement interface where we do not.

5.6.1 Interaction in Instrument Systems

The generalized instrument consists of a number of interconnected parts with both abstract and physical embodiments. An orifice *flow* metering system might consist of the following chain: An orifice plate converts the *flow* to a pressure drop; a diaphragm converts the pressure drop to a force; a spring (perhaps an unbonded strain-gage bridge) converts the force to a displacement; the strain gages convert the displacement to a resistance change; a bridge arrangement converts the resistance change to a differential voltage; and a galvanometer converts the voltage to a trace on paper. Figure 5.23 illustrates this chain.

In setting up this chain, the orifice is sized to minimize the *pressure drop* resulting from our *flow* measurement, and the diaphragm must be sized for minimum pumping volume during transients in pressure, or it will alter the *flow* reading in a transient sense. The diaphragm, however, has an output stiffness; the force it transmits decreases with increasing displacement, and it is driving the unbonded gages, converting force to displacement. The combined stiffness of the unbonded strain gages, which will be linear springs, must equal the average stiffness of the diaphragm[*] in order to ensure that the strain energy transfer is maximized for a given input energy from the fluid.

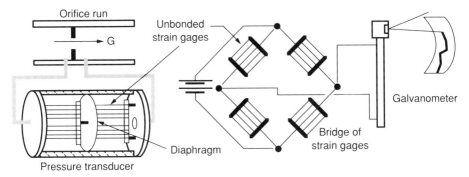

Fig. 5.23 The chain of a *flow* measurement.

[*] If the diaphragm were rigidly supported in the center, then the force on the support would be the pressure times the effective area of the diaphragm. If the support were not rigid, this output force would depend nonlinearly on the support displacement. The plot's negative slope of force transmitted vs. displacement is the output stiffness of the diaphragm, and pressure × area is the force source in this model.

The displacement-strain relationship is a definition. Further along the chain, however, this strain is converted to a differential voltage whose magnitude depends on the input impedance of the galvanometer, which is loading the bridge. For bridges constructed with the same resistance in each arm, whether the resistors are active gages or not, the input and output impedances are the same: the resistance of one arm, usually the gage resistance. If we ignore dynamic considerations (such as galvanometer damping), the optimal galvanometer will therefore have the same resistance for maximum deflection per unit strain as one gage in the bridge.

At the measurement interface, therefore, the objective is to mismatch the impedances between the measurement system and the measured system as much as possible. There is more to this than the selections of voltage measuring devices with higher impedance than the system in which voltage is being measured and current measuring devices with lower impedance than the circuits in which they are placed. The measurement and control engineer must always be aware of dynamic loading as well.

The output impedance of any piezoelectric device, for example, is almost purely capacitive and is typically only a few picofarads for small devices. The input impedance of most oscilloscopes is a parallel combination of one $M\Omega$ and 100 pF. Either or both of those would load a piezoelectric device to near uselessness, even though for most other purposes they are *high* impedances. The resistance would reduce the charge on the crystal much too quickly, and the capacitance would steal charge and reduce the voltage output drastically. An attempt to measure the pressure in a small volume with a transducer that has a large *swept* volume itself would meet with the same failure: the displacement of the transducer would alter the volume in which the pressure was measured. Holography has become popular in the study of the vibration of thin plates and shells because it does not *load* the structure by adding mass as an array of accelerometers would.

5.6.2   Dynamic Interactions in Instrument Systems

It is not only the steady-state *loading* of a measurement that is of concern; under many circumstances, the unsuspected dynamics of the instrument being used will lead to erroneous results. Consider the simple measurement interface shown in Fig. 5.24. The readout instrument, an oscilloscope for example, has both resistance and capacitance. The load impedance is therefore the parallel combination of these:

$$Z_{\text{inst}} = \frac{R_i \cdot \dfrac{1}{C_i s}}{R_i + \dfrac{1}{C_i s}} = \frac{R_i}{R_i C_i s + 1} = \frac{R_i}{\tau_i s + 1}, \quad \text{with } \tau_i = R_i C_i \tag{5.55}$$

Thus the load depends on the frequency of the voltage being measured. This might not concern us in the sense that we can predict it and compensate for the known phase shift which this will induce, but when the interaction of the two systems is considered, the problem becomes more obvious. The total impedance loading the measurement of V_s includes the source impedance Z_o. Suppose this is purely resistive (R_o):

$$Z_{\text{total}} = R_o + Z_{\text{inst}} = \frac{R_o R_i C_i s + (R_o + R_i)}{R_i C_i s + 1} \tag{5.56}$$

The readout instrument is sensitive only to the voltage it sees, which has been reduced by the voltage drop in R_o. In general terms, the measured voltage, V_{meas}, is the voltage across the instrument's input resistor and capacitor, R_i and C_i:

$$\frac{V_{\text{meas}}}{V_s} = \frac{Z_{\text{inst}}}{Z_{\text{total}}} = \frac{R_i}{R_o R_i C_i s + (R_o + R_i)} = \frac{1}{R_o C_i s + (R_o/R_i + 1)} \tag{5.57}$$

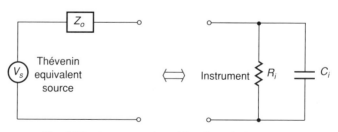

Fig. 5.24   A measurement with a dynamic instrument.

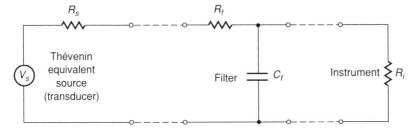

Fig. 5.25 Schematic of cascaded instrument system.

Unless we know the output impedance at the point of measurement, in this case R_o, we do not even know the time constant or break frequency germane to the measurement. In the event that Z_o is also complex (has reactive terms), this situation is more complicated, and if $Z_o = L_o s + R_o$ for example, the system could even be oscillatory.

Sometimes, frequency dependent impedances are intentionally introduced into a measurement system, most commonly in the form of passive filters. Figure 5.25[*] shows a first-order, low-pass filter being driven by a source with a nonzero output impedance (R_s) and being loaded by a readout instrument with finite input impedance (R_i).

The loading problem is potentially present at the interface of each stage indicated by the broken lines and terminals. What is not so obvious is that the dynamics of this filter are quite sensitive to its source and load. An unwary designer might choose the filter time constant ($\tau_f = R_f C_f$) with little regard for the values of R_f and C_f, and essentially design the filter under the simple assumption that source and load were ideal.

To the instrument, however, the filter is part of the source impedance, so the true source impedance combines R_s, R_f, and C_f, and the Thévenin equivalent source voltage (V_T) is no longer simply V_s. The Thévenin voltage (V_T) and the new output impedance (Z_T) with the filter included become:

$$V_T = \frac{V_s}{(R_s + R_f)C_f s + 1} \quad \text{and} \quad Z_T = \frac{R_s + R_f}{(R_s + R_f)C_f s + 1} \tag{5.58}$$

Figure 5.26 shows the new equivalent circuit for the cascaded instrument system.

The instrument is now loading this system, and the output (V_o) is equal to the voltage across R_i. Therefore, in terms of the original measured voltage (V_s):

$$\frac{V_o}{V_s} = \frac{\left\{ 1/(1 + (R_s + R_f)/R_i) \right\}}{\left\{ (1 + R_s/R_f)/(1 + (R_s + R_f)/R_i) \right\} R_f C_f s + 1} = \frac{K_f}{\tau s + 1} \tag{5.59}$$

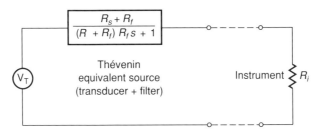

Fig. 5.26 Thévenin equivalent of cascade system.

[*] This example is drawn entirely from Nachtigal, "Dynamic Systems and Measurements," with the permission of Dr. C. L. Nachtigal. In this and the previous example, Fig. 5.24 and 5.25, the source voltage is referred to as a Thévenin equivalent source. In single-loop circuits the source voltage is by definition the Thévenin voltage, since removing one element from the loop causes it to become open-circuited. If the source is a sensor, for example, the magnitude of the voltage is governed by the value of the sensed variable and, of course, its own design parameters. The Thévenin, or open-circuit voltage, is specified on the sensor data sheet.

where

$$K_f = \frac{1}{1 + \lfloor R_s/R_i \rfloor + \lfloor R_f/R_i \rfloor} = \frac{1}{1 + \beta} \tag{5.60}$$

and

$$\tau = \left\{ \frac{1 + \lfloor R_s/R_f \rfloor}{1 + \lfloor R_s/R_i \rfloor + \lfloor R_f/R_i \rfloor} \right\} R_f C_f = \left\{ \frac{1 + \alpha}{1 + \beta} \right\} R_f C_f \tag{5.61}$$

Our filter has its expected characteristics, $\tau = \tau_f = R_f C_f$ and $K_f = 1$, only if both α and β approach zero, a circumstance that requires $R_s \to 0$ and $R_i \to \infty$ while R_f remains finite. As the instrument resistance (R_i) decreases, the static gain and time constant decrease as well, which means that the break frequency of the filter increases from the designed value because the instrument provides another path for the discharge of the capacitor in the filter. As the transducer resistance (R_s) increases from zero, the static sensitivity (K) again decreases, but this time the break frequency decreases as well.

This analysis shows us that source loading by a filter and filter loading by a readout instrument can cause significant changes in both the designed filter gain and break frequency. Only if α and β in Eqs. 5.60 and 5.61 remain equal does the filter break frequency remain unscathed by nonideal source and load impedances. This condition requires that

$$\frac{R_s}{R_f} = \frac{R_s + R_f}{R_i}$$

or

$$R_f^2 + R_s R_f - R_s R_i = 0 \tag{5.62}$$

If we divide the second of Eqs. 5.62 by R_s^2 and solve it for R_f/R_s it yields:

$$\frac{R_f}{R_s} = \frac{1}{2} \left\{ -1 \pm \sqrt{1 + 4\frac{R_i}{R_s}} \right\} \cong \sqrt{\frac{R_i}{R_s}} \tag{5.63}$$

since any realistic measurement system has $R_i/R_s \gg 1$, and since the resistances must be positive, the negative solution is discarded. This approximation is equivalent to saying that R_f should be chosen to be the geometric mean of the estimated or known values of R_i and R_s; that is, $R_f = \sqrt{R_i R_s}$ presuming only that $R_i \gg R_s$. For this choice of filter resistance, the resulting gain is given by

$$K = \frac{1}{1 + \sqrt{R_s/R_i} + R_s/R_i} = \frac{1}{1 + (R_s/R_i)^{3/2}} \cong 1, \quad \text{for} \quad R_i \gg R_s \tag{5.64}$$

If the filter fails these conditions, it must either be carefully designed for the task or it should be an active[*] filter with a high-input impedance and low-output impedance.

5.6.3 Null Instruments

Many instruments are servos, active systems designed to oppose the variable they measure so as to decrease their demand on the complementary energy or power variable to zero. In a steady-state sense, these instruments draw no power or energy from the measurement interface because these energetic requirements are provided by the instrument's power supply. These instruments therefore have infinite or zero input impedance in steady state. Examples abound: slidewire potentiometers for strain readout and thermocouple readout both measure voltage by servoing to zero current. Servo accelerometers avoid the problems with temperature inherent in the elasticity variations of metals and crystals by servoing the motion of the proof mass to zero and measuring the force required to do it. The DP-cells used in the process control industry measure differential pressure across a diaphragm or capsule while preventing the diaphragm or capsule (separating the two pressures) from moving. They thus avoid having to worry about the nonlinear elastic characteristics of the capsule and seals in the system.[†] Their output is either the current in a voice coil or the pressure in a bellows necessary

---

[*] Incorporating amplification, usually an operational amplifier.
[†] Dry friction would nonetheless be fatal to the instrument because it would be fatal to its servo and would lead to dynamic instabilities of the limit-cycle variety.

to oppose the motion. In all of these cases, however, there is a transient *displacement* until the servo zeros it out, and *some energy must be lost to transfer the information required by the servo in the instrument.* For this reason, null instruments are no better at measuring stored energy, the voltage on a small capacitor for example, than passive but high-impedance instruments, particularly if the energy consumed to reach null is not extremely low.

5.7 DISTRIBUTED SYSTEMS IN BRIEF

While a detailed study of the input-output relationships for distributed systems is beyond the scope of this chapter, a brief discussion can tie these in to the concepts already covered. All of the systems discussed to this point have been *lumped*, a label that implies physical dimensions are not of importance; the system parts can be considered as *point* objects. Considering a tank, for example, to be a point does not mean that the tank has no dimensions; it merely means that the internal pressure is considered to be the same everywhere within it; conditions in its interior are absolutely uniform. When studying lumped circuits, we are not concerned with the dimensions of the circuit elements or with their distances from each other on the circuit board. In reality, the nodes of the circuit have lengths, but they are ignored for the purposes of analyzing the lumped circuit.

In the mechanical domain, we consider that masses are rigid and behave as if the forces acting were applied at a point, and if we are interested in the distributed properties of an object, we consider only its moment of inertia. In each of these examples, changes in the physical variables of our model are assumed to propagate instantaneously, even though it is well known that all have finite propagation velocities. A system element must be considered to be distributed to have properties that vary with a physical dimension if that assumption is not true. This occurs whenever the dimensions of the object are large compared to the characteristic size of the *events* occurring.

Mechanical disturbances of all kinds propagate at the speed of sound in the medium involved, and electromagnetic disturbances propagate at speeds near the speed of light, depending again on the medium. A hammer blow to the end of a long, slender bar, for example, induces a strain pulse at the struck end which travels into the bar (*informing* the interior of the event) at the speed of sound in compression in the bar material ($c = \sqrt{E/\rho}$) where E is Young's modulus of the material, ρ is its mass density, and c is the propagation speed of compressive or tensile events. If the pulse duration is short, its physical length approaches that of the bar or may be shorter than the bar. Our simple lumped models of the bar's behavior ($F = M\ddot{y}$ and $\int F dt = Mv(t) - Mv(0)$) which treat it as a solid rigid object, are incorrect. Similarly, if a small explosion, a spark, for example, is initiated in a tank, the pressure in the tank will not remain uniform throughout. Instead, pressure waves will propagate within the tank at the speed of sound until damping takes its toll; we can no longer consider the tank to be a simple lumped capacitor, at least in the time scale of the spark event.

5.7.1 The Impedance of a Distributed System

Imagine a long, slender tank of water, a trough, open at the top, and perform the following *thought* experiment. If one end wall were moved inward suddenly, the level of water at that end would rise higher up than was required by the change in tank volume because the remaining water further along the tank would not change level instantaneously. Then, this wave coming down off the tank wall would travel to the other end where it would slosh up the far wall and return. This wave would continue to slosh back and forth at decreasing amplitude as viscosity took its toll, until finally the surface would be calm again at a new, slightly higher level. Each time the wave reached an end it would be returned in kind, that is with the same sign.

Now suppose that the far end of the tank opened into a large lake, so large that no level changes would take place when an end wall was moved, and again move the remaining end wall inward quickly. Then, when the first wave reached the opening to the lake, it would leave the tank and be lost in the lake. But that would involve water leaving the tank in excess of the volume change in the tank, and a negative wave would return from the open end to signal the new, lower level required by the loss. The closed end would reflect this wave as a rarefaction, and when that returned to the open end, lake water would spill back in as an upward wave. Eventually these alternating processes would return the level in the tank to that of the lake.

A closed end returns a wave of like kind, and an open end returns a wave of opposite kind. If there are no losses as the waves hit the ends of the tank, a wave of strength ' + 1' is reflected with strength ' + 1' from a closed end and with a strength '− 1' from an open end. This implies that there is an end condition somewhere between *closed* and *open* from which a wave will not be reflected at all. A suitably constructed porous wall, in this example, would simply absorb the wave completely by accepting and dissipating all of its energy. The impedance of this wave-matched wall is the wave impedance of the channel and is a characteristic of it, depending on the inductive and capacitive properties of the medium.

The "75 Ω" and "300 Ω" markings on the antenna connections of a television receiver imply two things: first the input impedances of those terminals are resistive at 75 Ω and 300 Ω respectively, and second, the coaxial cable and flat-lead antenna wiring are really wave guides whose wave impedances are 75 and 300 Ω. By matching the impedance of the cable at the receiver terminals, we are assured that all the incoming *wave energy* will be absorbed by the receiver, and none will be reflected back up the cable to the antenna and thus lost.

5.8 CONCLUDING REMARKS

This chapter has demonstrated an alternative viewpoint for the interaction of systems with each other. Control engineers are quite accustomed to transfer functions: relationships in the frequency (s or Laplace) domain between a variable at one point in a system and another at some other point, most often between inputs to and outputs from a controlled system. This chapter has dealt with a special class of these relationships between the complementary variables of power at a single point in a system. These special transfer functions are called driving-point impedances or admittances, and they determine how one subsystem will load or be loaded by another.

Admittances are the reciprocals of impedances, and both are unique properties of a system. The Laplace operator (s) expresses these properties as a polynomial ratio. The denominator polynomial (when set to zero, it becomes the characteristic equation of the system) is always the same, and the numerator polynomial is a function of the location of the point considered in the system. It was also shown that driving-point impedances are not a function of the controlled variables on any ideal sources the system contains. Instead, all *effort* sources may be replaced by solid connections, and all *flow* sources may be removed before the driving-point impedance is computed.

When two systems are connected together at a driving point, port, or pair of terminals, usually so that one can pass energy or information to the other, then there is a favorable relationship between the impedances of the two systems that depends on the objective of the connection. When it is desired to pass energy or power from one system to the other, then the output impedance of the driving system should match the input impedance of the driven system. If neither the driver nor the driven are adjustable, then a transducer, gyrator, or transformer is used to match them by selecting the modulus to achieve the match. Any impedance seen through a transformer, for example, appears to be increased or diminished by the square of the transformer ratio. In a chain of subsystems, it is not necessary to install the transformer at the driving point under consideration; the correct ratio can be determined no matter where it is placed within the chain because the square of the modulus will always appear in one of the driving-point impedances.

If the interconnection represents a measurement interface, then the most favorable relationship between the driving-point impedances is the largest possible mismatch consistent with obtaining the measurement. The ideal instruments for measuring *efforts* have infinite input impedance and the ideal instruments for measuring *flows* have infinite input admittances. Instruments that measure the integral of *flows*, such things as volume, charge, and displacements, should have very low compliance (should displace easily, have low volume themselves, or have small capacitances); while instruments that measure the integral of *efforts*, such things as flux linkage or momentum, must have low mass or inductance.

The operating point of a pair of coupled systems is at the intersection of their input and output characteristics in the power or energy plane. If one of these, for example the source or output characteristic, exists in the power plane, that is, is static, but the other is energetic (i.e., dynamic: massive, inductive, capacitive, etc.), then the source characteristic must enclose the trajectory of the load characteristic at the highest frequency of interest, and ideally, the source characteristic and load trajectory should be tangent at the maximum power point or should be made tangent there by suitable choice of system parameters.

The key issue in this chapter is this: whenever two dynamic systems are connected, an interaction occurs. If the connection is to meet its objectives, the nature of this interaction must be explored and controlled.

CHAPTER **6**

ELECTRONIC DEVICES AND DATA CONVERSION

STANLEY J. DOMANSKI

(formerly with Analog Devices, Inc.)
Wilmington, Massachusetts

JACKIE MARSH

SEMATECH
Austin, Texas

MARK A. McQUILKEN

Motorola, Inc.
Austin, Texas

LYNN A. KAUFFMAN

(formerly with Motorola, Inc.)
Austin, Texas

6.1 ANALOG OPERATIONAL AMPLIFIERS

Stanley J. Domanski

An operational amplifier (op amp) amplifies a voltage to produce a larger output voltage. The output voltage divided by the input voltage is known as voltage gain and is specified in volts per volt. The ideal operational amplifier is generally regarded as having an infinite voltage gain, infinite bandwidth, infinite input impedance, and zero output impedance. But as we shall see, the ideal does not exist, for these amplifiers have limitations bounded by the technologies used to fabricate them, such as modules that use discrete transistors, hybrids that use discrete active chips, and monolithic devices, where the entire operational amplifier is designed and fabricated on a single piece of semiconductor material.

6.1.1 Open-Loop Versus Closed-Loop Behavior

The open-loop gain of an op amp is quite high at frequencies below 10 Hz, and typical values of open-loop gain vary between 20,000 volts/volt at DC to values in excess of 1,200,000 volts/volt (see Figs. 6.1 and 6.2). In one case, the open-loop voltage gain is specified in volts per volt (Fig. 6.1)

117

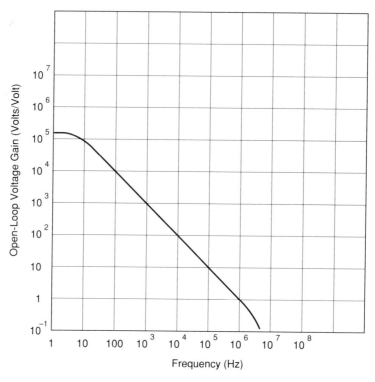

Fig. 6.1 Open-loop gain.

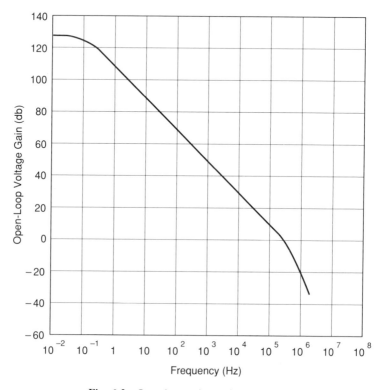

Fig. 6.2 Open-loop gain vs. frequency.

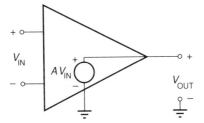

Fig. 6.3   Ideal op amp.

while in Fig. 6.2, the voltage gain is specified in db, an abbreviation for decibel. The decibel measure of a voltage gain is defined as $20 \log_{10} V_{out}/V_{in}$. For example, if $V_{out}/V_{in} = 10$, the decibel value is 20 db. Figure 6.3 shows the symbol for an op amp, which is a triangle with a plus and minus input terminal and internal voltage amplifier.

An important aspect of the open-loop response is the fact that as the frequency increases, the gain decreases and follows the general form

$$\frac{V_{out}}{V_{in}}(s) = \frac{K}{(s/\omega_p) + 1} \tag{6.1}$$

where K is the DC gain in volts/volt, s is the complex frequency $\sigma + j\omega$, and ω_p is the low frequency beyond which the open-loop gain begins to decrease at a rate of 20 db/decade. For every decade increase in signal frequency, the output voltage decreases by a factor of 10, assuming the input signal amplitude is held constant. This is commonly referred to as the -20 db/decade roll-off. Since the open-loop voltage is inversely proportional to frequency and can vary by plus or minus 20 percent, an op amp is seldom used in an open-loop configuration.

The more usual way for an op amp to be used is in a closed-loop configuration, in which various passive components like resistors and capacitors are connected to the op amp so as to establish the overall transfer function of the network or its frequency response.

Consider Fig. 6.4, which represents a generalized feedback circuit. G is the open-loop response, and H represents some feedback mechanism. The equation describing the response of V_{out}/V_{in} is

$$\frac{V_{out}}{V_{in}}(s) = \frac{G(s)}{1 + G(s)H(s)} \tag{6.2}$$

where both G and H are, in general, frequency dependent. If the product $G(s)H(s) \gg 1$, then $V_{out}/V_{in} \approx 1/H(s)$. In other words, the overall gain of the network, or closed-loop response, can be made independent of variations in $G(s)$. However, this is only true if the product of $G(s)$ and $H(s) \gg 1$. The most commonly used feedback components to synthesize $H(s)$ are resistors.

Table 6.1 lists the overall system gain $G/(1 + GH)$ for various values of G, the forward-loop gain. It should be noted that the larger the open-loop gain, the closer the overall system gain approaches $1/H$ or $10.0V/V$. Figure 6.5 compares the open-loop frequency response to the closed-loop frequency response for a feedback factor H of 0.1.

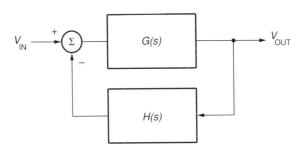

Fig. 6.4   Generalized feedback network.

TABLE 6.1   SYSTEM GAIN VS. FORWARD-LOOP GAIN

H (Feedback)	G (Forward-Loop Gain)	$1 + GH$	$\dfrac{G}{1 + GH}$ (System Gain)
.1	10	2	5
.1	100	11	9.091
.1	1,000	101	9.901
.1	10,000	1,001	9.990
.1	100,000	10,001	9.999

6.1.2   Op Amp Specifications

Offset voltage, input bias current, common-mode rejection ratio (CMRR), and noise specifications indicate how the real operational amplifier differs from the ideal model. It was shown in Section 6.1.1 that the op amp has a finite gain that varies inversely as frequency. Other non-ideal aspects of op amps will now be discussed. Figure 6.6 shows a model of a real-world op amp.

Input Imperfections

The actual characteristics of real op amps are considerably more complicated, of course. The inputs each have a DC current source (I_B) connected to them. A series DC voltage source (V_{OS}) appears in series with one input. An impedance Z_{DIFF} appears between the inputs, and another, Z_{CM}, appears between the inputs and ground. These impedances usually consist of a resistance and capacitance in parallel, and the finite Z_{CM} will introduce errors due to common-mode input voltages. In addition to the DC voltage and current sources, small AC sources representing the noise components must be included in the model.

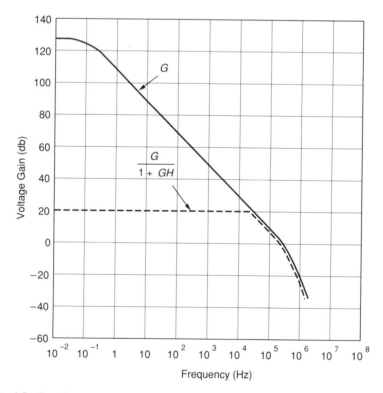

Fig. 6.5   Open-loop and closed-loop frequency response for a feedback factor $H = 0.1$.

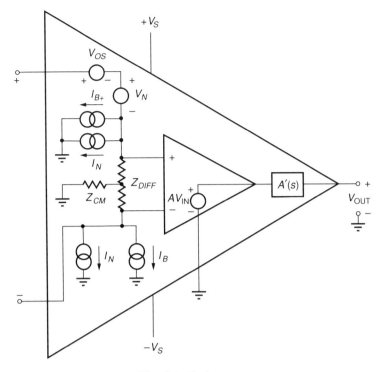

Fig. 6.6 Real op amp.

Output Obstacles

The output side of the model is also nonideal. First, an output impedance, R_0, is added in series with the voltage source. The "A" term (infinite in the ideal model) is both finite and a function of frequency in a real amplifier. It is also obvious that the output voltage and current capabilities of a real op amp are bounded. The real amplifier, thus, can be modeled as follows.

Offset Voltage

Each of these nonideal specifications should be examined in some detail. Consider first the DC errors. Offset voltage is the result of a mismatch in the base-emitter voltages of the differential-input transistors (or gate-source voltage mismatch in FET-input amplifiers). This offset voltage is indistinguishable from an input signal as far as the amplifier is concerned. This offset can usually be trimmed to zero by the user by means of an external potentiometer. This trim may be effective only at one temperature since offset voltage changes as a function of temperature.

Input Bias Current

Another DC error term is the input bias current. As a consequence of the practical characteristics of transistors, base current must be supplied to the input transistors to bias them into their active operating region. This current must also return to its originating point through some DC path. Thus, operational amplifiers cannot be used with input signal sources that are not referred to the same power source as the amplifier. Reducing errors induced by bias current is possible by providing a source (other than the signal path) to leak this current.

In many applications, the errors due to bias current are actually less annoying than the errors caused by the mismatch of the bias currents on the two inputs. This difference between the bias currents is called the input offset current and is usually specified in addition to the bias current.

Input currents, like input offset voltage, vary as a function of temperature. In the case of a bipolar-input operational amplifier, the bias currents are the input base currents to the input differential pair of transistors. As the temperature increases, so does the *beta* of the transistors. Since the sum of the emitter currents of the input differential pair is a constant base drive, the input bias currents decrease. In the case of an FET-input amplifier, the bias current is due to JFET gate leakage, which is in reality a reverse-biased junction leakage current. Such currents have the characteristic of doubling for every 10°C rise in junction temperature.

Common-Mode Rejection Ratio

The ideal operational amplifier is a pure differential amplifier and is insensitive to the absolute voltage on the inputs with respect to ground. The real amplifier has several nonideal characteristics associated with input levels. First, of course, is the allowable input voltage range. Most IC op amps will only operate when the voltages on the input terminals are within the range bounded by the supply voltages. The second, perhaps more subtle, characteristic is the common-mode rejection ratio (CMRR). CMRR is defined as the ratio of the change in common mode to the resulting change in input offset voltage. It is often convenient to specify this parameter logarithmically in db.

Noise

Two types of noise exist in circuits using op amps. The first type of noise will be referred to here as "interference." This is noise that originates from sources not related to the actual circuit. Such noise sources include ground and power-supply noise created by other circuitry in a system, stray electromagnetic pickup of line frequency energy (and the harmonics thereof), broadcasting stations, contact arcing in mechanical switches, and transients due to switching in reactive circuits. Even mechanical vibration can create noise in high impedance amplifier circuits. This external noise can often be minimized or even eliminated once the interfering source is identified and appropriate actions are taken.

The second type of noise is the inherent noise of the circuit itself. Unlike interference, it cannot be totally eliminated since it is caused by components in the actual circuit, such as resistors and sources within the amplifier. The best that can be accomplished is minimizing the noise in a specific bandwidth of interest. To do this effectively, it is important to construct reasonably accurate models of the noise sources in op amps. There are essentially three sources of noise to consider in op amps: the input noise voltage and the noise components of the bias current at the two inputs.

Types of Noise

Noise encountered in op amp circuits can be generally classified into a few specific categories:

Johnson Noise. Thermal agitation of electrons in resistive circuit elements results in random movement of charge through the resistive material. This flow of random current through a resistance creates a voltage. Johnson noise can be computed as

$$E_N(\text{rms}) = \sqrt{4kTRB} \qquad (6.3)$$

where
$$k = \text{Boltzmann's constant} (1.374 \times 10^{-23} \text{ J/K})$$

$$T = \text{temperature in Kelvin}$$

$$R = \text{resistance in ohms}$$

$$B = \text{bandwidth in hertz}$$

As a reference point, a 1 k ohm resistor produces 4 nV/$\sqrt{\text{Hz}}$ white noise, or 120 nV rms noise in a 1 kHz bandwidth. Johnson noise is generally less important inside the amplifier than it is in the external resistors in the circuit. In many circuits, the noise of external resistors is much higher than that generated by the amplifier.

"Flicker" or 1/f Noise. Flicker noise is usually the dominant noise source at frequencies below about 100 Hz. Noise with 1/f characteristic has a power spectrum that is inversely proportional to frequency; thus, the voltage noise will be inversely proportional to the square root of frequency. It is interesting to note the 1/f noise contains equal amounts of noise in any decade (or octave) of frequency. Noise with this spectral characteristic is also called "pink" noise.

"Shot" (or Schottky) Noise. Shot noise arises whenever current passes through a transistor junction. Shot noise is generally expressed as a current and follows the form

$$I_N = 5.7 \times 10^{-4} \sqrt{I_J B} \qquad (6.4)$$

where
$$I_N \quad \text{is noise current in picoamps rms}$$

$$I_J \quad \text{is junction current in picoamps}$$

$$B \quad \text{is bandwidth of interest (in hertz)}$$

As the current flows through various circuits' impedances inside the amplifier, it produces a noise voltage. Shot noise can be considered a "white" noise source and is the dominant contributor to amplifier noise at high frequencies.

Popcorn Noise. Occasionally, a transistor will exhibit an erratic 1 to 100 Hz fluctuation between two values of h_{fe} as a result of surface irregularities of contaminants. If such a transistor is used in the differential-input stage of an operational amplifier (and is biased at a constant I_E), the input bias current will jump between two values, creating a voltage noise as it flows through source and feedback impedances. Most amplifiers produced today are free from popcorn noise; however, occasional bad lots are processed, warranting periodic screening.

Noise Specifications

A complete data sheet for an operational amplifier should contain some noise specifications. Since it is not economically practical for a manufacturer to completely test all op amps for noise, most noise specifications quoted are typical. If noise is the most critical parameter in the application, many vendors are willing to screen amplifiers to a particular guaranteed maximum noise level (at additional cost).

Noise is often specified in terms of a graph of total noise up to a given frequency. Such curves are useful when the noise gain vs. frequency characteristics of a circuit is known. It is important to consider both the voltage and current noise curves (and the gain to each vs. frequency) in a comparison of total noise.

Slew Rate

The slew rate of an operational amplifier indicates how rapidly the output voltage of the op amp can respond to a step input voltage. Slew rate is specified in volts per microsecond. Slew rates for various op amps can vary from .1 volt per microsecond to hundreds of volts per microsecond.

6.1.3   Operational Amplifier Applications

While operational amplifiers were originally intended for use inside analog computational circuits, where feedback networks allowed them to perform analog operations such as integration and differentiation, op amps have become widely used as low-cost gain stages. The choice of op amps in such applications is dependent upon the characteristics of the signal source, the gain stage, and the desired accuracy.

Consider a preamplifier for a low-level signal as shown in Fig. 6.7. Assuming a noninverting gain stage is built with an ideal op amp and ideal resistors, the output will be equal to the input times one plus the resistor ratio.

Any input offset voltage in the op amp will appear as a voltage source in series with the signal (Figs. 6.8 and 6.9), and will be amplified along with the signal. Any DC bias current will flow into the amplifier input through the equivalent resistance seen "looking back" from its input.

Another useful set of specifications is a set of noise spectral density figures (at particular frequencies or plotted as a graph vs. frequency). Spectral noise density, e_n, at a given frequency is defined as the rms noise voltage in a 1 Hz band surrounding a specified value of frequency (see Fig. 6.10). It is usually expressed in nanovolts per root of frequency. The awkward units arise from the fact that it is noise power which is actually being characterized and e_n is proportional to the square root of the power spectral density.

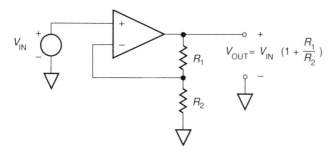

Fig. 6.7   Ideal noninverting gain stage.

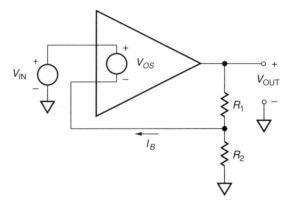

$$V_{OUT} = \left[V_{IN}\left(1 + \frac{R_1}{R_2}\right) + V_{OS}\left(1 + \frac{R_1}{R_2}\right) + I_B R_1 \right] = \left(1 + \frac{R_1}{R_2}\right)\left[V_{IN} + V_{OS} + I_B\left(\frac{R_1 R_2}{R_1 + R_2}\right) \right]$$

Fig. 6.8   Noninverting amplifier first-order errors.

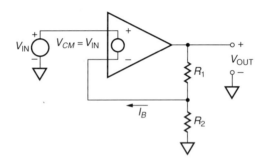

$$V_{OUT} = \left[V_{IN}\left(1 + \frac{R_1}{R_2}\right) + V_{OS}\left(1 + \frac{R_1}{R_2}\right) + I_B R_1 + \left(\frac{V_{IN}}{CMRR}\right)\left(1 + \frac{R_1}{R_2}\right) \right]$$

$$= \left(1 + \frac{R_1}{R_2}\right)\left[V_{IN} + V_{OS} + I_B + I_B\left(1 + \frac{R_1 R_2}{R_1 + R_2}\right) + \left(\frac{V_{IN}}{CMRR}\right) \right]$$

Fig. 6.9   Noninverting amplifier second-order errors.

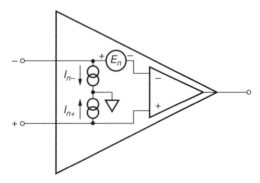

Fig. 6.10   Voltage and current noise model.

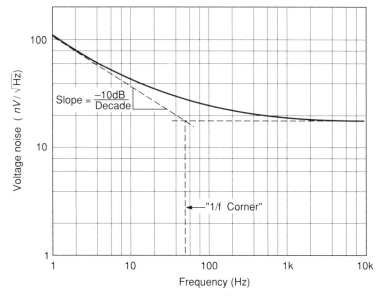

Fig. 6.11   Noise voltage plot as a function of frequency.

A typical plot of noise voltage spectral density is shown in Fig. 6.11. At high frequencies, the shot noise dominates the total noise, while at low frequencies, the $1/f$ noise dominates. It is possible to graphically determine a useful figure of merit called the "$1/f$ corner frequency from this curve. This is done in much the same way as a low-pass filter response is approximated. Rather than determine a pole frequency from the intersection of the extensions of the 10 db/decade roll-off and the midband gain, the intersection of the 10 db/decade noise roll-off with the midband white noise density yields the $1/f$ corner frequency. Generally, the lower the $1/f$ corner frequency, the better the amplifier.

Effect of Common-Mode Voltage

An additional error arises in noninverting amplifiers due to common-mode effects. Although an ideal op amp is insensitive to common-mode inputs, real amplifiers do respond to common-mode inputs (see Fig. 6.12). This error term can be modeled as an additional offset voltage equal to the common-mode voltage divided by CMRR (see Fig. 6.13). As an offset voltage, it is amplified by the same gain as the signal.

Finite open-loop gain adds yet another error. It can be shown that the actual gain of the circuit will be equal to the desired gain multiplied by

$$A_{\text{actual}} = \left(1 + \frac{1}{A_{\text{OL}}B}\right)A_{\text{desired}} \tag{6.5}$$

where A_{OL} is the amplifier's open-loop gain and B is the voltage feedback ratio (see Fig. 6.14).

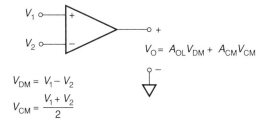

Fig. 6.12   Effect of common-mode voltage.

$$V_O = A_{OL}V_{DM} + A_{CM}V_{CM}$$

$$\text{Since } CMRR = \frac{A_{OL}}{A_{CM}}, \quad V_O = A_{OL}V_{DM} + \frac{A_{OL}}{CMRR}V_{CM}$$

$$= A_{OL}\left(V_{DM} + \frac{V_{CM}}{CMRR}\right)$$

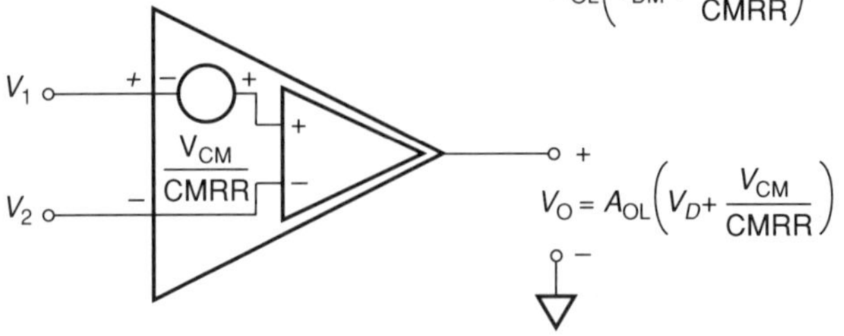

$$V_O = A_{OL}\left(V_D + \frac{V_{CM}}{CMRR}\right)$$

Fig. 6.13   Equivalent offset voltage induced by common mode.

Effect of Bias Current in Buffer Amplifier

Of course, the largest error is due to offset voltage. It may appear that since this error can be reduced or eliminated by trimming, it can be neglected. However, temperature-induced offset drifts are not so easily corrected. It is also possible to reduce errors induced by bias current by matching the impedances at each input of the op amp. The resultant error term is now I_{OS} times the source resistance. However, CMRR- and A_{OL}-induced errors cannot be compensated for because they are not constants. CMRR is often a nonlinear function of common-mode voltage. Although noise has been ignored thus far in this discussion, input noise also cannot be reduced.

Noninverting gain stages are often used to buffer high-output impedance sources (see Fig. 6.15). In these applications, it is possible that bias current will be the primary error source. Consider the case of a pH or other ion-selective electrode, where the desired signal voltage is a few hundred millivolts but the source impedance can exceed $10^8\Omega$. In these applications, a hybrid FET-input op amp is the only practical choice. For example, if the amplifier is connected as a simple follower, the actual output signal is as shown in Fig. 6.15.

The Inverting Amplifier.   So far, this discussion has been limited to noninverting amplifiers. Many circuits require a negative gain, and op amps are easily used. The normal inverting amplifier operates

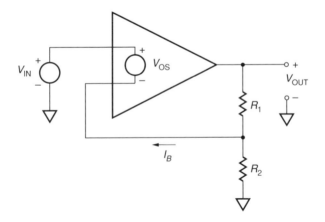

$$V_{OUT} = \left(\frac{R_1 + R_2}{R_1}\right)\left[V_{IN} + V_{OS} + I_B\left(\frac{R_1 R_2}{R_1 + R_2}\right) + \frac{V_{OS}}{CMRR}\right]\left[\frac{A_{OL}\dfrac{R_2}{R_1 + R_2}}{1 + A_{OL}\dfrac{R_2}{R_1 + R_2}}\right]$$

Fig. 6.14   Gain error introduced by finite A_{OL}.

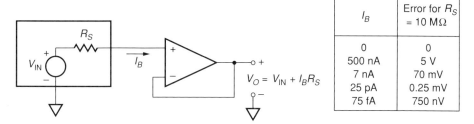

I_B	Error for R_S = 10 MΩ
0	0
500 nA	5 V
7 nA	70 mV
25 pA	0.25 mV
75 fA	750 nV

$V_O = V_{IN} + I_B R_S$

Fig. 6.15   Effect of bias current in buffer amplifier.

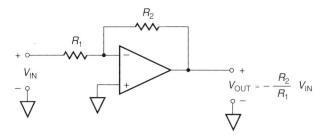

$V_{OUT} = -\dfrac{R_2}{R_1} V_{IN}$

Fig. 6.16   Ideal inverting amplifier.

by holding the noninverting input at 0 V, which causes the amplifier's output to drive the inverting input also to 0 V. If this node is at virtual ground, it is easy to prove that the current from the source is equal to V_{IN}/R_1 and the current from the output is V_{OUT}/R_2. Since these currents are equal and opposite (with an ideal op amp), the gain is simply $-R_2/R_1$ (see Fig. 6.16).

The error analysis is slightly different from the case of the noninverting amplifier. Errors due to offset are actually amplified by a greater gain than is the actual signal. Since the input offset voltage can be modeled as a voltage source in series with one of the inputs, it is clear that the gain to this voltage is actually $(1 + R_2/R_1)$. For circuits where the desired signal gain is low (including gains less than unity), this increased offset gain can be troublesome (Fig. 6.17).

In an inverting amplifier, $V_{CM} \sim 0$ so CMRR is not as important as it is in a noninverting amplifier, and "Noise gain" is equal to signal gain plus 1. Offset voltage will be amplified more than the signal.

Input bias current is another source of error. The bias current flows through a resistance equal to R_1 in parallel with R_2, creating an offset voltage that is amplified by a gain of $(1 + R_2/R_1)$, or a net output error of $I_b R_2$ (see Fig. 6.18). It is common practice to add a resistance R_3 (equal to R_1/R_2) in series with the noninverting input so that the $I_b R$ terms generate a small common-mode voltage that is rejected and an offset voltage of $I_{OS} R$ is the only error. Note that the application circuit must be capable of tolerating the common-mode signal; furthermore, it is worth noting that I_{OS} must be

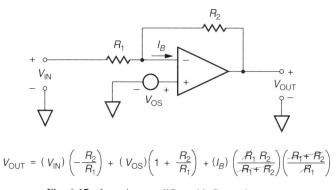

$$V_{OUT} = (V_{IN}) \left(-\frac{R_2}{R_1} \right) + (V_{OS}) \left(1 + \frac{R_2}{R_1} \right) + (I_B) \left(\frac{R_1 R_2}{R_1 + R_2} \right) \left(\frac{R_1 + R_2}{R_1} \right)$$

Fig. 6.17   Inverting amplifier with first-order errors.

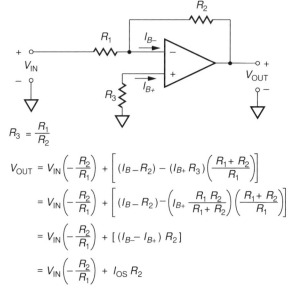

$$R_3 = \frac{R_1}{R_2}$$

$$V_{OUT} = V_{IN}\left(-\frac{R_2}{R_1}\right) + \left[(I_B - R_2) - (I_{B+} R_3)\left(\frac{R_1 + R_2}{R_1}\right)\right]$$

$$= V_{IN}\left(-\frac{R_2}{R_1}\right) + \left[(I_B - R_2) - \left(I_{B+} \frac{R_1 R_2}{R_1 + R_2}\right)\left(\frac{R_1 + R_2}{R_1}\right)\right]$$

$$= V_{IN}\left(-\frac{R_2}{R_1}\right) + \left[(I_{B} - I_{B+}) R_2\right]$$

$$= V_{IN}\left(-\frac{R_2}{R_1}\right) + I_{OS} R_2$$

Fig. 6.18 Use of bias current cancellation resistor.

significantly smaller than I_b for this technique to offer significant performance improvement. Since the ideal inverting amplifier operates at zero common-mode voltage, common-mode errors are zero.

The useful dynamic range of any preamplifier is limited by the random noise generated within the circuit. When analyzing the contribution of noise to the output of an amplifier, it is important to consider the gain applied to the noise signal as a function of frequency. In the case of a unity-gain inverting amplifier with an integrating capacitor as shown in Figs. 6.19 and 6.20, some problems arise from the difference in noise and signal gains (see Figs. 6.21 and 6.22).

Current-to-Voltage Converters. While some transducers deliver an output voltage proportional to some physical parameter, many transducers produce a current output. Such transducers include photodiodes, some temperature sensors, and a variety of biological probes. The currents produced are often very small—on the order of nanoamps or less. Generally, signal conditioning of some type will be needed before such a signal is at all useful or accurately measurable.

While a resistor is a genuine current-to-voltage converter, it is generally not practical to force the current directly through a resistor and then attempt to measure it differentially. Furthermore, most current sources have a limited range of compliance voltage.

A more useful technique for conditioning a current signal uses an op amp and a large feedback resistor as shown in Fig. 6.23. The input current source drives a constant terminal voltage (0 V), and the current flows through the resistor to the op amp output. The output voltage is then equal to the input current times the feedback resistor.

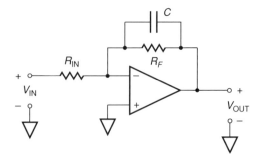

Fig. 6.19 Inverting amplifier with integrating capacitor.

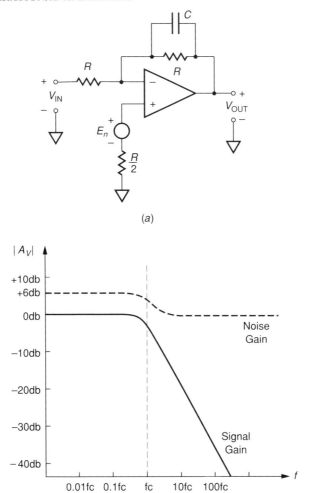

(a)

(b)

Fig. 6.20   Frequency response of unity-gain inverting amplifier.

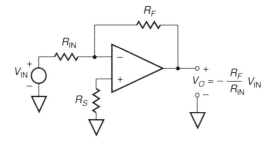

Fig. 6.21   Noiseless inverting amplifier.

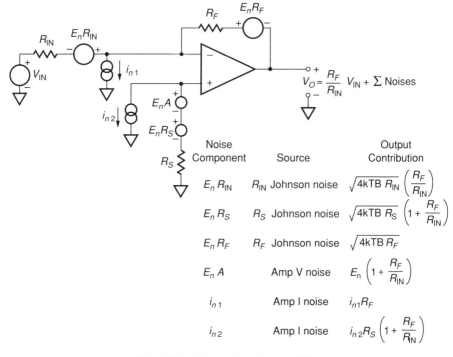

Noise Component	Source	Output Contribution
$E_n R_{IN}$	R_{IN} Johnson noise	$\sqrt{4kTB\ R_{IN}}\left(\dfrac{R_F}{R_{IN}}\right)$
$E_n R_S$	R_S Johnson noise	$\sqrt{4kTB\ R_S}\left(1+\dfrac{R_F}{R_{IN}}\right)$
$E_n R_F$	R_F Johnson noise	$\sqrt{4kTB\ R_F}$
$E_n A$	Amp V noise	$E_n\left(1+\dfrac{R_F}{R_{IN}}\right)$
i_{n1}	Amp I noise	$i_{n1}R_F$
i_{n2}	Amp I noise	$i_{n2}R_S\left(1+\dfrac{R_F}{R_N}\right)$

Fig. 6.22   Noise in inverting amplifier.

Current-to-Voltage Converter

Clearly, amplifier input bias current will cause an error. Thus, FET-input amplifiers are most often used in current-to-voltage converter applications since they exhibit the lowest bias current (at room temperature). The best performance is available from hybrid FET-input op amps since they can feature lower bias currents than presently manufacturable monolithic types.

6.1.4   Instrumentation Amplifiers

The Need for Gain Blocks

The real world is characterized by deviations from the ideal; practical transducers rarely exhibit zero output impedances and convenient output ranges. Furthermore, environmental conditions usually complicate the process of data acquisition.

In the simplest situation, a transducer is connected directly to the data processing system; more often some amplification is required. Under laboratory conditions, this amplification may be provided by a simple op amp and several resistors. Electrical interference, voltage drops caused by current through the resistance of leads from remote locations, nonlinear transducers, requirements

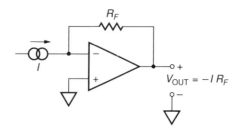

Fig. 6.23   Current-to-voltage converter.

for galvanic isolation, and fluctuating temperatures often complicate the task of providing accurate amplification.

In this section, we will discuss various means of providing gain in an environment hostile to precise measurements. Instrumentation amplifiers will often serve those applications where isolation is not required and extremely high common-mode voltages are not encountered. Isolation amplifiers are intended for use under more extreme conditions.

Why Gain Blocks Are Required

1. Inconvenient transducer output characteristics
- Format of output (capacitance, resistance, current, voltage, etc.)
- High output impedances
- Inconvenient voltage ranges
- Unbalanced outputs

2. Hostile environmental conditions
- Noise
- High common-mode voltage
- Remote locations
- Temperature variances

3. Requirements for isolation
- Safety
- Protection for circuitry
- Ground loops

Limitations of Operational Amplifiers

IC operational amplifiers are available with a wide variety of performance features. Costs are low for general-purpose devices while increased precision and/or speed is available at a slightly higher cost. Furthermore, the op amp is extremely versatile and can be configured to do more than simple amplification. Every analog circuit designer has at least a working knowledge of op amp techniques, increasing op amp usage. In less than ideal situations, though, op amps have several serious shortcomings.

Practical transducer applications usually involve differential connections, nonzero source impedances, and noise. A typical application is shown in Fig. 6.24.

The output impedance of such a circuit is nonzero and, in general, unbalanced. Lead resistance and noise pickup cannot be totally avoided. A differential amplifier is required to sense output voltage, but a single op amp in a differential configuration is not well suited to such nonideal applications.

Op Amp in Differential Connection

For balanced gain, $R_2 = R_2'$ and $R_1 = R_1'$ (Fig. 6.25). For balanced input impedance, $R_1' + R_2' = R_1$. To provide balanced impedance return paths for amplifier bias currents (to minimize offset voltage drift)

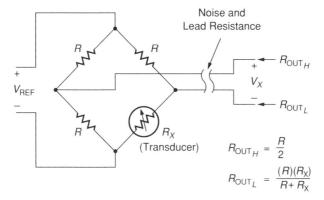

Fig. 6.24 Practical transducer circuit with differential output.

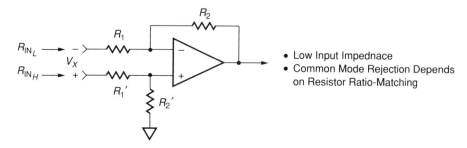

Fig. 6.25   Op amp in differential connection.

$$\frac{R_1 R_2}{R_1 + R_2} = \frac{R_1' R_2'}{R_1 + R_2} \tag{6.6}$$

These mutually exclusive conditions are only valid for ideal transducers. Finite input impedances, even if balanced, can disturb unbalanced transducers; lead resistance further aggravates the situation.

Op amp common-mode rejection is typically between 60 db and 90 db (some types may have up to 110 db). This may not be sufficient to reject common-mode noise. Nor is there galvanic isolation between input and output. Signals superimposed on common-mode voltages in excess of ± 10 volts cannot be handled, and if the common-mode input voltage exceeds supply voltage, the op amp might be destroyed.

Instrumentation Amplifiers

An instrumentation amplifier (IA) is a precision differential voltage gain device that is intended for use when acquisition of a useful signal is difficult. IAs are characterized by high input impedances, low bias currents, high common-mode rejection, balanced differential inputs, and stable, well-characterized specifications. Gain is determined by pin strapping with a user-selectable resistor or resistor pair. All other necessary precision components are internal, allowing the manufacturer to guarantee a specified level of performance. The output is single-ended; sense and output reference terminals are usually provided.

Instrumentation Amplifier Characteristics

1. High input impedances
2. Low bias currents
3. High common-mode rejection
4. Balanced differential inputs
5. Stable, well-characterized specifications
6. Gain determined by user-selectable resistor, resistor-pair, or pin strapping (internal)
7. Single-ended output

Specifications.   The performance of an IA is described by its specifications. Some specifications are self-explanatory and not unique to IA applications. Others are very significant when precision amplification in a real-world environment is required; a discussion of such specifications follows.

Gain equation. The gain of an IA is determined by pin strapping of internal resistors or externally applied resistors. The specification of the relationship between desired gain and IA gain is called the gain equation.

Gain range. This specification indicates the range of gains that the manufacturer recommends for proper operation. The IA may indeed function at higher or lower gains, but the manufacturer does not guarantee any particular level of performance. In practice, lower gains may compromise stability while higher gains may be impractical due to increased noise and drift.

Gain error. The number given by this specification describes maximum deviation from the gain equation, as shown in Fig. 6.26. The user can usually trim the gain or compensate elsewhere in the design. If data is eventually digitized and fed to an "intelligent system" (such as a microprocessor), gain error might be corrected by measuring a reference and multiplying by a constant.

Gain vs. temperature. These numbers give both maximum and typical deviations from the gain equation as a function of temperature. An intelligent system can correct for this with an "auto-gain" cycle (measure a reference and re-normalize).

Nonlinearity. Nonlinearity is defined as the deviation from a straight line on a plot of output

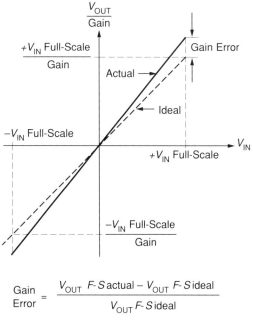

$$\frac{Gain}{Error} = \frac{V_{OUT}\,F\text{-}S\,actual - V_{OUT}\,F\text{-}S\,ideal}{V_{OUT}\,F\text{-}S\,ideal}$$

Fig. 6.26 Gain error.

versus input. Figure 6.27 shows the transfer function of a device with exaggerated nonlinearity. The magnitude of this error can be calculated thus:

$$\text{Nonlinearity} = \frac{(\text{Actual output}) - (\text{Calculated output})}{\text{Rated full scale output range}} \qquad (6.7)$$

This deviation can be specified relative to any straight line or to a specific straight line. There are two commonly used methods of specifying this straight line relative to the performance of a precision measurement device.

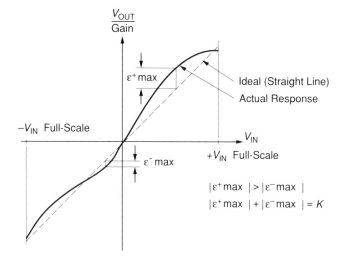

Fig. 6.27 Transfer function illustrating exaggerated nonlinearity.

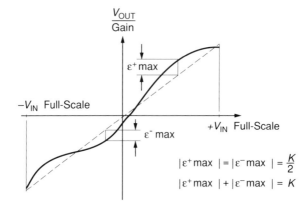

Fig. 6.28   Transfer function after calibration by best straight line method.

The "Best Straight Line" method of nonlinearity specification consists of measuring the peak positive and negative deviations and adjusting the slope of the device transfer function (by adjusting the gain and offset) so that these maximum positive and negative errors are equal (see Fig. 6.28). This method yields the best specifications but is difficult to implement in that it requires the user to examine the entire output signal range to determine these maximum positive and negative deviations.

The "Endpoint" method of specifying nonlinearity requires that the user perform offset and/or gain calibration at the extremes of the output range, as in Fig. 6.29. This is much easier to implement but may result in nonlinearity errors of up to twice those attained with best straight line techniques. This worst case will occur when the transfer function is bowed in one direction only.

Most linear devices, such as instrumentation amplifiers, are specified for best straight line linearity. The user must take this into consideration when evaluating the application error budget.

Nonlinearity errors are irreducible. In other words, these errors are neither fixed nor proportional to input or output voltage and cannot be reduced by adjustment. In some IAs, nonlinearity increases with gain.

Voltage Offset.   Initial voltage offset may be adjusted to zero, but shifts in offset voltage could cause errors. Intelligent systems can often correct with an auto-zero cycle, but there are many small-signal, high-gain applications without this capability. Voltage offset and offset drift are two components; input and output offset and offset drift. Input offset is that component of offset that is directly proportional to gain. Input offset, as measured at the output, with $G = 100$ is 100 times greater than with $G = 1$. Output offset is independent of gain. At low gains, output offset drift is dominant whereas input offset drift dominates at high gains. Therefore, the output offset voltage drift is normally specified as drift at $G = 1$ (where input effects are insignificant) while input offset

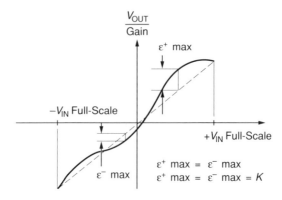

Fig. 6.29   Transfer function after calibration by end point method.

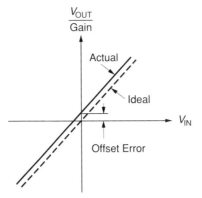

Fig. 6.30   Offset error.

voltage drift is given by a drift specification at a high gain (where output offset errors are negligible). All input-related numbers are referred to the input (RTI), which means that the effect on the output is G times larger. Voltage offset vs. power supply is also specified at one or more gain settings and is also RTI. See Fig. 6.30 for a plot of the following:

$$\text{Total Error RTI} = \text{Input Error} + \frac{\text{Output Error}}{\text{Gain}} \tag{6.8}$$

$$\text{Total Error RTO} = (\text{Input Error} \times \text{Gain}) + \text{Output Error} \tag{6.9}$$

Common-Mode Rejection Ratio.   The common-mode rejection ratio (CMRR) is a measure of the change in output voltage when both inputs are changed by equal amounts. These specifications are usually given for a full-range input voltage change and a specified source imbalance. CMRR is a ratio expression while "Common-Mode Rejection" (CMR) is the logarithm of that ratio (see Fig. 6.31). For example, a CMRR of 10,000 corresponds to a CMR of 80 db.

In most IAs, the CMRR increases with gain because most designs have a front-end configuration that does not amplify common-mode signals. Since the standard for CMRR specifications is referred to the output (RTO), a gain for differential signals in the total absence of gain for common-mode output signals will yield a one-to-one improvement of CMRR gain. This means that the common-mode output error signal will not increase with gain; it does not mean that it decreases with gain! At higher gains, however, amplifier bandwidth decreases. Since differences in phase-shift through the differential input stage will show up as a common-mode error, CMRR becomes more frequency-dependent at higher gains.

Settling Time.   Settling time is defined as the length of time required for the output voltage to approach and remain within a certain tolerance of its final value. It is usually specified for a fast full-scale input step and includes output slewing time. Since several factors contribute to the overall settling time, fast settling to 0.1% doesn't necessarily mean proportionally fast-settling to 0.01%. In addition, settling time is not necessarily a function of gain. Some of the contributing factors include slew rate limiting, under-damping (ringing), and thermal gradients ("long tails").

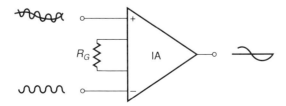

Fig. 6.31   Common-mode rejection.

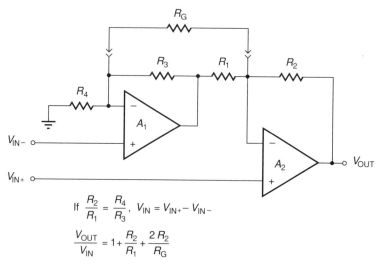

If $\dfrac{R_2}{R_1} = \dfrac{R_4}{R_3}$, $V_{IN} = V_{IN+} - V_{IN-}$

$$\frac{V_{OUT}}{V_{IN}} = 1 + \frac{R_2}{R_1} + \frac{2R_2}{R_G}$$

Fig. 6.32 Two-amplifier instrumentation amplifier.

Design Techniques. The most easily visualized IA configurations are based on IC op amps. The single-op amp differential stage previously shown is the simplest IA design, but it lacks the performance required for precision applications.

Figure 6.32 shows a two-amplifier IA design that overcomes some of the weaknesses in the one-amp approach. Input resistance is high, thus permitting sources to have unbalanced, nonzero output impedance. The major disadvantage of this design is that the common-mode voltage input range must be "traded-off" against gain range. A1 is called upon to amplify a common-mode signal by the ratio $(R_3 + R_4)/R_4$. If $R_3 > R_4$, saturation of A1 could occur, leaving no "headroom" to amplify the differential signal of interest. If $R_3 < R_4$, low gains cannot be realized.

The most popular configuration for op amp-based instrumentation amplifiers is shown in Fig. 6.33. The transfer function of this circuit is

$$V_{\text{out}} = (V_{\text{in}}^+ - V_{\text{in}}^-)\frac{(2R_1 + 1)}{R_G} \times \frac{R_3}{R_2} \qquad (6.10)$$

where $R_3 = R_3'$, $R_2 = R_2'$, and $R_1 = R_1'$.

In this configuration, gain accuracy and CMR depend on the ratio matching of R_2 and R_2', R_3 and R_3'. It can be shown, however, that CMR does not depend on the matching of R_1 and R_1'.

Within limits, the user may take as much gain in the front end as he wishes (as determined

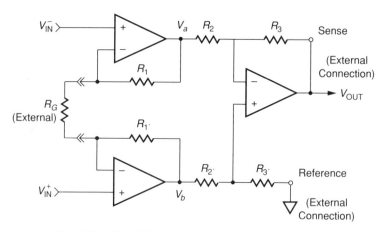

Fig. 6.33 "Classic" three-op-amp instrumentation amplifier.

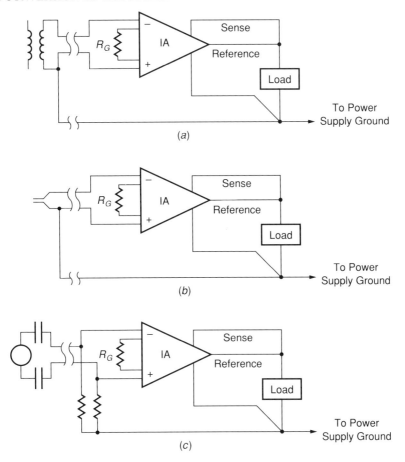

Fig. 6.34 Schemes for bias current return to ground: (*a*) transformer coupled; (*b*) thermocouple; (*c*) AC coupled.

by R_G) without increasing the common-mode error signal. Thus CMR will theoretically increase in direct proportion to gain, a very useful property. Furthermore, common-mode signals are only amplified by a factor of one, regardless of gain, since no common-mode voltage will appear across R_G and therefore, no common-mode current will flow in it because the input terminals of an op amp operating normally will have no significant potential difference between them. This means that large common-mode signals (within the op amp limits) may be handled independently of gain.

Finally, because of the symmetry of this configuration, first-order common-mode error sources in the input amplifiers, if they track, tend to be cancelled out by the output stage subtractor.

Indirect Ground Returns for Bias Currents

Although instrumentation amplifiers have differential inputs, there must be a return path for the bias currents. If this is not provided, those currents will charge stray capacitances, causing the output to drift uncontrollably or to saturate. Therefore, when amplifying "floating" input sources such as transformers and thermocouples, as well as AC-coupled sources, there must still be a DC path from each input to ground. Suitable connections are shown in Fig. 6.34. If it is not possible to provide a DC path for bias currents, an isolation amplifier will be required.

6.2   DATA CONVERSION TO THE DIGITAL DOMAIN

Stanley J. Domanski

Data conversion generally consists of transforming an analog signal, such as voltage or current, to some binary format or transforming a binary number to a voltage or current. The need for data conversion is evident when one considers that natural phenomena are basically analog in nature.

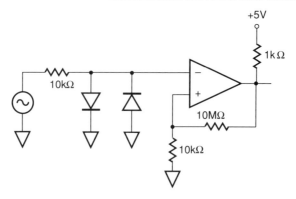

Fig. 6.35 Zero-crossing detector with hysteresis and clamping diodes.

To describe it in another way, natural occurrences may have an infinite variety of values. These occurrences could be temperature, pressure, humidity, luminosity, or the pH of a chemical reaction.

6.2.1 Comparators

Comparators are for the most part integrated circuit devices designed to detect signal voltages that exceed a given reference level. The reference threshold level generally has an arbitrary voltage. Comparators are similar to an operational amplifier being operated in the open-loop mode. They are generally characterized with high open-loop gain and fast response time, usually on the order of hundreds of nanoseconds. In general, the outputs of many comparators have open collectors that are pulled up to some supply voltage through an external resistor. Additionally, comparators can oscillate in a circuit unless care is taken in implementing circuit topology. Figure 6.35 shows a typical application of a comparator used as a zero crossing detector. Notice the use of the back-to-back clamping diodes and positive feedback to develop a hysteresis of 5 mV. Another important application of comparators is in the operation of successive approximation analog-to-digital converters.

6.2.2 Digital Codes

The best known code is natural binary. In a natural binary fractional code having N bits, the most significant bit (MSB) has a weight of $1/2$ (2^{-1}), the second bit has a weight of $1/4$ (2^{-2}), and so forth down to the least significant bit (LSB), which as a weight of 2^{-N}. The analog output value of the binary number is obtained by adding up the weights of all non-zero bits (see Fig. 6.36).

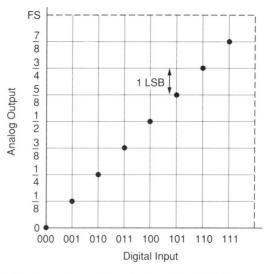

Fig. 6.36 Conversion relationship for an ideal 3-bit D/A converter.

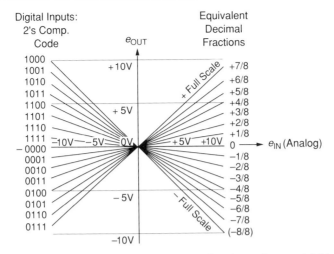

Fig. 6.37 D/A converter as four-quadrant multiplier of analog voltage and 3-bit-plus sign 2's complement digital number. Analog output vs. analog output as a function of digital input code.

The input-output relationship of an ideal 3-bit D/A converter shows exact (1/8) F-S steps (1 LSB) at each code change and no errors at either zero or full-scale. Such a converter would also switch between codes infinitely fast with no change in performance with temperature or time.

So far, the conversion relationships mentioned have been unipolar: the codes have represented numbers, which in turn represent the normalized magnitudes of analog variables without regard to polarity. A unipolar D/A converter will produce analog signals of only one polarity, and a D/A converter configured in the bipolar mode will produce signals of either polarity (see Fig. 6.37).

The D/A converter can be thought of as a digitally controlled potentiometer that produces an analog output that is a normalized fraction of its full-scale setting. The output voltage or current depends on the reference value chosen to determine full-scale output. If the reference varies in response to an analog signal, the output is proportional to the product of the digital number and the analog input.

The product's polarity depends on the analog signal polarity and the digital coding and conversion relationship. Four-quadrant multiplication is available if the D/A converter accepts reference signals of both positive and negative polarities and the digital input can represent either a positive or negative number. D/A converters, therefore, may be one-quadrant, two-quadrant (single polarity of either analog or digital variable) or four-quadrant. They may even be fractional-quadrant if the reference has a limited range of variation.

6.2.3   Analog-to-Digital Converters

The analog-to-digital converter is at the core of any data acquisition system designed to transform data in the form of continuous analog variables into a discrete binary code suitable for digital processing. A/D converters take on more varied forms than D/A converters, mainly because of the much wider variety of required characteristics. The discussion here will be limited to A/D converters implemented typically in integrated-circuit or hybrid form.

The ideal transfer curve for a 3-bit A/D converter is shown in Fig. 6.38 with the analog levels on the horizontal axis and the digital outputs corresponding to those input levels on the vertical axis. Note that a given digital output is valid over a small range of signal input, not just at a single point; this range is referred to as the "width" of the code. For ideal performance, each code width (except at the extremes) should be one LSB wide; acceptable performance will occur with codes 1/2 LSB to 1 1/2 LSB wide. If a code width becomes so narrow that it disappears, the A/D converter will never output that code—it will be a "missed" code. Integrating A/D converters and voltage-to-frequency converters have inherently uniform code widths which change significantly only over a large number of counts.

The specifications for A/D converters are similar to those for D/A converters (see Table 6.2). That is, the same concepts of gain error, offset error, and linearity apply with equal importance to A/D converters. The D/A converter output is an analog signal, and the input is a digital code that can only represent a finite number of input test conditions. The A/D converter, on the other hand, has an infinite number of input conditions but a finite number of digital output codes. Thus,

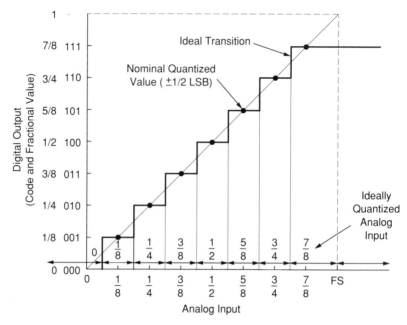

Fig. 6.38   Conversion relationship for an ideal A/D converter.

A/D converter errors are generally the difference between actual and ideal analog inputs that cause a particular code transition to occur.

In a D/A converter, differential linearity specifies the size of the analog output change response to any 1-LSB increment in the digital code. In the case of an A/D converter, differential linearity specifies the minimum and maximum limits on the width of any code.

The problem of missing codes in an A/D converter is most often caused by a non-monotonic conversion relationship on the part of the internal D/A converter. In the 3-bit example given in Fig. 6.39, the DAC's MSB is less than the sum of bits 1 and 2, hence the D/A conversion is non-monotonic. Since the MSB is always tried first, if it is less than the input value that is being compared with, it will foreclose the possibility of any lesser code being valid. Since it is turned on first and corresponds to a value less than the analog value of code 011, no output for code 011 can appear. Table 6.3 shows the results of trials at various levels in the vicinity of the non-monotonic bit. It provides verification that code 011 is indeed nowhere to be found—it is a missing code.

Missing codes in an A/D converter indicate differential nonlinearity more negative than -1 LSB. An A/D converter is specified for $\pm 1/2$ LSB and is at most $1 1/2$ LSB wide.

As with the D/A converter, it is important to verify that each of the specifications of a particular A/D converter remain within the system error budget over the full range of interest.

TABLE 6.2   COMPARISON OF D/A AND A/D SPECIFICATIONS

	D/A	A/D
Gain error	Error in slope of transfer curve.	Same.
Offset error	Output that occurs for input code that should produce zero output.	Input that causes first bit transition to occur (ideal is $1/2$ LSB).
Linearity error	Deviation of analog output from straight line.	Deviation of code midpoints from straight line.
Differential	Difference between actual output increments and ideal (1 LSB) steps.	Difference between actual width of code and ideal (1 LSB).

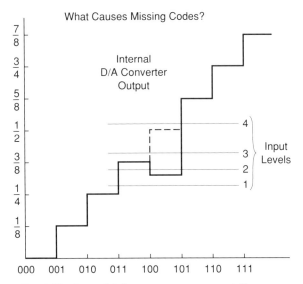

Fig. 6.39 Internal D/A converter output vs. A/D output.

In order to select the best ADC for a given application, the circuit designer should be familiar with the various types of converters available. In addition, familiarity with the manufacturing processes involved will often aid in selection of an ADC.

ADC Selection Factors

1. Resolution
2. Accuracy
    Initial (25°C)
    Drift
3. Speed
4. Power requirements
5. Reference
    Internal
    Ratiometric
6. Interfacing
7. Cost
8. Size
9. Availability

TABLE 6.3 SUCCESSIVE APPROXIMATIONS[a] (ACCEPT CODE IF LOWER THAN INPUT LEVEL)

Code	Input Level			
	1	2	3	4
100	No	Yes	Yes	Yes
010	Yes	—	—	—
011[b]	—	—	—	—
110	—	No	No	No
101	—	No	No	No
Answer	010	100	100	100
Correct Answer	010	010	011	100
Error	None	2 Bits	1 Bit	None

[a] Fig. 6.38 shows the internal D/A converter output.
[b] Missed Code = 011

The checklist above outlines the proper priorities for selecting an ADC. First and foremost, the required resolution of the converter must be determined. This will determine the number of recognizable quantization intervals in the ADC transfer function. Since all ADCs exhibit a fundamental quantization uncertainty of $\pm 1/2$ LSB, the designer must choose an ADC with sufficient resolution to reduce this "digitizing noise" to an acceptably low value. A useful rule of thumb is that each bit of resolution reduces quantization noise an additional 6 db. Thus, a 10-bit converter exhibits "best case" quantization noise approximately 60 db below full scale. Table 6.4 shows the quantization noise for converters of various resolutions.

Resolution, however, does not imply accuracy. Many A/D converters have been built with far higher resolution than accuracy. Specifications must be examined to ensure that the desired system accuracy can be achieved on a consistent basis. "Typical" accuracy specifications should be avoided whenever possible.

ADC Accuracy Requirements

1. Gain error
2. Offset error
3. Linearity
4. No missing codes
5. Evaluate all of above over expected temperature range

Of the specifications listed above, the most crucial is missed codes. Gain and offset errors are not as important because they can be trimmed in either hardware or software. Linearity errors will directly contribute harmonic distortion to a waveform that is digitized and subsequently reconstructed using an ideal DAC. Linearity can be improved in software by storing a calibration curve in a non-volatile memory, but such techniques are generally cumbersome and can cost more than a more accurate ADC. Missing codes, however, cannot be restored in either hardware or software. If a 10-bit ADC misses 10 codes in its transfer function, then it is reduced to a $\log_2$ (1014), or 9.98-bit converter.

Of course, all of these errors must be evaluated over the full temperature range of interest. Most manufacturers include temperature coefficients for accuracy specifications on their data sheets. It is important that any critical parameters be specified as minimum or maximum values, since "typical" specifications are usually not guaranteed or tested.

Conversion speed is another important selection factor since it defines the upper limit on system bandwidth. Generally speaking, the conversion speed requirements will dictate the type of converter selected. The total conversion speed includes both the ADC conversion time and the aperture and acquisition times of the sample/hold amplifier, if used.

The Nyquist theorem states that any signal sampled at twice the frequency of its highest frequency component will be sampled with no loss of information. This assumes the signal is ideally low-pass filtered. Since ideal filters are unrealizable, it is important to sample at a rate many times the cut-off frequency of the nonideal filter. This topic is covered in detail in Section 6.4, where sample/hold amplifiers are discussed.

System power limitations will often dictate the best ADC for the application. Available system supply voltages must be suitable for the converter chosen, and the total power dissipation should be held as low as possible in order to minimize heat generation and the attendant drift.

Applications that use a single ADC can best be satisfied by choosing an ADC with a built-in reference. By using such a converter, one less component must be procured, less circuit board space is required, and the error budget calculation becomes simpler. In systems where the reference must be variable, ratiometric operation is desired, or a master system reference is available, an ADC requiring an external reference can be used to advantage. Table 6.5 lists the advantages of having internal vs. external references in ADCs.

TABLE 6.4   QUANTIZATION NOISE VS. RESOLUTION

Resolution	1/2 LSB	P = P "Noise"*	"Noise," db$_{FS}$
8 Bits	0.19%	39 mV	−48.2 db
10 Bits	0.048%	9.8 mV	−60.2 db
12 Bits	0.012%	2.44 mV	−72.2 db
14 Bits	0.003%	610 μV	−84.3 db
16 Bits	0.00076%	152 μV	−96.3 db

* Assuming 10 V span

TABLE 6.5   INTERNAL VS. EXTERNAL REFERENCE

Internal Reference	External Reference
One less component	Ratiometric operation
Total error budget	Can be more accurate
Lower cost	Several converters can share

Since most ADCs are interfaced to microprocessors (with the notable exceptions of display-oriented converters and video speed converters), it logically follows that interface logic should be included on-chip. It is important to read manufacturers' claims of "microprocessor compatibility" with a great deal of caution. An ADC designed for simple interface to one particular processor may be nearly impossible to interface to a different machine.

Most newer ADCs also include at least a three-state buffer for bus interface. In most cases, direct bus interface is possible. However, high-speed data buses with significant activity during the conversion period may inject noise into the ADC, causing erratic or unstable output data. The solution to this problem (generally encountered only on ADCs with resolutions of 12 bits or higher) is to use an external three-state buffer. A/D converters designed for voltmeter applications generally provide either multiplexed seven-segment or BCD outputs.

ADC Output Interfacing

1. Three-state buffers
2. Multiplexed seven-segment or BCD
3. Serial

Selecting the Right ADC for the Application

We now address the process of choosing the optimum ADC for a particular application. In order to select an ADC, it is useful to understand exactly what types of ADCs are available and how they work (see Table 6.6).

This list of A/D converter types is certainly not complete, but it does include the most popular ones, especially those currently produced in integrated-circuit form. The integrating ADC design, one of the oldest implementations of this function, is still widely used in digital panel meters and digital laboratory instrumentation. It suffers from limited speed but can achieve high accuracy at relatively low cost. The successive-approximation A/D (which uses an internal D/A) is the most popular for data-acquisition systems because of its moderately high speed and resolution. We will discuss the implementation, application, and specification of this type of converter in considerable detail throughout the remaining discussions. The successive-approximation (S/A) converter is very versatile, and recent major cost reductions have made it suitable for more and more applications; it does, however, require some knowledge and experience for successful implementation. The tracking A/D and the voltage-to-frequency converter can be looked upon as variations of the successive-approximation and the integrator design techniques. In these types, the digital data is available on a virtually continuous basis.

The multi-comparator ladder (or parallel) design provides the highest speed at the expense of limited resolution. Most monolithic parallel converters are limited to 6- or 8-bit resolution.

TABLE 6.6   ANALOG-TO-DIGITAL CONVERTER TYPES

Type	Characteristic
Integrating	High Accuracy Low Speed Low Cost
Successive Approximation	High Speed Accuracy at Cost Flexibility
Tracking (Counter-Comparator)	High Speed Susceptible to Noise
Multi-Comparator "Flash"	Highest Speed High Resolution Expensive

Multi-Comparator "Flash" Converter

The multi-comparator "flash," or parallel converter consists of a tapped resistor network and an array of $2^N - 1$ comparators. Each comparator provides a logic output indicating whether the analog input is above or below a particular fraction of the reference voltage. These logic outputs are then decoded to a more useful code (binary or Gray code, for example). The total conversion time for this type of converter is equal to the sum of nanoseconds, translating into conversion rates of tens of megasamples per second.

The disadvantage of the parallel converter is the limited resolution available. Figure 6.40 shows that a 3-bit converter requires $2^3 - 1$ comparators. A 10-bit converter would, for example, require 1024 matched precision resistors, 1023 high-speed comparators and an extensive logic array. Integrating this complex a device remains currently beyond the practical state of the art.

Integrating Converter

The concept of the "dual-slope" integrating A/D is simple. A current proportional to the input signal charges a capacitor for a fixed length of time. The capacitor is then discharged by a current proportional to the reference until the starting point is crossed (See Fig. 6.41). The discharge time is then directly proportional to the average value of the input signal. Although components of good quality (especially the capacitor) most be used for reasonable accuracy, only the reference need be an expensive, high-quality component. Speed is an obvious limitation because of the long count time required (for example, you must count to 2,000 and do what amounts to 1,000 successive comparison tests to achieve 3-digit or 10-bit resolution).

Dual-slope integration has many advantages. Conversion accuracy is independent of both the capacitor value and the clock frequency because they affect both the up-ramp and the down-ramp in the same ratio. Differential linearity is excellent since all codes inherently exist, being generated by a clock and counter. Resolution is limited only by analog resolution, rather than by differential nonlinearity; hence, the excellent fine structure may be represented by more bits than would be needed to maintain a given level of scale-factor accuracy. The integration provides rejection of high-frequency noise and averaging of changes that occur during the sampling period. The fixed

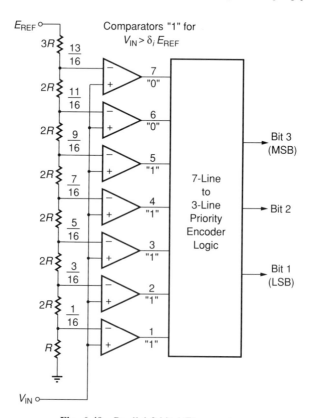

Fig. 6.40 Parallel 3-bit A/D converter.

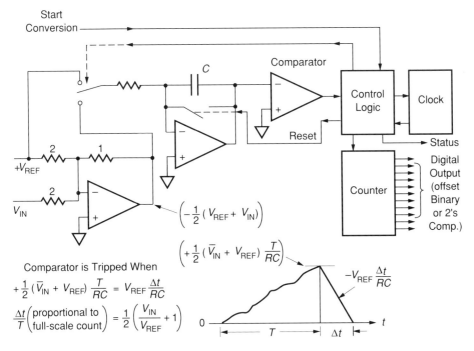

Fig. 6.41 Dual-ramp A/D converter for bipolar input.

averaging period also makes it possible to obtain "infinite" normal-mode rejection at frequencies that are integral multiples of $1/T$, as shown in Fig. 6.42.

Throughput rate of dual-slope converters is limited to somewhat less than $1/2T$ conversions per second. The sample time T is determined by the fundamental frequency to be rejected. For example, to reject 60 Hz and its harmonics, the minimum integrating time is $16\frac{2}{3}$ ms, and the maximum number of conversions is somewhat less than 30/s. Though too slow for fast data acquisition, dual-

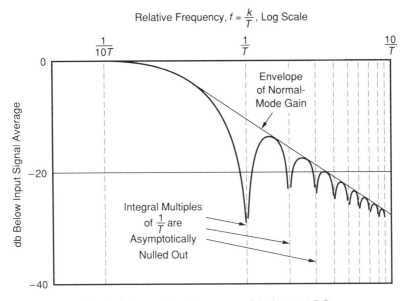

Fig. 6.42 Normal mode response of dual-ramp ADC.

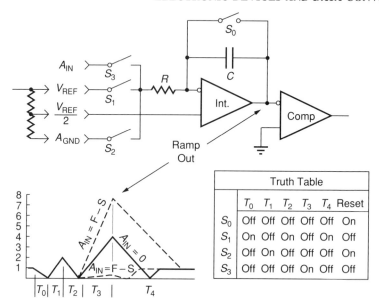

Fig. 6.43   Quad-slope sequence.

slope converters are quite adequate for transducers such as thermocouples and gas chromatographs, and they are the predominant circuit used in constructing digital voltmeters (DVMs). Since DVMs use sign-magnitude BCD coding, bipolar operation requires polarity sensing and reference-polarity switching rather than simple offsetting.

A shortcoming of conventional dual-slope converters is that errors at the input of the integrating amplifier or the comparator show up as errors in the digital word. Such errors are usually reduced by the introduction of a third portion of the cycle, at which time the capacitor is charged with zero-drift errors and then introduced in the opposite sense during the integration to nullify them.

Such error-nullifying techniques generally entail the addition of more logic. Since CMOS fabrication allows complex digital functions to be integrated without drastically increasing power consumption, it is the first choice for most integrating A/D converters.

The diagram in Fig. 6.43 shows the 13-bit ADC, which employs the integrating converter technique known as quad-slope. This technique digitally corrects for zero drift in the integrator and comparator and for long-term changes in oscillator frequency or integrating component values. The only external components required are three resistors, a capacitor, and a voltage reference. Typical conversion time to 13-bits is 40 milliseconds. The ADC is fully microprocessor compatible, with three-state controlled output data lines that will interface with an eight-bit data bus. The device will also operate in asynchronous serial binary mode. Bipolar operation with full accuracy is easily accomplished with a single voltage reference; ratiometric operation with a variable reference is equally simple. The fundamental limitation on the resolution of CMOS integrating A/D converters is the noise of the integrator amplifier and comparator. The inherently high $1/f$ noise of CMOS amplifiers can be avoided if bipolar input amplifiers are added externally.

Tracking Converter

One of the most important uses of D/A converters is in the core of certain types of A/D converters. In these A/D converters (see Fig. 6.44), the output of the internal DAC is compared to the input level; the logic section of the A/D converter uses the information from the comparator (internal level above or below input) to modify the code to the DAC until the two levels are brought as close together as possible (within the resolution limit of the DAC). In the simple tracking A/D converter shown in Fig. 6.44, an up/down counter drives the DAC in the proper direction to match the input in response to commands from the comparator. The accuracy of the D/A converter directly affects the accuracy of this system as an A/D converter.

If the output of the D/A converter is less than the analog input, the counter counts up. If the DAC analog input is constant, the counter output "hunts" back and forth between the two adjacent bit values. This converter can follow small changes quite rapidly (it will follow 1 LSB changes at the clock rate), but it will require the full count to acquire full-scale step changes. Since it seeks

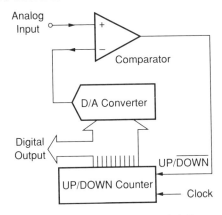

Fig. 6.44   Tracking (counter-comparator) A/D converter.

to "home in" on the analog value, the analogy to a servomechanism is quite evident. The tracking ADC is not often implemented in IC form, since a similar amount of complexity can yield a much more efficient converter—the successive-approximation type.

Successive-Approximation Converter

The successive-approximation ADC is popular for high-speed data aquisition systems since it is always able to complete a conversion with 8, 10, or 12 "counts" or comparisons (for 8-, 10-, or 12-bit resolution), whereas the time for a complete conversion with a tracking converter will vary directly with the input signal change. The performance requirements for a DAC inside a successive-approximation ADC will normally be more stringent than for one inside a tracking ADC.

The concept of the successive-approximation converter (Fig. 6.45) is directly analogous to that of a beam balance with a set of precision weights in successive binary ratio. One performs the measurement by placing the unknown on one side of the beam, then testing each known weight against the unknown, beginning with the largest (see Fig. 6.46). If a given weight tips the scale it is discarded; if not, it remains. This procedure continues down to the smallest weight. If the balance is true and the individual weights all exact, the unknown can be determined to a resolution equal to the smallest weight. If a weight equal to one-half the smallest weight is added to the unknown side, measurement is accurate to one-half the smallest weight.

The equivalent flow diagram for the successive-decision process and final result indicates that complex logic functions are required. However, the complex logic function can be done fairly rapidly

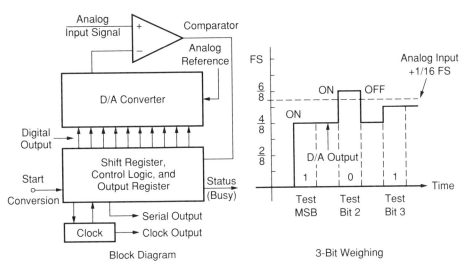

Fig. 6.45   Successive-approximation A/D converter.

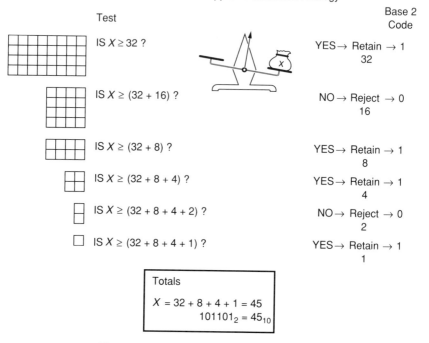

Fig. 6.46   Successive-approximation sequence.

in comparison with analog functions. Note that the successive-approximation ADC lies between the integrating converter and the parallel-comparator type in complexity, speed, and cost. More important, the complexity is not so great as to prevent its implementation in monolithic form.

The implementation of the successive-approximation A/D converter is basically very simple. One starts with a standard current-output D/A converter as the core. A successive-approximation register, which exercises the bit currents in proper sequence and stores the comparator decisions, is added, together with a comparator operating in a current-comparison mode. The successive-approximation register will then turn on the proper bits to produce a current that homes in on the input signal current (the input voltage divided by the input span resistance). A flow chart modeling the above process is shown in Fig. 6.47.

While it is possible to construct a successive-approximation ADC from the individual components, including the DAC, comparator reference, and successive-approximation logic or software, it is generally more cost effective to consider the ADC function as a component to be purchased. Medium- to low-cost converters are now readily available in a wide range of speeds and accuracies.

The construction of complete D/A converters with monolithic technology brought some insight to the problem of building complex, multi-function analog interface components. The construction complexity is extended one step further when the complete A/D converter is considered. In addition to all the componets required to implement a D/A converter, we must now add a fairly complex digital register and an accurate, high-speed comparator. Bipolar and CMOS technologies bring differing strengths to the solution of this fabrication problem, but both technologies have been brought to the point where complete A/D converters can be built on a single chip.

A/D Converter Additional Fabrication Requirements

1. Successive Approximation Register Logic or Counter
    High Speed
    Low Power

2. Comparator
    Accuracy
    Speed
    Low Noise

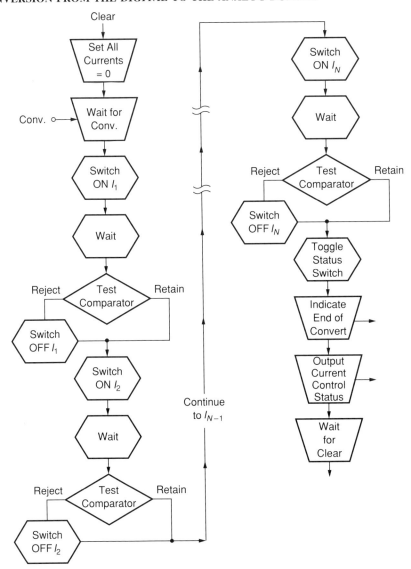

Fig. 6.47 Analog-to-digital flow chart.

CMOS technology allows the complex SAR (successive-approximation register) function to be implemented with very low power requirements and reasonable speed. The comparator function is somewhat more difficult to achieve with CMOS technology. CMOS comparators suitable for 8- and 10-bit resolution have been built successfully, but 12-bit resolution, though possible, compromises speed (see Figs. 6.48, 6.49, and 6.50).

6.3   CONVERSION FROM THE DIGITAL TO THE ANALOG DOMAIN

Stanley J. Domanski

In a multitude of applications it may be sufficient only to convert a signal from the analog domain to the digital domain and not back to the analog domain. This is true if the main purpose for the original conversion from analog to digital is not machine control.

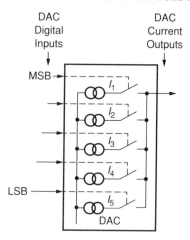

Fig. 6.48 Digital-to-analog conversion.

All physical phenomena are analog in nature, as in the case of current flow in a conductor. Ohm's law applies where $V = IR$ or $I = V/R$. For instance, a specific level of heat or light may be required, dictating an increase or decrease in current flow through a heating element or lamp. The amount of change in heat or light may have been the result of a computation performed by a digital computer. The result of the computation is a digital word whose numeric value is proportional to the physical value required. Hence a digital-to-analog converter is a device that transforms a digital representation of a physical parameter into a meaningful and useful form; this meaningful form is usually a voltage or current, for it is a voltage or current that is responsible for the generation of heat, light, and motion.

6.3.1 Digital-to-Analog Converters

The digital to analog function is to convert a digital code to an analog parameter by switching successive-ratio analog bit weights to a common summing point. Commonly, converted parameters

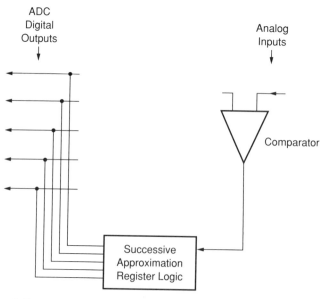

Fig. 6.49 Additional components to perform analog-to-digital conversion.

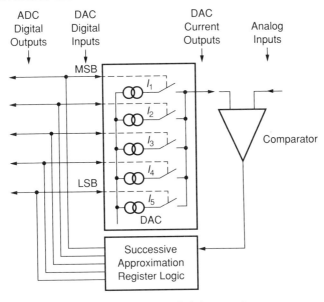

Fig. 6.50   Analog-to-digital conversion.

are digital to voltage, digital to current, and digital to gain or attenuation. The inherent resolution is limited by the number of bits of digital code.

Figure 6.51 shows the various components that make up a D/A converter. Not all of them are necessarily included in every device available on the market, but the parts shown inside the dotted line box are the minimum components needed. Digital switch control is necessary to convert standard logic levels (TTL, CMOS, etc.) to an analog switch-drive level. A reference amplifier is needed in many cases to properly relate the bit weights of the analog switches to an output amplifier, which changes the output current of the converter into a voltage. A class of converters includes storage registers and various other digital control functions in the same package. The reason for breaking down converter architecture so finely is that many integrated-circuit converter types have only a few of the above components, trading off complexity for chip size.

6.3.2   D/A Converter Specifications and Testing

The specifications for a D/A converter at first glance may seem a little overwhelming; however, they may be easily grouped so that a complete error budget can be done. The specifications can be

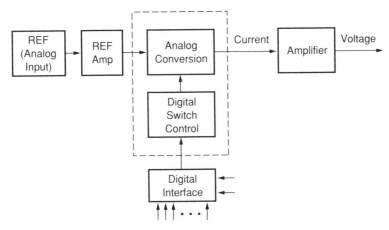

Fig. 6.51   D/A converter general functional diagram.

TABLE 6.7 ERROR TERMINOLOGY; 10 VOLT FULL SCALE

Resolution	LSB	mV	% of F-S	ppm
8-bit	1	39.1	0.391	3906
10-bit	1	9.77	0.1	977
	1/2	4.88	0.09	488
12-bit	1	2.44	0.024	244
	1/2	1.22	0.012	122
14-bit	1	0.61	0.006	61
	1/2	0.305	0.0031	31
16-bit	1	0.153	0.0015	15
	1/2	0.076	0.00076	8

grouped first by the static accuracy, second by temperature characteristics, and third by dynamic characteristics. Failure to consider any of these specifications may result in poor performance in a given application.

The parameters of D/A converters have error specifications which may have any one of four different units of measure associated with it (LSBs, % of FSR, error in volts, ppm). Table 6.7 lists error terms for converters of different resolutions, and their equivalents in each of the units of measure.

Static Accuracy Overview

The static accuracy of a D/A converter can be completely described by four error terms: offset, gain (full-scale), linearity (relative accuracy), and differential linearity (step size).

Four Basic D/A Converter Errors (Static, 25°C)

1. Zero (offset) The error when the digital input code calls for zero output. It affects all codes by the same additive amount (trimmable).

2. Gain (scale factor) An error in the slope of the transfer curve. It affects all codes by the same percentage amount (normally trimmed for calibration).

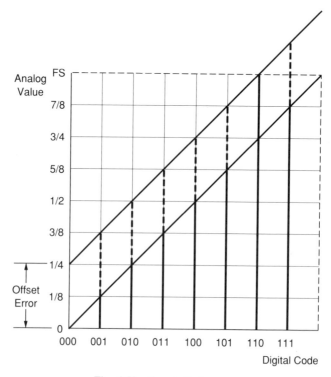

Fig. 6.52 Zero (offset) error.

3. Linearity (relative accuracy) A measure of the converter's transfer curve deviation from a straight line (not trimmable).

4. Differential linearity The difference between any two adjacent steps. If one step is exactly 1 LSB different from the previous step, the differential nonlinearity is zero (not trimmable).

Offset Error

The zero (offset) error is the deviation in the output when the digital input calls for a zero output. It affects all codes by the same additive amount, as shown in Fig. 6.52. In a voltage output D/A converter, it is the combination of DAC leakage current and the current-to-voltage converter's offset voltage. In a current output DAC the zero error is the DAC leakage current (sometimes just called leakage current).

The offset (bipolar zero) is the midpoint of a bipolar (e.g., ± 10 volt output) transfer characteristic that normally falls at zero output. For almost all bipolar converters (e.g., ± 10 volt output), instead of actually generating negative current to correspond to negative numbers, a unipolar DAC is used, and the output is offset by one-half full-scale (1 MSB). For best results, this offset voltage or current is derived from the same reference supply that determines the gain of the converter.

Gain Error

The gain of a converter is the analog scale factor that establishes the nominal (i.e., full-scale) conversion relationship. The gain error is an error in the slope of the transfer curve, as shown in Fig. 6.53. It affects all codes by the same percentage amount.

The sensitivity of a converter to changes in the power supplies is normally expressed in terms of percent change in analog value (D/A output, A/D input) for a 1% change in power supply (e.g., $0.05\%/\%\Delta V_S$). As a rule, the fractional change in scale factor for a good converter should be well below the equivalent of $\pm 1/2$ LSB for a 3% change in power-supply voltage. When power-supply voltage changes affect conversion accuracy excessively, the key culprit is often a marginal "constant-current-circuit" for the reference diode.

Linearity Error

Linearity error of a converter is the deviation of the analog values in a plot of the measured conversion relationship from a straight line, as shown in Fig. 6.54. The straight line can be either a "best straight line," determined empirically by manipulation of the gain and/or offset to equalize maximum positive and negative deviations of the actual transfer characteristic from this straight line, or a straight line passing through the end points of the transfer characteristic after the system is calibrated. Sometimes

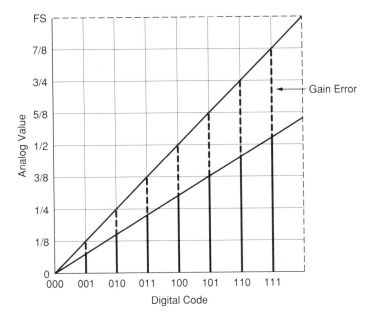

Fig. 6.53 Gain error.

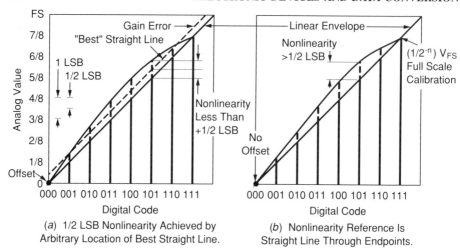

(a) 1/2 LSB Nonlinearity Achieved by
Arbitrary Location of Best Straight Line.

(b) Nonlinearity Reference Is
Straight Line Through Endpoints.

Fig. 6.54   Comparison of linearity criteria.

referred to as "endpoint" nonlinearity, the latter is the definition used by analog devices, both because it is a more conservative measure and because it is much easier to verify in actual practice. "Endpoint" nonlinearity is the same as relative accuracy.

Differential Linearity Error

In a converter, differential linearity error describes the variation in the analog value of transitions between adjacent pairs of digital numbers, over the full range of the digital input or output. If each transition is equal to its neighbors (i.e., 1 LSB), the *differential nonlinearity* is zero. If a transition differs from one of its neighbors by more than 1 LSB (e.g., if, at the transition 01111111 to 10000000, the MSB is low by 1.1 LSB), a D/A converter can be *non-monotonic*, and an A/D converter using it may miss one or more codes.

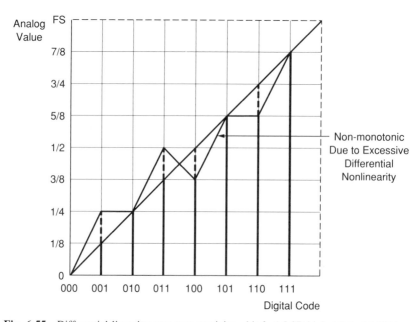

Fig. 6.55   Differential linearity: non-monotonicity—bit 3, 1 LSB high; bit 1, 1 LSB low.

Monotonicity

Monotonicity requires that a converter always give an increasing (in magnitude) analog value for increasing digital code. Non-monotonicity is a result of excess differential nonlinearity (±1 LSB). Monotonicity is essential for many control applications. Notice that a converter with a maximum differential nonlinearity specification of ±1/2 LSB is more tightly specified than one that is only guaranteed to be monotonic (see Fig. 6.55).

Relative Accuracy

It is important to draw a careful distinction between overall linearity (relative accuracy) and differential linearity (step size between codes). Relative accuracy error describes the deviation of the *points* along the transfer curve from the ideal true straight line. Differential linearity describes the equality of the analog *steps* between adjacent digital codes. If a given step between two codes is not exactly equal to one least significant bit (LSB = $V_{REF}/2n$, n = number of bits), that step has a differential linearity error. Note that, in this illustration, the differential linearity error is at least as bad as the linearity error at every code transition. Note that specifying only relative accuracy can allow very loose output performance.

Absolute Accuracy

Absolute accuracy of a D/A converter is the difference between the analog output that is expected when a given digital code is applied and the output that is actually measured with that code applied to the converter. Absolute accuracy error can be caused by gain error, zero error, linearity error, or any combination of the three. Absolute accuracy measurements should be made under a set of standard conditions with sources and meters traceable to an internationally accepted standard.

Settling Time (Fig. 6.56)

1. Settling time definition
   Error band definition
   Bits, Percent, or Millivolts
   Which transition?
   Zero to full-scale
   LSB
   MSB carry
   Reference input step without code change
2. Current settling vs. voltage settling
   Current—Fast, no capacitance charging
   Voltage—Slow, must charge capacitances

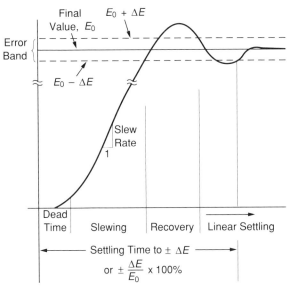

Fig. 6.56 Settling time measurement.

3. Measurement technique
4. Current-to-voltage conversion
5. More difficult than normal op amp settling

6.3.3   Temperature Effects in Digital-to-Analog Converters

The temperature performance of a D/A converter can be completely described by its offset, gain, and differential-nonlinearity temperature coefficients (T.C.). The gain T.C. of a multiplying D/A converter need only define the change with temperature of the ratio of the output to the external reference input and will normally seem very good since it depends only on matched-resistance temperature coefficients. However, if such a converter is used for generating absolute levels, its gain T.C. must be added to the T.C. of the external reference to determine the overall gain T.C. Converters with a self-contained reference are already specified for overall T.C.

Temperature Coefficients (Fig. 6.57)

1. **Zero temperature coefficient**   A zero shift due to temperature affects all output readings equally and is expressed as V/°C or ppm/°C of full-scale.
2. **Gain temperature coefficient**   A gain shift with temperature causes the slope to change and is expressed as ppm of reading/°C. (Notice if the reference is included.)
3. **Offset temperature coefficient**   Bipolar offset is affected by both zero and gain T.C.
4. **Differential linearity temperature coefficient**   This shows the relative change in bit weights with temperature and is a measure of when a converter can be expected to go non-monotonic. It is expressed as ppm of full-scale/°C.

Temperature coefficients of gain and offset are defined in terms of the "average" deviation over a range of temperature variation, for example, $(\Delta T_1 - \Delta T_2)/(T_1 - T_2)$. For specified temperature

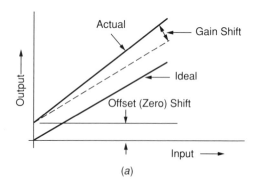

(a)

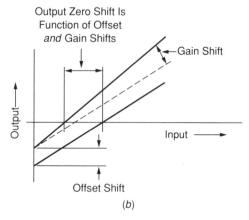

(b)

Fig. 6.57   Temperature coefficient effects: (a) unipolar (b) bipolar.

ranges that extend from below room temperature to above room temperature, the device is zeroed and calibrated at room temperature, and the temperature coefficients for the "high" range ($T_H - T_A$) and "low" range ($T_R - T_A$) are both compared with the specification; both must be better than specified. This is affected by the reference and the reference circuitry, including the reference amplifier and switches. The total gain (or scale factor) change is specified in ppm/°C.

The zero stability of a unipolar DAC is almost entirely governed by the output amplifier's zero stability. Since output amplifiers are usually employed as current-to-voltage converters, they operate at low values of closed-loop gain, and the zero T.C. is not greatly affected by the choice of programmable gain setting (i.e., 0–5 V or 0–10 V). Zero T.C. is usually expressed in μV/°C.

Converters that use offset-binary coding are "zero" set at the all-bits-off point, and their scale factor is set at either all-bits-on, or (for more precise zero) at the MSB transition. However, the zero T.C. is measured at the MSB transition (analog zero). It is affected by the reference T.C., the tracking of the offset reference, and the tracking of the bipolar-offset and gain-setting resistors.

The offset or zero T.C. will behave differently in unipolar and bipolar offset applications. In the unipolar mode, the zero T.C. depends only on leakage current terms or the offset voltage T.C. of the output amplifier; these are generally small. However, in the bipolar mode, the zero point occurs at the major carry (code 100 . . .). Both the unipolar offset T.C. and the gain T.C. of the device contribute to drift around the zero point. It is important to check this specification closely if bipolar operation will be used.

Monotonic behavior is achieved if the differential nonlinearity is less than 1 LSB at any temperature in the range of interest. The differential nonlinearity coefficient may be expressed as a ratio, as a maximum change over a temperature range, and/or implied by a statement that the device is monotonic over the specified temperature range.

Factors Contributing to Monotonicity

1. Initial accuracy
2. T.C. of differential nonlinearity
3. Temperature changes anticipated

Example: Assume 12 bit D/A converter with initial accuracy $= \frac{1}{4}$ LSB (0.006%, 60 ppm), including all errors at $+25$°C.

Differential nonlinearity T.C. $= 1$ ppm/°C.

Temperature range $= -55$°C to $+125$°C.

$\Delta T_1 = +80$°C, $\Delta T_2 = +100$°C.

Error $= 60$ ppm $+ 1$ ppm/°C $\times 100$°C $= 160$ ppm.

1 LSB $= 240$ ppm.

Monotonic D/A Converter

6.3.4 IC Technologies for D/A Converter Components

- Bipolar
- Bipolar/I^2L
- CMOS
- LC^2MOS
- BIMOS II

In order to manufacture high-resolution, high-accuracy data converters, the most important technology to have available is a high-accuracy resistor process. The technology required to produce well matched resistor ladder networks has advanced considerably in recent years.

Currently, using diffusion techniques, resistor ladder network linearities of 0.05% and differential linearities of 0.012% are achievable. With thin-film technology (silicon chromium), high resolution (16-bit) devices with linearities of 0.012% and differential linearities of 0.00076% are realizable on a production basis. Even though current processing allows fabrication of 12-bit ladder networks, the higher accuracy technologies do not permit low-pricing structures. Thus, some form of trimming after resistor definition is required to produce high-accuracy devices economically.

6.3.5 Basic D/A Conversion Techniques

The concept of D/A converter design, as discussed before, is to create a binary-weighted series of switchable currents, as shown in Fig. 6.58. Presumably, the binary resistor ratio could be continued indefinitely to build 8-, 10-, or 12-bit converters, but this is not feasible with IC fabrication techniques

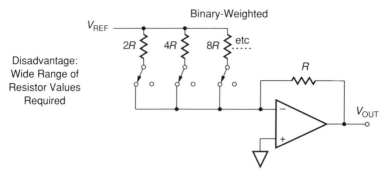

Fig. 6.58 Continuous binary-weighted.

since resistors cannot be well matched at ratios much greater than 20:1, thus setting an upper limit at five bits. The first design technique to circumvent this problem actually started in modular designs, with the use of the quad current switch and interquad dividers, as in Fig. 6.59. This design technique employs a group of identical 4-bit D/A converters strung together with appropriately sized attenuators between them to scale the output of each quad. Thus 8-, 12-, and even 16-bit D/A converters have been built without exceeding a 16:1 range of resistance values.

R-2R Ladder

However, even the successive-quad technique is not perfectly suited to IC fabrication—the large resistance ratios still present some difficulties in matching and also require a large amount of circuit area. Thus, newer designs tend to use the R-2R ladder network for optimum matching and area usage. Figure 6.60 shows the "standard" R-2R or voltage-switching converter. The inverted R-2R or current-switching ladder in Fig. 6.61 is being used in most new IC designs. If offers several performance and fabrication advantages. First, the resistance ladder itself always runs at the same bias; the switches simply steer the current either to ground or to the output through the virtual ground of the amplifier, thus achieving high speed. Second, there is a small (and constant) voltage drop across the switches, simplifying circuit design and giving stable, thermal-free operation. Finally, only two values of resistance need to be fabricated—ideal for monolithic technology.

The inverted R-2R ladder can be extended to a 16-bit DAC although this requires very tight resistor matching and tracking to guarantee monotonicity. In a 16-bit R-2R ladder DAC, the tolerance is most critical for the major carry where the 15 LSBs turn off and the most significant bits turn on. If the MSB is more than 0.0015% low, the converter will be non-monotonic.

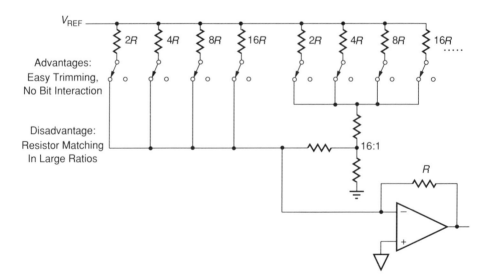

Fig. 6.59 Successively-weighted binary quads.

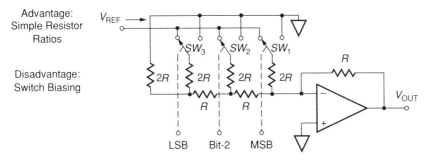

Fig. 6.60 Standard R-2R.

In many applications, very good differential linearity (monotonic behavior) is required, such as in a closed-loop servo-system, while linearity is not critical. Two segmentation techniques have been used to overcome the penalty inherent in the R-2R topology and therefore reduce the overall processing requirements. A current segmentation approach has been used on 12-bit DACs to achieve 12-bit monotonic performance without trimming.

Current-Segment DAC

In the current-segment approach (see Fig. 6.62), the two most significant bits are decoded into a switch pattern that directs each of the four equal current sources to the output, the complementary output, or to a current-dividing 4-bit DAC. As the digital code progresses higher, a given current source will be steered first into the complementary output, then into the 4-bit DAC (which steers from 0 to 31/32 of the current into the output), and then finally completely into the output. In this design, the monotonicity of the 4-bit DAC is critical to the monotonicity of the entire transfer function.

Voltage-Segment DAC

An extension of the current-segment DAC is the voltage-segment DAC. In a voltage-segment scheme, the most significant bits drive a decoder tree that selects the voltages from two adjacent taps of a series resistance ladder and applies them to the inputs of two precision buffers (see Fig. 6.63). These buffers then force the voltages across the terminals of a second series resistance ladder. The least significant data bits drive a second decoder tree that selects the voltage at one of the taps of this second ladder and routes it to the output. Since both ladders will be monotonic and the low-order bits are repeated four times, the entire structure will be monotonic.

The latches are controlled by the address inputs, A0–A3, and $\overline{CS}$ and $\overline{WR}$ inputs on the DAC. All control inputs are active low, consistent with general practice in microprocessor systems. The $\overline{CS}$ and $\overline{WR}$ inputs must both be low for any operation to occur. The four address lines each enable one of the four latches. All latches are level-triggered. This means that data present during the time

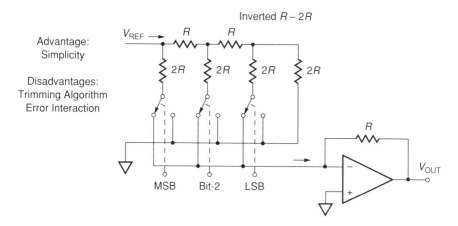

Fig. 6.61 Inverted R-2R.

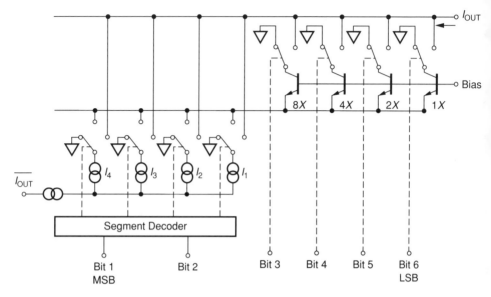

Fig. 6.62   Current-segment 6-bit DAC.

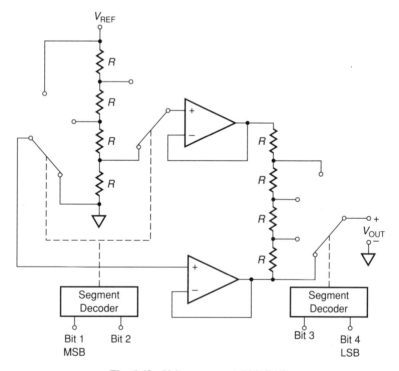

Fig. 6.63   Voltage-segment 4-bit DAC.

when the control signals are valid will enter the latch. When any one of the control signals returns high, the data are latched.

6.3.6 Microprocessor Interfacing

4-Bit Processor Interface

Many industrial control applications use 4-bit microprocessors but require 12-bit accuracy analog control voltages. Each DAC in Figs. 6.64 and 6.65 occupies four locations in a 4-bit microprocessor system. A single 74LS139 2-to-4 decoder (Fig. 6.66) is used to provide sequential addresses for the four DAC registers. CS is derived from an address decoder driven from the high-order address bits. The system WR is used for the WR input of the current output DAC.

8-Bit Processor Interface

Whenever a 12-bit DAC is loaded from an 8-bit bus, two bytes are required. If the program considers the data to be 12-bit binary fractions (between 0 and 4095/4096) and the data are left-justified, then the eight most significant bits comprise one byte, and the remaining bits make up the upper half of another byte. Right-justified data calls for the eight least significant bits to occupy one byte, with the four most significant bits residing in the lower half of another byte, simplifying integer arithmetic.

Left-Justified and Right-Justified Interface

Figure 6.67 shows an addressing scheme for use with a DAC set up for left-justified data in an 8-bit system. The base address is decoded from the high-order address bits and the resultant active-

$\overline{CS}$	A_3	A_2	A_2	A_1	Operation
1	X	X	X	X	No Operation
X	1	1	1	1	No Operation
0	1	1	1	0	Enable 4 LSBs of First Rank
0	1	1	0	1	Enable 4 Middle Bits of First Rank
0	1	0	1	1	Enable 4 MSBs of First Rank
0	0	1	1	1	Loads Second Rank from First Rank
0	0	0	0	0	All Latches Transparent

"X" = Don't Care

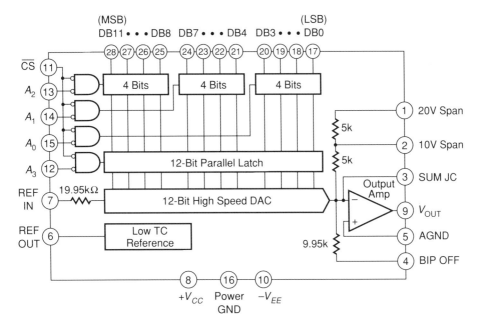

Fig. 6.64 Voltage-output DAC.

$\overline{CS}$	$\overline{WR}$	A_3	A_2	A_1	A_0	Operation
1	X	X	X	X	X	No Operation
X	1	X	X	X	X	No Operation
0	0	1	1	1	0	Enable 4 LSBs of First Rank
0	0	1	1	0	1	Enable 4 Middle Bits of First Rank
0	0	1	0	1	1	Enable 4 MSBs of First Rank
0	0	0	1	1	1	Loads Second Rank from First Rank
0	0	0	0	0	0	All Latches Transparent

"X" = Don't Care

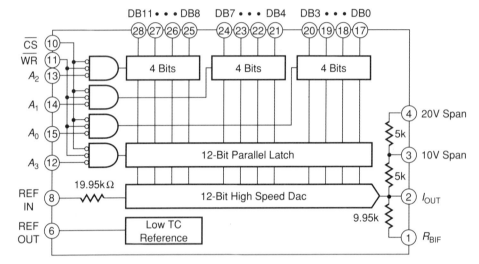

Fig. 6.65   Current-output DAC.

low signal is applied to $\overline{CS}$ (Fig. 6.68). The two LSBs of the address bus are connected as shown to the DAC address inputs. The latches now reside in two consecutive locations, with location X10 loading the four LSBs and location X10 loading the eight MSBs and updating the output.

Right-justified data can be similarly accommodated (Fig. 6.69). The overlapping of data lines is reversed, and the address connections are slightly different. The DAC still occupies two adjacent locations in the processor's memory map.

6.3.7   CMOS Digital-to-Analog Converters

Only CMOS D/A converters with a design similar to this function are *true* multiplying DACs. A DAC with this design multiplies any input signal, AC or DC, millivolts or volts, positive or negative, by exactly the digital code. The linearity of the device does not degrade with reduced signal, as will that of a bipolar DAC. In fact, a CMOS converter preforms in an analogous fashion to a potentiometer; as such it *really* is a digitally controlled attenuator (Fig. 6.70).

Attenuator

In a number of applications, when using a CMOS DAC, it is essential to connect the I_{out} terminal to a virtual ground. If that output point is not held at virtual ground, the linearity of the ladder network will be degraded, since the voltage drops across each leg of the ladder will be changed (Fig. 6.71). Typical unipolar and bipolar connections are shown in Fig. 6.72; note that these connections perform a net inversion of the input signal. The bipolar connection operates in full 4-quadrant mode in that it accepts both analog and digital signals of both polarities.

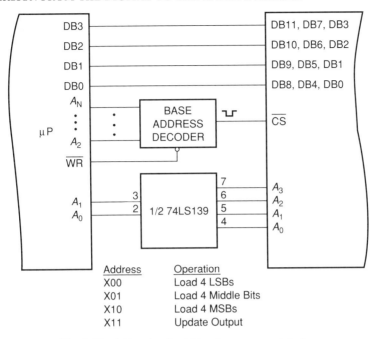

Fig. 6.66 Addressing for 4-bit microprocessor interface.

Gain Adjust

Figure 6.73 shows the feedback connection for a multiplying DAC to obtain the function

$$V_{out} = -V_{in}/N, \; 0 < N < 1 \tag{6.11}$$

where N is the fractional binary value of the input digital code. In the same way that the basic M-DAC circuit is analogous to an analog potentiometer unloaded by an op amp, this circuit is analogous to a follower, with the potentiometer connected in the feedback circuit. Examples of gains at various codes are given in the table in Fig. 6.77. As one might expect, the reduction of feedback causes amplifier noise and errors to be magnified at the higher gains. Since the incremental step changes of gain become increasingly large for small values of N, any errors in these gain steps are magnified correspondingly (e.g., at 00000001, a 10% error in the LSB value means a 10% gain error).

To limit the gain at all-zeros and to prevent the amplifier from "taking off," a large value of shunt feedback resistance may be used. This will have minimal effect on accuracy at the lower gains. For gains less than unity, use attenuation ahead of this circuit or connect resistance in series with R_F.

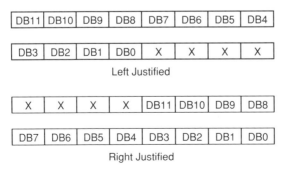

Fig. 6.67 Twelve-bit data formats for 8-bit systems.

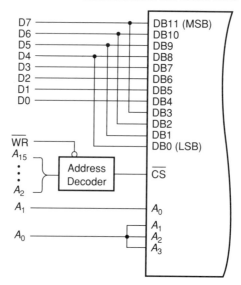

Fig. 6.68 Left-justified 8-bit bus interface.

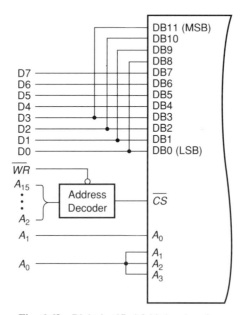

Fig. 6.69 Right-justified 8-bit bus interface.

CMOS Design

R-2R Ladder
Low R_{ON} Switches at Virtual Ground
Balanced Current Steering
Bi-Directional

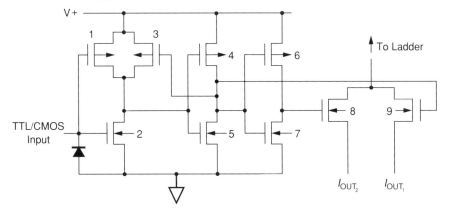

Fig. 6.70 CMOS switch.

6.4 DATA ACQUISITION SUBSYSTEMS

Stanley J. Domanski

A "Data Acquisition Subsystem," shown in Fig. 6.74, may be loosely defined as an ADC preceded by an analog multiplexer, buffer, or subtractor amplifiers, and a sample/hold amplifier.

Data Acquisition Subsystem Requirements

1. Multiple input channels
2. High input impedance
3. Single-ended or differential operation
4. Sample/hold function included
5. Maintain signal accuracy commensurate with ADC
6. Ease of use

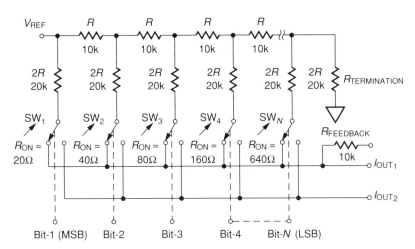

Fig. 6.71 R-2R ladder "bare bones" DAC.

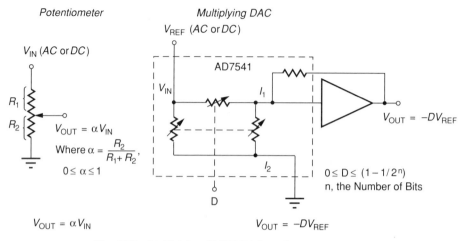

$V_{OUT} = \alpha V_{IN}$ $V_{OUT} = -D V_{REF}$

Fig. 6.72 Multiplying CMOS DAC used as an attentuator.

Each building block has certain performance characteristics. The multiplexer, for example, must have reasonably low "on" resistance, low leakage current, and a break-before-make operation. It is also desirable to include overvoltage protection, since many data acquisition subsystems will encounter input signals outside the expected range.

6.4.1 Sample Theory

In a data acquisition system (DAS), various physical parameters such as temperature, pressure, humidity, or vibration are monitored and usually converted to a digital format so that the data may be processed through the use of a digital computer. Some parameter, such as temperature, may vary slowly or vary rapidly. As an example, atmospheric temperatures may change at relatively slow rates—perhaps degrees per minute—whereas the temperature changes within the combustion chamber of a gasoline engine will vary drastically over short periods of time. Since many channels may be monitored, it becomes impractical to continuously monitor a given parameter. It is possible to simply sample the time-varying signal and reclaim the information.

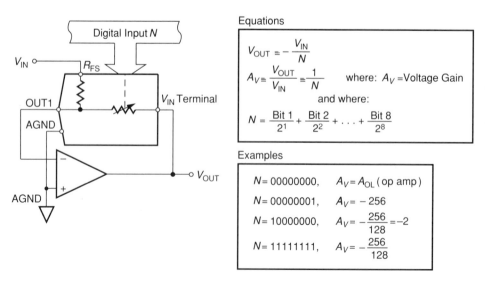

Fig. 6.73 CMOS multiplying DAC as a feedback element.

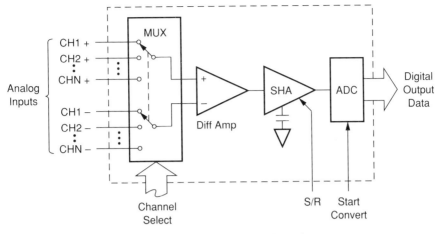

Fig. 6.74 Generalized data acquisition subsystem.

If the time-varying function has a base band frequency of f_b the requirement for reclaiming the information is related to the Nyquist criterion that the sampling frequency (f_s) must be greater than twice the frequency of the base band frequency. The sampling is usually accomplished with the aid of a sample/hold amplifier (SHA). Should f_s become less than twice f_b, a condition known as "aliasing" will occur.

Aliasing occurs when the sampling rate in a sampled data system is not sufficiently high. When a signal (or band of signals) is sampled at some frequency, the sum and difference frequencies appear in the reconstructed waveform. The "sum" frequencies are outside the band of interest and can be easily filtered. The difference frequencies can also be filtered, providing they do not fall within the band of interest.

In the case shown in Fig. 6.75, a signal at 25 kHz in a 25 kHz band-limited system is sampled and digitized at a 33.3 kHz rate. If the resulting points are used to reconstruct the signal in the analog domain, the output will obviously contain a frequency component at 8.33 kHz, which is inside the original band, even though the original signal contained no components at that frequency. The obvious solution to the problem is to increase the sampling rate by using a faster ADC, faster SHA, or both.

The primary use of sample/hold amplifiers is to maintain a stable voltage level at the input to an ACD during conversion. It is possible to use an ADC on some signals directly, but some limitations on signal bandwidth must be considered. Consider the following situation. A 12-bit, 35-microsecond ADC is being used to digitize a signal without a SHA. The input signal must then be limited to a maximum slewing rate that results in less than $\frac{1}{2}$ LSB (0.012% of full-scale) change during the conversion period. For a 10 V peak-to-peak sine wave, the maximum rate of change is 10π times the frequency. Solving for a 1.22 mV change in 35 microseconds yields a maximum input frequency of 1.1 Hz.

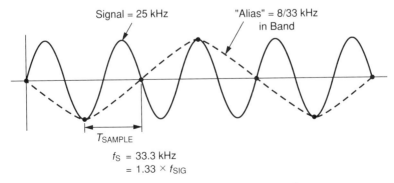

Fig. 6.75 In-band alias due to insufficient sampling rate.

A simplified formula for computing the maximum frequency that can be digitized without a SHA is

$$f_{max} = \frac{1}{(2^N + 1)\pi T_{CONV}} \tag{6.12}$$

where N is the number of bits of ADC resolution and T_{CONV} is the conversion time.

There are two limitations on the maximum signal frequency that can be digitalized by a combination of an ADC and SHA. One limit is related to the uncertainty or "jitter" in the SHA aperture time. If a sinusoidal input is sampled at the same point on the waveform, the sampling uncertainty will be equal to the maximum rate of change in a time period equal to the aperture uncertainty. Thus, the maximum sine wave frequency that can be sampled to N-bit accuracy is

$$f = \frac{1}{(2^N + 1)\pi T_{AU}} \tag{6.13}$$

where T_{AU} is the aperture uncertainty. For a SHA with one nanosecond aperture uncertainty in a 12-bit system, the maximum frequency is 383 kHz.

The sampling error due to aperture uncertainty is often not the limiting factor in sampled data systems. The Nyquist theorem states that a band-limited signal sampled at a rate at least twice the maximum signal frequency can be reconstructed without loss of information. This means that a sampled data system must sample, convert, and acquire the next point at a rate equal to twice the signal frequency. Thus the maximum input frequency when using a sample/hold amplifier is equal to

$$f_{i,max} = \frac{1}{2(T_{ACQ} + T_{CONV} + T_{AP})} \tag{6.14}$$

where T_{ACQ} is the acquisition time of the SHA, T_{AP} is the maximum aperture time (including jitter), and T_{CONV} is the conversion time of the ADC. For a combination of a 35 microsecond ADC and a 2.5 microsecond SHA (T_{AP} is generally small enough to be ignored), this means that the maximum signal frequency is 13.3 kHz.

6.4.2   Sample/Hold Amplifiers

The purpose of a sample/hold amplifier is to capture the value of the analog signal representative of a physical parameter and to hold this value stable during the time the A/D converter is doing a conversion. A sample/hold amplifier is often an essential component of the data acquisition system. These devices perform several valuable functions, as described below. As a front end to a successive-approximation A/D converter, a sample/hold amplifier "cleans up" and buffers the signal for the noise-sensitive, low-impedance converter. Sample/hold devices are essential for use in systems requiring conversion of rapidly changing signals with high accuracy. The sample/hold, in effect, improves the speed of the A/D converter so that it is as fast at catching transients as the aperture time of the sample/hold device. A sample/hold amplifier buffering a multiplexer output can hold a previous signal for conversion while the multiplexer is switching to acquire a new signal.

The sample/hold amplifier should have the characteristics of low offset voltage, high linearity, fast acquisition and aperture times, and low sample-to-hold offset. Some of these are illustrated in Fig. 6.76. Also, its associated hold capacitor should have sufficiently low dielectric absorption to prevent errors due to the "analog memory" effect.

DAS SHA Requirements

1. Low offset voltage
2. Low S/H offset
3. High speed
4. Low dielectric absorption capacitor

The ADC requirements are, of course, dictated by system performance goals. Conversion speed becomes important in large multi-channel systems, since only one channel can be converted at any given time. ADC accuracy is important in any system, and when the DAS under consideration is to be interfaced to a microprocessor, easy interface connections are desirable.

A sample/hold amplifier (SHA) can also be used to de-glitch a DAC output. The output of the DAC can be held during the time when the digital inputs change since this is the time during which unwanted transients or "glitches" occur. After the output is known to have reached a stable condition,

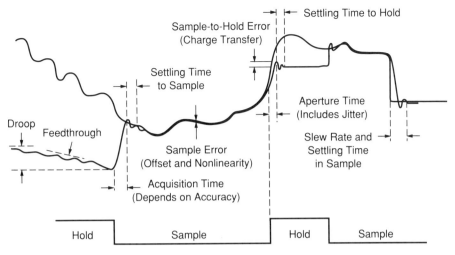

Fig. 6.76   Sample/hold timing and error diagram.

the SHA can be switched back to the sample mode, reducing the transient amplitude to the level of the SHA's glitch (which is usually smaller than that of the DAC).

One of the biggest problems involved in selecting sample/hold amplifiers involves the bewildering array of specifications. If one considers that an operational amplifier operates in only one mode—sample—and has several important specifications just for that, then the sample/hold specification list is clearly more formidable in that there are four modes—sample, hold, and the two transitions between them.

6.4.3   Multiplexers

The purpose of a multiplexer in a data acquisition system is to be able to arbitrarily monitor a particular parameter. (A multiplexer is often referred to as a MUX.) The parameter to be monitored is known, and it is a matter of selecting the appropriate channel. The channel to be monitored is selected by addressing the MUX with a binary word. The following list indicates some requirements for a MUX.

DAS MUX Requirements

1. Low R_{ON}
2. Low leakage current
3. Overvoltage protection
4. Break-before-make action

Generally MUXs are designed using CMOS technology, for these devices are unipolar and can accommodate both positive and negative signals (voltages).

6.5   DIGITAL INTEGRATED CIRCUITS*

Jackie Marsh

A digital integrated circuit uses electronic circuitry to perform various functions by converting or manipulating all forms of information as *binary code*, or *binary coded decimal (BCD)*. Digital signals have only two binary states—high and low, or logic "1" and "0" in positive logic systems. In negative logic systems, an electrical high is logic "0," and an electrical low is a logic "1."

6.5.1   Logic

Bipolar

Logic gates consist of resistors, capacitors, diodes, and transistors. The original bipolar family, introduced in the early 1960s, combined many discrete parts to make a resistor-transistor logic (RTL) device. The families that followed, the 7400 series, were developed in the mid-1960s. In diode-

---

* This material was written while the author was employed by Motorola, Inc., Austin, Texas.

transistor logic (DTL), the resistor was replaced by a diode. By the late 1960s, the diode gave way to a transistor, creating the TTL (transistor-transistor logic) family. In the early 1970s, a new process technique employing a Schottky transistor with TTL created a high-speed logic with the same power as TTL, but four times faster, labelled 74HS TTL. Low-power Schottky 74LS TTL devices, with one-fourth the power and approximately the same speed as TTL, became very popular. Variations such as ALS (Advanced Low-Power Schottky) and FAST TTL have also emerged. Emitter-Coupled Logic (ECL) is extremely fast, but it consumes high power and is difficult to design into a system, making it very specific for high speed applications such as very fast computers.

MOS

In 1973, RCA introduced the 4000 series family using MOS (metal-oxide-silicon) transistors with aluminum gate connections. These circuits, with complementary n-channel and p-channel substrates, were called CMOS (complementary MOS). They used very little power but were slow, sensitive, and susceptible to electrostatic discharge damage. Motorola, Inc., adapted the 4000 family into the 14000 family, implementing the functions of the TTL family in CMOS. Today's CMOS ICs are very fast, low in power consumption, and reliable. Recent HCMOS (also CHMOS) devices use silicon gates instead of metal and are gaining market share on the metal gate 14000 family as the cost curve is dropping dramatically.

Logic Gates

Integrated cicuits are composed of logic gates consisting of electronic digital circuitry elements that detect the presence or absence of discrete on/off pulses (bits). Voltage (V) is applied to a circuit, allowing the handling of a flow of pulses following a pattern prescribed by the logic elements (gates). The total number of inputs to a gate is designated as the "fan-in." The number of circuits that can be driven as a result of the output of a gate determines the "fan-out" rating. The possible input combinations and resulting outputs are exhibited in a truth table. Gates are represented by logic symbols.

AND Gate. Each input to an AND gate in Fig. 6.77 must have voltage applied to produce an output voltage. All inputs to an AND gate must be "terminated" to ground, the value of the "high" voltage level, or to a signal of pulses for the gate to operate properly. This applies to *all* logic elements.

OR Gate. Voltage applied to any or all of the inputs to an OR gate will result in output voltage (Fig. 6.78).

NOT Gate. When the voltage into the NOT gate (Fig. 6.79), also called an inverter, is high ("1"), the output voltage is low. If the input is low, the voltage output is high.

NAND Gate. A "high" voltage applied to all inputs of a NAND gate (Fig. 6.80) will result in an output near electrical ground. This gate is formed by combining the AND gate with the NOT gate, inverting the output of an AND gate. A NAND gate with all inputs tied together also functions as an inverter.

NOR Gate. A "high" applied to one or more inputs of a NOR gate (Fig. 6.81) results in a "low" output. This gate is formed by combining the OR gate with the NOT gate, inverting the output from the OR gate. A NOR gate with all leads tied together also functions as an inverter.

EXCLUSIVE OR (XOR) Gate. The EXCLUSIVE OR gate in Fig. 6.82 compares the input levels. If they are the same, a "0" output results. If they are different, a "1" output is produced.

Buffers

A buffer (Fig. 6.83) acts as two inverters in series to repeat and strengthen a digital signal without changing the state of the input signal. Combined with other gates, the buffer is used to supply and handle large currents and minimize delay time due to output capacitance loading. These low-impedance driver circuits have more fan-out or drive capabilities than standard gates.

6.5.2 Boolean Logic Notation

Logical statements may be written using "boolean algebra," or logical notation, as follows.

AND: $C(\text{output}) = A \cdot B$ or AB (boolean "multiply")
OR: $C = A + B$ (boolean "addition")
NOT: $C = \overline{A}$ (boolean "inversion")
XOR: $C = A \oplus B$

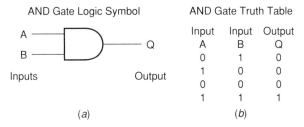

Fig. 6.77 AND gate: (*a*) logic symbol; (*b*) truth table.

AND Gate Logic Symbol

	Inputs		Output

AND Gate Truth Table

Input A	Input B	Output Q
0	1	0
1	0	0
0	0	0
1	1	1

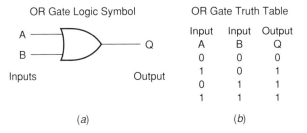

Fig. 6.78 "OR" Gate (*a*) logic symbol (*b*) truth table

OR Gate Logic Symbol

OR Gate Truth Table

Input A	Input B	Output Q
0	0	0
1	0	1
0	1	1
1	1	1

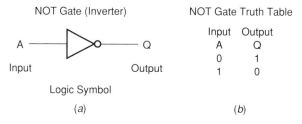

Fig. 6.79 NOT gate: (*a*) logic symbol; (*b*) truth table.

NOT Gate (Inverter)

NOT Gate Truth Table

Input A	Output Q
0	1
1	0

Logic Symbol

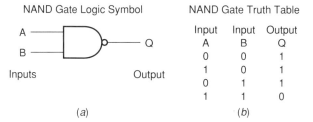

Fig. 6.80 NAND gate: (*a*) logic symbol; (*b*) truth table.

NAND Gate Logic Symbol

NAND Gate Truth Table

Input A	Input B	Output Q
0	0	1
1	0	1
0	1	1
1	1	0

NOR Gate Logic Symbol

NOR Gate Truth Table

Input A	Input B	Output Q
0	0	0
1	0	1
0	1	1
1	1	1

Fig. 6.81 NOR gate: (*a*) logic symbol; (*b*) truth table.

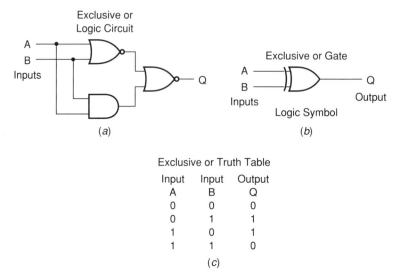

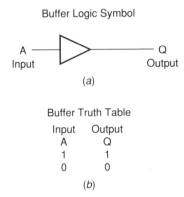

Exclusive or Truth Table

Input A	Input B	Output Q
0	0	0
0	1	1
1	0	1
1	1	0

(c)

Fig. 6.82 EXCLUSIVE OR gate: (a) logic symbol; (b) truth table; (c) logic circuit.

6.5.3 Flip-Flops

Flip-flop devices, also called bistable multivibrators, can store input levels, count pulses, control circuits, synchronize operations, and produce required wave shapes. All flip-flops have two output levels, Q and $\overline{Q}$, indicating complementary states. If Q is high, $\overline{Q}$ is low. There are several types of flip-flops presently in use, each with several variations. The most popular ones are explained below.

R-S (Set-Reset Latch) Flip-Flop

The R-S flip-flop (also termed reset-set or latch), consists of two 2-input NAND gates with cross-connected inputs and outputs (see Fig. 6.84). A pulse at the S input (set switch) of 1, results in an output with Q = 1 and $\overline{Q}$ = 0. The R switch resets the flip-flop to Q = 0. If both R and S inputs are high, Q will remain high for the duration of the pulse, and the resulting output will be indeterminate. Therefore, R and S are never applied simultaneously.

Clocked R-S Flip-Flop

The clocked, or synchronized, R-S flip-flop has a clock trigger in addition to the R and S inputs. In Fig. 6.85, an input to S or R will set the flip-flop if a clock pulse occurs. The output state is indeterminate if R and S are at 1 when a clock pulse occurs.

Buffer Logic Symbol

A — Input   Q — Output

(a)

Buffer Truth Table

Input A	Output Q
1	1
0	0

(b)

Fig. 6.83 Buffer: (a) buffer logic symbol; (b) buffer truth table.

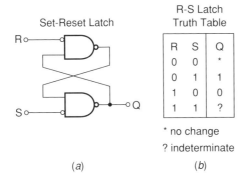

Set-Reset Latch

R-S Latch
Truth Table

R	S	Q
0	0	*
0	1	1
1	0	0
1	1	?

* no change

? indeterminate

(a)   (b)

Fig. 6.84   Set-reset latch: (a) circuit; (b) truth table.

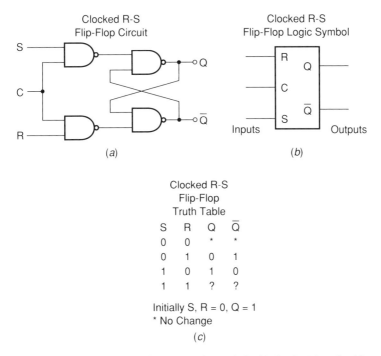

Clocked R-S
Flip-Flop Circuit

Clocked R-S
Flip-Flop Logic Symbol

Inputs   Outputs

(a)   (b)

Clocked R-S
Flip-Flop
Truth Table

S	R	Q	Q̄
0	0	*	*
0	1	0	1
1	0	1	0
1	1	?	?

Initially S, R = 0, Q = 1
* No Change

(c)

Fig. 6.85   Clocked R-S flip-flop: (a) logic symbol; (b) circuit; (c) truth table.

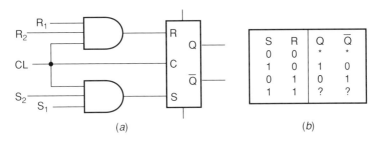

S	R	Q	Q̄
0	0	*	*
1	0	1	0
0	1	0	1
1	1	?	?

(a)   (b)

Fig. 6.86   R-S AND-gated latch: (a) circuit; (b) truth table. Assume preset and clear at H steady state.

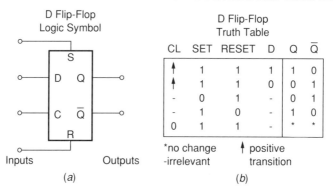

Fig. 6.87   D flip-flop: (*a*) logic symbol; (*b*) truth table.

R-S Latch (Gated Latch)

A gated latch (Fig. 6.86) uses AND gates to control the R and S states of the R-S flip-flops when multiple R and S inputs are used. While the clock is held high, data is retained in the gated latch. Outputs are not affected until the falling edge of the clock, when the data is transferred to the output.

Type-D Flip-Flop

Type D flip-flops respond to a positive edge trigger. The D flip-flop truth table in Fig. 6.87 indicates that input D is transferred to output Q, and $\overline{Q}$ assumes a logic value complementary to Q. Reset and Set override D and clock inputs.

J-K Flip-Flop

A versatile device, a J-K flip-flop like the one in Fig. 6.88 passes the J and K data to the outputs Q and $\overline{Q}$, respectively, except when the clock is pulsed while J and K are at 1 simultaneously. In

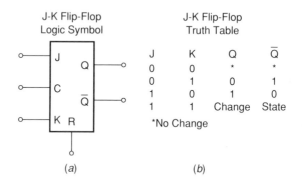

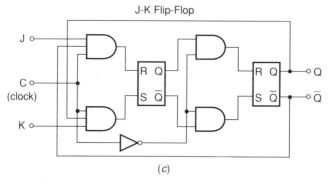

Fig. 6.88   J-K flip-flop: (*a*) logic symbol; (*b*) circuit; (*c*) truth table.

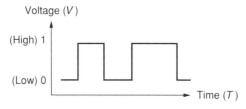

Fig. 6.89 Digital pulses.

that case, Q and $\overline{Q}$ toggle, or change states. This prevents an indeterminate output. If J and K are both low, there is no change in the outputs. Multiple J and K inputs to the leading AND gates are possible. The J-K flip-flop responds to an active clock transition, which may be positive or negative, depending on the technology and vendor involved. In the J-K flip-flop in Fig. 6.88, J and K input levels pass to the master on the positive edge of a clock pulse. They hold until the clock pulse falls, passing to the slave. The inverter blocks inputs to the slave to prevent resets of the slave flip-flop at an inappropriate time. Other variations of the J-K flip-flop require only a single active edge to effect the required output transitions.

6.5.4 Clocks

In a digital circuit (see Fig. 6.89), a pulse is a change of levels from low to high (0 to 1) and back to low (positive pulse) or high to low and back (negative pulse). Each pulse has a duration called a pulse width. When a series of pulses is used as a timing device, it is called a *clock*, and each pulse has a frequency of occurrence over time. A clock is used to synchronize two or more circuits that operate together or in parallel. Many operations may be combined and synchronized by a clock timing system. Pulse sources include multivibrators, oscillators, generators, triggers, differentiator and integrator networks, and clippers and clampers. In Fig. 6.90, the transition from 0 to 1, or low to high, is a positive transition, or the leading edge of a positive clock pulse. A negative transition from 1 to 0 is the trailing edge. The hold time t_h is the minimum stable time period after a positive transition to ensure continued recognition of a digital pulse. Clock pulse widths may be too short in duration to allow recognition of the pulse. Therefore, many digital devices specify acceptable operating dwell times, $t_w^{(H)}$ for the positive edge of a triggered flip-flop and $t_w^{(L)}$ for the trailing edge. A flip-flop sensitive to logic levels of a clock signal will latch at the trigger level as the clock switches between high and low. Time t_{su} is the setup time, the minimum time required to stabilize the correct logic level before the positive transition of the clock in order to ensure recognition of the clock transition from low to high. A Schmitt trigger can produce a square wave so that the setup time is negligible.

6.5.5 Counters

A counter is a sequential device that counts and stores the number of input pulses it receives. There are several types and many variations of each. Binary counters store the count in a series of flip-flops as a binary number. BCD counters store a decimal digit in each flip-flop. For example, in a binary counter the decimal 14 is stored in four flip-flops as 1110. In a BCD counter, two sets of

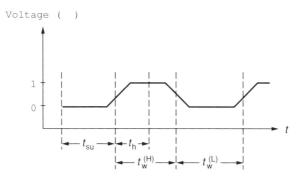

Fig. 6.90 AC waveform.

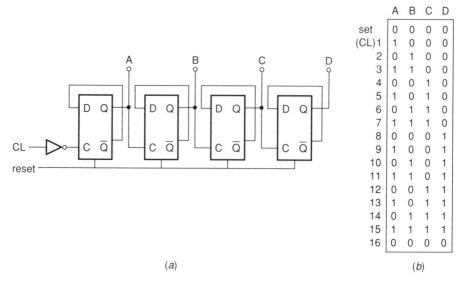

	A	B	C	D
set	0	0	0	0
(CL)1	1	0	0	0
2	0	1	0	0
3	1	1	0	0
4	0	0	1	0
5	1	0	1	0
6	0	1	1	0
7	1	1	1	0
8	0	0	0	1
9	1	0	0	1
10	0	1	0	1
11	1	1	0	1
12	0	0	1	1
13	1	0	1	1
14	0	1	1	1
15	1	1	1	1
16	0	0	0	0

(a)   (b)

Fig. 6.91 Four-bit ripple counter: (a) circuit; (b) clock sequence.

four flip-flops hold the binary four as the least significant digit and the binary number one as the most significant digit.

4-Bit Binary Serial Counter

Also called a ripple counter or asynchronous counter, a 4-bit binary counter consists of a series of four flip-flops, each capable of storing one binary bit. The output of a preceding flip-flop triggers the next successive flip-flop (see Fig. 6.91). Initially, all flip-flop outputs are cleared by a direct positive reset pulse. A negative clock pulse to the first flip-flop sets A to 1. The second pulse resets the first flip-flop to 0; A goes to 0, causing the second flip-flop to toggle and setting B to 1. C and D remain cleared. The third pulse resets flip-flop two to 0, toggles flip-flop one and three to 1, and results in A = 1, B = 1, and C = 0. D remains at 0. After the fourth pulse, A and B = 0, B resets to 0, C = 1, and D = 0. On the eighth pulse, D is set to 1. It takes 15 pulses to set all four counters to 1111, the binary decimal equivalent of 15. On the negative edge of the 16th pulse, all counters reset to 0 (see Fig. 6.91b). Thus, this counter can count 2^4 (16) pulses before full reset (see waveform in Fig. 6.92). Higher counts require more flip-flops and additional circuitry.

   If the output $\overline{Q}$ is used instead of Q, the counter descends from 15 to 0, and is called a *down* counter, as opposed to an *up* counter (0,1,2, . . . ,15). A counter can store either up counter inputs or down counter inputs, or both if additional circuitry is provided.

4-Bit Synchronous Counter

A 4-bit synchronous counter, or parallel counter (see Fig. 6.93), combines an inverter and NAND gates with J-K flip-flops to switch all flip-flops simultaneously. The waveform appears to be the

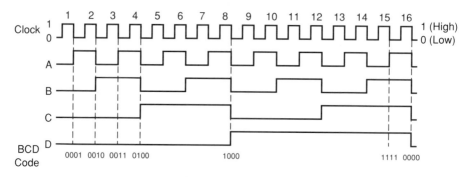

Fig. 6.92 Waveform for 4-bit counter.

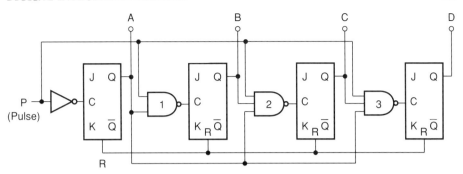

Fig. 6.93 Four-bit synchronous counter circuit. All J and K inputs tied to Logic 1 (toggle mode).

same as in Fig. 6.92, but the asynchronous counter has an inherent time delay toggling from one flip-flop to the next. The total time delay of a ripple counter is called the *settling time*. This delay is eliminated in the synchronous counter. For example, on the eighth clock pulse in Fig. 6.92, A, B, C, and D switch at the same time if the counter is synchronous.

Ring Counter

A simple ring counter sequentially sets each flip-flop, with the outputs of the last flip-flop in a chain fed to the J and K inputs of the first flip-flop to keep a data pattern circulating (see Fig. 6.94). All flip-flops are initially reset to 0, and then flip-flop A is set to 1. A clock pulse shifts the 1 to flip-flop B, the next to C, and so on to H. The outputs of the H flip-flop will toggle J and K of flip-flop A, continuing the sequence as long as the clock runs.

Decade Counter

A decade counter increments a counter every ten pulses. A decade counter may be asynchronous (ripple) and negative-edge triggered, or synchronous and positive-edge triggered. It may have serial load or parallel load. A counter may have many states (such as 4-bit binary, 6-bit binary, or divide-by-twelve), depending on the number of flip-flops and steering circuitry used. A 4-bit binary counter sequence and waveform is represented in Fig. 6.95.

6.5.6 Registers

Registers temporarily store binary information, acting as a link between main digital systems and input-output channels. Data can be manipulated within a register to increase functionality through selective utilization of flip-flops and steering circuitry.

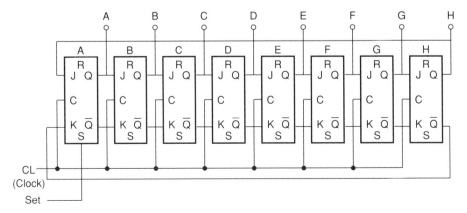

Fig. 6.94 Simple ring counter circuit.

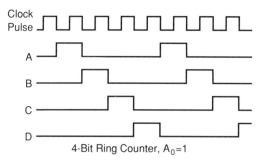

4-Bit Ring Counter, $A_0=1$

Fig. 6.95 Simple waveform.

Shift Registers

A shift register can enter a word into and out of a register one bit at a time (serial shift register), or simultaneously (parallel shift register). Words can be shifted left or right, performing arithmetic operations such as multiplication, division, and complementation. A one-bit right shift performs a division by two in base two, whereas a shift of one bit to the left multiplies the register contents by two. Conversion between parallel and serial data can occur, and input and output frequencies can differ.

Serial Shift Register

The serial shift register in Fig. 6.96 operates much like a ring counter without feedback from the last flip-flop to the first. The flip-flops are initially cleared (latched) by a reset input pulse. One clock pulse is required for each bit to be shifted into the register. To store a word, the clock pulse is

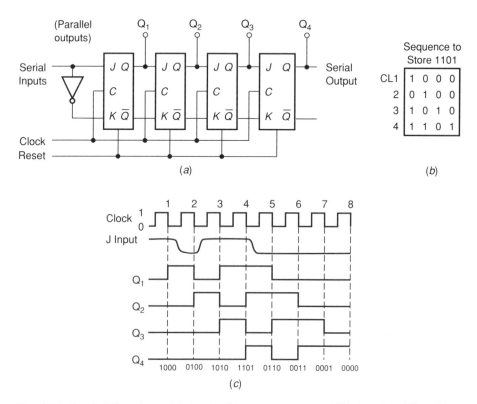

Fig. 6.96 Serial shift register: (a) circuit; (b) sequence to store 1101 in serial shift register; (c) waveform to shift 1101 through a 4-bit register.

stopped. Otherwise, the word will continue to shift until it is passed out of the register. Figure 6.96 shows the sequence followed to shift the binary word 1101 through a 4-bit register. It takes four clock pulses to store the word 1101 sequentially and four more to clear the register in the absence of additional data inputs.

Parallel Shift Register

Shift registers can be used to receive data serially for parallel read-out or for conversions from parallel to serial. Auxiliary gating can override asynchronous parallel entry.

Parallel Register

A group of flip-flops with a common clock can accept more than one bit at a time when data is transferred in parallel. A single clock pulse shifts a word into a parallel shift register after the register has been cleared. A parallel-load 8-bit memory register receives all inputs A, B, C, D,. . ., H simultaneously. If all outputs are also read at the same time, this is a parallel-input, parallel-output register.

6.5.7 Memories

Bipolar Memories

Bipolar ICs include transistor to transistor logic (TTL), emitter-coupled logic (ECL), and integrated injection logic (IIL). Bipolar devices offer very fast access times with low densities. All Schottky devices are bipolar. Static RAMs of ECL and TTL devices include 256 by 1, 256 by 4, 1024 by 1, and 4096 by 1 organizations, with address access times ranging from 10 ns to 70 ns. Bipolar ROMs have address access times of 45 ns to 90 ns, compared to an average of 200 ns to 400 ns for MOS ROMs. Because bipolar memories are most widely used in small-scale integration (SSI) and medium-scale integration (MSI) circuits, compatibility of TTL logic level and power supply is useful for the denser large-scale integration (LSI) NMOS devices.

PROM

PROMs, programmable read-only memories, are field programmable on special equipment by electrically blowing a fuse link to change the state for each bit selected. PROMs offer the convenience of "off the shelf" availability, combined with cost-saving avoidance of customized manufacturer masks and modifications. PROMs can be written only once, usually by a vendor or the user. Once programmed, a PROM is essentially a ROM. PROMs are used mainly in peripheral controllers, terminals, logic analyzers, and digital controllers.

Magnetic Bubbles

Magnetic bubbles are nonvolatile, high-density (as large as 4 megabits) read/write memories, with very slow serial access compared to RAMs. The 1M by 1 bit MBM2011A Motorola magnetic bubble device has an average address access time of 11.5 milliseconds per page. Bubbles store data as the presence or absence of locally polarized domains in a thin film of magnetic garnet material. The advantage of bubble memories is an inherent stability in hostile, rugged environments due to permanent magnets and magnetic shields. The disadvantage is that, being a relatively new technology, bubbles are expensive. Interface and control circuits required by bubbles make them less cost-effective than magnetic disks in their main market, high volume external memory for computer systems. Redundancy features require detection and correction circuitry. In addition, the new high-density DRAMs, with much faster random access, are taking the place of bubbles as an alternative to disks. Bubbles are most ideally suited to applications in avionics, communications systems, heavy industry, robotics, automated test equipment (ATE), defense systems, and portable equipment.

MOS Memories

Information in the form of binary digits is stored for future use in an array of elements called a memory. Memories are arranged in row and column addresses which are assigned a numerical code for access by the microprocessor or other peripheral devices. The processor "writes" to a specific address to store data, which can later be "read," or recalled, at the appropriate time. Most memories are designed to accommodate 1-bit, 4-bit, or 8-bit word requirements. Because some microprocessors, such as the M68000 family, can support data types of bits, bytes, words, long words, and binary coded decimal (BCD), memory can be accessed for operation on any bit field, from 1 to 32 bits, regardless of memory or register location. A 4K by 8 (32K) memory can store 4,096 bytes of eight bits each. Memory configurations have become so standardized that devices from different manufacturers are basically interchangeable within a system. This has made it easy to design these memories into a system and reduced the cost for those memories on the low end of the learning curve. Different memory types are available to perform the various functions required by a system or application.

ROM

Read-only memory (ROM) can be written only once during manufacturing using a memory pattern which is preprogrammed in the silicon. ROM is unalterable and nonvolatile, retaining all data in the absence of electrical power occurring when power is switched off or during power failures or system malfunctions. For this reason, ROM is useful in dedicated systems, control, and other fixed data applications. Access time is the same for all ROM address locations. The address locations may be randomly accessed, as opposed to disk, cassette, or tape storage methods involving additional setup time.

EPROM

A disadvantage of ROM is the difficulty and high cost associated with changing a customized mask program (mask charges). For nonvolatile small- to moderate-volume applications, or periodic program updating, erasable programmable read-only memories (EPROMs) have been developed. A transparent window on the package allows the memory contents to be erased with high intensity shortwave ultraviolet light and subsequently reprogrammed. EPROMs must be removed from a socket for erasure on special equipment, reprogrammed on a second machine, and inserted manually into the system. EPROMs may also be light sensitive, affecting reliability.

OTPs

One-time programmable read-only memories (OTPs), or "One-shot proms," are EPROMs without a transparent window for erasure by ultraviolet light. They are, therefore, field programmable one time only. Since 80% of the EPROM users only program the device once, the OTP offers a cost-effective solution in a less expensive plastic package.

EEPROM, EAROM

The high costs associated with EPROMs can be somewhat alleviated by using EEPROMs, electrically erasable programmable read-only memories (also called EAROMs, electrically alterable read-only memories). Most EEPROM processes use MOS materials and a derivative floating gate UV EPROM process. Some technologies use silicon nitride material in addition to MOS materials, as in MNOS (metal nitride oxide silicon), or SNOS (silicon nitride oxide silicon). The greatest advantage of EEPROMs is that they can be erased and programmed electrically in the system. This eliminates the need for specialized equipment, and the circuits can be housed in an inexpensive plastic package. They have erase and reprogram capabilities that require relatively high voltage. Some EEPROMs have on-chip charge pumps to reduce programming power supply requirements to TTL levels. The number of times an EEPROM can be erased and reprogrammed is limited by its ability to allow the passage of electrons through a thin silicon dioxide layer. EEPROMs are especially useful for prototype development, microprocessor systems employing adaptable visual or tactile sensors, self-adaptive intelligent systems, robotics, and systems with frequent basic program changes.

RAM

RAM has become the accepted usage for a read/write random access memory. A RAM can be written into and read more than once and has very fast write times. RAMs are used to store data that changes frequently during the operation of a program.

SRAMs

Static RAMs (SRAMs) require applied constant direct current to retain memory content, making them volatile memories. Static design eliminates the need for external clocks or timing strobes. CMOS circuitry has enabled designers to create SRAMs with very fast (sub-100 nanosecond) address access times while drawing low power supply current. For example, the MCM6168, a 4K by 4 SRAM, has maximum address access times of 45, 55, and 70 ns. Operation from a single 5V ($\pm 10\%$) power supply draws only 80 milliamps (mA) maximum active, 5 mA standby at TTL levels, and 2 mA at full-rail CMOS standby levels.

The pin assignment of a memory device corresponds to the address and data lines, power supply and ground connections, and chip enable lines from the microprocessor. In Fig. 6.97 the pin-out of the MCM6168 static RAM indicates 12 address lines, four data input/output lines, power supply and ground, chip write, and chip enable. Because this is a static RAM, chip enable (E) is not a clock, but a power-down feature to reduce system power requirements for standby mode.

The MCM6168, organized as 4096 words of 4 bits each, has a 128 bit row by 128 bit column matrix (see Fig. 6.98). The number of address lines (p) required for this device, with $N = 4096$ words, and where $N = 2^p$, is twelve. The matrix uses seven row address decoder lines for 2^7 rows and five column decoder lines for 2^5 columns of 2^2 bit words, for a total of 16,364 bits. Input/output lines share the same terminals, their nature dependent on the state of the read/write line.

DRAMs

Dynamic random access memories require periodic "refresh" (restored charge) to compensate for degradation of an electrical charge of a metal oxide semiconductor (MOS) capacitor. Dynamic RAMs

Pin Assignment

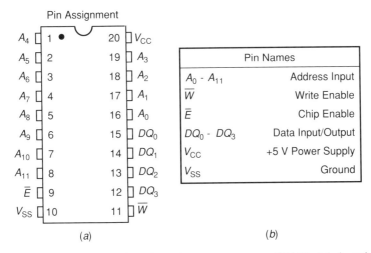

(a)

Pin Names	
A_0 - A_{11}	Address Input
$\overline{W}$	Write Enable
$\overline{E}$	Chip Enable
DQ_0 - DQ_3	Data Input/Output
V_{CC}	+5 V Power Supply
V_{SS}	Ground

(b)

Fig. 6.97 MCM6168 4,096 x 4-bit static random access memory (SRAM): (a) pin assignment for MCM6168 SRAM; (b) pin names.

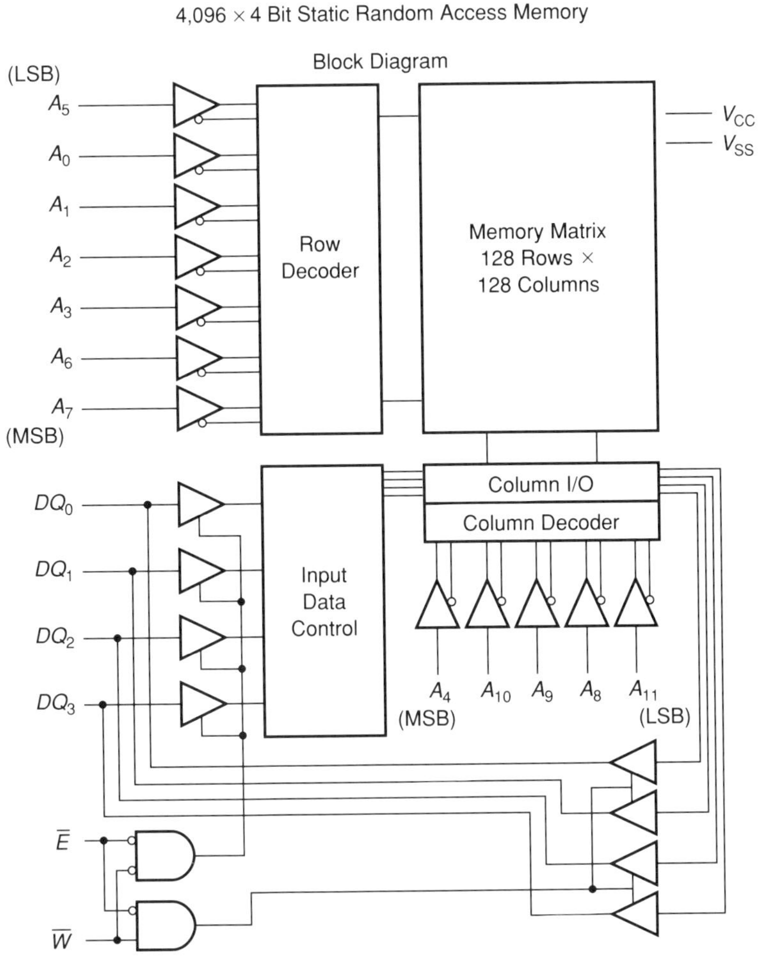

4,096 × 4 Bit Static Random Access Memory

Block Diagram

Fig. 6.98 Block diagram of MCM6168 SRAM.

Fig. 6.99   64K dram.

(DRAMs) can achieve very high density and maintain low power in the quiescent state. In addition to a standard external refresh mode, most DRAMs offer the option of using an on-chip address counter for automatic refresh. This allows operation from a single low-voltage power supply (in Motorola DRAMs, a single 5 V operation with ±10% tolerance). The basic cell of a dynamic RAM is smaller than a static cell, allowing greater density and high performance. Many DRAMs, such as the Motorola 256K DRAM, are available in page mode or nibble mode. The nibble mode allows the user to access up to 4 bits of data at a high data rate. The page mode feature allows column accesses of up to 512 bits within a selected row. DRAMs are useful in large, highly sophisticated systems requiring a deep memory and capable of supporting the higher initial system development costs associated with DRAMs.

Packaging for Memory Devices

A memory die is very regular, with a rectangular aspect ratio rather than a square one, to most easily accommodate the lead configuration, or pin-out, of the die (see photo of 64K DRAM, Fig. 6.99). The most popular package in use in memories today is the dual-in-line package (DIP), with an electrical lead line on either side of the package (see photo of DIP on wafer, Fig. 6.100). DIPs are available in plastic (DIP), ceramic (CERDIP), and side brazed. In CERDIP a ceramic "sandwich" encases the lead frame of a plastic DIP and is sealed with molten glass. It is used in approximately 14% of all DIP applications. Historically, plastic DIPs have not been hermetic, thus allowing the potential for moisture to destructively permeate the memory. Plastic DIPs also dissipate less heat than the CERDIP. However, plastic DIPs are durable, inexpensive, and extremely utilitarian in most applications and have captured about 85 percent of the DIP market. One percent of DIP usage is the more expensive ceramic side brazed package, with gold soldered to the top and side leads to hermetically seal the die cavity. With optimum tolerance for adverse temperatures, humidity, acids, and salt conditions, side braze packages can be compliant with military, space, telecommunications, and heavy industry specifications.

Surface Mount Packaging

Regular and uniform configurations in memory devices are conducive to surface mount techniques for high board density. Surface mount packaging offers many advantages over DIPs, and surface mount market share for all ICs is expected to increase from 0.4% in 1985 to over 8% in 1990. Surface mount packaging may be as high as 15% of the total 1990 memory market. Cost savings result

Fig. 6.100   Dual-in-line package (DIP).

from improved board yields, a reduced number of plated-through holes, and less expensive board rework techniques. Shorter lead lengths reduce parasitic capacitance, resistance, and inductance. In memories, plastic surface mount packages provide the cost savings, ease of assembly, and layout density required by memory applications. Some surface mount alternatives for memories are pictured in Fig. 6.101.

SO-ICs

A small-outline integrated circuit (SO-IC) is useful for low pin counts, from 8 to 28 pins. Pins with 150-mil and 300-mil widths are bent from the body in a gull wing shape. This design lends the SO-IC to surface mount techniques, occupying approximately one-fourth the board space required by a corresponding DIP. Larger pin counts cause a severe rectangular aspect ratio to the die, making SO-IC packages most popular for SSI (small-scale integration) and MSI (medium-scale integration) memory devices. Also, the leads, which are fashioned from the same leadframe material as the DIP, are stiff and non-compliant, resulting in problems during surface mounting of the package.

PLCCs

Plastic leaded chip carriers (PLCCs), offer space savings of 40% to 60% over DIPs, with board density and board level economies for high pin counts (18 to 124 pins). Forty- and fifty-mil centers compare favorably to 300- and 600-mil centers on DIPs. PLCCs have compliant J-shaped leads, reducing temperature coefficient of expansion problems and making adaptation to a variety of inexpensive substrate materials possible. Generally, PLCCs have four sides with an equal number of leads, the total of which, when divided by four, will result in an uneven number. For example, a 44-lead PLCC, when divided by four, results in eleven, an uneven number. In memories, the rule may deviate because of the rectangular aspect ratio. Thus, a memory PLCC may have more leads on two opposite sides than on the two ends. This low-cost package is gaining momentum in applications which involve interfacing with ICs having many input/outputs, severe board area restrictions, and very deep memories.

SOJs

The J-lead version of the small-outline SOJ offers a more efficient board layout than the PLCC but is also impractical for large pin counts. The gap in the center of the package for a one megabit DRAM in Fig. 6.101 provides efficiencies in wire bonding for a DRAM die.

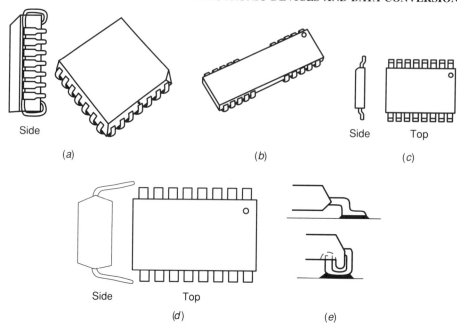

Fig. 6.101 Memory device package alternatives: (*a*) plastic leaded chip carrier (PLCC); (*b*) small outline J-lead (SOJ); (*c*) small outline integrated circuit (SOIC); (*d*) dual-in-line package (DIP); (*e*) gull-wing lead; J-lead.

JEDEC Standards in Memories

The Joint Electronics Device Engineering Council (JEDEC) was formed by about thirty systems and semiconductor vendor companies to set standardization specifications for packaging and pin configurations. Standard JEDEC pin and lead configurations for many of the memory packages are shown in Tables 6.8 to 6.13. A plastic 20-lead SOJ has been chosen as the JEDEC standard for the one megabit and four megabit DRAMs in surface mount packaging, with a JEDEC standard 18-pin DIP alternative.

6.6 INTRODUCTION TO MICROCOMPUTERS

Mark A. McQuilken

The microcomputer has become so commonplace in industry that many engineers, even those outside the electronics engineering realm, must acquire a good working knowledge of micros and their strengths, limitations, and operational requirements. In the last decade there has been an explosion in the number and types of microcomputer-based applications present in the engineering workplace. Microcomputers (micros) may now be found acquiring and manipulating data from flow sensors and thermocouples. Industrial micros decode and drive displays, manipulate stepper motors in industrial robots, and act as temperature controllers in wave solder machines and kilns. They provide the programmable mechanism of massive machine tools (referred to as numerical control), as well as perform some of the more mundane functions like set-point control, the monitoring and decoding of keyboards, and even the generation of alarm tones. In some applications, micros digitally manipulate signals to extract important information from noisy signals, decode telephone touch-tone signals, and even decode human speech. Other applications of micros include the generation of artificial human speech, signal modulation and demodulation, error correction and detection of important digital messages, and encryption and decryption of sensitive data. Increasingly micros find use in the implementation of many new broad spectrum discrete-time solutions to engineering problems.

As evidenced by these many applications, processes and controls previously the exclusive domain of analog-type devices are steadily being replaced by microcomputer-based equivalents. Of course, this increases the chance of contact between the micro and the engineer. Today's engineer must

TABLE 6.8   JEDEC STANDARD MEMORY DIP PIN CONFIGURATION (DRAMs)

PIN #	16K × 1 DRAM	256K × 1 DRAM	64K × 4 DRAM	1M × 1 DRAM	4M × 1 DRAM	256K × 4 DRAM	1M × 4 DRAM
1	N/C	A8	$\overline{G}$	D	D	DQ1	DQ1
2	D	D	DQ1	$\overline{W}$	$\overline{W}$	DQ2	DQ
3	$\overline{W}$	$\overline{W}$	DQ2	$\overline{RE}$	$\overline{RE}$	$\overline{W}$	$\overline{W}$
4	$\overline{RE}$	$\overline{RE}$	$\overline{W}$	A10	NC	$\overline{RE}$	$\overline{RE}$
5	A5	A0	$\overline{RE}$	A0	A0	NC	A9
6	A6	A2	A6	A1	A1	A0	A0
7	A7	A1	A5	A2	A2	A1	A1
8	V_{DD}	V_{DD}	A4	A3	A3	A2	A2
9	A9	A7	V_{DD}	V_{DD}	V_{DD}	A3	A3
10	A10	A5	A7	A4	A4	V_{DD}	V_{DD}
11	A11	A4	A3	A5	A5	A4	A4
12	A12	A3	A2	A6	A6	A5	A5
13	A13	A6	A1	A7	A7	A6	A6
14	Q	Q	A0	A8	A8	A7	A7
15	$\overline{CE}$	$\overline{CE}$	DQ3	$\overline{A9}$	$\overline{A9}$	A8	A8
16	V_{ss}	V_{ss}	$\overline{CE}$	$\overline{CE}$	$\overline{CE}$	$\overline{G}$	$\overline{G}$
17	—	—	DQ4	Q	Q	$\overline{CE}$	$\overline{CE}$
18	—	—	V_{ss}	V_{ss}	V_{ss}	DQ3	DQ3
19	—	—	—	—	—	DQ4	DQ4
20	—	—	—	—	—	V_{ss}	V_{ss}

Pin Names:

A0–A14	Address		**PGM**	Program
$\overline{E}$	Chip Enable		V_{PP}	Program Voltage
$\overline{G}$	Output Enable		$\overline{W}$	Write
Q0–Q7	Data Output		**DQ0–DQ7**	Data
V_{DD}	+5 V Power Supply		**N/C**	No Connect
V_{SS}	Ground		**S**	Chip Select
$\overline{RE}$	Row Address Strobe		$\overline{CE}$	Column Address Strobe
$\overline{DU}$	Don't Use		**UB**	Upper Byte
$\overline{LB}$	Lower Byte		**Opt**	Optional

Note: Chip Select (S) or Chip Enable ($\overline{E}$) is an optional function, exhibited by a particular device as determined by the manufacturer.

TABLE 6.9   JEDEC STANDARD MEMORY DIP PIN CONFIGURATIONS (ROMs, EPROMs)

PIN #	4K × 8 EPROM	4K × 8 EPROM	8K × 8 EPROM	2K × 8 ROM	4K × 8 ROM[a]	4K × 8 ROM[b]	8K × 8 ROM
1	A7	A7	A7	A7	A7	A7	A7
2	A6	A6	A6	A6	A6	A6	A6
3	A5	A5	A5	A5	A5	A5	A5
4	A4	A4	A4	A4	A4	A4	A4
5	A3	A3	A3	A3	A3	A3	A3
6	A2	A2	A2	A2	A2	A2	A2
7	A1	A1	A1	A1	A1	A1	A1
8	A0	A0	A0	A0	A0	A0	A0
9	DQ0	DQ0	DQ0	A0	Q0	Q0	Q0
10	DQ1	DQ1	DQ1	Q1	Q1	Q1	Q1
11	DQ2	DQ2	DQ2	Q2	Q2	Q2	Q2
12	V_{ss}	V_{ss}	V_{ss}	V_{ss}	V_{ss}	V_{ss}	V_{ss}
13	DQ3	DQ3	DQ3	Q3	Q3	Q3	Q3
14	DQ4	DQ4	DQ4	Q4	Q4	Q4	Q4
15	DQ5	DQ5	DQ5	Q5	Q5	Q5	Q5
16	DQ6	DQ6	DQ6	Q6	Q6	Q6	Q6
17	DQ7	DQ7	DQ7	Q7	Q7	Q7	Q7
18	A11	$\overline{E}$	A11	$\overline{S}$	A11	$\overline{EorS}$	A11
19	$\overline{A10}$	$\overline{A10}$	$\overline{A10}$	$\overline{A10}$	$\overline{A10}$	A10	$\overline{A10}$
20	$\overline{EorS}$	G/V_{PP}	$\overline{E}/V_{PP}$	$\overline{EorS}$	$\overline{EorS}$	S	$\overline{EorS}$
21	V_{PP}	A11	A12	S	S	A11	A12
22	A9	A9	A9	A9	A9	A9	A9
23	A8	A8	A8	A8	A8	A8	A8
24	V_{DD}	V_{DD}	V_{DD}	V_{DD}	V_{DD}	V_{DD}	V_{DD}

[a] Version B   [b] Version A

TABLE 6.10 JEDEC STANDARD MEMORY DIP PIN CONFIGURATIONS

PIN #	1K × 8 SRAM	1K × 8 SRAM	2K × 8 SRAM	64K × 1 SRAM	16K × 4 SRAM	8K × 8 SRAM	16K × 4 SRAM	8K × 16 SRAM	4K × 16 SRAM	16K × 16 SRAM	256K × SRAM
1	A6	A7	A7	A0	A0	NC	A1	V_{DD}	V_{DD}	V_{DD}	V_{DD}
2	A5	A6	A6	A1	A1	A12	A0	$\overline{E}$	$\overline{E}$	$\overline{E}$	$\overline{E}$
3	A4	A5	A5	A2	A2	A7	A2	DQ15	DQ15	DQ15	A7
4	A3	A4	A4	A3	A3	A6	A3	DQ14	DQ14	DQ14	A6
5	A2	A3	A3	A4	A4	A5	A4	DQ13	DQ13	DQ13	A5
6	A1	A2	A2	A5	A5	A4	A5	DQ12	DQ12	DQ12	A4
7	A0	A1	A1	A6	A6	A3	A6	DQ11	DQ11	DQ11	A3
8	DQ0	A0	A0	A7	A7	A2	A7	DQ10	DQ10	DQ10	A2
9	DQ1	DQ0	DQ0	Q	A8	A1	A8	DQ9	DQ9	DQ9	A1
10	DQ2	DQ1	DQ1	$\overline{W}$	$\overline{E}$	A0	$\overline{E}$	DQ8	DQ8	DQ8	A0
11	V_{SS}	DQ2	DQ2	V_{SS}	$\overline{G}$	DQ0	V_{SS}	V_{SS}	V_{SS}	V_{SS}	DQ0
12	DQ3	V_{SS}	V_{SS}	$\overline{E}$	V_{SS}	DQ1	$\overline{W}$	DQ7	DQ7	DQ7	DQ1
13	DQ4	DQ3	DQ3	D	$\overline{W}$	DQ2	DQ1	DQ6	DQ6	DQ6	DQ2
14	DQ5	DQ4	DQ4	A8	DQ1	V_{SS}	DQ2	DQ5	DQ5	DQ5	V_{SS}
15	DQ6	DQ5	DQ5	A9	DQ2	DQ3	DQ3	DQ4	DQ4	DQ4	DQ3
16	DQ7	DQ6	DQ6	A10	DQ3	DQ4	DQ4	DQ3	DQ3	DQ3	DQ4
17	$\overline{W}$	DQ7	DQ7	A11	DQ4	DQ5	A9	DQ2	DQ2	DQ2	DQ5
18	$\overline{E}$orS	$\overline{E}$orS	$\overline{E}$orS	A12	NC	DQ6	A10	DQ1	DQ1	DQ1	DQ6
19	A9	$\overline{S}$	A10	A13	A9	DQ7	A11	DQ0	DQ0	DQ0	DQ7
20	A8	$\overline{G}$	$\overline{G}$	A14	A10	E1	A12	$\overline{G}$	$\overline{G}$	$\overline{G}$	$\overline{E}$

Pin	1	2	3	4	5	6	7	8	9	10	11
21	A10	A0	A0	A0	A13	A10	A11	A15	$\overline{W}$	$\overline{W}$	A7
22	$\overline{G}$	A1	A1	A1	V_{DD}	$\overline{G}$	A12	V_{DD}	A9	A9	V_{DD}
23	A11	A2	A2	A2	—	A11	A13	—	A8	A8	—
24	A9	A3	A3	A3	—	A9	V_{DD}	—	V_{DD}	V_{DD}	—
25	A8	A4	A4	A4	—	A8	—	—	—	—	—
26	A13	A5	A5	A5	—	E2	—	—	—	—	—
27	$\overline{W}$	A6	A6	A6	—	$\overline{W}$	—	—	—	—	—
28	V_{DD}	A7	A7	A7	—	V_{DD}	—	—	—	—	—
29	—	A8	A8	A8	—	—	—	—	—	—	—
30	—	V_{SS}	V_{SS}	V_{SS}	—	—	—	—	—	—	—
31	—	A9	A9	A9	—	—	—	—	—	—	—
32	—	A10	A10	A10	—	—	—	—	—	—	—
33	—	A11	A11	A11	—	—	—	—	—	—	—
34	—	A12	NC	A12	—	—	—	—	—	—	—
35	—	A13	NC	NC	—	—	—	—	—	—	—
36	—	NC	NC	NC	—	—	—	—	—	—	—
37	—	$\overline{LB}$	$\overline{LB}$	$\overline{LB}$	—	—	—	—	—	—	—
38	—	$\overline{UB}$	$\overline{UB}$	$\overline{UB}$	—	—	—	—	—	—	—
39	—	$\overline{W}$	$\overline{W}$	$\overline{W}$	—	—	—	—	—	—	—
40	—	V_{DD}	V_{DD}	V_{DD}	—	—	—	—	—	—	—

TABLE 6.11 JEDEC STANDARD MEMORY SURFACE MOUNT CONFIGURATIONS (EPROMs, ROMs)

PIN #	2K × 8 EPROM	4K × 8 EPROM	8K × 8 EPROM	16K × 8 EPROM	32K × 8 EPROM	2K × 8 ROM	4K × 8 ROM	8K × 8 ROM	16K × 8 ROM	32K × 8 ROM
1	DU	DU	DU	DU	DU	DU	DU	DU	DU	DU
2	NC	NC	NC	NC	V_{PP}	NC	NC	NC	NC	NC
3	NC	NC	NC	NC	A12	NC	NC	NC	NC	A12
4	A7	A7	A7	A7	A7	A7	A7	A7	A7	A7
5	A6	A6	A6	A6	A6	A6	A6	A6	A6	A6
6	A5	A5	A5	A5	A5	A5	A5	A5	A5	A5
7	A4	A4	A4	A4	A4	A4	A4	A4	A4	A4
8	A3	A3	A3	A3	A3	A3	A3	A3	A3	A3
9	A2	A2	A2	A2	A2	A2	A2	A2	A2	A2
10	A1	A1	A1	A1	A1	A1	A1	A1	A1	A1
11	A0	A0	A0	A0	A0	A0	A0	A0	A0	A0
12	NC	NC	NC	NC	NC	NC	NC	NC	NC	NC
13	D0	D0	D0	D0	D0	D0	D0	D0	D0	D0
14	D1	D1	D1	D1	D1	D1	D1	D1	D1	D1
15	D2	D2	D2	D2	D2	D2	D2	D2	D2	D2
16	V_{ss}	V_{ss}	V_{ss}	V_{ss}	V_{ss}	V_{ss}	V_{ss}	V_{ss}	V_{ss}	V_{ss}
17	DC	DC	DC	DC	DC	DC	DC	DC	DC	DC
18	D3	D3	D3	D3	D3	D3	D3	D3	D3	D3
19	D4	D4	D4	D4	D4	D4	D4	D4	D4	D4
20	D5	D5	D5	D5	D5	D5	D5	D5	D5	D5
21	D6	D6	D6	D6	D6	D6	D6	D6	D6	D6
22	D7	D7	D7	D7	D7	D7	D7	D7	D7	D7
23	$\overline{E}$	$\overline{E}$	$\overline{E}$	$\overline{E}$	$\overline{E}$	$\overline{E}$	$\overline{E}$	$\overline{E}$	$\overline{E}$	$\overline{E}$
24	A10	A10	A10	A10	A10	A10	A10	A10	A10	A10
25	$\overline{G}$	V_{PP}	$\overline{G}$	$\overline{G}$	$\overline{G}$	$\overline{G}$	$\overline{G}$	$\overline{G}$	$\overline{G}$	$\overline{G}$
26	V_{PP}	NC	NC	NC	NC	NC	NC	NC	NC	NC
27	NC	A11	A11	A11	A11	NC	A11	A11	A11	A11
28	A9	A9	A9	A9	A9	A9	A9	A9	A9	A9
29	A8	A8	A8	A8	A8	A8	A8	A8	A8	A8
30	NC	NC	NC	NC	A13	NC	NC	NC	NC	A13
31	NC	NC	$\overline{PGM}$	$\overline{PGM}$	A14	NC	NC	NC	NC	A14
32	V_{DD}	V_{DD}	V_{DD}	V_{DD}	V_{DD}	V_{DD}	V_{DD}	V_{DD}	V_{DD}	V_{DD}

TABLE 6.12   JEDEC STANDARD MEMORY SURFACE MOUNT CONFIGURATIONS (SRAMs)

PIN #	64K × 1 SRAM	16K × 4 SRAM	1K × 8 SRAM	2K × 8 SRAM	4K × 8 SRAM	8K × 8 SRAM	16K × 8 SRAM	32K × 8 SRAM
1	A0	A0	DU	DU	DU	DU	DU	DU
2	A1	A1	NC	NC	NC	NC	NC	A14
3	A2	A2	NC	NC	NC	A12	A12	A12
4	A3	A3	A7	A7	A7	A7	A7	A7
5	A4	A4	A6	A6	A6	A6	A6	A6
6	A5	A5	A5	A5	A5	A5	A5	A5
7	A6	A6	A4	A4	A4	A4	A4	A4
8	A7	A7	A3	A3	A3	A3	A3	A3
9	Q	$\overline{A8}$	A2	A2	A2	A2	A2	A2
10	$\overline{W}$	$\overline{E}$	A1	A1	A1	A1	A1	A1
11	V_{ss}	V_{ss}	A0	A0	A0	A0	A0	A0
12	$\overline{E}$	$\overline{W}$	NC	NC	NC	NC	NC	NC
13	D	DQ1	DQ0	DQ0	DQ0	DQ0	DQ0	DQ0
14	A8	DQ2	DQ1	DQ1	DQ1	DQ1	DQ1	DQ1
15	A9	DQ3	DQ2	DQ2	DQ2	DQ2	DQ2	DQ2
16	A10	DQ4	V_{ss}	V_{ss}	V_{ss}	V_{ss}	V_{ss}	V_{ss}
17	A11	A9	DU	DU	DU	DU	DU	DU
18	A12	A10	DQ3	DQ3	DQ3	DQ3	DQ3	DQ3
19	A13	A11	DQ4	DQ4	DQ4	DQ4	DQ4	DQ4
20	A14	A12	DQ5	DQ5	DQ5	DQ5	DQ5	DQ5
21	A15	A13	DQ6	DQ6	DQ6	DQ6	DQ6	DQ6
22	V_{DD}	V_{DD}	DQ7	DQ7	DQ7	DQ7	DQ7	DQ7
23	–	–	$\overline{E}$	$\overline{E}$	$\overline{E}$	$\overline{E}$	$\overline{E}$	$\overline{E}$
24	–	–	NC	A10	A10	A10	A10	A10
25	–	–	$\overline{G}$	$\overline{G}$	$\overline{G}$	$\overline{G}$	$\overline{G}$	$\overline{G}$
26	–	–	NC	NC	NC	NC	NC	NC
27	–	–	NC	NC	A11	A11	A11	A11
28	–	–	A9	A9	A9	A9	A9	A9
29	–	–	A8	A8	A8	A8	A8	A8
30	–	–	NC	NC	NC	NC	A13	A13
31	–	–	$\overline{W}$	$\overline{W}$	$\overline{W}$	$\overline{W}$	$\overline{W}$	$\overline{W}$
32	–	–	V_{DD}	V_{DD}	V_{DD}	V_{DD}	V_{DD}	V_{DD}

now be a competent operator and often troubleshooter of microcomputer-based equipment. In many cases, the engineer is tasked to modify or even design new processes and controls using the newer microcomputer technology. Such active involvement with this technology requires the engineer to have, at the very least, a knowledge of microcomputer architecture, an understanding of the strengths and limitations of the microcomputer, and an awareness of current product offerings and trends in microcomputer development. The engineer must also have a working knowledge of the equipment required to develop and troubleshoot a microcomputer application.

There are many factors responsible for the microcomputer's increasing popularity in recent years. The most prominent contributing factor has been the increased level of integration. This means that the individual size of the many transistors that constitute the micro has decreased over the years while the number of transistors on a single chip has increased. The increase in integration level means that for a given area of silicon, one of the primary determinants in the cost of a microcomputer, more complex functions requiring more transistors may be constructed at very little or no extra expense. The result is that memory arrays become larger and the diversity of peripheral devices increases. Both resources become less expensive, permitting a wider range of application. An additional benefit of increased integration, which has also increased the breadth of microcomputer applications, is the increased operational bandwidth of the micro. The decrease in transistor size increases the speed of the individual transistors so that the computational speed of the micro has increased. This improvement in the microcomputer's bandwidth allows new applications, such as voice recognition or real-time process control, to be implemented. The micro is thus allowed to replace traditionally-executed analog functions like RMS-to-DC conversion and waveform generation.

TABLE 6.13 JEDEC STANDARD MEMORY SURFACE MOUNT CONFIGURATIONS (EEPROMs)

PIN #	1K × 8	2K × 8	4K × 8	8K × 8	16K × 8	32K × 8
1	DU	DU	DU	DU	DU	DU
2	OPT	OPT	OPT	OPT	OPT	A14
3	NC	NC	NC	A12	A12	A12
4	A7	A7	A7	A7	A7	A7
5	A6	A6	A6	A6	A6	A6
6	A5	A5	A5	A5	A5	A5
7	A4	A4	A4	A4	A4	A4
8	A3	A3	A3	A3	A3	A3
9	A2	A2	A2	A2	A2	A2
10	A1	A1	A1	A1	A1	A1
11	A0	A0	A0	A0	A0	A0
12	NC	NC	NC	NC	NC	NC
13	DQ0	DQ0	DQ0	DQ0	DQ0	DQ0
14	DQ1	DQ1	DQ1	DQ1	DQ1	DQ1
15	DQ2	DQ2	DQ2	DQ2	DQ2	DQ2
16	V_{ss}	V_{ss}	V_{ss}	V_{ss}	V_{ss}	V_{ss}
17	DU	DU	DU	DU	DU	DU
18	DQ3	DQ3	DQ3	DQ3	DQ3	DQ3
19	DQ4	DQ4	DQ4	DQ4	DQ4	DQ4
20	DQ5	DQ5	DQ5	DQ5	DQ5	DQ5
21	DQ6	DQ6	DQ6	DQ6	DQ6	DQ6
22	DQ7	DQ7	DQ7	DQ7	DQ7	DQ7
23	$\overline{E}$	$\overline{E}$	$\overline{E}$	$\overline{E}$	$\overline{E}$	$\overline{E}$
24	NC	A10	A10	A10	A10	A10
25	$\overline{G}$	$\overline{G}$	$\overline{G}$	$\overline{G}$	$\overline{G}$	$\overline{G}$
26	NC	NC	NC	NC	NC	NC
27	NC	NC	A11	A11	A11	A11
28	A9	A9	A9	A9	A9	A9
29	A8	A8	A8	A8	A8	A8
30	NC	NC	NC	NC	A13	A13
31	$\overline{W}$	$\overline{W}$	$\overline{W}$	$\overline{W}$	$\overline{W}$	$\overline{W}$
32	V_{DD}	V_{DD}	V_{DD}	V_{DD}	V_{DD}	V_{DD}

6.6.1 What Are Microcomputers?

It is intriguing to think that a single- or multi-chip "solution" can be applied to such a wide number of engineering situations. The microcomputer (MCU) has many unique and interesting qualities that enable its application to so many different types of problems. In fact, the most useful characteristics are those traits related directly to the second half of its name, specifically, the microcomputer's ability to *compute* by performing arithmetic and logical calculations. Basically, microcomputers may be defined as single- or multi-chip devices possessing all, or a reasonable subset, of the architectural elements of a computer but with a reduced physical size and often a designed limitation of some resource or capability.

Like its larger counterpart, the mainframe computer, the micro possesses the elements minimally necessary to classify it as a computer (see Fig. 6.102). The central processing unit (CPU) of the microcomputer contains both a control unit (CU) and an arithmetic/logic unit (ALU), which processes information, either data and/or program, and produces output based upon an operating program through a variety of devices. When it is available without peripheral devices such as memory, timers, and input/output (I/O) devices, the CPU portion of an MCU is called a microprocessor (MPU). For example, Motorola's MC6809, a mature yet sophisticated microprocessor (refer to Fig. 6.103), lacks data memory, program memory, or I/O devices. Still, because it has a control unit and an ALU, the MC6809 is an example of a microprocessor. Once some memory is added around the MC 6809, the system may then be considered a microcomputer. Devices that are added to such a system merely increase the size of the microcomputer. Other examples of microprocessors include Motorola's MC68000 family, which includes the MC68020 and MC68030. In recent years, MCUs on a single chip have been called microcontrollers. For purposes of the following discussion, the terms microcontroller and microcomputer will be used interchangeably.

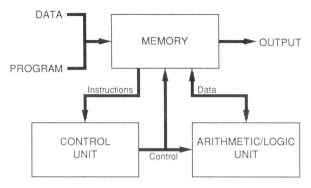

Fig. 6.102 Block diagram of a computer.

The given description of the microcomputer (MCU) is a general one which fails to express the variety and elegance of the practical MCU system. Technology has evolved to the point where whole microcomputer *systems* are now available in a single integrated circuit package (Fig. 6.104). These single-chip microcomputers couple the computational power of the CPU with on-chip memory (program and data) and on-chip input/output devices (i.e., peripherals). For example, one of the first microprocessors, the Motorola MC6800 (introduced in the mid seventies), required additional circuitry to operate (see Fig. 6.105). The MC6800 requires off-chip program/data memory, address decoding, and external peripheral devices including digital I/O. Even the MC6800's *clock*, a precise-period signal used to motivate and coordinate the actions of the CPU, has to be externally provided (see Fig. 6.106). Contrast this with one of Motorola's latest MCU offerings, the MC68HC11E9 microcomputer; only four resistors, one crystal used by the on-chip clock circuit to generate a precise timing reference, and three capacitors are minimally required to run the MC68HC11E9. This MCU comes with 12,000 bytes of on-chip program storage, 512 bytes of on-chip data storage, 512 bytes of on-chip *non-volatile* data storage, an 8-channel/8-bit A/D converter, a 16-bit timer subsystem,

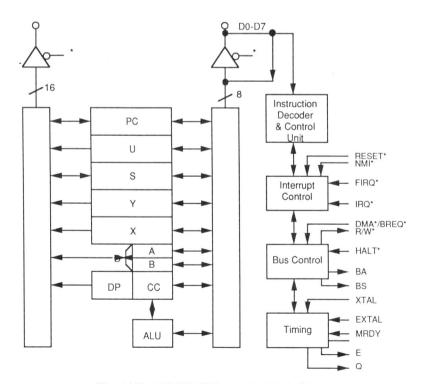

Fig. 6.103 MC6809 CPU expanded block diagram.

Fig. 6.104 Microprocessor packaging. Clockwise from top: plastic dual-in-line package (DIP), chip carrier in a socket, ceramic DIP (sidebrazed package), ceramic leadless chip carrier, ceramic pin grid array.

an asynchronous serial interface, and a high-speed synchronous serial interface for input/output expansion. In addition, it even possesses 31 digital input/output lines (Fig. 6.107). To build an equivalent microprocessor system based on the MC6800 would take approximately 10 additional integrated circuits (ICs) occupying over 6.5 square inches (41.9 cm^2), as opposed to less than 1 square inch (6.5 cm^2) for the MC68HC11E9! Additional examples of microcontrollers include Motorola's M6804 family, M6805 family, M68HC11 family, and the super high performance M68332, 32-bit microcontroller. All of these MCUs possess on-chip peripherals which make each MCU a virtually self-contained microcomputer system on a single silicon chip.

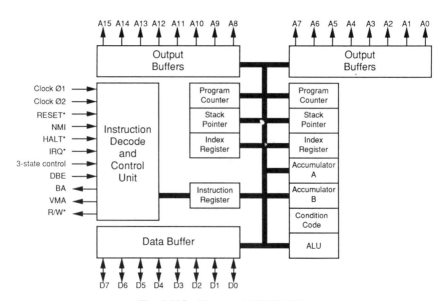

Fig. 6.105 Motorola MC6800 CPU.

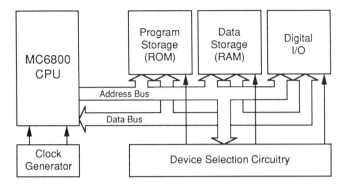

Fig. 6.106 MC6800 system implementation.

The MC6800 and the MC68HC11E9 represent a good sampling of the spectrum of devices available, ranging from the primitive MC6800 CPU to the considerably more elaborate MC68HC11E9 MCU. Despite the large differences in levels of integration between the MC6800 and MC68HC11, the CPU portions of both devices work similarly. Basically, the CPU of a computer performs conditional transfers and arithmetic/logic functions on data. For example, the first step in any CPU's operation is to fetch or *read* data from a sequence of instructions, called the *operating program* or program, for short. The CPU does this by specifying an address to a program storage device, along with a control signal which tells the program storage device (and other storage devices attached to the same address/data lines) whether the subsequent action is a read of the device (a transfer from memory to CPU) or a write to storage (a transfer from the CPU to the memory).

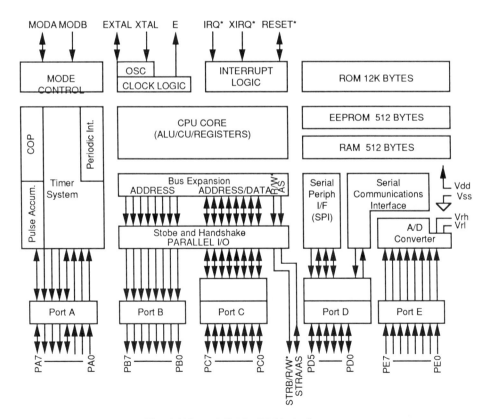

Fig. 6.107 MC68HC11E9 block diagram.

Direct STAA and STAB (Store Accumulators A or B into direct memory)

Cycle #	R/W* Line	Address Bus	Data Bus
1	1	Opcode Address	Opcode
2	1	Opcode Address + 1	Operand Address (low byte)
3	0	Operand Address	Data from Accumulator

Immediate CPD and CPY (Compare D Accumulator and Compare Y Register

Cycle #	R/W* Line	Address Bus	Data Bus
1	1	Opcode Address	Opcode (page select byte)
2	1	Opcode Address + 1	Opcode (second byte)
3	1	Operand Address + 2	Operand data (high byte)
4	1	Operand Address + 3	Operand data (low byte)
5	1	$FFFF	Irrelevant Data

Fig. 6.108   68HC11 Cycle-by-cycle count of some common instructions.

This first fetch, or read, of the program memory tells the CPU what subsequent sequence of actions to take. The data acquired first by the CPU, or instruction fetch, tells the internal *instruction decoder* what succeeding action, if any, is necessary to carry out the particulars of the instruction. For example, if the first instruction tells the CPU to load an internal CPU register (i.e., temporary storage location) with a piece of data stored at location "X" in memory, the CPU must perform a total of three data transfers: one for the opcode (instruction or a contraction of the phrase *operation code*), one for the address of the data to be fetched, and, finally, the transfer of the data of interest itself. A similar sequence would be used for the arithmetic addition of two pieces of data, one located in memory and one in a CPU register. The CPU would still perform three data transfers as just described, but this time one of the internal CPU registers would contain the sum of the data fetched from memory and the previous contents of the CPU register (i.e., the contents before execution of the add instruction). Figure 6.108 shows the execution details, cycle by cycle, of some common Motorola MC68HC11 opcodes.

Obviously, the fetch and write mechanisms of the CPU are useless without memory and peripheral devices. The selection of the type and size of memory required in a microcomputer system is critical to the success of the micro-based system. Traditionally, the medium most selected for the storage of the program in the microcomputer system has been *read only memory* (ROM). Two reasons for the selection of ROM as primary storage in micro-based systems have been the non-volatility of ROM, the ability of the ROM to retain data without power applied to its terminals, and the lower cost of the ROM when compared to the other choices for micro program/data storage. Contrast these two attributes with those of some alternatives. Magnetic core storage, although non-volatile, is very bulky and expensive. Mass storage devices, such as disk memory, require whole microcomputer systems just to control them. Any volatile memory device would have to acquire the microcomputer's operating program from a non-volatile source. Given the cost of the ROM alternatives, this adds a "chicken and egg" dilemma to the design of the MCU assembly without a ROM. The ROM is not a panacea however, since the process of producing a ROM is a costly one usually reserved for those microcomputer assemblies which will be produced in large quantities, usually 1,000 pieces or more. This expense tends to exclude many applications where only a few microcomputer assemblies will be produced.

A common variation of the ROM which is found frequently with the microcomputer is the *erasable-programmable read only memory* (EPROM). Although still non-volatile and inexpensive like the ROM, the EPROM is changeable "in the field" using simple apparatus such as an ultraviolet light (UV) source and a PROM programmer. This makes it ideal for those small-quantity applications not appropriate for the masked ROM. In fact, some external EPROMs have the same pin connections as masked ROM parts to allow development of the MCU's program with the EPROM (which will permit the multiple erase and program operation desirable during MCU-based product development). Functional pin-outs of some external ROM/EPROM devices are shown in Fig. 6.109.

In the description of the basics of CPU operation, it was mentioned that temporary storage is used during the performance of CPU calculations. The need for temporary volatile storage in a

Size:	32k x 8	16k x 8	8k x 8	4k x 8
P/N:	27256	27128	2764	2732
S I G N A L	Vpp	Vpp	Vpp	
	A12	A12	A12	
	A7	A7	A7	A7
	A6	A6	A6	A6
	A5	A5	A5	A5
	A4	A4	A4	A4
M N E M O N I C	A3	A3	A3	A3
	A2	A2	A2	A2
	A1	A1	A1	A1
	A0	A0	A0	A0
	O0	O0	O0	O0
	O1	O1	O1	O1
	O2	O2	O2	O2
	Gnd	Gnd	Gnd	Gnd

4k x 8	8k x 8	16k x 8	32k x 8	Size
2732	2764	27128	27256	P/N
	Vcc	Vcc	Vcc	S I G N A L
	PGM*	PGM*	A14	
Vcc	N.C.	A13	A13	
A8	A8	A8	A8	
A9	A9	A9	A9	
A11	A11	A11	A11	
OE*	OE*	OE*	OE*	M N E M O N I C
A10	A10	A10	A10	
CE*	CE*	CE*	CE*	
O7	O7	O7	O7	
O6	O6	O6	O6	
O5	O5	O5	O5	
O4	O4	O4	O4	
O3	O3	O3	O3	

Fig. 6.109 ROM/EPROM memory pinouts.

microcomputer in addition to the CPU's registers can be quite important. Many applications, like some signal processing techniques, require that digitized input signals be buffered. This may mean that thousands of input signal samples are to be held in temporary storage until the CPU is ready to use the data. To fill this requirement, the *random access memory* (RAM) memory device is used. RAM, a slightly more expensive device, differs from ROM primarily in that RAM may be modified by the CPU on the fly, while ROM or EPROM must be reprogrammed either by the semiconductor manufacturer (ROM) or in the field with the additional equipment (EPROM). Also, the contents of RAM are maintained only so long as power is applied to the terminals of the RAM, unlike the non-volatile ROM and EPROM.

Each memory type has its own particular strengths and weaknesses, as seen from the descriptions of the RAM and ROM. Other memory technologies are beginning to bridge the gap between the volatile and non-volatile memories. One such alternative is the *electronically erasable-programmable* ROM (EEPROM or E²PROM). This increasingly popular memory device permits non-volatile storage of data even under CPU control. This feature of the EEPROM permits useful applications not easily or economically possible with RAM or ROM. An example of such a unique application excerpted from *Motorola M68HC11 Reference Manual* employs the Motorola MC68HC11 microcontroller, which has 512 Bytes of EEPROM on-chip with the CPU, as the center of a sensor data acquisition system.

One of the most interesting uses for EEPROM in an MCU system is to implement self-adjusting or self-adapting systems. A fairly simple form of self-adaptation would be a system that can calibrate or recalibrate a sensor as it ages . . . The adaptation would be semipermanent so the modified behavior would be in effect the next time the system was activated (as if the system had originally been programmed that way).

Other interesting possibilities for the application of MCUs with EEPROM include storage of set-point and converter calibration information, logging of periodically measured data, and tailoring of MCU-based product behavior without hardware or mask ROM changes. Product development can be expedited by permitting MCU algorithms to be stored and run out of EEPROM without the costly erasure cycle associated with the EPROM, and then easily changed when necessary.

Even a microcomputer with EEPROM will provide minimum utility in most situations without adequate I/O devices. The potential value of the microcomputer is fulfilled only when it is given useful data to work on and outputs information, based upon that data, in a useful and usable form. Thus, another critical element in the design of the microcomputer system is to provide applicable I/O. As evidenced by Motorola's MC68HC11E9 and other such devices, many I/O choices are possible: asynchronous serial interfaces for communication with terminals and other data equipment, synchronous serial interfaces for digital I/O expansion, A/D converters for digitization of analog signals (the type of signals one would expect from a wide range of sensors), complex timers for period/pulse-width measurement of input signals, digital I/O lines to provide a source of digital device control under software direction, and even D/A converters to drive the myriad analog voltage controlled devices such as voltage-controlled oscillators (VCO) or voltage-controlled amplifiers (VCA). All of the peripherals and memories discussed constitute the contemporary microcomputer.

6.6.2   Why Use a Microcomputer?

Within the boundaries set by the miniature microcomputer and its memories, peripheral resources, and computing power, the primary determinant in the functionality and usefulness of the microcomputer is the amorphous software program. The performance of conditional transfer and logic/ arithmetic operations on data is perhaps the single strongest characteristic of the MCU. The same types of cycles occur within the CPU of the MCU, namely memory/peripheral reads and writes, regardless of the actual instructions executed. The implications are substantial. The same hardware may be programmed to perform functionally different tasks by merely changing the machine's program. This flexibility can best be realized with a simple illustration.

The Motorola MC68HC05B6 microcomputer has, among many on-chip features, an 8-channel/ 8-bit *analog-to-digital converter* (ADC). The ADC may be configured under software control to perform the conversion of a given input channel once every 16 microseconds. Initially, assume that a program has been written to calculate the average voltage of the input signal, which in this example, is a full-wave rectified version of a vibration sensor which is attached to a shaker table (see Fig. 6.110). After this micro-based product is put into service as a *dangerous level of vibration alarm* (DLVA for short), it is found that the error of the DLVA depends upon the type of modulation (sine, triangle, or square wave) to the vibration table. Rather than scrap the HC05B6-based DLVA, a slight modification may be made to the software that will enable calculation of the *root-mean-square* (RMS) of the input signal — a method of calculation that more accurately measures the energy delivered to the vibration table. In this example the importance of software to the micro is underscored.

This leads us to reason one for using a microcomputer: Within the boundaries set by the microcomputer's hardware resources and computing bandwidth (i.e., computing speed), the primary determinant in the functionality of the micro is its program. As implied in the previous illustration, changing the function of the micro through a few lines of program code may be a relatively painless exercise. This tends to make software modifications more cost effective than equivalent hardware changes. This may well be a corollary to reason one.

By definition, a microcomputer may include various I/O devices. As explained earlier, high levels of integration often imply that a given microcontroller will have more on-chip resources — some of which are obtained for little additional cost — than are really needed for a given application. Because of this, the micro has tremendous potential for growth, in features or size, in just about any application. The DLVA in the illustration may well have been less costly had discrete analog circuits been used to perform the averaging function. Of course, also as shown, the average detection failed to ultimately provide the correct alarm condition. This change would have required a redesign of the analog circuits; a replacement of analog circuits with more analog circuits. As the scenario might continue, when having to interface the DLVA to data logging equipment a year later, continuous replacement might require yet another major redesign. However, with the microcomputer solution, specifically the Motorola MC68HC05B6 and its built-in EEPROM and asynchronous serial port,

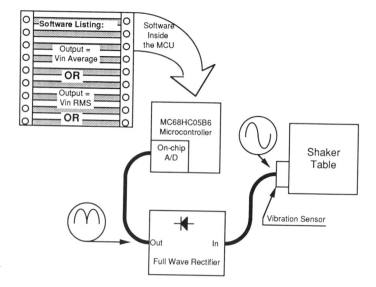

Fig. 6.110   Software flexibility example.

interfacing the DLVA to data collection equipment (such as an IBM-compatible personal computer programmed to collect data via RS232 COM1 port) is almost a trivial task. Thus, reason two for designing with microcomputers is the potential for growth in a micro-based application exceeds that obtainable with a dedicated hardware solution. Even though data logging of the vibration levels may eventually have been suggested, it is not hard to imagine that the suggestion was made because the DLVA was designed with a micro. Thus, a corollary to reason two is that given the software basing and peripheral types of a micro solution, the MCU-based assembly will help create new and useful uses for itself.

Finally, the most compelling reason for designing with an MCU is cost. Usually, an MCU-based assembly will have a higher performance/price ratio than alternate solutions if the following criteria are met.

1. The engineering staff must be able to apply the microcomputer and proper MCU development methods. Designing with the MCU requires a different way of looking at the world—a view which consists of the "big picture" and the details simultaneously. Above all, application and design must be done in an orderly, methodical fashion to reduce error.

2. The correct development tools for MCU applications must be available. The days of developing MPU/MCU applications by blowing EPROMs and then trying the EPROM in the circuit are gone. That is a dangerous business given the complexity of MCU systems today. Software validation is an almost impossible task for such a development environment. Thus, proper software development tools, such as macro-assemblers or compilers, and hardware emulators, such as Motorola Evaluation Modules (EVM), are necessary to allow systematic and quantifiable development of the application.

3. The nature of the problem must be reducible to a digital implementation. Where an applications bandwidth exceeds the current MCU product offerings, an MCU solution may not be possible. Other factors discouraging MCU solutions might include insufficient word resolution required for the task or too many additional peripherals required to make the MCU a cost effective solution in the given application. However, as the speed of the MCU continues to increase and more engineering problems are innovatively solved by the MCU, more and more processes will be implemented with MCUs.

4. The MCU selected must have features and resources that closely match the application. Despite reason two and its corollary, putting a comparatively expensive MC68332 32-bit microcontroller in an application where an MC68HC05C8 would suffice will guarantee cost *ineffectiveness*. A wide range of MCU types, such as Motorola's broad family of MCU offerings, is necessary to fit most effectively a given application.

5. The application must be a reasonably complex one which would benefit from the flexibility of a micro. Obviously if the end application is like a 1 kHz fixed-frequency square-wave oscillator with no intention on the designer's part to improve or grow the device beyond that primitive function, replacing the handful of resistors, capacitors, and SSI (small scale integration) gates may not be the most cost effective solution.

6.7 MICROCOMPUTER ARCHITECTURE

Lynn A. Kauffman

The architecture of a computer is the design of the internal structure, including the amount of memory required, the size and number of the buses, and the types of input-output elements. Key to the architecture of the CPU are the types of registers and the possible data transfers among them. The major design criterion of microcomputer architecture is the trade-off between hardware and software functions. One technique of computer architecture is to have the masked integrated circuitry placing the MPU, ROM, RAM, and I/O on one chip (a single-chip microcomputer). Another technique uses one IC containing the control unit, and several identical ICs containing 2-bit or 4-bit *slices* of the ALU and registers (a bit-slice MPU). In between the two extremes are microcomputers constructed from chip sets or families of integrated circuits, including the MPU with a fixed instruction set, and the necessary RAM, ROM, and I/O interface adapter chips.

The microcomputer system architecture is the collection of registers that make up the particular system and the possible data transfers among them. The MPU, ROM, RAM, and I/O ports can all be viewed as a collection of addressable registers. The microprocessor architecture includes the number and types of registers in the MPU and their possible data transfers. The registers inside the microprocessor are considered internal registers, and those in the memory and input-output sections are the external registers.

6.7.1 The Central Processing Unit

The central processing unit (CPU) of a computer includes the arithmetic logic unit (ALU) and the control unit (Fig. 6.111). In a microcomputer, the terms microprocessor (MPU) and CPU are interchangeable. The CPU (MPU) executes the program instructions and processes the data. In executing an instruction, the CPU must pull the instruction from memory (fetch), decode it, and then execute it. It transfers data to and from memory and I/O. The CPU also generates timing and control signals and recognizes external signals for interrupts. The ALU portion of the CPU performs the arithmetic operations for the computer (e.g., addition and subtraction) and makes logical decisions (e.g., logical AND, logical OR, and shift operations) based on the binary data it receives. The binary input for the ALU is supplied by the data registers and accumulators.

Registers

Registers are small memories which are part of the control unit. Registers, like other memory storage units, consist of binary cells and have unique addresses. Data can be saved in a register until a bus is ready to receive it or the program calls for it. Registers may be under program control, an advantage since the CPU may obtain data without a memory access. Internal buses connect the registers to each other, to the rest of the control section, and to external buses. Figure 6.112 shows the registers found in the MC6800 MPU that are visible to the programmer. Several common types of registers are described below.

Program Counters. The program counter (PC) contains the address of the next instruction in the program. The CPU places the contents of the PC on the address bus and fetches the first word of the instruction from memory. If the instruction occupies more than one word of memory, the CPU increments the PC. Unless a JUMP or BRANCH instruction is encountered, the CPU executes instructions sequentially. If the program breaks its sequence, the binary number stored in the PC is stored on the stack. The number for the repeated or out-of-sequence routine (subroutine) is now stored in the PC, and the program proceeds. The number on the stack is returned to the PC once the subroutine is completed.

Instruction Registers. The instruction register (IR) holds instructions until they are decoded. Pipelining makes use of two instruction registers. While one instruction is being executed, the other is being fetched and saved. The IR tells the decoder what is asked of the control unit. The decoder sets up the logic that controls the entire MPU, including the state of the control lines, the clock control, and bus access.

Memory Address Registers. Address registers contain the current addresses of the memory data. They are used to gain access to specific points in memory whenever the CPU calls for data transfers into or out of memory. This register is not software accessible.

Accumulators. Accumulators are used during calculations for temporary storage of one of the operands. Accumulators may also be used during logical operations and data transfers. They do dual duty in that they hold the binary numbers before going to the ALU and are usually used to hold the result of an arithmetic or logic operation. Generally, they are the most frequently used registers in the computer.

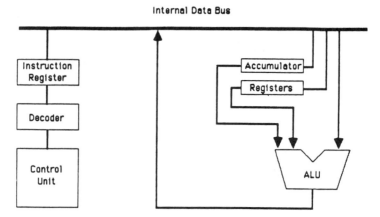

Fig. 6.111 The central processing unit including the control unit and the ALU.

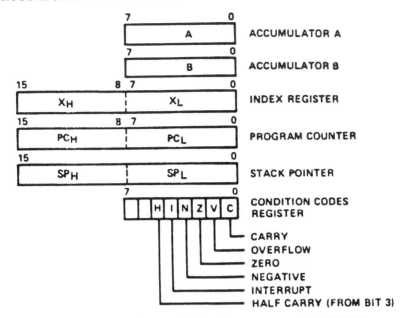

Fig. 6.112 The registers available to the programmer of the MC6800.

General Purpose Registers. General purpose registers may serve as temporary storage for data and addresses. They also may be assignable as accumulators or program counters.

Index Registers. Index registers (X) are used for addressing. Auto-indexing is the built-in hardware that automatically causes the index register to either increment or decrement each time it is used.

Condition Code Registers. Condition code registers are the status registers that hold 1-bit indicators or flags representing the conditions inside the CPU. Computer decision-making is based on these flags. Common flags are

Carry—set if the last operation generated a carry from the most significant bit.

Zero—set if the result of the last operation was zero.

Overflow—set if the last operation produced a two's complement overflow. This bit shows that the result of the arithmetic operation exceeded the word capacity.

Sign—set if the most significant bit of the result of the last operation was a "1" (a "1" indicates a negative two's complement number).

Parity—set if the number of "1"s in the result of the last operation was even (even parity) or odd (odd parity). This flag is useful in communication between the MPU and peripherals, and in character manipulation. It detects errors in transmission of data.

Half-Carry—set if the last operation generated a carry from the lower half-word. In 8-bit MPUs this flag is used to do BCD (binary coded decimal) arithmetic.

Interrupt Enable—set if an interrupt is allowed. The CPU may automatically disable interrupts during startup or service routines. During critical timing loops or multiword operations, the program may disable interrupts.

Stack Pointers. A stack pointer is the register containing the address of the top of the stack. Stacks are last-in first-out (LIFO) memories. Data can only be added to the top of the stack (PUSH) or taken from the top of the stack (PULL). The CPU adds an element to the stack by placing the element in the memory location addressed by the stack pointer. The stack pointer is then incremented. To remove an element from the stack, the stack pointer is decremented and the element is pulled from the memory location addressed by the stack pointer. The advantage of a stack is that data can be added without disturbing the data that is already present. A JUMP-TO-SUBROUTINE instruction moves a return address from the program counter (PC) to the stack. A RETURN instruction fetches the address from the stack and places it in the PC. Thus, the stack pointer is used to record the last program step prior to a subroutine.

Data Registers. Data registers are temporary storage for data coming from or going to the data bus.

Instruction Cycle

During one instruction cycle, the MPU performs several tasks. These tasks are considered to be part of either the fetch, decode, or execute stage of the instruction cycle (Fig. 6.113). First, the instruction address is placed on the memory address bus. The address is read by the address decoder, and the correct row in memory is read. The data stored at this location is placed on the data bus. The IR is gated on, and the data is latched into the IR. Then the instruction is decoded. A given binary word in the instruction register represents a given instruction. The control ROM, internal to the MPU, implements the sequence of the decoder. It is programmed to set up the proper control sequence to match the instruction code entered. By decoding the op code, the control section of the MPU fetches any addresses or data required by the instruction from either memory (external to the MPU) or the registers (internal). The MPU performs the operation specified by the instruction including arithmetic and logical operations, data transfers, and management tasks. The MPU must look for any control signal interrupts and respond. Finally, the MPU must generate any status, control, or timing signals required by the memory or I/O. The PC keeps track of every step of the program as the MPU executes it.

6.7.2 Bus Structures

The MCU bus structure is a means of transmitting information throughout the system. A bus is a set of conductors that transfers binary data ("1"s and "0"s). The system bus, also called an external bus (external to the MPU), connects the registers in the various subsystems of the MCU (RAM, ROM, I/O). The MPU moves information through the system via three buses: the address bus, the data bus, and the control bus. These three standard buses allow all of the units of the microcomputer to communicate with the microprocessing unit.

External Bus

The MPU communicates with the other elements in the system via the address bus. The address bus allows the MPU to select the origin or destination of the signals which are transmitted on another bus or line. It selects a register within one of the system devices where data is to be placed or found. The address bus originates from the program counter in the MPU. Several connections are made to each device, and the MPU addresses each device via the address lines. A unique address for each register is generated by charging some wires at a HIGH logic level and others at a LOW logic level. Traditionally, 8-bit microcomputers have used 16 lines for addresses, which allow 2^{16} or 64K locations to be addressed.

The bidirectional channel through which the MPU recieves and sends data to a device after it has been addressed is the data bus. A read/write line (one of the control bus lines) allows the MPU to control the direction of the data transfer. An 8-bit MPU requires an 8-bit data bus in order to be able to transmit eight parallel bits of data.

The control bus conductors control the sequence of events for the system. It generates the timing, synchronization, and isolation signals and controls the direction of data transfer for the memory and I/O devices. At least ten control lines are required by the MPU and generally more are actually used. Typical signals found on the control bus are READ, WRITE, INTERRUPT, RESET, and various acknowledgement signals.

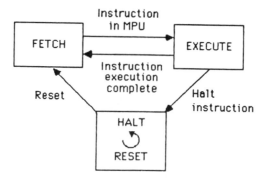

Fig. 6.113 The MPU instruction cycle.

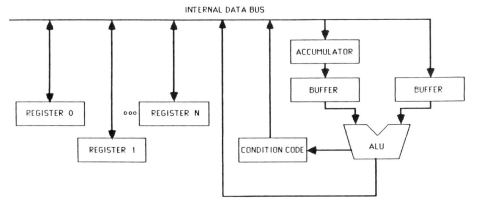

Fig. 6.114   Single bus architecture.

The actual communication lines of the system bus may be on the chip itself (laid out on the circuit), printed circuit traces on the printed circuit board, or cables or wires connecting the system units. An efficient bus design is crucial to the speed of the system. Each of the three buses in the MPU can carry up to one TTL load (typical capacitance is 100–130 picofarads, which translates to five to seven MOS packages). Therefore, in most systems it is necessary to add bus drivers on the data and address bus (and sometimes to the control bus).

Internal Bus

The microprocessor is said to have a single, dual, or triple bus architecture, depending on the number of internal data buses. The internal bus is the communication line between the ALU and the MPU registers. The bus arrangement determines the response time of the MPU. The simplest design is the single bus. The single bus system has only one set of conductors connecting the registers to the ALU (see Fig. 6.114). A common bus can handle only one condition at a time. Transfers of data must occur in steps, although many of the new RISC and DSP processors allow multiple operations per instruction by taking advantage of multiple on-chip buses. Buffer registers are necessary. The advantage of this structure is space conservation; execution of instructions, however, is slowed down considerably.

A double bus uses a single input bus or source bus connected between the registers and the ALU inputs. A separate result bus connects the ALU output to the register input.

The triple bus (Fig. 6.115) provides maximum performance. Two source buses are connected to the ALU inputs, and one bus is the destination (result) bus. The inputs no longer need to be

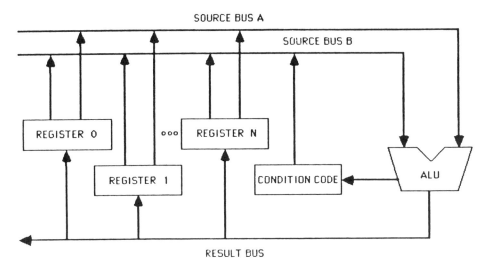

Fig. 6.115   Triple bus architecture.

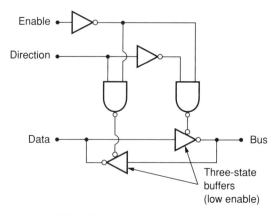

Fig. 6.116   Three-state bus control.

buffered. The triple bus eliminates time sharing with simultaneous data transfer on all three buses. However, this increases pin requirements on the MPU package (in an 8-bit MPU, an additional data bus internally means an additional eight pins externally).

Three-State Bus Control

One task of the bus configuration is to ensure that data is transferred to and from the microprocessor correctly. Proper design avoids bus contention, the situation in which two elements are both trying to control a bus. To avoid bus contention so that data is received or transmitted by one unique device, three-state devices are used on the data bus (Fig. 6.116). Three-state control isolates the signal paths and prevents erroneous data transfer. Three-state buffers have three output states. Two of the outputs are the standard TTL logic "1" and logic "0," and the third output is a high impedance state, which prevents the output from driving or loading any circuit connected to it. The high impedance output is said to be floating and causes the output to be electrically disconnected from the logic circuit to which it is physically attached.

6.7.3   Memory

Memory is indispensable in a microcomputer system. It contains the data and instructions necessary for using the data (i.e., the program). Within the microcomputer, there exists a memory hierarchy. The internal registers that are part of the MPU are the fastest level of data memory available. The internal registers can be accessed in less than 100 nanoseconds (ns), and they are generally few in number (between 8 and 64). The main memory or "the memory" of the system is implemented on one or more components, typically MOS LSI devices. A limited amount of memory may reside on the MPU chip, but the MPU generally requires external memory. The main memory has typical access times from 15 to 120 ns. The third level of memory is the mass memory or bulk storage. The mass memory is low-cost, high capacity storage consisting of magnetic tape, cassette, or disk. Mass storage capacity can be on the order of thousands of kilobytes. However, mass memory is very slow to access.

Some key features of the main memory are non-destructive read-out, volatile read/write memory, and single-chip forms. Unlike magnetic core memory, semiconductor memory may be read without changing the contents and without needing to rewrite the data to the memory. The RAM (random access memory) portion of memory is volatile, which means that the contents of the RAM will be lost if the power is lost. Therefore, RAM is used to store data or programs that are not critical in the event of a power failure or can be reloaded from the mass memory. The ROM (read only memory) is a non-volatile device and contains the permanent programs and data needed for the microcomputer. The RAM and ROM for the main memory are available in single packages with various word lengths and organizations. All the decoding and interfacing circuitry are part of the memory chip, and the memory device behaves like other digital ICs of the same technology.

Each word in memory has two parameters, its address and the data associated with that address. The address is the location in memory where the word resides. A memory access consists of a bit pattern specifying the address of the desired word being sent to the address bus. A decoder in the memory selects the word at the address specified by the pattern on the address bus. In response to a READ control signal, the word is sent from memory to the data bus. A WRITE signal causes the word to be written from the data bus into the specified memory location. After a certain access time, the data becomes available on the memory output connected to the data bus.

Addressing Modes

Each microprocessor has an instruction set which includes the codes that tell the machine what to do. The M6800 instruction set, shown in Fig. 6.117 includes 72 separate instructions. Although the MC6800 has been technically eclipsed by some of its successors (i.e., Motorola M68HC11), the balanced "orthogonal" instruction set is still an excellent example for a well-rounded MCU/MPU instruction set. A single instruction may have one to four different modes of addressing. The mode of addressing simply tells the MPU where to find the data. Since each of the M6800 instructions may have more than one addressing mode, there are actually 197 valid machine codes.

An opcode (operation code) is a set of binary digits that forms part of the MPU instruction set. It has two parts; the first indicates the addressing mode, and the second is the code for the operation. These instructions must be entered into memory in the proper sequence in order for the correct function to be performed. When the program runs, the opcodes flow from memory to the instruction register (IR). The IR output is decoded, and the proper response signals are placed on the control lines. The addressing mode tells the MPU how the instruction is to be handled. Multiple addressing modes add flexibility to the program execution, but complicates the MPU design. The addressing mode also affects the time required for the instruction to be executed. There are seven major addressing techniques.

Inherent Mode. In the inherent addressing mode, the data is stored in one of several internal registers. It is a one-word instruction (i.e., one computer word) which contains the code for a register

Executable instructions – alphabetic list.

ABA	Add Accumulators	INC	Increment
ADC	Add with Carry	INS	Increment Stack Pointer
ADD	Add	INX	Increment Index Register
AND	Logical AND	JMP	Jump
ASL	Arithmetic Shift Left	JSR	Jump to Subroutine
ASR	Arithmetic Shift Right	LDA	Load Accumulator
BCC	Branch if Carry Clear	LDS	Load Stack Pointer
BCS	Branch if Carry Set	LDX	Load Index Register
BEQ	Branch if Equal to Zero	LSR	Logical Shift Right
BGE	Branch if Greater or Equal Zero	NEG	Negate
BGT	Branch if Greater than Zero	NOP	No Operation
BHI	Branch if Higher	ORA	Inclusive OR Accumulator
BIT	Bit Test	PSH	Push Data
BLE	Branch if Less or Equal	PUL	Pull Data
BLS	Branch if Lower or Same	ROL	Rotate Left
BLT	Branch if Less than Zero	ROR	Rotate Right
BMI	Branch if Minus	RTI	Return from Interrupt
BNE	Branch if not Equal to Zero	RTS	Return from Subroutine
BPL	Branch if Plus	SBA	Subtract Accumulators
BRA	Branch Always	SBC	Subtract with Carry
BSR	Branch to Subroutine	SEC	Set Carry
BVC	Branch if Overflow Clear	SEI	Set Interrupt Mask
BVS	Branch if Overflow Set	SEV	Set Overflow
CBA	Compare Accumulators	STA	Store Accumulator
CLC	Clear Carry	STS	Store Stack Register
CLI	Clear Interrupt Mask	STX	Store Index Register
CLR	Clear	SUB	Subtract
CLV	Clear Overflow	SWI	Software Interrupt
CMP	Compare	TAB	Transfer Accumulators
COM	Complement	TAP	Transfer Accumulators to Condition Code Reg
CPX	Compare Index Register	TBA	Transfer Accumulators
DAA	Decimal Adjust	TPA	Transfer Condition Code Reg to Accumulator
DEC	Decrement	TST	Test
DES	Decrement Stack Pointer	TSX	Transfer Stack Pointer to Index Register
DEX	Decrement Index Register	TXS	Transfer Index Register to Stack Pointer
EOR	Exclusive OR	WAI	Wait for Interrupt

Fig. 6.117   The M6800 instruction set.

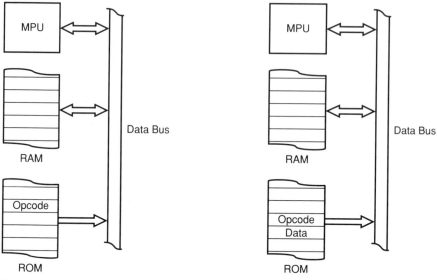

Fig. 6.118 Inherent addressing mode.   **Fig. 6.119** Immediate addressing mode.

(Fig. 6.118). It involves the transfer of data in the registers or the incrementing of a register. For example, the opcode ADD R would mean, "add register R to the accumulator and leave the result in the accumulator."

Immediate Mode. The immediate addressing mode gives the opcode plus the operand, the operand immediately following the opcode. The information might be data to be used by the processor or memory address information. In this mode the decoder is told how many bytes of data are required for the instruction. The immediate addressing mode is either a two- or three-word instruction (Fig. 6.119).

Direct Addressing Mode. The direct addressing mode is a two-word instruction (Fig. 6.120). It is used for branch instructions with a short (one-word) branch address specified after the opcode.

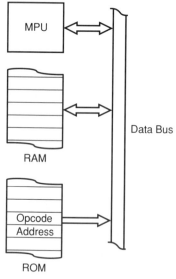

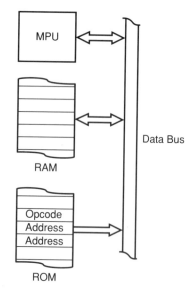

Fig. 6.120 Direct addressing mode.   **Fig. 6.121** Extended addressing mode.

Extended Addressing Mode. Extended addressing is very similar to direct addressing, except that it uses three-byte instructions (Fig. 6.121). In an 8-bit MPU, three bytes are required for a branch instruction; one byte for the instruction and two bytes for the target address. Extended addressing is used for normal branch or jump operations.

Indexed Addressing Mode. Indexed addressing is useful for accessing tabular information stored in memory. A table is a group of words stored in sequence in ROM (or RAM). It is necessary to refer to these entries in the table from the program. The indexed mode is a two-, sometimes three-byte instruction (Fig. 6.122). The starting address of the table is placed in the index register (X), and an offset (displacement) is included in the instruction. The offset is added to the contents of the specified index register.

Relative Addressing Mode. Relative addressing is designed to make jumps within a short distance easier. It is a two-word instruction, as shown in Fig. 6.123. The jump is relative to the current address in the program counter. The offset may be either positive or negative. With 8-bit words, +128 positions forward and −127 positions backward may be specified. The address generated is equal to the PC plus the offset. Relative addressing is important for arithmetic routines or block transfers of data.

Indirect Addressing Mode. Indirect addressing is used to share information among several programs or users. When different programs are executed, data resides at unpredictable locations in RAM. Any routine that makes use of a given data structure must look up the actual address of the structure in a pointer table that is in a given location in RAM. With indirect addressing, the data can easily be accessed. The instruction includes the address of the pointer, and the data at that location can be fetched. Figure 6.124 illustrates this addressing mode. Because shared data is generally stored in tables, indirect addressing coupled with indexed addressing can generate efficient programs.

Read Cycle

In either a read or write operation, the MPU must signal to the device that is being addressed by use of the chip select (CS) or chip enable (CE) pin. This selects the one memory chip from among the many connected to the address bus. The MPU places the contents of the PC on the address bus. This address determines the source of the instruction to be read. The timing signal required for the first step is produced by the processor clock. The clock, which is an oscillator, produces a regular series of pulses. The internal characteristics of the processor determine the required frequency and pulse width of the clock. The second step of the read cycle involves the MPU waiting for the data to be available from memory. The timing of the processor and the memory must be coordinated so the correct data is transferred. The delay of the MPU depends on the access time of the memory. The

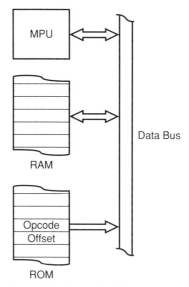

Fig. 6.122 Indexed addressing mode.

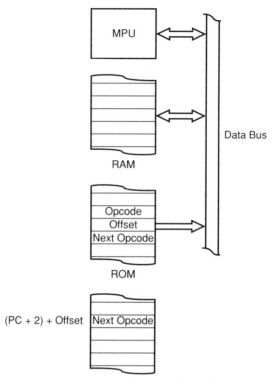

Fig. 6.123   Relative addressing mode.

simplest way to assure erroneous data is not received by the MPU is to design a memory section which has an access time less than one clock period. In the third step of the memory read cycle, the MPU places the data from the input bus into the IR. If the memory access time is less than one clock period, the timing signal to place the data input contents (an instruction in this case) in the IR can be the second cycle of the clock. In a data fetch from RAM, for example, the data from the input data bus would be placed in the data register.

Write Cycle

The memory write cycle occurs less frequently than a memory read cycle in most MPU systems. It is also more complex. First, the MPU places the contents of an address register on the address bus. The contents of the data register are placed on the data output bus, which requires a clocking signal. A selector can allow data to be obtained from several data registers. In the third step, the MPU sends the WRITE pulse to memory. In a write operation, the order in which the memory receives the signals is very important. Precise timing relationships must exist among the signals. A write operation, unlike a read operation, is destructive. Therefore, the address must reach the memory before the write pulse begins and data must remain constant after it ends. Finally, the MPU waits for the write operation to be completed. The simplest design has the write cycle completed in one clock period.

6.7.4   Input-Output Elements and Operations

A basic computer input-output system involves the communication between the computer (specifically the MPU) and the external devices (peripherals). The three main types of peripherals are input devices, output devices, and bulk-memory devices. Input-output (I/O) ports and peripheral interface chips are ways for the microcomputer to get data into and out of the computer system. A program directs the microprocessing unit either to read the status of input lines or to send data to the output lines. A write operation is the passing of data from the MPU to the peripheral. Passing data from the peripheral to the MPU is a read operation.

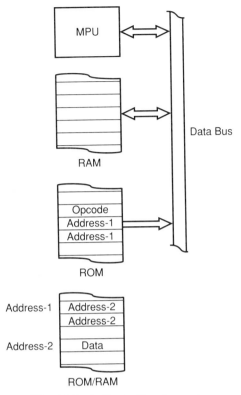

Fig. 6.124   Indirect addressing mode.

Inputs to the MCU system may include status from sensors, A/D (analog-to-digital) converters, and disks and tape units. Possible outputs include a printout of the results, D/A (digital-to-analog) converters, and electronic pulses which regulate a process. I/O operations are similar to memory access.

Some of the problems of I/O include the wide variety and types of available peripherals, the enormous range of speeds, the variety of signal types and levels, and the complexity of the signal structure. Peripherals can be mechanical, electromechanical, or electronic. They may use digital voltage signals, current signals, or analog signals. It may be necessary to hold or transform the signals before the MPU can accept them. Very few standards exist for I/O. Each peripheral poses a unique problem, unlike the memory chips that have the same signal levels as the MPU. The input-output section of an MCU can be the most expensive portion of the system, although with MCUs the overall cost of a system with I/O may be less than a CPU system with external I/O.

Interfacing Peripherals

Interfacing the various peripherals to the system is a difficult task. Interfacing includes translating the signals of the MCU and peripherals so they may communicate and providing proper timing and control signals. It involves both hardware and software design. One method of providing I/O functions is the parallel interface. A parallel interface has as many lines as the width of the data path. These lines are attached from the peripheral to the MPU and allow simultaneous data transfer. Another method of I/O is serial transmission. In serial transmission the bits of each word are sent over signal paths between the MCU and the peripherals. In simplex, the paths are one way (one path for each direction). In duplex, one path can transmit both ways. Serial transmission is an attractive method for interfacing devices that are some distance from the MPU, since at most only two transmission lines are needed. Serial transmission can occur either synchronously or asynchronously. In synchronous transmission, the data rate is locked into the system clocking. The reciever (MPU or I/O) and transmitter (I/O or MPU) are synchronized to the same clock rate. Asynchronous transmission means the peripherals are not locked into the system clocking. "Start" and "stop" bits must be added to the data word to inform the reciever where data transmission begins and ends.

Servicing Peripherals

Two methods of servicing the peripherals are polling and the interrupt routine. Polling is the technique of reading each peripheral periodically to see if it is ready to transfer data. The ports to be polled are read sequentially. An interrupt is a signal which informs the MPU that a peripheral device is requesting service. The interrupt signal causes the MPU to jump to a service routine for the requesting device. This gives control of the system to the external hardware.

Interrupts can also be used for other applications, such as (1) alarm inputs (e.g., a security system), (2) power fail warning, which allows the MPU to turn the necessary outputs off or on, save the data in a low-power memory, or switch to a backup power supply, (3) manual override, allowing external control of the system for maintenance, repair, testing, or debugging, (4) system debugging, allowing the insertion of corrections, or breakpoints, (5) hardware failure, (6) transmission error, and (7) control for direct memory access. The above is not an exhaustive list of interrupt uses. If the response time of the MPU must be much shorter than the average time between the occurrence of events, interrupt driven routines are useful. Otherwise, the processor could spend too much time searching for the event (polling) which would diminish the processor throughput.

The major disadvantage of interrupt routines is their random nature. The CPU is subject to the dictates of the external environment. Interrupt routines can be difficult to write and debug. Some CPUs require extra hardware (latches, flip-flops, gates, and encoders) as the number of interrupt sources increases. Rapid identification of the source of the interrupt is essential. Processors with several interrupt inputs must assign priority to each input and service the highest-priority interrupt first. Program execution becomes less formalized with interrupt routines. When I/O data rates are high, interrupts are less useful, as the CPU must still handle the data transfers and fetching and decoding of instructions. At high data rates, direct memory access (DMA), which substitutes hardware for software control, greatly increases the I/O capability.

General Input Procedure

The general input procedure is similar to a memory read cycle. The MPU places the address on the address bus, which selects a particular input port. The physical port may be of any bit length. However, it is most convenient if the port has the same word length as the processor. A single peripheral may require several ports if the length of the data word exceeds the word length of the MPU or if the control and status information must be transferred separately. The MPU waits for data to become available, and when it does, the MPU reads the data from the data bus and places it in a register.

Determining When the Data is Available. An alternative to the interrupt routine is to have the processor assume the data is always available. This assumption is adequate for slowly changing inputs such as data from mechanical switches and sensors that measure slowly changing physical quantities (e.g., pressure and temperature). During transition periods (dials being turned or sensor levels changing), erroneous data may be received by the MPU. A few possible solutions to avoid this are changing slow or irregular transition data into pulses by using monostable multivibrators or Schmitt triggers, using an additional input signal to inform the CPU that the data is ready, or, in software, having the MPU recheck and confirm the intial data reading after a delay.

Another method involves the MPU accepting data at a rate determined by an external clock. The input operation is performed at a regular rate. This method is adequate for moderate data rates and synchronous peripherals. Synchronous transfers are fastest. For transfers with higher data rates, DMA is the solution. The problem with synchronous transfer is the initial synchronization of the MPU with the external clock. A special control line or synchronization message is needed. A special code indicating end of transfer may also be used. This requires extra hardware and software.

Another method of servicing input ports is called handshaking. The peripheral generates a special signal (DATA READY) by using an extra bit, strobe, or code that has no other meaning. The MPU determines that the DATA READY signal is active. The MPU reads the data from the input port and then must provide an INPUT ACKNOWLEDGE signal to the peripheral to indicate that the transfer is complete. The input port must hold the data and control signals long enough to ensure that they have been received. This method is adequate for slow to moderate data rates and for synchronous peripherals. The MPU is alerted to the signal by an interrupt.

General Output Procedure

The general output procedure is similar to a memory write cycle. The MPU places the address on the address bus and selects an output port. The MPU places the data on the data bus and waits for the transfer to be successfully completed. A write pulse is generated, and the timing constraints are the same as those for a memory write cycle (Section 6.7.3). The MPU must determine if the peripheral is ready to accept the data. The output port must hold the data long enough for the peripheral to accept it.

The same methods of determining if data is available for the input procedure may be used to determine if the peripheral is ready for the output procedure. The processor may assume the peripheral is always ready. This method is adequate for slow peripherals which operate at mechanical or human speed. Some examples of slow output transfers are CRT displays, relays, and actuators. The MPU cannot exceed the response rate of these peripherals.

Handshaking with the peripheral is another method of determining if the peripheral is ready to receive the data. The peripheral sends an OUTPUT REQUEST or a PERIPHERAL READY signal to the output port of the microcomputer. The MPU determines that the PERIPHERAL READY signal is active. The MPU sends the OUTPUT READY signal to the peripheral. The next PERIPHERAL READY signal could inform the MPU that the transfer is complete. This method requires extra hardware, software, and time to ensure the transfer proceeds properly. Handshaking is adequate for low to moderate data rates and asynchronous peripherals.

For moderate data rates and synchronous peripherals, the MPU can transfer data to the output port at a rate determined by an external clock. A special synchronization message for "start" and "stop" is required.

Direct Memory Access

Direct memory access (DMA) is a hardware device which controls direct transfers between the I/O and memory, as illustrated in Fig. 6.125. With DMA, the data is not routed through the MPU registers. The MPU sets up the transfer of data. It must send the starting address and number of words to memory to initialize the transfer. The hardware of the DMA initiates and controls the actual data transfer. Blocks of data can be transferred at high speed using DMA.

Memory and I/O Addressing

Part of designing the input-output bus is determining how the MPU will distinguish between the external registers of the memory and those of the I/O. The two major methods of I/O addressing are called isolated or standard I/O and memory-mapped I/O. In isolated I/O, the I/O is assigned to

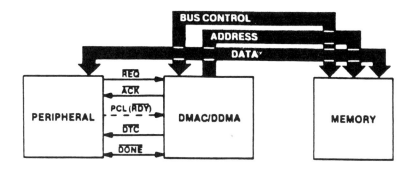

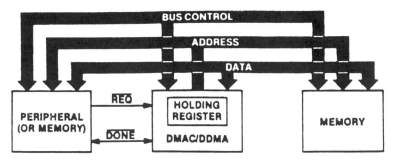

Fig. 6.125   Example of DMA control: (*a*) single address transfer; (*b*) dual address transfer.

an address space distinct from the main memory address space. In memory-mapped I/O, the I/O is assigned to the same address space as the memory.

Isolated I/O. In isolated I/O addressing, the input-output and memory are decoded separately. A SECTION SELECT signal distinguishes between memory and I/O. The advantages to this method are shorter addresses, clearer programs, and separate design of memory and I/O sections. With distinct memory and I/O addresses, the I/O port addresses are shorter, and the instructions are shorter. The decoding system is also simpler. Extra control signals can be developed for I/O transfers. The memories do not require strobes or handshaking signals. The programs are clearer because the I/O transfers can be distinguished from the memory accesses. Since the control structure of the I/O and memory are independent, the design of the two sections can be done separately. The disadvantage of this method is that extra decoding and instructions are required. A block diagram of a system using isolated I/O is shown in Fig. 6.126.

Memory-Mapped I/O. In memory-mapped I/O, the I/O ports are treated exactly the same as memory locations. The same instructions are used for both memory accesses and I/O transfers. The advantage of this method is threefold. Any instructions that operate on data in memory can operate on data at the I/O ports. By not having separate instructions, many tasks are simplified. Second, since no separate decoding or control structures are needed, the design uses fewer parts. Finally, LSI interfaces and special controllers can be incorporated directly, as these devices include storage elements and registers that must be programmed from the CPU. Some disadvantages are that I/O transfers are indistinguishable from other operations, the I/O ports take up address space, instructions are more difficult to understand, a more complex decoding system is required, and it is

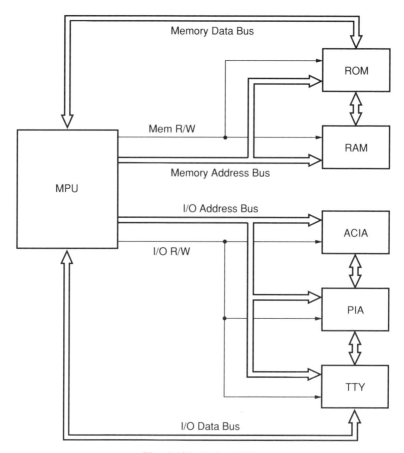

Fig. 6.126 Isolated I/O.

more difficult to program. Program debugging is more difficult because simple instructions must be used for complex I/O functions. Because of physical limitations of the I/O devices, the instructions are more difficult to understand. The I/O ports occupy less address than memory chips, making decoding more complex. The design must either extend the decoding system or waste memory space. The I/O ports seldom occupy more than a few memory addresses. The interface chips often need to generate additional signals under program control. Figure 6.127 shows a block diagram of a memory-mapped system.

Memory-mapped I/O is best suited for complex interface chips and for I/O ports that are LSI devices. Isolated I/O is best suited to SSI (small scale integration) and MSI (medium scale integration) integrated circuits where the I/O ports are TTL circuits.

6.7.5 Hardware and Software

An MPU is a hardware structure driven by a software program. Hardware is defined as the physical components of a system, from gates, flip-flops, and programmable LSI devices to the packaged subsystems such as floppy disks. The designer of a microprocessor-based structure must be familiar with the hardware, including the functions, operations, limitations, and performance characteristics of the components, as well as the software and the system integration possibilities. The software design at the assembly language level calls for a knowledge of the MPU instruction set and the ability to combine the instructions into a program. The program must be designed to carry out the desired function using the particular hardware configuration.

Programs for MCU systems are written in an environment different from that in which they are run. The hardware and interfaces are developed concurrently with the software. The problem of system integration is a difficult one. The system designer must understand how the components of the MCU interact. Hardware and software tasks are interrelated. Considerations include cost,

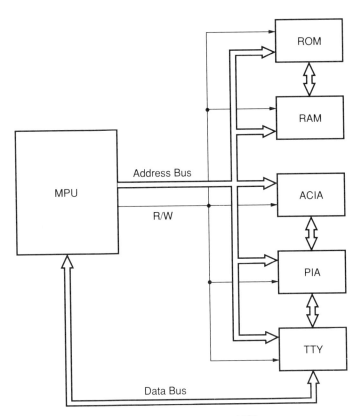

Fig. 6.127 Memory mapped I/O.

speed, and system timing. The MCU includes complex input/output operations with real-time constraints.

A programmer uses high-level languages to write a program, but the MPU executes a series of machine instructions. Machine instructions are represented as a sequence of bytes stored in memory. The program is stored as a pattern of "1"s and "0"s. Assembly language is the mnemonics used to represent the opcodes (e.g., ADD, SUB, JMP). Assemblers are the programs for the MPU which convert the assembly language programs into machine code. Resident or self-assemblers run on the MCU itself. Cross assemblers run on a mainframe or a minicomputer. Compilers are the programs that convert high-level languages such as PASCAL, FORTRAN, or the more popular "C" language into machine code. Compilers require a considerable amount of memory for execution.

Development system hardware and software are needed for troubleshooting the MCU design. Chip manufacturers generally have development systems for designing with their chip families. Good development tools include software editing and debugging aids. Also, ways to troubleshoot the hardware and monitor the interface between the MPU and peripherals are needed for system development.

6.7.6   Bit-Slice Architecture

In bit-slice architecture, the functions of the microcomputer are kept in discrete LSI blocks. Low-power Schottky integrated circuit technology is used to build custom-designed computers with outstanding performance advantages. The first advantage of this type of architecture is increased throughput (speed). The second advantage is the ability to customize the instruction sets to accommodate special requirements. The LSI blocks, called bit-slices, are specialized ALUs which manipulate a portion (typically four bits) of the computer word. Any number of bit-slices can be strung together to form words of any size. Control functions such as sequencing, addressing control, and timing must be provided by other LSI components. Most Schottky TTL microprocessors are 2- or 4-bit slices and require other devices and circuitry to make a microcomputer.

Microprogramming is required in most Schottky TTL processors. Microprogramming involves the writing of sequences of control instructions that decode and execute computer instructions. Several tasks are usually performed in a typical computer instruction. The more powerful the instruction, the more tasks that are required. If an instruction decoder is to perform several tasks simultaneously, it must have more inputs or more bits per control word. Each instruction usually consists of several microinstructions performed in sequence at very high speeds. These microinstructions determine the functions performed by each basic instruction; hence they determine the instruction set. In a microprogrammed computer, the instructions are stored in a ROM called the Microprogram Control Store (MCS).

A microprogrammed microprocessor is flexible because the instruction set is not hardware designed, and powerful, as special functions can be coded. However, the designer now has two levels of hardware/software interface to contend with. The macroinstruction set must first be designed in terms of microinstructions before any higher level of programming can be done.

Another MPU architecture which has gained popularity is RISC. RISC, which stands for "reduced instruction set computer," is a high-powered MPU which allows microprogramming of the CPU instruction decoder. By allowing programming of the instruction decoder, unique instructions streamlined for a special task may be developed. This in turn increases computational speed significantly. For example, if a given task required successive multiply and add instructions, an ordinary MPU would waste time doing two instruction fetches. By programming a single specific multiply and add instruction into a RISC machine, at the very least only one instruction fetch and decode must be done. RISC technology stands to rival many general purpose micro applications where high throughput is an operational requirement.

BIBLIOGRAPHY

Bennet, William S., and Carl F. Evert, Jr., *What Every Engineer Should Know About Microcomputers*, Marcel Dekker, New York, 1980.

Bishop, Ron, *Basic Microprocessors and the 6800*, Hayden Book Co., Inc., New Jersey, 1979.

"Eleventh Annual MCU/MPU Directory," *EDN*, November 1984.

Freedman, Alan, *The Computer Glossary for Everyone*, Prentice-Hall, Inc., New Jersey, 1983.

Getgen, Lawrence E., *Designing with Microprocessors*, Science Research Associates, Inc., Chicago, 1985.

Greenfield, Joseph D., *Using Microprocessors and Microcomputers: The 6800 Family*, John Wiley and Sons, New York, 1981.

"Hello, Mr. Chips," *Creative Computing*, June 1985, pp. 48–53.

"It's 16 Bits, but Is that Wide or What?" *Creative Computing*, June 1985, pp. 40–45.

Lawrence, O. R., *Computer Technology*, McGraw-Hill Ryerson Ltd., Toronto, 1984.

Levanthal, Lance A., *Introduction to Microprocessors: Software, Hardware, Programming*, Prentice-Hall, Inc., New Jersey, 1978.

"Macro Wonders from the Latest Micros," *Fortune*, December 10, 1984, pp. 115–128.

Sheingold, D. H. (ed), *Analog-Digital Handbook*, Prentice-Hall, Inc., New Jersey, 1986.

Short, Kenneth L., *Microprocessors and Programmed Logic*, Prentice-Hall, Inc., New Jersey, 1981.

Sippl, Charles J., *Microcomputer Handbook*, Van Nostrand Reinhold Co., New York, 1977.

Zaks, Rodney, *From Chips to Systems: An Introduction to Microprocessors*, Sybex, Berkeley, 1981.

CHAPTER 7

GROUNDING AND
CABLING TECHNIQUES

HENRY W. OTT

Henry Ott Consultants
Livingston, New Jersey

7.1   INTRODUCTION

Widespread use of electric and electronic circuits for communication, power distribution, automation, computation, and other purposes makes it necessary for diverse circuits to operate in close proximity. All too often these circuits affect each other adversely. Electromagnetic interference (EMI) has become a major problem for circuit designers, and it is likely to become more severe in the future. The large number of electronic devices in common use is partly responsible for this trend. In addition, the use of integrated circuits is reducing the size of electronic equipment. As circuitry becomes smaller and more sophisticated, more circuits are crowded into less space, thus increasing the probability of interference.

Today's equipment designers need to do more than just make their circuits operate under ideal conditions in the laboratory. Besides that obvious task, they must also make sure the equipment will work in the "real world," with other equipment nearby. This means the equipment should not be affected by external noise sources, and should not itself be a source of noise. Elimination—or really avoidance—of electromagnetic interference should be a major design objective.

Noise can be defined as any electrical signal present in a circuit other than the desired signal. An important exception to this definition is the distortion products produced in a circuit due to nonlinearities. These are really circuit design problems and not truly noise problems. Although these distortion products may be undesirable, they are not considered noise unless they get coupled into another part of the circuit. It follows from the definition of noise that a desired signal in one part of a circuit may be considered noise if inadvertently coupled into some other part of the circuit.

Noise sources can be grouped in three major categories. The first are the so-called intrinsic noise sources that arise from random fluctuations within physical systems. Examples of intrinsic noise are thermal and shot noise. Second are manufactured noise sources, such as motors, switches, and transmitters. The third category is noise due to natural disturbances, such as lightning and sun spots.

Interference can be defined as the undesirable effect of noise. If a noise voltage causes unsatisfactory operation of a circuit, it is interference. Usually noise cannot be eliminated but only reduced in magnitude until it no longer causes interference.

Susceptibility is the capability of a device or circuit to respond to unwanted electrical energy (noise). The susceptibility level of a circuit or device is the noise environment in which the equipment can operate satisfactorily.

Chapter 7 is reproduced, with permission, from Noise Reduction Techniques in Electronic Systems, by Henry W. Ott, published by John Wiley & Sons, NY, ©1976 by Bell Telephone Laboratories, Inc. The material was edited by Chester L. Nachtigal.

Fig. 7.1 Before noise can be a problem, there must be a noise source, a receiver that is susceptible to the noise, and a coupling channel that transmits the noise to the receiver.

7.1.1 Typical Noise Path

A block diagram of a typical noise path is shown in Fig. 7.1. As can be seen, three elements are necessary to produce a noise problem. First, there must be a noise source. Second, there must be a receiver circuit that is susceptible to the noise. Third, there must be a coupling channel to transmit the noise from the source to the receiver.

The first step in analyzing a noise problem is to define the problem. This is done by determining what the noise source is, what the receiver is, and how the source and receiver are coupled together. It follows that there are three ways to break the noise path: (1) the noise can be suppressed at the source, (2) the receiver can be made insensitive to the noise, or (3) the transmission through the coupling channel can be minimized. In some cases, noise suppression techniques must be applied to two or to all three parts of the noise path.

As an example, consider the circuit shown in Fig. 7.2. It shows a shielded dc motor connected to its motor-drive circuit. Motor noise is interfering with a low-level circuit in the same equipment. Commutator noise from the motor is conducted out of the shield on the leads going to the drive circuit. From the leads, noise is radiated to the low-level circuitry.

In this example, the noise source consists of the arcs between the brushes and the commutator. The coupling channel has two parts: conduction on the motor leads and radiation from the leads. The receiver is the low-level circuit. In this case, not much can be done about the source or the receiver. Therefore, the interference must be eliminated by breaking the coupling channel. Noise conduction out of the shield or radiation from the leads must be stopped, or both steps may be necessary.

7.1.2 Use of Network Theory

For the exact answer to the question of how any electric circuit behaves, Maxwell's equations must be solved. These equations are functions of three space variables (x, y, z) and of time (t). Solutions for any but the simplest problems are usually very complex. To avoid this complexity, an approximate analysis technique called "electric circuit analysis" is used during most design procedures.

Circuit analysis eliminates the spatial variables and provides approximate solutions as a function of time only. Circuit analysis assumes the following:

1. All electric fields are confined to the interiors of capacitors
2. All magnetic fields are confined to the interiors of inductors
3. Dimensions of the circuits are small compared to the wavelength(s) under consideration

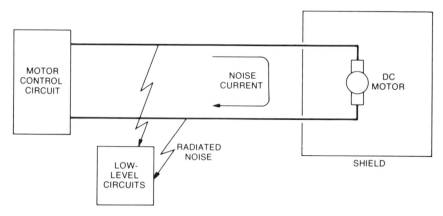

Fig. 7.2 In this example the noise source is the motor, and the receiver is the low-level circuit. The coupling channel consists of conduction on the motor supply leads and radiation from the leads.

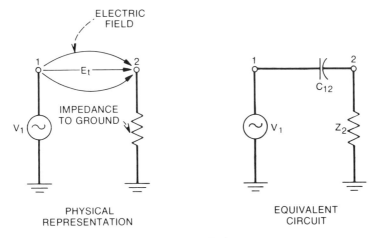

Fig. 7.3 When two circuits are coupled by an electric field, the coupling can be represented by a capacitor.

What is really implied is that external fields, even though actually present, can be neglected in the solution of the network. Yet, these external fields may not necessarily be neglected where their effect on other circuits is concerned.

For example, a 100 W power amplifier may radiate 100 mW of power. These 100 mW are completely negligible as far as the analysis of the power amplifier is concerned. However, if only a small percentage of this radiated power is picked up on the input of a sensitive amplifier, it may produce a large noise signal.

Whenever possible, noise-coupling channels are represented as equivalent lumped component networks. For instance, a time-varying electric field existing between two conductors can be represented by a capacitor connecting the two conductors (see Fig. 7.3). A time-varying magnetic field that couples two conductors can be represented by a mutual inductance between the two circuits (see Fig. 7.4).

For this approach to be valid, the physical dimensions of the circuits must be small compared to the wavelengths of the signals involved. This assumption is made throughout most of this book, and it is normally reasonable. For example, the wavelength of a 1 MHz signal is approximately 300 m. For a 300 MHz signal, it is 1 m. For most electronic circuits, the dimensions are smaller than this.

Even when the above assumption is not truly valid, the lumped component representation is still useful for the following reasons.

1. The solution of Maxwell's equations is not practical for most "real world" noise problems because of the complicated boundary conditions.

2. Lumped component representation, although it does not necessarily give the correct numerical answer, does clearly show how noise depends on the system parameters. On the other hand, the solution of Maxwell's equations, even if possible, does not show such dependence clearly.

In general the numerical values of the lumped components are extremely difficult to calculate with any precision, except for certain special geometries. One can conclude, however, that these components exist, and as will be shown, the results can be very useful even when the components are only defined in a qualitative sense.

7.1.3   Methods of Noise Coupling

Conductively Coupled Noise

One of the most obvious, but often overlooked, ways to couple noise into a circuit is on a conductor. A wire run through a noisy environment may pick up noise, and then conduct it to another circuit. There it causes interference. The solution is to prevent the wire from picking up the noise, or to remove the noise from it, by decoupling before it interferes with the susceptible circuit.

The major example in this category is noise conducted into a circuit on the power supply leads. If the designer of the circuit has no control over the power supply, or if other equipment is connected to the power supply, it becomes necessary to decouple the noise from the wires before they enter the circuit.

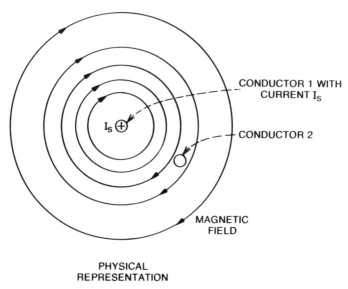

PHYSICAL
REPRESENTATION

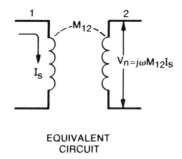

EQUIVALENT
CIRCUIT

Fig. 7.4   When two circuits are coupled by a magnetic field, the coupling can be represented as a mutual inductance.

Coupling through Common Impedance

Common impedance coupling occurs when currents from two different circuits flow through a common impedance. The voltage drop across the impedance seen by each circuit is influenced by the other. The classic example of this type of coupling is shown in Fig. 7.5. The ground currents 1 and 2 both flow through the common ground impedance. As far as circuit 1 is concerned, its ground potential is modulated by ground current 2 flowing in the common ground impedance. Some noise signal, therefore, is coupled from circuit 2 to circuit 1 through the common ground impedance.

Another example of this problem is illustrated in the power distribution circuit shown in Fig. 7.6. Any change in the supply current required by circuit 2 will affect the voltage at the terminals of circuit 1, due to the common impedances of the power supply lines and the internal source impedance of the power supply. Some improvement can be obtained by connecting the leads from circuit 2 closer to the power supply output terminals, thus decreasing the magnitude of the common line impedance. The coupling through the power supply's internal impedance still remains, however.

Electric and Magnetic Fields

Radiated electric and magnetic fields provide another means of noise coupling. All circuit elements including conductors radiate electromagnetic fields whenever charge is moved. In addition to this unintentional radiation there is the problem of intentional radiation from sources such as broadcast stations and radar transmitters. When the receiver is close to the source (near field), electric and

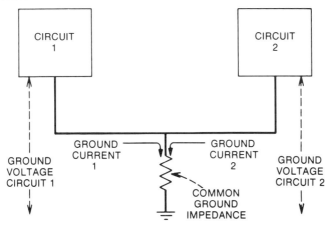

Fig. 7.5 When two circuits share a common ground, the ground voltage of each one is affected by the ground current of the other circuit.

magnetic fields are considered separately. When the receiver is far from the source (far field), the radiation is considered as combined electric and magnetic or electromagnetic radiation.[1]

7.1.4 Methods of Eliminating Interference

The primary methods available for combatting interference are:

1. Shielding
2. Grounding
3. Balancing
4. Filtering
5. Isolation
6. Separation and orientation
7. Circuit impedance level control
8. Cable design
9. Cancellation techniques (frequency or time domain).

This chapter covers items 1 and 2. Reference 1 covers all 9 items. Even with all these methods available, it should be remembered that noise usually cannot be eliminated; it can only be minimized to the point where it no longer causes interference.

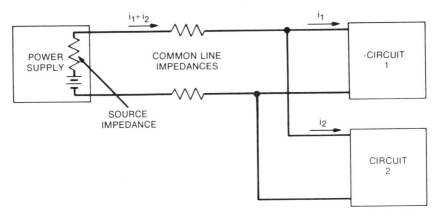

Fig. 7.6 When two circuits use a common power supply, current drawn by one circuit affects the voltage at the other circuit.

In all but the simplest cases, a single unique solution to the noise reduction problem may not exist. Compromises are generally required, and which of the many alternative solutions is the best can be the subject of considerable disagreement. Decisions on which techniques should be used in a specific case are things that must be determined by the system design engineer.

7.1.5 Summary

- Designing equipment that does not generate noise is as important as designing equipment that is not susceptible to noise.
- Noise suppression should be considered early in the design stage.
- Three items are necessary to produce a noise problem; there must be

    A noise source

    A coupling channel

    A susceptible receiver
- The three primary means of noise coupling are

    Conductive coupling

    Common impedance coupling

    Coupling by radiated electromagnetic fields
- A unique solution to most noise reduction problems does not always exist. There is usually more than one technique by which the noise objectives can be met.

7.2 SHIELDING OF CONDUCTORS

The two primary ways to minimize unwanted noise pickup are shielding and grounding. This section is devoted to the subject of shielding, and Section 7.3 covers grounding. The techniques of shielding and grounding are closely interrelated, however, and these two sections should be studied together as a single unit. In this section, for example, it is shown that a cable shield used to suppress electric fields should be grounded, but Section 7.3 explains where that ground should be made.

When properly used, shields can reduce the amount of noise coupling considerably. Shields may be placed around components, circuits, complete assemblies, or cables and transmission lines. This section is concerned with the shielding of conductors, although the same basic principles apply to any type of shielding. Reference 1 contains additional information on other types of shielding. In this section, the following three assumptions are made.

1. Shields are made of nonmagnetic materials and have a thickness much less than a skin depth at the frequency of interest.*
2. The receiver is not coupled so tightly to the source that it loads down the source.
3. Induced currents in the receiver signal circuit are small enough not to distort the original field. (This does not apply to a shield around the receiver circuit.)

To permit the problem of shielding to be studied, we shall represent the coupling between two circuits by lumped capacitance and inductance between the conductors. The circuit can then be analyzed by normal network theory.

Three types of coupling are considered. The first is capacitive, or electric, coupling, which is due to the interaction of electric fields between circuits. This type of coupling is commonly identified in the literature as electrostatic coupling, an obvious misnomer since the fields are not static. The second is inductive, or magnetic, coupling, which results from the interaction between the magnetic fields of two circuits. This type of coupling is commonly described as electromagnetic, again misleading terminology since no electric fields are involved. The third is a combination of electric and magnetic fields and is appropriately called electromagnetic coupling or radiation. The techniques developed to cope with electric coupling and magnetic coupling will normally, when used together, be appropriate for the electromagnetic case. For analysis in the near field, we normally consider the electric and magnetic fields separately, whereas the electromagnetic field case is considered when the problem is in the far field.† The circuit which is causing the interference is called the source, and the circuit being affected by the interference is called the receiver.

---

*If the shield is thicker than a skin depth, some additional shielding is present besides that calculated by methods in this section. This effect is discussed further in Ref. 1, Chapter 6.

† See Ref. 1, Chapter 6 for definitions of near and far fields.

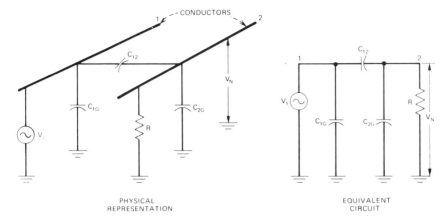

Fig. 7.7   Capacitive coupling between two conductors.

7.2.1   Capacitive Coupling

A simple representation of capacitive coupling between two conductors is shown in Fig. 7.7. Capacitance C_{12} is the stray capacitance between conductors 1 and 2. Capacitance C_{1G} is the capacitance between conductor 1 and ground, C_{2G} is the total capacitance between conductor 2 and ground, and R is the resistance of circuit 2 to ground. The resistance, R, results from the circuitry connected to conductor 2 and is not a stray component. Capacitance C_{2G} consists of both the stray capacitance of conductor 2 to ground and the effect of any circuitry connected to conductor 2.

The equivalent circuit of the coupling is also shown in Fig. 7.7. Consider the voltage V_1 on conductor 1 as the source of interference and conductor 2 as the affected circuit or receiver. Any capacitance connected directly across the source, such as C_{1G} in Fig. 7.7, can be neglected since it has no effect on the noise coupling. The noise voltage, V_N, produced between conductor 2 and ground can be expressed as

$$V_N = \frac{j\omega[C_{12}/(C_{12} + C_{2G})]}{j\omega + 1/R(C_{12} + C_{2G})} V_1 \tag{7.1}$$

Equation 7.1 does not show clearly how the pickup voltage depends on the various parameters. Equation 7.1 can be simplified for the case when R is a lower impedance than the impedance of the stray capacitance C_{12} plus C_{2G}. In most practical cases this will be true. Therefore, for

$$R \ll \frac{1}{j\omega(C_{12} + C_{2G})}$$

Eq. 7.1 can be reduced to

$$V_N = j\omega R C_{12} V_1 \tag{7.2}$$

This is the most important equation describing the capacitive coupling between two conductors, and it clearly shows how the pickup voltage depends on the parameters. Equation 7.2 shows that the noise voltage is directly proportional to the frequency ($\omega = 2\pi f$) of the noise source, the resistance, R, of the affected circuit to ground, the capacitance, C_{12}, between conductors 1 and 2, and the magnitude of the voltage, V_1.

Assuming that the voltage and frequency of the noise source cannot be changed, this leaves only two remaining parameters for reducing capacitive coupling. The receiver circuit can be operated at a lower resistance level, or capacitance C_{12} can be decreased. Capacitance C_{12} can be decreased by proper orientation of the conductors, by shielding (described in the next section), or by physically separating the conductors. If the conductors are moved farther apart, C_{12} decreases, thus decreasing the induced voltage on conductor 2.* The effect of conductor spacing on capacitive coupling is shown in Fig. 7.8. As a reference, 0 dB is the coupling when the conductors are separated by three

---

*The capacitance between two parallel conductors of diameter, d, and spaced D apart is $C_{12} = \pi\epsilon/\cosh^{-1}(D/d)$, (F/m). For $D/d > 3$, this reduces to $C_{12} = \pi\epsilon/\ln(2D/d)$, (F/m), where $\epsilon = 8.85 \times 10^{-12}$ farads per meter (F/m) for free space.

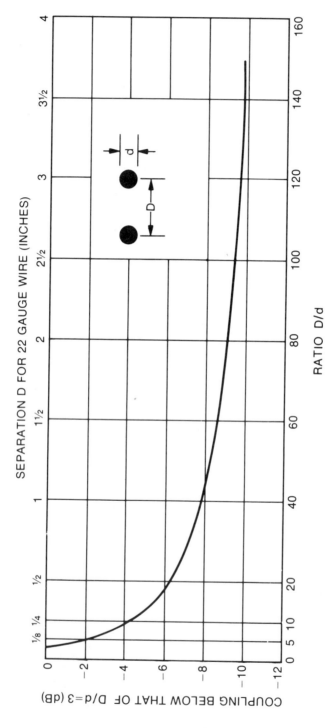

Fig. 7.8 Effect of conductor spacing on capacitive coupling. In the case of 22 gauge wire, most of the attenuation occurs in the first 25 mm of separation.

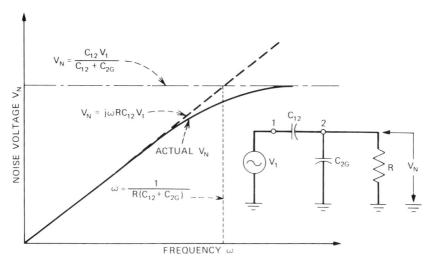

Fig. 7.9   Frequency response of capacitive coupled noise voltage.

times the conductor diameter. As can be seen in the figure, little additional attenuation is gained by spacing the conductors a distance greater than 40 times their diameter (25 mm or 1 in. in the case of 22 gauge wire).

If the resistance from conductor 2 to ground is large, such that

$$R \gg \frac{1}{j\omega(C_{12} + C_{2G})}$$

then Eq. 7.1 reduces to

$$V_N = \left(\frac{C_{12}}{C_{12} + C_{2G}}\right) V_1 \tag{7.3}$$

Under this condition, the noise voltage produced between conductor 2 and ground is due to the capacitive voltage divider C_{12} and C_{2G}. The noise voltage is independent of frequency and is of a larger magnitude than when R is small.

A plot of Eq. 7.1 versus ω is shown in Fig. 7.9. As can be seen, the maximum noise coupling is given by Eq. 7.3. The figure also shows that the actual noise voltage is always less than or equal to the value given by Eq. 7.2. At a frequency of

$$\omega = \frac{1}{R(C_{12} + C_{2G})} \tag{7.4}$$

Eq. 7.2 gives a value of noise that is 1.41 times the actual value. In almost all practical cases, the frequency is much less than this, and Eq. 7.2 applies.

7.2.2   Effect of Shield on Capacitive Coupling

First consider the case where the receiver (conductor 2) has infinite resistance to ground. If a shield is placed around conductor 2, the configuration becomes that of Fig. 7.10. An equivalent circuit of the capacitive coupling between conductors is included. The voltage picked up by the shield is

$$V_S = \left(\frac{C_{1S}}{C_{1S} + C_{SG}}\right) V_1 \tag{7.5}$$

Since there is no current flow through C_{2S} the voltage picked up by conductor 2 is

$$V_N = V_S \tag{7.6}$$

If the shield is grounded, the voltage $V_S = 0$, and the noise voltage V_N on conductor 2 is likewise reduced to zero. This case—where the center conductor does not extend beyond the shield—is an ideal situation and not typical.

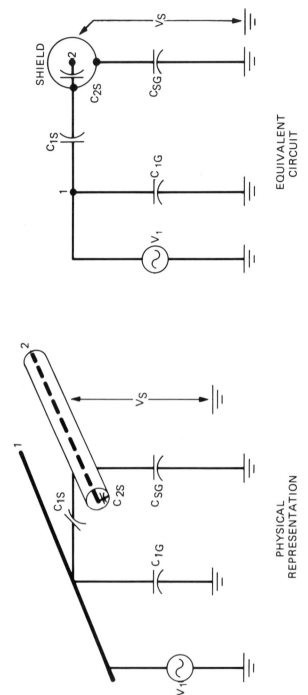

Fig. 7.10 Capacitive coupling with shield placed around receiver conductor.

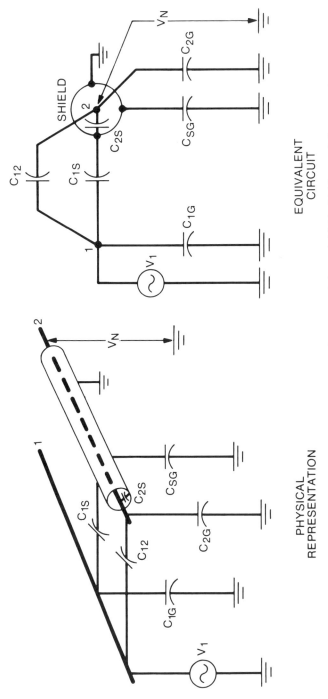

PHYSICAL REPRESENTATION

EQUIVALENT CIRCUIT

Fig. 7.11 Capacitive coupling when center conductor extends beyond shield; shield grounded at one point.

In practice, the center conductor normally does extend beyond the shield, and the situation becomes that of Fig. 7.11. There, C_{12} is the capacitance between conductor 1 and the shielded conductor 2, and C_{2G} is the capacitance between conductor 2 and ground. Both of these capacitances exist because the ends of conductor 2 extend beyond the shield. Even if the shield is grounded, there is a noise voltage coupled to conductor 2. Its magnitude is expressed as:

$$V_N = \frac{C_{12}}{C_{12} + C_{2G} + C_{2S}} V_1 \tag{7.7}$$

The value of C_{12}, and hence V_N in Eq. 7.7 depends on the length of conductor 2 that extends beyond the shield.

For good electric-field shielding, therefore, it is necessary to (1) *minimize the length of the center conductor that extends beyond the shield, and* (2) *provide a good ground on the shield.* A single ground connection makes a good shield ground, provided the cable is not longer than one-twentieth of a wavelength. On longer cables, multiple grounds may be necessary.

If in addition the receiving conductor has finite resistance to ground, the arrangement is that shown in Fig. 7.12. If the shield is grounded, the equivalent circuit can be simplified as shown in the figure. Any capacitance directly across the source can be neglected, since it has no effect on the noise coupling. The simplified equivalent circuit can be recognized as the same circuit analyzed in Fig. 7.7, provided C_{2G} is replaced by the sum of C_{2G} and C_{2S}. Therefore, if

$$R \ll \frac{1}{j\omega(C_{12} + C_{2G} + C_{2S})}$$

which is normally true, the noise voltage coupled to conductor 2 is

$$V_N = j\omega R C_{12} V_1 \tag{7.8}$$

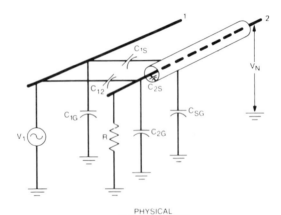

PHYSICAL
REPRESENTATION

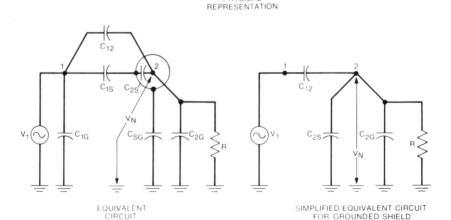

EQUIVALENT
CIRCUIT

SIMPLIFIED EQUIVALENT CIRCUIT
FOR GROUNDED SHIELD

Fig. 7.12 Capacitive coupling when receiving conductor has resistance to ground.

This is the same as Eq. 7.2, which is for an unshielded cable, except that C_{12} is greatly reduced by the presence of the shield. Capacitance C_{12} now consists primarily of the capacitance between conductor 1 and the unshielded portions of conductor 2. If the shield is braided, any capacitance that exists from conductor 1 to 2 through the holes in the shield must also be included in C_{12}.

7.2.3   Inductive Coupling

When a current, I, flows in a closed circuit, it produces a magnetic flux, ϕ, which is proportional to the current. The constant of proportionality is called the inductance, L. Hence we can write

$$\phi = LI \tag{7.9}$$

The inductance value depends on the geometry of the circuit and the magnetic properties of the medium containing the field. Inductance has meaning only for a closed circuit. However, at times we may talk about the inductance of only a portion of a circuit. In this case, we mean the contribution that a segment of the circuit makes to the total inductance of the closed circuit.

When current flow in one circuit produces a flux in a second circuit, there is a mutual inductance, M_{12}, between circuits 1 and 2 defined as

$$M_{12} = \frac{\phi_{12}}{I_1} \tag{7.10}$$

The symbol ϕ_{12} represents the flux in circuit 2 due to the current I_1 in circuit 1.

The voltage, V_N, induced in a closed loop of area $\overline{A}$ due to a magnetic field of flux density $\overline{B}$ can be shown[2] as

$$V_N = -\frac{d}{dt} \int_A \overline{B} \cdot \overline{A} \tag{7.11}$$

where $\overline{B}$ and $\overline{A}$ are vectors. If the closed loop is stationary and the flux density is sinusoidally varying with time but constant over the area of the loop, Eq. 7.11 reduces to

$$V_N = j\omega B A \cos \theta* \tag{7.12}$$

As shown in Fig. 7.13, A is the area of the closed loop, B is the rms value of the sinusoidally varying flux density of frequency ω radians per second, and V_N is the rms value of the induced voltage. This relationship can also be expressed in terms of the mutual inductance, M, between two circuits, as follows:

$$V_N = j\omega M I_1 = M \frac{di_1}{dt} \tag{7.13}$$

Equations 7.12 and 7.13 are the basic equations describing inductive coupling between two circuits. Figure 7.14 shows the inductive (magnetic) coupling between two circuits as described by Eq. 7.13. I_1 is the current in the interfering circuit, and M is the term that accounts for the

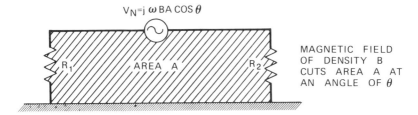

$V_N = j\omega BA \cos\theta$

AREA A

R_1 R_2

MAGNETIC   FIELD
OF   DENSITY   B
CUTS   AREA   A   AT
AN   ANGLE   OF θ

Fig. 7.13   Induced noise depends on the area enclosed by the disturbed circuit.

*Equation 7.12 is correct when the MKS system of units is being used. Flux density B is in webers per square meter (or tesla), and area A is in square meters. If B is expressed in gauss and A is in square centimeters (the CGS system), the right side of Eq. 7.12 must be multiplied by 10^{-8}.

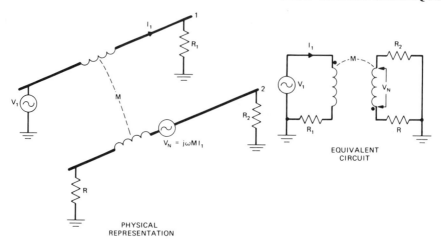

Fig. 7.14 Magnetic coupling between two circuits.

geometry and the magnetic properties of the medium between the two circuits. The presence of ω in Eqs. 7.12 and 7.13 indicates that the coupling is directly proportional to frequency. To reduce the noise voltage, B, A, or $\cos\theta$ must be reduced. The B term can be reduced by physical separation of the circuits or by twisting the source wires, provided the current flows in the twisted pair and not through the ground plane. The conditions necessary for this are covered in a later section. Under these conditions, twisting causes the B-fields from each of the wires to cancel. The area of the receiver circuit can be reduced by placing the conductor closer to the ground plane (if the return current is through the ground plane) or by using two conductors twisted together (if the return current is on one of the pair instead of the ground plane). The $\cos\theta$ term can be reduced by proper orientation of the source and receiver circuits.

It may be helpful to note some differences between magnetic and electric field coupling. First, reducing the impedance of the receiver circuit in a magnetically coupled situation does not decrease the pickup as it does in the case of electric field coupling. Second, in the case of magnetic field coupling, the noise voltage is produced in series with the receiver conductors, whereas in the case of electric field coupling, the noise voltage is produced between the receiver conductor and ground.

If an ungrounded and nonmagnetic shield is now placed around conductor 2, the circuit becomes that of Fig. 7.15, where M_{1S} is the mutual inductance between conductor 1 and the shield. Since the shield has no effect on the geometry or magnetic properties of the medium between circuits 1 and 2 it has no effect on the voltage induced into conductor 2. The shield does, however, pick up a voltage due to the current in conductor 1.

$$V_S = j\omega M_{1S} I_1 \qquad (7.14)$$

A ground connection on one end of the shield does not change the situation. *It can, therefore, be concluded that a shield placed around a conductor and grounded at one end has no effect on the magnetically induced voltage in that conductor.*

7.2.4 Magnetic Coupling between Shield and Inner Conductor

Before continuing the discussion of inductive coupling, it will be necessary to calculate the magnetic coupling between a hollow conducting tube and any conductors placed inside the tube. This concept is fundamental to a discussion of inductive shielding and will be needed later.

First consider the magnetic field produced by a tubular conductor carrying a uniform axial current, as shown in Fig. 7.16. If the hole in the tube is concentric with the outside of the tube, there is no magnetic field in the cavity and the total magnetic field is external to the tube.[3] Now let a conductor be placed inside the tube to form a coaxial cable, as shown in Fig. 7.17. All of the flux, ϕ, due to the current, I_S, in the shield tube encircles the inner conductor. The inductance of the shield is equal to

$$L_S = \frac{\phi}{I_S} \qquad (7.15)$$

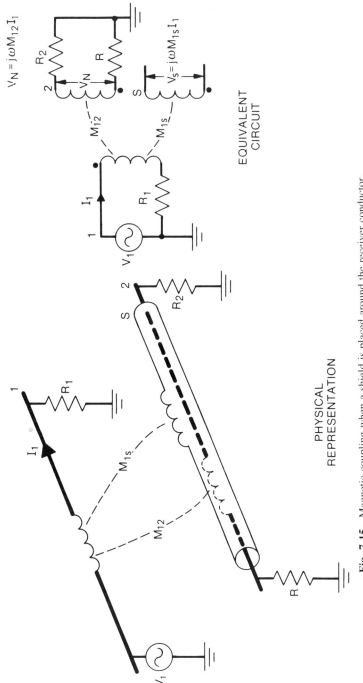

Fig. 7.15 Magnetic coupling when a shield is placed around the receiver conductor.

229

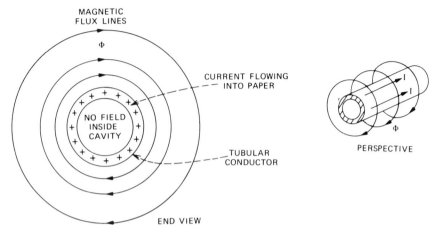

Fig. 7.16 Magnetic field produced by current in a tubular conductor.

The mutual inductance between the shield and the inner conductor is equal to

$$M = \frac{\phi}{I_S} \qquad (7.16)$$

Since all the flux produced by the shield current encircles the center conductor, the flux in these two equations is the same. The mutual inductance between the shield and center conductor is therefore equal to the self inductance of the shield

$$M = L_S \qquad (7.17)$$

Equation 7.17 is a most important result and one that we will often have occasion to refer to. It was derived to show that the mutual inductance between the shield and the center conductor is equal to the shield inductance. Based on the reciprocity of mutual inductance,[2] the inverse must also be true. That is, the mutual inductance between the center conductor and the shield is equal to the shield inductance.

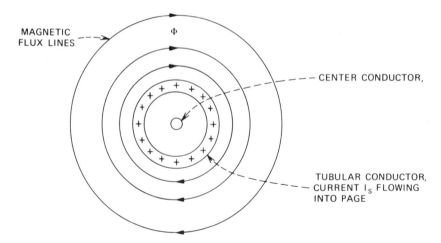

Fig. 7.17 Coaxial cable with shield current flowing.

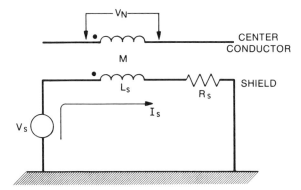

Fig. 7.18 Equivalent circuit of shielded conductor.

The validity of Eq. 7.17 depends only on the fact that there is no magnetic field in the cavity of the tube due to shield current. The requirements for this to be true are that the tube be cylindrical and the current density be uniform around the circumference of the tube. Equation 7.17 applies regardless of the position of the center conductor within the tube. In other words, the two conductors do not have to be coaxial.

The voltage, V_N, induced into the center conductor due to a current, I_S, in the shield can now be calculated. Assume that the shield current is produced by a voltage, V_S, induced into the shield from some other circuit. Figure 7.18 shows the circuit being considered; L_S and R_S are the inductance and resistance of the shield. The voltage, V_N, is equal to

$$V_N = j\omega M I_S \tag{7.18}$$

The current, I_S, is equal to

$$I_S = \frac{V_S}{L_S}\left(\frac{1}{j\omega + R_S/L_S}\right) \tag{7.19}$$

Therefore

$$V_N = \left(\frac{j\omega M V_S}{L_S}\right)\left(\frac{1}{j\omega + R_S/L_S}\right) \tag{7.20}$$

Since $L_S = M$ (from Eq. 7.17),

$$V_N = \left(\frac{j\omega}{j\omega + R_S/L_S}\right)V_S \tag{7.21}$$

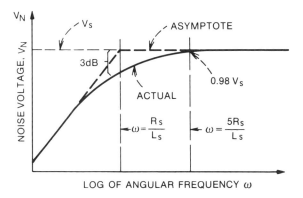

Fig. 7.19 Noise voltage in center conductor of coaxial cable due to shield current.

TABLE 7.1   MEASURED VALUES OF SHIELD CUTOFF FREQUENCY f_c

Cable	Impedance (Ω)	Cutoff frequency (kHz)	Five times cutoff frequency (kHz)	Remarks
Coaxial cable				
RG-6A	75	0.6	3.0	Double-shielded
RG-213	50	0.7	3.5	
RG-214	50	0.7	3.5	Double-shielded
RG-62A	93	1.5	7.5	
RG-59C	75	1.6	8.0	
RG-58C	50	2.0	10.0	
Shielded twisted pair				
754E	125	0.8	4.0	Double-shielded
24 Ga.	—	2.2	11.0	
22 Ga.[a]	—	7.0	35.0	Aluminum-foil shield
Shielded single				
24 Ga.	—	4.0	20.0	

[a]One pair out of an 11 pair cable (Belden 8775)[4].

A plot of Eq. 7.21 is shown in Fig. 7.19. The break frequency for this curve is defined as the shield cutoff frequency (ω_c) and occurs at

$$\omega_c = \frac{R_S}{L_S} \quad \text{or} \quad f_c = \frac{R_S}{2\pi L_S} \tag{7.22}$$

The noise voltage induced into the center conductor is zero at dc and increases to almost V_S at a frequency of 5 R_S/L_S rad/s. Therefore, if shield current is allowed to flow, a voltage is induced into the center conductor that nearly equals the shield voltage at frequencies greater than five times the shield cutoff frequency.

This is a very important property of a conductor inside a shield. Measured values of the shield cutoff frequency and five times this frequency are tabulated in Table 7.1 for various cables. For most cables, five times the shield cutoff frequency is near the high end of the audio-frequency band. The aluminum-foil-shielded cable listed has a much higher shield cutoff frequency than any other. This is due to the increased resistance of its thin aluminum-foil shield.

7.2.5   Shielding to Prevent Magnetic Radiation

To prevent radiation, the source of the interference may be shielded. Figure 7.20 shows the electric and magnetic fields surrounding a current-carrying conductor located in free space. If a shield grounded at one point is placed around the conductor, the electric field lines will terminate on the shield but there will be very little effect on the magnetic field. This is shown in Fig. 7.21. If a shield current equal and opposite to that in the center conductor is made to flow on the shield, it generates

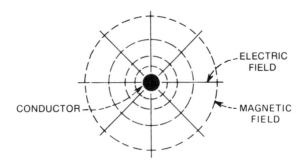

Fig. 7.20   Fields around a current-carrying conductor.

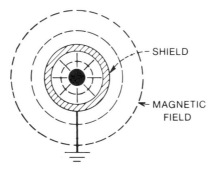

Fig. 7.21 Fields around shielded conductor; shield grounded at one point.

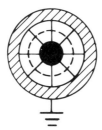

Fig. 7.22 Fields around shielded conductor; shield grounded and carrying a current equal to the conductor current, but in the opposite direction.

an equal but opposite external magnetic field. This field cancels the magnetic field caused by the center conductor external to the shield. This results in the condition shown in Fig. 7.22, with no fields external to the shield.

Figure 7.23 shows a circuit which is grounded at both ends and carries a current I_1. To prevent magnetic field radiation from this circuit the shield must be grounded at both ends and the return current must flow from A to B in the shield (I_S in the figure) instead of in the ground plane (I_G in the figure). But why should the current return from point A to B through the shield instead of through the zero-resistance ground plane? The equivalent circuit can be used to analyze this configuration. By writing a mesh equation around the ground loop ($A - R_S - L_S - B - A$) the shield current I_S can be determined:

$$0 = I_S (j\omega L_S + R_S) - I_1(j\omega M) \tag{7.23}$$

where M is the mutual inductance between the shield and center conductor and as previously shown (Eq. 7.17), $M = L_S$. Making this substitution and rearranging produces this expression for I_S.

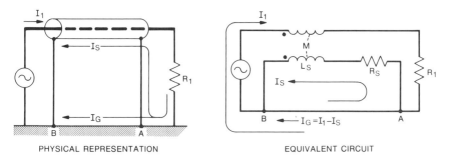

PHYSICAL REPRESENTATION   EQUIVALENT CIRCUIT

Fig. 7.23 Division of current between shield and ground plane.

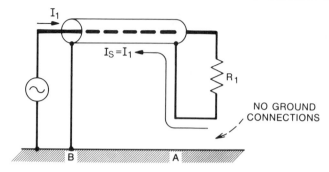

Fig. 7.24 Without ground at far end, all return current flows through shield.

$$I_S = I_1 \left(\frac{j\omega}{j\omega + R_S/L_S} \right) = I_1 \left(\frac{j\omega}{j\omega + \omega_c} \right) \tag{7.24}$$

As can be seen from the above equation, if the frequency is much above the shield cutoff frequency, ω_c, the shield current approaches the center conductor current. Because of the mutual inductance between the shield and center conductor, the shield provides a return path with lower total circuit inductance than the ground plane at high frequency. As the frequency decreases below $5\omega_c$, the cable provides less and less magnetic shielding since more of the current returns via the ground plane.

To prevent radiation of a magnetic field from a conductor grounded at both ends, it should be shielded and the shield should be grounded at both ends. This provides good magnetic field shielding at frequencies considerably above the shield cutoff frequency. This reduction in the radiated magnetic field is not because of the magnetic shielding properties of the shield as such. Rather, the return current on the shield generates a field that cancels the conductor's field.

If the ground is removed from one end of the circuit, as shown in Fig. 7.24, then the shield should not be grounded at that end since the return current must now all flow on the shield. This is true especially at frequencies less than the shield cutoff frequency. Grounding both ends of the shield, in this case, reduces the shielding since some current would return via the ground plane.

7.2.6   Shielding a Receiver against Magnetic Fields

The best way to protect against magnetic fields at the receiver is to decrease the area of the receiver loop. The area of interest is the total area enclosed by current flow in the receiver circuit. An important consideration is the path taken by the current in returning to the source. Quite often the current returns by a path other than the one intended by the designer and, therefore, the area of the loop changes. If a nonmagnetic shield placed around a conductor causes the current to return over a path that encloses a smaller area, then some protection against magnetic fields will have been provided by the shield. This protection, however, is due to the reduced loop area and not to any magnetic shielding properties of the shield.

Figure 7.25 illustrates the effect of a shield on the loop area of a circuit. In Fig. 7.25*a*, the source, V_S, is connected to the load, R_L, by a single conductor, using a ground return path. The area enclosed by the current is the rectangle between the conductor and the ground plane. In Fig. 7.25*b*, a shield is placed around the conductor and grounded at both ends. If the current returns through the shield rather than the ground plane, the area of the loop is decreased and a degree of magnetic protection is provided. The current returns through the shield if the frequency is greater than five times the shield cutoff frequency as previously shown. A shield placed around the conductor and grounded at one end only as shown in Fig. 7.25*c* does not change the loop area, and therefore, provides no magnetic protection.

The arrangement of Fig. 7.25*b* does not protect against magnetic fields at frequencies below the shield cutoff frequency since then most of the current returns through the ground plane and not through the shield. This circuit should be avoided at low frequencies for two other reasons: (1) since the shield is one of the circuit conductors, any noise current in it will produce an IR drop in the shield and appear to the circuit as a noise voltage, and (2) if there is a difference in ground potential between the two ends of the shield, this too will show up as a noise voltage in the circuit.*

---

* See Section 7.3.11 for further discussion of a shielded cable grounded at both ends.

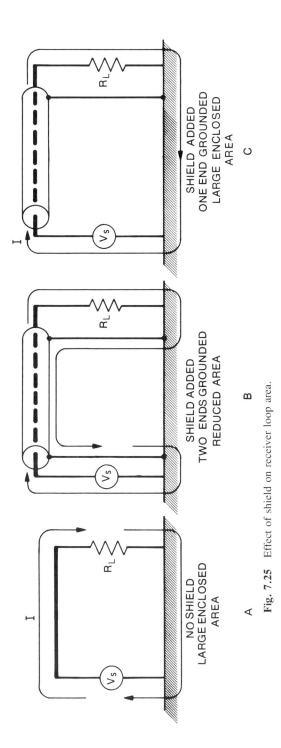

Fig. 7.25 Effect of shield on receiver loop area.

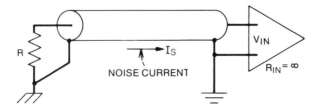

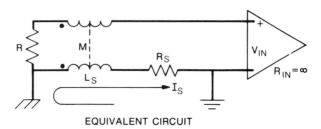

PHYSICAL REPRESENTATION

EQUIVALENT CIRCUIT

Fig. 7.26 Effect of noise current flowing in the shield of a coaxial cable.

    Whenever a circuit is grounded at both ends, only a limited amount of magnetic field protection is possible because of the large noise current induced in the ground loop. Since this current flows through the signal conductor, a noise voltage is produced in the shield, equal to the shield current times the shield resistance. This is shown in Fig. 7.26. The current, I_S, is the noise current in the shield due to a ground differential or to external noise coupling. If voltages are added around the input loop, the following expression is obtained:

$$V_{IN} = -j\omega M I_S + j\omega L_S I_S + R_S I_S \tag{7.25}$$

Since $L_S = M$, as was previously shown

$$V_{IN} = R_S I_S \tag{7.26}$$

Whenever shield current flows a noise voltage is produced in the shield due to the $I_S R_S$ voltage drop.

    Even if the shield is grounded at only one end, shield noise currents may still flow due to capacitive coupling to the shield. *Therefore, for maximum noise protection at low frequencies, the shield should not be one of the signal conductors, and one end of the circuit must be isolated from ground.*

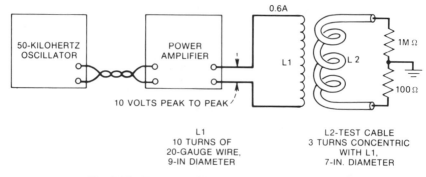

Fig. 7.27 Test setup of inductive coupling experiment.

7.2.7 Experimental Data

The magnetic field shielding properties of various cable configurations were measured and compared. The test setup is shown in Fig. 7.27, and the test results are tabulated in Figs. 7.28 and 7.29. The frequency (50 kHz) is greater than five times the shield cutoff frequency for all the cables tested. The cables shown in Figs. 7.28 and 7.29 represent test cables shown as $L2$ in Fig. 7.27.

In circuits (a) through (f) (Fig. 7.28), both ends of the circuit are grounded. They provide much less magnetic field attenuation than do circuits (g) through (k) (Fig. 7.29), where only one end is grounded.

Circuit (a) in Fig. 7.28 provides essentially no magnetic field shielding. The actual noise voltage measured across the one megaohm resister in this case was 0.8 V. The pickup in configuration (a) is used as a reference, and is called 0 dB, to compare the performance of all the other circuits. In circuit (b), the shield is grounded at one end; this has no effect on the magnetic shielding. Grounding the shield at both ends as in configuration (c) provides some magnetic field protection because the

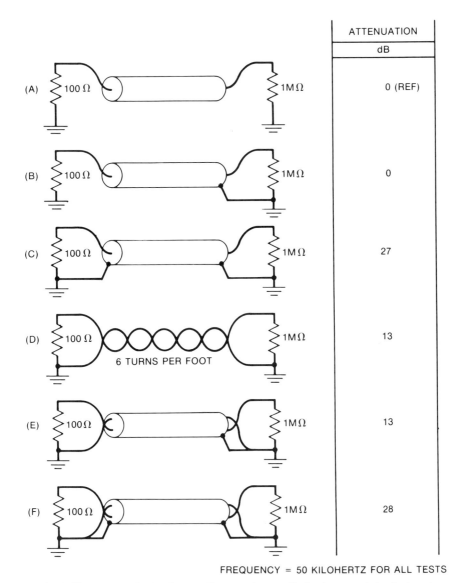

FREQUENCY = 50 KILOHERTZ FOR ALL TESTS

Fig. 7.28 Results of inductive coupling experiment; all circuits grounded at both ends.

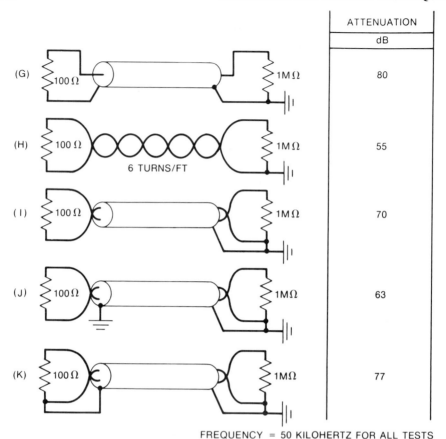

FREQUENCY = 50 KILOHERTZ FOR ALL TESTS

Fig. 7.29   Results of inductive coupling experiment; all circuits are grounded at one end only.

frequency is above the shield cutoff frequency. The protection would be even greater if it were not for the ground loop formed by grounding both ends of the circuit. The magnetic field induces a large noise current into the low impedance ground loop consisting of the cable shield and the two ground points. The shield noise current then produces a noise voltage in the shield, as was shown in the preceding section.

Use of a twisted pair as in circuit (d) should provide much greater magnetic field noise reduction, but its effect is defeated by the ground loop formed by circuit grounds at both ends. This can clearly be seen by comparing the attenuation of circuit (h) to that of circuit (d). Adding a shield with one end grounded to the twisted pair as in (e) has no effect. Grounding the shield at both ends as in (f) provides additional protection, since the low-impedance shield shunts some of the magnetically induced ground-loop current away from the signal conductors. In general, however, none of the circuit configurations in Fig. 7.28 provide good magnetic field protection because of the ground loops. If the circuit must be grounded at both ends, configurations (c) or (f) should be used.

Circuit (g) shows a significant improvement in magnetic field shielding. This is due to the very small loop area formed by the coaxial cable and the fact that there is no ground loop to defeat the shielding. The coax provides a very small loop area since the shield can be represented by an equivalent conductor located on its center axis. This effectively locates the shield at or very near the axis of the center conductor.

It was expected that the twisted pair of circuit (h) would provide considerably more shielding than the 55 dB shown. The reduced shielding is due to the fact that some electric-field coupling is now beginning to show up. This can be seen in circuit (i), where attenuation increases to 70 dB by placing a shield around the twisted pair. The fact that attenuation in circuit (g) is better than (i) indicates that in this case the particular coaxial cable presents a smaller loop area to the magnetic field than does the twisted pair. This, however, is not necessarily true in general. Increasing the number of turns per foot for either of the twisted pairs ((h) or (i)) would reduce the pickup. In general, circuit (*i*) is preferred over circuit (*g*) for low-frequency magnetic shielding since in (*i*) the shield is not also one of the signal conductors.

Grounding both ends of the shield as in circuit (*j*) decreases the shielding slightly. This is probably due to the high shield current in the ground loop formed by the shield inducing unequal voltages in the two center conductors. Circuit (*k*) provides more shielding than (*i*) since it combines the features of the coax (*g*) with those of the twisted pair (*i*). Circuit (*k*) is not normally desirable since any noise voltages or currents that do get on the shield can flow down the signal conductor. It is almost always better to connect the shield and signal conductors together at just one point. That point should be such that noise current from the shield does not have to flow down the signal conductor to get to ground.

7.2.8 Coaxial Cable Versus Shielded Twisted Pair

When comparing coaxial cable with a shielded twisted pair, it is important to recognize the usefulness of both types of cable from a propagation point of view, irrespective of their shielding characteristics. This is shown in Fig. 7.30. Shielded twisted pairs are very useful at frequencies below 100 kHz. In some applications, the frequency may reach as high as 10 MHz. Above 1 MHz, the losses in the shielded twisted pair increase considerably.

On the other hand, coaxial cable has a more uniform characteristic impedance with lower losses. It is useful, therefore, from zero frequency (DC) up to VHF frequencies, with some applications extending up to UHF. Above a few hundred megahertz, the losses in coaxial cable become large, and waveguides become more practical. A shielded twisted pair has more capacitance than a coaxial cable and, therefore, is not as useful at high frequencies or in high-impedance circuits.

A coaxial cable grounded at one point provides a good degree of protection from capacitive pickup. But if a noise current flows in the shield, a noise voltage is produced. Its magnitude is equal to the shield current times the shield resistance. Since the shield is part of the signal path, this noise voltage appears as noise in series with the input signal. A double-shielded, or triaxial, cable with insulation between the two shields can eliminate the noise produced by the shield resistance. The noise current flows in the outer shield, and the signal current flows in the inner shield. The two currents (signal and noise), therefore, do not flow through a common impedance.

Unfortunately, triaxial cables are expensive and awkward to use. A coaxial cable at high frequencies, however, acts as a triaxial cable due to skin effect. For a typical shielded cable, skin effect becomes important at about 1 MHz. The noise current flows on the outside surface of the shield while the signal current flows on the inside surface. For this reason a coaxial cable is better for use at high frequencies.

A shielded twisted pair has characteristics similar to a triaxial cable and is not as expensive or awkward. The signal current flows in the two inner conductors, and any noise currents flow in the shield. Common-resistance coupling is eliminated. In addition, any shield current is coupled equally into both inner conductors by mutual inductance, and the voltages therefore cancel.

An unshielded twisted pair, unless it is balanced, provides very little protection against capacitive pickup, but it is very good for protection against magnetic pickup. The shielded twisted pair provides the best shielding for low-frequency signals, in which magnetic pickup is the major problem. The effectiveness of twisting increases as the number of twists per unit length increases.

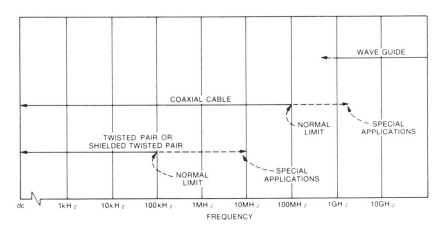

Fig. 7.30 Useful frequency range for various transmission lines.

7.2.9 Braided Shields

Most cables are actually shielded with braid rather than with a solid conductor. The advantages of braid are flexibility, durability, strength, and long flex life. Braids, however, typically provide only 60–90% coverage and are less effective as shields than solid conductors. Braided shields usually provide just slightly reduced electric field shielding (except at UHF frequencies) but greatly reduced magnetic field shielding. The reason is that braid distorts the uniformity of the shield current. A braid is typically from 5 to 30 dB less effective than a solid shield for protecting against magnetic fields.

At higher frequencies, the effectiveness of the braid decreases further. This is because the braid holes become larger compared to a wavelength as the frequency increases. Multiple shields offer more protection, but with higher cost and less flexibility. Cables with double or even triple shields are used in some critical applications. Recently, cables with solid aluminum foil shields have become available. These shields provide almost 100% coverage and more effective shielding. They are not as strong as braid, however, and usually have a higher shield cutoff frequency due to their higher shield resistance.

7.2.10 Uniformity of Shield Current

The magnetic shielding previously discussed depends on a uniform distribution of the longitudinal shield current around the shield circumference. Solid shields such as aluminum foil produce the most uniform shield current distribution and, therefore, provide the best magnetic shielding if the frequency is above the shield cutoff frequency. Braided shields are considerably less effective for magnetic shielding since their current distribution is less uniform than that of a solid shield. The braid can be plated—typically with solder or silver—and current flow is more uniform due to better conductor-to-conductor contact. Unplated shields tend to build up an oxide coating and have poor electrical contact between individual braid wires.

The magnetic shielding effectiveness near the ends of the cable depends on the way the braid is terminated. A pigtail connection, Fig. 7.31, causes the shield current to be concentrated on one side of the shield. For maximum protection, the shield should be terminated uniformly around its cross section. This can be accomplished by using a coaxial connector such as the BNC, UHF, or Type N connectors. Such a connector provides 360° electrical contact to the shield. A coaxial termination also provides complete coverage of the inner conductor, preserving the integrity of electric field shielding.

7.2.11 Summary

- Electric fields are much easier to guard against than magnetic fields.
- The use of nonmagnetic shields around conductors provides no magnetic shielding per se.
- A shield grounded at one or more points shields against electric fields.
- The key to magnetic shielding is to decrease the area of the loop. To do that use twisted pair, or use coaxial cable if the current return is through the shield instead of in the ground plane.
- For a coaxial cable grounded at both ends, virtually all of the return current flows in the shield at frequencies above five times the shield cutoff frequency.
- To prevent radiation from a conductor, a shield grounded at both ends is useful above the shield cutoff frequency.

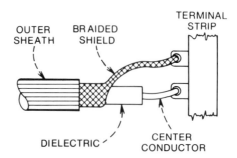

Fig. 7.31 Pigtail shield connection concentrates current on one side of shield.

- Only a limited amount of magnetic shielding is possible in a receiver circuit that is grounded at both ends, due to the ground loop formed.
- Any shield in which noise currents flow should not be part of the signal path. Use shielded twisted pair or triaxial cable at low frequencies.
- At high frequencies, a coaxial cable acts as a triaxial cable due to skin effect.
- The shielding effectiveness of a twisted pair increases as the number of twists per unit length increases.
- The magnetic shielding effects listed previously require a cylindrical shield with uniform distribution of shield current over the circumference of the shield.

7.3   GROUNDING

Grounding is one of the primary ways to minimize unwanted noise and pickup. Proper use of grounding and shielding, in combination, can solve a large percentage of all noise problems. A good grounding system must be designed just like the rest of the circuit. It is sometimes difficult to convince one's self that expensive engineering time should be spent on the minute details of where every circuit should be grounded. However in the long run, time and money are usually saved by not having to solve mysterious interference problems that turn up when the equipment is later built and tested.

The grounding principles covered here are just as applicable to large complex electronic systems as they are to individual circuits on a single printed wiring board. There are two basic objectives involved in designing good grounding systems. The first is to *minimize the noise voltage generated by currents from two or more circuits flowing through a common ground impedance*. The second is to *avoid creating ground loops which are susceptible to magnetic fields and differences in ground potential*. Grounding, if done improperly however, can become a primary means of noise coupling.

In the most general sense a ground can be defined as an equipotential* point or plane which serves as a reference voltage for a circuit or system. It may or may not be at earth potential. If the ground is connected to the earth through a low impedance path, it can then be called an earth ground. There are two common reasons for grounding a circuit: (1) for safety, and (2) to provide an equipotential reference for signal voltages. Safety grounds are always at earth potential, whereas signal grounds are usually but not necessarily at earth potential. In many cases, a safety ground is required at a point which is unsuitable for a signal ground, and this may complicate the noise problem.

7.3.1   Safety Grounds

Safety considerations require the chassis or enclosure for electric equipment to be grounded. Why this is so can be seen in Fig. 7.32. In the left-hand diagram, Z_1 is the stray impedance between a point at potential V_1 and the chassis, and Z_2 is the stray impedance between the chassis and ground. The potential of the chassis is determined by impedances Z_1 and Z_2 acting as a voltage divider. The chassis potential is

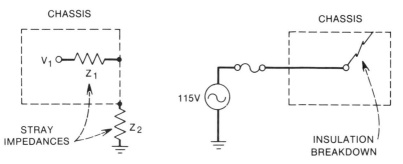

Fig. 7.32 Chassis should be grounded for safety. Otherwise, it may reach a dangerous voltage level through stray impedances (left) or insulation breakdown (right).

*A point where the voltage does not change regardless of the amount of current supplied to it or drawn from it.

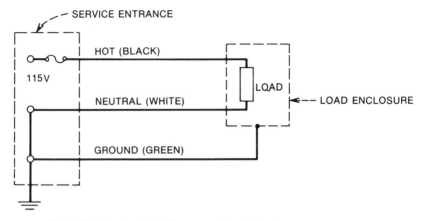

Fig. 7.33   Standard 115 V_{AC} power distribution circuit has three leads.

$$V_{\text{chassis}} = \left(\frac{Z_2}{Z_1 + Z_2} \right) V_1 \qquad (7.27)$$

The chassis could be a relatively high potential and be a shock hazard, since its potential is determined by the relative values of the stray impedances over which there is very little control. If the chassis is grounded, however, its potential is zero since Z_2 becomes zero.

The right-hand diagram of Fig. 7.32 shows a second and far more dangerous situation: a fused AC line entering an enclosure. If there should be an insulation breakdown such that the AC line comes in contact with the chassis, the chassis would then be capable of delivering the full current capacity of the fused circuit. Anyone coming in contact with the chassis and ground would be connected directly across the AC power line. If the chassis is grounded, however, such an insulation breakdown will draw a large current from the AC line and cause the fuse to blow, thus removing the voltage from the chassis.

In the United States, AC power distribution and wiring standards are contained in the National Electrical Code.[5] One requirement of this code specifies that 115 V_{AC} power distribution in homes and buildings must be a three-wire system, as shown in Fig. 7.33. Load current flows through the hot (black) wire, which is fused, and returns through the neutral (white) wire. In addition, a safety ground (green) wire must be connected to all equipment enclosures and hardware. The only time the green wire carries current is during a fault, and then only momentarily until the fuse or breaker opens the circuit. Since no load current flows in the safety ground, it has no IR drop and the

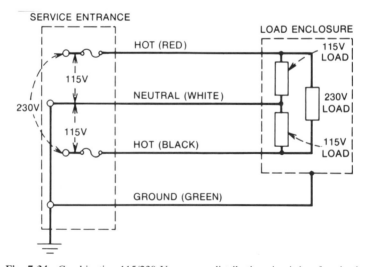

Fig. 7.34   Combination 115/230 V_{AC} power distribution circuit has four leads.

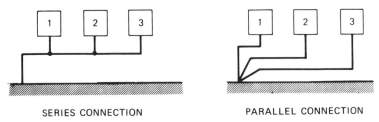

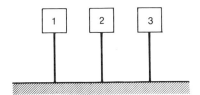

Fig. 7.35   Two types of single-point grounding connections.

Fig. 7.36   Multipoint grounding connections.

enclosures connected to it are always at ground potential. The National Electrical Code[6] specifies that the neutral and safety ground shall be connected together at only one point, and this point shall be at the main service entrance. To do otherwise would allow some of the neutral current to return on the ground conductor. A combination 115/230 V system is similar, except an additional hot wire (red) is added, as shown in Fig. 7.34. If the load requires only 230 V, the neutral (white) wire shown in Fig. 7.34 is not required.

7.3.2   Signal Grounds

Signal grounds generally fall into one of two classes: (1) single point grounds, and (2) multipoint grounds. These schemes are shown in Figs. 7.35 and 7.36. There are two subclasses of single point grounds—those with series connections and those with parallel connections. The series connection is also called a common ground system, and the parallel connection is called a separate ground system.
In the following discussion of grounding techniques, two key points should be kept in mind:

1. All conductors have a finite impedance, generally consisting of both resistance and inductance. At 11 kHz, a straight length of 22-gauge wire one inch above a ground plane has more inductive reactance than resistance.

2. Two physically separated ground points are seldom at the same potential.

The AC power ground is of little practical value as a signal ground. The voltage measured between two points on the power ground is typically hundreds of millivolts, and in some cases many volts. This is excessive for low-level signal circuits. A single-point connection to the power ground is usually required for safety, however.

7.3.3   Single-Point Ground Systems

From a noise point of view, the most undesirable ground system is the common ground system shown in Fig. 7.37. This is a series connection of all the individual circuit grounds. The resistances shown represent the impedance of the ground conductors and I_1, I_2, and I_3 are the ground currents of circuits 1, 2, and 3, respectively. Point A is not at zero potential but is at a potential of

$$V_A = (I_1 + I_2 + I_3)R_1 \tag{7.28}$$

and point C is at a potential of

$$V_C = (I_1 + I_2 + I_3)R_1 + (I_2 + I_3)R_2 + I_3R_3 \tag{7.29}$$

Although this circuit is the least desirable grounding system, it is probably the most widely used because of its simplicity. For noncritical circuits it may be perfectly satisfactory. This system should

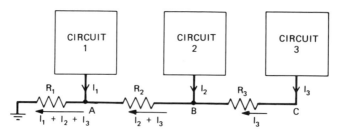

Fig. 7.37 Common ground system is a series ground connection and is undesirable from a noise standpoint but has the advantage of simple wiring.

not be used between circuits operating at widely different power levels, since the high-level stages produce large ground currents which in turn adversely affect the low-level stage. When this system is used, the most critical stage should be the one nearest the primary ground point. Note that point A in Fig. 7.37 is at a lower potential than point B or C.

The separate ground system (parallel connection) shown in Fig. 7.38 is the most desirable at low frequencies. That is because there is no cross coupling between ground currents from different circuits. The potentials at points A and C, for example, are as follows:

$$V_A = I_1 R_1 \tag{7.30}$$

$$V_C = I_3 R_3 \tag{7.31}$$

The ground potential of a circuit is now a function of the ground current and impedance of that circuit only. This system is mechanically cumbersome, however, since in a large system an unreasonable amount of wire is necessary.

A second limitation of the separate ground system occurs at high frequencies, where the inductances of the ground conductors increase the ground impedance and also produce inductive coupling between the ground leads. Parasitic capacitance between the ground conductors also allows coupling between the grounds. At still higher frequencies the impedance of the ground wires can be very high if the length coincides with odd multiples of a quarter wavelength. Not only will these grounds have large impedance, but they will also act as antennas and radiate noise. Ground leads should always be kept shorter than one-twentieth of a wavelength to prevent radiation and to maintain a low impedance.

7.3.4 Multipoint Ground Systems

The multipoint ground system is used at high frequencies to minimize the ground impedance. In this system, shown in Fig. 7.39, circuits are connected to the nearest available low-impedance ground plane, usually the chassis. The low ground impedance is due primarily to the lower inductance

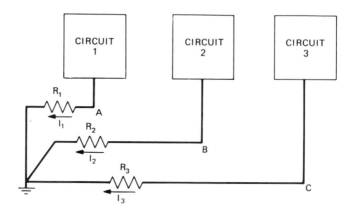

Fig. 7.38 Separate ground system is a parallel ground connection and provides good low-frequency grounding, but is mechanically cumbersome.

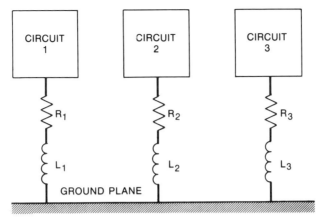

Fig. 7.39 Multipoint ground system is good choice at frequencies above 10 MHz. Impedances $R_1 - R_3$ and $L_1 - L_3$ should be minimized.

of the ground plane. The connections between each circuit and the ground plane should be kept as short as possible to minimize their impedance. In very high frequency circuits the length of these ground leads must be kept to a small fraction of an inch. Multipoint grounds should be avoided at low frequencies since ground currents from all circuits flow through a common ground impedance — the ground plane. At high frequencies, the common impedance of the ground plane can be reduced by silver-plating the surface. Increasing the thickness of the ground plane has no effect on its high-frequency impedance, since current flows only on the surface due to skin effect.

Normally at frequencies below one megahertz a single-point ground system is preferable; above 10 MHz, a multipoint ground system is best. Between 1 and 10 MHz a single-point ground can usually be used provided the length of the longest ground conductor is less than one-twentieth of a wavelength. If it is greater than one-twentieth of a wavelength a multipoint ground system should be used.

7.3.5 Practical Low-Frequency Grounding

Most practical grounding systems at low frequencies are a combination of the series and parallel single-point ground. Such a combination is a compromise between the need to meet the electrical

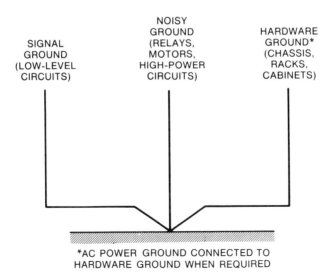

Fig. 7.40 These three classes of grounding connections should be kept separate to avoid noise coupling.

noise criteria and the goal of avoiding more wiring complexity than necessary. The key to balancing these factors successfully is to group ground leads selectively, so that circuits of widely varying power and noise levels do not share the same ground return wire. Thus, several low-level circuits may share a common ground return, while other high-level circuits share a different ground return conductor.

Most systems require a minimum of three separate ground returns, as shown in Fig. 7.40. The signal ground used for low-level electronic circuits should be separated from the "noisy" ground used for circuits such as relays and motors. A third "hardware" ground should be used for mechanical enclosures, chassis, racks, and so on. If AC power is distributed throughout the system, the power ground (green wire) should be connected to the hardware ground. The three separate ground return circuits should be connected together at only one point. Use of this basic grounding configuration in all equipment would greatly minimize grounding problems.

An illustration of how these grounding principles might be applied to a nine-track digital tape recorder is shown in Fig. 7.41. There are three signal grounds, one noisy ground, and one hardware ground. The most sensitive circuits, the nine read amplifiers, are grounded by using two separate ground returns. Five amplifiers are connected to one, and four are connected to the other. The nine write amplifiers, which operate at a much higher level than the read amplifiers, and the interface and control logic are connected to a third ground return. The three DC motors and their control circuits, the relays, and the solenoids are connected to the noisy ground. Of these elements, the capstan motor control circuit is the most sensitive; it is properly connected closest to the primary ground point. The hardware ground provides the ground for the enclosure and housing. The signal grounds, noisy ground, and hardware ground should be connected together only at the source of primary power, that is, the power supply.

When designing the grounding system for a piece of equipment, a block diagram similar to Fig. 7.41 can be very useful in determining the proper interconnection of the various circuit grounds.

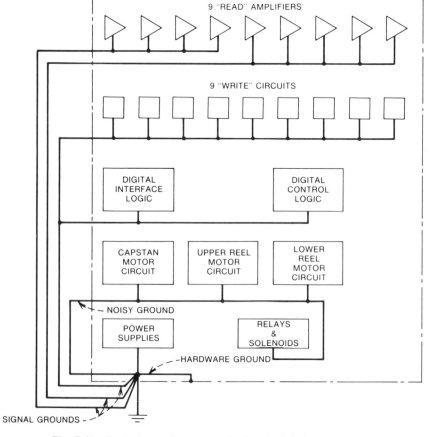

Fig. 7.41 Typical grounding system for 9-track digital tape recorder.

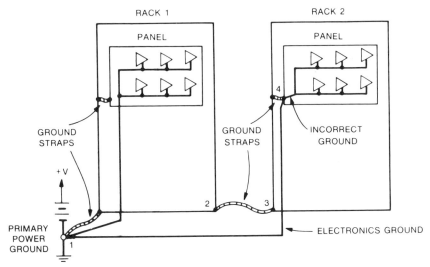

Fig. 7.42 Electronic circuits mounted in equipment racks should have separate ground connections. Rack 1 shows correct grounding; rack 2 shows incorrect grounding.

7.3.6 Hardware Grounds

Electronic circuits for any large system are usually mounted in relay racks or cabinets. These racks and cabinets must be grounded for safety. In some systems such as electromechanical telephone offices, the racks serve as the return conductor for relay switching circuits. The rack ground is often very noisy, and it may have fairly high resistance due to joints and seams in the rack or in pull-out drawers.

Figure 7.42 shows a typical system consisting of sets of electronics mounted on panels which are then mounted to two relay racks. Rack number 1, on the left, shows correct grounding. The panel is strapped to the rack to provide a good ground and the racks are strapped together and tied to ground at the primary power source. The electronics circuit ground does not make contact with the panel or rack. In this way noise currents on the rack cannot return to ground through the electronics ground. At high frequencies, some of the rack noise current can return on the electronics ground due to capacitive coupling between the rack and electronics. This capacitance should, therefore, be kept as small as possible. Rack number 2, on the right, shows an incorrect installation in which the circuit ground is connected to the rack ground. Noise currents on the rack can now return on the electronics ground, and there is a ground loop between points 1, 2, 3, 4, 1.

If the installation does not provide a good ground connection to the rack or panel, it is best to eliminate the questionable ground, and then to provide a definite ground by some other means, or be sure that there is no ground at all. Do not depend on sliding drawers, hinges, and so on, to provide a reliable ground connection. When the ground is of a questionable nature, performance may vary from system to system or time to time, depending on whether or not the ground is made.

Hardware grounds produced by intimate contact, such as welding, brazing or soldering, are better than those made by screws and bolts. When joining dissimilar metals for grounding, care must be taken to prevent galvanic corrosion and to assure that galvanic voltages are not troublesome. Improperly made ground connections may perform perfectly well on new equipment but may be the source of mysterious trouble later.

When electrical connections are to be made to a metallic surface, such as a chassis, the metal should be protected from corrosion with a conductive coating. For example, finish aluminum with a conductive alodine or chromate finish instead of the nonconductive anodized finish. If chassis are to be used as ground planes, careful attention must also be paid to the electrical properties of seams, joints, and openings.

7.3.7 Single Ground Reference for a Circuit

Since two ground points are seldom at the same potential, the difference in ground potential will couple into a circuit if it is grounded at more than one point. This condition is illustrated in Fig. 7.43; a signal source is grounded at point A and an amplifier is grounded at point B. Note that in this discussion, an amplifier is generally mentioned as the load. The amplifier is simply a convenient example, however, and the grounding methods apply to any type of load. Voltage V_G represents the

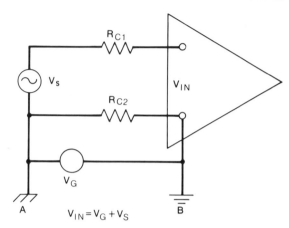

Fig. 7.43 Noise voltage V_G will couple into the amplifier if the circuit is grounded at more than one point.

difference in ground potential between points A and B. In Fig. 7.43 and subsequent illustrations, two different ground symbols are used to emphasize that two physically separated grounds are not usually at the same potential. Resistors R_{C1} and R_{C2} represent the resistance of the conductors connecting the source to the amplifier.

In Fig. 7.43 the input voltage to the amplifier is equal to $V_s + V_G$. To eliminate the noise, one of the ground connections must be removed. Elimination of the ground connection at B means the amplifier must operate from an ungrounded power supply. A differential amplifier could also be used as discussed later in this chapter. It is usually easier, however, to eliminate ground connection A at the source.

The effect of isolating the source from ground can be determined by considering a low-level transducer connected to an amplifier, as shown in Fig. 7.44. Both the source and one side of the amplifier input are grounded.

For the case where $R_{C2} \ll R_s + R_{C1} + R_L$, the noise voltage, V_N, at the amplifier terminals is equal to

$$V_N = \left[\frac{R_L}{R_L + R_{C1} + R_s} \right] \left[\frac{R_{C2}}{R_{C2} + R_G} \right] V_G \tag{7.32}$$

Example 7.1. Consider the case where the ground potential in Fig. 7.44 is equal to 100 mV, a value equivalent to 10 A of ground current flowing through a ground resistance of 0.01 Ω. If

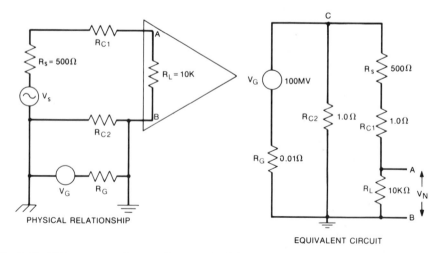

Fig. 7.44 With two ground connections, much of the ground-potential difference appears across the load as noise.

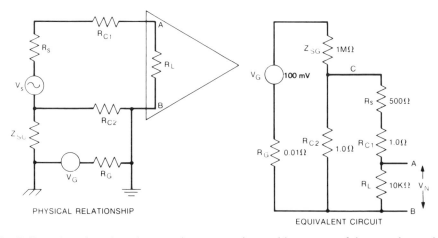

PHYSICAL RELATIONSHIP

EQUIVALENT CIRCUIT

Fig. 7.45 A large impedance between the source and ground keeps most of the ground-potential difference away from the load and reduces noise.

R_s = 500 Ω, R_{C1} = R_{C2} = 1 Ω, and R_L = 10 kΩ, then from Eq. 7.32 the noise voltage at the amplifier terminals is 95 mV. Thus, almost all of the 100 mV ground differential voltage is coupled into the amplifier.

The source can be isolated from ground by adding the impedance Z_{SG}, as shown in Fig. 7.45. Ideally, the impedance Z_{SG} would be infinite, but due to leakage resistance and capacitance it has some large finite value. For the case where $R_{C2} \ll R_s + R_{C1} + R_L$, and $Z_{SG} \gg R_{C2} + R_G$, the noise voltage V_N at the amplifier terminals is

$$V_N = \left[\frac{R_L}{R_L + R_{C1} + R_s} \right] \left[\frac{R_{C2}}{Z_{SG}} \right] V_G \qquad (7.33)$$

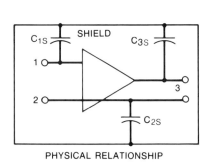

PHYSICAL RELATIONSHIP

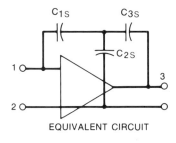

EQUIVALENT CIRCUIT

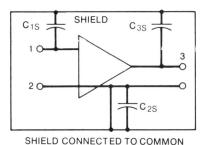

SHIELD CONNECTED TO COMMON

Fig. 7.46 Amplifier shield should be connected to the amplifier common.

Most of the noise reduction obtained by isolating the source is due to the second term of Eq. 7.33. If Z_{SG} is infinite, there is no noise voltage coupled into the amplifier. If the impedance Z_{SG} from source to ground is 1 MΩ and all other values are the same as in the previous example, the noise voltage at the amplifier terminals is, from Eq. 7.33, now only 0.095 μV. This is a reduction of 120 dB from the previous case where the source was grounded.

7.3.8 Amplifier Shields

High-gain amplifiers are often enclosed in a metallic shield to provide protection from electric fields. The question then arises as to where the shield should be grounded. Figure 7.46 shows the parasitic capacitance that exists between the amplifier and the shield. From the equivalent circuit, it can be seen that the stray capacitances C_{3S} and C_{1S} provide a feedback path from output to input. If this feedback is not eliminated, the amplifier may oscillate. *The only shield connection that will eliminate the unwanted feedback path is the one shown at the bottom of Fig. 7.46 where the shield is connected to the amplifier common terminal.* By connecting the shield to the amplifier common, capacitance C_{2S} is short circuited and the feedback is eliminated. This shield connection should be made even if the common is not at earth ground.

7.3.9 Grounding of Cable Shields

Shields on cables used for low-frequency signals should be grounded at only one point when the signal circuit has a single point ground. If the shield is grounded at more than one point, noise current will flow. In the case of a shielded twisted pair, the shield currents may inductively couple unequal voltages into the signal cable and be a source of noise. In the case of coaxial cable, the shield current generates a noise voltage by causing an IR drop in the shield resistance, as was shown in Fig. 7.26. But if the shield is to be grounded at only one point, where should that point be? The top drawing in Fig. 7.47 shows an amplifier and the input signal leads with an ungrounded source.

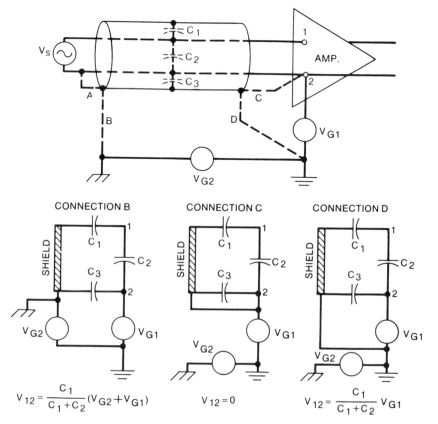

Fig. 7.47 When amplifier is grounded, the best shield connection is (C), with shield connected to amplifier common.

Generator V_{G1} represents the potential of the amplifier common terminal above earth ground, and generator V_{G2} represents the difference in ground potential between the two ground points.

Since the shield has only one ground, it is the capacitance between the input leads and the shield that provides the noise coupling. The input shield may be grounded at any one of four possible points through the dotted connections labeled A, B, C, and D. Connection A is obviously not desirable, since it allows shield noise current to flow in one of the signal leads. This noise current flowing through the impedance of the signal lead produces a noise voltage in series with the signal.

The three lower drawings in Fig. 7.47 are equivalent circuits for grounding connections B, C, and D. Any extraneous voltage generated between the amplifier input terminals (points 1 and 2) is a noise voltage. With grounding arrangement B, a voltage is generated across the amplifier input terminals due to the generators V_{G2} and V_{G1} and the capacitive voltage divider formed by C_1 and C_2. This connection, too, is unsatisfactory. For ground connection C, there is no voltage V_{12} regardless of the value of generators V_{G1} or V_{G2}. With ground connection D, a voltage is generated across the amplifier input terminals due to generator V_{G1} and the capacitive voltage divider C_1 and C_2. The only connection which precludes a noise voltage V_{12} is connection C. *Thus, for a circuit with an ungrounded source and a grounded amplifier, the input shield should always be connected to the amplifier common terminal, even if this point is not at earth ground.*

The case of an ungrounded amplifier connected to a grounded source is shown in Fig. 7.48. Generator V_{G1} represents the potential of the source common terminal above the actual ground at its location. The four possible connections for the input cable shield are again shown as the dashed lines labeled A, B, C, and D. Connection C is obviously not desirable since it allows shield noise currents to flow in one of the signal conductors in order to reach ground. Equivalent circuits are shown at the bottom of Fig. 7.48 for shield connections A, B, and D. As can be seen, only connection A produces no noise voltage between the amplifier input terminals. *Therefore, for the case of a grounded source and ungrounded amplifier, the input should be connected to the source common terminal, even if this point is not at earth ground.*

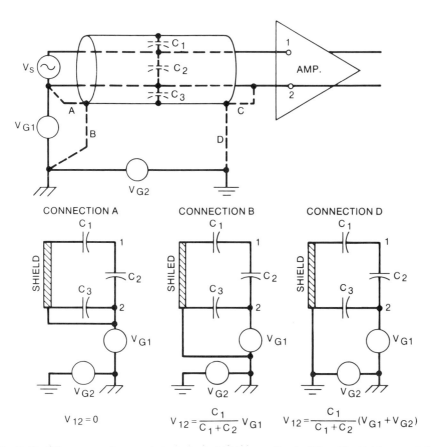

Fig. 7.48 When source is grounded, the best shield connection is (A), with shield connected to the source common. The configuration can also be used with a differential amplifier.

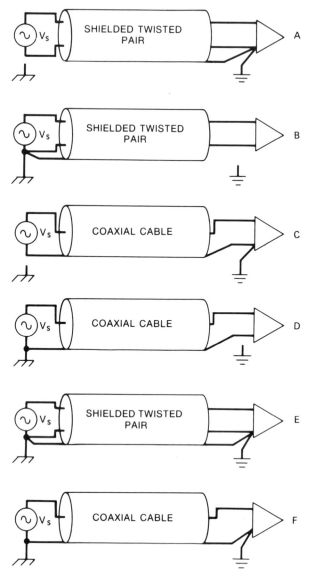

Fig. 7.49 Preferred grounded schemes for shielded, twisted pairs and coaxial cable at low frequency.

Preferred low-frequency shield grounding schemes for both shielded twisted pair and coaxial cable are shown in Fig. 7.49. Circuits (*a*) through (*d*) are grounded at the amplifier or the source, but not at both ends.

When the signal circuit is grounded at both ends, the amount of noise reduction possible is limited by the difference in ground potential and the susceptibility of the ground loop to magnetic fields. The preferred shield ground configurations for cases where the signal circuit is grounded at both ends are shown in circuits (*e*) and (*f*) of Fig. 7.49. In circuit (*f*), the shield of the coaxial cable is grounded at both ends to force some ground-loop current to flow through the lower-impedance shield, rather than the center conductor. In the case of circuit (*e*) the shielded twisted pair is also grounded at both ends to shunt some of the ground-loop current from the signal conductors. If additional noise immunity is required, the ground loop must be broken. This can be done by using transformers, optical couplers, or a differential amplifier.

An indication of the type of performance to be expected from the configurations shown in Fig. 7.49 can be obtained by referring to the results of the magnetic coupling experiment presented in Figs. 7.28 and 7.29.

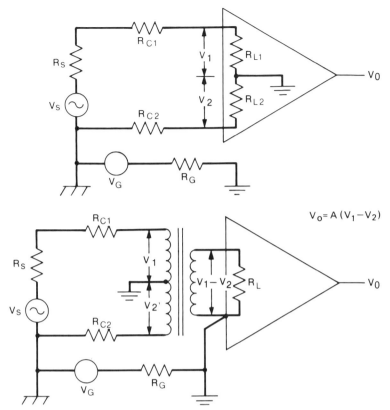

Fig. 7.50 A differential amplifier—or a single-ended amplifier with transformer—can be used to reduce the effects of a common-mode noise voltage.

7.3.10 Differential Amplifiers

A differential (or balanced-input) amplifier may be used to decrease the effect of a longitudinal (common-mode) noise voltage. This is shown in the upper drawing of Fig. 7.50, where V_G is the longitudinal voltage. The differential amplifier has two input voltages, V_1 and V_2, and the output voltage is equal to the amplifier gain (A) times the difference in the two input voltages, $V_0 = A(V_1 - V_2)$.

The lower drawing of Fig. 7.50 shows how a single-ended (or unbalanced) amplifier can be used to simulate the performance of a true balanced amplifier. The transformer primary has a grounded center tap and the voltages across the two halves are V_1 and V_2. The secondary voltage (assuming a 1:1 turns ratio) is equal to $V_1 - V_2$. Amplifier output again is equal to the gain times this voltage difference, duplicating the balanced amplifier output.

The response of either circuit in Fig. 7.50 to the noise voltage can be determined from the equivalent circuit shown in Fig. 7.51. For resistance R_{L2} much larger than R_G the input voltage to the amplifier due to common-mode noise voltage, V_G, is

$$V_N = V_1 - V_2 = \left(\frac{R_{L1}}{R_{L1} + R_{C1} + R_s} - \frac{R_{L2}}{R_{L2} + R_{C2}} \right) V_G \qquad (7.34)$$

Example 7.2. If in Fig. 7.51, $V_G = 100$ mV, $R_G = .01\ \Omega$, $R_s = 500\ \Omega$, $R_{C1} = R_{C2} = 1\ \Omega$, and $R_{L1} = R_{L2} = 10\ \text{k}\Omega$, then from Eq. 7.34, $V_N = 4.6$ mV. If, however, R_{L1} and R_{L2} were $100\ \text{k}\Omega$ instead of $10\ \text{k}\Omega$, then $V_N = 0.5$ mV. This represents an almost 20 dB decrease in the input noise voltage.

From the above example, it is obvious that increasing the input impedance (R_{L1} and R_{L2}) of the differential amplifier decreases the noise voltage coupled into the amplifier due to V_G. From

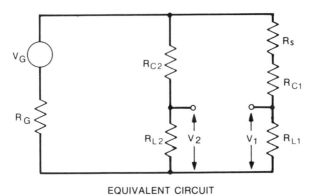

EQUIVALENT CIRCUIT

Fig. 7.51 Equivalent circuit for analysis of differential-amplifier circuit.

Eq. 7.34, it can be seen that decreasing the source resistance R_s also decreases the noise voltage coupled into the amplifier. Figure 7.52 shows a way to modify the circuits of Fig. 7.50 to increase the input impedance of the amplifier to the longitudinal voltage V_G without increasing the input impedance to the signal voltage V_1. This is done by adding resistor R into the ground lead as shown. When using a high-impedance differential amplifier, both the input cable shield and the source common should be grounded at the source as was shown in Fig. 7.49b.

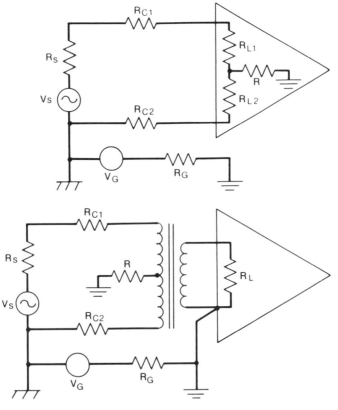

Fig. 7.52 Insertion of resistance R into ground lead decreases the noise voltage.

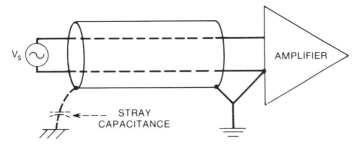

Fig. 7.53 At high frequencies, stray capacitance tends to complete the ground loop.

7.3.11 Shield Grounding at High Frequencies

At frequencies less than 1 MHz, shields should normally be grounded at one end only. Otherwise, as previously explained, large power-frequency currents can flow in the shield and introduce noise into the signal circuit. The single-point ground also eliminates the shield ground loop and its associated magnetic pickup.

At frequencies above 1 MHz or where cable length exceeds one-twentieth of a wavelength it is often necessary to ground a shield at more than one point to guarantee that it remains at ground potential. Another problem develops at high frequencies; stray capacitive coupling tends to complete the ground loop, as shown in Fig. 7.53. This makes it difficult or impossible to maintain isolation at the ungrounded end of the shield.

It is therefore common practice at high frequencies to ground cable shields at both ends. For long cables, ground may be required every one-tenth of a wavelength. The noise voltage due to a difference in ground potential that couples into the circuit (primarily at power frequencies and its harmonics) can usually be filtered out, because there is a large frequency difference between the noise and the signal frequency. At frequencies above one megahertz the skin effect reduces the coupling due to signal and noise current flowing on the shield. This skin effect causes the noise current to flow on the outside surface of the shield and the signal current to flow on the inside surface of the shield. The multiple ground also provides a degree of magnetic shielding at higher frequencies when coaxial cable is used.

The characteristics of the circuit shown in Fig. 7.53 can be put to advantage by replacing the stray capacitance with a small capacitor, thus forming a combination or hybrid ground. At low frequencies a single point ground exists since the impedance of the capacitor is large. However, at high frequencies the capacitor becomes a low impedance, thus converting the circuit to one having a multiple ground. Such a ground configuration is often useful for circuits that must operate over a very wide frequency range.

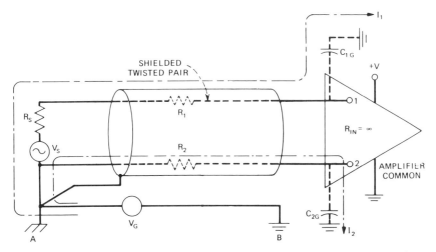

Fig. 7.54 Amplifier and a grounded source are connected by a shielded twisted pair.

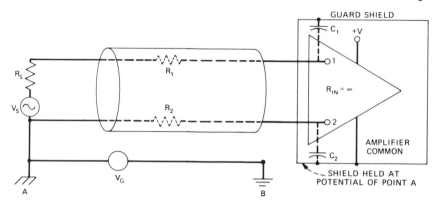

Fig. 7.55 Guard shield at potential of point A eliminates noise currents.

7.3.12 Guard Shields

Noise reduction greater than that obtainable with a differential amplifier can be obtained by using an amplifier with a guard shield. A guard shield is placed around the amplifier and held at a potential which prevents current flow in the unbalanced source impedance. The effect of a guard shield can best be explained by considering an example in which a guard shield is used to cancel the effects of a difference in ground potential.

Figure 7.54 shows an amplifier connected by a shielded twisted pair to a grounded source. V_G is a common-mode (longitudinal) voltage due to a difference in ground potentials. V_s and R_s are the differential signal voltage and source resistance, respectively. R_{IN} is the input impedance to the amplifier. C_{1G} and C_{2G} are stray capacitances between the amplifier input terminals and ground, including the cable capacitance. There are two undesirable currents flowing as a result of voltage, V_G. Current I_1 flows through resistors R_s and R_1, and capacitance C_{1G}. Current I_2 flows through resistor R_2 and C_{2G}. If each current does not flow through the same total impedance there will be a differential input voltage to the amplifier. If, however, a guard shield is placed around the amplifier, as shown in Fig. 7.55, and the shield is held at the same potential as point A, currents I_1 and I_2 both become zero because both ends of the path are at the same potential. Capacitances C_1 and C_2 now appear between the input terminals and the shield.

The shield accomplishes the objective of eliminating the differential input noise voltage. Unmentioned, however, has been the problem of how to hold the shield at the potential of point A. One way to do this is shown in Fig. 7.56, where the guard shield is held at the potential of point A by connecting it to the cable shield. The other end of the cable shield is then grounded at point A. This discussion assumes that the source common (lower) terminal is at the same potential as point A. That is, there is no noise voltage generated between point A and the source common. If there is any possibility of a noise voltage being generated between the common terminal of V_s and ground point A, the guard shield should be connected to the source common as illustrated, instead of directly to point A.

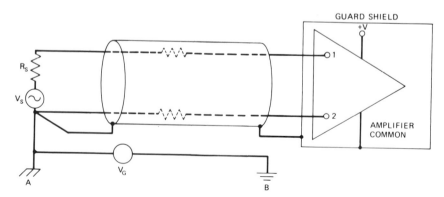

Fig. 7.56 Guard shield is connected to point A through the cable shield.

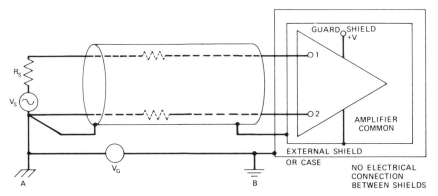

Fig. 7.57 Practical circuit often has a second shield around the guard shield.

Notice that the amplifier and shield connections of Fig. 7.56 do not violate any of the previously described rules. The cable shield is grounded at only one point (point *A*). The input cable shield is connected to the amplifier common. The shield around the amplifier is also connected to the amplifier common terminal.

In the guarded amplifier of Fig. 7.56, any ground point at potential *B* inside the amplifier guard shield increases the capacitance from the input leads to ground (unguarded capacitance). For the scheme to work, therefore, it means the amplifier must be powered by self-contained batteries, or else power must be brought in through an electrostatically shielded transformer. No point of the guard shield can come in contact with ground *B* without nullifying its effectiveness. A practical circuit, therefore, has a second shield placed around the guard shield to guarantee the guard's integrity, as shown in Fig. 7.57. This second or external shield is grounded to the local ground, point *B*, and satisfies the safety requirements.

A guard shield is usually only required when extremely-low-level signals are being measured, or when very large common-mode voltages are present and all other noise reduction techniques have also been applied to reduce the noise pick-up to an absolute minimum. A guard shield may be placed around a single-ended amplifier as well as a differential amplifier.

Example 7.3. Consider a numerical example, as illustrated in Fig. 7.58, where $R_1 = R_2 = 0$, $R_s = 2.6$ kΩ, $C_{1G} = C_{2G} = 100$ pF and $V_G = 100$ mV at 60 Hz. The reactance of 100 pF is 26 MΩ at 60 Hz. The differential input noise voltage across the amplifier input terminals without a guard shield can be written as

$$V_N = \left(\frac{R_s + R_1}{R_s + R_1 + Z_{1G}} - \frac{R_2}{R_2 + Z_{2G}} \right) V_G \qquad (7.35)$$

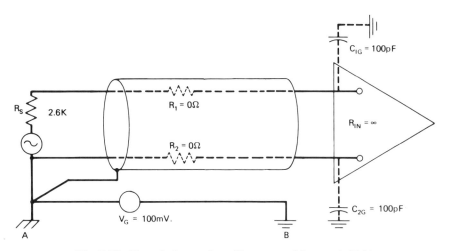

Fig. 7.58 Numerical example to illustrate need for guard shield.

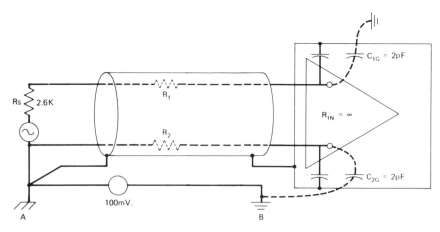

Fig. 7.59 Guard shield reduces line capacitance to ground and, therefore, noise voltage.

where Z_{1G} and Z_{2G} are the impedance of capacitance C_{1G} and C_{2G}, respectively. Substituting numerical values into Eq. 7.35, the input noise voltage without the guard shield is 10 μV. If the use of the guard shield reduces each line's capacitance to ground to 2 pF, as shown in Fig. 7.59, the differential input noise voltage across the amplifier input terminals with the guard shield in place can still be written as shown in Eq. 7.35, but the input noise voltage is now reduced to 0.2 μV, a 34 dB improvement. The 2 pF capacitance to ground is due to the fact that the guard shield is not perfect. If it were perfect, there would be no capacitance to ground and the noise voltage would be zero. It should be noted that the noise voltage coupled into the amplifier increases as the frequency of the noise source is increased, since the impedances of C_{1G} and C_{2G} decrease as the frequency is increased.

7.3.13 Guarded Meters

Even for those who do not intend to design equipment using a guard shield, there is still a good reason to understand the operating principles. Many new measuring instruments are being manufactured

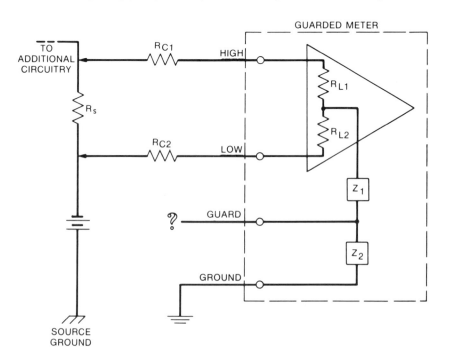

Fig. 7.60 When a guarded meter is used, a common problem is how to connect the guard terminal.

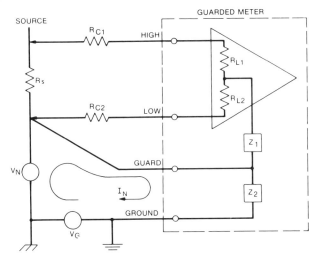

Fig. 7.61 When measuring voltage across R_s, best connection for guard is to the low-impedance side of R_s; noise current does not affect amplifier.

with a guard shield (see Fig. 7.60). It is up to the user to connect the guard shield to the proper place in the circuit being measured. When a user does not understand the purpose of a guard shield, he is likely to leave it open or connect it to the meter ground; neither of these connections produces optimum results. To take maximum advantage of the guard shield, the following rule should be followed: *The guard shield should always be connected such that no common-mode current can flow through any of the input resistances.* This normally means connecting the guard to the low-impedance terminal of the source.

Example 7.4. Refer to Fig. 7.60. The problem is to measure the voltage across resistor R_s, neither end of which is grounded, with a guarded digital voltmeter. What is the best connection for the guard shield? Five possible ways to connect the guard shield are shown in Figs. 7.61 through 7.65. Voltage V_G is the ground differential voltage and V_N is the battery noise voltage. Figure 7.61 shows the best connection, with the guard connected to the low-impedance terminal of the source. Under this condition no noise current flows through the input circuit of the meter.

The connection shown in Fig. 7.62, where the guard is connected to ground at the source is not as good as the previous connection. Here, the noise current from the generator V_G is no problem, but noise current from V_N flows through impedances R_{C2}, R_{L2}, and Z_1 and causes a noise voltage to

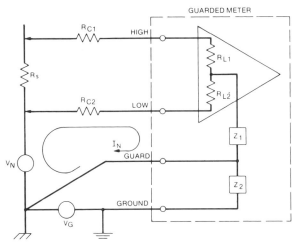

PROTECTION AGAINST V_G ONLY.

Fig. 7.62 Guard connected to source ground gives no protection against V_N.

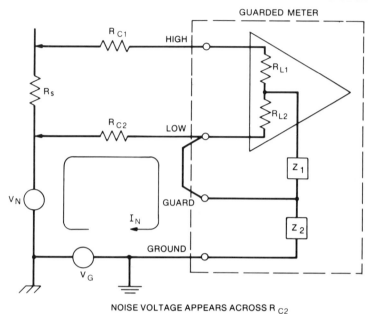

Fig. 7.63 Guard connected to low side of meter allows noise current to flow in line resistance R_{C2}.

be coupled into the amplifier. The connections of Figs. 7.63, 7.64, and 7.65 all allow noise current to flow through the meter input circuit and are, therefore, undesirable.

7.3.14 Cables and Connectors

Inadvertent generation of ground loops and poor shielding practices are likely to occur in system cabling, especially when different design groups are responsible for different sides of the interface. Good cabling requires design; it is not something that just happens.

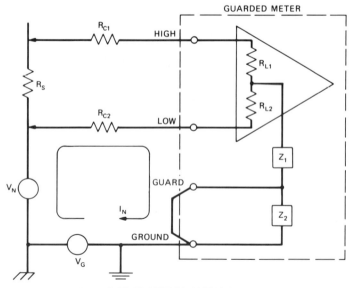

Fig. 7.64 Guard connected to local ground is ineffective; noise current flows through R_{C2}, R_{L2}, and Z_1.

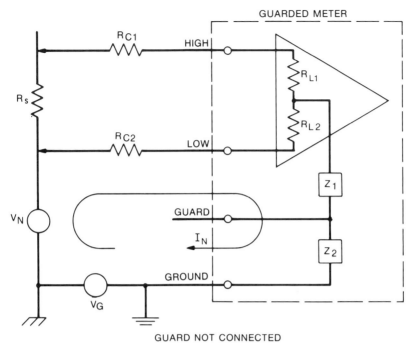

Fig. 7.65 Guard not connected; noise currents due to V_N and V_G flow through R_{C2}, R_{L2}, Z_1, and Z_2.

When possible, high-level and low-level leads should not be put in the same cable. If they must be in one cable, the high-level leads should be grouped and put in a shield. Normal precautions should be taken with the low-level leads.

Low- and high-level leads should be run through separate connectors where possible. If high-and low-level leads must be in one connector, they should be placed on pins which are physically separated. Ground leads should be placed on the intervening pins, as shown in Fig. 7.66. If all pins in the connector are not used, the spare one(s) should be in the middle, separating the high- and low-level leads.

Shielding integrity should be maintained when cables are run between systems. Cable shields should be carried through connectors. When more than one shielded cable goes through a connector, each shield should be carried through on a separate pin. Connecting all the shields to a single pin produces ground loops and allows shield currents to flow between individual shields.

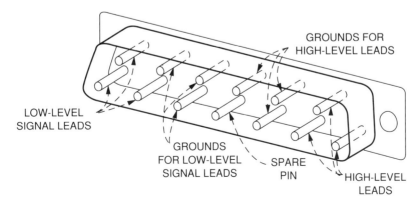

Fig. 7.66 When assigning pins in a connector, high-level and low-level leads should be physically separated with ground leads in between.

Where low-level signal cables require shielding and are grounded at one point only, insulation is necessary over the shield. This prevents the shield from inadvertently touching ground at some other point.

7.3.15 Summary

- At low frequencies, a single-point ground system should be used.
- At high frequencies, a multipoint ground system should be used.
- A low-frequency system should have a minimum of three separate ground returns. These should be
  - signal ground,
  - noisy ground,
  - hardware ground.
- The basic objectives of a good ground system are
  - minimize noise voltage from two ground currents flowing through a common impedance,
  - avoid generating ground loops.
- For the case of a grounded amplifier with an ungrounded source, the input cable shield should be connected to the amplifier common terminal.
- For the case of a grounded source with an ungrounded amplifier, the input cable shield should be connected to the source common terminal.
- A shield around a high-gain amplifier should be connected to the amplifier common.

REFERENCES

1. H. W. Ott, *Noise Reduction Techniques in Electronic Systems*, John Wiley & Sons, New York, 1976.
2. W. H. Hayt, Jr., *Engineering Electromagnetics*, 3rd ed., McGraw-Hill, New York, 1974.
3. Smythe, W. R., *Static and Dynamic Electricity*, McGraw-Hill, New York, 1924.
4. Belden Engineering Staff, *Electronics Cable Handbook*, Howard W. Sams & Co., New York, 1966.
5. ITT, *Reference Data for Radio Engineers*, 5th ed., Howard W. Sams & Co., New York, 1968.
6. *National Electrical Code*, National Fire Protection Association, Boston, MA, 1975. (This code is normally reissued every three years.)

BIBLIOGRAPHY

Section 7.1

Bell Laboratories, *Physical Design of Electronic Systems*, vol. 2, Chapter 5 (Electrochemistry and Protection of Surfaces), Prentice-Hall, Englewood Cliffs, NJ, 1970.

Cohen, T. J., and McCoy, L. G., "RFI—A New Look at an Old Problem," *QST*, March, 1975.

White, D. R. J., *Electromagnetic Interference and Compatibility*, vol. 1 (Electrical Noise and EMI Specifications), Don White Consultants, 1971.

White, D. R. J., *Electromagnetic Interference and Compatibility*, vol. 2 (EMI Test Methods and Procedures), Don White Consultants, 1974.

Section 7.2

Bell Laboratories, *Physical Design of Electronic Systems*, vol. 1, Chapter 10 (Electrical Interference), Prentice-Hall, Englewood Cliffs, NJ, 1970.

Buchman, A. S., "Noise Control in Low Level Data Systems," *Electromechanical Design*, September, 1962.

Ficchi, R. O., *Electrical Interference*, Hayden Book Co., New York, 1964.

Ficchi, R. O., *Practical Design For Electromagnetic Compatibility*, Hayden Book Co., New York, 1971.

Frederick Research Corp., *Handbook on Radio Frequency Interference*, vol. 3, (Methods of Electromagnetic Interference Suppression), Frederick Research Corp., Wheaton, Maryland, 1962.

Morrison, R., *Grounding and Shielding Techniques in Instrumentation*, John Wiley & Sons, New York, 1967.

Nalle, D., "Elimination of Noise in Low Level Circuits," *ISA Journal*, vol. 12, August, 1965.

Timmons, F., "Wire or Cable Has Many Faces, Part 2," *EDN*, March, 1970.

Section 7.3

Ady, R., "Applying Opto-Isolators," *Electronic Products*, June 17, 1974.

Bell Laboratories, *Physical Design of Electronic Systems*, vol. 1, Chapter 10 (Electrical Interference); Prentice-Hall, Englewood Cliffs, NJ, 1970.

Brown, H., "Don't Leave System Grounding to Chance," *EDN/EEE*, January 15, 1972.

Buchman, A. S., "Noise Control in Low Level Data Systems," *Electromechanical Design*, September, 1962.

Cushman, R. H., "Designer's Guide to Optical Couplers," *EDN*, July 20, 1973.

Ficchi, R. O., *Electrical Interference*, Hayden Book Co., New York 1964.

Ficchi, R. O., *Practical Design For Electromagnetic Compatibility*, Hayden Book Co., New York 1971.

Frederick Research Corp., *Handbook on Radio Frequency Interference*, vol. 3 (Methods of Electromagnetic Interference Suppression) Frederick Research Corp., Wheaton, Maryland 1962.

Hewlett-Packard, *Floating Measurement and Grounding*, application note 123, 1970.

Morrison, R., *Grounding and Shielding Techniques in Instrumentation*, John Wiley & Sons, New York, 1967.

Nalle, D., "Elimination of Noise in Low Level Circuits," *ISA Journal*, vol. 12, August, 1965.

White, D. R. J., *Electromagnetic Interference and Compatibility*, vol. 3 (EMI Control Methods and Techniques), Don White Consultants, Germantown, Maryland 1973.

CHAPTER **8**
BRIDGE TRANSDUCERS

PATRICK L. WALTER

Sandia National Laboratories
Albuquerque, NM

8.1   TERMINOLOGY

A telemetry system responding to a measurand consists of four basic parts: the transducer, the transmitting system, the receiving system, and the data output or display system.

Telemetry—The transmission of information about a measurand.

Measurand—The object of a measurement. The process to be defined.

Transducer—A component in the telemetry system which provides information about a process and, as a by-product, transfers energy from the process. Typical bridge transducers convert physical quantities such as force, pressure, displacement, velocity, acceleration, temperature, humidity, etc. into electrical quantities for input to the transmitting system.

Transmitting system—The transmitting system typically consists of some or all of the following devices: cable, amplifier, subcarrier oscillator, filter, analog-to-digital (A/D) converter, transmitter, and antenna.

Receiving system—The receiving system typically consists of some or all of the following devices: antenna, preamplifier, multicoupler, receiver, tape recorder, discriminator, decommutator, digital-to-analog (D/A) converter, and output filter.

Data output and display system—The data output and display system typically consist of some or all of the following devices: oscilloscope, analog meter, digital meter, graphic display, and digital printer. These devices may either be connected directly to the output of the receiving system or a computer may process the data from the receiving system before display.

8.2   FLEXURAL DEVICES IN MEASUREMENT SYSTEMS

Bridge transducers depend on a measurand to directly modify some electrical or magnetic property of a conductive element. For example, the thermal coefficient of impedance can result in a change in impedance of a conductive element proportional to temperature (e.g., resistance thermometer). Similarly, hygroscopic materials can have their impedance change in a deterministic fashion due to humidity (e.g., humidity sensor). Most bridge transducers, however, depend on the displacement of a flexure to vary the impedance of a conductive element, resulting in an electrical signal proportional to the measurand. Advantage is taken of either the strain pattern on the surface of the flexure or the motion of this surface. Among the gamut of flexure elements associated with bridge transducers are cantilever beams, Bourdon tubes, and clamped diaphragms.

8.2.1   Cantilever Beams

Cantilever beams are routinely designed into bridge transducers. Strain near the clamped end of the beam can be correlated to displacement of the free beam end, force or torque applied to the free beam end, dynamic pressure associated with fluid flow acting over the beam surface, etc. The compliance of a cantilever beam is defined as

$$\frac{y}{F} = \frac{L^3}{3EI} \tag{8.1}$$

where y is deflection of the beam free end, F is the force applied to this end, L is the beam length, E is the modulus of elasticity of the beam material, and I is the beam area moment of inertia. A compliant flexure will result in a bridge transducer with a large electrical signal output. Equation 8.1 indicates a compliant flexure design can be achieved by a long, thin, narrow beam of low modulus material. The penalty attached to such a design in application could be a transducer which is bulky, displays undesirable response to physical inputs orthogonal to its sensing direction, and has poor dynamic response.

8.2.2   Bourdon Tubes

Bourdon tubes are one of the most widely used flexures for sensing pressure. The original patent for this device was granted to Eugene Bourdon in 1852. Bourdon tubes are hollow tubes that are twisted or curved along their length. The application of pressure deforms the tube wall which, depending on tube shape, causes it to untwist or unwind. Motion of the tube is typically used to modify the AC impedance of bridge transducers. Bourdon tubes can be integrated into transducers to achieve extremely high accuracies and have been manufactured from perfectly elastic materials such as quartz.

   Transducers employing Bourdon tubes tend to be physically large and easily damaged by environmental inputs such as acceleration. In addition, the tubes themselves afford poor frequency response to time-varying pressure.

8.2.3   Clamped Diaphragms

Clamped diaphragms are another flexure used to transform a measurand into a strain or displacement proportional to applied pressure. A small, flat, circular diaphragm can be made simply, and it can be placed flush against surfaces whose flow dynamics are being studied. This type of diaphragm is typically designed to deflect in accord with theory associated with clamped circular plates. Corrugated diaphragms provide extensibility over a greater linear operating range than do flat diaphragms. A catenary diaphragm consists of a flexurally weak seal diaphragm bearing against a thin cylinder whose motion is measured. The compliance of a flat, clamped circular diaphragm is defined as

$$\frac{y}{P} = \frac{3R_0^4(1 - \nu^2)}{16t^3E} \tag{8.2}$$

where y is the deflection of the center of the diaphragm, P is the applied pressure, R_0 is the diaphragm radius, ν is Poisson's ratio, t is the diaphragm thickness, and E is the modulus of elasticity of the diaphragm material. Somewhat analogous to the cantilever beam, a compliant diaphragm will have a large radius, be thin, and be made of a low modulus material. Equation 8.2 holds for deflections no greater than t.

   Figure 8.1 shows the radial and tangential strain distribution in a flat, clamped, circular diaphragm. The radial and tangential strains at the center of the diaphragm are identical. The tangential strain decreases to zero at the periphery while the radial strain becomes negative. Figure 8.2 describes a strain gage pattern designed to take advantage of this strain distribution. The central sensing elements measure the higher tangential strain while the radial sensing elements measure the high radial strains near the periphery. Resistance strain gages are discussed beginning in Section 8.3.

8.2.4   Error Contributions from the Flexure Properties

When flexures are designed for bridge transducers, the final transducer may have to possess an accuracy over its operating temperature range of from a few to a fractional percent. Knowledge of

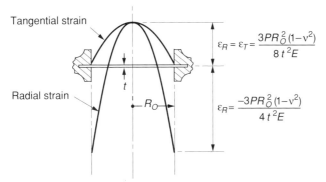

$$\varepsilon_R = \varepsilon_T = \frac{3PR_O^2(1-v^2)}{8t^2E}$$

$$\varepsilon_R = \frac{-3PR_O^2(1-v^2)}{4t^2E}$$

Fig. 8.1 Radial and tangential strain distribution in a flat, clamped, circular diaphragm (courtesy of Measurements Group, Inc., Raleigh, NC, USA).

the inelasticities and metallurgical behaviors of flexural elements must be considered in transducer design. Metals under a constant load experience a minute deformation with time, called creep. Differences between the loading and unloading curve of a flexure, due to energy absorbed by the material as internal friction, introduce another effect, known as hysteresis. The modulus of elasticity of materials changes with temperature. Corrosion resistance, machinability, magnetics, fatigue effects, thermal conductivity, and thermal expansion are other properties of flexural materials to consider in design application. The 300 series stainless steels are useful flexural materials due to their corrosion resistance, desirable low temperature properties, and good creep properties at elevated temperatures. Inconel is a good flexural material in corrosive salt water environments. These materials and others are discussed in an extremely good article on transducer flexures in Chapter 11 of Reference 1.

8.3 THE RESISTANCE STRAIN GAGE

Strain gages are used to measure the strain pattern on the surface of the flexure in bridge transducers. In 1938, Simmons, at the California Institute of Technology, and Ruge, at the Massachusetts Institute of Technology, discovered independently that fine wire bonded directly to a surface being studied would respond to surface strain. Dr. Ruge's original gage was made by unwinding a constantan wire-wound vitrified resistor, gluing a portion of this wire with Duco cement to a celluloid bar, attaching brass shim stock as terminals, and interfacing the completed assembly to a galvanometer. The first strain gage manufacturer established in the United States was the Baldwin Lima Hamilton Corporation (now BLH Electronics). BLH gages operating on the principle discovered by Simmons and Ruge have been designated the SR-4 gage to include the initials of both men. The evolution of the bonded resistance wire strain gage occurred during the early 1940s. The first practical bridge transducer load cell was built by Baldwin Lima Hamilton in 1941. By the mid-1950s, Baldwin Lima Hamilton remained the only major strain gage manufacturer and the foil strain gage was beginning to appear. Subsequent work by W. P. Mason and R. N. Thurston (reported in the *Journal of the Acoustical Society of America*, vol. 29, 1957) resulted in the introduction of the commercial semiconductor strain gage. Continued maturation of the bonded resistance strain gage has enabled a transducer industry centered around this technology to develop.

Other sources provide the derivation of all the equations dealing with ensuing topics in this chapter (References 2–5).

8.3.1 Strain Gage Types and Fabrication

Paper-backed wire strain gages typically consist of a grid of resistance wire to which a paper backing has been attached with nitrocellulose cement. The wire is manufactured by drawing the selected alloy through progressive forming dies. To protect it during handling and assembly, the wire is usually sandwiched between two thin layers of paper. Typical grid wires are 0.02 mm in diameter.

Foil strain gages are essentially small printed circuits. Art work for a master gage pattern is first prepared. This pattern is then photographically reduced, and multiple images are placed on a photographic plate. A sheet of foil (typically 0.003 to 0.005 mm thick) of the appropriate alloy has a light-sensitive emulsion applied, is exposed to the photographic plate, and then undergoes a

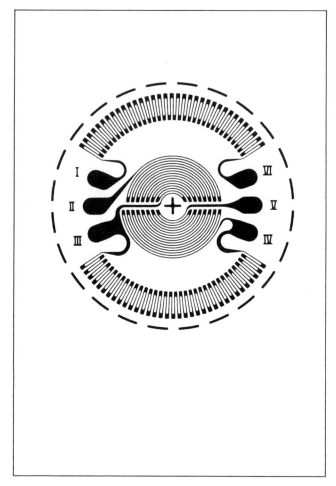

Fig. 8.2 Micro-Measurements' "JB" pattern strain gage for circular diaphragm pressure transducers (courtesy of Measurements Group, Inc., Raleigh, NC, USA).

development process. Chemical etching removes all but the grid material. The resultant grid cross section is square as opposed to round for wire. Advantages inherent in foil gages include better strain transmission due to improved bonding of the grid to the backing, a better thermal path for dissipation of electrically generated heat, and a grid that can more readily be configured to minimize sensitivity to transverse strains.

The total combination of wire and foil gages span grid lengths from 0.2 mm to more than 250 mm. Foil gages satisfy the smaller of these requirements. Both wire and foil gages have associated with them a variety of ohmic values, such as 120, 175, 350, 1000, and so on. Standard values which have historically evolved are 120 and 350 Ω. These values are carry-overs from impedance-matching requirements for galvanometers which were formally used for strain recording. Figure 8.3 displays numerous configurations of wire and foil strain gages.

In the manufacture of bridge transducers using metallic strain gages, vacuum deposition of the gages is an alternate technique to bonding individual gages to the transducer flexure. The flexure is coated with aluminum oxide and then the metal gages are selectively deposited. This process yields the closest match of thermal and electrical characteristics for each bridge element.

The manufacture of semiconductor strain gages starts with a single high-purity silicon crystal. Atoms such as phosphorus (n type) or boron (p type) are doped into the material to lower its resistivity. The parent crystal is sliced into wafers before the dopant is added in a furnace at high temperature (>1000°C). The wafer is masked and etched to produce a suitable grid pattern (usually

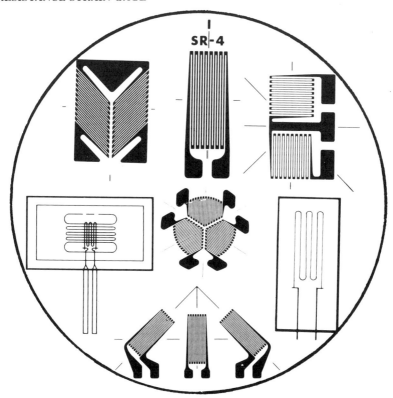

Fig. 8.3 Numerous configurations of wire and foil strain gages—not to scale (courtesy of BLH Electronics, a Bofors Company).

either a straight element or U-shape). This grid can remain unbacked or can be mounted on a suitable carrier.

Alternately, in diffused semiconductor transducers, the transducer flexure itself may be made of silicon. Its surface can be passivated, etched, and doped to form gage elements integral to the flexure. Similar to vacuum deposition of metal gages, diffused semiconductor transducers offer a more nearly optimum match of thermal and electrical properties for each bridge element. Problems with slippage associated with the bond of a gage carrier or backing is nonexistent.

Bridge transducers using semiconductor gages typically possess poorer thermal and linearity specifications than those using metal gages. However, the sensitivity of semiconductor gages to surface strain is much greater than metal gages. This allows them to be used in transducers providing more signal output (typically 100 to 500 mV versus 30 mV) and on stiffer flexures, resulting in smaller transducer size, higher frequency response, and increased ruggedness. Although not strictly correct, by convention it has become equivalent to refer to semiconductor-based transducers as either piezoresistive or solid-state transducers.

8.3.2 Gage Factor

The gage factor F for a strain gage is defined as

$$F = \frac{\Delta R}{R\epsilon} \tag{8.3}$$

where R is resistance and ϵ is strain equal to $\Delta l/l$ (Δl is the change in length of l_0). Equation 8.3 may be redefined as

$$F = 1 + 2\nu + \frac{d\rho}{\rho\epsilon} \tag{8.4}$$

where ν is Poisson's ratio and ρ is the resistivity of the grid material. Most metal gages have a gage factor between 2 and 4.5. For strain gages made from a semiconductor material, the change in resistivity with applied stress is the dominant factor and values as high as 170 are possible. Table 8.1 lists properties and gage factors for various grid materials.

8.3.3   Mechanical Aspects of Gage Operation

To build effective bridge transducers, one must be aware of the interaction between the gage and the surface of the transducer flexure to which it is mounted. Mechanical aspects of this interaction include the influences of temperature, backing material, size, orientation, transverse sensitivity, distance from the surface, bonding and installation, and gage frequency response.

Temperature

A qualitative discussion of temperature effects on bonded strain gages indicates the effects to be attributable to three principal causes: (1) the transducer flexure to which the gage is attached expands or contracts, (2) the strain gage resistance changes with temperature, and (3) the strain gage grid expands or contracts. With some gages (particularly semiconductors), the change in gage factor with temperature is also extremely significant. These temperature effects are accounted for by temperature-strain calibration, self-temperature-compensated gages where combined effects 1, 2, and 3 above are minimized over a given temperature range for a given combination of grid and flexure material, and a dummy gage integrated into a bridge circuit (discussed later) to electrically subtract temperature-induced strain.

Backing Material

The purpose of the backing material used in constructing strain gages is to provide support, dimensional stability, and mechanical protection for the grid element. The backing material of the gage element(s) acts as a spring in parallel with the flexure to which it is attached and can potentially modify flexure mechanical behavior. In addition, the temperature operating range of the gage can be constrained by its backing material. Most backings are epoxies or glass fiber–reinforced epoxies. Some gages are encapsulated for chemical and mechanical protection as well as extended fatigue life. For high temperature applications, some gages have strippable backings for mounting with ceramic adhesives. Still other metal gages can be welded. The frequency response of welded gages, due to uncertainties in dynamic response, is a subject area still requiring investigation.

TABLE 8.1   GRID MATERIAL COMPOSITION, TRADE NAME, PROPERTIES, AND GAGE FACTOR

	Composition	Trade Name	Properties	Gage Factor
1.	Copper/nickel (57%/43%)	Constantan	Strain sensivity relatively independent of level Strain sensitivity relatively independent of temperature Used to 200°C High resistivity applicable to small grids Measures strains to 20 percent in annealed form	2.0
2.	Nickel/chromium/iron/ molybdenum (36%/8%/55.5%/0.5%)	Isoelastic	High gage factor High fatigue life Used to 200°C High temperature coefficient of resistance Nonlinear at strain levels above 5 percent	3.5
3.	Nickel/chromium (80%/20%) Nickel/chromium (75%/20%) plus iron and aluminum	Nichrome V Karma	Good fatigue life Stable High resistance applicable to small grids Used to 400°C	2.2

Note: Nickel alloy gages are susceptible to magnetic fields.

Size

The major factors to be considered in determining the size of strain gage to use are available space for gage mounting, strain gradient at the test location, and character of the material under test. The strain gage must be small enough to be compatible with mounting location and concentrated strain field. It must be large enough so that, on metals with large grain size, it measures average strain as opposed to local effects. Grid elements greater than 3 mm generally have greater fatigue resistance.

Transverse Sensitivity and Orientation

Strain gage transverse sensitivity and mounting orientation are concurrent considerations. Transverse sensitivity in strain gages is important due to the fact that part of the geometry of the gage grid is oriented in directions other than parallel to the principal gage sensing direction. Values of transverse sensitivities are provided with individual gages but typically vary between fractional and several percent. The position of the strain gage axis relative to the numerically larger principal strain on the surface to which it is mounted will have an influence on indicated strain.

Distance from the Surface

The grid element of a strain gage is separated from the transducer flexure by its backing material and cement. The grid then responds to strain at a location removed from the flexure surface. The strain on flexures such as thin plates in bending can vary considerably from that measured by the strain gage.

Bonding Adhesives

Resistance strain gage performance is entirely dependent on the bond attaching it to the transducer flexure. The grid element must have the strain transmitted to it undiminished by the bonding adhesive. The elimination of this bond is one of the principal advantages of vacuum-deposited metallic and diffused semiconductor bridge transducers. Typical adhesives are

> *Epoxy adhesives*—Epoxy adhesives are useful over a temperature range of –270°C to +320°C. The two classes are either room temperature curing or thermal setting type; both are available with various organic fillers to optimize performance for individual test requirements.

> *Phenolic adhesives*—Bakelite, or phenolic adhesive, requires high bonding pressure and long curing cycles. It is used in some transducer applications because of long-term stability under load. The maximum operating temperature for static loads is 180°C.

> *Polyimide adhesives*—Polyimide adhesives are used to install gages backed by polyimide carriers or high temperature epoxies. They are a one-part thermal setting resin and are used from –200°C to +400°C.

Ceramic cements (applicable from –270°C to +550°C) and welding are other mounting techniques.

Frequency Response

The frequency response of bridge transducers cannot be addressed without considering the frequency response of the strain gage as well. It is assumed that the transducer is used in such a manner that mounting variables do not influence its frequency response.

Piping in front of pressure transducer diaphragms and mounting blocks under accelerometers are two examples of variables which can violate this assumption. Transducers, particularly those which measure force, pressure, and acceleration, typically are dynamically modeled as single degree-of-freedom systems characterized by a linear second-order differential equation with constant mass, damping, and stiffness coefficients. In reality, transducers possess multiple resonant frequencies associated with their flexure and their case. Figure 8.4 presents the actual frequency response of a bridge-type accelerometer; the response indicates this single degree-of-freedom model to be adequate through the first major transducer resonance. Such devices have a frequency response usable (constant within 4% referenced to their DC response) to one-fifth of the value of this major resonance. The strain gage itself acts as a spatial averaging device whose frequency response is a function of both its gage length and the sound velocity of the material on which it is mounted. Reference 6 discusses this relationship from which Fig. 8.5 is extracted. Figure 8.5 contains curves for three different length gages. Its abscissa must be multiplied by a specific sound velocity. For most bridge transducers, the structural resonance of the flexure constrains its frequency response.

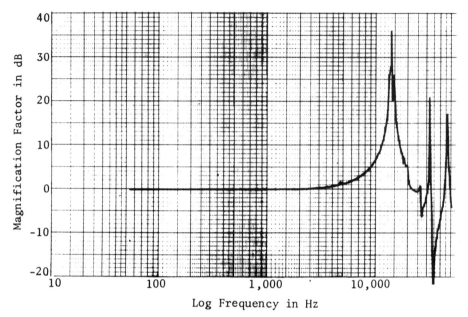

Fig. 8.4 Magnitude of transfer function of piezoresistive accelerometer.

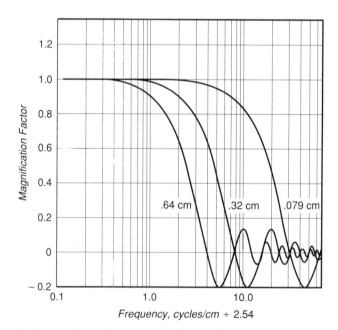

Fig. 8.5 Transfer function for strain gages of varying lengths when analyzed as spatial averaging transducers. (©Instrument Society of America, 1980; reprinted with permission from *ISA Transactions*, Vol. 19, Issue 3.)

8.3.4 Electrical Aspects of Gage Operation

The resistance strain gage, which manifests a change in resistance proportional to strain, must form part of an electrical circuit such that a current passed through the gage transforms this change in resistance into a current, voltage, or power change to be measured. The electrical aspects of gage operation to be considered include current in the gage, resistance to ground, and shielding.

Strain gages are seldom damaged by excitation voltages in excess of proper values, but performance degrades. The voltage applied to a strain gage bridge creates a power loss in each arm which must be dissipated in the form of heat. By its basic design, all of the power input to the bridge is dissipated in the bridge with none available to the output circuit. The sensing grid of every strain gage then operates at a higher temperature than the transducer flexure to which it is bonded. The heat generated within the gage must be transferred by conduction to the flexure. Heat flow into the flexure causes a temperature rise which is a function of its heat sink capacity and gage power level. The optimum excitation level for strain gage applications is a function of the strain gage grid area, gage resistance, heat sink properties of the mounting surface, environmental operating temperature range of the gage installation, required operational specifications, and installation and wiring techniques. Rigid operating requirements for precision transducers require performance verification of the optimum excitation level. Zero-shift versus load and stability under load at the maximum operating temperature are the performance tests most sensitive to excessive excitation voltage.

Table 8.2 and Figs. 8.6, 8.7, and 8.8 allow a first approximation at optimizing bridge excitation levels. Table 8.2 defines the suitability of various structural materials for providing an adequate heat sink for gage mounting dependent on both accuracy requirements and static or dynamic measurements. Figures 8.6 to 8.8 define the recommended excitation voltage for specific gages as a function of the power density capability of the heat sink and gage grid area.

Resistance-to-ground is an important parameter in strain gage mounting since insulation leakage paths produce shunting of the gage resistance between the gage and metal structure to which it is bonded, producing false compressive strain readings. The ingress of fluids typically leads to this breakdown in resistance-to-ground value and can also change the mechanical properties of the adhesive. A minimum gage-to-mounting-surface resistance-to-ground value of 50 megohms is recommended.

Since signals of interest from strain gage bridges are typically on the order of a few millivolts, shielding of the bridge from stray pickup is important. Gage leads should also be shielded and proper grounding procedures followed. Stray pickup may be introduced by 60 Hz line voltage associated with other electronic equipment, electrical noise from motors, radio frequency interference, etc. Note

TABLE 8.2 SUITABILITY OF VARIOUS MATERIALS AS A HEAT SINK FOR STRAIN GAGE MOUNTING (Units are watts/in^2 on top, kilowatts/m^2 on bottom)

Accuracy Requirements	EXCELLENT Heavy Aluminum or Copper Specimen	GOOD Thick Steel	FAIR Thin Stainless Steel or Titanium	POOR Filled Plastic such as Fiberglass/ Epoxy	VERY POOR Unfilled Plastic such as Acrylic or Polystyrene
Static					
High	2–5 *3.1–7.8*	1–2 *1.6–3.1*	0.5–1 *0.78–1.6*	0.1–0.2 *0.16–0.31*	0.01–0.02 *0.016–0.031*
Moderate	5–10 *7.8–16*	2–5 *3.1–7.8*	1–2 *1.6–3.1*	0.2–0.5 *0.31–0.78*	0.02–0.05 *0.031–0.078*
Low	10–20 *16–31*	5–10 *7.8–16*	2–5 *3.1–7.8*	0.5–1 *0.78–1.6*	0.05–0.1 *0.078–0.16*
Dynamic					
High	5–10 *7.8–16*	5–10 *7.8–16*	2–5 *3.1–7.8*	0.5–1 *0.78–1.6*	0.01–0.05 *0.016–0.078*
Moderate	10–20 *16–31*	10–20 *16–31*	5–10 *7.8–16*	1–2 *1.6–3.1*	0.05–0.2 *0.078–0.31*
Low	20–50 *31–78*	20–50 *31–78*	10–20 *16–31*	2–5 *3.1–7.8*	0.2–0.5 *0.31–0.78*

Courtesy of Measurement Group, Inc., Raleigh, NC, USA.

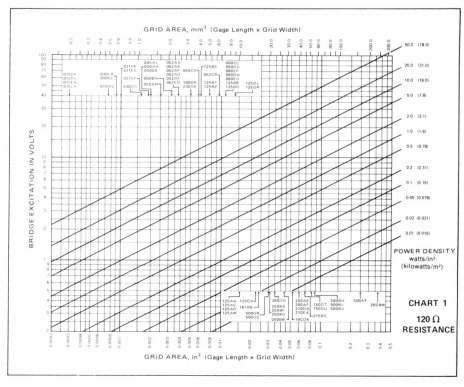

Fig. 8.6 Bridge excitation versus grid area for various power densities and 120 Ω gages (courtesy of Measurements Group, Inc., Raleigh, NC, USA).

that shielding materials for electrical fields are different from those for magnetic fields. Nickel alloy strain gages are particularly susceptible to magnetic fields.

8.3.5   Technical Societies and Strain Gage Manufacturers

In concluding a discussion of the resistance strain gage, it is appropriate to identify some of the technical societies dealing with strain gages and some of the manufacturers of strain gages. In 1956, to accelerate the development of the resistance strain gage, BLH Electronics established a users group to accomplish this purpose and to further the state of the art in strain gage technology in general. This users group was formed primarily of various aircraft companies in the western United States and is entitled the Western Regional Strain Gage Committee (WRSGC). For 15 years, the WRSGC was an autonomous organization financed by BLH Electronics. Since 1971, WRSGC has operated under the auspices of the Technical Committee on Strain Gages (TCSG) of the Society for Experimental Mechanics (SEM). The SEM is the premier organization in the United States involved with strain gages and experimental mechanics in general. The SEM (formerly Society for Experimental Stress Analysis) was founded by Dr. William Murray at the Massachusetts Institute of Technology. Publications of this society include *Experimental Mechanics* and *Experimental Techniques*. A similar European organization is the Joint British Committee for Stress Analysis whose publication is *The Journal of Strain Analysis for Engineering Design*.

A list of strain gage manufacturers includes the following.

> *BLH Electronics*—BLH manufactures wire, foil, and semiconductor strain gages. BLH also manufactures a wide variety of bridge transducers, particularly load cells, and associated signal conditioning equipment.

BLH Electronics
42 Fourth Avenue
Waltham, MA 02154

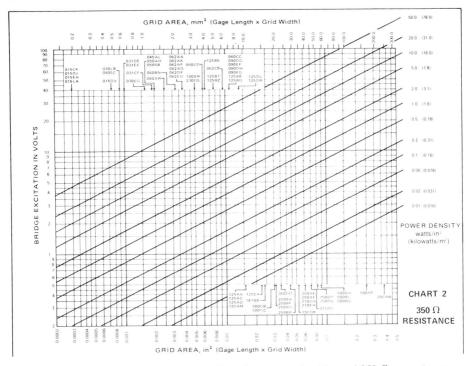

Fig. 8.7 Bridge excitation versus grid area for various power densities and 350 Ω gages (courtesy of Measurements Group, Inc., Raleigh, NC, USA).

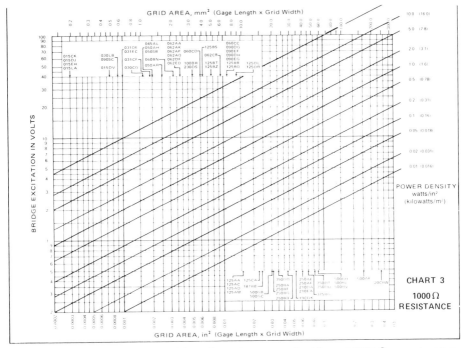

Fig. 8.8 Bridge excitation versus grid area for various power densities and 1000 Ω gages (courtesy of Measurements Group, Inc., Raleigh, NC, USA).

Hottinger Baldwin Measurements (HBM)—HBM manufactures foil gages only. Principally a German company, it also has a significant United States work force. They too manufacture a wide variety of bridge transducers and associated signal conditioners, with load cells a specialty.

Hottinger Baldwin Measurements, Inc.
139 Newburg Street
Framingham, MA 01701

HITEC Corporation—HITEC manufactures weldable strain gages. The gage element is a bonded, heat-cured, factory-installed foil strain gage. HITEC also makes a capacitive strain gage which operates to 840°C.

HITEC Corporation
65 Powers Road
Westford, MA 01886

Micro Engineering II—Micro Engineering II manufactures foil and semiconductor gages. This company purchased the Magnaflux line of foil gages.

Micro Engineering II
14 North Benson Avenue
Upland, CA 91786

Micro Measurements—Micro-Measurements produces a broad spectrum of foil strain gages. Micro-Measurements is part of the *Measurements Group, Inc.*, which emphasizes all aspects of experimental stress analysis.

Micro-Measurements
P. O. Box 27777
Raleigh, NC 27611

Precision Foil Technology—Precision Foil manufactures foil strain gages under the same parent company as *Micro Engineering II*. They specialize in O.E.M. requirements.

Precision Foil Technology
28 N. Benson Avenue
Upland, CA 91786

8.4 THE WHEATSTONE BRIDGE

Best transducer performance can be achieved by minimizing the strain level in the transducer flexure. Lower strains allow increased safety without mechanical overload protection. Effective overload stops are usually troublesome to design and an added expense to make. Reduced strain levels almost always produce an improvement in linearity accompanied by a reduction in the hysteresis originating in the transducer flexure material.

Small strains result in small impedance changes in resistive strain gage elements. Electromechanical transducers use a Wheatstone Bridge circuit to detect a small change in impedance to a high degree of accuracy.

8.4.1 Bridge Equations

The circuit most often used with strain gages is a four-arm bridge with a constant voltage power supply. Figure 8.9 shows a basic bridge configuration. The supply voltage E_{ex} can be either AC or DC, but for now we assume it is DC so equations can be written in terms of resistance R rather than a complex impedance. The condition for a balanced bridge with e_0 equal to zero is

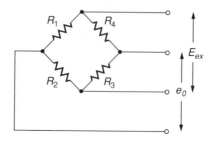

Fig. 8.9 Four-arm bridge with constant voltage (E_{ex}) power supply.

$$\frac{R_1}{R_2} = \frac{R_4}{R_3} \tag{8.5}$$

Next, an expression is presented for e_o due to *small* changes in R_1, R_2, R_3, and R_4.

$$e_o = \left[-\frac{R_3 dR_4}{(R_3 + R_4)^2} + \frac{R_4 dR_3}{(R_3 + R_4)^2} - \frac{R_1 dR_2}{(R_1 + R_2)^2} + \frac{R_2 dR_1}{(R_1 + R_2)^2} \right] E_{ex} \tag{8.6}$$

In many cases, the bridge circuit is made up of equal resistances. Substituting for individual resistances, a strain gage resistance R, and using the definition of the gage factor from Eq. 8.3, Eq. 8.6 becomes

$$e_o = \frac{F E_{ex}}{4} \left(-\epsilon_4 + \epsilon_3 - \epsilon_2 + \epsilon_1 \right) \tag{8.7}$$

The unbalance of the bridge is seen to be proportional to the sum of the strain (or resistance changes) in opposite arms and to the difference of strain (or resistance changes) in adjacent arms.

Equations 8.6 and 8.7 indicate one technique to compensate strain gage circuits to minimize the influence of temperature-induced strain. This was referred to in Section 8.3 as the dummy gage method.

Assume that we have a bridge circuit with one active arm and arbitrarily let this arm be number 4. Equation 8.7 becomes

$$e_o = \frac{F E_{ex}}{4} \left(-\epsilon_4 \right) \tag{8.8}$$

Arm 4 responds to the total strain induced in it, which is comprised of both thermal (t) and mechanical (m) strain

$$\epsilon_4 = \epsilon_m + \epsilon_t \tag{8.9}$$

A problem arises if it is desired to isolate the mechanical strain component. One solution is to take another strain gage (the dummy gage) and mount it on a strain-isolated piece of the same material as that on which gage 4 is mounted. If placed in the same thermal environment as gage 4, the output from the dummy gage becomes simply ϵ_t. If the dummy gage is wired in an adjacent bridge arm to 4 (1 or 3), Eq. 8.7 becomes

$$e_o = \frac{F E_{ex}}{4} \left(-\epsilon_m - \epsilon_t + \epsilon_t \right) \tag{8.10}$$

Equation 8.10 indicates that thermal strain effects are canceled. Similarly, in Fig. 8.2, four gages were shown mounted on a transducer diaphragm. Equation 8.7 indicates that thermal strain effects from this circuit should be canceled.

In reality, perfect temperature compensation is not achieved since no two strain gages from a lot track one another identically. However, compensation adequate for many applications can be accomplished.

The biggest thermal problem with bridge transducers occurs in transient situations, such as explosive or combustion environments. Here, due to individual physical locations, gages in a bridge are not in the same time-varying temperature, and compensation cannot be achieved. The only technique which can be used in this situation is either to cool the transducer by circulating water or gas around it or to delay the thermal transient until the measurement is complete.

The alternating signs in Eq. 8.7 are useful in isolating various strain components when using bridge circuits containing strain gages. Figure 8.10 shows a beam flexure used in an accelerometer. Four gages are mounted on the beam—two on the top and two on the bottom. Notches are placed in the beam to intensify the strain field under the gages. Due to symmetry, the tension gages see the same strain as do the compression gages. If the tension gages occupy two adjacent arms of the bridge and the compression gages the other two, Eq. 8.7 indicates that the net bridge output will be zero. However, if the tension and compression gages are in opposite arms, Eq. 8.7 indicates that a bending strain signal four times that of an individual gage will be produced with temperature compensation also achieved.

Equation 8.6 presented the generalized form of the bridge equation for four active arms. If only one arm (e.g., arm 4) is active, this equation becomes

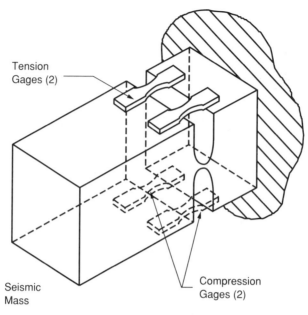

Tension
Gages (2)

Seismic
Mass

Compression
Gages (2)

Fig. 8.10   Strain gaged beam flexure used in accelerometer (courtesy of Endevco).

$$e_o = \left[\frac{-R_3 dR_4}{(R_3 + R_4)^2} \right] E_{ex} \tag{8.11}$$

This equation was specifically presented for *small* changes in resistance, such as those associated with metallic strain gages. If the change in resistance in arm 4 is large, Eq. 8.11 is better expressed as

$$e_o = \frac{(R_4 + \Delta R_4) E_{ex}}{[(R_4 + \Delta R_4) + R_3]} - \frac{R_4 E_{ex}}{R_4 + R_3} \tag{8.12}$$

For an equal-arm bridge, this becomes

$$e_o = \frac{\Delta R E_{ex}}{4R + 2\Delta R} = \frac{F E_{ex} \epsilon}{4 + 2F \epsilon} \tag{8.13}$$

For an equal-arm bridge, Eq. 8.11 becomes

$$e_o = \frac{dR E_{ex}}{4R} = \frac{F E_{ex} \epsilon}{4} \tag{8.14}$$

The difference between Eq. 8.14 and Eq. 8.13 is that Eq. 8.14 describes a linear process while Eq. 8.13 describes a nonlinear one. Semiconductor gages, because of their large gage factor, require analysis using Eq. 8.13.

Semiconductor gages may be used in constant-voltage four-arm bridge circuits when two or four gages are used in adjacent arms and strained so that their outputs are additive. Analysis of the bridge equations for this situation will show that if gages in adjacent arms are subjected to equal but opposite values of ΔR, the output signal is doubled and the nonlinearity in the bridge output is eliminated. Another approach to eliminating this nonlinearity is to design a circuit where the current through the strain gage remains constant.

Table 8.3 provides generalized bridge equations for 1, 2, and 4 equal active arm bridges of various configurations. The dimensionless bridge output is presented in mV/V for a constant voltage power supply. Strain is presented in microstrain. No small strain assumption is built into these equations. For large strains with semiconductor gages, F may not be a constant and this correction also has to be built into the equations. In this table, the Poisson gage is one which measures the lateral

TABLE 8.3   EQUATIONS FOR 1, 2, AND 4 EQUAL ACTIVE ARM BRIDGES

BRIDGE/STRAIN ARRANGEMENT	DESCRIPTION	OUTPUT EQUATION $-e_o/E_{ex}$ IN mV/V
	Single active gage in uniaxial tension or compression.	$$\frac{e_o}{E_{ex}} = \frac{F\epsilon \times 10^{-3}}{4 + 2F\epsilon \times 10^{-6}}$$
	Two active gages in uniaxial stress field—one aligned with max. principal strain, one "Poisson" gage.	$$\frac{e_o}{E_{ex}} = \frac{F\epsilon(1+\nu) \times 10^{-3}}{4 + 2F\epsilon(1-\nu) \times 10^{-6}}$$
	Two active gages with equal & opposite strains—typical of bending-beam arrangement.	$$\frac{e_o}{E_{ex}} = \frac{F\epsilon}{2} \times 10^{-3}$$
	Two active gages with equal strains of same sign—used on opposite sides of column with low temperature gradient (bending cancellation, for instance).	$$\frac{e_o}{E_{ex}} = \frac{F\epsilon \times 10^{-3}}{2 + F\epsilon \times 10^{-6}}$$
	Four active gages in uniaxial stress field–two aligned wth max. principal strain, two "Poisson" gages (column).	$$\frac{e_o}{E_{ex}} = \frac{F\epsilon(1+\nu) \times 10^{-3}}{2 + F\epsilon(1-\nu) \times 10^{-6}}$$
	Four active gages in uniaxial stress field–two aligned wth max. principal strain, two "Poisson" gages (column).	$$\frac{e_o}{E_{ex}} = \frac{F\epsilon(1+\nu) \times 10^{-3}}{2}$$
	Four active gages with pairs subjected to equal and opposite strains (beam in bending or shaft in torsion).	$$\frac{e_o}{E_{ex}} = F\epsilon \times 10^{-3}$$

Courtesy of Measurements Group Inc., Raleigh, NC, USA.

compressive strain accompanying an axial tension strain. As noted earlier, only for two adjacent active gages with equal and opposite strains, or for four active gages with pairs subjected to equal and opposite strains, is the bridge output a linear function of strain.

8.4.2   Lead Wire Effects

There has been a historical lack of agreement between manufacturers of strain gages as to color codes and wiring designations. This is particularly true in bridge transducers. Figures 8.11 and 8.12 are suggested industry standards which have assisted in lessening this confusion. Figure 8.11 covers the situation where all bridge elements are remote from the power supply, while Fig. 8.12 covers the situation where only one bridge arm is remote from the power supply. The bridge balance network and shunt calibration are discussed in Sections 8.5 and 8.6, respectively. Table 8.4 presents guidelines for multi-conductor strain gage cable.

The previous discussion has assumed that the only resistive elements in the circuits are the gages themselves. Resistance of circuit lead wires also is a consideration.

One possible need for remote recording occurs when the bridge power supply and the readout instrumentation are at one location and the bridge transducer is at a remote location. In this situation, the resistance R_L of each lead wire between the bridge and the power supply or readout must be accounted for. Most readout instruments have very high input impedances, so the effect of R_L in series with them can be ignored. The significant effect of lead-wire resistance is to modify the

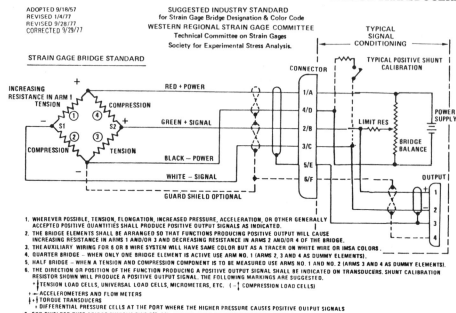

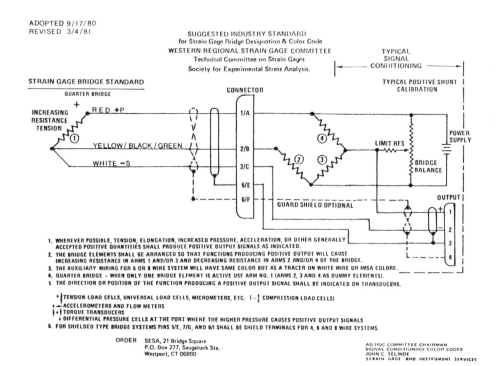

Fig. 8.11 Color code and wiring designation, four-arm bridge (courtesy of Western Regional Strain Gage Committee).

Fig. 8.12 Color code and wiring designation, single-arm bridge (courtesy of Western Regional Strain Gage Committee).

TABLE 8.4 MULTICONDUCTOR STRAIN GAGE CABLE GUIDELINE, WESTERN REGIONAL STRAIN GAGE COMMITTEE

A need exists for low millivolt signal levels to be transmitted by better quality multiple conductor cables of considerable length.

I.	Conductors:	Three through eight twisted, stranded conductors of tinned copper per ASTM-B-174, AWG 20-16/32, or AWG 18-16/30
II.	Color Code:	Jacket: orange, grey, white, or black.
		Conductors: Per ISA-S37.3, ANSI-MC6.2-1975, and WRSGC/SESA 5-6-1960
III.	Insulation:	Superior to the PVC materials currently in use. The dielectric material should be nonhygroscopic and approach zero water absorption and zero water permeability. Target jacket thickness of .016 in. or less and conductor insulation of .012 in. or less. Target resistance values should be constant as practical humid and wet environments and as high as possible (target value of 500 megohms per 1,000 feet). The breakdown level of the dielectric materials shall be greater than 150 volts DC.
IV.	Construction:	The cable shield shall be aluminized polyester tape with 100% coverage of all conductors. A 22-AWG drain wire shall be in intimate contact with the shield throughout the entire cable length. The cable shall have as small a diameter as practical and flexible enough to have a bend radius less than six (6) cable diameters. Overall cable strength sufficiently high to be pulled through conduits.

Courtesy of Society for Experimental Stress Analysis.

resistance in series with the power supply from R_{bridge} to $R_{bridge} + 2R_L$. For example, a lead-wire resistance of 3 Ω and a bridge resistance of 120 Ω will produce loading effects which, if not corrected, will result in a 5 percent error in bridge transducer output.

There are at least three simple techniques to eliminate this error source.

1. The bridge transducer can be calibrated with the long length of cable with which it will operate.
2. The excitation voltage E_{ex} can be measured at the bridge itself instead of at the power supply and appropriate values substituted in Eq. 8.6 or equivalent versions of it.
3. The bridge voltage E_{ex} can be determined by measuring the current to the bridge (I_{bridge}) and calculating E_{ex} as the product of $I_{bridge} \times R_{bridge}$.

Another possible need for remote recording occurs when two gages (either both active or one active and one for temperature compensation) are at the test site. The other two bridge completion resistors are in parallel with the power supply and located adjacent to it. In Fig. 8.9, assume R_3 and R_4 are the two remote active arms. In Eq. 8.6, the last two terms would be zero since these arms are not active. R_3 and R_4 in this equation would become respectively $R_3 + R_L$ and $R_4 + R_L$. If the strain gages in both arms are identical, Eq. 8.6 reduces to

$$e_o = \frac{FE_{ex}R}{4(R + R_L)}(-\epsilon_4 + \epsilon_3) \qquad (8.15)$$

Other situations can be investigated by substituting appropriate values for the resistance in each bridge arm (including lead-wire resistance) in the governing bridge equation. In addition, Sec. 8.6 will show that shunt calibration is one technique that can be used to compensate the system for the effects of lead-wire resistance.

8.4.3 Temperature Compensation

Before leaving the analysis of the Wheatstone Bridge circuit, temperature compensation of bridge type transducers should receive additional discussion. An ideal transducer would yield an output voltage which is a constant calibration factor times its mechanical input, independent of other environmental factors. Ambient temperature variations are one of the major error sources in precision transducers. Even when using self-temperature-compensated strain gages, and taking advantage of the ability of the Wheatstone Bridge circuit to subtract in the dummy gage method, some residual error remains. These remaining errors are of two types.

First, the transducer zero-output can change with temperature. Unequal mechanical expansion of transducer members can cause this effect. Second, the calibration factor, span, or sensitivity also can change with temperature. This can be caused, for example, by a change in the stiffness of the transducer flexure with temperature.

The following discussion provides one compensation scheme for each type (metallic and semi-conductor) of bridge transducer. References 7 and 8 are sources of more detailed information. An equal-arm bridge transducer operating with a constant-voltage supply is assumed. Metallic strain gages are discussed first.

Figure 8.13 shows one scheme for compensating for transducer zero-shift. A corner of the bridge is brought out to terminals, and a temperature-sensitive resistor, r, is placed in one side of the bridge. Typically, a wire resistor such as Balco, nickel, or copper with a positive temperature coefficient is used.

The transducer must first be temperature calibrated and the change in zero reading for a given temperature range determined. This can be characterized in volts of output change per volt of input. Definitions:

b = output voltage change per degree per input volt
a = temperature coefficient of resistance of r
R = bridge arm resistance
T = temperature change from reference temperature

If the bridge supply voltage is E_{ex}, and R is changed a small amount by the addition of r, the bridge output is

$$e_o = \frac{E_{ex}r}{4R} \tag{8.16}$$

Equation 8.16 can further be expressed as

$$e_o = \frac{E_{ex}r_o(1 + aT)}{4R} \tag{8.17}$$

or

$$e_o = constant + E_{ex}bT \tag{8.18}$$

where r_o is the value of r at the reference temperature. The effect of the constant term is eliminated by a temperature-insensitive trim resistor in an adjacent arm. The above equations indicate that at the reference temperature r_o should be selected equal to $4Rb/a$. If the transducer is properly designed, b is very small compared to a, keeping the compensating resistor small in value. The compensating resistor should be located in an arm causing a voltage change of opposite sign to the zero drift with increasing temperature.

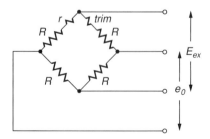

Fig. 8.13 Transducer bridge compensation for zero shift, metal gages.

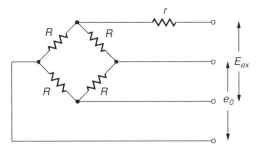

Fig. 8.14   Transducer bridge compensation for span, metal gages.

After zero-shift is compensated, the calibration or span factor remains to be compensated. Most metal strain gage transducers give larger outputs with increasing temperature, so the temperature coefficient of the calibration scale factor, K, is positive. The trick in span compensation is to hold the transducer supply voltage constant while automatically varying the bridge current, causing it to decrease with increasing temperature. In this discussion, r is identified to be a positive series resistor (Fig. 8.14). Definitions:

$r = r_o(1 + aT)$
a = positive temperature coefficient of r
T = temperature difference from reference temperature
c = temperature coefficient of the calibration factor K, so $K = K_o(1 + cT)$
E_{ex} = transducer supply voltage

The voltage on the transducer at the reference temperature is $RE_{ex}/(R + r_o)$ and at temperature T is $RE_{ex}/(R + r)$. The ratio by which it changes is $(R + r_o)/(R + r)$, which is used to correct for the variation in K. This variation is corrected for when $K_o(1 + cT)(R + r_o)/[R + r_o(1 + aT)] =$ constant. The value of r_o which satisfies this requirement can be shown to be

$$r_o = \frac{cR}{(a - c)} \tag{8.19}$$

Note that in span and zero compensation as discussed thus far, the compensating resistors must be at the same temperature as the transducer. Usually, this is accomplished by mounting the resistors inside the transducer.

Figure 8.15 shows one technique for correcting for zero-shift due to temperature in semiconductor bridges. Temperature compensation is performed by adding non-temperature-sensitive resistors in series and parallel to the gage having the highest resistance change with temperature. The objective of this method is to achieve both zero balance and temperature compensation together. Since the compensation resistors are non-temperature-sensitive, they can be added wherever convenient in the circuit.

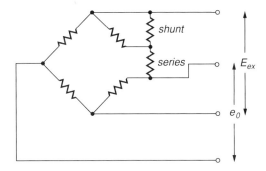

Fig. 8.15   Transducer bridge compensation for zero shift, semiconductor gages.

The bridge is first balanced using a series resistor at ambient room temperature. Next, the transducer is cycled over the temperature extremes. A parallel resistor is installed across the gage having the greatest resistance change. The bridge is then rebalanced, and the procedure repeated until satisfactory performance is achieved.

Semiconductor bridge transducers are typically compensated for calibration or span factor with a circuit as in Fig 8.14. However, r for this situation is a non-temperature-sensitive resistor. For P-type silicon gages, the strain sensitivity drops with temperature while the resistance rises. The increase in resistance occurs at a greater rate than does the decrease in sensitivity. Figure 8.14 shows that the effect of an increase in resistance R, with r constant, is to increase the voltage applied to the bridge, offsetting the decrease in strain sensitivity. Alternately, in Fig. 8.14, r can be replaced by a thermistor instead of a fixed dropping resistor. The thermistor is generally a more efficient method of compensation but must be in the same thermal environment as the bridge network.

When balancing Wheatstone Bridges, it must be determined that the balancing circuit does not significantly alter the thermal compensation network. Balancing methods are discussed next.

8.5   RESISTANCE BRIDGE BALANCE METHODS

Even when a best attempt is made at matching resistors, the output from a bridge transducer with zero measurand applied is always something other than zero volts. With microprocessors and scanners, this is of little consequence. The initial bridge output can be acquired and stored in the memory of the microprocessor and then subtracted from all subsequent readings. Frequently, however, it is desired to initialize a bridge circuit such that a zero value of measurand corresponds to zero voltage. For example, assume it is desired to acquire a vibration measurement on a space vehicle using a bridge transducer. Assume the channel is to be calibrated for $\pm\,20$ g and the accelerometer has a sensitivity of $1mV/g$ ($g=$ standard acceleration of gravity). If the data channel range were ±20 mV, and the accelerometer acquiring the measurement had a zero offset of 5 mV, the channel could transmit only in the range of $+15$ g to -25 g as opposed to ±20 g. Balancing the bridge would solve this problem.

Equation 8.5 presented the requirement for a balanced bridge. Basically, the resistance ratio of any two adjacent bridge arms must be equal to the resistance ratio of the other two arms. Any bridge balancing network must then have as its objective the satisfying of this criterion. The two main types of zero balancing methods are those which manipulate one arm of a transducer bridge to bring its output to the desired condition and those which manipulate two adjacent arms of the transducer bridge.

Figure 8.16 presents the most common circuit for manipulating a single bridge arm. A variable resistor R_B is placed across one of the resistors (say R_4) whose value needs to be lessened such that $R_1/R_4 = R_2/R_3$. The effect of R_B in parallel with R_4 is to lessen the value of the bridge arm from R_4 to some new value R_T.

The overall combination of R_B in parallel with R_4 must be variable over a range at least equal to the maximum possible initial unbalance of the bridge. Selecting this range, other than by trial and error, requires knowledge of the strain gage resistance, R, its tolerance in percentage, m, and the number of active gages, n, in the bridge. The range of the balancing circuit should be

$$\frac{2Rmn}{100} \tag{8.20}$$

Note that the presence of the variable resistor R_B desensitizes the bridge network since $\Delta R_4/R_4$ is not equal to $\Delta R_T/R_T$. If the strain gages are initially closely matched, the influence of this effect is small since R_B will remain large and R_T will closely approximate R_4. For optimum precision, the best method to minimize the influence of the variable resistor is to calibrate the transducer once

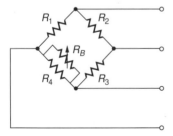

Fig. 8.16   Circuit for manipulating a single bridge arm.

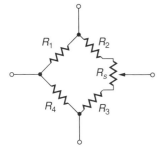

Fig. 8.17 Circuit for series manipulation of two adjacent bridge arms.

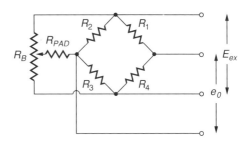

Fig. 8.18 Circuit for parallel manipulation of two adjacent bridge arms.

the bridge is balanced. Of course, if less than four arms of the bridge are active and balancing is performed across a dummy completion resistor, no desensitizing of the bridge occurs.

Two techniques are available to manipulate two adjacent arms in a bridge. Again, the rationale for this manipulation is to satisfy Eq. 8.5. The first technique is series manipulation which assumes the bridge is open such that the variable resistance may be inserted in series with two arms of the bridge. This technique is not applicable to a closed bridge.

Figure 8.17 shows a variable series resistor R_S inserted in one corner of the bridge. The insertion of R_S, which is typically quite small, allows adjustment of the ratio of R_2 to R_3 to achieve balance. Reference 2 provides the best discussion of bridge balance networks and indicates that minimum bridge desensitization occurs when bridge power is applied across the vertical terminals of Fig. 8.17 as opposed to the horizontal.

The second technique discussed (and the one typically used) is parallel manipulation of two adjacent bridge arms. Figure 8.18 illustrates this technique. R_B is a parallel variable resistor which allows the ratio of R_2/R_3 to be adjusted. A pad resistor, R_{PAD}, serves simply to keep the individual bridge arms from being shorted out at the end of travel of R_B. Again, the secret is to keep the combination of R_B and R_{PAD} as high as possible to avoid bridge desensitization. If no other guidance is available, start out with a pad resistor about 100 times the bridge resistance and a variable resistor about 20 times the bridge resistance. Again, maximum accuracy is achieved when the bridge transducer is calibrated with the balance network with which it will be used.

As alluded to earlier, the addition of a balance network to a bridge transducer may react unfavorably with temperature compensation resistors placed in the transducer's circuitry by its manufacturer. Temperature compensation can be severely modified by the presence of this balancing network. The prerequisite to insertion of a balance network should be an exact knowledge of the circuit of the transducer. For this reason, and for reasons associated with desensitization of the transducer, balance networks should be avoided unless required.

8.6   RESISTANCE BRIDGE TRANSDUCER MEASUREMENT SYSTEM CALIBRATION

A basic component in any measurement system is the transducer. The measurement system can be as simple as a transducer, a cable from the transducer, and a recorder. Alternately, the measuring system can contain many more of the elements of the transmitting and receiving system defined in Section 8.1. Cables, amplifiers, filters, digitizers, tape recorders, and so on all have the capability, when inserted into a measurement system, to modify both the amplitude and spectral content of the signal from the transducer defining the measurand. The response of these components may also drift with time.

To obtain measurements of the highest possible quality, one must accurately and carefully calibrate the entire measurement system as near to the time of actual measurement as possible. The calibrations may be conducted prior to, immediately after, or even during the time of actual measurement. Such calibration of an entire measurement system is referred to as "end-to-end" calibration. This calibration ordinarily does not replace the evaluation of individual components of the measurement system.

One group concerned with "end-to-end" calibration of measurement systems is the Telemetry Group/Range Commanders Council whose Secretariat is headquartered at White Sands Missile Range, New Mexico. The Transducer Committee of the Telemetry Group coordinated the writing of Chapter

2 of Ref. 9, entitled "Test Methods for Transducer-Based System Calibrations." The following information is largely extracted from that chapter which also deals with piezoelectric transducers, servo transducers, capacitive and inductive transducers, and thermoelectric transducers.

A preferred calibration procedure is one in which a known value of the measurand is applied directly to the transducer in the measurement system (transmitting, receiving, and display) in which it will be used. This procedure permits the output display to be read directly in terms of units of the measurand.

8.6.1 Static Calibration

The basic equipment for the static calibration of transducer systems consists of a measurand source supplying accurately known and precisely repeatable values of the measurand and an output-indicating or recording system. The combined errors or uncertainties of the calibration system should be sufficiently smaller than the permissible tolerance of the system performance characteristic under evaluation so as to result in meaningful calibration values. All calibration system components should be periodically checked against standards. Environmental conditions during calibrations should be constant and specified to permit corrections to the data, as required.

The procedures which will be specified are based on the assumption that the measuring system is linear. For systems that will ultimately measure dynamic data, linearity is a prerequisite.

The static calibration sequence consists of the following steps:

1. Zero measurand output verification
2. Sensitivity verification
3. Linearity and hysteresis verification
4. Repeatability verification

Zero measurand output verification starts with a measurement of system output with zero measurand applied to the trandsducer. Zero measurand is an important measurement for several reasons:

1. In many measurement systems, the transducers are exposed to zero measurand before the test begins.
2. For many transducers, zero measurand means that no external input to the transducer is required, thus greatly simplifying the test procedure.
3. Measurement system malfunctions, including drift, will frequently appear when the system output is monitored over some reasonable time period with zero measurand applied.
4. In some measurement systems with no external measurand input, the ambient environment will furnish an important reference point for the system calibration. For example, for an absolute pressure measuring system, the ambient atmospheric pressure provides this reference. Similarly, for certain accelerometer systems, the measurable attitude of the test vehicle prior to launch represents a known component of the earth's gravitational field as input to the accelerometer.

For a measurement system with a linear response, the slope of the line characterizing the input measurand versus system output represents the sensitivity of the system. There are a number of straight lines which may be chosen to provide this sensitivity verification. These include:

1. End point line—The straight line between the outputs at the specified upper and lower limits of the range.
2. Best straight line—The line midway between the two parallel straight lines closest together that enclose all output versus measurand values on a calibration curve.
3. Terminal line—The straight line for which the end points are 0 percent and 100 percent of both measurand and output.
4. Theoretical slope line—The straight line connecting the specified points between which a specified-theoretical curve has been established.
5. Least squares line—The straight line for which the sum of the squares of the residuals is minimized for all calibration points.

Procedures used in the verification of sensitivity will depend on specific accuracy, calibration time, and expense trade-off choices for each system. For unidirectional transducers, it is typical to calibrate from zero to full-scale and back again in 10 percent of full-scale increments (21 points). For bidirectional transducers, a 21-point calibration cycles the transducer from negative full-scale to positive full-scale and back again in 20 percent of full-scale increments.

Data extracted from these calibrations are typically linearity and hysteresis. Linearity is the closeness of a calibration curve to the specified straight line, expressed as the maximum deviation of any calibration point from that line during any one calibration cycle. Hysteresis is the maximum difference in output at any measured value within the specified range when the value is approached first with increasing and then with decreasing measurand.

The reference straight line selected is often the linear least-squares line. This line is based on the following principle: the most probable value of an observed quantity is such that the sum of the squares of the deviations of the observations from this value is a minimum. This is based on the fact that most measurements of physical quantities show a normal distribution with both positive and negative deviations from the mean probable and very large deviations less likely than small deviations. The line can be defined unequivocally in terms of the quantities measured. The line also is statistically significant, and standard deviations can be assigned to estimates of the slope, intercept, and other parameters derived from it.

An additional parameter describing the performance of the measurement system is obtained by repeating the static calibration of the system. A minimum of two, but preferably three, consecutive static calibrations yield data from which the "repeatability" of the system is verified. Repeatability is the ability of the measurement system to reproduce output readings when the same measurand value is applied to it consecutively under the same conditions and in the same direction. It is expressed as the maximum difference between corresponding values from at least two consecutive calibrations. Although there is no universal agreement as to the particular values selected, a value close to full-scale output is commonly used.

If the bridge transducer will be used to acquire time-varying measurements, the measuring system must be both dynamically and statically calibrated. The dynamic response of any system is described by a frequency response function which is a complex function of frequency. The frequency response function relates system output to system input in the frequency domain. For measurement systems, this frequency response function is typically represented by Bode plots which are log amplitude and phase versus log frequency.

8.6.2 Dynamic Calibration

Dynamic calibrations are inherently more difficult to perform than are static calibrations and usually require specialized equipment. Dynamic calibrations can be performed using several types of well-defined input signals, such as applications of sinusoids, transients, or broad-band noise. The principal requirement that the input must satisfy is that it must contain significant energy at frequencies over the range of the frequency response function of interest.

The dynamic calibration sequence consists of the following steps:

1. Dynamic sensitivity determination
2. Dynamic amplitude linearity determination
3. Amplitude-frequency verification
4. Phase-frequency verification

If the measuring system does not have zero frequency response, its end-to-end calibration is of necessity made by dynamic methods. The simplest approach to dynamically determining system sensitivity is the application of a sinusoidally varying measurand to the transducer. The amplitude of the measurand should be equal to the range of the transducer. For unidirectional transducers, this test involves biasing the transducer to its half-range point. At the test frequency, it is possible to relate the peak amplitude of the system response to the amplitude of the measurand and determine system dynamic sensitivity.

It is further desired to acquire dynamic amplitude linearity by performing tests equivalent to dynamic sensitivity determination at several levels of the measurand (usually levels of 25, 50, 75, and 100 percent of full-scale are adequate). This testing should be performed at several different frequencies over the range of the frequency response function of interest. If the measurement system cannot be verified to be linear, it should not be used to acquire time-varying measurements.

The concept of a measuring system having a unique amplitude-frequency and phase-frequency response is only meaningful for systems which have been verified to be dynamically linear. Amplitude-frequency response tests consist essentially of a series of dynamic sensitivity determinations at a number of frequencies within the bandwidth of the system. Three is the *minimum* number of test frequencies. One test should be performed close to the upper limit of the frequency band where the response has not been affected by the high frequency roll-off characteristic of the system. The second frequency should be sufficiently higher than the first to provide some indication of the roll-off rate of the system. The third frequency should be about halfway between zero frequency and the first test frequency to verify a flat response to the upper band edge. More improved definition obviously can be provided by increasing the number of test frequencies.

Phase-frequency response characteristics of a measuring system can often be acquired simultaneously with the amplitude-frequency response. An output recording device is required with two identically responding channels. The system output is recorded on one channel. The second channel records the measurand, which is typically acquired by a previously calibrated monitoring transducer whose amplitude-frequency and phase-frequency response characteristics are well established. A time correlation between the system output and this monitoring transducer can establish measuring system phase-frequency response. For systems measuring signals whose time history is important, a linear phase response with frequency is required. For those signals about which only statistical information is to be acquired (e.g., random vibration), phase response is not an important system characteristic.

With today's technology, frequency response functions can also be characterized by transient or random system excitation. Dual channel spectrum analyzers can ratio input-to-output measuring system Fourier transforms in near real time. Recall that the system input stimulus must contain significant signal content at all frequencies of interest.

8.6.3 Electrical Substitution Techniques

If actual values of the measurand cannot be used to calibrate resistance bridge transducers, electrical substitution techniques can be used. Test equipment required includes a precision voltage source, precision resistors or decade box, and a signal generator. The techniques include shunt calibration, series calibration, and bridge substitution. Shunt calibration techniques are discussed first.

Inserting a resistor of known value in parallel with one arm of a strain gage bridge is single-shunt calibration. The calibration resistor is inserted across the arm opposite the strain gage conditioning system. The conditioning system may contain a balance potentiometer, a limit or pad resistor, modulus resistors, and temperature compensation resistors. Standard practice is to insert the shunt resistor between the negative input (excitation) and the negative output (Fig. 8.19). This reduces errors caused by shunting some of the bridge conditioning resistors.

The value of the shunt resistor R_C is determined by first applying a value of the measurand to the transducer and monitoring the voltage change at the transducer output terminals (Fig. 8.19). With the measurand removed, a decade box is substituted for R_C and its resistance adjusted until a voltage change results with a magnitude equal to that caused by the measurand. For subsequent calibrations, a fixed resistor R_C can be substituted for the decade box. When the switch in series with R_C is closed, it will produce a step voltage through the measuring system of amplitude equal to that produced by the measurand. When shunting one arm of the bridge, the resistance change produced in that arm is $-R^2/(R_c + R)$.

In the calibration laboratory, the small lead length associated with the transducer introduces no error in establishing R_C. The application of the bridge transducer in the field can require significant lengths of cable with significant transmission line resistance. Figure 8.20 illustrates the situation where R_C must be applied remotely. If R_C were applied directly at the bridge, loading errors introduced by transmission lines would be accounted for. If (as in Fig. 8.20) R_C is applied at a remote location, the effect of the transmission line resistance $2R_L$ in series with R_C must be considered.

Bipolar shunting is used when the physical loading creates both positive- and negative-going signals and it is desired to create positive and negative calibration outputs. The calibration resistor is alternately inserted across the two arms opposite the bridge-conditioning network. If line resistance is significant, it must be considered as in Fig. 8.21.

Series calibration of bridge transducers is considerably different from shunt calibration. Figure

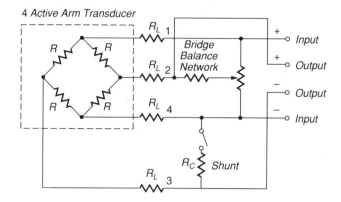

Fig. 8.19 Single-shunt calibration of bridge transducer.

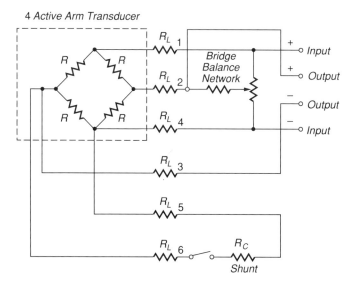

Fig. 8.20   Remote location single-shunt calibration of bridge transducer.

8.22 describes this process. Series calibration consists of two distinct calibration phases. In the zero calibration phase, excitation is removed from the bridge. The sensitivity resistor (R_{SENS}) is concurrently placed across the bridge input terminals, simulating the power supply impedance to result in the same overall system impedance encountered in the data circuit. Zero bridge transducer output is recorded.

The next calibration phase is the series phase. In the series calibration mode the two switches in Fig. 8.22 are closed, introducing R_{c1} and R_{c2} into the circuit. This removes power from one corner of the bridge and puts a calibration resistor R_{c1} in series with the sensitivity resistor and one side of the bridge output. The second calibration resistor becomes intermediate between the R_{SENS} to R_{c1} connection and excitation return. This second resistor, R_{c2}, is selected to maintain the approximate equivalent bridge impedance across the excitation. The calibration circuit then electrically simulates the bridge transducer. The value of R_{c1} is determined experimentally, corresponding to some measurand equivalent.

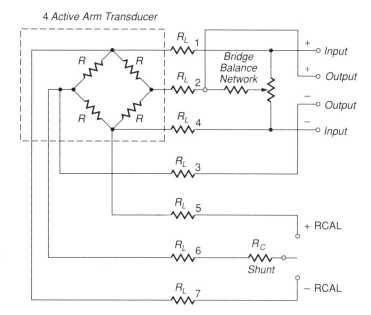

Fig. 8.21   Remote location double-shunt calibration of bridge transducer.

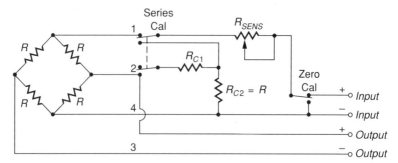

Fig. 8.22   Series calibration of bridge transducer.

Series calibration overcomes a serious shortcoming of shunt calibration. During application of a shunt resistor, the transducer can still respond to mechanical input. The calibration step is superimposed upon any mechanically induced signal present. If the mechanical input is static and of sufficient magnitude, over-ranging will invalidate the calibration step. If the mechanical input is dynamic, it may be impossible to accurately measure the magnitude of the calibration step. The magnitude of the series calibration step is significantly more independent of this mechanical input. As in all calibration, transmission line resistance must be considered where significant. Similarly, a change in sensitivity resistance modifies the effect of the series calibration resistance. However, the typical error incurred is negligible.

The final electrical substitution technique discussed is bridge substitution. This technique involves substitution of a model for the bridge transducer itself. Figure 8.23 represents a typical low-level bridge system.

An accurate bridge transducer model has the same terminal impedance as the transducer and provides a fast and simple method of generating a static and dynamic output equivalent to that generated by the transducer for a given physical load. It also provides a convenient method for verifying the calibration resistor's measurand equivalency for shunt and series systems. The two types of bridge transducer models employed for system calibrations are the shunt resistor adapter and the shunt resistor bridge.

Figure 8.24 describes the shunt resistor adapter which is simple, inexpensive to construct, and an exact model since it is used in conjunction with the actual transducer. The adapter is inserted between the transducer and the rest of the measurement system. It performs three primary functions.

1. It supplies the stimulus for performance of system end-to-end calibrations. Shunting the arms of a transducer bridge with the appropriate resistors produces an unbalance in the bridge equivalent to that produced by a given measurand. The adapter provides a convenient method of applying these shunt resistors directly to the bridge with negligible line loss (S_1 and S_2).

2. It performs a system frequency response test. A convenient system frequency response can be performed by selecting the appropriate shunt resistor and sweeping the adapter's AC power supply over the desired range. Figure 8.25 shows a typical oscillograph display of the results.

3. It provides a convenient check of the system's calibration resistors (R_c) and equivalents. The system's R_c equivalent will differ from the values established by the laboratory calibration as a function of line resistance, calibration resistor tolerance, etc. Since the adapter shunt resistors are precision resistors that are applied directly to the bridge, the equivalency of the adapter shunt resistors will not be affected by lead resistance and other variables.

Although the shunt resistor adapter model is a very powerful and simple calibration tool, it has two undesirable characteristics. The least desirable characteristic is that the system calibration and calibration resistor equivalents generated by the adapter are incremental values superimposed on the transducer output resulting from the physical stimulus acting at the time of the test. Also, the adapter does not provide a fixed independent reference since it is used in conjunction with the transducer.

The undesirable features of the shunt resistor adapter can be eliminated by replacing the actual transducer with a bridge model (Fig. 8.26). Since the shunt resistor bridge (bridge model plus shunt resistor adapter) is a stable, complete model of the transducer, it can be used to perform an absolute end-to-end system calibration and can be a valuable tool in troubleshooting.

Several disadvantages are encountered when using the shunt resistor bridge as a calibration tool. Since some transducers are hard to model, it is difficult to insure that the bridge is a representative model of the transducer under all conditions. Furthermore, a different bridge model is required for each major transducer design.

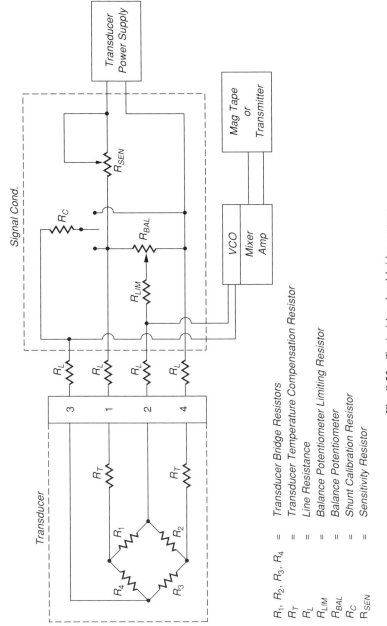

R_1, R_2, R_3, R_4 = Transducer Bridge Resistors
R_T = Transducer Temperature Compensation Resistor
R_L = Line Resistance
R_{LIM} = Balance Potentiometer Limiting Resistor
R_{BAL} = Balance Potentiometer
R_C = Shunt Calibration Resistor
R_{SEN} = Sensitivity Resistor

Fig. 8.23 Typical low-level bridge system.

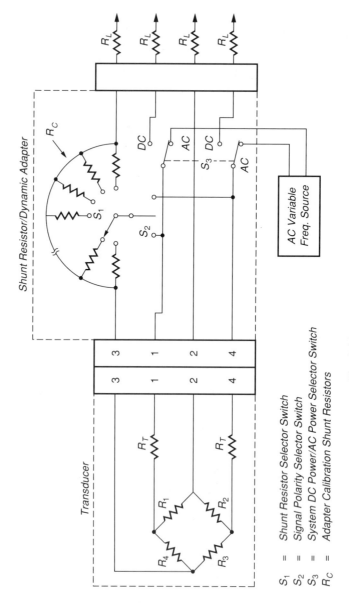

S_1 = Shunt Resistor Selector Switch
S_2 = Signal Polarity Selector Switch
S_3 = System DC Power/AC Power Selector Switch
R_C = Adapter Calibration Shunt Resistors

Fig. 8.24 Shunt resistor adapter.

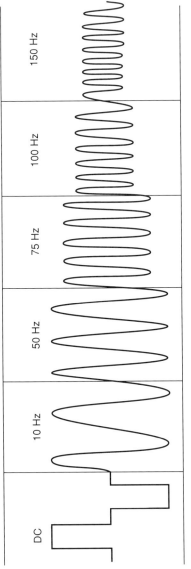

Fig. 8.25 Oscillograph frequency response display.

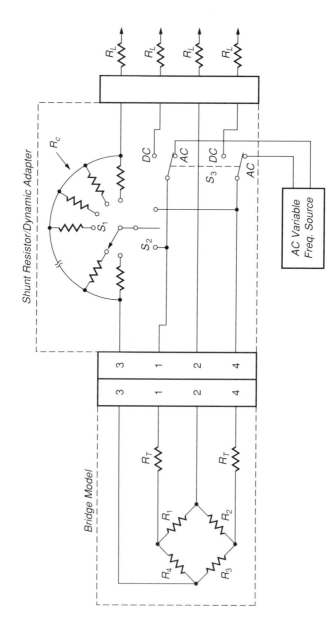

Fig. 8.26 Shunt resistor bridge.

As a final note, remember that the resistance of semiconductor bridge transducers is strongly a function of temperature. When using shunt or series calibration techniques on semiconductor bridges, ambient temperature changes should be taken into account.

8.7 RESISTANCE BRIDGE TRANSDUCER MEASUREMENT SYSTEM CONSIDERATIONS

8.7.1 Bridge Excitation

When amplifiers and power supplies were formally designed around vacuum tubes, component drift was a problem in bridge transducer measurement systems. AC power supplies in bridge circuits eliminated many of these problems by operating at frequencies above DC. Most bridge transducer power supplies today are DC. When comparing DC supplies with AC, the following advantages are associated with DC:

1. Simpler circuitry
2. Wider resultant instrumentation system frequency response
3. No cable capacitive or inductive effects due to the excitation
4. Shunt calibration and bridge balance circuitry are simpler

Independent of type of supply, the power level selected has to take account of all variables which affect the measurement. These include gage resistance, gage grid area, thermal conductivity of flexure to which gage is mounted, flexure mass, ambient test temperature, whether used on a static or dynamic test, accuracy requirements, and long- or short-term measurement. These variables account for the fact that a strain gage is a resistance which has to dissipate heat when current passes through it. Most of the heat is conducted away from the gage grid to the transducer flexure. The result of inadequate heat conduction is gage drift.

For transient measurements, a steady transducer zero-reference is not as important as for static measurements. Bridge power can be significantly elevated to increase measurement system signal-to-noise ratio.

The following specifications define key performance parameters of DC output instrumentation power supplies. Input supply can be either AC or DC.

1. Warmup time—The time necessary for the power supply to deliver nominal output voltage at full-rated load. It is usually specified over the range of operating temperatures.
2. Line regulation—The change in steady-state DC output voltage resulting from an input voltage change over the specified range.
3. Load regulation—The change in steady-state DC output voltage resulting from a full-range load change.
4. Efficiency—The ratio of the output power to the input power.
5. Load transient recovery—The time required for the output DC voltage to recover and stay within a specified band following a step change in load.
6. Periodic and random deviation—The AC ripple and the noise of the DC output voltage over a specified bandwidth with all other parameters held constant.
7. Stability (drift)—The deviation in the DC output voltage from DC to an upper limit which coincides with the lower limit as specified above in (6).
8. Temperature coefficient—The change in output voltage per degree change in ambient temperature.

Reference 10 defines test procedures for these specifications.

8.7.2 Signal Amplification

In addition to providing a precision power source to bridge transducers, the resultant millivolt signals from these transducers often require amplification. This amplification is usually performed by a differential DC amplifier. A differential DC amplifier is an electronic circuit whose input lines are conductively isolated from the output lines, power and chassis ground and whose output voltage is proportional to the differential input signal voltage. Ideally, both input lines have equal impedance and transfer characteristics with respect to the amplifier ground structure. The amplifier has a frequency response from 0 Hz to a value determined by the bandwidth of the amplifier.

Selecting amplifiers can be difficult because specification terminology is not universally standardized. Amplifier specifications are either referred-to-input (RTI) or referred-to-output (RTO). Discussing these specifications can lead to an understanding of the amplifiers themselves.

1. Input impedance—The minimum impedance the amplifier will present when operated within its specification. It is the impedance seen between the two ungrounded input lines of the amplifier.

2. Source current—The bias current flowing through the circuit comprised of the amplifier input terminals closed through the source resistance. The amplifier input transistors act as constant current generators in series with the input terminals. This current can result in both offset voltage and common mode voltage.

3. Common mode rejection—The measure of the conversion of common mode voltage to normal differential signal. The common mode input voltage is the voltage common with both inputs to the amplifier. A common mode rejection of 60 dB implies that a 10 volt signal applied simultaneously to both inputs produces an error signal RTI of 10 millivolts.

4. Linearity—The maximum deviation from the least-squares straight line established through the output voltage versus differential input voltage characteristic. In evaluating linearity, it is usually sufficient to test at the highest and lowest gains, since linearity will be worst at these settings.

5. Gain range—Gain is defined as the slope of the least-squares straight line established through the output voltage versus the differential input voltage characteristic of the amplifier. The gain range is the maximum and minimum values of gain available from the amplifier without causing any degradation in performance beyond the limits of the specification.

6. Gain stability with temperature—Gain stability with temperature is the change in amplifier gain as a function of ambient temperature for any gain in the specified gain range.

7. Zero stability with temperature—Zero stability with temperature is defined as the change in output voltage with temperature. It must be specified as RTI or RTO, and this test is typically performed with the amplifier input leads terminated with the maximum source impedance and no signal applied. A warm-up period is usually specified for both this test and gain stability with temperature.

8. Frequency response—Frequency response is defined as the minimum frequency range over which the amplifier gain is within ± 3 dB of the DC level for all specified gains for any output signal amplitude within the linear output voltage range. In writing specifications, it is not uncommon for a user to also specify the desired phase characteristics over the frequency range of interest and the number of filter poles.

9. Slew rate—Slew rate is defined as the maximum rate at which the amplifier can change output voltage from the minimum to the maximum limit of linear output voltage range. It is expressed in volts per microsecond with a large amplitude step voltage applied to the input of the amplifier and the amplifier driving a specified capacitive load. The usual source of slew rate difficulty is current limiting, and this specification (a nonlinear process) should not be confused with rise time (a linear process).

10. Settling time—The time following the application of a step voltage input for the amplifier output voltage to settle to within a specified percentage of its final value.

11. Overload recovery—The time required for the amplifier to recover from a specified differential input signal overload. It is specified as the number of microseconds from the end of the input overload to the time that the amplifier DC output voltage recovers to within the linear output voltage range. Amplifier gain must be specified.

12. Noise—Noise is divided into two components, i.e., RTI and RTO. RTI noise is that component of noise that varies directly with gain. It is measured with the amplifier input leads terminated in the maximum source impedance and no signal applied. RTO noise is that component of noise which remains fixed with gain.

13. Harmonic distortion—Harmonic distortion is the maximum harmonic content for any amplifier frequency or output amplitude within the specified limits.

14. Output impedance—The maximum impedance the amplifier will present when it is operated anywhere within its specification. This specification is important in resistive loading ratings or in determining the amount of capacitance which can be connected across the output without causing instability.

Reference 10 describes test procedures for these specifications and discusses them further. Figure 8.27 presents the basic DC amplifier circuit. Reference 11 provides additional discussion directed towards understanding DC instrumentation amplifiers.

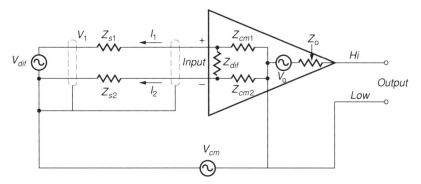

Fig. 8.27   Basic DC amplifier circuit.

8.7.3   Slip Rings

In many measurement applications, it is necessary to acquire data from rotating machinery. Turbines, rate tables, and centrifuges are examples of such machinery. If it is necessary to measure strain, pressure, torque, force, etc. on the rotating machine component, signals from bridge transducers must be coupled from this component to a stationary instrumentation system. Instrumentation slip rings accomplish this function.

In their simplest form, slip rings consist of a metal ring on the rotating machine component against which a brush attached to the stationary machine portion is spring-loaded to make ohmic contact. Precious metals are generally used for mating surfaces to minimize contact resistance.

Slip rings came into existence in the 1940s with initial application in the aircraft industry. In the 1950s, mercury slip rings came into existence. These latter rings, which first found application at Rolls Royce in England, use mercury as the signal transfer medium. The mercury is entrapped between the rotor and the stator of the ring assembly. Today, slip rings are capable of operating from very low rpm to tens of thousands of rpm.

Noise induced in slip rings is of the ohmic contact type, i.e., it is roughly proportional to current. A high brush pressure reduces noise at the expense of increased brush wear. Brush wear is a function of the brush pressure, material, finish (usually microinch), and flatness. One technique for lowering contact noise is to mount several brushes in parallel on the same ring.

Because ohmic changes in the slip rings can be of the same order of magnitude as resistance changes in the bridge transducer, full bridges are almost always used on the rotating part to avoid inserting slip rings within the bridge itself. Slip rings in the output circuits of bridge transducers using voltage monitoring do not create significant problems because any small resistance changes in the rings are in series with the large input impedance of the voltage measuring device and are effectively ignored. Slip rings in the input circuits of bridge transducers operating from a constant voltage source can create problems if they cause fluctuating voltage drops in series with the transducer. For this reason, constant current sources are preferred when using slip rings.

Other techniques for extracting data from rotating machinery have evolved over the years. These include rotary transformers, light modulation, and rf telemetry. Of these schemes, rf telemetry has displayed the most promise with commercially available low power transmitters capable of operating up to 30,000 g's.

8.7.4   Noise Considerations

Many other sources besides slip rings can induce unwanted spurious signals in these transducers. Since the unamplified output from bridge transducers is typically ones or tens of millivolts, and never more than a few hundred, they are easily influenced by noise sources. The following discussion defines noise, documents how to verify its existence (or hopefully nonexistence), and provides some hints as to how to suppress noise in bridge transducer measuring systems. Reference 12 provides a basis for this discussion.

The output of measuring system components represents combinations of responses to environments. These environments can be divided into two categories: desired and all others (undesired). For example, consider a bridge pressure transducer in a hostile explosive environment. Its desired environment is pressure. Other undesired environments it encounters are temperature, acceleration, ionized gas, etc. Ideally, the transducer would respond to pressure alone. In practice, an additional response is elicited from the transducer due to the other environments; usually, but not always, the response is small compared to the pressure response.

Two response types exist for a bridge transducer: self-generating and non-self-generating. Non-self-generating responses are due to changes in the material properties or geometries within a transducer. Power has to be applied to the transducer to elicit a non-self-generating response. For example, pressure applied to the diaphragm of a pressure transducer with bridge electrical power supplied modifies the impedance of the strain gage circuit and results in a millivolt output (non-self-generating). Self-generating responses are those attributable to various measurands applied to bridge transducers without electrical power supplied. Examples of these responses include thermoelectric, photoelectric, pyroelectric, magnetoelectric, etc., induced voltages within the bridge circuit. Thus, there exist four environment-response combinations in bridge transducers. The non-self-generating response to the desired environment is defined as signal. The non-self-generating response to the undesired environment, as well as the self-generating response to both the desired and undesired environment, is noise. Figure 8.28 illustrates the paths associated with these four combinations, with path 4 being signal and paths 1, 2, and 3 being noise.

The quantifications of paths 1 through 4 can be accomplished by switching. If at some time during the test bridge power is switched off, Fig. 8.28 indicates that only paths 1 and 3 will exist. Since these paths are both noise, the bridge transducer response ideally should approach zero. Similarly, if the desired environment can be switched off for some time period, only paths 1 and 2 remain. If path 1 was verified as being noise-free when bridge power was removed, path 2 also becomes quantified. If paths 1, 2, and 3 are all shown to provide negligible signal level, transducer output becomes attributable to path 4, which has been defined as signal. In summary:

Remove bridge power: document paths 1, 3

Reapply power, remove desired environment: document paths 1, 2

If the documented signal paths are of sufficiently small magnitude to be considered inconsequential during test, the non-self-generating response to the desired environment (signal) is recorded.

Some question may arise as to how to implement these procedures, particularly in transient measurement situations. For example, assume a bridge accelerometer is to be used to measure a transient acceleration event. Three accelerometers can be fielded in close physical proximity. The first can be mounted without power applied to document paths 1 and 3. The second can have power applied but be mounted in a piece of foam to isolate it from the acceleration environment, resulting in documentation of paths 1 and 2. The third can be powered and properly mounted to measure the acceleration environment. If the first two accelerometers produce no output, then the output from the third is the noise-free signal.

If noise is present in measuring systems containing bridge transducers, noise-control efforts are dictated by the specific noise type. Electric and magnetic fields can be shielded, noise components at

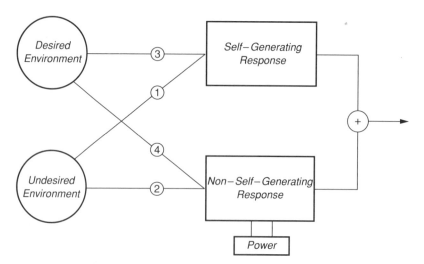

Fig. 8.28 Bridge transducer model for noise hunting and documentation (adapted with permission from "Information as a 'Noise Suppression' Method," by Peter Stein, Stein Engineering Services, Inc., Phoenix, AZ, LR/MSE Publ. 66, 1975).

specific frequencies can be filtered, thermal transient effects can be absorbed or delayed, steady-state temperature effects can be compensated, etc. The prerequisite to any noise control is documentation of its presence.

As noted earlier, most modern resistance bridge transducers use DC power supplies as opposed to AC power supplies. AC power supplies, however, still have an important role to play with resistance bridge transducers. AC power supplies accomplish noise suppression by separating the self-generating responses from the non-self-generating responses in a transducer by moving the frequency content of a signal into some new range of the frequency spectrum. This procedure is known as amplitude modulation. Reference 13 forms the basis for the following discussion.

Referring to Fig. 8.28, amplitude modulation can eliminate paths 1 and 3 (noise) from the net output signal. Thus, self-generating emf's, such as thermoelectric and electromagnetic ones, can be separated from emf's attributable to resistance changes in a bridge transducer.

A design procedure is presented for an AC powered bridge where the signal input to the self-generating response extends to frequency ω_1 and the input to the non-self-generating response extends to ω_2. Power supplied is at frequency ω_3. The design method developed requires a knowledge of these frequencies. A procedure to determine these frequencies involves:

1. Performing a frequency analysis of the signal recorded with no power supplied to determine ω_1
2. Applying power under normal operating conditions and comparing signal frequency content to the results of (1) to determine ω_2

An example follows where noise is present as a self-generating response and the bridge is powered first by a DC supply and then by an AC supply.

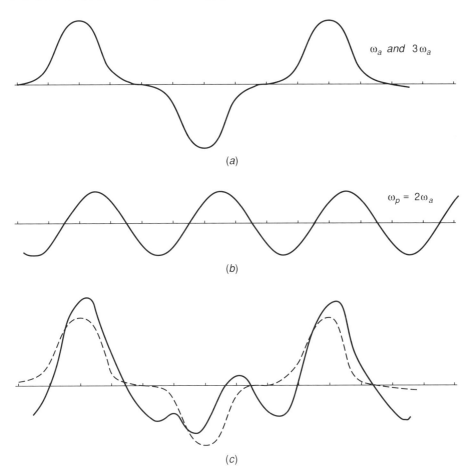

(a)

(b)

(c)

Fig. 8.29 Signal waveshapes DC bridge (adapted with permission from "Information as a 'Noise Suppression' Method," by Peter Stein, Stein Engineering Services, Inc., Phoenix, AZ, LR/MSE Publ. 66, 1975).

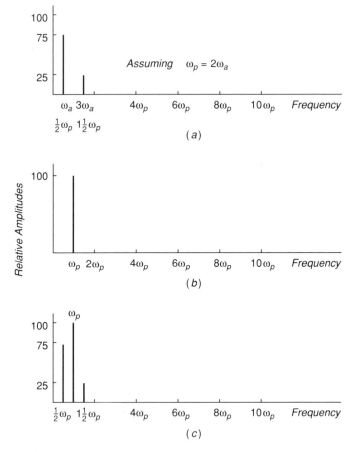

Fig. 8.30 Signal frequency contents DC bridge (adapted with permission from "Information as a 'Noise Suppression' Method," by Peter Stein, Stein Engineering Services, Inc., Phoenix, AZ, LR/MSE Publ. 66, 1975).

In this example, the non-self-generating response (signal) occurs at frequency ω_p and the self-generating response (noise) occurs at two frequencies bracketing ω_p ($\omega_a = \omega_p/2$ and $3\omega_a$). Figure 8.29 illustrates the waveshapes assumed for the self-generating and non-self-generating responses with DC power supplied ($\omega_3 = 0$). Figure 8.29a represents both self-generating response input and output, Fig. 8.29b represents non-self-generating response input and output, and Fig. 8.29c represents the total transducer output (summation of 8.29a and 8.29b). It is seen that the two responses are hopelessly intermingled and that the signal cannot be separated from the noise. Figure 8.30 shows the frequency content of the waveshapes, with Fig. 8.30a corresponding to 8.29a, Fig. 8.30b corresponding to 8.29b, and Fig. 8.30c corresponding to 8.29c. The AC-powered bridge will be presented as a solution to measurement problems such as this. ω_3 is typically selected as 10 times ω_p.

Figure 8.31 describes this situation for the AC bridge. Figure 8.31a describes bridge power. Figure 8.31b describes the output from the non-self-generating response which now contains frequencies at ($\omega_c - \omega_p$) and ($\omega_c + \omega_p$). ω_c here is defined to be the carrier frequency, ω_3. Figure 8.31c describes the net transducer output which is a summation of Fig. 8.29a and Fig. 8.31b. Figure 8.32 describes the frequency content associated with Fig. 8.31 respectively. The frequency content in Fig. 8.32c associated with the time history of Fig. 8.31c shows conclusively that the non-self-generating information has been moved from its original frequency, ω_p, to occupy a new frequency range, ($\omega_c - \omega_p$) to ($\omega_c + \omega_p$), while the self-generating response is left at ω_a and $3\omega_a$. The non-self-generating response is then in that part of the frequency spectrum where no appreciable noise exists and can be separated by band pass filtering.

$\omega_3 = 10\omega_p$

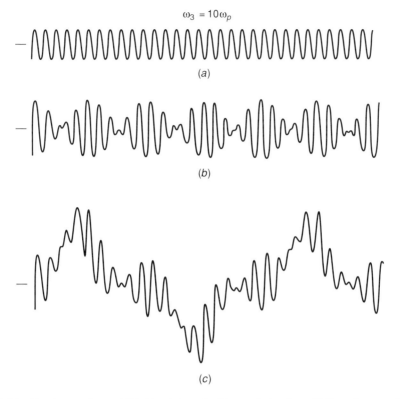

(a)

(b)

(c)

Fig. 8.31 Signal waveshapes AC bridge (adapted with permission from "Information as a 'Noise Suppression' Method," by Peter Stein, Stein Engineering Services, Inc., Phoenix, AZ, LR/MSE Publ. 66, 1975).

After band passing, a problem still remains in phase sensing. Figure 8.33 describes this problem. Figure 8.33 illustrates an amplitude-modulated signal after band passing to remove the effects of any self-generating response which may be present. This amplitude-modulated signal is ambiguous in that it could correspond to any of the lower four signal inputs to the non-self-generating response. The problem of phase sensing associated with a modulated signal is that of determining which portion of the modulated waveshape is positive and which is negative.

If a modulated signal emerges from a measuring system which is initially balanced (zero output for zero input), phase sensing must be done in a phase-sensitive manner. The general principle for all phase sensors is as follows:

If the system output is of the same sign at the same time as the time-varying supply power, the measurand must have been positive. If the signs are opposite, the measurand must have been negative.

Phase sensing is accomplished by a phase-sensitive demodulator. A reference signal is fed from the bridge supply power to the phase-sensitive demodulator. This signal is compared with the amplitude-modulated signal for phase determination. A half-wave rectifier with transformer coupled reference and amplitude-modulated signals forms one basis for a demodulator.

After phase sensing, a low-pass filter is required to separate the non-self-generating analog signal proportional to the measurand from the other frequencies which appear as sidebands around harmonics of power supply frequency. Final selection of a power supply frequency is a trade-off between the maximum frequency content in the measurand and the low-pass filter roll-off characteristics. While a 10:1 ratio is typical, the power supply frequency may vary between 3–20 times the maximum non-self-generating signal frequency.

Note that introduction of an AC power supply requires a bridge balancing network incorporating complex impedance in the balance controls.

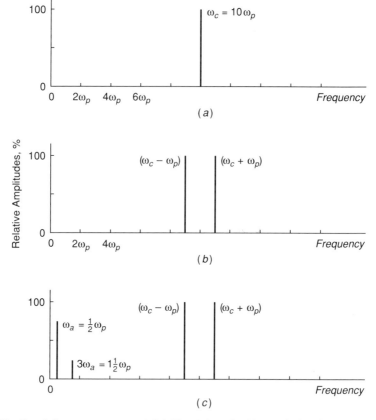

Fig. 8.32 Signal frequency contents AC bridge (adapted with permission from "Information as a 'Noise Suppression' Method," by Peter Stein, Stein Engineering Services, Inc., Phoenix, AZ, LR/MSE Publ. 66, 1975).

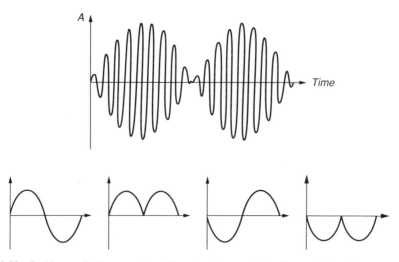

Fig. 8.33 Problems of phase sensing (adapted with permission from "Information as a 'Noise Suppression' Method," by Peter Stein, Stein Engineering Services, Inc., Phoenix, AZ, LR/MSE Publ. 66, 1975).

8.8 AC IMPEDANCE BRIDGE TRANSDUCERS

Having discussed bridge transducers that use resistive sensing elements, and AC power supplies (amplitude modulation), a logical question is whether bridge transducer sensing elements can be capacitive or inductive. In practice, many bridge transducers do employ capacitive or inductive elements. While resistance bridge type transducers typically possess resonant frequencies in the tens or hundreds of kilohertz range, AC impedance bridge transducers typically possess resonant frequencies of less than 10 kHz. The larger physical size of AC impedance bridge transducers makes them environmentally more fragile but improves their performance by increasing their sensitivity to low-level measurands.

8.8.1 Inductive Bridges

Figure 8.34 shows an example of how variable reluctance sensing elements can be incorporated into an AC bridge transducer. A differential pressure transducer containing a magnetically-permeable stainless steel diaphragm as the mechanical flexure is portrayed. This diaphragm is clamped between

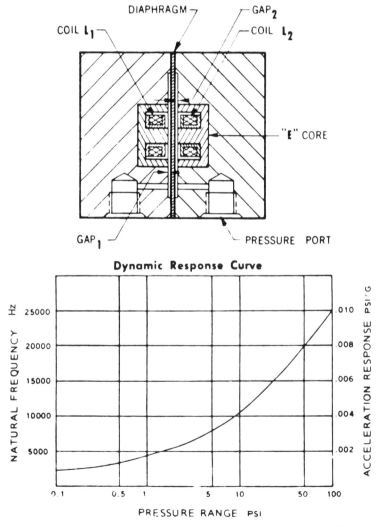

Fig. 8.34 Variable reluctance sensing in a bridge transducer (courtesy of Validyne Engineering Corporation).

two blocks and deflects when a pressure difference is created across it through the two ports shown. An E-core and coil assembly is embedded in each block. A small gap exists in front of each E-core. When the diaphragm is undeflected, a condition of equal inductance exists in each coil. When the diaphragm does deflect, an increase of gap in the magnetic flux path of one core occurs, with a resultant decrease in the gap in the magnetic flux path of the other. Magnetic reluctance varies with gap, determining the inductance value. The diaphragm motion then changes the inductance of the two coils, one increasing and one decreasing. These two coils can be placed in adjacent arms of an AC-powered bridge. Resistive elements can be used to complete the bridge. Once the bridge is balanced, an amplitude-modulated signal results when a differential pressure is applied across the ports of the transducer. When the resultant signal is properly demodulated, the applied pressure can be quantified.

Eddy current inductive displacement measuring systems are another example of the use of impedance as opposed to resistive bridges. Placing a coil with an AC current flowing in it a nominal distance from a metal target induces a current flow on the surface and within the target. The induced current produces a secondary magnetic field that opposes and reduces the intensity of the first field. Changes in the impedance of the exciting coil provide information about the target. The target can be the diaphragm of a pressure transducer, the seismic mass of an accelerometer, the flexure of a load cell, etc. The coil is one leg of a balanced bridge network. Unbalanced bridge conditions are sensed and converted into a signal directly proportional to the distance between coil and target. Figure 8.35 schematically illustrates this conversion.

The electrical parameters of resistivity and permeability in the target material influence performance of eddy current transducers. For a specific material, displacement sensitivity is influenced by coil geometry and operating frequency.

Target thickness is generally not a limiting factor. At one "skin depth," the eddy current density is only 36 percent of the maximum encountered on the target surface; at two "skin depths," it is 13 percent. Figure 8.36 defines skin depth as a function of target resistivity and permeability.

Target shape and alignment also should be considered in application. A flat, circular target equal to the coil diameter appears as an infinite plane. Smaller target diameters produce smaller voltage unbalances in the impedance bridge. Since the transducer senses the average distance to the target, the non-parallelism effect is small up to 15 degrees.

A differential transformer also is briefly mentioned here although it does not operate in an impedance bridge. A linear variable differential transformer (LVDT) consists of three symmetrically spaced coils wound onto an insulated bobbin. A magnetic core moving through the bobbin provides a path for magnetic flux linkage between coils. The center coil is the primary and has an AC voltage applied. The two secondary coils are wired in a series-opposing circuit. When the core is centered between two secondary coils, the voltages in the two coils cancel. As the core is displaced, the phase-referenced and demodulated output signal provides a linear voltage output with displacement.

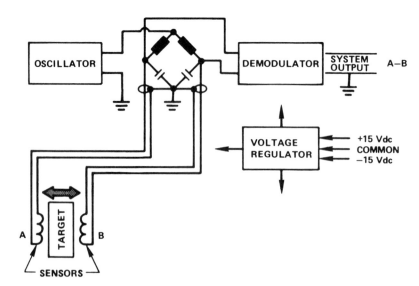

Fig. 8.35 Eddy current inductive displacement measuring system (courtesy of Kaman Instrumentation Corporation, Measurement Systems Group, Colorado Springs, CO).

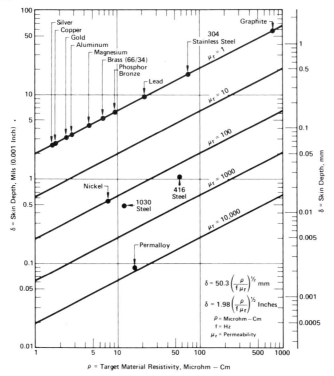

Fig. 8.36 Skin depth versus target resistivity and permeability at 1 Mhz (courtesy of Kaman Instrumentation Corporation, Measurement Systems Group, Colorado Springs, CO).

8.8.2 Capacitive Bridges

Capacitance sensors can be integrated into bridge transducers. The capacitance between two metal plates separated by an air gap is $C = kK A/h$ where C is capacitance, K is the dielectric constant for the material between the plates, A is the plate overlapping area, h is the gap thickness between the two plates, and k is a proportionality constant. The range of a capacitance sensor can be shown to remain linear with changes in area but become nonlinear when the change in gap displacement becomes a significant portion of the original gap.

Again, it should be pointed out that advantages of small size and enhanced dynamic response are to be found with resistance bridge transducers. Increased sensitivity will be displayed by impedance bridge transducers. Both impedance bridges and AC-powered resistance bridges offer noise suppression through separating non-self-generating responses from self-generating responses.

REFERENCES

1. P. K. Stein, *Measurement Engineering*, Stein Engineering Services, Inc., Phoenix, 1964.

2. W. M. Murray and P. K. Stein, *Strain Gage Techniques*, Engineering Extension UCLA and Society for Experimental Stress Analysis, 1958.

3. C. C. Perry and H. R. Lissner, *The Strain Gage Primer*, 2nd ed., McGraw-Hill, New York, 1962.

4. R. C. Dove and P. H. Adams, *Experimental Stress Analysis and Motion Measurements*, Charles E. Merrill Books, Columbus, OH, 1964.

5. A. L. Window and G. S. Hollister, *Strain Gauge Technology*, Applied Science Publishers, London and New York, 1982.

6. P. L. Walter, "Deriving the Transfer Function of Spatial Averaging Transducers," *ISA Transactions*, Vol. 19, No. 3, 1980.

7. *Statham Instrument Notes Number 5*, "Temperature Compensation of Bridge Type Transducers," Gould Inc., Oxnard, CA, August 1951.

8. TECHNICAL DATA TD4354-1, "When and How—Semiconductor Strain Gages," BLH Electronics, Waltham, MA, June 1975.

9. Document 118-79, "End-to-End Test Methods for Telemetry Systems," *Test Methods for Telemetry Systems and Subsystems*, Vol. 1, Secretariat, Range Commanders Council, White Sands Missile Range, NM, 1979.

10. Document 118-82, "Test Methods for Vehicle Telemetry Systems," *Test Methods for Telemetry Systems and Subsystems,"* Vol. 5, Secretariat, Range Commanders Council, White Sands Missile Range, NM, 1982.

11. J. W. Jaquay, "Understanding DC Instrumentation Amplifiers," *Instruments and Controls*, September, 1972.

12. P. K. Stein, "A Unified Approach to Handling of Noise in Measuring Systems," AGARD LS-50, NATO, Nevilly-sur-Seine, France, September 1972.

13. P. K. Stein, "Information Conversion as a Noise Suppression Method," Lf/MSE Publication 66, Stein Engineering Services, Phoenix, January 1975.

CHAPTER 9

POSITION, VELOCITY, AND ACCELERATION MEASUREMENT

AIRPAX CORPORATION

Engineering Staff
Ft. Lauderdale, Florida

WILLIAM BRENNER

Temposonics Division of MTS
  Systems Corporation
Plainview, New York

JOHN W. G. BUDD

Schaevitz Engineering
Pennsauken, New Jersey

DALE R. DUVALL

Scantek, Incorporated
Irving, Texas

HOWARD JASLOW

Digital Signal Corporation
Bohemia, New York

ROBERT W. LALLY

PCB Piezotronics, Incorporated
Depew, New York

GEORG F. MAUER

Department of Mechanical and
  Civil Engineering
University of Nevada,
Las Vegas, Nevada

CHESTER L. NACHTIGAL

Kistler-Morse Corporation
Redmond, Washington

R. J. PFEIFER

Combustion Engineering Inc.
Columbus, Ohio

GEORGE A. QUINN

Farrand Controls
Valhalla, New York

FRANK S. RUHLE

Farrand Controls
Valhalla, New York

JEFFREY F. TONN

Heart Technology, Inc.
Bellevue, Washington

9.1 INTRODUCTION

Chester L. Nachtigal

9.1.1 Terminology

In its broadest context, a "transducer" is a device that accepts an input variable and produces an output variable of a different nature. An electrical transformer is a transducer in that it accepts a voltage input of a given amplitude and frequency and produces a voltage at a different amplitude. But here we are interested in measurement transducers only—those devices that are in communication with a variable of interest and that convert that variable to another, more amenable form for subsequent operations. This other form is invariably a voltage or a current.

The distinction between transducers in general and measurement transducers in particular is that those in the latter category are designed to extract information only and operate in parallel with the main stream of activity of the system that produces the variable of interest. Referring to the example of the electrical transformer, if the transformer were removed from the system, the main stream activity would come to a halt. A measurement transducer may well be connected to either the primary or the secondary terminals of the transformer for the purpose of monitoring and recording these voltage amplitudes. If the measurement transducers were removed, the system itself would be unaffected. So the measurement transducer in this case is not in the mainstream of activity.

The terms "sensor" and "transducer" are often used interchangeably, but we shall try to maintain these two as separate entities for purpose of clarity of thought. A sensor or sensing element is that part of the transducer that communicates directly with the variable to be measured, called the "measurand," and responds in a way that can be converted to an electrical form. In summary, a measurement transducer includes the two functions of sensing and converting to another domain. This definition differs from that of an entire measurement system in that the output of a transducer, while electrical in nature, is not yet a signal that may be conveniently manipulated. It is likely a very low level signal that requires special handling, such as needing a high input impedance and amplification.

A second component, called a "signal conditioner," is generally required to mate with a transducer in order to complete the task of generating a useful output signal. The functions most often provided by the signal conditioner are: a) supplying AC or DC excitation voltage or current to the transducer; b) balancing or nulling the transducer output signal; c) converting a pair of push-pull transducer output signals to a single-ended signal; d) amplifying the response; e) demodulating the transducer output if it is AC-excited; and f) filtering out unwanted frequency components (including the zero-frequency or DC component). As more and more measurement systems embrace the digital computer, the function of the signal conditioner may include conversion of the signal from the analog, or continuous-event domain, to the digital, or discrete-event domain. Section 9.7 explores this area further in the context of digital transducers—transducers that include some type of conversion to the digital domain.

The "nonlinearity" of a measurement system is of great concern largely because most readout devices are capable of accurately displaying only "linear" signals. By this term we mean that if the true value of the measurand doubles, the output from the measurement system also doubles in magnitude. As measurement systems get "smarter," this linearity requirement will become less and less important because the system's intelligence will be able to cope with some degree of instrument nonlinearity without a similar degree of error. Section 9.8 carries this discussion further.

Just as "sensor" and "transducer" are often used interchangeably, so too are "transducer" and "signal conditioner" thought of as a single device. Here again we will attempt to preserve the separate identities of these two components even though they are frequently housed in the same jacket.

It has been stated that the initial function of a transducer is communication with the variable of interest and sensing or extracting information concerning the state of that measurand. This extraction process must be carefully executed. Ideally, only information but no energy is to be withdrawn from the system of interest. If some energy, however small an amount, is required to actuate the sensing mechanism of the transducer, the original system without the transducer has been altered by the energy extraction. This degrading effect is termed loading the signal source and is discussed in Chapter 5. It is doubtful whether there exists a transducer that qualifies as an ideal measurement device in terms of no energy extraction, but the point is only a philosophical one if considerable care is taken in minimizing the loading effect and/or accounting for the effect of the loading.

If the initial function of a transducer is one of primary sensing, then it may be asserted that the final function is performing the conversion function in a timely fashion. We generally are interested in producing a signal that is *statically* related to the incoming variable. By this we mean that the output is instantaneously related to the input. A more sophisticated version of this statement is that the input-output relationship is independent of the rate of change of the input variable. Conversely, a *dynamic* system is one whose output to input ratio *is* a function of the rate of change of the input.

Note that this definition of "static" and "dynamic" is independent of whether the system in fact is in motion or at rest, that is, steady state.

When one carefully probes the chain of events that make up the operation of any transducer, the static relationship becomes an idealization that is impossible to achieve in practice. Obviously the transducer can only respond after and not in anticipation of the occurrence of the input variable. Therefore the question of quantifying the degree to which the transducer is static must be answered in terms of the range of frequency components in the measurand and the extent to which the transducer departs from an ideal, static relationship. Rise time, 3 db frequency, time constant, resonant frequency, phase shift, and delay time are all terms that provide the user with some measure of departure from an ideal, instantaneous, input-output relationship. Put simply, these terms also characterize a transducer's speed of response.

The majority of transducers have their mode of operation based on one of a few well-known fundamentals, such as modulation of an electrical resistance, inductance, or capacitance. We will illustrate this in the subsequent sections, but here it is pointed out that once these principles of sensor and transducer design are grasped, one may apply them in the selection of a transducer for measuring variables of a widely diverse nature. In particular, the area of bridge transducers is of general interest in instrumentation because so many measurement transducers are based on this principle. Chapter 8 deals with this topic in considerable detail. Examples of transducers that will be discussed in this chapter and in Chapters 10 and 11 include the following variable measurement capability: position, geometric profile, velocity, acceleration, force, torque, pressure, strain, temperature, and fluid flow rate.

9.1.2 Principles of Transducer Design

One approach to the characterization of transducers is by their principle of operation. By far the greatest majority of measurement systems are comprised of transduction of the signal of interest to the electrical domain. Further, the majority of electrical domain measurement systems have their output signals in the form of voltages. Most signal processing components such as filters, readout equipment, and data acquisition systems are designed to operate on voltage signals as opposed to current signals. The method of current-mode signal transmission is important for long distances and is presented in Section 10.6.

The most fundamental question to be asked of the transducer manufacturer is how a given transducer performs its conversion from the nonelectrical to the electrical domain. This conversion process should be as direct as possible since the dynamic performance of a transducer is directly related to the type, number, and complexity of these intermediate conversions. For example, a transducer designed to sense and convert a mechanical motion such as time-varying linear displacement should not be designed to convert the incoming variable to another intermediate mechanical variable such as rotary shaft motion before generating its final response signal in the electrical domain. Mechanical components invariably have problems with wear, lowered frequency response, nonlinear friction, hysteresis, and resolution.

In the electrical domain there are three and only three constitutive elements: the resistor, the inductor, and the capacitor. Many measurement transducers are based on modulating one of these three elements by the physical variable to be measured. But no output signal is available if no excitation source is applied because they are passive elements and therefore carry no energy supply. There are, however, some types of transducers which are based on still other principles of operation which do not require an external power source in order to generate an output signal. Consequently, these transducers must convert a portion of the energy available from the system that supplies the measured variable to provide the output power. We shall call this type self-generating. Because they require no power source, they are also called passive.

As an example of the limitation generally imposed by a passive transducer, consider a measurement application that requires information about both constant as well as time-varying components of a mechanical displacement variable. If the transducer is passive, the instrument cannot yield information about a constant component of the measurand because there is no power available in the stationary portion of the input. Even if the output signal were connected to an infinite input impedance device, charge would eventually dissipate through the atmosphere, thus causing the output voltage to decay to zero. This charge leakage is a current to ground, and since current times voltage is power, there is a flow of power from the output terminal of the transducer. But the constant portion of the variable can supply no power, and therefore the transducer is incapable of producing indefinitely a correct measure of the constant portion of the variable.

In summarizing this section, two methods of grouping the vast array of transducers are employed in this handbook. In this introductory section they are grouped by their principle of conversion to the electrical domain, including: 1) change in resistance; 2) change in inductance; 3) change in capacitance; and 4) self-generating. In subsequent portions of this chapter and in Chapter 10 the method of grouping transducers is by the type of variable to be measured, such as position, velocity, or applied force.

9.2 POSITION TRANSDUCERS

Chester L. Nachtigal (9.2.1–9.2.3)
George A. Quinn and Frank S. Ruhle (9.2.4)
Georg F. Mauer (9.2.5)
Howard Jaslow (9.2.6)
Jeffrey F. Tonn (9.2.7)
William Brenner (9.2.8)

This section and subsequent sections of this chapter will focus on commercially available transducers and systems to measure variables of interest, starting with position measurement. The words "displacement" and "position" are taken to be completely interchangeable in this book. Both may be used to describe stationary or time-varying points in space. On the other hand, "motion" is a much broader term which includes time-varying position, velocity, acceleration, or any other time derivative of position.

9.2.1 Potentiometric and Strain Gage Position Transducers

Potentiometers

The most readily available and simplest position transducer is a linear (straight-line) or rotary potentiometer excited by a DC source such as a battery. It may be hooked up to deliver an output voltage that is essentially proportional to a straight-line or a rotary position varying between zero and a maximum. Alternatively, a potentiometer may be hooked up to deliver an output varying between a negative and a positive voltage in proportion to a mechanical displacement that also varies between a maximum negative and a maximum positive value relative to a defined null position. This schematic is shown in Fig. 9.1. A particular feature that rotary instrumentation potentiometers have that is usually not available in voltage control potentiometers is their ability to rotate through a full 360° without mechanical stops. The type of rotary potentiometer used for position measurement is called a servopotentiometer. It is specially designed to withstand many rotational cycles.

The fact that there must be a rubbing contact between the wiper of the potentiometer and the conducting element connected to the reference voltage is the principal shortcoming of potentiometers used as position transducers. Electrical noise and mechanical failure are direct consequences of this rubbing action.

When using a potentiometer to make a position measurement, care must be taken to minimize the electrical loading of the pot output by a subsequent meter input impedance. The degree of loading may be evaluated rather conveniently by the use of Thèvenin's Theorem (see Chapter 5). If we let ρ be the distance of wiper travel normalized with respect to the allowable wiper travel, we have for Fig. 9.1:

$$R_T = \frac{(1 - \rho)\frac{R_p}{2}(1 + \rho)\frac{R_p}{2}}{(1 - \rho)\frac{R_p}{2} + (1 + \rho)\frac{R_p}{2}} = (1 - \rho^2)\frac{R_p}{4} \tag{9.1}$$

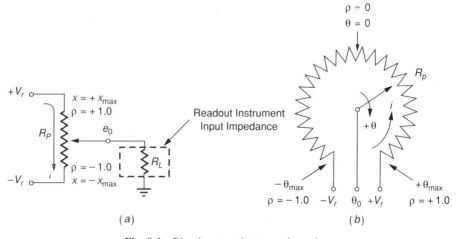

(a)   (b)

Fig. 9.1 Bi-polar potentiometer schematics.

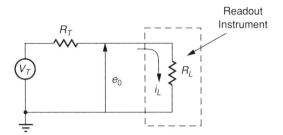

Fig. 9.2 Equivalent circuit of a potentiometer connected to a readout instrument.

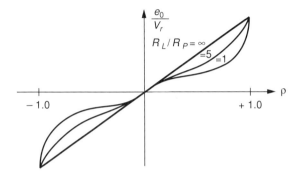

Fig. 9.3 Bi-polar potentiometer response.

The Thèvenin voltage is the ideal output voltage under open-circuit conditions and is

$$V_T = e_{0_{\text{o.c.}}} = \rho \frac{R_p}{2} i_{\text{o.c.}} = \rho \frac{R_p}{2} \frac{E_{ex}}{R_p} = \rho V_r \qquad (9.2)$$

Now let R_L be the input impedance of the readout instrument, connected between the e_0 terminal and ground and R_T is the Thèvenin resistance as shown in Fig. 9.2:

$$e_0 = i_L R_L = \frac{V_T}{R_T + R_L} R_L$$

$$= \frac{1}{1 + R_T/R_L} V_T = \frac{\rho V_r}{1 + (1 - \rho^2) R_p/4 R_L} \qquad (9.3)$$

The normalized output, e_0/V_r, is shown in Fig. 9.3. See also Section 9.2.6 for information about rotary potentiometer manufacturers and Chapter 5 for Thèvenin analysis.

Strain Gage Position Transducers

Strain gages can withstand only relatively small deflections. Therefore, this type of displacement transducer is not widely used. However, for applications with deflection ranges within the limits of a strain gage, this method offers the advantages of accuracy and simplicity.

A position transducer may consist of an elastic element and one or more attached strain gages. Ideally, a position transducer would require no force to deflect the elastic element. Obviously this is impossible due to the elastic element that carries the strain gages. So the question becomes one of determining how much inaccuracy is introduced (system loading) by the force needed to deflect the elastic element. Yet another question is the possibility of significant dynamic interactions between the system undergoing measurement and the transducer. A commercially available solid-state strain gage position transducer is shown in Fig. 9.4.[1] It produces a high-level output signal (± 1.0 V for 12 V DC excitation with no amplification at a full-scale displacement of ± 0.38 mm (0.015 in)). The electrical analysis of this transducer follows that of a two-active-arm strain gage bridge network in every respect.

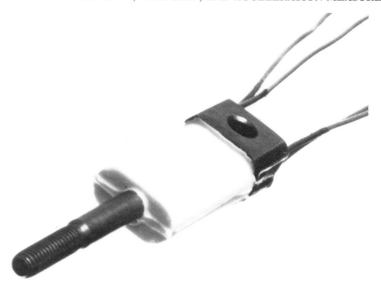

Fig. 9.4 Solid-state strain gage position transducer (courtesy of Kistler-Morse Corporation, Redmond, WA).

9.2.2 LVDT and RVDT Transducers and Signal Conditioners

LVDT Transducer

The Linear Variable Differential Transformer, or LVDT, is one of the most popular and optimum methods of measuring mechanical position in use today. As such it has found use as a primary transducer in the measurement of a variety of variables such as force, pressure, torque, and acceleration. In each of these transducers the position of an elastic member undergoing deflection in response to an applied force is measured by an LVDT. The value of the measured variable is then inferred from the position signal. Variable differential transformers have been used as electrical control devices since the turn of the century.[2] But it was not until the late 1920s and early 1930s that the use of this concept in measurement transducers became more widespread. This adaptation received its impetus when chemical process control manufacturers sought an electrical telemetering system for remote indication of process variables. The onset of World War II saw the incorporation of the LVDT into many mechanical position control applications aboard aircraft and submarines. Since the war it has been developed and refined to a remarkable degree.

The word "linear" appears in the name of the LVDT to denote straight-line motion as opposed to a linear relationship between the input and output. Its construction and principle of operation are deceptively simple. Three coils of electrically conducting wire are wound on an insulating form as shown in Fig. 9.5. By the principle of mutual inductance an AC voltage across the terminals of the primary coil induces a voltage of the same frequency in each of the two secondary coils. If the movable ferromagnetic core is centered, the two secondary voltages are of the same amplitude. For a positive displacement of the core (solid response curves in Fig. 9.6b), the voltage appearing across the number 1 secondary is greater in amplitude than at the null condition, while the amplitude across the number 2 secondary coil is less. The black dots in Fig. 9.6a indicate relative polarity of the three AC voltages.

When the two secondary coils are connected in series opposition as shown in Fig. 9.7a, a single response voltage is generated whose amplitude is closely related to the core displacement history (Fig. 9.7b, c). This type of output signal is not very convenient, however. We would like a signal directly proportional to the core position—a positive voltage when the core is positively displaced and a negative voltage when the core is negatively displaced. This requires that the amplitude-modulated secondary voltages are demodulated to yield such a DC signal.* We shall analyze a demodulator circuit later under *Signal Conditioner*.

---

* In this chapter the use of "DC" is especially confusing. When an input excitation voltage is described as DC, this means that it is a constant voltage. When the output is referred to as DC, it is a voltage proportional to the input variable even though it varies with time.

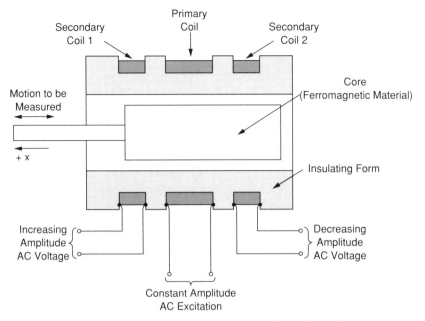

Fig. 9.5 Cross-section of a linear variable differential transformer.

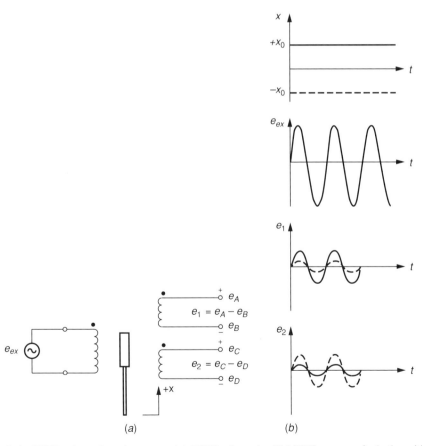

Fig. 9.6 LVDT schematic and response: (*a*) LVDT schematic; (*b*) LVDT response for both positive and negative displacement.

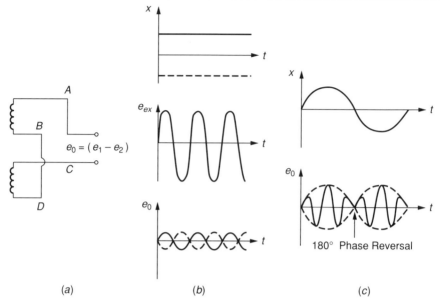

Fig. 9.7 LVDT with series opposition (subtraction) hookup of the secondaries: (*a*) series opposition hookup; (*b*) LVDT response voltage; (*c*) time-varying input and response.

With reference to the series opposition hook-up as shown in Fig. 9.7*a*, the input and output voltages are shown exactly in phase with each other or exactly 180° out of phase, depending on whether the core is positively or negatively displaced from its null position. In practice there is some phase difference between e_o and e_{ex} which is a function of the excitation frequency. The output voltage of the series opposition hook-up circuit may be quickly obtained by writing

$$e_o = e_A - e_C = e_A - e_B + e_B - e_C$$

$$= (e_A - e_B) - (e_C - e_D) \tag{9.4}$$

In all cases where an LVDT is used to measure position, the excitation applied to the transformer primary coil is AC. This fact is mentioned here to guard against some potentially confusing terminology in the industry. An "AC LVDT" or an "AC-operated" LVDT should be interpreted as the three-coil transformer and core assembly. A "DC LVDT" or a "DC-operated" LVDT is an LVDT that contains within its case all the electronics necessary to generate a sinusoidal carrier signal from an input DC excitation and to demodulate the response voltages. A DC power supply is required to perform these functions, and hence the name "DC-operated." In some cases an operational amplifier is also included in the electronics so that the output is a ±10 volt, low output impedance signal. These DC input, DC output measurement systems usually have a special name, such as "DCDT"[3] or "DC-DC LVDT."[4]

One of the most advantageous attributes of the LVDT is the fact that the core does not touch the internal bore of the transformer. Therefore, the problems of mechanical wear, friction loading of the input displacement source, and electrical noise created as a result of rubbing action are all eliminated. Most LVDTs, except for the gage head design, do not include a means of guiding the core with respect to the centerline of the transformer housing. (See Fig. 9.8.) This task is performed by either providing a bearing system for the core and/or its actuating rod or depending on the nature of the system undergoing motion. For example, it is necessary to measure the longitudinal position of a piston within a cylinder in electrohydraulic control systems. An LVDT is well-suited for this task because the lateral motion of the core is limited by the lateral motion of the piston relative to the cylinder. See Fig. 9.9, which shows an electrohydraulic actuator used in an aircraft. For purposes of redundancy four LVDTs are employed to measure the piston displacement.

In one sense an LVDT is a noncontact displacement sensing device but the core must be secured to the part of the system to be measured. This may be inconvenient if not impossible, as in the case of measuring the transverse vibration of a rotating shaft. If the core cannot be secured to the moving element that is to be sensed but may be in contact with it, a gage head LVDT may be selected. This type of LVDT, illustrated in Fig. 9.10, is similar to a conventional LVDT but has two additional features: a means of guiding the core along the axis of the transformer and a mechanism

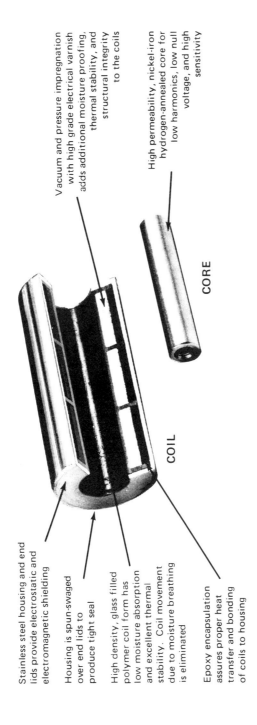

Stainless steel housing and end lids provide electrostatic and electromagnetic shielding

Housing is spun-swaged over end lids to produce tight seal

High density, glass filled polymer coil form has low moisture absorption and excellent thermal stability. Coil movement due to moisture breathing is eliminated

Epoxy encapsulation assures proper heat transfer and bonding of coils to housing

Vacuum and pressure impregnation with high grade electrical varnish adds additional moisture proofing, thermal stability, and structural integrity to the coils

High permeability, nickel-iron hydrogen-annealed core for low harmonics, low null voltage, and high sensitivity

COIL

CORE

Fig. 9.8 Cutaway view of an AC Schaevitz LVDT (courtesy of Schaevitz Engineering, Pennsauken, NJ).

Fig. 9.9   Electrohydraulic servoactuator with quad-redundant LVDT position transducers (courtesy of G. L. Collins Corporation, Long Beach, CA).

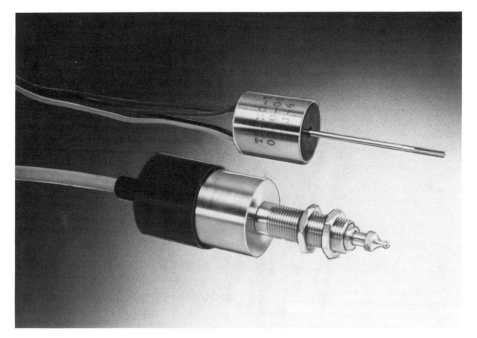

Fig. 9.10   DC-DC gage head position transducer (courtesy of Trans-Tek Incorporated, Ellington, CT). This photograph shows a gage-head LVDT below the DC LVDT which is used inside of it.

for maintaining contact between the tip and the moving system. In some cases the friction in the core guidance bearing causes a slight repeatability error. Two methods for maintaining contact between the tip and the moving element are presently in use: a simple coil spring or pneumatic actuation.[2] The latter method is ordinarily used only with ultraprecision gage-head LVDTs.

The outer diameter of a typical LVDT transformer is approximately 1.9 cm and the length varies from 4 cm for a ±0.13 cm stroke unit to 50 cm for a ±13 cm stroke unit. Even longer range units are available. Because of the frictionless operation of the LVDT and its basis on the induction principle, the resolution of an LVDT is said to be infinite. In fact, the low-level signal processing capability of the external electronics is the limiting factor on resolution.

RVDT

The principle of the LVDT has also been adapted for angular position measurement in a device known as the Rotary Variable Differential Transformer, or RVDT. A typical RVDT is shown in Fig. 9.11. A cardioid-shaped cam of magnetic material is used as a core, the shape of which is carefully chosen to produce a highly linear output over a specified range of rotation. Figure 9.12 presents the output voltage for a complete revolution of the input shaft. The linearity of a typical RVDT[2] is better than 0.5% of full scale for a ±40F operating range. Although the core is free to rotate through many revolutions, one would not ordinarily use this type of device beyond its useful linear range.

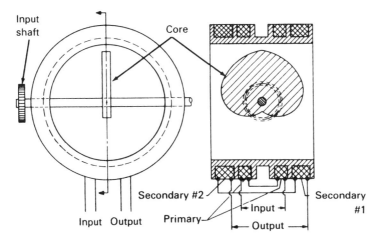

Fig. 9.11 Rotary variable differential transformer (RVDT) (taken by permission from Fig. 3.4, *Handbook of Instrumentation and Control* by E. E. Herceg, Schaevitz Engineering, Pennsauken, NJ, 1986).

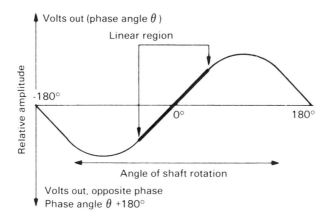

Fig. 9.12 Output characteristic of an RVDT (taken by permission from Fig. 3.5, *Handbook of Instrumentation and Control* by E. E. Herceg, Schaevitz Engineering, Pennsauken, NJ, 1986).

Signal Conditioner

We have seen that the output of an LVDT may be reduced to a single voltage referred to ground consisting of an amplitude-modulated carrier signal as in Fig. 9.7c. Since the envelope of the output signal contains information about both the instantaneous value of the core position as well as the negative of the core position, the system of Fig. 9.7 is not yet complete. Ideally we would like to have a DC output proportional to $x(t)$. Extracting such a signal from the modulated carrier signal is called "demodulating" the LVDT output. Frequently, the tasks of generating a fixed-frequency sinusoid to excite the LVDT primary coil and demodulating, converting the two secondary voltages to a single-ended signal, filtering, and voltage gaining are all performed by a single unit called the signal conditioner.

An LVDT demodulating circuit is simply a rectifying and low-pass filtering device. Figure 9.13 shows a representative sample of such a circuit together with the LVDT secondary coils and a gain stage. The overall operation of this circuit may be understood by considering the individual voltages. Let us first assume that the capacitor is disconnected temporarily. On the positive half-cycles of $(e_A - e_B)$ and $(e_C - e_D)$, the diodes may be replaced by a short circuit. During the negative half-cycles, the diodes may be open-circuited. Since the loops consisting of the secondary coil, the

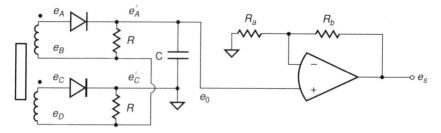

Fig. 9.13 LVDT demodulator circuit with gain.

diode, and the resistor R are equivalently open-circuited, no current can flow through either resistor R and thus the voltages $(e'_A - e_B)$ and $(e'_C - e_D)$ must be zero during the carrier negative half-cycle. Thus the wave forms are as shown in Fig. 9.14a. Because the two secondary coils are connected in series opposition we have

$$e_0 = e'_A - e'_C$$

$$= e'_A - e_B + e_B - e'_C$$

$$= (e'_A - e_B) - (e'_C - e_D) \tag{9.5}$$

Therefore, the output voltage is simply the difference between the two half-wave secondary voltages as shown in Fig. 9.14. Adding the capacitor has the effect of preventing the output voltage e_0 from

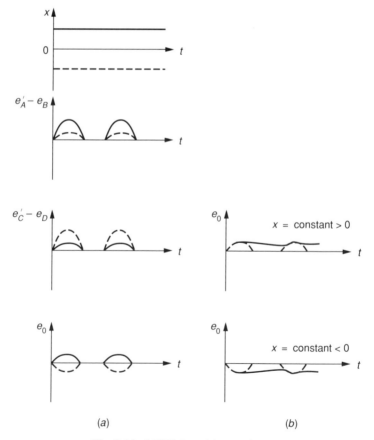

(a)   (b)

Fig. 9.14 LVDT demodulator voltages.

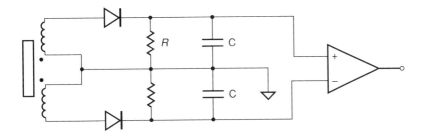

Fig. 9.15 Differential to single-ended output LVDT demodulator.

falling to zero instantaneously during the time that $(e'_A - e_B)$ and $(e'_C - e_D)$ would otherwise be zero. Thus the total response with the filter capacitor in place is as shown in Fig. 9.14b. When the voltages $(e'_A - e_B)$ and $(e'_C - e_D)$ are decreasing from their local peak values in any one cycle of the carrier frequency, the capacitor begins discharging through the two series resistors $R + R$. Therefore, the time constant of this discharge process is $\tau = 2RC$. The value of this time constant is chosen to minimize the discharge between successive peaks of the carrier on the one hand and to follow the modulating envelope of the carrier wave on the other hand. The modulation frequency is that of the input position imposed on the core. Let us assume a carrier frequency of 10,000 Hz and a signal frequency of 100 Hz. Thus the time per carrier cycle is 10^{-4} s and the time per signal cycle is 10^{-2} s. A discharge time constant of $\tau = 10^{-3}$ s would meet both objectives. Let $R = 5000\Omega$ and $C = 0.1$ μF. If the resistance is increased and the capacitance is decreased from these values, the demodulator circuit would have a larger output impedance. Increasing the capacitance and decreasing the resistance would lower the output impedance but would require a larger charging current to charge the capacitor and, therefore, create a possible loading error due to the output impedance of the transformer secondary coils.

A second demodulator circuit designed to interface with a differential amplifier is shown in Fig. 9.15. This circuit would be especially useful if common-mode noise were a problem. Still other circuits that have additional stages of filtering are in existence. In general they are all based on a diode rectification principle and RC filtering of the half-wave- or full-wave-rectified signal to eliminate the carrier frequency. This filtering requirement is the most significant factor that limits the dynamic frequency response performance of the LVDT system.

9.2.3 Inductive Proximity Probe Displacement Measurement Systems

A truly noncontact method of measuring small-amplitude displacements is embodied in an inductive proximity probe. This transducer is used primarily where the range of displacement is no greater than 2.5 mm, although some units are capable of measuring up to 125 mm of displacement. A typical frequency response is 0 to 30 kHz. A proximity probe is a cylindrical jacket with a multiturn coil of wire implanted very close to the surface of the sensing face, which is perpendicular to the centerline of the cylindrical body. The probe is mounted near—but not touching—a vibrating surface that must be an electrically conducting material. This type of position transducer is widely used in rotating machinery diagnostics.[5,6]

A few of the plethora of applications include shaft vibration, axial and radial shaft runout, startup and shutdown, machine monitoring, normal machine vibration "signature" identification, abnormal machine vibration detection for avoiding failure, bearing and shaft eccentricity, rotating machinery balancing, maintenance scheduling, sheet thickness, shaft diamcter, machine parts alignment, and oil film thickness. All of these applications make use of continuously varying displacement or gap width between the probe face and the sensed machine part or object. This class of applications is not to be confused with the use of a simpler type of proximity probe that senses only the presence or absence of material within the sensing range of the probe. This latter application of proximity probes is not a position-measuring instrument in the fullest sense and therefore is not included in this section.

Several different variable inductance methods of sensing the air gap between probe face and target surface are cited in technical and manufacturers' literature. Kaman Instruments[5] uses two different principles of operation in its line of noncontact displacement transducers. Most of their instruments are based on an inductive bridge concept. Figure 9.16 shows a block diagram of such a system. Slight changes in inductance of the sensor coil caused by changes in the electromagnetic coupling between the sensor and the target result in large shifts in the output of the bridge.

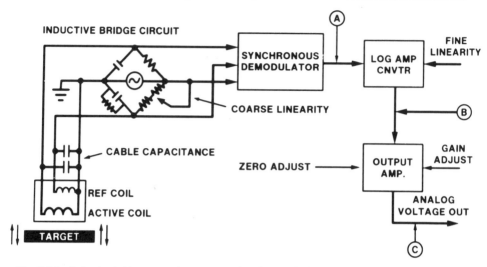

Fig. 9.16 Inductive bridge measuring system (taken by permission from p. 13 of "Application of a Precision Measurement Technology," Kaman Instrumentation Corp., Colorado Springs, CO, 1986).

Figure 9.17 shows a pair of transducers and their signal conditioners. These units are modularized and can be mechanically and electrically interconnected for a variety of signal processing functions. RMS-to-DC, peak-to-peak, and 10-hour battery pack are some of the choices. Figure 9.18a shows a typical input-output calibration characteristic for the system shown in Fig. 9.17. Figure 9.18b shows that there is a trade-off in linearity and range; that is, an improvement in linearity may be obtained

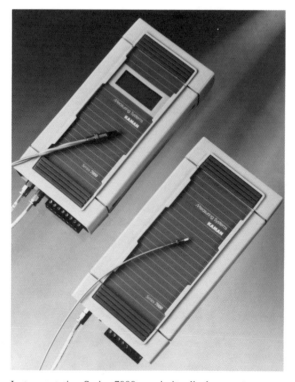

Fig. 9.17 Kaman Instrumentation Series 7000 proximity displacement measurement system (courtesy of Kaman Instrumentation Corp., Colorado Springs, CO).

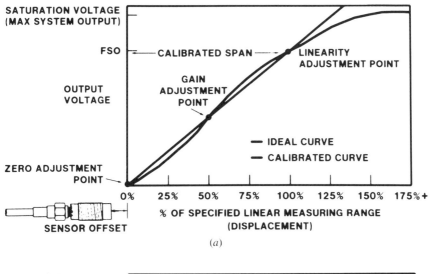

(a)

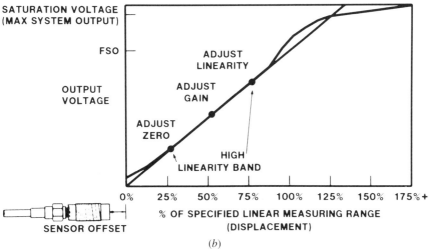

(b)

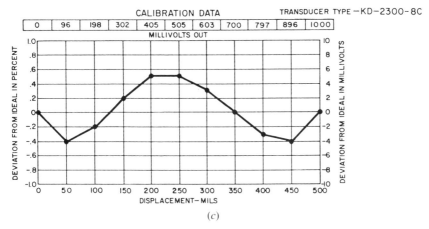

(c)

Fig. 9.18 Calibration procedures for a Series 2300 proximity displacement transducer: (a) full span calibration; (b) high linearity band; (c) error in percent of full-scale versus input displacement (taken by permission from p. 35 of "Application of a Precision Measurement Technology," Kaman Instrumentation Corp., Colorado Springs, CO, 1986).

321

at the expense of range. In order to examine deviations from an ideal linear calibration curve, it is customary to plot error versus input:

$$\text{Error (\% of F-S output)} = \frac{\text{Actual} - \text{Ideal}}{\text{Ideal}} \times 100$$

$$= \frac{\text{Actual} - \text{slope} \times \text{input}}{\text{slope} \times \text{input}} \times 100 \qquad (9.6)$$

Figure 9.18c is a plot of Fig. 9.18a on error versus input coordinates.

Another position-measuring product family by the instrument manufacturer of Reference 5 has its principle of operation based on the traditional Colpitts oscillator or eddy current effect. In this case the sensor is the resonating coil for the oscillator. As the proximity of the target changes, the oscillator frequency and modulation amplitude change also. This amplitude-modulated carrier signal is then half-wave rectified and filtered to produce a DC output. (See Section 9.2.2 for an example of this demodulation process.) Although proximity-measuring systems based on the eddy current effect are less expensive than those based on the inductive bridge concept, they tend to have a more limited input range and exhibit a greater nonlinearity of their entire range.[5]

A second manufacturer[6] offers a proximity probe system based on the eddy current effect. A radio frequency signal is radiated through the probe sensing face into the observed surface. Eddy currents are generated in the surface of the conducting target, and the loss in signal strength is detected by the signal conditioner. Figure 9.19 shows a typical proximity probe and Proximitor offered by Bently Nevada. A typical frequency response is 0 Hz to 10 kHz. Figure 9.20 illustrates the linearity of this system near its midrange and the extent to which it is nonlinear at the extremes of its range. Also,

Fig. 9.19 Displacement probe and mating carrier-demodulator unit (courtesy of Bently Nevada, Minden, NV).

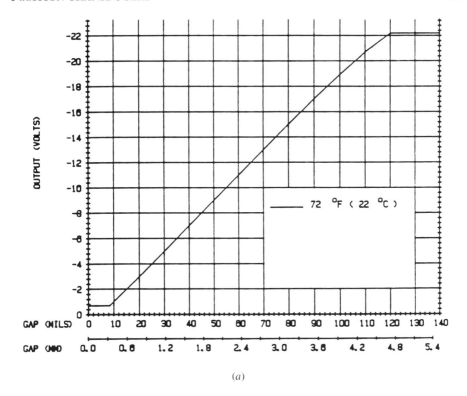

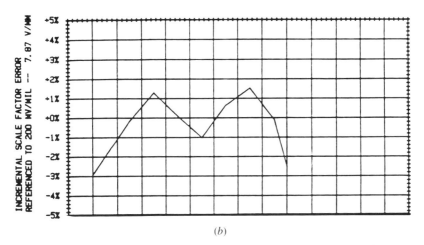

(b)

Fig. 9.20 Static calibration curve for a Series 7200 proximity displacement transducer: (*a*) output voltage versus input displacement; (*b*) error in percent of full-scale versus input displacement (courtesy of Bently Nevada, Minden, NV).

note in Fig. 9.20*b* that the shape of the output error is directly opposite that of Fig. 9.18*c*, namely, m-shaped instead of w-shaped.

Reference 2 describes the operating principle of a variable reluctance proximity transducer as a coupling effect between a primary and a secondary coil. Both coils are located in a proximity probe mounted near a vibrating surface as illustrated in Fig. 9.21. The material in motion must have a ferromagnetic surface. The windings have close magnetic coupling except at the sensitive

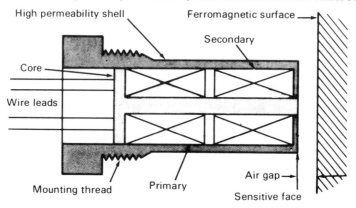

Fig. 9.21 Cross-section of a variable reluctance proximity transducer (taken by permission from Fig. 9.7 of *Handbook of Measurement and Control*, by E. E. Herceg, Schaevitz Engineering, Pennsauken, NJ, 1986).

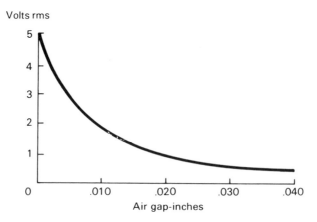

Fig. 9.22 Output voltage versus air gap for a typical variable reluctance proximity transducer (taken by permission from Fig. 9.8 of *Handbook of Measurement and Control*, by E. E. Herceg, Schaevitz Engineering, Pennsauken, NJ, 1986).

face, where the proximity of the ferromagnetic target serves as a part of the flux path. Its primary winding is excited by an AC carrier voltage having a frequency in the range of 400 to 5000 Hz. The mutual inductance of the secondary with the primary is a function of the reluctance of the flux path at the sensitive face. As a ferromagnetic material is brought closer to the secondary winding the air gap reluctance is lowered, thus increasing the coupling between primary and secondary coils. Consequently the voltage induced in the secondary coil is increased in amplitude. This voltage may be rectified and filtered to yield a DC output related to the air gap spacing. The output rms voltage is approximately inversely proportional to the square root of the distance from the sensitive face as given by Eq. 9.7.

$$e_O = \frac{K}{\sqrt{\text{air gap}}} \tag{9.7}$$

This equation is illustrated by Fig. 9.22.

Reference 5 offers a summary of a variety of noncontacting gaging devices, shown in Table 9.1. Perhaps not unexpectedly, proximity probe systems based on the inductive bridge principle of operation have the fewest disadvantages according to this manufacturer's view.

TABLE 9.1 NONCONTACTING MEASUREMENT DEVICES

There are a number of noncontacting measurement devices on the market, each with advantages and disadvantages. Here is a brief overview.

Device	Advantages	Disadvantages
Air gaging	It works	Costly (requires a clean air supply system, pressure transducer, and electronic), very limited range, user sensitive, environmentally sensitive
Capacitance, stray and grounded	Stray capacitance is relatively target insensitive	Costly, difficult to linearize, environmentally sensitive (need a constant dielectric and clean surface)
Infrared or photonic, and fiber optics	Reasonable cost, small spot size, will work with most target materials	Environmentally sensitive (requires a clean reflective target), limited range, poor linearity
Laser	Greater range, excellent resolution and linearity	Very expensive, difficult to set up and use, environmentally sensitive
X-Ray and Beta backscatter	Effective density measuring device, handles a wide pass line variation, effective with most materials	Expensive and hazardous to use, limited to thickness only
Ultrasonic	Long measuring ranges (in feet), good linearity	Limited applications and frequency response, poor resolution
Hall effect	Inexpensive	Needs a magnet for a target, poor linearity, temperature sensitive
Eddy current (Colpitts)	Inexpensive, unaffected by nonconductive intervening material, easy to use, prefers highly resistive targets, long term stability	Requires conductive target, limited range, poor linearity over full range, temperature sensitive, no zero offset, cannot synchronize, limited sensor cable length, high output impedance, poor frequency response, high noise floor
Inductive bridge	Inexpensive, excellent linearity, easy to calibrate, temperature stable, unaffected by intervening nonconductive material, good frequency response, very high resolution, easy to use, long term stability, low impedance low noise floor, can be synchronized, zero offset capability	Requires a conductive target

9.2.4 Linear and Rotary INDUCTOSYN® Position Transducers*

Introduction

Current military/aerospace control systems are demanding higher positioning accuracies for gyros, gimbal platforms, optical sighting devices, and antennas. Linear and rotary INDUCTOSYN® position transducers, obtainable in absolute and incremental types, are among the most accurate transducers available for the measurement and control of linear and angular displacement. Accuracies on the order of ±1 arc second for the rotary and ±0.0025 mm (±100 μin) for the linear are typically achieved.

Currently used in military, aerospace, satellite, radar, and navigational systems, and in machine tools, their electrical output signals drive numerical readout displays, generate computer input data, and provide servo feedback signals. They can be designed to be unaffected by dust, oil films,

* This section is excerpted from a paper by G. A. Quinn and F. S. Ruhle (see Reference 7) and was edited by C. L. Nachtigal.

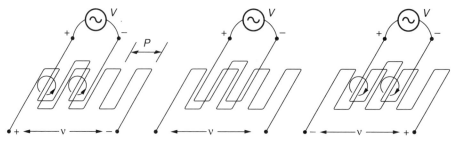

Fig. 9.23 Inductive coupling between precision windings.

vapors, seawater, light, radiation, extreme pressure, and temperature. INDUCTOSYN® transducers have operated at full accuracy for extended periods in space, immersed in liquid nitrogen, in strong magnetic fields, and immersed in 200 meters of seawater.

Principles of Operation

INDUCTOSYN® position transducers consist of printed circuit conductor patterns deposited or laminated on a pair of flat substrates of steel, other metals, or stable nonmetallic materials. This pair of transducer elements contains no moving parts, and is usually mounted directly on the user's fixed and movable machine parts. Since the two transducer elements are not in contact, they are not subject to wear, and mechanical life is indefinite. The elements are simple and rugged, permitting the selection of alternate materials that can withstand nearly any environmental extreme. The operating principle used is the variation of inductive coupling between the conductor patterns on the two elements as they move relative to one another.

Attached to fixed and movable machine parts with the hairpin loops parallel, the two elements are separated by a small air gap. An alternating current flowing in one conductor will induce a voltage in the other that depends on the relative position of the conductors. The induced voltage is maximum when the loops are facing. Passing through zero midway, it rises to a negative maximum at the next facing location. See Fig. 9.23.

INDUCTOSYN® elements behave as transformer windings with a coupling constant k. Calling

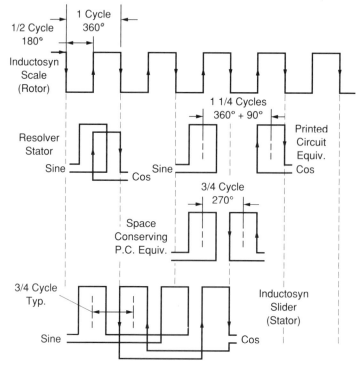

Fig. 9.24 Magnetic field coupling.

the pitch P and the input voltage V, the relation between induced voltage, v_a, and displacement, x is

$$v_a = kV\cos 2\pi \frac{x}{P} \tag{9.8}$$

A second output winding with the same pitch, P, may be located adjacent to the first, and displaced $P/4$ from it. The voltage v_b induced in this conductor is

$$v_b = kV\sin 2\pi \frac{x}{P} \tag{9.9}$$

Only the amplitudes of v_a and v_b change. Phase remains constant while amplitude is a function of relative displacement x/P. There is a unique pair of induced voltages v_a and v_b for every location within one cycle of the pitch P. Thus by measuring these voltages, it is possible to subdivide the accurately known pitch interval with high precision.

As indicated, patterns for inductive coupling consist of precisely spaced and repeated hairpin turns that are connected externally as coils. One element contains a single conductor pattern, while the other element contains two patterns in space quadrature. Figure 9.24 illustrates the coupling between INDUCTOSYN® windings.

Extremely high accuracy is achieved by averaging the measurement over many cycles of the scale pattern. Linear accuracies to 0.005 mm (0.0002 in) (± 0.0025 mm, or ± 0.0001 in), and angular accuracies to 2 arc seconds (± 1 arc second) are standard. With selected units, 0.002 mm (0.000080 in) (± 0.001 mm, or ± 0.000040 in) and 1 arc second (± 0.5 arc second) can be achieved. Repeatability is 10 times better than rated accuracy in most models.

Linear Position Transducers

The linear INDUCTOSYN® position transducer consists of two magnetically coupled parts. One part, the scale, is fixed to the axis along which measurement is to take place (e.g., the machine tool bed). The other part, the slider, is arranged so that it can move along the scale in association with the device to be positioned (e.g., the machine tool carrier). See Fig. 9.25.

The scale consists of a base material such as steel, stainless steel, aluminum, etc., covered by an insulating layer. Bonded to this is a printed circuit track forming a continuous rectangular waveform. (In actual fact the scale is usually made up of 250 mm (10 in) sections that have to be mounted end to end.) The cyclic pitch of the waveform is usually 0.1 in, 0.2 in, or 2 mm and is formed from two conductive poles.

The slider is normally about 100 mm (4 in) in length and has two separate, identical, printed circuit tracks bonded to it on the surface that faces the scale. These two tracks are formed from a waveform of exactly the same cyclic pitch as on the scale but one track is shifted $\frac{1}{4}$ of a cyclic pitch from the other, that is, 90°.

The slider and the scale are separated by a gap of about 0.125 mm (0.005 in) and an electrostatic screen is placed between them. A diagram of the relationship between slider and scale is shown in Fig. 9.26.

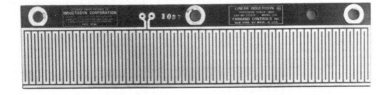

Fig. 9.25 Linear INDUCTOSYN® transducer scale and slider.

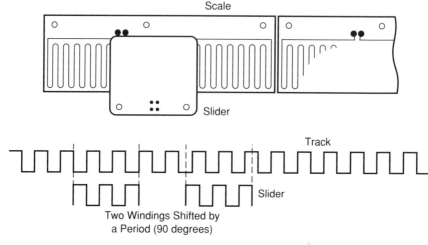

Fig. 9.26   The relationship between the linear INDUCTOSYN® scale and slider.

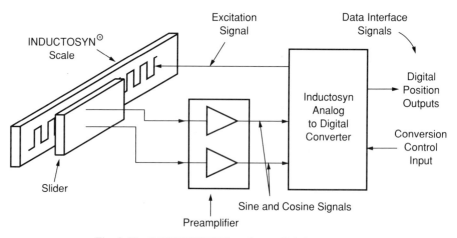

Fig. 9.27   INDUCTOSYN® analog to digital converter.

The principle of operation is similar to that of a resolver. If the scale is excited by an AC voltage, $V\sin\omega t$ (normally between 5kHz and 10 kHz), then the outputs from the slider windings will be

$$V \sin \omega t \cos\frac{2\pi x}{P}$$

$$v \sin \omega t \sin\frac{2\pi x}{P} \tag{9.10}$$

where x is the linear displacement of the slider and P is the cyclic length or pitch.

Therefore, the slider voltages are proportional to the sine and cosine of the distance moved through any one pitch of the scale. The INDUCTOSYN® signal repeats itself. The output signals derived from the slider are the result of averaging a number of poles, and therefore the effects of any small residual errors in conductor spacing are averaged. This is one of the reasons why such a high accuracy can be achieved. The magnetic coupling between the slider and the scale is not nearly as high as the coupling between rotor and stator in a resolver, and for this reason the transformation ratio from input to output is very low, giving rise to relatively small output signals.

INDUCTOSYN® analog to digital converters are analog to digital converters that match the INDUCTOSYN® transducer output characteristics, and transform their accurate analog position signals to digital signals that meet user requirements. See Fig. 9.27. Virtually all such converters subdivide each INDUCTOSYN® cycle length into a number of equal intervals, such as 1000 or

1024. Thus, used with a 2.5 mm (0.1 in) linear INDUCTOSYN® cycle length, the resolution per bit can be 0.0025 mm (0.0001 in). By a proper selection of cycle length and cycle division ratio, the digital output can be direct reading in any units desired, and resolution can be commensurate with the inherent accuracy of the INDUCTOSYN® transducer itself.

Rotary Position Transducers

Rotary INDUCTOSYN® scale elements consist of a pair of discs called the rotor and stator. They are mounted directly on the user's shaft and frame. The analogous linear INDUCTOSYN® scale elements are called the scale and slider, respectively. Rotary coupling transformers are built into the discs in some models, permitting continuous rotation without the use of slip rings and brushes. See Fig. 9.28.

Rotary INDUCTOSYN® position transducers were developed to provide the highest possible accuracy in applications typically served by resolvers. Certain terminology has thus been carried over from resolver practice.

The simplest possible resolver has two poles, and its output signal repeats itself once for each full revolution. To increase resolution, step-up gears are introduced between the resolver and the shaft to be measured, and the resolver is said to operate at a "speed" numerically equal to the gear ratio. For example, a resolver that makes 18 turns for each turn of the measured shaft is called "18 speed." In practice, this would be combined with a "single speed" resolver that simultaneously provided absolute angle information at lower resolution.

Since the output of a rotary INDUCTOSYN® transducer repeats itself once each cycle length, it is equivalent to a simple two-pole resolver with a gear ratio numerically equal to the number of INDUCTOSYN® cycle lengths per revolution. Thus a rotary INDUCTOSYN® transducer with a 1° cycle length is called "720 pole" or "360 speed." One cycle length equals two poles. The length of one complete cycle of the pattern is called the pitch or cycle length. One complete INDUCTOSYN® cycle length corresponds to one complete revolution of our two-pole resolver example, and the variations of inductive coupling as the patterns slide relatively past each other produce output signals identical to resolver signals.

The stator of a rotary INDUCTOSYN® has the two separate rectangular printed track waveforms arranged radially on a disc. The sine track is made up of a number of sections that alternate with the cosine track. In this way the whole of the stator disc is covered in track and any errors in spacing will be averaged out. This gives the rotary INDUCTOSYN® its exceptionally high accuracy.

The rotor of the device is a disc with a complete track of near rectangular printed track waveform. The coupling from the rotor to the outside world can be either by sliprings and brushes or by a rotating transformer as in the case of an electromagnetic brushless resolver.

Rotary INDUCTOSYNS® come in diameters of 75 mm, 100 mm, 175 mm, and 300 mm (3, 4, 7, and 12 in) and have pole counts in the range of 2–2160.

The units can be supplied as separate stator and rotor discs or as a completely mounted and assembled unit.

When the rotor of the rotary INDUCTOSYN® is excited by the AC voltage, $V\sin\omega t$, (normally 5 kHZ to 10 kHz), the stator voltages will be

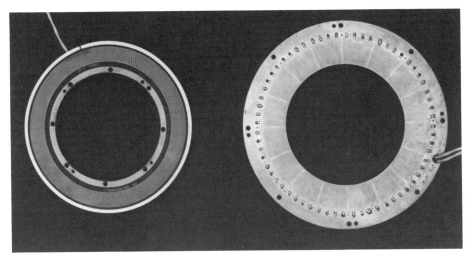

Fig. 9.28 Rotary transducer rotor and stator.

$$V \sin \omega t \, \sin \frac{N\theta}{2}$$

$$\tag{9.11}$$

$$V \sin \omega t \, \cos \frac{N\theta}{2}$$

where θ is the angle of rotation of the rotor with respect to the stator and N is the number of poles of the rotor.

9.2.5 Capacitive Position Transducers

Capacitive position transducers are noncontacting sensors. Their sensitivity is inversely proportional to the distance to be measured; therefore they are especially suitable for the measurement of small distances in the millimeter to micrometer range. In many applications, they are easier to install and less costly than optical equipment. The capacitive sensor geometry can be adapted to the location of measurement; round and rectangular probe shapes are commonly used. The sensors can be built small in size and can be used for distance measurements in locations that would be hard to reach with other gaging methods, such as distance measurements inside a hole. Like many noncontacting sensors, they can be used for accurate and rapid scanning of small distance variations, for example, measuring the runout of rotating shafts and computer disk drives.

Capacitive Sensor Theory

When two parallel plates of area A are placed at a distance d from each other, the capacitance C according to Faraday's Law is defined by

$$C = \frac{\epsilon_r \epsilon_o A}{d} \tag{9.12}$$

where ϵ_o = dielectric constant in vacuum

              ϵ_r = relative dielectric constant of the medium between the plates

If the relative dielectric constant and the area are known, the distance between the plates can be determined by measuring the capacitance.[8,9] A practical capacitive position transducer (Fig. 9.29) has a probe tip with defined surface area A, which represents one capacitor plate. The other plate, typically much larger than the probe tip area, is represented by the object surface, which must be electrically conductive. The requirements for surface conductivity are quite relaxed and are met by all metals as well as many other conductive materials. Metal coating of target surfaces is an option when the target surface is an insulator. Numerous types of electronic circuits are in use to measure the

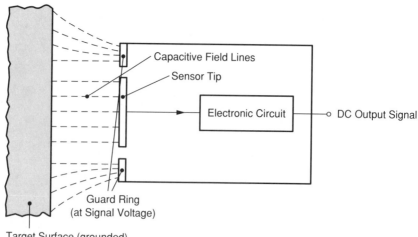

Fig. 9.29 Capacitive displacement transducer.

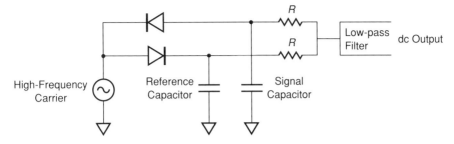

Fig. 9.30 Passive network for capacitance detection (Lion, 1964).

TABLE 9.2 CHARACTERISTICS OF CAPACITIVE DISTANCE SENSORS

Vendor	Lion Precision	Hitec Corp.
Measuring range	±0.03 to ±3 mm	0.6 to 6.0 mm
Sensor tip diameter	1.5 to 5.6 mm	1.0 to 10.0 mm
Linearity (full scale)	±0.5%	±0.2%
Operating Temperature	up to 370°C with air cooled probe	−70 to 150°C (to 810°C in special versions)

Source: Lion Precision Corp., Newton, MA and Hitec Corp., Westford, MA.

capacitance between the two plates.[10,11,12] One common approach is to make the sensor capacitance part of a passive network that is driven by a constant frequency oscillator (Fig. 9.30). Capacitance variations change the amplitude of the passive network's oscillations. After filtering a DC voltage signal is obtained which is a measure of the distance between object and probe tip. As is evident from Eq. 9.12, the relationship between distance and capacitance is highly nonlinear. The distance measurement is most accurate when the distances to be measured are small compared to the probe tip diameter. The accuracy of the measurement is also influenced by the inevitable presence of stray capacitances and by nonlinearities at the plate edges. By using a grounded guard ring, most of the edge nonlinearities are eliminated.[13] Variations of the dielectric constant are negligible when the medium between the plates is air at room temperature and ambient pressure.

Capacitive Position Transducers

Capacitive displacement transducers are available for measurement ranges from about 0.03 to 6 mm with sensor resolution typically 0.1% of range. The sensors listed in Table 9.2 are equipped with circular probe tips and guard rings. For special applications, such as measurement in locations with limited access, flat probe designs with rectangular probe tips may also be used.

Capacitive Position Encoder

The INDUCTOSYN® capacitive transducer[7] employs a noncontacting capacitive pickup moving over an incremental encoder (Fig. 9.31). As the pickup sensor travels over the comb pattern of the encoder, a cyclical variation of capacitance is electronically detected and counted. Capacitive coupling is advantageous for operation in strong magnetic fields, and capacitively coupled units are easily interfaced with MOS circuits. Operation is based on the capacitive coupling between two closely spaced elements that are directly attached to fixed and moving parts.

Linear units have interleaved comb-shaped precision conductor patterns applied to suitable mechanical supports. See Figure 9.31. Center-to-center spacing of the teeth of the comb is called the pitch. Two such sets of patterns are mounted on fixed and moving parts, respectively, so that the teeth of the comb patterns pass over each other as the parts move. The capacity between the comb patterns thus varies cyclically, with one complete cycle for each pitch distance traveled. A second set of patterns may be added, in space quadrature with the first. This permits sensing the relative direction of motion, and subdividing the pitch interval with great accuracy.

Auxiliary patterns with only one or a few elements may also be placed on the substrate. These may serve as an absolute index of position, or may signal control circuitry that the moving part is passing through one or more special locations. Track location signals on digital disc recorders are a typical example.

The pitch of the comb pattern may be chosen to precisely match the desired dimensions or motions of the mechanism. Recorder track spacing is again a good example. The pitch may also be exactly 0.01 in, 0.5 mm, or some other convenient interval for measuring instrument applications. Pitches of typical units now in production are 0.01, 0.0202125, and 0.2 in, and 0.4 mm. Rotary units are made in a comparable way, with the teeth of the pattern extending radially. The two elements are mounted concentrically, and

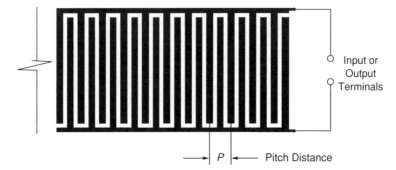

Fig. 9.31 Capacitive INDUCTOSYN® conductor pattern (Farrand Controls).

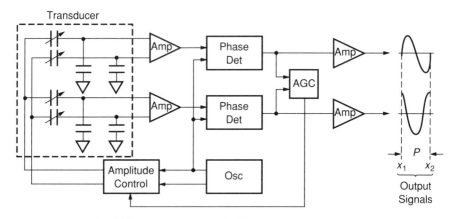

Fig. 9.32 Capacitive INDUCTOSYN® transducer operation.

TABLE 9.3 TYPICAL SPECIFICATIONS FOR CAPACITIVE ROTARY AND LINEAR INDUCTOSYN® POSITION TRANSDUCERS

Type of motion	Linear (straight line)
	Rotary—disc type
	Rotary—flexible tape wrap
Substrate material	Aluminum, steel, G-10 fiberglass,
	Flexible or rigid
Pattern material	Copper (other materials on special order)
Pattern pitch	0.01, 0.02, 0.10, 0.20 in
	1.0, 2.0 mm
Accuracy	2.5 μm TIR for short travel, linear
	$\pm$15 arc seconds for 94 mm OD, rotary
Excitation frequency	10 to 400 kHz
Voltage transfer ratio (VTR)	(V_{in}/V_{out}) 12–2000
Protective finish	Varnish, epoxy, polyurethane

the pitch distance is measured angularly. Auxiliary signals are still possible. The capacitive analog of a multispeed resolver can be supplied by concentric patterns with coarse and fine pitch.

Figure 9.32 is a block diagram showing typical capacitive INDUCTOSYN® transducer operation. Table 9.3 lists typical specifications for capacitive rotary and linear INDUCTOSYN® position transducers.

9.2.6 Rotary Position Encoders[*]

Rotary position encoders provide either analog or digital signal outputs, the latter being used for processing in feedback control systems or for use in digital readout instruments. Analog position

[*] This material was written while the author was employed by ILC Data Device Corporation, Bohemia, New York.

outputs may be used directly or amplified to drive instruments or other devices as in synchronous or slaved motion. The electronics associated with each encoder are generally required for filtering, conversion, or logic operations.

Five unique types of rotary position encoders are available.

1. Potentiometers
2. Synchros and Resolvers
3. Encoders (optical, brush, or magnetic)
4. RVDT
5. Rotary INDUCTOSYN®

The RVDT and INDUCTOSYN® (Farrand Controls, Inc.) are discussed in detail in Sections 9.2.2 and 9.2.4, respectively. Although the potentiometer is presented in Section 9.2.1, it is also discussed in this section with respect to rotary position.

As a general observation, INDUCTOSYNS® are the highest accuracy position transducers followed by resolvers, synchros and optical encoders on a comparable level, with the least accurate being potentiometers. In descending order of cost, taking into account the associated electronics, there is the optical encoder, followed by resolvers and synchros, and then the least costly potentiometers. Resolvers and synchros are truly the most reliable; then come optical encoders, which are more sensitive to environmental conditions; the least reliable are the potentiometers, which are affected by contact wear.

Each of these encoders is readily available from many manufacturers. A list of those manufacturers that supply position encoders is presented at the end of each section.

Rotary Potentiometers

Rotary potentiometers operate on the same basic principles as the linear potentiometers, details of which are given in Section 9.2.1.

The rotary potentiometer consists of a resistive element and a movable brush (Fig. 9.33). As the rotational position changes, the brush sweeps along a variable resistor resulting in a potential that is proportional to displacement. The output is an analog signal varying from zero at one end to a maximum at the other end. The proportionality is critically dependent on having a stable DC reference voltage. (Induction potentiometers are excited by an AC reference voltage.) Since the signal is modulated by the brush movement over a wire surface, noise is inherent in this type of transducer and must be filtered in the interface electronics. In addition, if the position information is to be processed digitally, then an A/D converter is required.

The potentiometer is the simplest and perhaps the oldest of position transducers. But its accuracy and linearity depend on brush-to-resistor contact and quality of material (wire, conductive film, or ceramic).

Characteristics. As would be expected, potentiometers are generally the lowest in cost among position transducers, but are limited in life and reliability due to wear. Accuracy can be affected by the stability of the reference voltage in addition to the influence of stiction between the brush

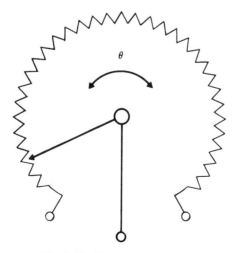

Fig. 9.33 Rotary potentiometer.

TABLE 9.4   POTENTIOMETER MANUFACTURERS

ASI Astrosystems, Inc.
C.J. Enterprises
CTS Corporation
Carter Manufacturing Corp.
Celesco Transducer Products, Inc.
ETI Division Polaris Ind. Enterprises
Farrand Industries, Inc.
Humphrey, Inc.
Johnson Controls, Inc.
Klinger Scientific
Magnetic Power Systems, Inc.
Maurey Instrument Corporation
Micro Switch Division, Honeywell
Micron Instrument Corp.
New England Instrument Co.
Samarius, Inc.
Scientific-Atlanta, Inc.
Vernitech Division, Vernitron Corporation
Waters Manufacturing, Inc.

and resistive element. For wirewound potentiometers, resolution is limited by wire diameter and wire-to-wire spacing, and with linearity, precision potentiometers with less than 0.025% nonlinearity become expensive enough to make the other position transducers more cost-effective. On the positive side in addition to cost, some wirewound precision potentiometers (e.g., those using nichrome) can withstand extreme temperatures. They are also very rugged and have a high survivability rate to shock and vibration. The brush-to-resistor contact can be adversely affected when operating in a severe shock and vibration mode, however.

Availability.   Table 9.4 presents a list of manufacturers of potentiometers.[14]

Synchros and Resolvers

Synchros and resolvers are rotary transformers with variable coupling; the coupling varies as a function of rotor position. Voltage outputs are proportional to the sine and cosine of the shaft angle, or, conversely, voltage inputs rotate the shaft to a specific angle.

In appearance, synchros and resolvers resemble small electric motors. The basic construction and operating principles of synchros and resolvers are similar, each consisting of a stator housing and a rotor core. Both stator and core have their own windings (Fig. 9.34). For the synchro, the stator has three sets of windings 120° apart, whereas the resolver stator has two sets of windings 90° apart.

The rotor is excited with a reference AC voltage ($A \sin \omega t$). This input excitation induces a voltage in the stator. The ideal outputs for the synchro and resolver are as follows:

synchro:

$$V_{1-3} = A \sin \omega t \sin \theta$$

$$V_{2-3} = A \sin \omega t \sin(\theta + 120°) \qquad (9.13)$$

$$V_{2-1} = A \sin \omega t \sin(\theta - 120°)$$

resolver:

$$V_{1-3} = A \sin \omega t \sin \theta$$

$$V_{2-4} = A \sin \omega t \cos \theta \qquad (9.14)$$

By using ratios of these voltages, fluctuations in reference amplitude and frequency cancel under ideal conditions and are minimized to a tolerable level for nonideal operation. These voltages, or their ratios, can be used directly in an analog servo system. With modern-day digital techniques, synchros (and resolvers) can be combined with synchro data converters for digital control and monitoring. By the nature of their construction, synchros (and resolvers) are very accurate transducers and with 12- to 16-bit synchro data converters currently available for the industrial market, these transducers are finding their way into many robotic applications, especially since they are both rugged in construction and can be subjected to harsh environments.

Ideally, synchros and resolvers are infinite resolution devices; practically, they are available in accuracies down to 10 arc seconds. For higher accuracies, multispeed synchros are available that can enhance the accuracy by as much as 10:1 and 36:1. Synchros and resolvers, in addition to being very rugged mechanical devices, can survive in temperatures of several hundred degrees Celsius.

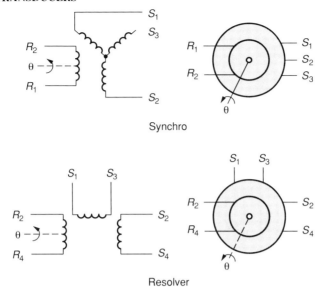

Fig. 9.34 Synchro and resolver windings.

Brushless synchros and resolvers are usually immune to the effects of normal contaminants since they do not have contacting electrical surfaces.

Applications. Synchros were developed over 40 years ago and were used as torque transmitting devices that could read position and drive a remote dial on a position indicator. The resolver is a direct outgrowth of the synchro for use in analog computing chains to manipulate angular position, such as in coordinate transformations. The original resolvers and synchros were brush type. Today many brushless versions are on the market.

Applications of synchros and resolvers cover both the analog and digital worlds. Analog signals can be used directly or amplified as inputs or outputs to synchros, or they may be converted to digital signals for processing and then reconverted to analog. The latter requires additional electronics for analog synchro (resolver)-to-digital conversion (SDC, RDC) and for digital-to-synchro (resolver) conversion (DSC, DRC). Data conversion is discussed in the subsection entitled Data Converters.

Analog. Synchro systems consist of more than one synchro. In its simplest form, two synchros are used; one serving as a transmitter, the other as a receiver. The shaft of the transmitter synchro can be operated manually (e.g., angle setting) or it can be used to measure a shaft angle (e.g., as a master shaft controlling slaved units). By connecting like terminals (i.e., two rotor terminals and three stator terminals) of one synchro to another synchro, the second synchro, which is the receiver, is driven to the rotational position of the first. A synchro system such as this provides remote control, synchronous rotations through slaved units, and measurements. An example of the latter is an aircraft control surface indicator, where the shaft of the control surface is connected to the transmitter synchro and the shaft of the readout dial is driven by the receiver synchro.

A slightly more complex synchro system can be used to detect the error of an actual shaft rotation relative to a desired shaft rotation. This is done with three synchros: a synchro transmitter, a synchro differential, and a synchro receiver. The synchro differential has as its inputs the actual and desired shaft positions, and its output is the difference or error that is transmitted to the synchro receiver. This synchro system has applications in automatic control systems. By changing the terminal-to-terminal connections of these three synchro units, the sum of two shaft angles is transmitted rather than the difference.

Synchro systems are used to drive shafts directly only if the loads are light; that is, a low torque system. These systems use specifically designed torque synchros. When heavy loads are encountered (i.e., high torque), however, torque synchros are inadequate and lose accuracy as the torque increases. For applications requiring heavy loads, such as in driving heavy radar antennas, control synchros are used. These synchros are used as components of servo systems. The synchros provide the control, the servos provide the power.

Resolvers have applications unique to their operation which synchros cannot perform as simply. These applications take advantage of the fact that the output signals are in quadrature and can therefore provide simple sine and cosine computations. Some applications include:

Rotation of rectangular coordinates

Phase shifters

Radar sweep resolution

Data transmission

Digital. Conversion to digital signals permits use of the manifold capabilities of digital signal processing. Programmable control with digital feedback provides the designer with a multitude of versatility. In addition, with high resolution converters (16 bits), accurate velocity computations can be made from successive position data. Typical digital control applications include:

Digital adaptive navigation

Antenna positioning

Multiaxis tool positioning control

Punch-press synchronizing control

Motion control

Adaptive control

Synchro Types. Synchros may be conveniently classified as shown in Table 9.5. Note that resolvers are included as a special type of synchro. Each conventional symbol is included in this table.

There are two basic types of synchros, torque and control, which are further subdivided by function. Torque synchros are employed in direct drive systems in which low torques are sufficient; for example, instruments. Control synchros are high precision devices that require signal amplification for high torque drive systems.

Figure 9.35 illustrates the functional interactions of the different types of synchros and Fig. 9.36 shows the conventional symbols for each functional type.

The "transmitter synchro" (torque, TX, or control, CX in Table 9.5) is a device that converts mechanical rotation into voltage outputs for transmission to differential transmitters or receivers. It accepts an AC reference excitation at its rotor terminals and develops, at its stator terminals, a three-wire AC output at the reference frequency. The amplitude ratios of the output voltages define the angular position of the shaft with respect to a reference shaft position.

The "differential transmitter" (torque, TDX, or control, CDX in Table 9.5) is a device that adds or subtracts its rotor angular position from the electrical angle received from a synchro transmitter. This algebraic sum is converted to a voltage output for transmission to other differential transmitters, receivers, or differential receivers.

The "torque receiver" (TR) rotates to the angular position defined by the electrical input to its stator from either a torque transmitter or a torque differential transmitter. For the synchros to perform correctly, they must each operate off the same power source.

The "torque differential receiver" (TDR) rotates to the angular position defined by the sum (or difference) of the electrical angles transmitted from two torque transmitters, two torque differential transmitters, or a combination of one torque transmitter with one torque differential transmitter.

The "control transformer" (CT) produces an electrical output that is proportional to the sine of the difference between its own shaft angle and that of the angular position defined by the electrical input to its stator from a control transmitter or a control differential transmitter.

The "resolver" (RS) produces a pair of electrical outputs that are proportional to the sine and cosine, respectively, of its shaft rotational position. Resolving the shaft angle into the sine and cosine components has many applications where coordinate transformations are required. In addition to providing the designer with component information not available from a synchro, the resolver also has the advantage that it can be used in control systems with greater accuracy due to its

TABLE 9.5 SYNCHRO CLASSIFICATION

Function	Type	
	Torque	Control
Transmitters	TX	CX
Differential Transmitters	TDX	CDX
Torque Receivers	TR	—
Torque Differential Receivers	TDR	—
Control Transformers	—	CT
Resolvers	—	RS
Transolvers	—	TY

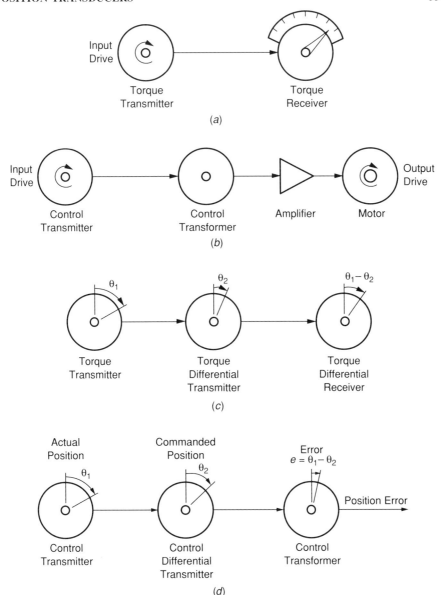

Fig. 9.35 Synchro functional interaction: (*a*) direct torque transfer; (*b*) control system; (*c*) differential positioning; (*d*) position error feedback.

better transmission characteristics. When used in control systems, resolvers are produced as resolver transmitters, resolver differentials, and resolver control transformers. As a resolver differential, it has a four-wire input and a four-wire output to accommodate the sines and cosines of the input and output angles.

The "transolver" (TY) converts signals from synchro to resolver format or resolver to synchro format. It is a bidirectional device (i.e., either rotor or stator may be used as input) in which the rotor windings are in three-wire synchro format but whose stator windings are in four-wire resolver format. It can convert signals from synchro to resolver format, and can be used as a control transformer (ignoring one stator winding) or as a control transmitter (ignoring the other stator winding). By rotating the shaft, the device can rotate the reference axis of (i.e., add to or subtract from) the angle that is being converted from synchro to resolver (or resolver to synchro) format.

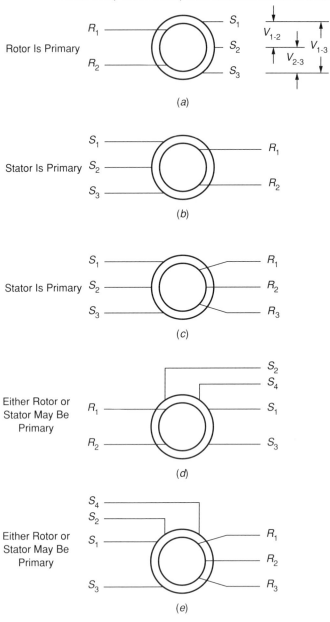

Fig. 9.36 Synchro symbols: (*a*) transmitters and receivers; (*b*) control transformer; (*c*) differential transmitters and differential receivers; (*d*) resolver; (*e*) transolver.

Characteristics. Synchros come in both conventional brush type and brushless. The latter has the same advantages as are found in brushless motors, three significant factors being

increased life
increased reliability
less hazardous in explosive environments

Synchro sizes are identified by a number corresponding to 10 times its nominal outside diameter (O.D.) in inches. For example, a size 11 is approximately 1.1 in O.D. The more common synchro sizes range from 8 to 23 with the 8 corresponding to 19 mm (0.750 in) and the 23 corresponding to 57 mm (2.250 in). Length-to-diameter ratios can vary from 0.8 to 2.5 with an average being about 1.6. A useful rule of thumb for weight-to-volume ratio (for a simple cylindrical volume) is 4500 kg/m^3 (2.6 oz/in^3). These numbers are useful for initial sizing estimates, but updated manufacturers' specifications should be consulted.

Synchros generally operate at excitation frequencies of 60 Hz and 400 Hz, and rated voltages of 11.8 V, 26 V, 90 V, and 115 V.

Data Converters. To use synchros and resolvers for digital measurement and control, data converters are required. For measurements with digital readout or for digital feedback and processing in control systems, the analog signals must be converted to digital format. Conversely, to operate a synchro from digitally acquired information, the digital signal must be converted back to analog format. To effect these transformations, there are many hybrid converters on the market today that provide the following:

synchro-to-digital conversion (SDC)

resolver-to-digital conversion (RDC)

digital-to-synchro conversion (DSC)

digital-to-resolver conversion (DRC)

Table 9.6 presents typical characteristics for both single-speed and multispeed data converters. They are available in standard military and industrial grades with resolutions of 10, 12, 14, 16; accuracies to within one arc minute; and sizes from 79 mm × 66 mm × 20 mm (3.1 in × 2.6 in × 0.8 in) down to 51 mm × 28 mm × 5 mm (2.0 in × 1.1 in × 0.2 in). Newer models available today have analog velocity signal outputs in addition to the digital position outputs (see Section 9.5.2). These versions are particularly well suited for tachometer applications.

Availability. Tables 9.7 and 9.8 each present a list of manufacturers[14,15] of synchros and resolvers and of synchro and resolver data converters, respectively.

TABLE 9.6   POSITION DATA CONVERTER CHARACTERISTICS

Converter*	Form Factor	Features
SDC	Encapsulated Module 79 mm × 66 mm × 10 mm (3.1 in × 2.6 in × 0.4 in)	Resolution: 10, 12, 14 bits Accuracy: ±21, ±8.5, ±4 min
SDC	36 Pin DDIP Hybrid	Resolution: 16 Bits Accuracy Options: ±5.3, ±2.6, ±1.3 min 1.25 rps tracking converter
SDC/RDC	Industrial Hybrid 53 mm × 53 mm × 5 mm (2.1 in × 2.1 in × 0.2 in)	Resolution: 10, 12, 14, 16 bits Accuracy: ±21, ±8.5, ±5.3, ±2.6 min Outputs: 3-state parallel data, direction count, DC analog velocity
RDC	Industrial Hybrid 51 mm × 28 mm × 5 mm (2.0 in × 1.1 in × 0.2 in)	Programmable Resolution: 10 to 16 bits Accuracy: 8, 4, 2, 1 min Scalable analog velocity output
SDC/RDC	Module 79 mm × 66 mm × 20 mm (3.1 in × 2.6 in × 0.8 in)	Two Speed Converter Resolution: 16 bits Accuracy: ±20 s
DSC	Encapsulated Module 79 mm × 66 mm × 20 mm (3.1 in × 2.6 in × 0.8 in)	Resolution: 14 bits Low power Output: 90V synchro @ 400 Hz and 60 Hz
DRC	32-Pin TDIP Hybrid	Resolution: 14 bits High power
DSC/DRC	36-Pin DDIP Hybrid	Resolution: 14 bits Accuracy: ±4 min

*SDC = Synchro-to-Digital Converter
 RDC = Resolver-to-Digital Converter
 DSC = Digital-to-Synchro Converter
 DRC = Digital-to-Resolver Converter

TABLE 9.7 SYNCHRO AND RESOLVER MANUFACTURERS

Manufacturer	Synchro	Resolver
ASI Astrosystems		X
Analite, Inc.		X
Autotech Corporation		X
Clifton Precision	X	X
Electronic Counters and Controls		X
GAP Instrument	X	X
Harowe Servo Controls	X	X
IMC Magnetics Corporation	X	X
Micron Instrument Corporation		X
Portescap Transicoil	X	X
Singer Co., Kearfott Division	X	X
Transmagnetics	X	X
Vernitron Corporation		X

TABLE 9.8 SYNCHRO AND RESOLVER DATA CONVERTER MANUFACTURERS

Analog Devices, Inc.
Autech Corporation
Computer Conversions Corporation
Control Sciences, Inc.
Control Technology Co.
EG & G Torque Systems
ILC Data Device Corporation
Natel Engineering Co., Inc.
Northern Precision Labs.
Ragen Data Systems, Inc.
Transmagnetics

Encoder

An encoder is a device used to generate digital position data. It consists of an encoding mask with incremental or absolute codes excited by optical, electrical, or magnetic energy. The basic operation of electrical and magnetic encoders is similar to the optical encoder. The electrical encoder (often referred to as a brush or contact encoder) has essentially the same limitations as a potentiometer, using a brush to make contact with a conductive coded mask. The magnetic encoder is relatively new and has the advantage over the optical encoder in that there is no light source that may decay or fail; the magnetic encoder also has good resistance to shock and vibration. Optical encoders have to be well sealed to prevent contamination from interfering with the light path. Nevertheless, optical encoders are the most widely used.

Operation. The basic operation of the rotary optical encoder is shown in Fig. 9.37 for both the incremental and absolute encoders. Each encoder consists of a light source, a coded disc, and a light detector. The light source can be an LED or an incandescent lamp. LEDs can survive mechanical vibration better than lamps. However, lamps are less sensitive to temperatures and have greater output than LEDs. The light detector, which can be a solar cell or a photo-transistor, generates an output when light from the source passes through the mask. This output is generally in the low millivolt range and has to be amplified for transmission and then processed. Included in the electronics processing package are filters, buffers, and registers, plus up-down counting logic for the incremental encoder.

Incremental Encoder. The coded disc for an incremental encoder, shown in its simplest form in Figure 9.38, consists of two concentric sets of equally spaced lines, each set in quadrature with the other (i.e., 90° phase difference). One set of lines provides equal pulsing for constant shaft velocity and the number of pulses (count) is directly proportional to angular displacement. The pulsing of one set relative to the other (i.e., which precedes which) defines the direction of rotation.

Incremental encoders require the use of counters to count pulses. The number of pulses defines the angular position. The number of pulses in a given time defines the shaft velocity. The resolution,

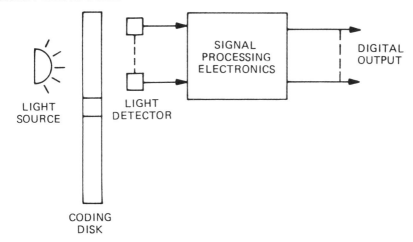

Fig. 9.37 Rotary optical encoder.

Fig. 9.38 Incremental encoder disc.

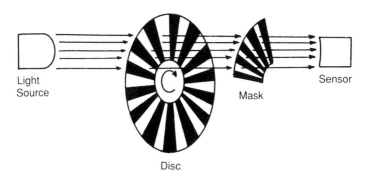

Fig. 9.39 High resolution encoder.

or number of lines per shaft revolution, is critical to the particular application at hand. Although resolution is limited by physical constraints (number of lines that can be placed on a disc of a given diameter) and light transmission quality, it can be increased four-fold by decoding both quadrature signals. For standard sized incremental encoders, as many as 10,000 lines per revolution can be placed on a disc of 89 mm (3.5 in) diameter. With quadrature decoding, this is equivalent to 40,000 lines per revolution.

High resolution encoders use a coded disc and mask together, each of the same pitch (Fig. 9.39), to enable better light transmission with more efficient light collection. When the disc and mask are aligned, light passes through; when they are out of phase, light is blocked.

Incremental encoders are less expensive than absolute encoders. However, if there is a loss in power, position information is lost with incremental encoders. To avoid this problem, absolute encoders can be used in place of, or in conjunction with, incremental encoders.

Fig. 9.40   Absolute encoder disc.

Absolute Encoder.   An absolute encoder disc can be coded to provide absolute position data in natural binary, gray code, or binary-coded decimal (BCD). Each of these codes has its own application: natural binary for direct digital processing, gray code to prevent glitches at major transitions, and BCD for position displays.

An N-bit absolute encoder has N sets of concentrically spaced lines; 2^{N-1} equally spaced lines on one concentric circle for the least significant bit (LSB) down to one 180° line for the most significant bit (MSB). This is illustrated in Fig. 9.40 for the natural binary code. In addition, each bit requires its own detector to provide a binary output.

Characteristics.   The accuracy and resolution of the encoder depend on the spacing of the lines (incremental encoder), the number of concentric circles (absolute encoder), and how accurately the code is deposited on the disc. These parameters, in turn, depend on the disc diameter and material (glass, plastic, or metal). The larger the diameter, the more concentric circles and, thus, greater resolution. But large diameters have greater inertias and are more sensitive to shock and vibration. However, heavy duty encoders are available in a shock-isolated package that can take up to 200 G.

Incremental encoder sizes vary from miniature 13 mm (0.5 in) diameter up to 89 mm (3.5 in) diameter. Absolute encoders are generally larger than the incremental encoders because of the need to have many concentric coded circles (one per bit); up to 230 mm (9 in) to 300 mm (12 in) diameter for high resolution encoders. Absolute encoders are available in resolutions from 6 to 16 bits; 6 bits is equivalent to a resolution of 5.6°, whereas 16 bits is equivalent to a resolution of 20 arc seconds.

Availability.   Table 9.9 shows a list of encoder manufacturers.[14,15]

9.2.7   Ultrasonic Ranging Transducers

Introduction

The technique of ultrasonic ranging utilizes pulses of sound to measure distances, analogous to the way pulses of electromagnetic energy are used in radar. As shown schematically in Fig. 9.41, pulses of acoustic energy are transmitted, usually at frequencies above the audible, reflections off a desired target object detected, and the resulting time of flight used to determine distance. Ultrasonic ranging requires only that the target produce a sufficient acoustic reflection. This noncontact means of distance measurement thus lends itself to a wide range of applications: targets can be non-magnetic or nonconducting, solid, powder, or liquid, at ranges from a few centimeters up to 60 m or more.[16]

Piezoelectric

Ultrasonic transducers used for ranging in air are commonly of two types, piezoelectric or capacitive. Piezoelectric transducers utilize a ceramic such as barium titanate or lead zirconate as the active element.[17] Often the ceramic is sandwiched between metal plates to produce a half-wavelength oscillator of suitable frequency, while optimizing the use of relatively expensive ceramic. One problem with this type of design is the high acoustic density of the ceramic and metal compared to air, which results in poor coupling between the transducer and air. Matching layers of cork, plastic, foam, or other materials of intermediate acoustic density are often applied to the front of the transducer to improve the coupling. Piezoelectric transducers are typically rugged devices suitable for hostile environments. Many are approved for explosive environments, a result of their solid

TABLE 9.9   ENCODER MANUFACTURERS

ASI Astrosystems, Inc.	Kessler-Ellis Products Co.
Airflyte Electronics Co.	Klinger Scientific
Allen-Bradley Industrial Control	Kollmorgen Corporation
Analite, Inc.	Librascope Division of Singer
Autotech Corporation	Litton Encoder Division
BEI Electronics, Inc., Industrial Encoder	McGill Manufacturing Co., Inc.
Division	Measurement Systems, Inc.
Bausch & Lomb	Micro Mo Electronics, Inc.
Buckminster Corporation	Micron Instrument Corp.
Burleigh Instruments, Inc.	Motion Control Devices, Inc.
CTS Corporation	North Atlantic Industries
C-Tek, Inc.	Northern Precision Labs.
Celesco Transducer Products, Inc.	Oriel Corporation
Clifton Precision	Opto Technology
Comptrol, Inc.	Pepperi & Fuchs, Inc.
Data Technology	Photronics Labs
Datametrics-Dresser Industries	Portescap Transicoil
Disc Instruments Division, Honeywell	PMI Motors, Kollmorgen Corp.
Duncan Electronics	Real Time Systems
Dynamics Research Corporation	Renco Corporation
Dynapar Corporation	Sensor Tech
Electro-Craft	Sequential Info Systems
Electro Sensors, Inc.	Servo-Tek Products, Inc.
Electromatic Components Ltd.	Sigma Instruments
Electronic Counters & Controls	Singer Co., Kearfott Division
Encoder Products	Stocker & Yale, Inc.
Fork Standards, Inc.	Teledyne–Gurley
General Equipment & Mfg. Co.	Theta Instrument Corp.
HH Controls Co., Inc.	Trans Kinetic Systems
Hewlett Packard Co.	Trumeter Co.
Hyde Park Electronics, Inc.	U.S. Digital
Innovatek Microsystems, Inc.	Vernitech Division, Vernitron Corporation
International Sensing	Waters Manufacturing, Inc.
Itek Measurement Systems	Xercon, Inc.

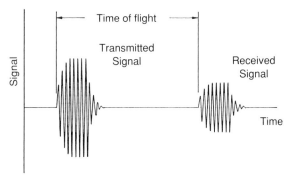

Fig. 9.41 The distance of a reflecting object is proportional to the time between when a burst of sound is transmitted and when the echo is received.

construction with no internal voids where explosive vapors can collect. Piezoelectric transducers are temperature limited by the requirement of staying well under the curie temperature of the ceramic, limiting them to maximum temperatures from 82°C to 121°C (180°F to 250°F).

Capacitive

Capacitive transducers employ a thin plastic film driven electrostatically to transmit and detect sound. The most familiar capacitive sensor is one developed by the Polaroid Corporation for use in automatic focusing cameras.[18] The low force, large displacement drive of capacitive transducers results in good acoustic coupling with air. While piezoelectric transducers are resonant devices with narrow bandwidths, capacitive sensors are inherently broad band, capable of functioning over a wide range of frequencies. The low mass of a plastic film means high frequencies are possible, up to 200 kHz or more. The use of an exposed plastic thin film does imply limited ruggedness when compared to piezoelectric transducers. At the same time they are usually lighter and more compact than a comparable piezoelectric transducer, while their high frequency and wide bandwidth makes them well suited for precision measurements at short to moderate distances.

Speed of Sound

The use of time of flight to calculate distances assumes that the speed of sound is accurately known. The speed of sound in dry air at 20°C is 343.2 m/s (1125 ft/s).[19] Contrary to common belief, the speed of sound does not vary significantly with pressure. Humidity can have a small effect on the speed of sound, up to about 0.35% at 20°C. The speed of sound varies proportionally with the square root of the absolute temperature:

$$c = c_o \sqrt{\frac{T(°K)}{273.15}}$$

$$= c_o \sqrt{1 + \frac{T(°C)}{273.15}} \tag{9.15}$$

where c_0 is the speed of sound at 0° C (323.3 m/s). Near room temperature the speed of sound varies 0.18% per °C, or about 1.0% per 10 °F.

Applications where the temperature varies significantly require compensation to achieve accuracies exceeding a few percent. This can be done by measuring the temperature and adjusting the assumed value for the speed of sound accordingly. It should be noted that it is the air temperature that is relevant, not the transducer temperature. Alternatively, temperature compensation can be done using a fixed target at a known distance as a reference for calculating distance.

Operating Frequencies

Operating frequencies of ultrasonic transducers vary from below 8 kHz to over 200 kHz, or in wavelength from 4.3 cm to 0.18 cm. The lower frequencies are used in long-range, high-power applications where absorption in the air is significant. Frequencies below 14 kHz have the disadvantage of being clearly audible. Higher frequency devices are more compact, produce a narrower beam and higher resolution, and are better suited for shorter range applications where absorption is not a problem. In general, acoustic interference from other sources is less likely to be a problem at higher frequencies. Resolution tends to be proportional to wavelength, but is also a function of the sophistication of the associated electronics. Most commercial transducers have no trouble resolving under half a wavelength, and resolution as small as 0.0025 cm has been achieved with a sensor operating at 200 kHz.[20]

Beam divergence angles also vary with frequency. Because the wavelength is relatively long compared to the dimensions of the transducer, ultrasonic energy does not propagate in a well-defined beam like light from a spotlight, but rather in a diffuse cone, more intense at the center and dropping off at the edges. The most common practice is to quote a 3 db included angle, that is, the included angle at which the amplitude of the received signal has dropped to 0.707 of the maximum. The 3 db included angle may be calculated for sensors radiating uniformly across a circular face using the relation:

$$\theta = 2 \arcsin\left(\frac{1.6c}{f \pi D}\right) \tag{9.16}$$

where c is the speed of sound, f is the operating frequency, and D is the effective diameter of the transducer. Ultrasonic transducers also emit small amounts of energy in side modes, which occasionally may result in spurious echoes.

The most suitable targets for ultrasonic transducers are flat, perpendicular surfaces, either solid or liquid. These reflect the acoustic energy directly back to the transducer, while surfaces at an angle reflect sound away from the transducer. Flat targets with surfaces stepped approximately one-half wavelength can produce poor echoes due to destructive interference. Rough surfaces can be thought of as reflecting the acoustic energy that strikes them uniformly in all directions. For this reason a coarse bulk solid may often produce a stronger echo than a fine solid with the same angle of repose. The most challenging targets are liquids with large foam layers or diffuse solids that pack poorly; these tend to absorb rather than reflect sound.

TABLE 9.10   ABSORPTION IN AIR (DB/M)

Frequency (Hz)	Percent Relative Humidity		
	0	40	100
10,000	0.03	0.15	0.06
20,000	0.08	0.48	0.21
40,000	0.26	1.10	0.83

Applications

The fact that the speed of sound varies for different gases generally precludes its use in applications where the gas mixture varies. Unsuitable applications include tanks containing volatile liquids or where gases such as carbon dioxide or sulfur dioxide are being evolved. Ultrasonic transducers have been used to measure gas ratios, one example being the control of a helium/air mixture by observing the time of flight over a known distance and calculating the speed of sound, which is a function of the helium/air ratio.

Absorption varies strongly with frequency. As a rule of thumb, reducing the frequency by half reduces the absorption by half. For this reason long-range transducers utilize low frequencies. Moisture also increases the absorption. At room temperature the highest absorption occurs around 37% relative humidity. Table 9.10 shows absorption in db per meter for different frequencies and humidities. The presence of dust or water droplets in the air will increase absorption. Placement of the transducer should always be done to minimize condensation or dust on the face of the transducer, as this can severely degrade the signal.

One characteristic of ultrasonic transducers is the existence of a dead band at close distances. This is a consequence of using the same instrument as both a transmitter and a receiver, because the sensor must ring down after the transmit pulse before a receive pulse can be detected. Piezoelectric transducers, being narrow bandwidth resonant devices, usually display larger dead bands than capacitive transducers. "Smart" electronics that reduce the transmit pulse amplitude and duration at short ranges reduce this problem. In practice it is usually possible to mount ultrasonic transducers so their operating range does not include distances below their minimum range.

Excessive ringing may be caused by interaction between a transducer and mechanical resonances in its mounting. This can be controlled by use of a vibration isolator of acoustically dead material such as rubber. When a transducer is mounted on a round flange or plate, mounting it off-center will reduce the coupling to external mechanical resonances. Excessive ringing may also result from standing wave effects, particularly when the transducer is mounted in a long, narrow standpipe, as is sometimes done to place the minimum range outside the dead band. This problem can be reduced by lining the standpipe with acoustically absorbent foam.

Mention should be made of the use of ultrasonic ranging in liquids. The high acoustic density of liquids compared to air means it is easier to couple acoustic energy from an appropriately designed sensor. The speed of sound is higher in liquids, 1482 m/s in water at room temperature versus 343 m/s in air. The variation of the speed of sound with temperature in liquids is more complex than for gases, so that empirical temperature compensation is usually used.

9.2.8   Magnetostrictive/Ultrasonic Displacement and Velocity Transducers

Introduction

This section describes a transducer that employs unique magnetostrictive and ultrasonic techniques. The transducer is used to measure displacement and velocity and can interface with either analog or digital systems. Units have been built to measure up to 10 m (30 ft) in length. (See photograph Fig. 9.42.) Before the availability of the Temposonics™ device the prior art offered a number of different methods for making these measurements. These products include:

Potentiometers
LVDTs (Linear Variable Differential Transformers)
INDUCTOSYNS®
Encoders (Optical Scales)
Ultrasonic Transmitters & Receivers
Magnetic Reed Switches
Magnescales

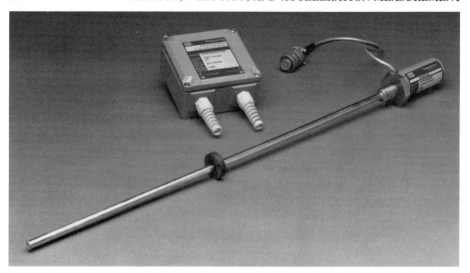

Fig. 9.42 A magnetostrictive/ultrasonic displacement and velocity transducer (courtesy of MTS Systems Corp., Sensors Division, Research Triangle Park, NC).

The listed devices have various attributes and limitations which determine their suitability for different applications in which they might be used. The characteristics of interest include accuracy, repeatability, resolution, reliability, ruggedization, temperature coefficient, and cost. We will review where this magnetostrictive device fits with respect to these important characteristics.

Principle of Operation

The principle of operation is as follows. Initially a current pulse is applied to the conductor within the waveguide (see Fig. 9.43.). This sets up a magnetic field circumferentially, around the waveguide over its entire length. There is another magnetic field generated by the permanent magnet that exists only where the magnet is located. This field has a longitudinal component. (See Fig. 9.44.) These two fields join vectorially to form a helical field near the magnet which in turn causes the waveguide to experience a minute torsional strain or twist only at the location of the magnet. This twist effect is known as the Wiedemann Effect. This torsional strain pulse propagates along the waveguide at the speed of sound in this material, which is about .36 μs/mm (9.1 μs/in) or 2.8 mm/μs (.11 in/μs). When this torsional pulse arrives at the tapes in the head (see Fig. 9.45) it is converted into a dynamic longitudinal pulse injected into the tapes. The tapes are made of a special nickel material whose reluctance changes with strain. The longitudinal pulses cause the tapes to experience a momentary change in reluctance. This change in reluctance due to dynamic impulse is known as the Villari Effect. Two coils coupling these tapes mounted in the field of two bias magnets will generate a momentary electrical pulse caused by the change in reluctance in the tapes. In order to extract the useful position information we measure the time between when we launch the initial current pulse and the time we receive the signal from the output coils. This time is a very precise function of the position of the moving magnet because we are able to control the velocity of the torsional strain pulse in the waveguide accurately and repeatably.

To provide an analog output, simply use the pulse width to develop a pulse-width-modulated wave train that is averaged to provide an analog output proportional to position. (See Fig. 9.46). When a digital output is desired we use the pulse width information to gate a counter and we can typically generate a 16-bit natural binary or BCD digital output proportional to position. This digital output is absolute rather than incremental, which is important for many applications. Referring to the analog output, we also have the ability to generate velocity output information by differentiating the position output. These two outputs, both position and velocity, are available simultaneously from the same transducer.

It may be convenient at this time to describe the Temposonics device by examining a photograph of the system. Figure 9.42 shows a long tube that contains the magnetostrictive elements mounted within the tube and the processing electronics located in the head and in the electronic box. A movable magnet is shown on the tube. It is the position of this magnet attached to the object of interest that is detected by the system. The magnetostrictive element or waveguide is suspended in the tube or protective element. (See Fig. 9.45.) The waveguide is fixed to the base on one end and

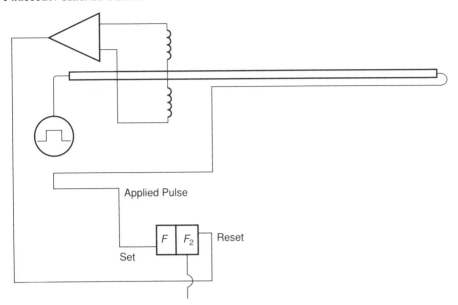

Fig. 9.43 A simplified analog circuit.

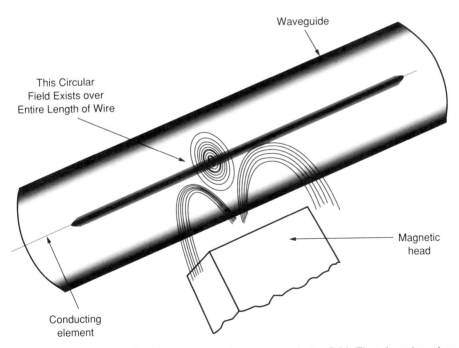

Fig. 9.44 External magnetic field interacting with central conductor field. The tube twists where an external magnetic field interacts with a field induced by the central conductor.

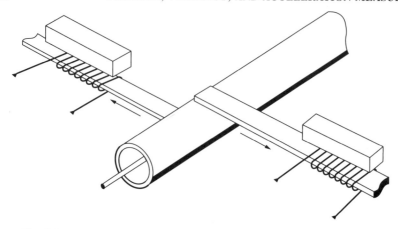

Fig. 9.45   A pair of tapes used to detect the transmission of a torsional strain pulse.

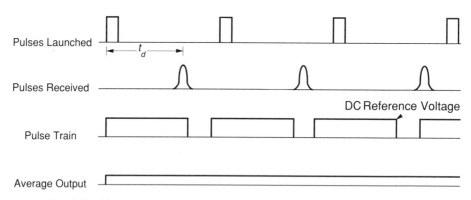

Fig. 9.46   Signal conditions representative of those encountered in an analog unit.

spring-loaded at the other end. Special tape elements are attached to the waveguide near the head end to detect the sonic pulses as they are received.

In the digital version resolutions on the order of 0.001 in or 0.1 mm are typical. It is possible to produce resolutions much finer by using special electronic techniques. One simple approach is to count at higher frequencies. For example a 55 mHz clock provides resolution of 0.05 mm (0.002 in). Using a 110 mHz clock will make the resolution 0.025 mm (0.001 in). It is difficult to work with these high frequencies so another approach was developed with the digital unit where the output signal is used to retrigger the input pulse a finite number of times, thereby extending the measurement time. This technique offers resolutions of microinches; however, one must evaluate the fact that this fine resolution is obtained at the sacrifice of update time which might be typically from 1 ms to 4 ms. For the shorter strokes, update times of 300 mμ are available.

Performance Specifications

When all of these principles are put together we are able to build transducers that offer a unique combination of valuable performance characteristics.

Linearity.   We define linearity as the departure of our output from the theoretical best straight line. We are able to offer ±0.05% of the total stroke length. We find that as the stroke lengths become longer, say 1 m and longer, we can provide better linearity to ±0.03% total stroke.

Repeatability.   These devices have extremely good repeatability. We can offer 0.001% of full scale or 0.0001 in, whichever is greater. With some special stabilizing techniques we have developed the ability to produce units with good long-term and short-term stability. There is virtually no warmup time and units in the field have shown high stability over many years.

Resolution. In the analog version the output has a resolution that approaches infinity in that it is stepless. We are simply averaging a pulse-width-modulated wave train.

Reliability. This transducer is completely solid state. There need not be any contacting or wearing parts. As a result the reliability of the transducer is very high. An MTBF of 4 million hours is obtained using the standard reliability handbooks for arriving at this figure of merit.

Temperature Coefficient. The temperature performance of the rod portion of the transducer is on the order of 3 ppm/°F. This low change with temperature is achieved by selecting a material for the waveguide with this characteristic and with special processing techniques in the manufacture of the device. The temperature performance of the analog systems is affected to a greater degree by the limitations of the analog electronics.

Optional Features

Velocity. In the analog version a velocity output is offered. The position output is clean enough to differentiate it and provide a velocity output from the position information simultaneously. This capability exists for velocities from a fraction of an inch per second to over 500 in/s. There are lags in the information at the higher speeds due to the filters and the finite speed of sound in the waveguide, but these are limited and are predictable.

Multiple Magnet Output. Units have been built where up to five magnets have been mounted on a single transducer rod simultaneously. It is possible to detect the output from each magnet individually.

Differential Output. Special electronics have been offered to manufacture units where the output is proportional to the distance between two magnets mounted and moveable on the rod.

Flexible Design. It is possible to manufacture the transducer in a flexible rod form so that the unit can be placed in a 1 meter (3 ft) diameter container for shipment. This is of great value when the transducers are made for very long measurements, say, 10 m (30 ft) or so. Another use of the flexible unit is where the measurement of a curved motion is required.

Applications

In view of the many advantages of the magnetostriction/ultrasonics technology combined in the standard transducer and the various options available, this position transducer has found acceptance in many different applications. Table 9.11 shows some of these applications. A few of these areas are highlighted here.

**TABLE 9.11   APPLICATION AREAS FOR
A MAGNETOSTRICTION/ULTRASONICS
DISPLACEMENT TRANSDUCER**

Hydraulic Cylinders and Actuators
Injection Molding Machines
Die Casting Machines
Industrial Robots
Forest Industry Machinery
Atomic Reactor Rod Position Control
Oil Tank Farms
Underground Gasoline Station Tank Monitoring
Flight Simulators
Underground Mine Fault Detection System
Presses
Rolling Mills
Glass Plate Manufacture
Wavemaking Machines
Underground Caverns (Liquid Level)
Forges
Programmable Controllers
Motion Controllers
Transfer Machines
Packaging Machines
Intrinsic Safety Applications (Paint Sprayers)

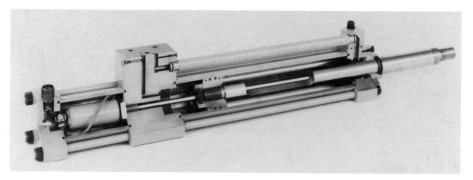

Fig. 9.47   A magnetostrictive/ultrasonic displacement transducer installed inside a hydraulic actuator.

Hydraulic cylinders and actuators.   This transducer design lends itself to be installed inside hydraulic cylinders. The thick wall used for the rod enables it to withstand oil pressures of about 5000 psi so that the transducer can be mounted within the cylinder. See Fig. 9.47. This permits an elegant design since the transducer is internal to the cylinder and coaxial with it.

Injection Molding Machines.   Until recently many of the functions on injection molding machines were controlled by potentiometers or cam-operated limit switches. These machines use linear displacement on three different functions on the machine: mold clamp position; injection of the molding material; and ejection of the molded part.

Die Casting Machines.   A magnetostrictive/ultrasonics transducer is being used to monitor both position and velocity of the injection process.

Industrial Robots.   The very high repeatability feature of this transducer makes it very attractive for this application. One manufacturer is working with six axes where one transducer handles two axes with dual magnets on the same rod.

9.3   SOLID-STATE IMAGERS

Dale R. DuVall

Solid-state image sensors are gaining widespread use in scanning applications where durability, reliability, and consistent performance are required. Solid-state camera systems provide numerous performance advantages over tube-type systems in many applications. These include parts inspection, process control, document scanning, laboratory instrumentation, and television.

Solid-state imagers are well suited as position transducers due to their precise geometrical accuracy. In addition, they exhibit no image lag, minimize blooming, and provide long life. They are inherently stable, requiring no periodic adjustments for image registration, and are relatively free from the effects of stray magnetic fields. These factors, combined with the benefits of solid-state ruggedness and small size, make them ideal candidates for most industrial imaging applications.

Development has been in both linear (one-dimensional) arrays for line scan applications and two-dimensional arrays, directed primarily toward television applications. Many configurations have appeared, some addressing unique applications. The topics addressed in this section deal primarily with imagers and imaging systems in configurations that are applicable to general usage in both the laboratory and industrial environment. Usage in astronomy, spectroscopy, and other applications where arrays (often uniquely designed) may be operated near their theoretical limits are not included.

Linear devices typically range in size from 64 photoelements (commonly referred to as picture elements, shortened to pixels or pels) to more than 5000 photoelements, fulfilling a wide range of requirements. High-resolution document scanning has been a major influence on the development of longer linear imagers. Two-dimensional devices can be found in various configurations, from small formats such as 14 pixels by 40 pixels to large formats as large as 1000 pixels by 1000 pixels. The smaller two-dimensional devices have found use in inspection applications and point-of-sale terminals, but the main thrust today is the development of superior performance in television applications. Progress continues to be made, including improvements in sensitivity, resolution, and noise reduction, as well as the integration of ancillary circuitry on the device itself. Recently, color filters have been deposited in various configurations on the photosites, providing the capability of color scanning. The size of the element spacing continues to be reduced; some linear devices have pixel spacings as small as 7 μm, while a larger element spacing of up to 40 to 50 μm is found in two-dimensional devices.

9.3.1 Imager Architecture

A variety of semiconductor technologies are used to fabricate these devices, with each having its own advantages and disadvantages. The more common are the charge-coupled device (CDD),[21] the photodiode detector with MOS switches, and the charge injection device (CID). Many newer devices utilize photodiodes as detectors and CCD shift registers as the interrogation means.

In the early stages of semiconductor development, each photodiode had its own output, requiring a large amount of support electronics. Today's devices incorporate either analog shift registers or a series of switches that are sequentially activated to sample the individual elements. Current products are available in a variety of interconnections providing single or multiple outputs. The single-output devices obviously minimize the external signal conditioning required, but multiple-output devices provide maximum effective operational speed without putting difficult bandwidth demands on any single channel. Typical maximum operating speeds per channel range from 2 to 10 MHz, although some devices can be operated at even greater speeds. The configuration required is determined by each unique application.

The photodetector array consists of three main elements. The first is the photosites themselves, which measure the incident light. The second is the interrogation circuitry, which samples the photosites and sequentially outputs the information (usually through a charge to voltage converter). The third is the control circuitry which buffers and/or generates the timing necessary to properly sequence the device.

Photosites

Two types of photosites are employed, either the p-n junction photodiode or the MOS capacitor (CCD structure providing a field-induced junction). Both devices provide the means for separation of the photon-generated electron-hole pair. A disadvantage to the CCD structure is the electrodes that lie above the cell and act as an optical interference filter, reducing the efficiency of the photosite. This effect is a function of the wavelength of the incident light.

The photosite behaves as an integrator of photoelectrons (or depletion of stored charge utilized in some devices). This provides a signal very linear with exposure, which is the product of incident light intensity and integration time (Fig. 9.48). In most devices, the resetting of each photosite is usually at the time its output is sampled. CIDs can be sampled without being reset, providing an advantage in some unique applications. Several newer devices allow simultaneous resetting of all

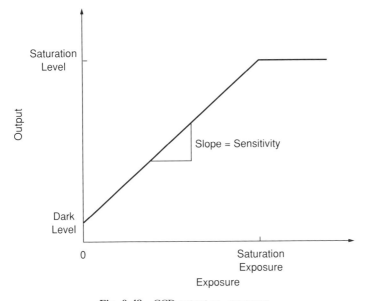

Fig. 9.48 CCD output vs. exposure.

photosites, allowing controlled exposures at intervals less than the frame rate. The complete resetting of the photosites, by whatever scheme, eliminates any residual image from frame to frame, referred to as lag that is so common in vidicon systems.

The point at which the photosite has stored its maximum capacity is referred to as its saturation charge or saturation level. Beyond this point the detector is no longer linear, and severe oversaturation can affect other sections of the array. The "spillover" of this excess charge can fill nearby elements or otherwise impact their output, resulting in the phenomenon referred to as blooming. This excess charge can also spill over into the interrogation register, affecting complete rows of data. Newer devices have made great improvements in eliminating this problem, and several types can withstand overexposures of one or more orders of magnitude with minimum degradation of the performance of neighboring photosites.

Interrogation (Sampling)

One method of interrogation is accomplished by draining the charge from the photosite into an analog shift register, either of the CCD or MOS bucket brigade type (Fig. 9.49). The shift register is then clocked to provide a sequential output of the video data. The efficiency at which the charge is transferred initially had detrimental effects on the performance of these devices, but current efficiencies of 0.99999 or better have minimized this concern.

The second technique uses MOS transistors to sequentially sample the photosites. One side of a sampling transistor is connected to its respective photosite, the other being tied to a common output line. A digital shift register sequentially gates the transistors "on" to provide a sampled stream of analog data on the common-line representative of the video (Fig. 9.50).

Linear arrays usually have two interrogation registers associated with the row of photosites, each connected to alternate photosites. In low- and moderate-speed applications, the two channels are multiplexed onto a common line before being output as a single video signal. In high-speed

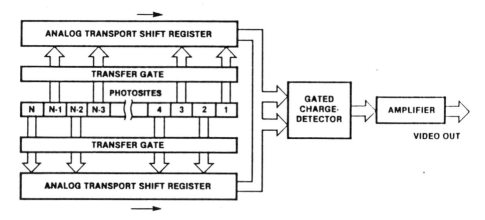

Fig. 9.49 Simplified block diagram of a line array using CCD (analog) shift register sampling.

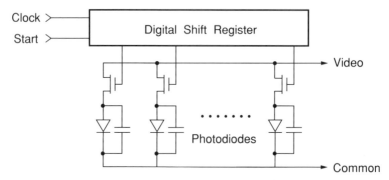

Fig. 9.50 Simplified block diagram of a line array using digital shift register sampling.

applications, the odd and even channels are brought out separately. Two-dimensional devices have a variety of interconnect schemes, with some two-dimensional devices employing a full-frame video buffer.[22]

Initially CCD photosites were used with CCD registers, while photodiodes were used with the MOS gating technique. Currently the use of photodiodes as sensing elements combined with CCD shift register interrogators provide a high level of performance.

The output of each photosite is in the form of charge, but most devices contain some type of charge to voltage converter that simplifies the usage of the outputs. Several imagers include a resettable integrator as part of this conversion, providing a "boxcar" type output voltage that greatly simplifies the electronics interface to the ancillary video circuits. In the highest-speed applications, this type circuitry can limit the frequency, so some designs provide integrators that can be disabled if desired.

Control Circuitry

All devices require some sort of control circuitry to generate the required sampling sequence. Some devices require multiple-phase clocks that must be carefully phased to ensure proper operation, while others internally generate the multiple clock phases from a single input. Suppliers of arrays requiring multiple clocks often have a separate clock generator circuit available to the systems designer.

9.3.2   Array Specifications

Many manufacturers differ in the way they specify their devices, and proper interpretation of those specifications is required to understand the limits of any meaningful data. The more significant considerations are described below.

Responsivity

The responsivity of silicon imagers has steadily improved over the last 10 years. Two factors affect the useful responsivity. The first is the quantum efficiency (the efficiency of the conversion of photons into accumulated charge). The second is the improvement of the signal to noise ratio so that low light levels can generate usable signals. A typical plot of photodetector responsivity is illustrated in Fig. 9.51.

Many photodetectors have a spectral responsivity typical of silicon photodiodes, whose range is from approximately 400 nm to 1100 nm, peaking at approximately 900 nm. Some devices have their responses enhanced at the shorter wavelengths, providing good sensitivity in the blue region, while others have minimized their sensitivity to the near infrared.

The sensitivity of the imagers as a function of wavelength is specified in several ways, depending upon the design and the manufacturer. Some specify the response in volts per unit of incident energy density (volts/joule/cm^2 or volts/lux), which includes the charge to voltage transform of the on-board amplifier. When defined in this manner, the spectral characteristics of the reference source must be provided or the responsivity must include the response per unit wavelength.

Others, especially those whose outputs are provided as currents, specify the more traditional amps per watt per unit wavelength common with discrete devices. While amps/watt/wavelength is

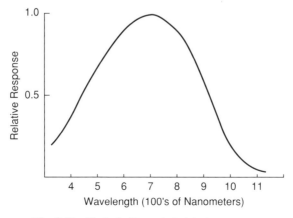

Fig. 9.51   Typical silicon photodetector response.

the fundamental measure of conversion efficiency, this value requires the area of the photosensitive element to be known to calculate the power.[23] In either case, the saturation level of the device, the spectral characteristics of the incoming energy, and the integration time must be known before the output of the imager can be estimated.

The output of the device is the integral of the product of its responsivity and the incident light as a function of wavelength, expressed as

$$\text{OUTPUT} = \int_{\lambda_1}^{\lambda_2} R(\lambda)H(\lambda)d\lambda \tag{9.17}$$

where $R(\lambda)$ is the responsivity of the device per incremental wavelength, $H(\lambda)$ is the spectral characteristic of the incident light, and $d(\lambda)$ is incremental wavelength. Responsivity variations among photosites within an array can vary as much as $\pm 10\%$, but 3 to 5% variations are often achieved in many devices. The effects of this variation, along with other array specifications, are discussed in the applications section.

Usually the one or two photosites at either end of the array are less consistent in uniformity than the interior elements, so it is generally best not to use these elements unless absolutely required.

Dark Current

The primary limitation of the allowable integration (exposure) time is thermally generated dark current. This current is accumulated at the photosites and is indistinguishable from photogenerated signals. This dark current signal is linearly proportional to exposure time and is also a function of temperature, doubling approximately every 8°C (Fig. 9.52). In typical devices operated at room temperature, the dark current can represent approximately 1 to 4% of the saturation exposure with an integration time of 40 ms. In extremely low light applications, the photodetectors are cooled so that integration times can be extended.

Dynamic Range

The maximum useful range of any one photodetector is referred to as the dynamic range. This is usually defined as the ratio of the saturation level to the RMS noise level of the detector. Peak noise levels can be five times as great. While this range has continually improved to ratios better than 2500:1, it does not reflect the typical useful range in general applications.

Odd/Even Noise

In arrays employing multiple interrogation registers, characteristics of the individual channels can differ somewhat. These variations can affect both the base levels (dark levels) and the apparent responsivity (light sensitivity). Since the most common interlacing scheme uses two channels, one connected to the even-numbered pixels, the other connected to the odd-numbered ones, this difference is referred to as odd/even noise. Since this noise appears as a systematic error, it can generally be corrected.

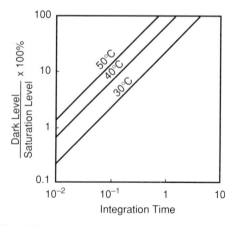

Fig. 9.52 Dark signal level versus integration time.

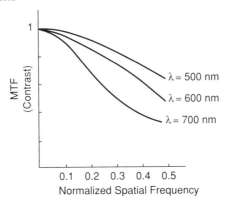

Fig. 9.53 Resolution versus wavelength.

Resolution

The photodetector arrays provide very high levels of performance in regard to resolution. The plot in Fig. 9.53 illustrates the MTF* of a typical device. At the Nyquist limit (where the pixel center-to-center spacing is twice the spatial frequency of the image on the device), contrasts of better than 50% can be achieved. At this limit, the actual output is a function of the position of the image on the device, so this degree of contrast should not be expected (in fact, the contrast can approach 0, depending on phase).

Window Effects

Almost all scanning devices include a glass window as part of their packaging for protection of the array itself. These windows can greatly deteriorate device performance in some applications. The more nearly parallel the incident light, the greater the influence of imperfections on and in the glass, as well as accumulated dust and dirt. These imperfections produce shadows and light scattering that make the apparent uniformity of sensitivity much worse than it otherwise would be. For this reason, the use of the array with small lens apertures or in shadowing applications where the light is nearly parallel must be approached with caution. When monochromatic light such as a laser is used for direct illumination of the array, the window can act as a glass wedge, producing interference patterns that can severely affect the data.

The window material used often reduces the spectral response in the blue region due to its absorption at wavelengths less than 400 nm. If response is required in this region, quartz windows should be used. These are available on some devices.

9.3.3   Scanning System Considerations

Scanning Techniques

Solid state imagers function well as measurement transducers, with each photosite effectively being a single-channel position detector. With a known system magnification, the pixel location where image edges (or other distinguishing features) fall is a precise indication of their position. Each unique application dictates the scanner configuration. In many cases, a one-dimensional scan will provide all the data required, and a linear scanner is the obvious choice. When two-dimensional images are necessary, either linear or two-dimensional arrays can be considered (Fig. 9.54).

Two-dimensional imagers are very useful in two-dimensional applications that do not require high resolution. A typical two-dimensional array, for example, may contain 577 pixels in one axis, with up to 581 pixels in the other. If limited resolution is adequate, then a two-dimensional array may be a good choice.

In high resolution systems requiring two-dimensional images, longer linear arrays have generally been used to provide scanning in one axis, with the scanning in the other axis being accomplished

---

*MTF is the modulation transfer function, defined in this application as the maximum contrast obtainable as a function of the spatial frequency incident on the device. The MTF is also a function of the wavelength of the incident light. Because longer wavelengths yield lower contrast than the shorter wavelengths, many manufacturers specify their devices with IR filters included.

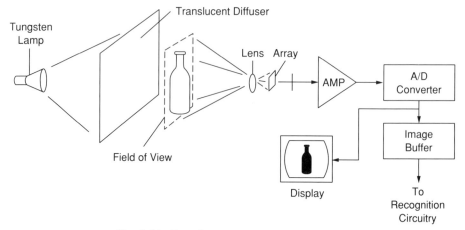

Fig. 9.54 Two-dimensional scanner application.

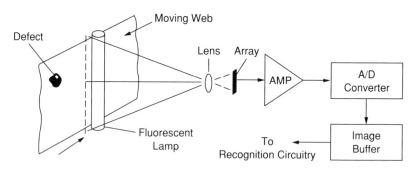

Fig. 9.55 Line scanner application.

by apparent relative motion between the detector and the object (Fig. 9.55). In many cases, such as parts moving down an assembly line at a constant velocity, this configuration is ideal. In other configurations the detector and lens are moved, or a rotating mirror imparts apparent motion on the object. A rotating mirror system has the disadvantage of the image distance changing if the object is in a plane. For small angles, this may not be a problem.

Using linear devices in two-dimensional applications also provides flexibility in selecting different resolutions in orthogonal axes, while two-dimensional devices usually have resolution ratios from 1:1 to 3:4.

The scan rate of linear devices is often the limiting factor in this mode. The resolution required in the axis of motion sets the integration time (scan rate) of the device, so that high resolutions of rapidly moving objects may be impractical. For example, a 1024 element, linear CCD array operating at a scan period of 100 μs (or a pixel rate in excess of 10 MHz) yields a limiting resolution in the axis of motion equal to the target's displacement during the scan period. Motion during the scan period induces an "image smear," which degrades the resolution. The effects of this can be calculated by convolving the moving image on the array with the pixel aperture, but details of this procedure are beyond the scope of this discussion.

To minimize this smearing effect, short duration strobes are sometimes used, especially with two-dimensional devices. The strobe duration becomes the useful integration period, instead of the frame rate of the device itself.

Practical Utilization

The utilization of the output of the photodetector device is totally dependent on the application. In systems where the target is of high contrast and the resolution requirements are not stringent, a simple thresholding of the video data will provide an adequate binary image. This might be the case, for example, where the target is back illuminated (providing a high contrast image) and the required resolution is less than the pixel pitch in object space.

On the other hand, a system used to detect material flaws, for example, might require the ability to detect changes in contrast of less than 5% with the defect position being required to within a pixel. In this system, multiple shades of gray are required, and corrections in responsivity nonuniformity are required. For precise linear measurements, the system geometry must be controlled and the lens system must have limited geometric distortion. Otherwise calibration targets, such as accurately ruled grid patterns, must be scanned to allow proper corrections.

In many imaging applications, the data from a photodetector array are utilized collectively, with the outputs of the photosites relative to their neighbors (or all other photosites in the array), often having at least as much significance as their absolute outputs. Because of this, the variations in responsivity and the offsets of each photodetector relative to its neighbors are of key importance. With variations in responsivity as much as 10% of saturation exposure and offsets between neighbors being as much as 3% of saturation, comparison of outputs among neighboring devices cannot be accurate unless these output variations are measured and compensated. In addition, the optics configuration, including the illumination characteristics on the object, can be as important as the detector responsivity itself to achieve satisfactory imaging results.

To correct the array variations, the characteristics of each photosite can be recorded and applied as a calibration correction. Since each output is linear, the use of the equation of a straight line correction works well; the offset being the y-intercept and the responsivity being the slope (or gain) value.

This type calibration is also applicable for correcting the systematic errors of the optics system, including nonuniformity of scene illumination and lens effects. For this type calibration, the base level of each pixel is obtained by scanning the array with the field dark or the optics path otherwise blocked. Then the field is scanned with a target of uniform reflectance covering the field. The outputs from each of these two scans are then used to determine the base level and gains of each pixel, respectively. Periodic recalibration of this type will compensate for long-term system variations such as lamp aging, and so forth.

Selection of Integration Times

The system designer generally has a variety of parameters that can be manipulated in order to arrive at an effective solution to the scanning requirement. One of the key parameters is the selection of the integration time. It is desirable to have a scan rate low enough to make signal conditioning and analog-to-digital conversion simple, and provide a data rate sufficient to allow the data to be routed into a computer or other processing device with minimal buffering and storage. On the other hand, the scan times should not be made so long that the dark signal levels are significant. The upper limit of the integration time is the serial speed at which pixels can be output from each port of the device. Maximum speeds from 1 to 10 MHz per channel are usually specified, but with care, some devices can be operated at twice these rates with acceptable performance in many applications.

Synchronization with the illumination source may also be desirable if it is driven from an AC source (see illumination considerations below). Fast scan rates require high illumination levels, which can be a significant problem in itself. In many cases, however, the spatial sampling requirements of a moving target eliminate options otherwise available.

Optic Considerations

Determination of the basic configuration of the imaging system can be adequately estimated using the simple lens formulas. These are found in numerous elementary texts and are not included here. (Reference 24 provides a good introduction to practical methods.)

Imaging Lens Selection

Many factors must be considered in selecting an appropriate imaging lens. Key factors include the required system magnification, the lens speed, the field angle, the depth of field required, the acceptable distortion, and the desired resolution. Final selection of the imaging lens is usually a series of compromises and cannot be treated in proper detail here; however, some considerations are mentioned in what follows.

When selecting a lens, a typical first step is to determine the magnification of the system. In most applications, the array is smaller than the object, so that the magnification is less than unity. With the magnification known, the focal length is then estimated from the track length (object to image distance) desired. Generally this estimated length is adjusted to fit a lens of standard focal length, and then the final track length is calculated. Select a focal length lens that keeps field angles small, minimizing the illumination falloff at the edges of the array due to the cosine 4th effect. Vignetting is generally not acceptable.

To minimize illumination requirements, a fast (small f-number) lens is best suited, but use of the wide aperture settings reduces the depth of field. The resolution of the lens is very important for sharp image detail, as the system MTF is the product of the array MTF and lens MTF. For example, a lens MTF of 70% at the pixel pitch and a limiting array MTF of 50% yield a maximum system MTF of 35% at the pixel spacing.

Geometric distortion can be a significant factor when the lens is used for precise dimensional measurements, introducing errors in excess of several pixels at the edges of the field of view. These errors can be reduced by proper calibration or by keeping the measurement point near the optical axis. Small changes in object to image distance alter the system magnification, introducing significant dimensional errors. Keeping the measurement near an axis or using a telecentric optics configuration minimizes this effect.

Usually a commercially available lens will function satisfactorily as an imaging lens. At short, finite, conjugate distances, enlarger lenses can function well. They typically range in focal length from 50 mm to 300 mm and are designed for maximum performance at reductions from 2 to 20. In addition, they usually provide no vignetting with small arrays and the field geometric distortion is usually well below 1%. For longer conjugate distances, camera lenses function well, but distortion is usually more significant.

Illumination

The illumination reaching the array is a function of the lens transmission and speed, the target illumination, and the reflectivity of the target. A relationship relating the irradiance (incident illumination) reaching the array, H_i, and the irradiance upon the target, H_o, is provided:

$$H_o = \frac{4H_i(f\#)^2(1 + m)^2}{TR} \tag{9.18}$$

where

$f\#$ = infinity f-number of the lens

m = magnification

T = lens transmission

R = target reflectance

In actuality, this expression is valid only for a true Lambertian reflector; however, it provides a reasonable estimate for most diffuse targets.

For objects containing the information required on the surface, front illumination is required. Illumination angles vary with application. For flat targets, illumination angles are not critical, but should be at least 10° off the optical imaging axis to avoid spurious specular reflections from many objects. The illumination configuration on contoured surfaces is unique to each application; in one instance it may be desirable to minimize unwanted shadows; in another, angles may be optimized to enhance the surface contrast.

Where only the outline or profile of an object, or holes completely through an object require detection, rear illumination has the advantage. In this configuration, contrast can be optimized so that simple threshold electronics can be used to generate black/white images. The background lighting can be provided by a front-lit, diffuse, reflecting panel or a back-illuminated, diffuse, transmitting panel.

Illumination Sources

Any type of illumination that emits radiation within the sensitivity of the detectors may be used, determined by the application. Several of the more common are mentioned in what follows.

Fluorescent lamps are well suited for applications when the information lies in the visible spectrum. These lamps, because of their large radiating surfaces, cannot brightly illuminate an object and are therefore not candidates when high illumination levels are required. With fluorescent lamps, there is significant output ripple (up to 50%) when operated at low frequency. In 50 or 60 Hz applications, the scan rates of the detector must be synchronized with the power frequency to avoid amplitude variations between successive scans. Operating the lamps at frequencies of 10 to 20 kilohertz or above eliminates the ripple. The lamps can be operated at DC but only with a significant loss of useful life.

Tungsten halogen lamps are very stable light sources that can provide very high power levels required for short integration times. These lamps typically have 75% of their output in the IR range above the response range of the photodetectors. If it is undesirable for the object to receive considerable heating from this lamp, an IR filter should be used between the source and object. Tungsten sources exhibit ripple from AC operation, but the thermal inertia of the filament typically limits the ripple from 3 to 8% at 60 Hz. If this level of ripple is unacceptable, the lamps can be operated at DC or electronic compensation can be provided in the video circuit.

In many applications, the contrast of the information desired is present only in visible light, and use of significant IR energy may destroy the contrast so apparent to the eye. Because of this, several manufacturers produce a filter that alters the spectral response of silicon to that nearly matching the responsivity of the eye. This response is referred to as photopic. With a 3200° K tungsten halogen source, only about 10% of its output energy lies within this spectral profile.

Light-emitting diodes can be used in many applications, especially where the field of view is small. They can be switched on and off at high frequencies, effectively providing a strobe to freeze action or control exposure. Lasers are sometimes used to provide flying spot scanners in one or two dimensions. Their use is usually limited to scanning diffuse targets due to the interference problems that arise with window interactions mentioned earlier.

Electronic strobes can be used when periodic sampling of high-speed targets is required. Strobe durations of a few microseconds are readily obtained, and the sampling rate is usually limited by the ability to recharge the energy storage capacitors of the strobe.

Since there are so many variations of contrast dependent on color, and most imagers behave as black/white detectors, it is necessary to understand the spectral characteristics of the source, the object, and the detector to predict true performance. The system designer must exercise his or her freedom to change sources and manipulate the responsivity of the system with the appropriate filter selection; the penalty being the overall drop in detector output when narrow-band filtering is required.

Video Electronics

The output of the photodetector array is a serial analog signal representative of the amount of light incident upon the array during the previous integration period. This serial stream of data must be sampled at the pixel rate in order to achieve maximum available resolution. The means to accomplish this is dependent on the device type and the accuracy of the quantization required. If the output is proportional to the charge, an integrating-type charge amplifier is well suited to provide a voltage proportional to the incident light. If the output is from an internal buffer inside the array, voltage amplification and offset correction are generally required. The signal from the array generally is from 1 to 2 V swing at saturation to dark levels, with a DC bias level of several volts. Removal of this bias level generally makes additional signal processing easier.

Almost all the devices have some sort of clock noise superimposed on the video signal. This noise is coherent, that is, consistent in amplitude and phase with respect to the data, and can therefore be ignored by proper sampling of the signal. Some arrays provide a dummy output containing this clock noise, but with no video signal. This line is usually routed to a difference amplifier along with the video, allowing easy removal of the majority of the clock noise as a common mode signal.

In almost all applications, the video signal requires some degree of digital processing before it is useful. The simplest case is the setting of a threshold level, which provides the means of generating a binary black/white image. This can be accomplished with a high-speed comparator sampled at the pixel rate.

Usually more processing is required, which generates the requirement for digitized shades of gray. A parallel (or flash) converter strobed at the pixel rate works well at high frequencies, eliminating the need for a sample and hold network required by slower conversion techniques.

Accurate signal level measurement beyond 1% of full scale is difficult to realize without very precise video calibrations and careful signal conditioning, with measurements approaching the dynamic range of the device being restricted to a laboratory environment.

Dimensional Measurements

The measurement of the edge of an object is limited to within ±1 pixel when amplitude thresholding is used, either on the analog signal or a multiple-level digital signal. When the video signal is converted to multiple-level digital information, edge detection can be made repeatable within a fractional pixel size by using linear interpolation between adjacent pixels and calculating the point on the line where a threshold is crossed. This technique still requires the considerations of geometric distortion and varying magnification mentioned earlier.

9.4 GAGING TRANSDUCERS

R. J. Pfeifer

9.4.1 Definitions and Scope

In his 1942 treatise *Gages and Their Use in Inspection*, F. H. Colvin defines a gage as "a measuring instrument which is taken to the work."[25] As the complexity and throughput of manufacturing processes has increased and electronic technology has developed, measuring instruments have been not only "taken to the work," but integrated into the manufacturing process to the extent that one particularly significant connotation of gages has developed: that of being specialized online devices for the measurement of critical parameters of high throughput manufacturing processes for the purposes of quality and process control. As computer technology has developed, such gages have provided the basic inputs for process automation and information systems. This section, then, emphasizes the technologies suitable for online automatic gaging for high throughput piece part and continuous manufacturing processes.

The terminology "gaging transducer" loosely defines the physical subsystem that senses the appropriate parameter (the measurand) and converts or "transduces" the basic sensor response into another form, typically electronic signals, which are further processed into measurement outputs in a signal conversion subsystem.[26] Schematically, the subsystems of the measurement system are defined in Fig. 9.56. Throughout the remainder of this section, the terms "gaging transducer" and "gage" are used interchangeably. Gaging transducers are described which sense not only position and/or dimensional parameters, but also those dependent parameters, mass and density, which are related to dimensions by

$$m = \rho \prod_{i=1}^{3} l_i \qquad (19.19)$$

where m is the mass, ρ the density, and l_i one of three orthogonal dimensions of the objects being measured.

The gages are considered in terms of the fundamental physical principles of measurement. These comprise:

1. Electromechanical
2. Pneumatic/fluidic
3. X-ray
4. Isotopic
5. Infrared
6. Microwave
7. Optoelectronic (discussed in Section 9.3)
8. Acoustic/ultrasonic (discussed in Section 9.2)

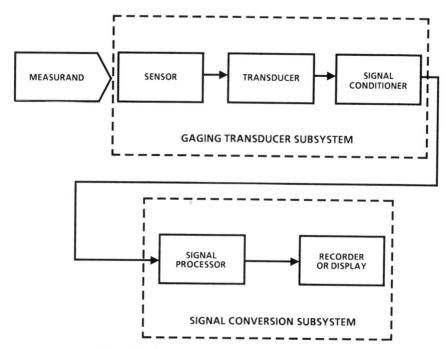

Fig. 9.56 Generalized measurement system functions.

The fundamental characteristics of all gaging transducers and systems are first addressed: range, accuracy, response time, and calibration. The principles of operation of the various gages are described and performance properties in terms of these fundamental characteristics are discussed.

9.4.2 Fundamental Characteristics of Gaging Transducers

The fundamental characteristics of gaging transducers and systems include:

1. Range—extent of measurand variations that can be measured with adequate accuracy.
2. Responsivity—temporal characteristics of gage response to variations in the parameter measured.
3. Accuracy—the degree to which the gage measures truly the measurand.
4. Calibration—the establishment of an algorithm that relates basic transducer response to the measurand in quantitative engineering units.

Many secondary characteristics of gaging transducers and measurement systems are also of significance, but these secondary characteristics typically comprise subsets of the fundamental characteristics.[27]

The range, or extent of parameter variations that can be measured with adequate accuracy, is determined by inherent properties of the measurand and sensor and their interaction. Typically, the structure of the sensor and transducer are optimized for a specific range of interest, so the gage cannot operate outside that range, as in the case of a feeler gage whose very dimensions determine the application range. Furthermore, the structure of the gage may result in changes in sensitivity to the measurand that ultimately limit or determine the range.

The responsivity of the gage is defined by the transfer function that relates the transducer output to the measurand dynamics. Again, the structure of the sensor and transducer, and the interaction with the measurand, determine this characteristic. Gage responsivity is frequently described in terms of the time constant, rise and fall times, cutoff frequency, delay time, and damping characteristics. The temporal and frequency characteristics of gages are discussed in the comprehensive dynamic systems analysis of Chapter 3.

Accuracy of gaging transducers, the degree to which a gage measures truly the measurand, is described in terms of

1. Repeatability—the degree to which the gage measures identically the same measurand.
2. Resolution—the smallest change in the measurand that can be detected.
3. Calibration accuracy—the degree to which the gage output can be transformed into quantitative units of the measurand.

Degradation of accuracy is described in terms of noise, drift, backlash, hysteresis, and so forth, and inaccurate measurement can be caused by interferences to which the gage may respond, more or less, as it does to the measurand. Thus temperature changes can interfere with precision-dimensional measurements by causing changes in the sensor structure that are indistinguishable from changes in the measurand.

Calibration, the establishment of an algorithm that relates the gage response to the measurand in quantitative units, is determined typically from gage response to a set of standard measurands. In principle, the gage need only have a single-valued correspondence with the measurand so that the calibration curve can be determined. Departures of a gage output from its calibration curve are often described in terms of nonlinearity (a term whose origin in piecewise segmenting of the calibration curve is historical). As a result of long-term inherent changes within a gage, the calibration is often restored by a process of zeroing, standardizing, or normalizing, all of which involve the measurement of a limited set of standards or base conditions and appropriate modification of the gage output or algorithm for variations from the true values of the standards.

One of the most challenging aspects of gaging is the establishment of adequate standards and correlation procedures with respect to the standards to determine calibration. With much of the mechanical gaging impetus coming from go/no-go gaging of piece parts, samples of such parts representing the go and no-go limits are frequently used as standards for many pass/fail gages.

9.4.3 Electromechanical Gaging Transducers

Electromechanical gaging transducers are those gages that directly sense and convert mechanical displacements to electrical responses for measurement of position or displacement. Typically, the gage consists of a sensor that tracks the item being measured, a mechanical linkage, a position-to-electronic converter, and an appropriate signal conditioner system. The classes of position-to-electronic converters are described in Section 9.2.

The unique characteristics of electromechanical gaging transducers are usually embodied in the structure of the mechanical sensing element that tracks the item that is measured and by the mechanical linkage, including the means whereby the gaging is freed from or compensated for typical interfering effects. In automatic piece-part gaging applications, prime consideration must be given to the means whereby the parts are transferred to the gaging station and presented to the gage. Typically, key points on the part are located, and the gage is referenced to these points.

Part motion and surface characteristics are of primary concern in the design of the tracking or feeler element of the gage. If the part velocity is substantial, then the tracking element must include a contacting stylus or anvil of appropriate shape, hardness, and lubricity to accurately track the surface without marking the part or wearing itself out. In addition, the tracking mechanism must be designed so that its mass and damping are appropriately matched to the dynamics of the item measured.

Mechanical gaging of continuous webs, such as steel and paper, where gage-to-process velocities may exceed 1000 m/min, requires special designs to match the sensor dynamics to the process characteristics. For relatively crude measurements, the rollers in the process itself may be instrumented to track the thickness of the sheet material.[28] Certain precise thickness gages in the steel industry use a diamond stylus to contact the sheet; in the paper industry, sapphire buttons have been used.[29] In these cases, the shape of the contacting anvil and mechanical properties of the tracking arm must be so designed to follow the surface, not skip over the high spots.

The tracking arm is typically connected to a mechanical-to-electronic transducer through a linkage that modifies the displacements of the tracking arm, amplifying them if fine structure measurement is required. Linkages include simple and compound levers, racks, pinions, gears, springs, reeds, and so forth.[30]

Optical elements, such as light sources, mirrors, lenses, prisms, and detectors may be included to reduce internal friction and inertia, and take advantage of amplifications of mechanical motion by means of optical levers.

The appropriately amplified mechanical motion is typically converted to an electronic output by means of a position transducer optimized for the application. Particularly suitable are LVDT, split core, piezoelectric, and capacitive transducers.

In certain applications, no tracking arm or feeler structure is required; the position or dimensional changes of the item being measured are directly sensed by the transducer, which typically provides an electronic output signal. Magnetic, eddy current, or capacitive interactions between the sensor-transducer and the item measured form the basis for most of these measurements. In the case of the Schuster Gage, an interesting intermediate variation is provided; here the thickness of nonmagnetic material passing over a steel calender roll is measured by means of a split-core transducer that rides on the exposed surface of the sheet as it passes over the roll.

The range of mechanical gaging transducers is quite variable and depends largely on the structure of the tracking arm. Typical automatic gaging resolutions and accuracies are 1 to 100 μm (0.04 to 4 mils) over ranges from 100 to 1000 times the resolution. Dynamic response is limited due to the mass of the structures involved, but can reach 1000 Hz. In general, the gages are versatile, reliable, and accurate.

9.4.4 Pneumatic Gaging Transducers

Pneumatic gaging transducers, based fundamentally on the variable obstruction to flow in a pneumatic circuit which is caused by the proximate presence of a piece part, provide great versatility and potential freedom from wear of the sensor elements, which need not be in physical contact with the part being measured. In general, the flow obstruction can be determined by flow or back pressure measurements in the pneumatic circuit. These flow characteristics are critically determined by the geometrical structure of the sensor/part geometry, and undergo substantial changes for relatively minute variations in part dimensions, thus rendering the technique quite sensitive for gaging. Since the pressure and flow changes can be readily transduced to mechanical displacements, the pneumatic response can be converted into electronic response and the gaging technique can be readily applied to automatic gaging applications.

The fundamental principles involved in pneumatic gaging can be described from Fig. 9.57. In this simplified schematic, regulated air enters the measuring device, passes through an orifice into a plenum and out an exit aperture to impinge on a surface (typically a workpiece) located at distance x from the exit aperture. The flow from the exit aperture and pressure in the plenum are related to the distance x; as the distance increases, the flow increases and the backpressure decreases. Typically, the gage is calibrated by use of go/no-go standards. By incorporating pneumatic nozzles and orifices within a contacting feeler-type mechanical gage, an intermediate mechanical-to-pneumatic transduction can be accomplished for the purpose of amplifying sensitivity to minute changes or replacing part of the mechanical linkage of typical mechanical gages.

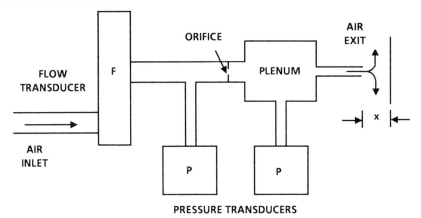

Fig. 9.57 Schematic of pneumatic gage.

Pneumatic gages can be classified as[30]

1. Direct Flow
2. Direct Backpressure
3. Differential Backpressure

In the first two types of gages, measurement is made directly of the flow or backpressure developed. Flow meters of various types are in common use, with the conical tube/float and rotating vane measurement/readout devices replaced by electronic transducers for automatic gaging applications. Similarly, various types of pressure indicators are in common use, such as Bourdon tubes, which can be replaced by electronic transducers (discussed in Chapter 10) for automatic gaging applications. For the most accurate gaging, small backpressure changes can be detected by various differential pressure devices that contain some arrangement of internal chambers and passages connected to a sensitive differential pressure detector. Typical differential detection elements include diaphragms, bellows, and venturis in clever arrangements to amplify the minute backpressure changes and thus increase the gage sensitivity.

In pneumatic systems, the gaging sensors, defined as those parts of the pneumatic circuit comprising the exit orifices or nozzles and any mechanical linkage used in positioning the nozzles with respect to the workpiece, can be classified as

1. Direct
2. Noncontacting
3. Contacting

In the direct sensor class, the exit nozzle is essentially the part being gaged, as in the case of carburetor jets. In the noncontacting class, the pneumatic sensor nozzle is inserted without contact into an orifice of the workpiece or is brought into close proximity with the part, so that the sensor nozzle and part form a larger, complex nozzle system whose flow is determined by the proximity of the part to the sensor and, therefore, by the precise dimensions of the part being measured. Such sensors are typically used for gaging the diameters of holes and simple external dimensions. In the contacting class, an integral element of the sensor nozzle, such as a piston, ball, or lever, is brought into contact with the workpiece and the position of the element determines the flow characteristics in the pneumatic circuit.

Pneumatic gages offer tremendous flexibility in application through ingenious arrangements of the pneumatic nozzles and gaging circuitry. Through the use of various differential backpressure devices, high sensitivity to variations can be achieved. Applications include systems to measure hole depths, diameters, tapers, concentricity, and straightness, as well as external diameters and shapes, including the precise contours of complex parts such as turbine blades. The low wear of pneumatic gages is a particular advantage of the genre, though the requirement for clean instrument air is demanding, because the lodging of even a small particle of dirt in a critical orifice or nozzle can cause significant errors. In general, the pneumatic gages are very responsive (100 to 1000 Hz) and accurate to 1 to 100 μm (0.04 to 4 mils). The basic range of the pneumatic sensor is limited, due to flow characteristics, to a few millimeters, but this limitation is overcome by appropriate mechanical positioning of either the workpiece or sensor for the gage range.

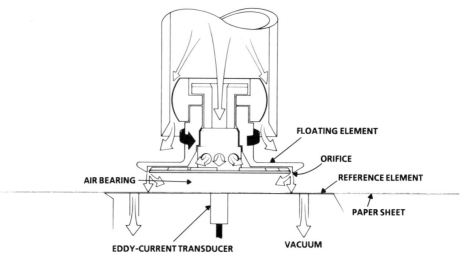

Fig. 9.58   Pneumatic gage for paper caliper.

One particularly ingenious pneumatic gage, used for the online measurement of caliper in paper,[31] is shown in Fig. 9.58. In this gage, an air bearing of 200 μm is established between a floating and reference element of the gage, with the thickness of the bearing determined and held constant by orifices in the neck and flat of the floating element. The separation of the floating and reference elements is detected by an eddy-current transducer flush-mounted in the reference element. When the moving paper web is present in the gap, it is drawn into contact with the reference element by a vacuum, so that the separation between reference and floating elements is determined exactly by the sum of the paper caliper and the air bearing thickness. The floating element orifice is specially designed to maintain a constant air bearing thickness, independent of the characteristics of the moving sheet. As a result, the device is capable of measurement accuracy of 1 to 3 μm (0.04 to 0.12 mils).

9.4.5   Nucleonic Gaging Transducers

Nucleonic gaging transducers comprise those gages utilizing high-energy radiation for measurement purposes. Typically, the high-energy radiation is created by an X-ray tube or radioisotope, so that the radiation is a beam of X-rays, gamma rays, or beta particles. These radiation beams are used for

1. Mass gaging
2. Image formation

Only the mass gaging application is addressed in this section; the image formation application is an important aspect of nondestructive testing (discussed briefly in Section 9.4.7).

As pointed out in Section 9.4.1, the mass of an object is related to the dimensions of the object. Where the density of a material is well known and constant, measurement of the mass is then an indirect measure of the dimensionality. Nucleonic gages are basically sensitive to the mass of a material, as the interaction between the high energy radiation and the material is at an atomic level, where incident radiation is either absorbed or scattered by the nucleus or electrons of the atoms constituting the material.

Nucleonic gaging techniques are utilized extensively in the gaging of sheet materials, particularly in online, automatic gaging applications.[29,32] Gage configuration is typically a two-sided "transmission" or one-sided "reflection" geometry, as shown in Fig. 9.59. In the transmission configuration, the amount of radiation detected decreases as the mass increases:

$$I = I_0 \exp(-\mu \rho t) \tag{9.20}$$

and in the reflection configuration, the amount increases as the mass increases:

$$I = I_{max}(1 - \exp(-s \rho t)) \tag{9.21}$$

where ρ is the density of the material, t the thickness, I the amount of detected radiation, I_0 the amount transmitted with no material interposed between the source and detector, I_{max} the amount

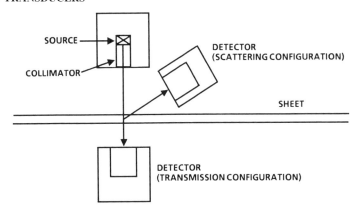

Fig. 9.59   Schematic of nucleonic gage.

reflected with a very thick $(\exp(-s\rho t) = 0)$ material interposed, μ the attenuation coefficient, and s the scattering coefficient. The coefficients are functions of the material and radiation.

For metals, the composition and density are well known and controlled so that transmission nucleonic measurements are commonly used online to determine the thickness of the material. Many different types of radiation are used, optimized in penetrating power for the thickness and material being measured. Low-energy X-rays (<10 kV) are used for measurement of aluminum foil; Sr-90 beta rays, Am-241 gamma rays, and medium energy X-rays (<100 kV) are used for sheets up to 4 cm (1.5 in) of aluminum and 13 mm (0.5 in) of steel. For thick materials, such as slabs of steel, high-energy gamma radiation from Cs-137 or Co-60 is used.

Transmission nucleonic mass gaging is also used for other sheet materials where the density is more variable, and where thickness is less important than the mass/area of the sheet. In these cases, characteristic of plastic sheets and paper, Kr-85 and Pm-147 beta radiation are typically used for measurement.

The scattering configuration is used for the measurement of certain coatings. A hybrid reflection configuration, in which both the source and detector are located opposite a reflecting material such as a calender roll, is used for measurements such as the thickness of rubber fabric for tires. In this configuration, the radiation is reflected from the roll, being partially absorbed in both incident and reflected beams by the calendered material.

The range of nucleonic gages restricts their online use to sheet materials, where the various source/detector combinations provide the capability to measure virtually any sheet material. Accuracy of the method is up to 0.1% of the thickness being gaged. Responsivity of the measurement is typically on the order of 1 to 100 Hz, with X-ray gages generally somewhat faster but less accurate than nucleonic gages.

9.4.6   Infrared and Microwave Gages

Gages using infrared or microwave energy are primarily utilized to measure the mass/area of materials that have significant absorption at appropriate infrared or microwave energies. The fundamental absorption phenomena result from excitation of resonant molecular vibrational or rotational modes by the incident energy, with the dependence on the concentration of material exponential, as in nucleonic gages.

Since the energy required to cause a resonant excitation is critically dependent on the structure of the molecule that is excited, infrared absorption techniques for measurement can be selective for specific constituents, depending on the energy used. Thus, the thickness of a particular constituent in a sandwich of sheets can be measured effectively. As a result, infrared gaging is typically utilized for measurement of coextruded plastics and for coatings where excitation of various organic bonds can be stimulated by radiation in the near infrared.[33] The capacitive behavior of plastics and similar materials at low microwave frequencies is the basis for microwave thickness measurement of these materials.

The range of thickness measurement of these techniques depends greatly on the extent of the radiation absorption characteristic of a particular material at the energy level selected. Typically, measurements of 1% accuracy over a range of 10 to 1000 μm (0.4 to 40 mils) can be made. The measurements are typically fast, limited primarily by detector response to about 100 Hz.

9.4.7   Related Technologies

A number of measurements are closely related to the position/dimension gaging discussed in Section 9.4. A survey of these technologies is made here for completeness.

The surface smoothness or finish of materials is a characteristic of the microstructure or fine-scale dimensionality of the surface of the materials. Optical measurements, in which the scatter of incident light from the surface is determined, are typically utilized to measure such microstructure effects. Surface smoothness can also be detected by use of a hardened stylus that tracks the surface with a very light, responsive arm.

Displacement of a lever or some element in a mechanism is frequently utilized as a transducer to convert to mechanical motion the change in a nondimensional measurand. A simple example of such a transducer is a bimetallic strip, whose deformation as temperature changes is a common indicator of temperature.

The nucleonic measurements frequently used for thickness determinations are also used for direct measurement of the density of materials. Configurations for such density measurements typically involve radiation transmission through a medium in a constant geometry so that density, rather than thickness, is determined. Common applications include measurement of the density of slurries in pipes.

The broad field of nondestructive testing includes numerous technologies applied to the detection and location of the position of defects in materials. Common applications include use of X-ray and gamma ray systems to radiograph castings and welds for voids. Some contemporary systems include fluorescent screens that are viewed with cameras, and sophisticated image processing such as tomography for automatic analysis. Ultrasonic approaches are utilized for detection and location of defects that provide interfaces in the material, such as microcracks and similar boundaries.

Position/dimension gaging utilizing optical techniques, especially imaging techniques, is discussed in Section 9.3. Computer analysis of the images renders these techniques appropriate for automatic control. Acoustic and, especially, ultrasonic techniques for position determination are discussed in Section 9.2.

9.5   VELOCITY TRANSDUCERS

Chester L. Nachtigal (Introduction, 9.5.1, 9.5.4)
Howard Jaslow (9.5.2)
Airpax Corporation (9.5.3)

Measurement of velocity occurs rather infrequently by comparison with position or acceleration measurement. Unlike displacement, when the motion of a system comes to rest, the velocity goes to zero. This suggests that a passive operating principle may be a viable approach to velocity transducer design. In fact, most velocity transducers do not require an external power source but rely instead on the power supplied by the system whose velocity is to be measured. These transducers are known as self-generating transducers.

Any transducer that does not rely on an external source of power must be treated especially carefully—loading the output port (see also Chapter 5) by drawing power from it also affects the input port. More power must be delivered by the system to be measured. This draw of additional power at the input to the transducer may cause the measured variable to change, and thus the original system with its added transducer is no longer the same as the system without the transducer.

This property of a transducer whereby loading at its output port also causes a loading effect at its input port classifies it as a passive system.[34] Contrast this with a strain gage bridge network, where no amount of electrical loading at its output terminals could in any way affect the input mechanical deflection of the structure to which the strain gages are attached. Thus, a strain gage bridge transducer together with its excitation source is an active transducer.

9.5.1   Linear Velocity Transducers and Signal Conditioners

Transducer

An example of a self-generating velocity transducer is a Linear Velocity Transducer (LVT), very similar to an LVDT in external appearance. (See Fig. 9.60a for a representative set of LVTs.) Its principle of operation is quite simple: an Alnico magnet core is driven along the axis of two surrounding coils by the system whose velocity is to be measured. The magnetic field surrounding the coils cuts the constant flux lines, inducing a voltage in each turn of the coils, a schematic of which is shown in Fig. 9.60(b). The dotted lines represent the magnetic flux lines emanating from the north pole and returning to the south pole of the magnet. The voltage in each set of coils may be found by the line integral:

$$\mathrm{emf}_{1,2} = \oint (\mathbf{v} \times \mathbf{B}) \cdot \mathbf{dL}$$

$$= k\,\mathbf{vBL} \qquad\qquad (9.22)$$

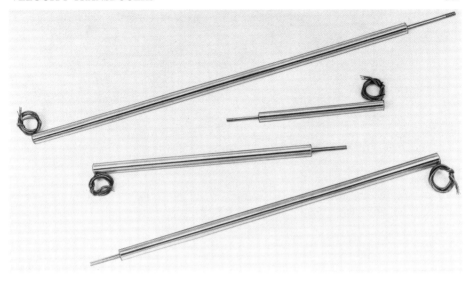

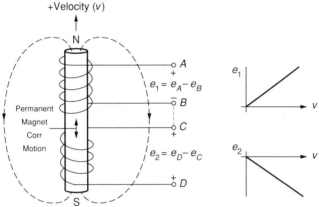

Fig. 9.60 Linear velocity transducer: (top) several LVTs for different displacement amplitudes (courtesy of Trans-Tek Incorporated, Ellington, CT); (bottom) schematic of a linear velocity transducer.

where $\mathbf{B}$ is the flux density, $\mathbf{L}$ is the length of each of the two multiturn coils, $\mathbf{v}$ is the velocity of the magnet, and k is a proportionality constant. The two sets of coils are wound in series opposition so that the voltage $e_D - e_C$ is negative going when $e_A - e_B$ is positive going. These voltage polarities may be readily determined by application of the right-hand rule. Thus the transducer has a push-pull response if the two terminals B and C are connected to signal ground. If points B and C are tied together but not to ground, the output voltage is given by

$$e_o = e_A - e_D = (e_A - e_B) - (e_D - e_C)$$
$$= e_1 - e_2 \qquad (9.23)$$

It is customary to ground point D and amplify the transducer output voltage by connecting point A to the input of a single-ended amplifier. Because this type of transducer is a passive instrument, its output impedance should be of concern. In fact, there is a direct trade-off between transducer sensitivity and output impedance—the greater the number of turns, the greater the output resistance due to the additional wire length. The inductive output impedance due to the self-inductance may also be a significant factor in limiting the dynamic response. These two components of the output impedance will be examined in more detail in the context of linear velocity transducer signal conditioning later in this section. Suffice it to say that an output impedance of 20,000 Ω and a sensitivity of 200 mV per cm/s are representative values for this type of transducer. Calibration of this transducer is difficult because of the relatively short distance for which the output is linearly related to the input velocity. Therefore, it is generally calibrated in conjunction with small-amplitude periodic vibrations.

DC-Excited AC-LVDT

A second type of linear velocity transducer is obtained by exciting the primary of an LVDT with a constant current. The secondary output is a function of the level of excitation and the velocity of the core.[2] However, this approach has two significant disadvantages. The DC primary voltage must be kept at a low value to avoid heating of the winding and a consequent increase in coil resistance. This in turn would cause a decrease in current and thereby cause the response to decrease in magnitude. A second disadvantage is the relatively small magnetomotive force developed by the necessarily small primary current. Thus, the LVDT core is loosely coupled to the secondaries and a small sensitivity results, typically of the order of a fraction of a millivolt per cm/s. The output linearity is also rather poor and, therefore, the user must generate a calibration curve over the entire range of intended use.

LVT Signal Conditioner

It was pointed out earlier in this section that velocity transducers are frequently passive devices with a nonnegligible output impedance. If this is the case one must exercise great caution to avoid loading down the signal source. Because the principle of operation of a linear velocity transducer is one of self-inductance, the output impedance of this device has both an inductive and a resistive component. Therefore, output loading may result in significant phase shifting as well as amplitude attenuation. Specification sheets frequently provide sufficient information to perform some simple Thèvenin-type analyses aimed at quantifying the loading effects. (See Chapter 5 for Thèvenin analysis.) These studies may help guide the design of an appropriate signal conditioner to minimize the loading problems.

For example, assume that a linear velocity transducer is chosen for measuring the velocity of a sinusoidal motion. (Model 0112-0000 in Ref. 35.) Its sensitivity at no-load conditions (infinite input impedance of the readout) is 0.197 V/cm/s. The output impedance is 19,000 Ω resistive and 2.9 H inductive. Let us further assume that the frequency of the velocity input ranges from 1 Hz to 100 Hz with a maximum displacement amplitude of 0.025 mm. We wish to design an operational amplifier signal conditioner that has a maximum output in the 1 to 10 V range. The maximum velocity is given by $\dot{x}_{max} = (100 \times 6.28)(0.025 \text{ mm}) = 15.7 \text{ mm/s} = 1.57 \text{ cm/s}$. The maximum open-circuit transducer voltage is $(1.57 \text{ cm/s}) (0.197 \text{ V/cm/s}) = 0.309 \text{ V}$. A signal conditioner gain of 10 would satisfy the conditions stated above. As a first iteration choice of a signal conditioner, let us use an operational amplifier inverter having a gain of 10 configuration with a 20,000 Ω input resistor and a 200,000 Ω feedback resistor. (See Section 6.1 for operational amplifier terminology and practice.) The transducer and amplifier equivalent circuit is illustrated in Fig. 9.61. The input impedance of this signal conditioner is R_i. Therefore, the transducer output voltage is given by

$$E_o = \frac{R_i}{Lj\omega + R + R_i} V_s = \frac{1/(1 + R/R_i)}{Lj\omega/(R + R_i) + 1} V_s \qquad (9.24)$$

where the uppercase variables are the complex amplitudes of corresponding lowercase time-domain variables. The signal conditioner output voltage is given by

$$E_s = -\frac{R_f}{R_i} E_o = -\frac{(R_f/R_i)/(1 + R/R_i)}{Lj\omega/(R + R_i) + 1} V_s \qquad (9.25)$$

The loading effects are threefold: (1) a loss in overall sensitivity due to attenuation of the static gain, $(R_f/R_i)/(1 + R/R_i) = 10/1.95 = 5.13$ instead of the desired gain of 10; (2) a loss in overall sensitivity due to dynamic attenuation, $|1/(\tau j\omega + 1)|\omega_{max} = 0.999$ ($\tau = 0.074 \times 10^{-3}$s); and (3) a phase shift, $\phi_{max} = -tan^{-1}\tau\omega_{max} = -2.67°$.

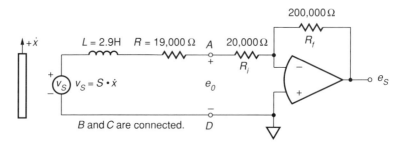

Fig. 9.61 Linear velocity transducer and op amp inverter signal conditioner.

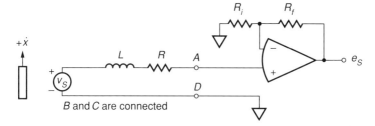

Fig. 9.62 Linear velocity transducer and op amp follower-with-gain signal conditioner.

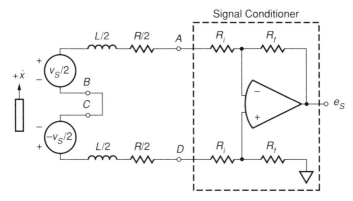

Fig. 9.63 Linear velocity transducer and differential-voltage op amp signal conditioner.

Referring to Eq. 9.25, we note that all of these loading problems may be eliminated by using an infinite input impedance signal conditioner. This suggests a follower-with-gain configuration as shown in Fig. 9.62. The static gain for this circuit is $1 + R_f/R_i$. Since the noninverting input of an FET op amp is very large (10^{11} MΩ), there is negligible current flowing from the transducer and hence no voltage drop across L and R. We may therefore write

$$e_s = (1 + R_f/R_i)V_s$$

$$= (1 + R_f/R_i)S\dot{x} \qquad (9.26)$$

where S is the open-circuit static sensitivity of the transducer. An added benefit of this signal conditioner configuration is the possibility of relatively low values of R_i and R_f as opposed to those in Fig. 9.61.

Now let us add an additional requirement, the need for rejection of noise superposed on the transducer leads, perhaps because of a remote location relative to the signal conditioner. A first approach might be to use a differential voltage amplifier configuration as shown in Fig. 9.63. Notice that the velocity transducer is now represented in push-pull form. But if the phase shift and static and dynamic loading are of concern, this signal conditioner would not be acceptable for the same reasons as the op amp inverter with gain as in Fig. 9.61. A possibility is immediately suggested: why not use a follower in each of the two transducer leads marked A and D together with a differential amplifier to convert the push-pull follower amplifier signals into a single-ended response? This solution is acceptable provided we tie the terminals B and C to signal ground, resulting in the circuit of Fig. 9.64. The reason for the need to tie B and C to ground lies in the fact that the signal source, $e_A - e_D$, as shown in Fig. 9.63, is known as a differential floating signal source since no terminal of the transducer is tied to signal ground. If terminals B and C were not referenced to ground, one or the other of the outputs of the buffer amplifiers in Fig. 9.64 would drift and saturate rather quickly.

9.5.2 Rotary Velocity Transducers*

There are many basic types of rotary velocity transducers, or tachometers, each operating from different physical principles and each providing different levels of accuracy for different applications.

*This material was written while the author was employed by ILC Data Device Corporation, Bohemia, New York.

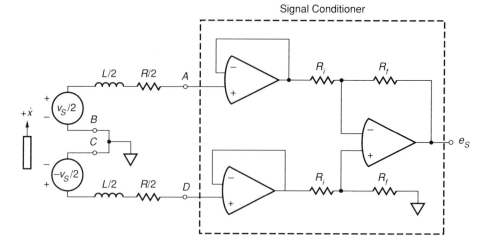

Fig. 9.64   Linear velocity transducer and buffered differential-voltage signal conditioner.

In general, the lower accuracy tachometers are used for coarse instruments, such as automobile speedometers, whereas high accuracy tachometers are required for high precision, stable servo operation.

Counting-type tachometers, which count revolutions, provide average measurements of relatively low accuracy, whereas those that count pulses at small angle increments, such as high resolution optical encoders (Section 9.2.6, Rotary Position Encoders) can provide reasonably good accuracy for many applications. In general, most tachometers based on mechanical principles, such as centrifugal force and velocity head (hydraulic tachometers), are not as accurate as the tachometer generator which is based on electromagnetic principles.

The tachometer generator or tachogenerator, as it is often called (or, more often, simply referred to as tachometer), generates a voltage that is proportional to the speed at which a current-carrying wire moves through a magnetic field. Since the current carrying wire is a coil, the rotational velocity is proportional to the voltage output. This voltage output can be used to either drive an instrument or provide velocity feedback in a motion control system.

There is a new type of tachometer available today that is an outgrowth of resolver converters. (See Section 9.2.6.) Use of a resolver with a resolver-to-digital converter provides both analog velocity data and digital position data. The velocity and position data can be used separately or together in motion control systems having velocity and/or position feedback.

Tachometer Generator

The tachometer generator (tachogenerator) is an electromagnetic generator that produces an output voltage proportional to the rotational speed of a mechanically driven shaft. The polarity of the output voltage defines the direction of rotation. The tachogenerator provides velocity feedback for motion control and for velocity damping in position servo systems.

In its operation, the tachogenerator has the same basic construction as a motor. But instead of supplying the device with a current and producing mechanical motion, the tachogenerator converts input mechanical shaft power into electrical power (output voltage and current). Rather than performing like an ordinary generator, the tachogenerator has certain demanding requirements, namely (1) output voltage must be proportional to shaft speed within a controlled linearity, (2) output voltage must be free from rotation-dependent voltage ripple and random voltage changes (noise), and (3) voltage gradient (ratio of output voltage to input shaft speed) must be stable over the operating temperature range. Based on temperature stability, tachogenerators are grouped into four main classes[36] determined by their construction (Table 9.12).

Tachogenerators are very accurate velocity transducers that are capable of attaining linearity to within 0.2%. Although tachogenerators are generally used as separate transducers, in many servo applications they can be obtained as an integral part of the motor—a motor-tachogenerator combination. In addition, tachogenerators come in both DC and AC versions; the DC version itself is available in both conventional brush types and brushless, the latter having the same advantages as are found in brushless motors: increased life, increased reliability, and less hazard in explosive environments.

Tachometer generators are available as frameless or housed units. Length-to-diameter ratios can

TABLE 9.12   TACHOGENERATOR CLASS

Class	Temperature Error
Ultrastable	< 0.01% per °C
Stable	< 0.02% per °C
Compensated	< 0.05% per °C
Uncompensated	< 0.20% per °C

TABLE 9.13   TACHOMETER GENERATOR MANUFACTURERS

Aerotech, Inc.
Airpax Corporation
Electro-Craft Corporation
Harowe Servo Controls
Inland Motor
Servo-Tek Products, Inc.
Transducer Systems, Inc.

vary from 1/8 to 2-1/2. Weight-to-volume ratios (for a simple cylindrical volume) range from 1700 to 6000 kg/m^3 (1.0 to 3.5 oz/in^3). Rotor inertias normalized with respect to the total tachometer value of WD^2/g (W = weight, D = diameter, g = gravitational acceleration) generally vary from 0.01 to 0.10. These ranges of values are useful for initial sizing estimates only; updated manufacturers' specifications should be consulted for specific design values. Table 9.13 presents a list of manufacturers of tachometer generators.[15]

Resolver with Resolver-to-Digital Converter

A resolver used together with a resolver-to-digital converter (RDC) provides digital, rotational position data. Typical characteristics of these converters are presented in Section 9.2.6, Rotary Position Encoders. RDCs are now available [37] with analog velocity voltage output signals in addition to digital position data. The state of the art of currently available RDCs is rapidly approaching tachometer generator accuracies. Today's RDCs can achieve linearity to within less than 1%.

With a resolver and an RDC, both digital position (θ) and analog velocity ($\dot{\theta}$) signals can be obtained simultaneously for shaft rotation. θ and $\dot{\theta}$ are used together in a number of control law systems, such as the proportional-integral-derivative (PID) technique.[38] Use of both θ and $\dot{\theta}$ greatly enhances tight motion control.

RDCs are either of the successive approximation type converters or tracking type converters. The tracking converter in itself is unique. By virtue of its tracking operation, it inherently generates an analog velocity signal. Since these converters use a type II tracking loop to null position error, there is no lag error in velocity. A block diagram of the operation of this type of converter is presented in Fig. 9.65.

Resolver-to-digital converters are available in small hybrid packages. The resolution for digital position data is logic programmable to 10, 12, 14, or 16 bits with accuracies of 8, 4, 2, or 1 arc minute, respectively. In addition, the bandwidth is logic programmable (approximately 200 Hz and 800 Hz). The high bandwidth (800 Hz) is applicable to servo systems requiring tight controls, whereas the low bandwidth (200 Hz) is useful for noisy environment applications where filtering is required.

Wide tachometer output dynamic range of 250,000:1 is available along with high tracking rates (800 rps) and scalable velocity gradients. Current models are available as industrial hybrids measuring approximately 51 mm × 28 mm × 5 mm (2.0 in × 1.1 in × 0.2 in). Table 9.14 lists the manufacturers of resolver-to-digital converters having analog velocity output. A list of manufacturers of resolvers is presented in Section 9.2.6.

TABLE 9.14   RESOLVER DATA CONVERTER MANUFACTURERS HAVING VELOCITY OUTPUT

Analog Devices, Inc.
Control Sciences, Inc.
ILC Data Device Corporation
Natel Engineering Co., Inc.

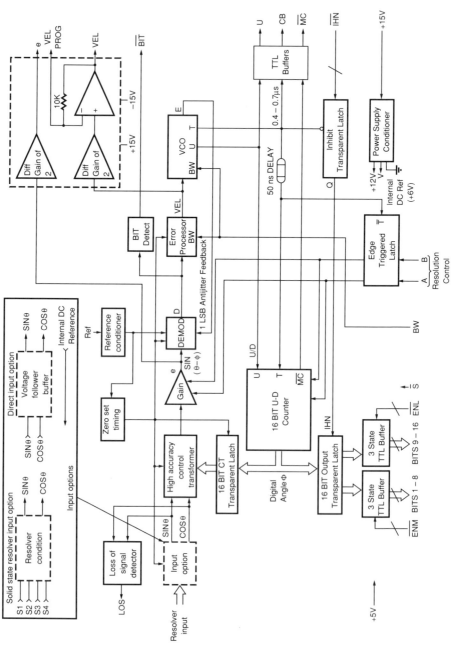

Fig. 9.65 Resolver-to-digital converter block diagram

9.5.3 Rotary Velocity Pickups*

Rotational speeds of pumps, computer peripherals, electric motors, automated machinery, engines, and similar equipment are routinely measured in industrial plants. Traditionally, DC tachometers have been used for this purpose. When attached to the rotating shaft of the machine whose speed is being measured, the instruments generate a DC output voltage that is directly proportional to the rotational speed of the shaft. For example, if 10 rpm produces 1 V, 20 rpm yields 2 V.

Shortcomings of conventional DC tachometers include:

Excessive size and bulkiness

A portion of the rotational power of the equipment being measured is required to turn the tachometer, thus contributing to measuring inaccuracies

The DC output voltage is nonlinear near zero speed and usually flattens at high speed

Signals are susceptible to noise pickup, especially if routed near power lines

Generator brushes and commutators require maintenance and repair considerations

Digital Systems

Recently developed AC (digital) tachometer systems eliminate the shortcomings of DC tachometers. The three key components of such systems are the exciter, magnetic pickup sensor, and electronic tachometer. The exciter (usually a gear) divides one complete revolution into a precise number of discrete parts, typically 60. The magnetic pickup sensor is an electromechanical transducer that translates exciter movement into an AC or digital signal, which has a one-to-one correspondence to the sensed discrete parts. The electronic tachometer either converts the signal to DC (so it can be displayed on an analog meter or used in servomechanism systems) or retains and processes it in its digital format (so it can be displayed in a digital readout or be sent to a digital computer interface).

Exciters (or activators) can be any ferrous device that has precisely spaced discontinuities such as sprockets, keyways, cams, bolt heads, rivet heads, splines, holes in a shaft, toothed wheels, or ordinary gears. Gears are normally used because they are manufactured to uniform industrial codes.

The largest practical gear-tooth size for the application should be selected to achieve a sufficient air gap and to compensate for runout tolerances. If the speed range is low, a gear with a large number of teeth helps ensure maximum frequency sensing ability and measuring accuracy. The minimum gear face width should be 13 mm (1/2 in) to provide sufficient magnetic flux to the pickup.

Magnetic Pickups

These devices are the most critical part of AC tachometer systems. The two major types of magnetic pickups are passive and active. Proper selection and mounting are important to ensure optimum performance. General guidelines are shown in the accompanying Table 9.15.

TABLE 9.15 SELECTION AND MOUNTING GUILDINES FOR ACTIVE AND PASSIVE MAGNETIC SPEED SENSORS

Requirements	Active Pickup	Passive Pickup
Detect zero speed	Zero velocity	No
Detect direction	Bidirectional	No
Constant signal levels	Yes	No
Direct logic interface	Yes	No
Large air gaps	Yes	No
Self-powered	No	Yes
Very low cost	No	Yes
Detect high frequency signal polarity	DOT	Yes
Indicates approach of metal	DOT	Yes
High temperature	No	Yes
Large tooth gears	Select large sensor	Select large pole piece
Small tooth gears	Select small sensor	Select small pole piece
Very slow rpm	Select many tooth gear for accurate readings	Select high voltage model
Very high rpm	Select DOT	Select high voltage model
Best mounting material	Nonmagnetic	Nonmagnetic
Worst mounting material	Magnetic conducting	Magnetic conducting
Big air gap	Use large tooth gears	Use large tooth gears

*This section is excerpted from Reference 39 and was edited by C.L. Nachtigal.

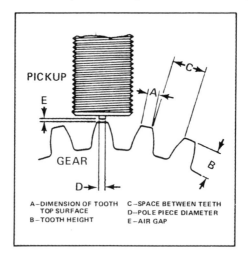

Fig. 9.66 Gear teeth geometry. Gears are normally used as excitors or activators with AC tachometer systems. Signal generated from pickup is a sine wave. These dimensional ratios are desirable in selecting passive pickups and gear combinations that will produce the maximum output at the lowest speeds: (A) equal to or more than (D), (B) equal to or more than (C), (C) approximately three times (D), (E) as small as possible—0.125 mm (0.005 in.) being typical. Gear width should be equal to or more than (D).

Magnetic pickups convert mechanical motion to an ac voltage signal and eliminate the need for mechanical linkage. The resulting pulses are then transmitted to speed indicators, speed transmitters and converters, or custom electronic measuring and controlling devices. There are no moving parts or bearings to wear, so periodic maintenance is eliminated. The environment can be dirty, oily, dusty, caustic, hot, cold, or explosive.

The magnetic pickup derives its name from the fact that it contains a magnet. The pickup points at the gear face and the flux return path is through the air. But the air gap between the pickup and gear must be relatively small for the pickup to work properly (Fig. 9.66).

Passive Pickups

These devices contain a magnet, a coil wound of fine wire, and a pole piece that touches the magnet, extends through the coil, and points at the sensed gear (Fig. 9.67).

Operating principle is based on the fact that the generated voltage is directly related to the rate of change of flux with respect to time. The generated signal is a sine wave when a gear tooth is used. As the gear speed increases, the generated voltage increases. However, at very low speeds, the voltage is typically in the millivolt range, which makes it practically useless for speed measuring purposes.

Advantages of passive pickups include low cost (often under $10) and simple two-wire connections. Wire connections are interchangeable. Shorting one lead to the other will not damage

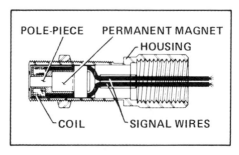

Fig. 9.67 Passive magnetic pickup. Typical passive, magnetic pickup contains magnet, coil, and pole piece. Pole piece touches magnet, extends through coil, and points at sensed gear.

the pickup. The nonelectronic, rugged design of passive pickups ensures a practical, infinite service life. Frequencies in excess of 20,000 Hz can be sensed. Some units provide signal current as well as voltage, permitting self-powered tachometer systems.

A major disadvantage of passive pickups is their inability to sense very low frequencies or zero speed. The pickup's output signal voltage amplitude varies with the exciter speed. Thus, the voltage may either be too low to be measured, or it may be excessive to the point of burning out the tachometer's circuitry at high exciter speeds.

Output signals must be wave shaped to be used with most electronic circuitry. Because the slopes of sine waves vary with their amplitude, variable trigger points may be created in the wave shapes that can cause timing errors in some applications. The usable air gap in passive pickups ranges from 0.125 mm to 1.0 mm (0.005 in to 0.040 in). This relatively small gap makes it impractical to use these sensors in applications where gear runout or vibration is great.

Active Pickups

These pickups are electrically powered devices that are categorized as either nonzero or zero-velocity sensing units.

Nonzero velocity sensing pickups are called Digital Output Transducers (DOT). In general, they are hybrids between passive and active zero-velocity pickups. The internal construction of DOTs is virtually identical to that of passive pickups. The main difference is an electronic microcircuit that amplifies and converts the coil's sine wave signal into a constant-amplitude, digital, square-wave signal.

The microcircuit is powered by an external voltage. The amplifier is limited to a predetermined gain. The same limitations that apply to passive pickups apply to nonzero velocity pickups.

Basically, the signal-conditioning circuits found in the input channel of a tachometer have been miniaturized and included in the pickup housing (Fig. 9.68). The signal on the system wiring is a large voltage level digital signal that can be directly interfaced with most digital circuits without additional wave conditioning. The absence of low-voltage signals minimizes noise problems.

Zero-velocity magnetic pickups are less common than nonzero units because their use is more specialized. Zero-velocity pickups are capable of detecting zero motion and movements as slow as one gear tooth per year. The units use magnetically sensitive resistors whose ohmic value changes with the strength of the magnetic field.

The resistors are installed in a bridge circuit configuration and the pickup must be oriented in a directional position. Bridge unbalance permits the circuit to respond to differences as low as one ohm (Fig. 9.69).

In normal operation, the unbalanced bridge voltage is amplified, wave shaped, and converted to a digital output signal. The resistors are available in two different sizes so that very fine-pitched gears (30 pitch and smaller), as well as standard-sized gears (20 pitch and larger), may be sensed.

Zero-velocity pickups generally overcome the shortcomings of passive pickups. However, because of the use of more complicated internal electronic circuits, zero-velocity units cost at least $50 more than passive pickups.

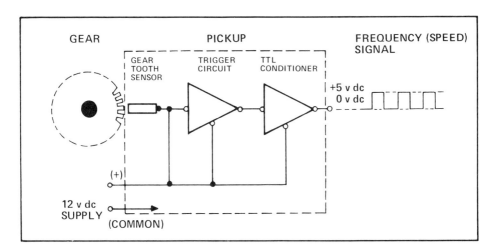

Fig. 9.68 Signal conditioning circuit for active magnetic pickups. Miniaturized circuits in active magnetic pickup convert sine-wave signal into a square-wave signal. Pickup is powered by external 12 V DC supply.

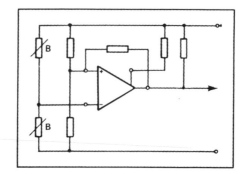

Fig. 9.69 Magnetic resistors in a bridge circuit. Magnetic resistors permit bridge circuit configuration to be used in zero-velocity magnetic pickups. Bridge unbalance allows circuit to respond to differences as low as 1 ohm.

The tachometer that connects to the pickup contains an adequate DC electrical supply for the pickup. The pickup detects gear-tooth edge changes, not the presence of ferrous metal or air. Gear-tooth edges are an ideal location for tachometry measurements because they are a more precise location than other targets, such as the flat areas on top of the gear teeth.

Maximum frequency sensing ability for zero-velocity pickups is 10,000 to 20,000 Hz; 95% of all applications fall within the sensing range of these units.

Bidirectional Pickups

In many applications, it is important to determine whether a motor or shaft is turning clockwise or counterclockwise, in addition to its rotational speed. This task can be readily accomplished by using dual sensors in an active magnetic pickup housing, (Fig. 9.70). The arrangement provides the pickup with five leads: one power, one ground, two gear-tooth signals, and one directional. The use of dual sensors also eliminates the mechanical problems frequently associated with other bidirectional measurement techniques.

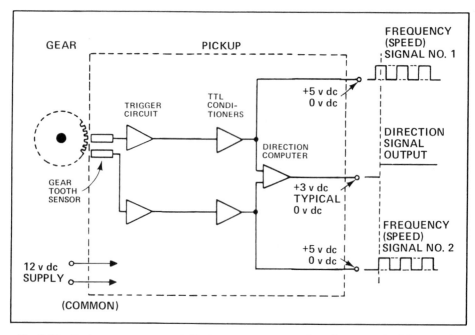

Fig. 9.70 Dual sensor arrangement. Dual sensor arrangement in active magnetic pickup permits bidirectional sensing to be performed.

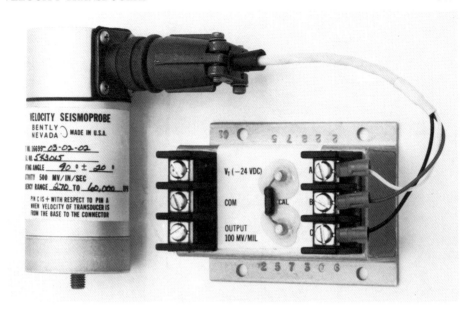

Fig. 9.71   Seismic velocity transducer (courtesy of Bently Nevada, Minden, NV).

9.5.4   Seismic Velocity Transducer

A transducer system that measures the absolute velocity of a vibratory structure is shown in Fig. 9.71 and is described in Reference 40. Its principle of operation consists of a generating coil mounted on a seismic mass. The transducer case is rigidly attached to a magnet that moves inside the coil, consequently generating a voltage proportional to the relative velocity between coil and magnet. If the seismic mass, and hence the coil, remains stationary, the only movement is the absolute velocity of the case and magnet. This occurs for a range of frequencies above the natural frequency of the mass on its sprung mounting.

A seismically suspended mass is shown in Fig. 9.73. The transfer function relating the relative position x_r and the case acceleration is

$$\frac{X_r}{\ddot{X}}(s) = \frac{1/\omega_n^2}{(s^2/\omega_n^2) + (2\zeta/\omega_n)s + 1} \tag{9.27}$$

For this discussion the transducer shown in Fig. 9.73 which measures the relative motion x_r is responsive to velocity $\dot{x}_r$, i.e., $e_0 = s_v \cdot \dot{x}_r$ and the overall velocity transducer is designed to measure the case velocity $\dot{x}$. We have

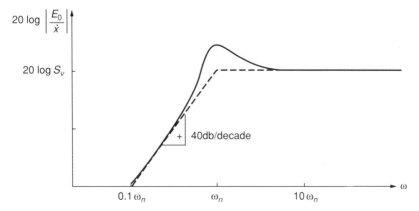

Fig. 9.72   Magnitude frequency response for a seismic velocity transducer.

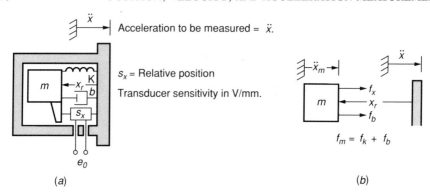

Fig. 9.73 Basic accelerometer structure.

$$\dot{X}(s) = \ddot{X}(s)/s$$

$$\dot{X}_r(s) = sX_r(s)$$

$$E_o(s) = S_v\dot{X}_r(s) \tag{9.28}$$

where the uppercase notation designates Laplace transform versions of corresponding lowercase symbols in Fig. 9.73. Inserting Eqs. 9.28 into Eq. 9.27

$$\frac{X_r}{\ddot{X}}(s) = \frac{s^2/\omega_n^2}{(s^2/\omega_n^2) + (2\zeta/\omega_n)s + 1} \tag{9.29}$$

$$\frac{E_o}{\ddot{X}}(s) = \frac{S_v(s^2/\omega_n^2)}{(s^2/\omega_n^2) + (2\zeta/\omega_n)s + 1} \tag{9.30}$$

The magnitude frequency response of Eq. 9.30 is shown in Fig. 9.72. If this transducer is to measure the velocity of the structure to which it is bolted, the natural frequency of the sprung mass must be quite small, and the frequency components in the case velocity which are accurately measured by the seismic velocity probe are larger than ω_n, say $\omega > 5\omega_n$.

9.6   ACCELERATION TRANSDUCERS

Robert W. Lally (9.6.1, 9.6.2)
John W. G. Budd (9.6.3)
Georg F. Mauer (9.6.4)
Chester L. Nachtigal (9.6.5, 9.6.6)

9.6.1   Basics of Acceleration Measurement—Seismic Device

Acceleration-sensing transducers, called accelerometers, implement Newton's Laws of Motion with instrumented, classical, spring-mass "seismic" structures. Such accelerometers measure the acceleration aspect of the absolute motion of an object or point in inertial space. By contrast, displacement transducers, employed inside accelerometers to measure the deflection of a spring, sense the relative motion between two points.

Absolute acceleration measurements play a vital role in many feedback control systems and processes. In autonavigators (inertial guidance systems) accelerometers sense the general motion of spacecraft, airplanes, and missiles. In active dampers, they supply stabilizing feedback signals. In environmental and simulation tests, they help control the level and spectral content of motion. In testing and perfecting the structural behavior and performance of products, accelerometers measure the shock, vibratory, and general motion experienced. Accelerometers also furnish valuable information for monitoring the health, testing the behavior, and checking the quality of machines and structures.

If the mass of an object and the forces acting on it are known, its acceleration and motion can be readily computed. It does not have to be measured. However, if the forces are unknown or if the mass is changing, which is usually the case, a known parasitic (seismic) mass is taken along for the ride and the forces on it measured as an indication of the motion being experienced. The resulting acceleration signals are often integrated to provide velocity or displacement information. Thus, with an elastic spring and a displacement-sensing element or a combination thereof, an accelerometer measures the force required to automatically give its seismic mass the same motion as the test object.

The basic structure and operation of a linear, translational accelerometer may be graphically modeled as a spring-mass-dashpot seismic structure as shown in Fig. 9.73. The displacement of the seismic mass in response to an input case acceleration may be modeled as a single degree of freedom, second-order system as follows:

$$m\ddot{x}_m = f_k + f_b = Kx_r + b\dot{x}_r \tag{9.31}$$

$$\ddot{x}_m = \ddot{x} - \ddot{x}_r \tag{9.32}$$

Combining Eqs. 9.31 and 9.32 yields:

$$m\ddot{x}_r + b\dot{x}_r + Kx_r = m\ddot{x}$$

$$\frac{\ddot{x}_r}{\omega_n^2} + \frac{2\zeta\dot{x}_r}{\omega_n} + x_r = \ddot{x} \tag{9.33}$$

The appropriate transfer functions in the Laplace domain are:

$$\frac{X_r}{\ddot{X}}(s) = \frac{1}{s^2/\omega_n^2 + 2\zeta\omega_n/s + 1}$$

$$\frac{E_o}{\ddot{X}}(s) = \frac{S_x}{s^2/\omega_n^2 + 2\zeta\omega_n/s + 1} \tag{9.34}$$

The magnitude ratio of $E_o/\ddot{X}$, derived by substituting the frequency operator $j\omega$ for s (see Chapter 3), is shown in Fig. 9.74. From this plot it is clear that an accelerometer accurately responds to all frequency components in the case acceleration $\ddot{X}$ (the acceleration that the transducer is designed to measure) up to approximately $\omega = 0.5\omega_n$. From a physical point of view the spring forces the mass to follow the motion of the test object that is being measured over the bandwidth from 0 rad/s to $0.5\omega_n$ rad/s. The force required to impart the test object's acceleration to the seismic mass also deflects the spring, thereby imposing a change in displacement across the position transducer S_x.

The function, structure, and behavior of accelerometers are similar to that of pressure and force transducers. A universal stress-strain transducer is modeled in Fig. 9.75. Accelerometers contain an exaggerated mass. Force is applied directly to a stress-distributing mass. Pressure sensor diaphragms sum the applied pressure into a force on the elastic spring.

The elastic sensing element, which functions to transfer deflection into an electrical signal, can be made of a variety of materials exhibiting resistive, piezoresistive, piezoelectric, inductive, capacitive, or optical effects. In piezoelectric accelerometers, the crystal sensing elements perform a

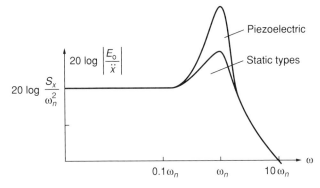

Fig. 9.74 Magnitude frequency response of a second-order system.

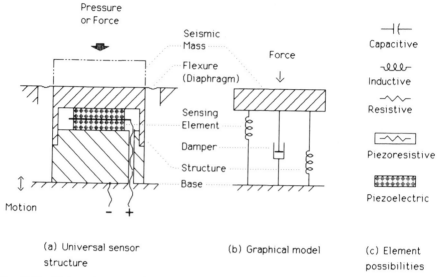

Fig. 9.75 Universal stress-strain gage transducer (PCB Piezotronics, Inc., Depew, NY): (*a*) universal sensor structure; (*b*) graphical model; (*c*) element possibilities.

dual function. They act both as a precision spring and as a sensing element, generating an electrical signal proportional to their deflection. Servo accelerometers employ an internal, closed-loop feedback control system that behaves like a mechanical spring. Vibratory velocity sensors use a seismic structure above its first resonance and below subsequent structural resonances, where the seismic mass stands still and the case moves about it.

Rotary accelerometers for measuring the rotational components of motion operate on similar principles. They employ either an instrumented seismic mass having a moment of inertia and a rotational degree of freedom or use two opposite polarity linear accelerometers spaced apart on a diameter to measure the tangential components of acceleration. However, efforts to implement rotational accelerometers along these conventional lines have not been very successful due to sensitivity to the usually much higher translational components of motion, which obscure the rotational signals. New array sensors, based on a profound spatial averaging idea, recently reported in Reference 41, promise to solve the rotational dilemma as well as provide a practical omnidirectional instrument to measure the six components of motion simultaneously. On the forefront of technology, these new array accelerometers and associated array calibrators incorporate smart sensors, intelligent amplifiers, and a sophisticated computer, all intimately interacting.

9.6.2 Piezoelectric Accelerometer and Its Signal Conditioner

Self-generating, piezoelectric, stress-gage accelerometers are stiff, elastic structures that function to transfer deflection caused by inertial forces into electrical signals, more convenient for recording or analyzing.

Piezoelectric accelerometers sense the absolute motion of an object or point in inertial space. They measure the acceleration aspect of shock and vibratory motion relative to an initial or average level, usually zero. But electrostatic-type piezoelectric instruments cannot measure constant accelerations or absolute levels for more than a few seconds of time, except under closely controlled, constant temperature laboratory conditions. Acceleration signals integrate into velocity and displacement information.

Rugged, all solid state, electrostatic piezoelectric accelerometers excel in dynamic applications involving testing, modifying, monitoring, and controlling the vibratory and related acoustic behavior of almost all things. Principal applications include the measurement of shock and vibratory motion and short duration, transient accelerations involved in environmental testing, structural behavior testing, performance testing, experimental modal analysis, machinery health monitoring, resonance searching, crash studies, blast hardening, earthquake simulation, oil exploration, fusing, knock detection, and acoustic noise abatement. These activities involve the testing of vehicles, cars, trucks, airplanes, rockets, spacecraft, weapons, projectiles, ships, submarines, torpedoes, trains, helmets, people, machines, pumps, fans, turbines, compressors, engines, pile drivers, tools, buildings, bridges, electronic equipment, and so on.

Electrostatic piezoelectric accelerometers come in a wide variety of sizes, shapes, materials, and behavior characteristics. Selection involves paying attention to many details, including performance, behavior, accuracy, interaction, configuration, mounting, maintenance, environment, noise, cabling, conditioning, calibration, cost-effectiveness, and cost per channel.

There exists another speciality type of piezoelectric accelerometer worth mentioning, a resonant, vibrating crystal type, that does indeed accurately measure constant accelerations and calibrates statically. It is presently being used in crash recorders because of the simplicity and stability of the system. Its frequency-modulated output is directly compatible with tape recorders, without converting or conditioning. In this relatively new instrument, a mass-loaded, elastic, vibrating quartz crystal changes frequency when subjected to inertial acceleration forces. Pressure and force-measuring vibrating crystal type transducers are also commercially available and gaining in popularity. The material in this and subsequent piezoelectric sections applies mainly to electrostatic-type instruments. The word electrostatic, somewhat of a misnomer, refers to the type of signal and mode of operation, not the response.

The Piezoelectric Effect

Commercial accelerometers employ a number of different natural, cultured, and artificial crystals exhibiting a phenomenon known as the piezoelectric effect, a law of nature. This ability to generate electricity by squeezing crystals, which mystified early humanity, was first scientifically discovered by the Curie brothers during the late 1800s. Since then, a number of famous scientists and engineers have helped put the piezoelectric effect into the service of humankind in a variety of ways.

The piezoelectric effect, illustrated in Fig. 9.76, states that when certain asymmetrical, elastic crystals are deformed by force, an electrical potential developed within the distorted crystal lattice causes electrical current to flow in external circuits. In piezoelectric transducer systems, the electrical charge generated (Q) is collected in a capacitor (C), which converts it into a voltage (E) according to the law of electrostatics: $E = Q/C$, illustrated in Fig. 9.77. Thus, both the charge and voltage signals are proportional to the applied force (F) according to the relationship $Q = KF$, where K is the constant of proportionality.

The charge signal stored in the capacitor will leak off through the insulation resistance (R) of the components or through a special ultrahigh megohm resistor installed for this purpose. In response to a step function input, if the charge signal continued to leak off at its initial rate, it would reach zero in one discharge time constant (τ), where $\tau = RC$. With a typical capacitance of 20 pF and a resistance of $10^{11}\,\Omega$, the discharge time constant is 2 s.

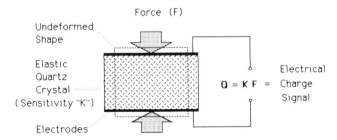

Fig. 9.76 The piezoelectric effect.

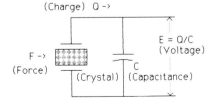

Fig. 9.77 The law of electrostatics.

Quartz and Tourmaline

Crystalline quartz and tourmaline are the two most popular natural crystals employed in transducers. Both are single pure crystals exhibiting many modes of piezoelectric, pyroelectric, and resonant vibratory behavior. Quartz is extremely high in insulation resistance and low in thermal sensitivity, permitting electrostatic operation down to very low frequencies, much below 1 Hz. Tourmaline is uniquely hydrostatically sensitive, which adapts it for underwater, shock wave, and calibration applications. The high thermal sensitivity of tourmaline, known as the pyroelectric effect, limits its use in other areas. Widely used, cultured quartz crystals, which in many ways are superior to natural crystals, are regrown on a seed from natural materials. Starting with the two most abundant elements on earth, silicon and oxygen, and in an autoclave under high pressure and temperature, people with nature's help can now duplicate in a month's time a process that perhaps took nature alone millions of years. The United States government sponsored the research and development of cultured quartz crystals to meet critical military communications, radar, and navigational needs. Cultured quartz is now readily available and reasonably priced.

Ceramic Crystals

Manufactured, artificially polarized, polycrystalline, ferroelectric materials are also used extensively today as crystal sensing elements in piezoelectric accelerometers. Typical compositions include barium titanate, lead zirconate titanate, and other exotic materials. Sensor manufacturers generally keep their formulas secret. Although not as stable or linear as natural crystals, these versatile ceramic crystals can be formulated and physically tailored to fit the application. The surfaces of the resulting crystal disks, slabs, and cylinders are metal-plated with electrodes to collect the electrical charge signals. More recently, metallized plastic piezoelectric films are finding application in specialized transducers such as microphones and hydrophones.

Structures

Typical electrostatic-type, piezoelectric accelerometer structures are illustrated in Fig. 9.78. Common configurations include compression, inverted compression, and shear mode structures. After decades of evolutionary development, the three designs are competitive in almost all respects. Inverted and shear mode models may have an edge in low base-strain sensitivity and thermal response, whereas the stiffer basic compression mode units have a higher first resonant frequency. With recent innovative improvements, segmented crystals and wrap-around masses, quartz compression mode accelerometers rival the other types in low strain and low temperature sensitivities.

In crystal accelerometers the crystal sensing elements perform a dual function. They act as a precision spring opposing the applied force and generate an electrical charge-type signal proportional to their deflection. Single-ended, compression mode structures generally employ a central preload stud or an external preload sleeve to prestress the crystal sensing elements. Prestressing beyond the specified range of the instrument imparts a capability to measure in either direction. It also forces imperfect interface surfaces into intimate contact, ensuring high rigidity, which results in high accuracy and performance. Metal or metallized electrodes in contact with the major surfaces of the crystal collect the electrical charge generated.

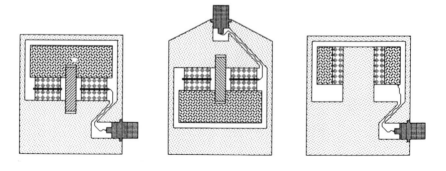

(a) Compression mode   (b) Inverted compression   (c) Shear mode

Fig. 9.78   Typical piezoelectric accelerometer structures (PCB Piezotronics, Inc., Depew, NY): (a) compression mode; (b) inverted compression; (c) shear mode.

Other innovative structures not shown in Fig. 9.78, such as differential bimorph beams, adhesive-bonded structures, and built-in, self-calibrating elements, meet specialized needs. For automotive active damping and structural behavior testing applications, relatively low-cost, limited-performance models are now appearing on the market.

Mounting

The external shape of accelerometer housings is designed to facilitate mounting, either directly or with accessory hardware such as studs, magnets, or adhesive mounting pads. During operation in precision applications, motion is generally transmitted from the test object to the accelerometer through precision, flat mounting surfaces, usually drawn into intimate contact with a flexible stud. The compliant elastic stud clamps the rigid, elastic, imperfect, mounting surfaces into intimate contact so that forces are transmitted through the interface surfaces and not through the stud. A thin coating of silicone grease or other lubricant on the mounting surfaces improves the transmissibility at high frequencies. Accelerometers can also be attached temporarily to test objects with adhesives, wax, putty, tape, magnets, and so forth. Since these added interface mounting structures affect the frequency response of the system, their behavior should be tested, calibrated, and corrected for if necessary. Some accelerometers and mounting accessories have built-in insulators to electrically isolate the instrument from the local ground. Other specialized mounting pads may incorporate mechanical filters or water cooling to reduce environmental effects. Low-profile, ring, or washer-shaped accelerometers are sometimes easier to mount in restricted spaces.

Behavior

Piezoelectric accelerometers, because of the exceptional linearity and repeatability of their rigid crystal sensing elements, exhibit near ideal, straight-line behavior over extremely wide amplitude and frequency ranges. They cover a dynamic range greater than 10,000 to 1, which accounts for their popularity in structural behavior testing. When switching exciter, amplifier, or recorder ranges, it is usually not necessary to change transducers. However, the electrostatic nature of piezoelectric accelerometers, which allows the charge signal to leak away (exponentially decay), precludes static measurements and conventional static calibration.

The high frequency response and related rise time of piezoelectric sensors are governed by the resonant behavior of the sensor's mechanical structure, or by built-in, electronic, low-pass filters. The low frequency response and related transient decay time of piezoelectric sensors depends on the behavior of the sensor discharge circuit, discussed previously. The shorter the discharge time constant, the faster the signal decay and low frequency rolloff. Because of the automatic elimination of static signal components by this discharge circuit, electrostatic piezoelectric sensors measure amplitudes relative to the average or initial level or to some other point in the signal.

Basic piezoelectric accelerometers with very little internal damping introduce practically no time delay or phase shift into the measurements. However, the lightly damped resonances of piezoelectric accelerometers can cause considerable trouble if they are excited during measuring transactions. Signal conditioning electronics must have sufficient range to accommodate the huge resonant signals without overloading. This means operating the system at very low sensitivity, or employing a charge-driven filter ahead of any electronics.

Under shock loading, small zero shifts caused by changes in residual stress levels plague highly preloaded piezoelectric as well as most other sensors and generally preclude integrating shock-motion signals to obtain velocity or displacement information. This application requires special sensor structures with bonded elements and no prestress.

Signal Conditioning

In a piezoelectric sensor system, signal-conditioning charge or voltage amplifiers match the output signal of the sensor to the input of the recorder or analyzer. The signal conditioners are essentially ultrahigh-input-impedance, very-low-output-impedance isolation amplifiers that reduce the interaction and effect valid communicating transactions. If directly connected, the charge signal from the crystals would quickly leak off through the input resistance of the readout instrument, invalidating the measurement.

Figure 9.79 compares the two most popular piezoelectric sensor systems. As illustrated, both charge- and voltage-type amplifiers employ a capacitor to convert charge into voltage according to the law of electrostatics. They differ primarily in the location of the capacitor—at the input of voltage amplifiers and in the feedback path of charge amplifiers. Versatile external charge amplifiers offer control of the range and the discharge time constant, but require special expensive, low-noise, coaxial cables and expose ultrahigh impedance circuits to the environment. Operating over two wires, ICP (integrated circuit piezoelectric) systems with built-in, voltage mode, isolation amplifiers hermetically seal the potentially troublesome circuits inside their cases. A complete ICP measuring system is diagrammed in Fig. 9.80.

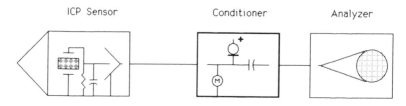

(a) Modern ICP voltage-mode system
(integrated-circuit-piezoelectric)

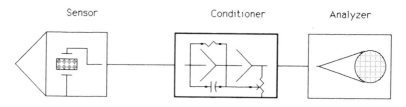

(b) Conventional charge-mode system
with dial-sensitivity charge amplifier

Fig. 9.79 Piezoelectric sensor systems (PCB Piezotronics, Inc., Depew, NY): (a) modern ICP voltage-mode system (integrated-circuit-piezoelectric); (b) conventional charge-mode system with dial-sensitivity charge amplifier.

Signal conditioners, sometimes called modifying transducers, also function to amplify, attenuate, standardize, filter, calibrate, limit, debias, clamp, check, isolate, convert, digitize, or monitor the signal. Basic ICP power units power the sensor amplifier over the signal lead, debias the output signal, and indicate normal or faulty operation: open or short circuits. Some commercial analyzers now feature built-in ICP conditioners.

Another specialty type of signal conditioner not shown, a classical integrating operational amplifier, functions to transfer the electrical current generated by the crystals into a voltage signal proportional to pressure rate, force rate, or jerk. Sophisticated differential-mode charge or voltage amplifiers with high common-mode rejection reduce noise in adverse electrical environments.

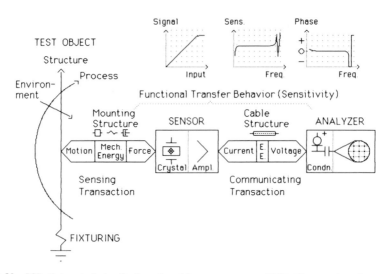

Fig. 9.80 ICP (integrated-circuit-piezoelectric) sensor system (PCB Piezotronics, Inc., Depew, NY).

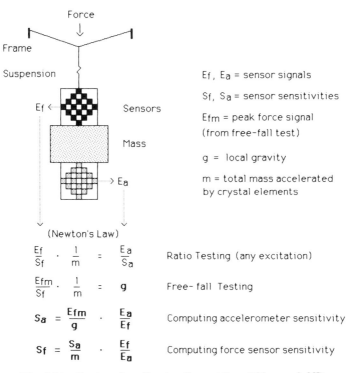

Force

Frame

Suspension

Ef ←   Sensors

Mass

→ Ea

(Newton's Law)

Ef, Ea = sensor signals

Sf, Sa = sensor sensitivities

Efm = peak force signal
(from free-fall test)

g = local gravity

m = total mass accelerated
by crystal elements

$$\frac{Ef}{Sf} \cdot \frac{1}{m} = \frac{Ea}{Sa} \qquad \text{Ratio Testing (any excitation)}$$

$$\frac{Efm}{Sf} \cdot \frac{1}{m} = g \qquad \text{Free-fall Testing}$$

$$Sa = \frac{Efm}{g} \cdot \frac{Ea}{Ef} \qquad \text{Computing accelerometer sensitivity}$$

$$Sf = \frac{Sa}{m} \cdot \frac{Ef}{Ea} \qquad \text{Computing force sensor sensitivity}$$

Fig. 9.81   Gravimetric calibration (Instruct Inc., Ridgewood, NJ).

Calibration

Several dynamic methods of calibrating piezoelectric accelerometers include reciprocity, laser-interferometer, comparison, and gravimetric. The latter two methods adapt to industrial laboratory and field use. Comparison calibration involves applying the same input motion to the accelerometer being calibrated and a reference standard accelerometer, and comparing the output signals. The reference standard is shuttled to a primary standards laboratory for periodic recalibration. Gravimetric ratio calibration, modeled in Fig. 9.81, is an absolute method utilizing gravity and structural technology. Implementing Newton's Laws, it involves simply exciting an instrumented unknown mass, the simplest of structures, with any kind of excitation, measuring the ratio of the resulting acceleration and force signals or their frequency components, and multiplying by a scaling factor. The scaling factor is the peak force signal experienced during a free-fall test, which effectively weighs the mass on the force sensor. Gravimetric calibration extends accepted and approved static tilt and comparison methods to dynamic instruments and to higher levels. Dividing by the local value of gravity readily converts the results to SI or English units.

9.6.3   Strain Gage Accelerometer

The general arrangement for displacement-type, linear accelerometers is shown in Fig. 9.73. This illustrates a single-degree-of-freedom system with viscous damping, having an unforced equation of motion

$$m\ddot{x}_r + b\dot{x}_r + Kx_r = 0 \qquad (9.35)$$

where $x_r = 0$ is the equilibrium position of the mass

$$m = \text{mass}$$

$$b = \text{damping constant}$$

$$K = \text{spring constant}$$

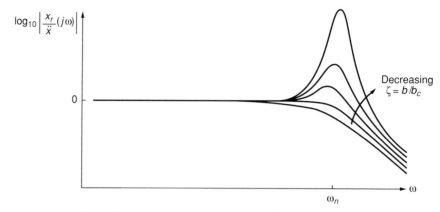

Fig. 9.82   Second-order system magnitude frequency response.

the transient solution of which depends on the degree of damping. Such a system can be underdamped, overdamped or critically damped.

The critical damping factor is

$$b_c = 2\sqrt{Km} = 2m\omega_n \qquad (9.36)$$

where ω_n = natural frequency in rad/s. If the system damping, b, is less than critical, i.e.,

$$\zeta = \frac{b}{b_c} < 1$$

where ζ is called the damping ratio, the solution of Eq. 9.35 is

$$x_r = Ce^{-\zeta\omega_n t}\sin(\omega_d t + \theta) \qquad (9.37)$$

where $\omega_d = \sqrt{1 - \zeta^2}\omega_n$, implying an oscillation with a damped natural frequency ω_d and phase angle θ. If the system damping, b, equals critical, i.e.,

$$\zeta = \frac{b}{b_c} = 1$$

the solution of Eq. 9.35 is

$$x_r = (A + Bt)e^{-\zeta\omega_n t} \qquad (9.38)$$

and there is no oscillation.

When the system damping, b, is greater than critical, i.e., $b/b_c > 1$, the solution of Eq. 9.35 is

$$x_r = e^{-\zeta\omega_n t}(Ae^{\omega_n \sqrt{\zeta^2 - 1}t} + Be^{-\omega_n \sqrt{\zeta^2 - 1}t}) \qquad (9.39)$$

which is also nonoscillatory.

The magnitude frequency response curves associated with these equations are shown in Fig. 9.82,

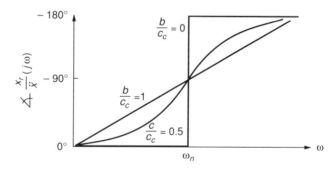

Fig. 9.83   Second-order system phase angle frequency response.

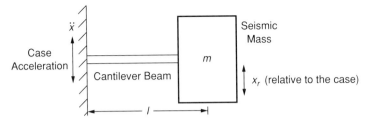

Fig. 9.84   Strain gage accelerometer mechanical configuration.

and the phase response in Fig. 9.83. These responses are for an input acceleration of the case, $\ddot{x}$. See also Eqs. 9.34. The foregoing applies to any single-degree-of-freedom system with viscous damping. A transducer must be employed to sense the motion of the mass, m, due to an applied case acceleration. This transducer can take many forms, including strain gage sensing.

The mechanical configuration for an accelerometer can take many forms, and probably the most widely used is the cantilever beam—seismic mass system (Fig. 9.84). Strain gages are bonded to the beam to sense the induced strain due to the relative motion of the mass m. From Newton's Second Law of motion the force, f, is equal to the mass, m, times its acceleration, $\ddot{x}_m$. The strain comes from the bending of the cantilever beam, length l.

One, two, or four gages can be bonded to the beam, but no matter how many sensing elements are used they are normally connected in a Wheatstone bridge circuit (Fig. 9.85).

Consider the fully active bridge, part (c). Two strain gages are bonded to each side of the cantilever beam of length, l, which is loaded at one end with mass, m. If each strain gage has a bonded resistance, R, then the bridge is balanced and $V_{out} = 0$.

The beam is deflected by application of a force due to acceleration, $\ddot{x}$, and each gage experiences a change of resistance ΔR due to the change in length Δl of the beam at its upper and lower surfaces. Since the two gages are bonded to each side of the beam, deflection will result in the gages on one side of the beam going into tension and the gages on the opposite side of the beam going into compression. The tension gages are wired into opposite arms of the bridge and their resistances become $R + \Delta R$. The compression gages are wired into the other arms of the bridge and their resistances become $R - \Delta R$.

The sensitivity for the Wheatstone bridge sensor is derived as follows:

$$V_{out} = 1/4 \times (\text{No. of active arms}) \times (\text{Gage factor}) \times (\text{Strain}) \times (\text{Excitation voltage}) \qquad (9.40)$$

In the case of the fully active four-arm bridge, with gage factor $GF = 100$, strain $\epsilon = 200\mu\epsilon$ and 15 V excitation

$$V_{out} = (1/4)(4)(100)(200)(10^{-6})(15)$$

$$= 300 \ \mu V = 0.3 \ V = 300 \ mV$$

A piezoresistive accelerometer employing semiconductor strain gages is thermally sensitive. With

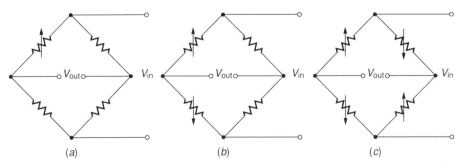

Fig. 9.85   Strain gage bridge networks: (a) one active element; (b) two active elements (half-bridge network); (c) four active elements (full bridge network).

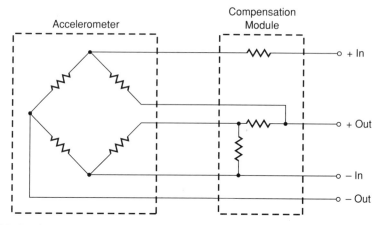

Fig. 9.86 Passive circuit temperature compensation for a Piezoresistive strain gage bridge network.

change of temperature two phenomena occur: thermal sensitivity shift, and thermal zero shift. Both of these changes must be compensated for. For semiconductor strain gage accelerometers, sensitivity goes down with increasing temperature, so if bridge voltage can be increased at an equal rate, compensation will occur. This can be accomplished using an active circuit such as a transistor or thermistor, but can also be accomplished using a passive resistive circuit with a constant voltage input. Zero shift can be either positive or negative with increasing temperature, but can also be compensated using a passive resistive circuit (Fig. 9.86), with constant voltage input.

The advantage of using this passive circuit is that it is temperature insensitive and does not have to be at the same temperature as the accelerometer. It can therefore be placed in a remote location, making it possible to miniaturize the accelerometer to a greater extent.

Piezoresistive accelerometers have inherently high output, as we have seen, so it is possible to use them to drive, directly, DVMs, strip chart recorders, tape recorders, and other readout devices. Sometimes it is necessary, however, to produce a high level output such as 0–5 V, 0–10 V, 4–20 mA or 10–50 mA. In this case, use of an amplifier or signal conditioner is necessary. A word of caution is in order when considering signal conditioners for use with semiconductor strain gage sensors. Conventional strain gage signal conditioners may not be suitable for two reasons:

1. Because they are designed for use with foil gage devices with low sensitivity, they often have a minimum gain of 100, resulting in saturation.

2. Zero adjust circuits are often placed in the input of the signal conditioner, which modifies the accelerometer temperature compensation circuit, rendering it ineffective.

A prospective user of semiconductor accelerometers should always contact the sensor manufacturer to determine compatibility.

9.6.4 Capacitive Accelerometers

A capacitive accelerometer consists of an elastic element, a small seismic mass, and a capacitive displacement transducing mechanism. External accelerations move the seismic mass relative to the sensor housing. This relative displacement is a measure of the acceleration. (See Section 9.6.1.) As discussed in Section 9.2.5, capacitive transducers are particularly well suited to detect very small distance variations. The high sensitivity of the capacitive pickup allows the designer to make the elastic element comparatively stiff, providing higher natural frequencies than other mechanical accelerometers. Capacitive accelerometers are rugged and can withstand large overloads. In comparison to piezoaccelerometers, their simpler electronic circuitry makes them easier to install and cost-effective for low to medium frequency applications.

The concept of a capacitive accelerometer is shown in Fig. 9.87. The capacitive sensor of the Setra accelerometer[42] consists of a thin, stiff disk and elastic elements that are sandwiched between two rigid capacitor plates. Displacement of the disk relative to the fixed plates is a measure of the acceleration. The electronic circuitry is integrated into the sensor. This accelerometer is quite rugged and can withstand overloads up to 2000 g. Full-scale operating ranges of these accelerometers vary from ±2 g to ±600 g, with measurement accuracy of 1% of full scale. The linear operating range (±3 db margins) extends to frequencies up to 3000 Hz.

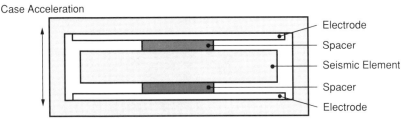

Fig. 9.87   Concept of a capacitive accelerometer.

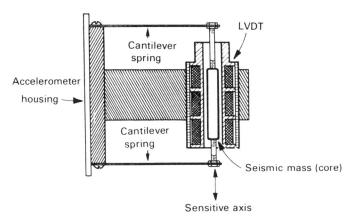

Fig. 9.88   LVDT accelerometer (taken by permission from Fig. 12.2 of *Handbook of Measurement and Control*, by E. E. Herceg, Schaevitz Engineering, Pennsauken, NJ, 1986).

9.6.5   LVDT Accelerometer

The basic elements of any accelerometer are a seismic mass and a position measuring transducer. The acceleration measurement range is from zero frequency to approximately 20% of the seismic mass natural frequency, depending on the required measurement accuracy. Figure 9.88 shows how an LVDT may be used to construct an LVDT accelerometer. This method of measuring acceleration is convenient because the necessary seismic mass is already available—the core and its attached rod. The core is suspended from the case by means of two leaf springs. The linearity is typically 1% of full scale and their sensitivity to cross-axis acceleration is less that 5%. Open-loop LVDT accelerometers were formerly much more popular than they are today.[2] Their application is mostly in areas where a high acceleration range is required (100 to 1000 g). Owing to its method of construction, an LVDT accelerometer could readily sustain large overloads that would be damaging to more fragile units.

9.6.6   Other Types of Accelerometers

Inductive Sensing

In the instrument world the manufacturer of an accurate and reliable position transducer often incorporates the same concept into other transducers as well. An inductive proximity transducer when combined with a second-order system as shown in Fig. 9.73 could be the basis for an inductive accelerometer. Because the basic transducer is in fact a proximity displacement transducer, the signal conditioner for this accelerometer would be identical to that for the displacement transducer. For further description of inductive proximity devices the reader is referred to the position transducer section of this chapter.

Servo Accelerometers

Closed-loop transducers and instruments generally offer superior performance relative to their open-loop counterparts. For example, the linearity of typical servo accelerometers[2] is 0.05% of full scale,

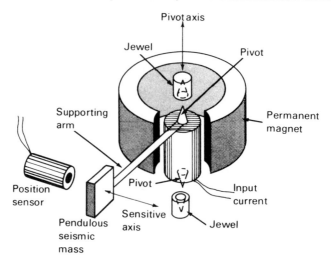

Fig. 9.89 Simplified mechanism of a servo accelerometer (taken by permission from Fig. 12.6 of *Handbook of Measurement and Control*, by E. E. Herceg, Schaevitz Engineering, Pennsauken, NJ, 1986).

or about 20 times better than most open-loop acceleration transducers. Resolution is about 0.0001% of full scale and repeatability exceeds 0.01% of full scale. Typical cross-axis sensitivity is less than 0.005 g per g, which means that the response to cross-axis acceleration is only 1/200 of the response to the same acceleration in the sensitive axis.

A closed-loop or servo accelerometer is a measurement system whereby an input acceleration causes a seismic mass to exhibit a displacement response much the same as an open-loop accelerometer. The servo accelerometer includes a displacement transducer as does the open-loop transducer. But here the similarity ends. On the one hand, the latter transducer simply converts this position response to an output voltage for readout. On the other hand, the closed-loop accelerometer uses this position response signal to apply a restoring force so that the seismic mass position returns to zero. The output signal that represents the measured acceleration, that of the case, is obtained by measuring the restoring force.

Figure 9.89 illustrates a typical servo accelerometer. The position sensor measures the position response of the pendulous mass and a current is set up that creates an opposing torque on the armature of the D'arsonval movement. The current that creates the torque is passed through a small resistor and is thereby converted to a voltage signal for purposes of readout. A block diagram of a linear accelerometer such as the one in Fig. 9.89 is shown in Fig. 9.90. Because the feedback torque returns the pendulous mass to zero, the linearity of the position transducer is of little or no consequence. This factor is largely responsible for the superior linearity specifications of this type of accelerometer. The same configuration as that shown in Fig. 9.89 is also applicable for the case of an angular servo accelerometer. However, the mass is balanced in an angular acceleration transducer. Some versions of servo accelerometers use flexural pivots instead of jewel-supported pivots.

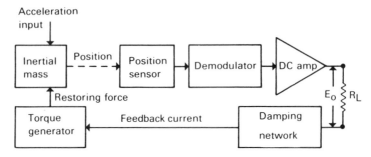

Fig. 9.90 Block diagram of a servo accelerometer (taken by permission from Fig. 12.5 of *Handbook of Measurement and Control*, by E. E. Herceg, Schaevitz Engineering, Pennsauken, NJ, 1986).

9.7   DIGITAL TRANSDUCERS

Chester L. Nachtigal

Most of the transducers mentioned in Chapters 8, 9, 10, and 11 are analog transducers. A transducer in this class has the property that its output may take on a continuous set of values ranging from a minimum to a maximum. The variable being measured by the transducer is encoded in the magnitude of the transducer output signal. Modulated-carrier transducers fit this description as well as DC-output devices. The output of a digital transducer on the other hand may take on only discrete values. A number system in which only integer values have meaning is an example of a discrete set of values. A transducer that can produce only integer values does not have the capability of quantifying the portion of an input variable lying between integer values. It is therefore frequently necessary to question the resolution capability of a digital transducer; much more so than that of an analog transducer.

With the advent of low-cost digital computers, there is a growing incentive to develop transducers that respond to an analog variable by delivering a digital output. Although our engineered world will become more and more dominated by digital data processing, nature is and will remain primarily analog. Hence the continuing need for transducers that have an analog front end. At the same time, the need for a digital output will grow stronger and stronger.

Reference 43 provides some advantages of using digital signals in ordinary instrumentation systems, even if a computer is not used in the complete installation. These are

ease of generating, manipulating, and storing digital signals

need for high measurement accuracy and discrimination

relative immunity of a high-level digital signal to external disturbances (noise), especially in transmitting signals over long distances

ergonomic advantages in simplified data presentation (for example, digital readout avoids interpretation errors in reading scales or needle movements)

logistic advantages concerning maintenance and spares compared with analog or hybrid systems

Three types of digital transducers are summarized in this section: (a) those with A/D converters; (b) frequency domain transducers where the magnitude of the measured variable modulates the frequency of a sinusoid or a square wave, and (c) direct digital, where the output is an inherently discrete response to the analog input variable.

9.7.1   Transducers with Analog-to-Digital Conversion

In the first group of digital transducers there is the transducer that generates a high-level analog signal that is then converted to digital format by a conventional analog-to-digital converter. This converter resides in the same package as the signal conditioner of the analog portion of the transducer and the sensing element itself. An advantage of this type is that the digital processor receiving this format does not need to dedicate any additional conversion time to the incoming signal. A significant disadvantage is that the transducer output consists of n individual wires, where n is the number of bits required to encode the analog signal to the required resolution. Values of $n = 12$, 14, or even 16 are quite common.

Resolvers (see Section 9.2.6) are a special case of this type of digital transducer when it is coupled with its own special converter, called a resolver-to-digital converter, or RDC. The resolver produces a pair of output voltages proportional to the sine and the cosine of the input shaft angle. Since these two signals are continuous functions, the resolver itself is an analog transducer.

9.7.2   Frequency Domain Transducers

The output of a frequency domain transducer is typically a sine wave, but other repetitive signals are also used, for example, a square wave. An example of this type of transducer is the vibrating wire technique for measuring strains and small motion, which has been known for about 100 years. (See Section 10.4.5.) Pressure transducers using the vibrating wire principle of operation have been available since the mid-1950s when their durability and long-term stability began to be recognized.

A more recent frequency domain transducer (reference 44) makes use of a crystalline-quartz resonator. It detects changes in pressure-induced stress by changes in the oscillation frequency of the quartz crystal. The relationship between the input pressure and the variable frequency response signal can be expressed in terms of either frequency or period. Since the signal period can be measured by a simple gated counting circuit, the transducer input-output transfer function is commonly expressed by relating pressure and period. Note that the resolution of this type of digital transducer is limited only by the resolution of the counting circuit.

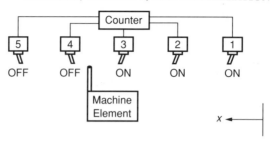

Fig. 9.91   Direct digital position transducer.

Turbine flowmeters fall into this transducer category in that the output is a pulse train created by passing the blades close to a velocity pickup of the type described in Section 9.5.3. Increasing the fluid volume flow rate through the turbine causes it to spin faster, thereby decreasing the time spacing of the output pulse train. A counter can be used to measure the period of the output signal.

9.7.3   Direct Digital Transducers

Direct digital transducers* share the feature that nowhere is there a continuous signal indicative of the input variable. Consider a series of on-off switches arranged to sense the position of a machine component that travels in a straight line as shown in Fig. 9.91. As it passes each successive switch, it turns the switch on and a simple counter circuit produces an output. Note that the resolution of this output is the distance between two adjacent switches. Notice also that any velocity changes of the machine component can be detected no earlier than the time of arrival at the next switch position.

This dead time is characteristic of digital measurement systems and can be troublesome, especially if the transducer is part of a closed-loop system. (See Chapters 14 and 15 for further discussion of this point.)

There are numerous examples of direct digital transducers at work today. A mechanical GO-NO GO gage is a simple one-bit digital transducer. Linear and optic arrays are one-dimensional, multibit position sensors. Computer-aided design digitizing tablets are two-dimensional digital transducers. A TV reticon tube with its pixels across the screen and its lines down the screen, or a solid-state TV screen consisting of charge-coupled elements (Section 9.4) may be used as two-dimensional digital position transducers. Addition of gray-scale imaging creates a three-dimensional transducer.

Perhaps the most widely used direct digital transducer is a rotary encoder (Section 9.2.6). Modern encoders[43] offer higher resolution, greater reliability, and greater accuracy than any currently available analog transducer of comparable size. Resolution as high as one millionth of a revolution or one second of arc is achieved in encoders providing a 20- or 21-bit binary output. This level of performance is achieved by using optical or photoelectric techniques.

9.8   SMART SENSORS, INTELLIGENT TRANSDUCERS AND TRANSMITTERS

Chester L. Nachtigal

The concept of "sensor" distinguished from "transducer" as described at the beginning of this chapter collides with the popular usage of "sensor" as in "smart sensor." Furthermore, our present understanding of "smart" and "intelligent" with reference to current instrumentation capabilities will surely be vastly superseded in less than a decade given the present rate of growth and improvement of the computer and specifically the microcomputer industry. But for purpose of this section, the generally accepted terms "smart sensor," "intelligent transducer," and "intelligent transmitter" are used. Substitution of "transmitter" for "sensor" or "transducer" (Ref. 45, 46, and 47) is unavoidable given that the sensed information must be delivered to a central computer or data

---

*Reference 43 defines a direct digital transducer as one where a parallel digital signal is generated, each bit weighted according to some code. That author further points out that direct digital transducers are few and far between due to the scarcity of natural phenomena that detect analog changes and respond in a discrete manner. This author contends that a transducer can be designed to respond in a discrete fashion in order to qualify as a digital transducer. In other words, it need not produce a parallel digital output signal naturally.

acquisition system located some distance from the location where the transducer actually performs the measurement function. Reference 45 makes the point that the intelligence of the measurement system most likely resides in the transmitter portion and not in the sensor or transducer portions of the device.

9.8.1   Definition of Smart Sensors, Intelligent Transducers and Transmitters

Articles dealing with smart sensors invariably discuss integrated sensors as well.[48,49] Functionally they are separate concepts, but the same technology that has led the way for integrated sensors has also enabled and driven the development of smart sensors. Reference 48 defines integrated sensors as sensors that have electronics on the sensing and transduction element itself. An example of a high-volume commercially successful transducer that has greatly benefited from integration is the silicon piezoresistive pressure transducer. Because piezoresistive strain sensing elements are heavily temperature dependent, the need for close-coupled compensation electronics has fueled this transducer's integration development.

Two definitions for smart sensors are offered. Reference 45 states that the basic characteristics of a smart sensor are:

direct computation of the physical magnitude of a measurement at the transducer level after correcting for parasitic effects instead of using calibration coefficients at the central computer level

a digital instead of an analog signal

identification of the sensor

not just simply transmission to, but also two-way communication with, the network or the central computer

integration of additional functions into the transmitter part, thus relieving the central computer of those tasks

possibilities of self-diagnosis for proper operation and preparation of data for maintenance assistance

Reference 48 defines smart sensors as those transducers that can do one of the following:

perform a logic function

perform two-way communication

make a decision

Reference 48 continues by stating that smart sensors are enhancing the following applications:

self-calibration and diagnostics

computation

communication

multisensing

In addition to these application areas, real-time compensation is an exceedingly important issue, not only for closed-loop control systems but also for measurement and data acquisition systems. This subject is discussed in Section 9.8.6.

9.8.2   Self-Calibration and Diagnostics

Calibration normally requires two parameter adjustments—offset or zero adjustment and span adjustment. Smart sensors have the built-in ability to make corrections upon receiving a message that a known standard input signal is currently being applied to the transducer, such as a zero value or a full-scale value. If span and offset are known functions of temperature, an intelligent transducer could also make corrections if it receives information about its ambient temperature at the time it updates its calibration. Real-time compensation for temperature inputs is described in Section 9.8.6.

9.8.3   Computation

A transducer can be improved[48] by fabricating a number of supposedly identical sensors operating in parallel and applying statistics to the multiplicity of sensor outputs. Thus the average, the variance, and the standard deviation can be computed for the set of simultaneous measurements. If an individual sensor output is determined not to be a member of the set (e.g., if it exceeds $\pm$ 3 standard deviations), it can be discarded.

9.8.4   Communication

Communication is a necessity for all transducers if for no other purpose than to deliver measurement information to the data collection system. A microprocessor resident in the transducer itself increases the flexibility of the type and frequency of the communication instances between the intelligent transmitter and the host system. Thus the host computer can be used more efficiently if, for example, the transducer information exchange occurs on an as-needed basis rather than at regular intervals. As stated by Reference 45, the concept of two-way communication with the host computer is an important attribute for improving performance by the use of smart sensors.

9.8.5   Multisensing

The ability of a single transducer to measure more than one physical variable affords numerous performance improvements. An example is the case of piezoresistive pressure transducers, which are quite temperature sensitive. If the transducer also carries an on-board temperature sensor for compensating the pressure measurement, performance can be significantly upgraded.

9.8.6   Compensation

The most common forms of sensor defects that benefit by compensation are:[50]

parameter drift; e.g., temperature-induced zero shift
sensor nonlinearities
cross-axis sensitivity
noise
limited time or frequency response

Temperature Compensation

Figure 9.92 shows the structure of an intelligent transmitter with compensation for changes due to temperature changes. The specifics of the compensation strategy are, of course, unique to a particular sensor, but the two parameters that normally are the most susceptible to temperature variations are span and zero offset. The power of the microprocessor in this application lies in its capacity to store information about the precise relationship between a temperature input and the sensor's unwanted response. With the enhancements afforded by the transmitter's two-way communication, the relationship between temperature input and sensor response can be readily updated from a central computer. Reference 50 classifies those compensation schemes requiring a second sensor as monitored compensation.

Nonlinear Compensation

Sensor nonlinearities occur in many, many different forms. Perhaps the most commonly encountered form is that of a polynomial instead of a straight-line relationship between output and input, as shown

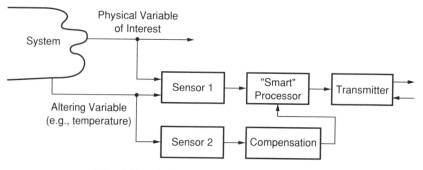

Fig. 9.92   Monitored temperature compensation.

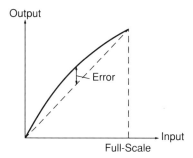

Fig. 9.93 Polynomial nonlinearity.

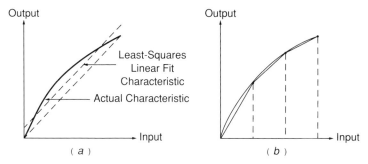

Fig. 9.94 Fitting a straight line to a polynomial characteristic: (*a*) least-squares linear fit; (*b*) piecewise linearization.

in Fig. 9.93. Two common methods of compensating this type of nonlinearity are shown in Fig. 9.94. The least-squares method requires an offline study of the sensor response and computation to find the least-squares straight-line (see Chapter 4), which is then used for the online measurement process. The intelligence of the transducer would greatly enhance subsequent recalibration of the transducer. A piecewise linearization method of reducing the error of Fig. 9.93 also requires offline study of the transducer. Once the input-output characteristic is obtained, the online processing of the data is by means of a table lookup procedure as shown in Fig. 9.94*b*. The input-output straight-line relationship changes depending on the magnitude of the input.

A third approach to compensating the nonlinearity shown in Fig. 9.93 consists of programming into the software of the intelligent transducer a polynomial function that approximates the inverse of the sensor nonlinearity. This structure is shown in Fig. 9.95. The intent is that the input-output relationship of the cascaded combination of sensor plus compensation is nearly linear. Other forms

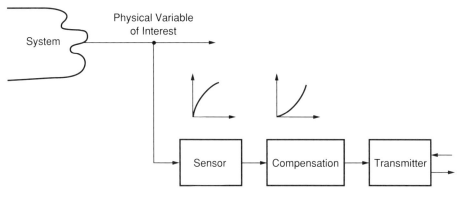

Fig. 9.95 Polynomial nonlinearity compensation.

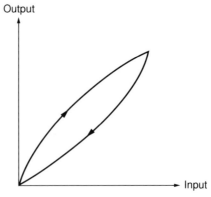

Fig. 9.96   Hysteresis.

of the compensation input-output relationship may also be programmed, such as base 10 logarithmic and base *e* natural log functional forms.

Hysteretic input-output relationships (see Fig. 9.96 for one example) are generally more difficult to process because of their double-valued nature. An algorithm for compensating this nonlinearity would have to take into account the direction of change of the input variable and perhaps its rate of change as well.

Deduced Measurement Compensation

Occasionally information is required about a subsystem or a component that is inaccessible to a transducer. As a case in point the designer of internal combustion engines would like to measure the temperature of the internal cylinder walls. Clearly it is difficult if not impossible to install a temperature transducer at that location because of mechanical interference with the piston and because of the environment. A temperature measurement from the cooling-jacket surface of the cylinder would have to be corrected or compensated, perhaps statically with a nonlinear relationship and most certainly by some dynamic transfer function. Figure 9.97 shows the structure of this type of compensation. Reference 50 categorizes this approach as "deductive" compensation.

The major difficulty in this type of compensated measuring concept is the need for modeling information about the intervening structure between the measurable location and the location of the physical variable of interest. Any parameter drift in the portion of the system between the actual measurement location and the desired measurement location will create errors unless the parameters of the model also change in unison with the real system. But such a parameter identifier is highly computation-intensive.

A common example where deduced compensation is used regularly with very little error is in a load cell. Measurement of force or weight is accomplished by applying the force or weight to a stiff structure generally made of steel and sensing the deflection of the structure with a strain sensor or other means (see Sections 10.1 and 10.2). The readout of the instrument relies on the fact that the relationship between force and displacement of the load cell structure is linear and static (no frequency response effects), or that the bandwidth of interest in the measurand is below the range of frequencies where the load cell causes phase shifting. Calibrating the load cell is the experimental process of finding the stiffness of the load cell structure and the sensitivity of the embedded position transducer. The online measurement process may be modeled by the block diagram of Fig. 9.95 except that the sensor and compensation characteristics are linear.

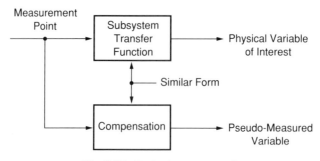

Fig. 9.97   Deductive compensation.

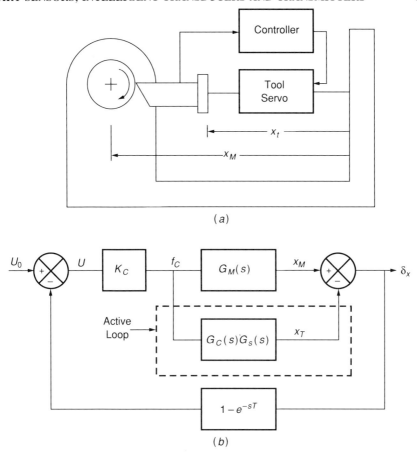

Fig. 9.98 Deductive compensation of machine-tool chatter: (*a*) controlled machine-tool schematic; (*b*) closed-loop control block diagram.

An example of deduced compensation of a dynamic nature is the case of controlling machine-tool chatter,[51] which is essentially a time-varying error in position of the tool relative to the workpiece. Ideally the control of this position error would include measuring the distance between the tip of the cutting tool and the workpiece. However, the cutting geometry and its environmental conditions preclude installing a position sensor at this location. A relatively simple measurement is the cutting force, which is the cause of the position error. Figure 9.98 illustrates the overall machine-tool and closed-loop control system. In Fig. 9.98b, $G_m(s)$ represents the machine-tool dynamic transfer function, $G_c(s)$ is the compensator that takes the form of $G_m(s)$, and $G_s(s)$ is the unavoidable band-limited transfer function of the tool servo.

Although the work reported in Reference 51 was performed entirely in the analog domain, a commercialization of this scheme would greatly benefit from two-way communication between a central computer and the local processor of an intelligent transmitter. The central processor would need to download the compensation transfer function form and/or the parameter values of each new workpiece shape. It is conceivable that the compensation transfer function itself would be implemented in the analog domain in order to achieve the required speeds of response.

Cross-Axis Sensitivity

It is exceedingly difficult to build a sensor responsive to one and only one sensing direction. The fact that the sensing element occupies a finite area is itself a source of error in some measurement applications. Some types of motion sensors, such as accelerometers, are especially prone to cross-axis errors. A possible solution is to use two or more single-axis sensors deployed in the anticipated directions of motion of the system to be measured. Depending on the nature of the coupling between the two or more axes, the compensation may resemble the block diagram shown in Fig. 9.92. The algorithm of the compensation block may simply be to subtract from Sensor 1's total response a signal coming from Sensor 2 and its series compensation which is Sensor 1's ascertained response to variable 2 alone.

Noise and Limited Frequency Response

Nearly all measurement systems are affected by noise to some extent. The process of low-pass filtering to attenuate unwanted noise is described in Sections 12.1 through 12.3 for analog filtering and 12.4 through 12.8 for digital filtering. A series-cascaded compensation structure as shown in Fig. 9.95 would be applicable for synthesizing a system with noise filtration. Instead of a nonlinear function, the compensator would be a linear transfer function.

Measurement systems with time or frequency response limitations can be compensated somewhat by a series-cascaded transfer function with a numerator polynomial that resembles the denominator of the transducers's transfer function. If the transducer has a pole at, say, 100 Hz, the compensator would have a zero at 100 Hz. This approach is called lead-function compensation and is discussed in Section 17.3. The structure of this type of compensation would also be similar to the structure shown in Fig. 9.95. As in the previous example, the nonlinear compensator would be replaced by a linear transfer function.

REFERENCES

1. DS Data Sheet, Kistler-Morse Corporation, Redmond, WA.

2. E. E. Herceg, *Handbook for Instrumentation and Control*, Schaevitz Engineering, Pennsauken, NJ, 1986.

3. "LVDT and RVDT Linear and Angular Displacement Transducers," Technical Bulletin 1002A, Schaevitz Engineering, Pennsauken, NJ.

4. "Displacement Transducer DC-DC: Series 200," Bulletin S012-0029 IHD, Trans-Tek Incorporated, Ellington, CT.

5. "Application of a Precision Measuring Technology," Seminar Workbook SWB 865, Kaman Instrumentation Corp., Colorado Springs, CO, 1986.

6. "Rotating Machinery Information Systems and Services," Technical Notes, Bently Nevada, Minden, NV, 1987.

7. George A. Quinn, "High Accuracy Rotary and Linear INDUCTOSYN® Position Transducers," Farrand Controls, Division of Farrand Industries, Inc., Valhalla, NY, 1985.

8. T. L. Foldvari and K. S. Lion, "Capacitive Transducers," *Instrum. & Control Syst.*, November 1964, pp. 77–85.

9. K. S. Lion, "Capacitive Transducers," *Instrum. & Control Syst.*, June 1966, pp. 157–159.

10. K. S. Lion, "Nonlinear Twin-T Network for Capacitance Transducers," *Rev. Sci. Instrum.*, Vol. 15, No. 3, March 1964.

11. S. Y. Lee, "Variable Capacitance Signal Transduction and the Comparison with other Transduction Schemes," Tutorial Proceedings of the Sixteenth International Aerospace Instrumentation Symposium, ISA Pittsburgh, Seattle, WA, May 1970.

12. W. H. Ko, B. X. Shao, C. D. Fung, W. J. Shen, and G. J. Yeh, "Capacitive Pressure Transducers with Integrated Circuits," *Sensors and Actuators*, Vol. 4, 1983, pp. 403–411.

13. L. Michelson, "Greater Precision for Noncontact Sensors," *Machine Design*, December 6, 1979, pp. 117–121.

14. John Hall, "Position Sensor and Proximity Switch," *I&CS — The Industrial and Process Control Magazine*, July 1984.

15. Jim McDermott, "Digital Motion Control," *EDN*, Vol. 29, No. 1, January 12, 1984.

16. Kneal Hollander, "Ultrasonic Level-Sensing Technology," *Powder and Bulk Engineering*, March 1988.

17. A. J. Pointon, "Piezoelectric Devices," *IEE Proceedings*, Vol. 129, No. 5, July 1982.

18. C. Biber, S. Ellin, and E. Shenk, "The Polaroid Ultrasonic Ranging System," *Audio Eng. Soc. Preprints*, No. 1696(A-8), 1980.

19. L. E. Kinsler, A. R. Frey, A. B. Coppens, and J. V. Sanders, *Fundamentals of Acoustics*, 3rd ed., Wiley, New York, 1982.

20. D. Campbell, "Ultrasonic Noncontact Dimensional Measurement," *Sensors*, Vol. 3, No. 7, July 1986.

21. M. J. Howes and D. V. Morgan, "Charge-Coupled Devices and Systems," Wiley, Chichester, England, 1979.

22. Robert Lewis, "Solid-State Cameras," *Electronic Imaging*, Vol. 4, No. 1, January 1985, pp. 44–52.

23. Application Note 121, "Spectral Response of Reticon Linear Photodiode Arrays," EG&G Reticon, Sunnyvale, CA.

24. Warren Smith, "Modern Optical Engineering," McGraw-Hill, New York, 1966.

25. F. H. Colvin, *Gages and Their Use in Inspection*, McGraw-Hill, New York, 1942.

26. J. W. Dally, W. F. Riley, and K. G. McConnell, *Instrumentation for Engineering Measurements*, Wiley, New York, 1984.

27. D. M. Considine, ed., *Encyclopedia of Instrumentation and Control*, McGraw-Hill, New York, 1971.

28. D. M. Considine, ed., *Handbook of Applied Instrumentation*, McGraw-Hill, New York, 1964.

29. J. Goss, "On-Line Sensors for Paper Systems," *Southern Pulp and Paper Manufacturer*, Vol. 42, No. 3, 1979.

30. W. Grohe, *Precision Measurement and Gaging Techniques*, Chemical Publishing Co., New York, 1960.

31. P. H. VanMunn, "On-Line Caliper Measurement and Control," *TAPPI*, Vol. 53, No. 5, 1970.

32. R. J. Pfeifer, "A Review of the Development of On-Line Paper Machine Sensors," *Pulp and Paper*, Vol. 55, No. 2, 1981; "On-Line Paper Machine Sensors—Review of Current Technology," *Pulp and Paper*, Vol. 55, No. 9, 1981.

33. G. Burk, "Infrared Sensors Control Thickness of Film Coextrusion Plies," *Modern Plastics*, January 1984.

34. Personal conversation with Prof. Adam C. Bell, Nova Scotia Technical Institute, Halifax, Nova Scotia, March 1989.

35. "Linear Velocity Transducer: Series 100," Bulletin S012-0028 IHE, Trans-Tek Incorporated, Ellington, CT.

36. Engineering Staff, *DC Motors, Speed Controls, Servo Systems*, Electro-Craft Corp., 1980.

37. Seymour Lanton, "Resolver-Based Motion Control Sensors," *Conference on Applied Motion Control Proceedings*, 1985.

38. Mike Glass, "Choosing the Right Encoder Simplifies Motion Control," *Computer Design*, March 1984.

39. Engineering Staff, "Understanding Magnetic Sensors for Digital Tachometer Systems," *Plant Engineering*, Cahners Publishing Company, Des Plaines, IL, April 26, 1984.

40. "Velocity Transducer System," Bulletin L5005-00, Bently Nevada, Minden, NV, 1986.

41. A. J. Severyn, R. W. Lally, R. Zimmerman, and D. L. Brown, "Six-Axis Dynamic Calibration of Accelerometers," *Proceedings of the 5th International Modal Analysis Conference*, Vol. II, Imperial College of Science & Technology, London, England, April 6–9, 1987.

42. Model 141 Accelerometer Data Sheet, Setra Systems, Inc., Acton, MA.

43. G. A. Woolvet, *Transducers in Digital Systems*, Peter Peregrinus, Ltd., Southgate House, Stevenhage, England, 1977.

44. Donald W. Busse, "Quartz Transducers for Precision Under Pressure," *Mechanical Engineering*, Vol. 109, No. 5, May 1987.

45. J. M. Favennee, "Smart Sensors in Industry." *Journal of Physics E*, Vol. 20, No. 9, September 1987.

46. Jerome M. Paros, "Use of Intelligent Transducers Increases," *Measurements & Control*, Vol. 124, September 1987.

47. Digiquartz® Intelligent Transmitters, Data Sheet, Paroscientific, Redmond, WA.

48. Joseph M. Giachino, "Smart Sensors," *Sensors and Actuators*, Vol. 10, No. 3–4, November–December 1986.

49. S. Middlehoek and S. A. Audet, "Silicon Sensors: Full of Promises and Pitfalls," *Journal of Physics E*, Vol. 20, No. 9, September 1987.

50. J. E. Brignell, "Digital Compensation of Sensors," *Journal of Physics E*, Vol. 20, No. 9, September 1987.

51. C. L. Nachtigal and N. H. Cook, "Active Control of Machine-Tool Chatter," *Journal of Basic Engineering*, Transactions of the ASME, Series D, Vol. 92, No. 2, June 1970.

BIBLIOGRAPHY

Beckwith, T. G., N. L. Back, and R. D. Marangoni, *Mechanical Measurement*, Addison-Wesley, Reading, MA, 1982.

Bolton, W., *Engineering Instrumentation and Control*, Butterworths, London, 1980.

Cerni, R. H., and L. E. Foster, *Instrumentation for Engineering Measurement*, Wiley, New York, 1962.

Considine, Douglas M., *Encyclopedia of Instrumentation and Control*, McGraw-Hill, New York, 1971.

Cullum, William, and Harry Kratzer, "Measuring Shaft Position by Applying Synchro or Resolver Transducers," *Control Engineering*, 1982.

Dally, J., W. Riley, and K. McConnell, *Instrumentation for Engineering Measurements*, Wiley, New York, 1984.

Davidson, J. W., III, "Using Linear and Rotary Optical Encoders," *I&CS*, November 1984.

Doebelin, Ernest O., *Measurement Systems: Application and Design*, McGraw-Hill, New York, 1975.

Engineering Staff, *DC Motors, Speed Controls, Servo Systems*, Electro-Craft Corp., 1980.

Engineering Staff, *Synchro Conversion Handbook*, ILC Data Device Corp., 1982.

Ewins, D. J., *Modal Testing: Theory and Practice*, Wiley, New York, 1984.

Glass, Mike, "Choosing the Right Encoder Simplifies Motion Control," *Computer Design*, March 1984.

Glass, Mike, "Synchro and Microprocessor Combine for Versatile Multiturn Position Sensing," *Control Engineering*, Vol. 30, No. 5, May 1983.

Graeme, Jerald G., Gene E. Tobey, and Lawrence P. Huelsman, eds., *Operational Amplifiers: Design and Applications*, McGraw-Hill, New York, 1971.

Harvey, Glenn F., ed., *Standards and Practices for Instrumentation*, Instrumentation Society of America, 4th ed., Pittsburgh, 1974.

Herceg, Edward E., *Handbook of Measurement and Control*, Schaevitz Engineering, 1976.

Hordeski, Michael F., *The Design of Microprocessor, Sensor, and Control Systems*, Reston, VA, 1985.

Hordeski, Michael F., *Transducers for Automation*, Van Nostrand Reinhold, New York, 1987.

Jaslow, Howard, "Position and Velocity Sensing for Robot Control," *Robots 9 Proceedings*, 1985.

Lanton, Seymour, "LSI Chips Shrink Synchro-to-Digital Converter Hybrids," *Electronics*, June 1981.

Lanton, Seymour, and Bart Sakasai, "Resolver-Based Position Encoders Satisfy Demands of Robotics," *Design News*, March 28, 1983.

Mansfield, P. H., *Electrical Transducers for Industrial Measurement*, Butterworths, London, 1973.

Martinez, Hugo, "Measuring the Position of Mechanical Shafts," *Powerconversion International*, October 1983.

Martinez, Hugo, and Seymour Lanton, "Position Sensing for Robot Control," *Robotics World*, May 1983.

Muth, Steve, "Consider Basic Parameters When Choosing S/D Converters," *EDN*, June 23, 1982.

Muth, Steve, "Getting Reliable Digital Data from Shaft Position Sensors," *Application Note*, ILC Data Device Corp., July 1981.

Norton, Harry N., *Handbook of Transducers for Electronic Measuring Systems*, Prentice-Hall, Englewood Cliffs, NJ, 1969.

Norton, Harry N., *Sensor and Analyzer Handbook*, Prentice-Hall, Englewood Cliffs, NJ, 1982.

Ohr, Stephan, "Data Converters for Robotics Applications," *Electronic Products*, September 7, 1982.

Oliver, Frank J., *Practical Instrumentation Transducers*, Hayden, New York, 1971.

Sakasai, Bart, "Shaft Encoders—Absolute Position Shaft Encoders for Control Systems," *Measurement and Control*, September 1982.

Sakasai, Bart, and William Cullum, "Shaft Position Encoders for Industrial Application," *Electronic Products*, May 12, 1983.

Smith, John I., *Modern Operational Circuit Design*, Wiley Interscience, New York, 1971.

CHAPTER 10

FORCE, TORQUE, AND PRESSURE MEASUREMENT

Albert E. Brendel

Sensor Developments Inc.
Lake Orion, Michigan

David A. Hullender

Department of Mechanical Engineering
University of Texas at Arlington
Arlington, Texas

Robert W. Lally

Orchard Park, New York

Georg F. Mauer

Department of Civil and Mechanical
    Engineering
University of Nevada, Las Vegas
Las Vegas, Nevada

Chester L. Nachtigal

Kistler-Morse Corporation
Redmond, Washington

J. Barrie Sellers

Geokon, Inc.
West Lebanon, New Hampshire

Bernard H. Shapiro

Dynisco
Norwood, Massachusetts

Anthony D. Wang

Burr-Brown Corporation
Tucson, Arizona

10.1 TERMINOLOGY

Albert E. Brendel

A transducer transforms one form of energy into a different form. We will restrict this chapter to those transducers that convert a physical input into a form of energy that can be transmitted to a remote location. We will further restrict the discussion to those transducers that produce an electrical signal in response to a mechanical input. However, an infinite variety of sensing and conversion devices exist which use other transmission media, such as sound, light, hydraulic pressure, radio waves, or magnetic waves.

While this chapter will discuss only the more common types of these transducers, most of the concepts discussed are directly applicable to the other types. Therefore, the reader should keep an open mind when choosing the optimum sensor.

Virtually all physical input transducers are based upon the principle that a *structure*, when subjected to a force, will *deform* in proportion to the applied force. The method used to *sense* this deformation takes two basic forms:

1. Sensing the change in physical properties of the structure material itself, such as its change in magnetic properties (in the case of ferromagnetic materials) or piezoelectric properties (in the case of crystalline structures).
2. Sensing the physical deformation indirectly with a secondary transducer, such as a strain gage or Linear Variable Differential Transformer (LVDT).

These two categories can be further subdivided into *self-generating* and *interrogated* transducers in which a second energy source is required to *readout* the deformation.

The Piezoelectric transducers are the most common type of *self-generating* transducers. They rely upon changes in the structure itself to generate an electrical voltage proportional to the deformation of the primary load structure. They are normally characterized by extremely high spring constants (permitting high frequency responses), but they suffer the penalty of poor long-term stability, which generally limits their uses to dynamic or short-term static measurements.

The strain gage transducer exemplifies the *interrogated* type of transducer in which the change of resistance of a *bonded-on* element is monitored. While the spring constant of this type of transducer is slightly lower than that of an equivalent capacity piezoelectric device. The ability to interrogate with a separate energy source permits this type transducer to exhibit reasonably high frequency response and long-term stability. In general, the strain gage can be considered a *modulator* of the excitation source, and the electrical signal produced can be thus transformed into a form that is more tolerant of transmission in a high *noise* environment. An example of this technique is ac or *chopped dc* excitation in a *carrier system* in which the combined strain gage/excitation is later demodulated to extract the signal information.

10.2   FORCE TRANSDUCERS AND SIGNAL CONDITIONERS

Albert E. Brendel (10.2.1, 10.2.5)
Robert W. Lally (10.2.2)
Chester L. Nachtigal (10.2.3)
Georg F. Mauer (10.2.4)

10.2.1   Strain-Gage Load Cell and Its Signal Conditioner

Load Cell Basics

The strain-gage load cell consists of a structure that deforms when subjected to a force and a strain-gage network that produces an electrical signal proportional to this deformation.

The principal structures are the cantilever beam, the shear beam, the diaphragm, the proving ring, and the column. The choice of structural element is primarily based upon physical size constraints of the sensor itself. For reasonably sized sensors, the structural elements can be determined by the maximum force that will be applied to the transducer.

Force N (lbf)	Structure Type
< 500 N (< 100 lbf)	Cantilever beams, Proving rings
500–10,000 N (100–2000 lbf)	Multiple beams, Diaphragms
10,000–50,000 N (2000–10,000 lbf)	Columns, Shear beams
> 50,000 N (> 10,000 lbf)	Multiple columns, Multiple shear beams

These guidelines are rules of thumb since in practice there is considerable overlap of the structures due to secondary considerations, such as extraneous loads, which are nonmeasured but must be supported by the structure.

The simple cantilever beam (Fig. 10.1) can be used as a force sensor, but generally suffers because it senses a bending moment produced by an applied force rather than the force itself. The strain produced at the gage location is

$$\epsilon = \frac{M}{E(I/c)} = \frac{FL}{E(I/c)} \tag{10.1}$$

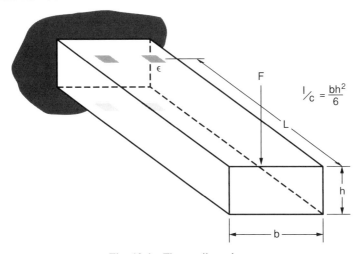

Fig. 10.1 The cantilever beam.

where ϵ is the unit deflection (dimensionless), M is the bending moment, E is the modulus of the elasticity, I/c is the section modulus of the structure at the gage location, F is the applied force, and L is the distance between the applied force and the gaged section (all units must be consistent).

Generally, four gages are wired in a Wheatstone bridge network (Section 8.4) such that the system senses the difference in strains between the upper and lower surfaces of the beam. These strains are only proportional to the bending moment, thereby canceling the effects of end loads or common temperature effects.

The sensitivity to force position can be removed easily since the Wheatstone bridge is a computational circuit, which can add or subtract signals produced by the individual strain gages. If the simple cantilever is gaged to sense moments produced in the beam at two locations, the gage system can be wired to produce a signal proportional to the difference in moments between the two locations, which forms a signal proportional to the *shear* force in the beam (see Fig. 10.2).

Even though the force position changes, the signal produced is proportional only to the distance between the two gage locations ($L_1 - L_2$). While this technique is often used, it suffers from the high stress at the cantilever base. Great care is required to prevent structural failure.

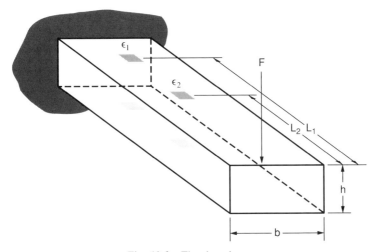

Fig. 10.2 The shear beam.

$$\epsilon_1 = \frac{FL_1}{E(I/c)} \tag{10.2}$$

$$\epsilon_2 = \frac{FL_2}{E(I/c)} \tag{10.3}$$

$$\epsilon_1 - \epsilon_2 = \frac{F(L_1 - L_2)}{E(I/c)} \tag{10.4}$$

If we change the load application point (Fig. 10.3), the stress in the structure is equalized. Variations on this structural concept form the basis of most high-performance beam-type transducers. The common S-shaped and parallelogram beam sensors are variations on this theme (Figs. 10.4–10.5).

The parallelogram sensor consists of two (or more) parallel beam systems in which each beam shares or equally carries the applied load. It is analyzed like the simple shear and cantilever beam. It has the added advantage, however, of being able to isolate force-produced bending moments more effectively. A free-body diagram of a parallelogram beam system (Fig. 10.6) shows that the bending moment produced by an applied force is supported in the structure by a force *couple* generating simple tension and compression forces in the two separated beams. These axial forces are canceled (not sensed) by the strain-gage network. They can be reduced to relatively low values if the beam systems are widely separated, which allows higher measurement accuracy.

As the force magnitude increases, required dimensional sizes of the beam system increase until it becomes impractical to use even multiple beam systems for the deformable structure. At these higher forces, it is sometimes possible to extend the use of beams if we sense shear deflections directly, applying gages to the beam's sides at 45° to the beam axis. This structural concept allows high forces to be measured, and the shear strains at the beam's neutral axis are proportional to the cross-sectional area rather than to the beam's moment of inertia.

For a rectangular beam,

$$\epsilon = \frac{3}{2} \frac{F}{AE} \tag{10.5}$$

where F, E, and ϵ are previously defined, and A is the cross-sectional area of the beam at the gage location.

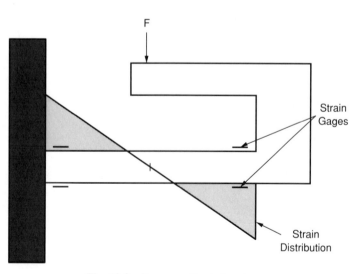

Fig. 10.3　Stress equalized shear beam.

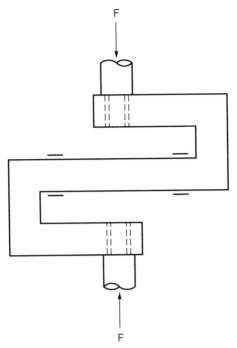

Fig. 10.4 The "s" load cell.

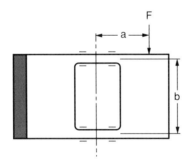

Fig. 10.5 Parallelogram beam sensor.

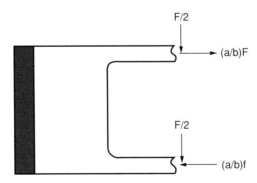

Fig. 10.6 Free-body diagram of parallelogram beam system showing shear force isolation from moment forces.

Because of the additive effect of the Wheatstone bridge and Poisson strains in the beam, the signals produced are 1.3 times higher than in a simple cantilever beam system. Unfortunately, the yield stress in shear of most materials is less than the yield stress in tension, and therefore little emphasis should be placed upon the increased output from the shear beam systems.

Shear beam load cells generally are constructed of multiple beams arranged in a radial configuration (Fig. 10.7). Their obvious advantage is in the low profile possible in this configuration, while their main weakness is in the inability of the hub and outer rim structure to support properly the beams without distorting themselves. Distortion in either of these elements tends to change severely the strain distribution in the shear beam itself, leading to potential inaccuracies.

The simple column is often used for high-force measurements. The strain is equal to

$$\epsilon = \frac{F}{AE} \tag{10.6}$$

where the definitions are previously given. As the column is loaded, Poisson strains are generated normal to the applied force axis, which also can be measured, allowing a semi-active Wheatstone bridge to be used. However, since these Poisson strains are approximately one-third the magnitude of the axial strains, the effective bridge output is reduced by a factor of 1.3, and higher stresses must be used to produce signals similar to those obtained from beam or shear beam structures. Generally the column must be at least twice as high as it is wide, because of strain gradients along the column. Since the width is proportional to the area, as the force capacity is increased, the column structure becomes higher. Also, bending moments from either extraneous forces or measured forces applied parallel to the measurement axis can produce significant bending stresses in the column, which again are not measured but can lead to structural failure.

To overcome bending moments in column structures, an auxiliary structure, parallel diaphragms at the input end, is often added, which effectively removes these moments by the same technique described for parallel beam systems. Figure 10.8 shows a free-body diagram of a diaphragm-supported column.

Multiple column structures reduce bending moment stresses and at the same time allow the reduction of sensor height. In these structures, wide spacing between the columns allows force couples, which increase or decrease the axial stress in each column, to support bending moments. All columns are individually *gaged*, and the Wheatstone bridge extracts the average column load. Since each column carries only a fixed proportion of the load (determined by the number of columns used), each column may be made shorter, thereby reducing the overall height of the sensor. Multiple-column load cells have been made with four to sixteen columns with all gages wired in a Wheatstone bridge configuration to average the strains produced in each column.

The diaphragm structure may be considered an extension of the simple cantilever beam sensor because its beam width is extended to wrap around on itself. Its advantage is in its structural simplicity, while its disadvantage is in its difficulty of adequately controlling the supporting structure's rigidity, much like the limitation of shear beam structures. As a diaphragm structure deflects, the radial tensile forces generated in the diaphragm affect the strain distribution, resulting in non-repeatability and non-linearity of the structure's force responses.

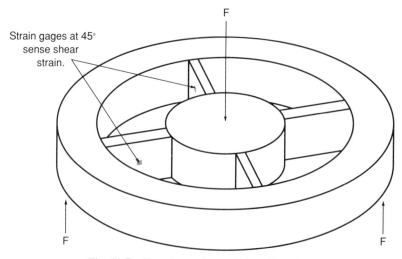

Fig. 10.7   Shear beams in a radial configuration.

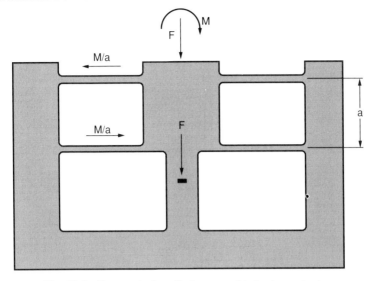

Fig. 10.8 Force paths in a diaphragm-assisted column structure.

The proving ring can be considered another extension of the cantilever beam system with individual beams arranged in a series/parallel combination. In this arrangement, gage placement can produce systems that are linear. The proving ring structures suffers, however, from relatively large deflections under load and its inability to support extraneous loads and/or bending moments without the use of an auxiliary support structure, such as the parallel diaphragm system discussed under column structures.

Signal Conditioning

A 350 Ω strain gage will increase (or decrease) its resistance approximately 0.7 Ω when subjected to normal design-strain levels. While the resistance change of a single strain gage could be measured with an ohmmeter, the more common approach is to utilize gages in multiples of four arranged in a Wheatstone bridge configuration (Fig. 10.9). In this configuration, the transducer can be considered a modulator of the excitation source, having typical transfer functions with units as mV/V/load.

If a DC voltage source excites the bridge, the sensor output is a low-level DC voltage proportional to the applied force. Generally, the sensor bridge is balanced to produce zero output voltage at zero load. It is important to recognize that most commercially available transducers use additional resistive networks in series with the strain-gage bridge to compensate for temperature. These networks

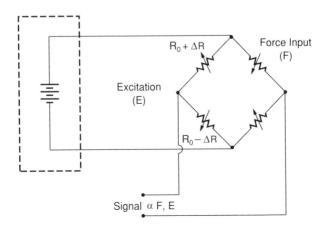

Fig. 10.9 DC-excited strain-gage transducer.

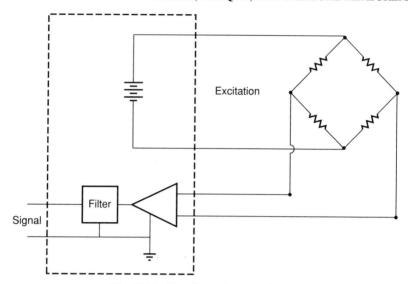

Fig. 10.10   Transducer signal conditioner.

located inside the transducer are designed to function with constant voltage excitation. They may be inadvertently defeated by using other types of excitation sources, such as constant current.

In a DC-excited system, the excitation voltage level is limited by the individual strain gages' and structure's ability to dissipate the i^2R heating. Excitation levels are therefore limited to about 20 V with 5–10 V excitation systems being common. These excitation levels produce full-scale transducer outputs between 10 mV and 40 mV and generally require amplification for subsequent display or control applications.

Commercial *transducer signal conditioners* excite the transducer, balance the bridge, amplify the low-level signal, and add low-pass filtering with the cut-off frequency determined by the use of the amplified signal. Figure 10.10 shows a block diagram of this simple system. Figure 10.11 shows a *transducer signal indicator*, which utilizes the same components and, in addition furnishes a digital or analog display capable of being scaled to engineering units.

These signal conditioners and indicators are available with a wide range of options, such as peak level detection, setable limits with relay closures, and analog outputs. Conditioners using

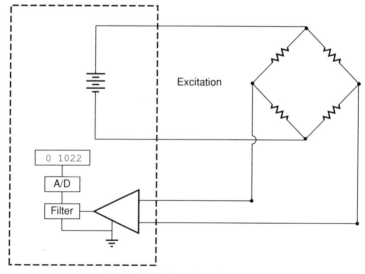

Fig. 10.11   Transducer indicator.

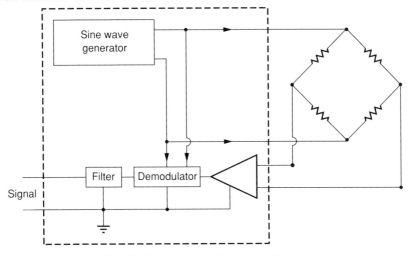

Fig. 10.12 Carrier-based signal conditioner.

microprocessors allow even more flexibility, with the displayed data relating to the input data by software programs rather than by simple fixed, scale factors.

While constant voltage excitation systems are the most common, other systems can sometimes be useful in special measurement applications. High frequency (3 kHz) sine wave excitation coupled with synchronous demodulation of the transducer output signal forms the basis for *carrier amplifier* signal conditioners (Fig. 10.12). These are more complex than constant excitation voltage systems but are capable of extracting small signals from high noise environments. Square-wave excitation systems are used similarly, but they are used principally to allow for amplifier drift suppression, much like a chopper-stabilized amplifier system.

Many transducer signal conditioners and indicators allow for *remote sensing* of the excitation voltage (Fig. 10.13). This technique allows relatively long cable runs between the transducer and the indicator without inadvertent signal loss due to the voltage drop in excitation lead wires. Resistance of the signal leads is unimportant when high-input impedance amplification is used in the indicator.

High voltage-pulsed systems produce high output signals on a sampled basis while they prevent excessive heating of the gage systems. The disadvantages of this technique are (1) the high voltages can lead to possible safety hazards and (2) the sampling rate limits the system's frequency response.

Most transducer signal conditioners and indicators incorporate some type of balancing networks to allow for tare loads, which are to be zeroed from the transducer output signal. The T balance network (Fig. 10.14) is often used but can degrade the transducer performance because of the unequal shunting of the strain-gage network. A better zero balancing technique (Fig. 10.15) uses an offset

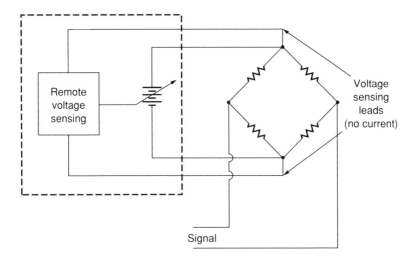

Fig. 10.13 Remote excitation sensing to eliminate voltage drop of lead wires.

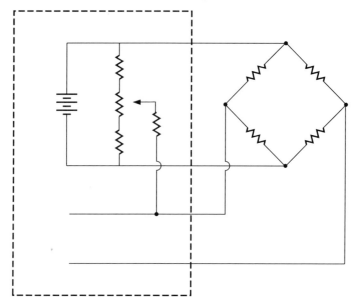

Fig. 10.14   "T" balance network (not recommended).

voltage injection technique with a high-impedance amplifier, which prevents any of the sensor's bridge arms from being shunted.

Many signal conditioners and indicators also contain provision for *shunt calibration* of the transducer (Fig. 10.16). An external resistor can be temporarily placed across one bridge arm, which produces a change in the bridge balance appearing as a simulated load on the transducer. The shunt resistor's effect is often supplied with commercial transducers, or it's easily determined by calibration and constitutes a rapid and simple method for subsequent calibration of the force-measurement system. This calibration method relies on the relationship between input load and strain-gage resistance changes. Therefore it is an indirect method of calibration and should be augmented with an occasional direct calibration method, such as placing a known load directly on the load cell.

Many attempts have been made to establish wiring standards for strain-gage bridge transducers. While there still exist some exceptions, the Western Regional Strain Gage Committee established the standard wiring codes most recognized (Fig. 10.17).

While the above systems constitute the majority of transducer signal conditioners, we should remember that the strain-gage transducer contains simple resistive elements that can be utilized in circuits other than the Wheatstone bridge. For example, DC-excited systems can utilize AC-coupled amplification, which allows a strain-gage-based transducer to sense only the change in force level. Half-bridge sensors are ideally suited for this type of signal conditioning. By using constant-current

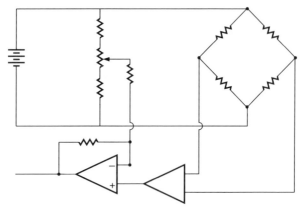

Fig. 10.15   Voltage injection balance network (recommended).

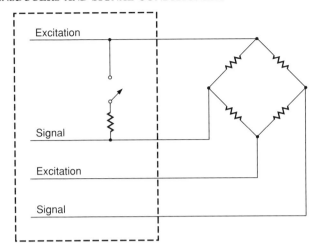

Fig. 10.16   Shunt calibration system.

excitation and AC-coupled amplification, a single strain gage can measure dynamic force (Figs. 10.18–10.20).

10.2.2   Piezoelectric Force Transducer and Its Signal Conditioner

Self-generating piezoelectric stress-gage force transducers are stiff, elastic structures that convert deflection caused by force into electrical signals more convenient for recording or processing. The reader is referred to Section 9.6.2, which presents the basics of piezoelectricity.

Piezoelectric force transducers, dynamic instruments, respond to repetitive and transient forces. They measure a force relative to an initial, average, or known level in the process and signal. They cannot measure, however, static forces or absolute levels for more than a few seconds, except under carefully controlled, constant-temperature, laboratory conditions. Tare weights and other static signal components are automatically eliminated, thus tiny fluctuations in force at any level may be measured: piezoelectric force sensors operate over a dynamic range greater than 10,000 to 1.

Piezoelectric force transducers, because of exceptional ruggedness, rigidity, and stability, excel in punishing applications, such as punch and tablet presses, machine-tools, tool dynamometers, recoil mechanisms, hammers, bolts, studs, impactors, printers, vibrators, targets, switches, actuators, exciters, test rigs, calibrators, and phonograph pickups. Like resistive strain-gage installations, piezoelectric stress-gage force sensors are often installed integrally in machine parts or structures, and piezoelectric force sensing elements are an integral part of many accelerometer and pressure transducer structures.

Piezoelectric force transducers come in a variety of sizes and shapes: cells, rings, links, and beams (Fig. 10.21). Selection usually involves choosing one that conveniently installs in the test object structure, with a range to accommodate sufficiently the forces involved.

Specialty type force sensors include multi-component force cells and driving-point sensors. Driving-point sensors, formerly called mechanical-impedance transducers, combine a force sensor and an accelerometer to measure both the force and motion at the point where an exciter is connected to the test object. Such instruments often explore nonnodal attachment locations that excite all resonances of interest.

Structures

Most piezoelectric force transducers employ quartz sensing elements because the low thermal and low frequency response of quartz permit simple static calibration. A typical force transducer structure (Fig. 10.22) can be viewed and modeled as a classical spring-mass seismic system. It basically consists of a stress-distributing, structural member (seismic end mass) preloaded against a crystal spring sandwiched between the seismic mass and a rigid base mass. The crystal elements perform a dual function: they act as a precision spring to oppose the applied force, and they generate an electrical charge signal proportional to their displacement. The elastic crystals, serving as a structural element, experience both stress and strain during the measuring transaction. The preload, exerted by a central stud or external sleeve, forces the imperfect interface surfaces into intimate, rigid contact and allows both tension and compression to be measured. The undercut in the seismic end mass tends to distribute point forces into an even stress pattern over the crystal elements, enhancing accuracy and reducing uncertainties.

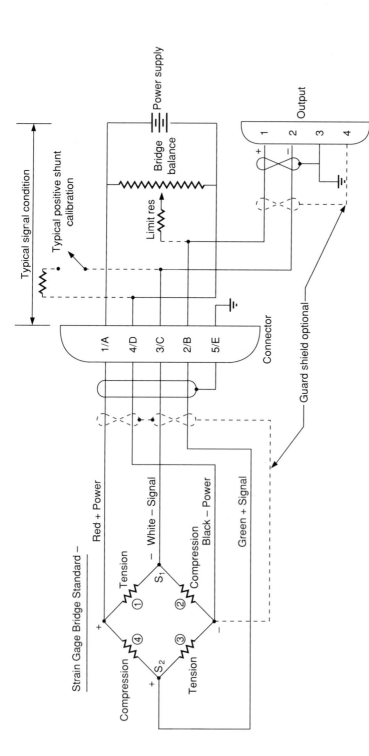

Fig. 10.17 Suggested industry standard—for strain-gage bridge designation and color code. (1) Wherever possible, tension, elongation, increased pressure, acceleration, or other generally accepted positive quantitites shall produce positive output signals as indicated. (2) The bridge elements shall be arranged so that functions producing positive output will cause increasing resistance in arms 1 and/or 3 and decreasing resistance in arms 2 and/or 4 of the bridge. (3) The auxiliary wiring for 6 or 3 wire system will have same color but as a tracer on white wire or IMSA colors. (4) Quarter bridge—when only the one bridge element is active, use arm No. 1 (arms 2, 3, and 4 as dummy elements). (5) Half bridge—when a tension and compression component is to be measured, use arms No. 1 and No. 2 (arms 3 and 4 as dummy elements). (6) The direction of position of the function producing a positive output signal shall be indicated on transducers. Shunt calibration resistor shown will produce a positive output signal. The following markings are suggested: (+|) tension load cells, universal load cells, micrometers, etc; (+ →) accelerometers and flow meters; (|+|) torque transducers; (+) differential pressure cells at the part where the higher pressure causes positive output signals. (7) For shielded type bridge systems pins 5/E, 7/G, and 9/I shall be shield terminals for 4, 6, and 8 wire systems. (*Source:* Western Regional Strain Gage Committee, Regional Committee of Technical Committee on Strain Gages of the Society for Experimental Stress Analysis.)

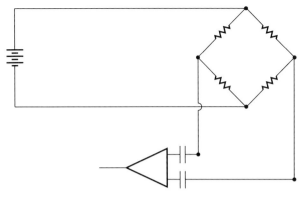

Fig. 10.18   AC-coupled amplifier system.

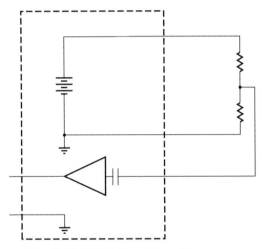

Fig. 10.19   Half-bridge AC amplifier system.

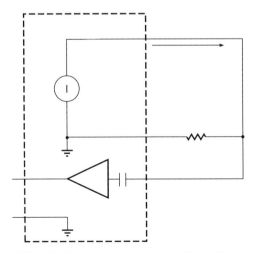

Fig. 10.20   Single gage/constant current/AC amplifier system.

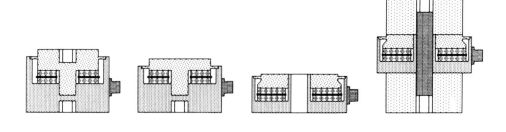

(a) Force cell   (b) Impact Sensor   (c) Force ring   (d) Force link

Fig. 10.21   Typical force transducer structures (PCB Piezotronics, Inc., Depew, NY).

Behavior

Force sensors introduce additional elasticity into instrumented machine parts and add their seismic end mass to objects being tested. Also, the test object adds mass to the force sensor, lowering the resonant frequency of the combined assembly. The interaction that results from these measuring transactions sometimes precludes valid measurements, especially at high frequencies. The solution to this dilemma is to select relatively stiff transducers with little seismic end mass or to test and compensate or correct for the mass loading and reduced elasticity effects.

Installation

Force transducers usually stud mount to the test object and support structure. A compliant elastic stud, similar to internal preload mechanisms, clamps the imperfect, stiff, elastic, interface surfaces together into intimate, rigid contact so that forces are transmitted primarily through the interface surfaces and not through the stud. Silicone grease or a similar lubricant on the surfaces increases the transmissibility at high frequencies. With force rings, the central preload element carries about ten percent of the load, necessitating in-place or assembled calibration. Force links can tolerate less precise mounting surfaces. High-precision measurements often require force application through a ball or spherical joint to ensure central loading. Off center, transverse, and moment loads cause errors. In structural behavior testing applications, a flexible link, a stinger, couples the vibrator to the force sensor mounted on the test object, thus reducing transverse force inputs and protecting the vibrator armature.

Calibration

Piezoelectric quartz force sensors calibrate by quick, conventional static methods. Quasi-static calibration involves quickly applying or removing a known force, with weights or through a proving ring, and measuring the corresponding transducer output signal. A simple free-fall calibrator for applying small forces is shown in Fig. 10.23.

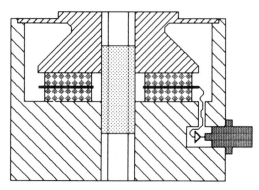

Fig. 10.22   ICP force transducer structure with built-in isolation amplifier and central preload stud (PCB Piezotronics, Inc., Depew, NY).

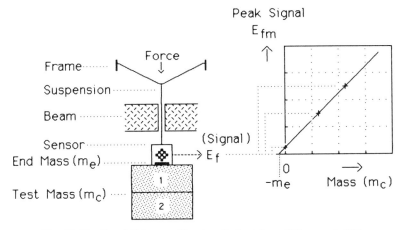

Fig. 10.23 Free-fall force calibration (Instruct Inc., Ridgewood, NJ).

Dynamic ratio calibration implementing Newton's Laws of Motion actuates a known mass. Measured indirectly, the force input is the known mass multiplied by the acceleration, which is measured with a calibrated accelerometer. Test hammers are similarly calibrated with a pendulously supported known mass. Because of the accelerometer's inertial nature, it doesn't sense the pendulum's free oscillations. For testing the behavior of structures, ratio testing with a known mass supplies the ratio of the accelerometer and force sensor (or hammer) sensitivities, which is the scaling factor as a frequency function for the signals processed in the analyzer.

Signal conditioning

Signal conditioners match the output signal of force sensors to the input of recorders or analyzers. To communicate effectively, signal conditioning of piezoelectric force sensors generally involves an isolation amplifier of the voltage or charge type located between the crystal sensing element and the recorder or analyzer. Otherwise, the charge signal from the crystals would quickly leak out through the input resistance of the readout instrument. Static calibration and press-monitoring applications often require direct-coupled amplifier or power units. Figure 10.24 shows a typical system that monitors and controls the forces forming in a tablet press. If a significant deviation from a normal pattern is detected, which indicates a faulty product, the faulty tablet is automatically rejected or the machine is shut down for repair. This system also incorporates a clamping circuit to rezero automatically the signal level between events, thus eliminating zero drift.

The sophisticated controller sets high and low limits for different events or parts of the cycle. With the potentially troublesome ultra-high-impedance circuits and components safely hermetically sealed inside their case, ICP (Integrated-Circuit-Piezoelectric) type sensors with built-in isolation amplifiers offer an almost unlimited, trouble-free life in this type of application.

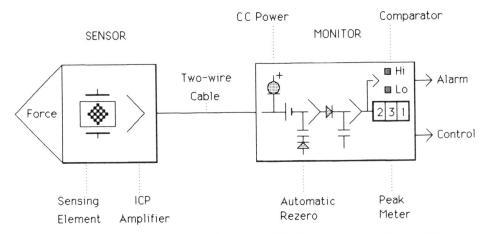

Fig. 10.24 ICP tablet press monitoring system (PCB Piezotronics, Inc., Depew, NY).

10.2.3 LVDT Load Cell

Since a load cell is essentially a position transducer connected across an elastic member, it is not surprising to find the LVDT as an integral part of a force-measurement system. Figure 10.25 shows examples of an LVDT combined with an elastic member. Proving rings and cascaded beam configurations (Fig. 10.25c,d) are the elastic beams commonly found in commercial LVDT load cells. Since the physical size of a proving ring LVDT load cell is somewhat larger than that of a comparable force range transducer using a different type of elastic member, the popularity of proving ring LVDT load cells is declining.[1]

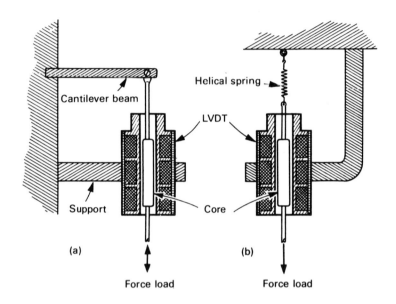

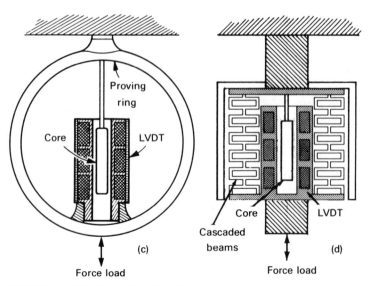

Fig. 10.25 LVDT Load Cells (Taken from *Handbook of Measurement and Control*, Herceg, E.E., Schaevitz Engineering, 1976, p. 10-2).

Fig. 10.26   Capacitive force transducer (reproduced with permission of Kavlico Corp., Chatsworth, CA).

10.2.4   Capacitive Force Transducer

Like other load cells, the force to be measured deflects an elastic element. Since the capacitive transducer can detect very small deflections accurately (see also Section 9.2.5), a highly sensitive force transducer can be built. The force to be measured (see Fig. 10.26) is directed onto a membrane whose elastic deflection is detected by a capacitance variation. An electronic circuit converts the capacitance variations into DC-voltage variations. The capacitive force transducer (Fig. 10.26) is built with full-scale operating ranges from 1 to 300 N; accuracy is better than 0.5 percent of range. Both the diaphragm and the substrate are made from ceramic material, with the electronic circuit mounted on the back of the substrate. The sensor is compact, with an overall diameter of less than 43 mm and length below 60 mm.

10.2.5   Other Types of Load Cells

As mentioned before in this chapter, there exist many methods to measure force. The principal mechanism of a force sensor's operation is to sense the deflection of a deformable structure.

Vibrating wire load cells utilize the change in natural frequency of a taut wire as the wire is shortened or lengthened. Magnetic reluctance sensors, by using the change in magnetic field, sense the change in magnetic properties of a material subjected to stress. Gyroscopic force sensors are built with the rate of precession proportional to the applied load. A simple dial indicator with its internal or external spring structure produces a visual signal proportional to applied load. A valve can be actuated by a force, and the pressure drop in the pneumatic circuit, measured at a remote location. A mirror can be rotated, and optical lever techniques can amplify force-related signals. The simple carbon microphone is a force sensor in which the electrical resistance of the structure itself is the sensing system. Conductive closed-cell foams extend this technique in which entrapped gas reduces the spring constant.

One of the earliest strain gages was a simple rubber tube containing mercury, which changed its diameter and thereby changed the resistance of the mercury column. When translucent materials change their optical properties (this can be observed at a distance) it can form the basis for photoelastic stress analysis. Timing of transmitted sonic waves can measure accurately the length of a structure and has formed the basis of at least one bolt tension measurement system.

The electrical output sensor is one of the most popular sensors today because it is relatively simple to utilize and condition the electrical signal with amplifiers and other signal conditioners. An infinite variety of electrical readout devices, microprocessors, and electrically controlled actuators are available for this sensor. With all of the possibilities available, the reason for the measurement itself should indicate the appropriate sensor.

In any measurement system the deflection of the sensor (or the amount of activation energy required) directly affects the measurement itself and therefore low-deflection systems, such as piezoelectric, strain gage, or vibrating wire systems, have a considerable advantage over other possible systems. Since deflection of a force measurement system directly relates to the frequency response characteristics, dynamic force measurement requires sensors with low deflections.

10.3 TORQUE TRANSDUCERS

Albert E. Brendel (10.3.1, 10.3.2, 10.3.4)
Chester L. Nachtigal (10.3.3)

Torque transducers, like their load cell counterparts, consist of a mechanical structure that undergoes a torsional deflection when subjected to an applied torsional load, and a means to sense this deflection.

Since most torque transducers are used to determine either the input torque level to drive a machine or the output torque of a motor or engine, they are often used in rotating systems. Systems operating at constant speeds obey the laws of statics. It is possible to measure a rotating torque by noting that the reaction torque restraining the motor's housing (or an absorber's housing) is equal to the rotating torque. Therefore, a choice exists as to where the torque measurement system can be placed, whether in an in-line rotating component or a stationary location.

10.3.1 Rotating vs. Nonrotating Transducers

In a rotating torque transducer, the sensor is directly in series with the rotating shaft, rotates with the shaft, and is complicated because the electrical signals must be brought out to the nonrotating world through slip rings, rotating transformers, or telemetry systems.

Reaction sensors supporting the housing of devices, on the other hand, are not rotating and therefore do not have this restriction.

The choice between the two approaches is largely dictated by the ultimate function of the measurement, the relative magnitude of the torque to be sensed, and the inertia of the components of the system. In massive systems, a reaction torque sensor generally must not only sense the torque transmitted through the housing, but also support the weight of the housing itself. Structural constraints on the sensor often limit the sensitivity of these systems because the structure must support these tare loads and therefore, either an auxiliary support structure (such as a bearing system) must be provided or the sensor moved to the rotating shaft where these conditions are not present. Parasitic load paths such as electrical cables and water or hydraulic lines also can present problems in reaction torque systems since they also can produce deflection-related torque reactions.

In rotating systems, where either the torque is fluctuating or the rotational speed is not constant, the deflection characteristics and the supported torsional masses create additional acceleration torques ($T = I\alpha$), which are added to or subtracted from the reaction torque but not the transmitted torque.

Reaction torque systems therefore are largely limited to the average measurement of a steady-state torque. Torque measurement of actual transmitted torque in accelerating or massive systems is generally made with rotating sensors, using various schemes to transmit the signal to a stationary vantage point.

10.3.2 Strain-Gage Torque Transducers

A torque sensor is the rotary equivalent of a load cell (see Section 10.2.1). There are three principal structures in a torque sensor:

1. The cruciform beam system.
2. The radial beam system.
3. The torsion rod.

The choice of structure, like load cells, is dictated by size constraints and the magnitude of the torque.

Low torques (below 100 N·m[1000 (lbf)·in]) generally dictate the use of a cruciform or radial beam structure (see Figs. 10.27 and 10.28). In these cases, the torque is carried by *force couples* in which each beam measures the force of the torque acting at a distance from the torque measurement axis (see Fig. 10.29). These structures are force sensors, which are discussed in Section 10.2.1.

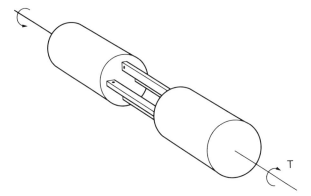

Fig. 10.27   Cruciform beam.

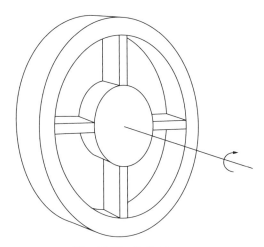

Fig. 10.28   Radial beam.

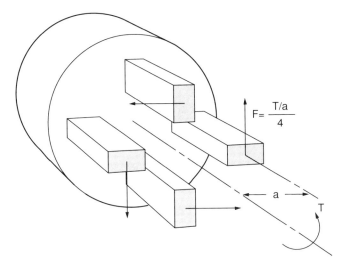

Fig. 10.29   Force paths in a cruciform structure.

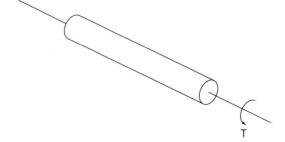

Fig. 10.30   The torsion rod.

For higher torque levels [> 100 N·m(1000 (lbf)·in)], an alternate structural element, the torsion rod, becomes feasible (see Fig. 10.30).

In the torsion rod, the shear strain at a section is

$$\epsilon = \frac{T}{G(J/P)} \tag{10.7}$$

where ϵ is the strain at 45° to the measurement axis (dimensionless); T is the torque; G is the shear modulus of elasticity; and (J/P) is the polar section modulus.

For a round section

$$(J/P) = \frac{\pi d^3}{16} \tag{10.8}$$

where d is the shaft diameter.

Similar to the discussion for shear beams in Section 10.2.1, the Wheatstone bridge and Poisson strain effects result in an effective strain multiplication of 1.3 with the same allowable stress reduction applying to this type of torsional transducer as in the case of load cells.

10.3.3   LVDT Torque Transducers

A second type of torque transducer is an LVDT that measures the deflection due to an applied torque. The torque shaft is somewhat more complicated than a strain-gage torque transducer. Figure 10.31 illustrates three different arrangements that incorporate a variable differential transformer. In Fig. 10.31a the torque sensing transducer is actually an LVDT load cell with its own elastic element. The measured force multiplied by the radial length of the arm that bears on the load cell is the input torque. Figure 10.31b shows a subtle variation of the previous case where the connecting arm between the input shaft and the LVDT is the elastic member. The LVDT response measures the deflection of the bending beam due to the applied torque. In the third method (Fig. 10.31c), an RVDT measures the relative angular deflection of two plates coupled by coil springs. The case or stator of the RVDT is connected to the input shaft in this figure while the rotor or shaft is connected to the output shaft.

All three of these applications don't indicate provisions for connecting the input excitation or the output signal. Thus if the torque transducer is used in a multi-revolution system, the LVDT connections must be made through a set of slip rings or by telemetering the output signal. DC excitation could be provided by a battery mounted on the rotating torque transducer.

At least one commercially available torque transducer[2] incorporates a variable differential transformer in such a way that no slip rings are required. The output voltage is produced in the stationary, housing-mounted secondary coils, while the excitation is applied to a stationary, housing-mounted primary coil. This transducer may be purchased with a magnetic pickup mounted on one end for speed measurement. The manufacturer offers a family of this Torsional Variable Differential Transformer (TVDT) type of torque transducer ranging in capacity from 22.6 N·cm to 338. 940 N·cm and ranging in stiffness from 790.85 N·cm/rad to 15.9×10^6 N·cm/rad.

10.3.4   Other Types of Torque Transducers

The operating principles of most torque sensors are related to force transducers. Sections 10.1, 10.2.1, and 10.2.4 are directly related to this section.

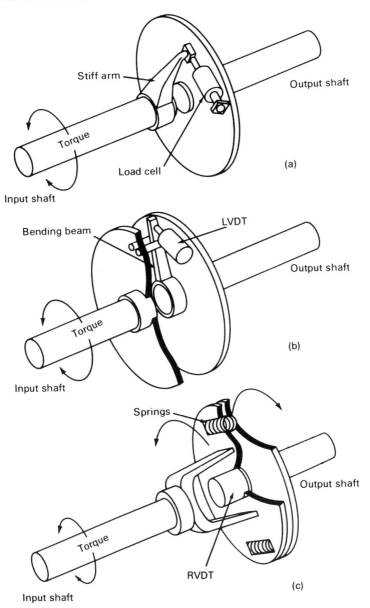

Fig. 10.31 Application of variable differential transformers in the measurement of torque (figure taken from *Handbook of Measurement and Control*, E.E. Herceg, Shaevitz Engineering, Pennsauken NJ, p. 10-4).

Once a torsionally deformable structure has been developed, the measurement of torque becomes a problem of merely measuring angular deflection.

Two toothed wheels installed on opposite ends of a shaft can be interrogated with a light source and a photoelectric receiver, using an optical shutter effect.

Magnetic position sensors generating individual pulsed signals from these same wheels also can be used. The phase of the two signals is proportional to the windup, which in turn is proportional to torque.

The magnetic reluctance of a rotating ferromagnetic shaft can be sensed with stationary magnetic coils. This reluctance forms the basis of many torque measurement systems used in large ship drive shafts.

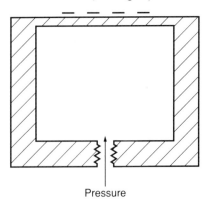

Pressure

Fig. 10.32 Diaphragm pressure transducer.

The electric current driving an electric motor is also proportional to output torque and, while less accurate than direct measurement techniques, can often satisfy measurement requirements.

Indirect torque measurements can be deduced by noting the acceleration upon an inertial mass or the precession of a gyroscope.

In a differential gear system, the reaction torque restraining the ring gear is proportional to the transmitted torque. It has the advantage of measuring rotary torque from a stationary reference point (so long as reduction ratios and/or changes in direction of the drive system are acceptable).

10.4   PRESSURE TRANSDUCERS

Albert E. Brendel (10.4.1)
Bernard H. Shapiro (10.4.2)
Robert W. Lally (10.4.3)
Georg F. Maur (10.4.4)
J. Barrie Sellers (10.4.5)
Chester L. Nachtigal (10.4.6)

10.4.1   Basics of Pressure Sensing

Pressure transducers measure the distortion produced by a pressure acting upon a deformable member. The two most common structures used are the diaphragm and the Bourdon tube. In the first diaphragm (see Figs. 10.32 and 10.33), strain gages may be applied to the surface of the diaphragm itself to

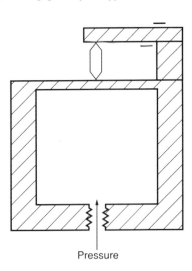

Pressure

Fig. 10.33 Diaphragm pressure transducer with secondary measurement structure.

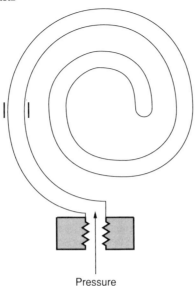

Pressure

Fig. 10.34 Directly gaged bourdon tube.

measure the diaphragm strains, or a secondary deflection or a force measurement structure, such as a cantilever beam, may be used to measure the deflection of the diaphragm, which now can be considered as a pressure to force converter.

The Bourdon tube systems can be similarly gaged directly or equipped with a secondary measurement structure (see Figs. 10.34 and 10.35). Size constraints and the magnitude of the pressure being measured dictates the structure choice and the use of auxiliary displacement or force sensing structures.

Low-pressure measurements generally favor the diaphragm (auxiliary) sensor system and use relatively large-area diaphragms to amplify the produced force. As pressure increases, the actual diaphragm strains become significant, and direct strain-gage applications become possible.

Bourdon tube systems are generally more complex but produce a high-deflection system that can often drive less sensitive displacement sensors, such as LVDTs or potentiometers. A modification of the Bourdon tube is the straight tube in which the hoop strain is sensed (see Fig. 10.36). Figure

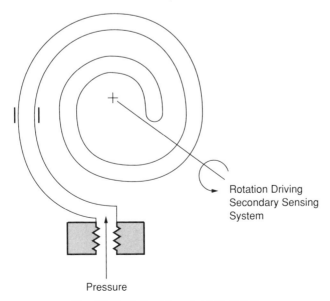

Rotation Driving
Secondary Sensing
System

Pressure

Fig. 10.35 Indirect bourdon tube sensor.

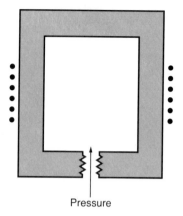

Fig. 10.36   Hoop strain pressure sensor.

10.37 shows a flow-through pressure transducer, which uses the same principle but can be inserted directly in the flow path of a fluid system. These latter sensors are generally applicable to high-pressure systems since the hoop strain is generated without the multiplying effect of the diaphragm or the Bourdon tube.

10.4.2   Strain-Gage Pressure Transducers

Introduction

The origin of the strain-gage pressure transducer is lost in history. However, we do have a few landmarks that provide some glimpses into the past. In 1908, Dr. S. Lindeck of Charlottenburg, Germany suggested a concept that might have been the first bonded strain-gage pressure transducer. He was working with precision resistors, which consisted of fine manganin wire wound on thin-wall brass tubes. Manganin is a resistance alloy whose composition is 4% nickel, 12% manganese and 84% copper. It has a specific resistance of 290 ohms per Circular Mil Foot (Ω/CMF) and a Gage Factor of $+0.47$. The Gage Factor is well below that of present strain gages, which cover the span from $+2$ for metal gages to $+200$ for semiconductors.

Dr. Lindeck coated the wire with shellac and wound it carefully around the tube. He then baked the assembly to bond firmly the wire to the tube in one continuous filament. The closed-end tube was pressurized internally to about 800 PSI in discrete pressure steps. The purpose of this test was to study the factors affecting stability of such resistors.

In doing this, he found that the coil changed resistance linearly with pressure. His report stated, "A method of measuring high pressure can possibly be based hereon." How right he was! In 1944, U.S. Patent Number 2,365,015 was issued to Mr. E. E. Simmons for a bonded strain-gage pressure transducer, which incorporated some concepts of Dr. Lindeck's wire-wound tube. Since that time the strain-gage pressure transducer has been changed considerably.

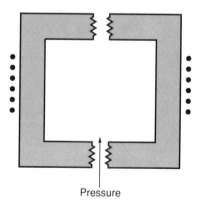

Fig. 10.37   Typical piezoelectric pressure transducer structures (PCB Piezotronics, Inc., Depew, NY).

Transducers

Strain-gage transducers consist of two basic elements, one mechanical and the other electrical. The mechanical, or force-summing element, converts the applied pressure into a deflection or displacement that is proportional to the pressure. The mechanical displacement is transmitted to the strain-sensing element, thus changing electrical resistance of the sensing element. The sensing elements can be any of the five strain-gage types. Regardless of type, each sensor experiences a change in resistance that is directly proportional to the applied pressure.

Just as there are five different strain-gage sensors, there are also a number of mechanical devices for converting the applied pressure to displacement. The most frequently used ones are diaphragms— flat, spherical, or convoluted. Other configurations are flattened tubes, thick and thin-wall circular tubes, bellows, and pressure capsules. The choice of force-summing element is based on pressure range, performance, size and weight requirements, ability to interface readily with the strain gage, and cost.

A strain-gage pressure transducer is a complete system, accepting an input pressure at one end and providing an electrical output at the other. This system includes the force-summing element, the strain-sensitive electrical element, compensation components, protective covers, and if required an amplifier. It is this combination of system components that have a direct bearing on overall transducer performance.

Strain Gages

Strain-gage pressure transducers are classified according to their electrical sensing elements. The sensing element can be any one of the five types. In most transducers four strain gages are wired into a Wheatstone Bridge, and each bridge arm has a single strain gage. Regardless of strain gage type, four strain gages are the preferred configuration, providing optimum electrical output and accuracy.

By applying an excitation voltage to the bridge, a voltage output is obtained. Depending on the material used for the strain gages, a 10 V excitation can produce an electrical output from 10 to 500 mV at rated pressure. An amplifier can be included in the transducer package to provide higher outputs of 5 V, 10 V, or 4–20 mA.

Strain gages used in pressure transducers can be classified as follows:

Metallic strain gages
>   Unbonded wire
>   Bonded wire
>   Bonded foil
>   Thin film

Semiconductor strain gages (also called "solid-state" or "piezoresistive")
>   Bonded Bar
>   Diffused

Unbonded Wire.   The unbonded strain-gage transducer employs a fine gage wire or wires as the strain-sensing element. The wires are stretched between insulated posts and are not supported or bonded between attachment points. The wire is approximately 0.0003 inches (0.0076 mm) in diameter and is a platinum-tungsten alloy. The displacement of the strain gage is generally 0.0005 to 0.001 inches (0.0127 to 0.0254 mm) at rated full-scale pressure.

Bonded Foil.   Bonded foil strain gages are applied over their entire length to the transducer's mechanical element, using special epoxy cements. The cement is cured at specific clamping pressures and temperature levels to assure a strong, stable system. The strain-gage foil, approximately 0.0002 inches (0.0051 mm) thick, is made of copper-nickel or nickel-chrome alloy and bonded to a thin substrate. The substrate, which serves as an insulator, is about 0.001 inches (0.0254 mm) thick and made of polyimide or epoxy film.

Bonded Wire.   The bonded wire gage is the forerunner of the bonded foil strain gage. Instead of foil, the strain-gage filament is made of wire, approximately 0.0005 inches to 0.001 inches (0.0127 to 0.0254 mm) in diameter. This type of gage has been replaced by the foil version in almost all transducers. However, some pressure transducers utilize a wire-wound strain tube as the sensing element. In this instance, the strain-gage wire is wound around a closed-end metal tube, cemented, and then cured in place.

Thin Film.   Thin film pressure transducers are made by vapor deposition or sputtering of both insulation and strain-gage filaments directly on the pressure-sensing element. First, an electrical insulation film, such as silicon monoxide, is deposited on the polished element. Next, the metal strain-gage elements are deposited. The strain gages are usually nickel-chrome or chrome-silicon alloys.

Bonded Bar.   The Bonded Bar semiconductor transducer is similar to the bonded foil transducer. The gages are applied to the pressure-sensing element and bonded in place with a special epoxy cement, which is temperature cured. It holds the gage securely in place and provides an insulation barrier. For most transducer applications, P-type semiconductors are used, providing an increase in gage resistance under tensile strains.

Diffused.   Diffused gage transducers are made from a pure silicon wafer, utilizing the same technology developed for transistors and integrated circuits. The strain-gage elements are diffused into a silicon substrate that in most cases is also the transducer pressure-sensing element. The gages are P-type silicon diffused into an N-type silicone substrate. Resistance values, gage factor, and gage performance are similar to the Bonded Bar semiconductor.

Reference Pressure

The conventional units for defining pressure are Pounds per Square Inch, PSI. However, in all pressure transducers, a reference pressure has to be established and the PSI be qualified to indicate the reference datum. The suffixes "G," "A," "S," and "D" are used to identify the zero pressure reference as follows:

PSIG—Gage pressure
PSIA—Absolute pressure
PSIS—Sealed pressure
PSID—Differential pressure

Gage Pressure.   The transducer measures pressure with respect to atmospheric pressure. To accomplish this, the transducer case is vented to the atmosphere. When the pressure port is exposed to atmospheric pressure, the transducer will indicate 0 PSIG. This occurs because the pressure on both sides of the sensing element is the same and there is no net output.

Absolute Pressure.   This transducer measures pressure with respect to a vacuum. Typically, the interior of the transducer case is evacuated and hermetically sealed. When the pressure port is exposed to atmospheric pressure, the transducer will indicate actual atmospheric pressure. This occurs because there is a vacuum on one side of the diaphragm and atmospheric pressure on the other. The net output represents the difference, which is atmospheric pressure.

Sealed Pressure.   A sealed pressure transducer measures pressure with respect to a fixed atmospheric pressure sealed within the case. This design is used to prevent ambient media from entering the inside of the transducer. When the pressure port is exposed to the atmosphere, the transducer will indicate approximately 0 PSIS. This occurs because there is a fixed atmospheric pressure on one side of the diaphragm and the ambient pressure on the other. If they happen to be the same, the net output is 0 PSIS. If not, then the net output will be a reading other than 0 PSIS.

Differential Pressure.   A differential pressure transducer has two pressure ports rather than one. This design permits the application of pressure to both sides of the strain-gage-sensing element. By doing this, small pressure differences at high line pressures can be measured with accuracies superior to those obtained by measuring each pressure independently. Differential transducers are available in either bidirectional or unidirectional pressure-sensing capability. A bidirectional, or plus/minus, unit monitors pressure changes in which the higher of the two pressures is unknown or changing. In the unidirectional transducer, the higher of the two pressures must always be assigned to the same pressure port and maintained accordingly.

Electrical Output

Strain gage pressure transducers are available in a variety of low-level and high-level electrical outputs. This selection is necessary to accommodate the overall requirements of the measuring system. Cable runs between components, local noise, quality of signal required, and environmental factors are considerations in selecting the appropriate signal level.

Millivolt (mV) signal levels are the basic output obtained from the strain-gage-sensing element. This is the electrical output from the strain-gage Wheatstone Bridge when rated pressure is applied to the transducer. Voltage level and 4–20 mA outputs depend on integral amplifiers to raise the millivolt output to higher levels. Selection of a transducer with an integral amplifier is generally made because of severe operating noise, limited installation space for external amplifiers, and compatibility with system components, such as recorders, controllers, and data gathering systems.

Knowledge of electrical noise at the transducer location is an important consideration in looking at the electrical output. Low-level signals from a millivolt unit, operating at the low end of its pressure range, may be masked by electrical interference.

Basic signal levels depend on the type of sensing-element employed, and standardized outputs have evolved over the years. At an input voltage of 10 VDC, the standardized electrical output at rated pressure for each type of strain-gage-sensing element is as follows:

Unbonded Wire—40 mV (4 mV/V)

Bonded Wire—30 mV (3 mV/V)

Bonded Foil—30 mV (3 mV/V)

Thin Film—30 mV (3 mV/V)

Bonded Bar Semiconductor—100 mV (10 mV/V)

Diffused Semiconductor—100 mV (10 mV/V)

By using amplification and associated circuitry, all of the above are also available in standardized outputs of 5V, 10V, or 4–20 mA.

The data provided for the low-level outputs is provided in mV as well as mV/V based on 10 V excitation. Also, the low-level units are available in other mV levels spanning the spectrum from 10 to 500 mV.

Industry Standards

The Instrument Society of America (ISA) has prepared certain standards whose goals are to achieve uniformity in the instrumentation field. In the field of strain-gage pressure transducers, two such standards have been written and are used as guides by manufacturers and users of transducers.

ISA Standard S37.1—Electrical Transducer Nomenclature and Terminology.

ISA Standard S37.3—Specifications and Tests for Strain Gage Pressure Transducers.

Both documents are excellent tools in developing an understanding of strain-gage pressure transducers. In particular, S37.3 provides important information on transducer acceptance tests and calibration procedures. Because these standards are universal, the information applies to all strain-gage-sensing elements.

Calibration Techniques

Calibration and test procedures used by manufacturers of pressure transducers follow generally the ISA recommendations. The manufacturer calibrates and compensates each transducer to meet published specifications, including all parameters related to performance at room temperature as well as at upper- and lower-limits over the compensated temperature range.

Test procedures are performed under static conditions. All pressure inputs are steady state, stable, and done at uniform, controlled temperatures. Before any data is taken, all temperature and pressure inputs are stabilized. This is done to ensure reproducibility of the test data no matter where the calibration is performed, providing a reference point for the manufacturer and end-user.

Transducer Selection

To select the right pressure transducer requires knowledge of both the application and characteristics of transducers that make the measurements. Generally, all strain-gage pressure transducers share many common characteristics and are adequate for the majority of test measurements. However, there are some differences in performance, and actual specifications must be reviewed to ensure that a proper choice has been made.

In determining the appropriate transducer, a systems approach may reveal considerations beyond those directly associated with the transducer. Power supplies, amplifiers, signal conditioners, data acquisition equipment, and displays usually dictate transducer output levels and input excitation. Pressure fittings, electrical terminations, size, and weight will dictate the mechanical requirements, which may override other criteria. Total system cost can have a direct bearing on the funds allocated for the transducer, which may govern final selection. An overall look at present and future requirements for the installation may also have an influence on the selection of the transducer type.

10.4.3 Piezoelectric Pressure Transducers

Self-generating, piezoelectric, stress-gage pressure transducers are stiff, elastic structures that transfer deflection caused by force into electrical signals more convenient for recording or processing. Pressure acting on a flush diaphragm creates the force. The piezoelectric-sensing elements generate charge-type signals, which are converted into voltages with charge-mode or voltage-mode, signal-conditioning amplifiers.

Piezoelectric pressure sensors respond to repetitive and transient pressures. They measure relative pressure (PSIR)—relative to an initial, average, or known level at some point in the process and signal. But they cannot measure static pressures or absolute levels for more than a few seconds, except under carefully controlled, constant-temperature laboratory conditions. Some higher-range models employing quartz or tourmaline crystals can be statically calibrated by quick conventional methods. Operating over a dynamic range greater than 10,000 to 1, a piezoelectric pressure transducer covers the equivalent of several ordinary ranges. By automatically eliminating static signal components, piezoelectric sensors measure tiny variations in pressure at any level. Sophisticated models employ various means of behavior modification, isolating, compensating, and filtering to reduce sensitivity to environmental inputs other than the measurand.

Employing a variety of both natural and man-made crystals, piezoelectric pressure and sound pressure (microphone) transducers are available in many shapes and sizes. Many standard and special mounting adaptors facilitate installation and isolate the sensing elements from severe environments.

Applications are legion. Piezoelectric pressure sensors help test the behavior and monitor the health of acoustic, hydraulic, pneumatic, and fluidic structures and associated processes. They are involved in testing, modifying, and controlling the behavior of machines, tools, cars, airplanes, spacecrafts, ships, rocket motors, engines, control systems, valves, mufflers, baffles, silencers, dampers, ammunition, guns, shells, bombs, explosives, targets, imploders, combustors, fuel injectors, actuators, containers, manifolds, sound absorbers, pumps, compressors, pipes, nozzles, flow meters, bearings, jet printers, and fluidic computers. In these things, piezoelectric transducers measure dynamic phenomena, such as pulses, pulsations, oscillations, variations, changes, waves, shocks, blasts, explosions, detonations, and sonic booms.

Structure

The internal structures of typical piezoelectric pressure transducers are shown in Fig. 10.38. These solid-state structures are quite rigid, imparting very high frequency response. Since the sensing elements directly experience most of the force transmitted, they behave essentially like stress-gage sensors. Sophisticated models also employ accelerometer elements internally to cancel out vibration sensitivity and embody tuned absorbers to reduce amplification at resonance. The latter feature, frequency-tailoring, results in nearly nonresonant behavior.

Most piezoelectric pressure sensors employ crystalline quartz-sensing elements because of their little temperature sensitivity and short-term static response. In Fig. 10.38, a thin-wall sleeve contains and preloads the quartz crystal elements. Hydrostatically-sensitive tourmaline uniquely qualifies it for valid underwater blast, shock wave, and calibration reference measurements. Versatile, man-made ceramic elements readily adapt to special shapes and reduce costs significantly in applications such as microphones. For shock and detonation wave applications, which excite structural resonances, nonpreloaded tourmaline sensors experience very little zero shift due to changes in residual stress patterns.

Installation

Generally flush and near flush diaphragm models install, directly in precision, threaded ports in the test object structure and are secured by integral threads or separate clamp nuts. Available mounting adaptors include straight thread, pipe thread, water or air-cooled, helium-bleed, and spark plug. Optional, conformal coating of a pressure probe with a hard plastic electrically isolates it from the local ground.

Behavior

Piezoelectric pressure sensors generally excel in ruggedness, stability, repeatability, linearity, and wide operating ranges. The high frequency response of piezoelectric pressure sensors is determined by the resonant behavior of the sensor's mechanical structure or by electronic filters, located internally or externally. The high frequency response of basic models is quite flat, within 20 percent of the first structural resonance. In some models, electronic filters extend this flat operating range, but introduce phase shift (time delay).

Electronic discharge and coupling circuits, operating in series, govern the low-frequency behavior of piezoelectric systems. These circuits attenuate very low-frequency signal components, which roll off the low-frequency response and cause transient signals to exponentially decay or to undershoot in response to step-function and pulse inputs.

Signal conditioning

Mainly for practical reasons, voltage-mode, quartz, ICP (Integrated-Circuit-Piezoelectric) pressure sensors with built-in, microelectronic isolation amplifiers are quite popular today. ICP sensors

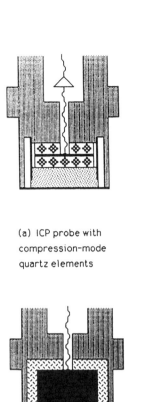

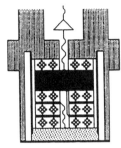

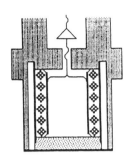

(a) ICP probe with
compression-mode
quartz elements

(b) Quartz probe with
acceleration-compensation
and frequency-tailoring

(c) Quartz probe with
transverse-thickness-
mode elements

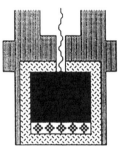

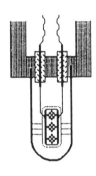

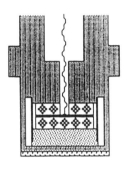

(d) Blast transducer
with plastic encased,
ceramic element

(e) Hydrostatic-mode,
freely-suspended
tourmaline element

(f) Ceramic-coated,
integral diaphragm
ballistics transducer

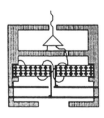

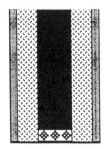

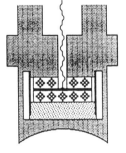

(g) Sound pressure sensor
with bimorph ceramic element
and accel. compensation

(h) Sub-microsecond
response tourmaline
pressure bar

(i) Conformal quartz
sensor for testing
unmodified ammunition

Fig. 10.38 High-sensitivity piezoelectric pressure transducer concept.

TABLE 10.1   CHARACTERISTICS OF SOME CAPACITIVE PRESSURE GAGES

	Kavlico	Setra
Full Scale Measuring Ranges	7×10^3 to 35×10^5 Pa (1 to 500 psi) absolute or differential pressure	10 to 6.9×10^7 Pa (0.02 to 10^4 psi) absolute or differential pressure
Linearity at 25°C	$\pm$ 0.3% of span typical	$< \pm$ 0.1% of span
Operating Temperature Range	-10 to 60°C	0 to 80°C
Pressure Overload	Up to 20 times rated pressure	Up to 2 times rated pressure

compromise and trade off static calibration capability for near trouble-free, drift-free, unattended, dynamic operation. But charge-mode, charge-amplifier systems are extensively used in high-temperature systems and in high-pressure ballistics applications, where in the past, static calibration was the only feasible method.

Calibration

For highest accuracy measurements, piezoelectric pressure transducers ought to be calibrated over the specific range or partial range being used.

Charge-mode quartz and tourmaline models calibrate by quick, conventional static means. Ceramic and ICP types require dynamic calibration by step-function pulse, comparison, or piston-phone methods. Comparison methods generally employ precision, statically calibrated quartz, or tourmaline sensors as reference standards. Pulse generators apply a known step change in pressure, with about a 1 ms rise time. Shock tubes generate submicrosecond pressure steps of calculated amplitude. With precision instruments, shock and pulse results agree within a couple percent with static calibrations. Perfected, pulse-comparison fixtures employing free, tourmaline reference sensors now offer an alternative for high-pressure calibrations.

10.4.4   Capacitive Pressure Transducers

Capacitive pressure transducers are diaphragm-type gages used mostly for low to medium high-pressure measurement.[3,4] Since capacitive sensors can detect very small diaphragm displacements, the diaphragm can be made comparatively stiff and rugged; some sensors withstand overloads up to 20 times their rated pressure without damage (see Table 10.1). The membrane area required is small, diameters from 4 to 30 mm are commonly used, allowing the construction of compact sensors. In a typical transducer an electronic circuit converts the capacitance variations to a DC voltage output signal. By carefully matching the capacitor and membrane characteristics, both of which are nonlinear, with the electronic circuit design, the output voltage of most capacitive pressure transducers is a linear function of the pressure signal.

High sensitivity can be achieved by using a differential capacitor in a bridge arrangement (Fig. 10.39). The pressure sensor is excited by an AC carrier. The demodulated signal is proportional to the pressure difference between the two pressure ports. This concept is applied in the Barocel system (Dresser Industries, Stratford, Connecticut), which is offered with measurement ranges from 1.3×10^{-3} to 7×10^5 Pa (10^{-6} to 100 psi) with an accuracy of 0.05 percent. Special versions of the transducer operate at temperatures up to 450°C. Other capacitive pressure transducers use a flat ceramic diaphragm mounted over a ceramic substrate (Fig. 10.40). The electronic circuit is mounted onto the back of the substrate. Several versions are available to perform gage, absolute, or differential pressure measurements. Other sensors are designed with a metal diaphragm (Setra

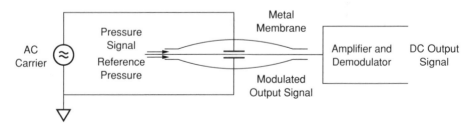

Fig. 10.39 High-sensitivity differential pressure transducer.

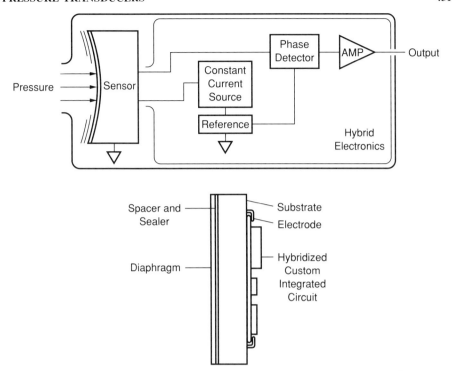

Fig. 10.40   Membrane type pressure transducer (Kavlico Corp., Chatsworth, CA).

Systems, Acton, Massachusetts). The electronic circuit, which is integrated into the transducer, gives a DC output proportional to the pressure signal.[5]

10.4.5   Vibrating Wire Pressure Transducer and Its Signal Conditioner

The vibrating wire technique for measuring strains and small movements has been around for about 100 years. The first recorded commercial use was in France where, in 1928, Andre Coyne patented a vibrating wire strain gage. This technique applied to the measurement of water pressures began in Germany and France in the mid-1950s, where the requirement arose for a pressure transducer with remote readout capability and good long-term stability to monitor pressures inside dam embankments. It is reported[6] that of approximately 3000 vibrating wire transducers installed in the 1950s, 90% were still functioning accurately some 30 years later. Further improvements in miniaturization and sealing of the transducers were made in the 1980s by McRae and Sellers[7] utilizing a wire grip technique developed first by the Norwegian Geotechnical Institute and later modified by Hawkes.[8]

The vibrational frequency of a tensioned steel wire is governed by the following equation:

$$f = \frac{1}{2l}\sqrt{\frac{Tg}{w}} \tag{10.9}$$

where l is the length of the wire, w/g is the mass per unit length, and T is the tension in the wire.

Transducer

In the vibrating wire pressure transducer the pressure is sensed by means of a diaphragm. The wire is attached to the back side of this diaphragm and is held under tension (Fig. 10.41). Pressure acting on the diaphragm causes it to deflect, which in turn causes a change in the wire tension and hence a change in the vibrational frequency of the wire. The steel wire is vibrated by an electromagnetic coil, and this or a second coil is used to measure the vibrational frequency.

The range and sensitivity of the vibrating wire pressure transducer are controlled by the length of the vibrating wire and by the diameter and thickness of the diaphragm according to the following relationship:

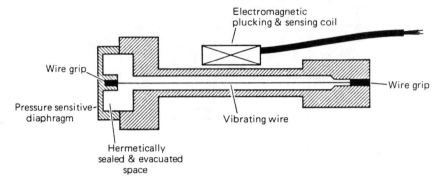

Fig. 10.41   Idealized cross-section of a vibrating wire pressure transducer.

$$P = \frac{kl^3t^3}{r^4}(f_0{}^2 - f^2)$$ (10.10)

where f_0 is the wire frequency when the pressure equals zero, f is the wire frequency at pressure p, l is the length of the wire, t is the thickness of the diaphragm, and r is the radius of the diaphragm. k is a constant for a wire of a given diameter and material composition. The normal range of frequencies is dependent on the wire length and wire diameter. For wire diameters in the range 0.15–0.25 mm the approximate operating frequency ranges are in the range $60.000 / l - 150,000 / l$, where l is the wire length in mm. For example, a wire of 50 mm length will have a normal operating frequency range of 1200–3000 Hz. For stability reasons, the wire tension should never be allowed to stress the wire beyond 20% of the yield stress. Nor should the wire be allowed to go too slack. Given these constraints, the pressure ranges of the transducer may be from as low as 0 to 30 KPa to as high as 0 to 50 MPa, and the resolution is typically .025%.

Temperature effects due to different expansion rates between the wire material and the transducer body material can be avoided either by making the two out of the same material or, in the case where the body is hermetically sealed and the electromagnetic coil is outside the body, by judicious choice of nonmagnetic materials that make up the various components of the housing assembly. Temperature effects caused by expansion and contraction of any gas trapped within the transducer can be avoided either by evacuating this space completely or by venting the space to atmosphere. If the latter is done, care should be taken to prevent the migration of moisture and other contaminants back along the vent line into the inside of the transducer. Internal venting is necessary when transducers are required to measure pressures referenced to ambient air pressure.

Long-term zero stability is greatly improved by heat treatment of the wire element before and after assembly. (A typical heat cycle holds the wire at 50% of the yield load for a span of 10 hours at 200°C.)

Signal Conditioner

Signal conditioning can be accomplished in two ways:

a. The *pluck and read* method in which an intermittent voltage pulse is applied to the plucking coil, causing the wire to be *plucked*, i.e., drawn to one side and then released, which allows the subsequent vibrations of the wire to decay naturally until the next voltage pulse occurs. When the coil is not plucking the wire, it is used as a sensing coil in conjunction with a permanent magnet located next to the wire. A sinusoidal voltage is induced in the sensing coil with a frequency equal to that of the wire's vibrational frequency. The frequency is usually measured by measuring the elapsed time of 100 cycles of the decaying vibration, using a digital quartz crystal oscillator. This method of frequency measurement greatly enhances the accuracy and resolution of the output signal. Various plucking techniques can be used, including single voltage pulses, 15 V–40 V, or a train of pulses with lower voltages, 1 V–5 V. The pulse train has a swept frequency band in the range of the characteristic transducer frequencies. This latter plucking method is used in explosive atmospheres (coal mines), where the readout instrument is permissible.

b. The *auto resonant* method in which the wire is vibrated continuously by one coil, using a feedback loop to synchronize the plucking frequency with the resonant frequency of the wire. A second coil then provides a continuous frequency output, which can be read by means of a frequency counter as before, or it is converted into a continuous voltage, or into a 4–20 mA output so that the output signal can actuate conventional control or recording devices. This second method can follow pressure fluctuations with frequencies of up to 30 Hz. Fluctuations more rapid than this cannot be followed accurately.

Since the relationship between pressure on the diaphragm and vibrational frequency is nonlinear, it is normal to square the measured frequency or period before either displaying it or converting it into a voltage or current. This squaring is done by microcomputer circuitry and linearizes the output signal with respect to the input pressure.

The main advantages of the vibrating wire transducer lie first in the zero stability inherent in a mechanical system (a vibrating wire) and secondly in its frequency output signal, which can be transmitted over long distances without loss of accuracy and whose essentially digital format is suitable for direct input into computer-driven control and data acquisition systems.

10.4.6 Other Types of Pressure Transducers

Pressure transducers are named after the manner in which the final conversion to a voltage is carried out; for example, a strain-gage pressure transducer uses strain gages to accomplish this task. A wide variety of such conversions are in use today. In each case, a primary sensor interfaces with the pressurized medium and deflects in a proportionate manner. In some transducers, the primary sensor is also the elastic member which not only senses the pressure but also displaces the conversion transducer, while in others the primary sensor couples with an elastic secondary sensor, which is attached to the conversion device.

The primary sensor is also called the pressure force-summing device since it sums the incoming pressure force, the elastic force transmitted to the conversion device, and its own elastic force. A number of different primary sensors are incorporated into modern pressure transducers, largely dictated by the force-displacement characteristics of the conversion transducer.

LVDT

The force-summing device for a *Linear Variable Differential Transformer* (LVDT) pressure transducer is frequently a diaphragm but may also be a bellows or a Bourdon tube (See Fig. 10.42). A push rod links the pressure-sensing element to the core of the LVDT.[9] The frequency response of this type of transducer is considerably less than other types. Typical LVDT pressure transducers have a natural frequency between 160 and 2000 Hz, depending on the selected pressure range. Because of

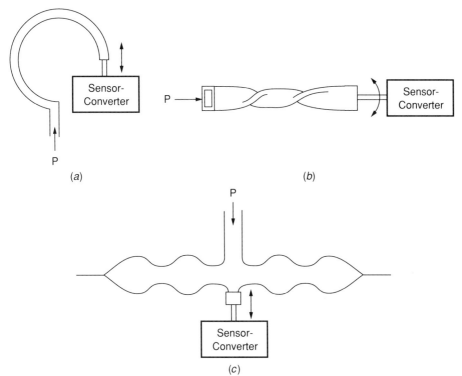

Fig. 10.42 Pressure force-summing devices: (*a*) "C" bourdon tube; (*b*) spiral bourdon tube; (*c*) diaphragm capsule.

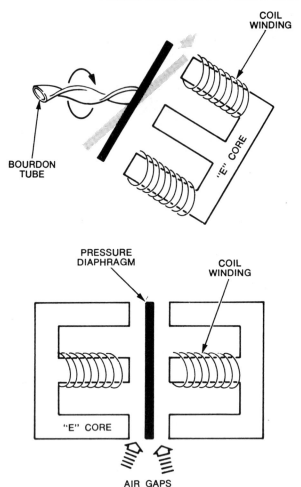

Fig. 10.43 Variable reluctance pressure transduction method (taken from *Pressure Transducer Handbook*, CEC Instruments, Pasadena CA).

their relatively massive primary-sensing and conversion elements (e.g., diaphragm or Bourdon tube and LVDT core), this type of pressure transducer is susceptible to errors in a vibratory environment. The acceleration forces can cause a significant response unrelated to pressure.

Inductive

The principle of inductive displacement transducers is discussed in Chapter 9. Another version of this sensing technique is to use an "E" core (see Fig. 10.43). In its adaptation to pressure measurement, two variations have evolved. In Fig. 10.43*a*, a spiral Bourdon tube actuates an armature relative to a single "E" core with two windings. In Fig. 10.43*b*, a pressure diaphragm is positioned between two "E" cores, each having a single winding. In both cases the position change of a ferromagnetic material alters the reluctance in an inductive loop, thereby causing a change in the coil voltage. This approach is readily capable of producing a voltage change of around 40 mV/V of ac excitation. Thus a 25 V carrier signal would produce an output level of 1 V without amplification.

10.5 EFFECTS OF FLUID TRANSMISSION LINES IN PRESSURE MEASUREMENT

David A. Hullender

Installing a pressure transducer at the end of a connection tube, often referred to as a fluid transmission line, may result in a significant distortion in pressure measurement. If the pressure is constant and if there is no flow through the line, then there will be no distortion. For a transient or oscillatory pressure, however, the line causes a pure time delay; a phase shift, which is a function of frequency; a frequency-dependent amplification or attenuation in the signal magnitude; and a dispersion in the signal shape with time. The significance of these events depends on the length and diameter of the line, the fluid properties, the frequency of the signal, and the internal fluid capacitance associated with the inlet to the transducer.

10.5.1   Transfer Function

For circular and noncircular[10] fluid lines considering laminar viscous friction and heat transfer effects, the distributed parameter model transfer function relating the transducer pressure, P_T, to the actual pressure, P_a, is given by[11,12]

$$\frac{P_T(\bar{s})}{P_a(\bar{s})} = \frac{1}{\cosh(\rho) + \overline{C}\overline{Z}\bar{s}\sinh(\rho)} \tag{10.11}$$

where $\bar{s}$ is the normalized Laplace variable defined by $\bar{s} = ls/c_0$, l is the length of the line, and c_0 is the speed of sound in the fluid. The normalized, line characteristic impedance, $\overline{Z}$, and the propagation operator, ρ, are functions of D_n and zero- and first-order Bessel functions of the first kind.[13] The dissipation number, D_n, represents the effective damping of the line and is defined by $D_n = 4\nu l/c_0 d^2$ where d is the line hydraulic diameter and ν is the kinematic viscosity. For values of $D_n < 10^{-4}$, the line is essentially lossless[12] with very large resonant characteristics, representing a pure time delay. In this case, $\overline{Z} \cong 1$ and $\rho \cong \bar{s}$.

10.5.2   Capacitance

The capacitance ratio, $\overline{C}$, is the fluid capacitance at the transducer's inlet divided by the capacitance of the transmission line. The capacitance of the line is the volume of the line, V_l, divided by the bulk modulus, β, of the fluid in the line. The capacitance at the transducer is due to the dead volume, V_T, in the transducer, which is due to the mechanical compliance of the transducer, $\Delta V/\Delta P$, and in the case of liquids, may also be due to entrained gas bubbles. In general, the equation for $\overline{C}$ is[11]

$$\overline{C} = \frac{V_T}{V_l} + \frac{\Delta V/\Delta P}{V_l/\beta} + \frac{V_0 P_0 \beta}{\gamma V_l P^2} \tag{10.12}$$

where V_0 is the volume of the gas bubbles at one atmosphere, P_0; P is the mean, operating absolute pressure: and γ is the specific heat ratio for the gas. The mechanical compliance term, $\Delta V/\Delta P$, may or may not be significant, depending on the deflection of the transducer-sensing element with pressure variations.

10.5.3   Frequency Response

The frequency response of the transfer function in Eq. 10.11 is shown in Fig. 10.44 for various line dissipation numbers, D_n, in the case of a zero capacitance ratio, $\overline{C}$. Lightly damped lines corresponding to relatively small values of D_n exhibit resonant peak conditions. As the value of D_n increases, the degree of magnitude attenuation and phase lag also increases.

The effects of the transducer capacitance are shown in Fig. 10.45. Depending on the value of $\overline{C}$, the reduction in the effective bandwidth of the line and transducer may be significant. Since the effective bandwidth, ω_b, represents the maximum pressure signal frequency that can be measured without significant magnitude distortion ($\pm$ 3 dB), it is beneficial to be able to estimate ω_b without having to generate the frequency response in Eq. 10.11. For a broad range of values of $\overline{C}$ and $D_n \leq 0.01$, $\omega_b \cong 0.5 \, \omega_r$ where ω_r is the first mode resonant frequency in Fig. 10.45.

For values of $\overline{C} \leq 1$, ω_r can be computed with less than 5% error using the following equations[14,15]

$$\omega_r(\text{rad/sec}) = \begin{cases} \dfrac{c_0}{l} \dfrac{a}{\sqrt{1 + 2\overline{C}}} & D_n \leq 0.001 & (10.13) \\[3ex] \dfrac{c_0}{l} \dfrac{b}{\sqrt{1 + 2\overline{C}}}(D_n)^{-d} & 0.001 < D_n \leq 0.01 & (10.14) \end{cases}$$

For liquids, a, b, and d are $\pi/2$, 1.33, and 0.0178 respectively; for air, the appropriate values are $\pi/2$, 1.18, and 0.0352. For $D_n \leq 0.001$ and $\overline{C} > 5$, the first mode resonant frequency is approximated by

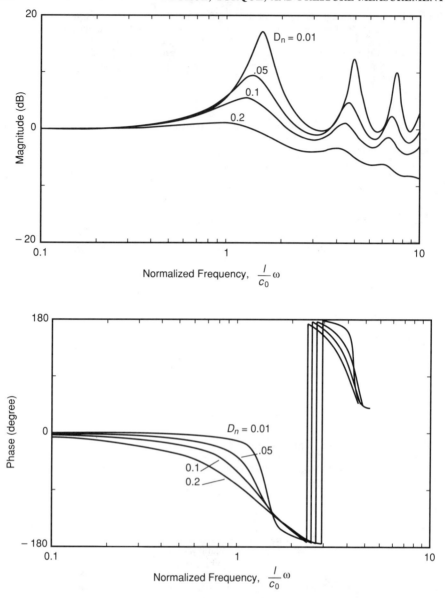

Fig. 10.44 Theoretical frequency response of a blocked pneumatic transmission line with $C = 0$.

$$\omega_r = \frac{c_0}{l} \frac{1}{\sqrt{C}} \qquad (10.15)$$

One additional consideration is the detrimental effects of source resistance, R_s, due to a valve or other restriction at the entrance to the transmission line.[2] To minimize the reduction in bandwidth, $R_s \ll 4\rho c_0/\pi d^2$ where ρ is the fluid density.

10.5.4   Example

To demonstrate the use of these equations, assume that a pressure transducer is connected to the end of a 6 mm (1/4 in) diameter hydraulic line with a length of 100 mm (4 in). For $\nu = 2 \times$

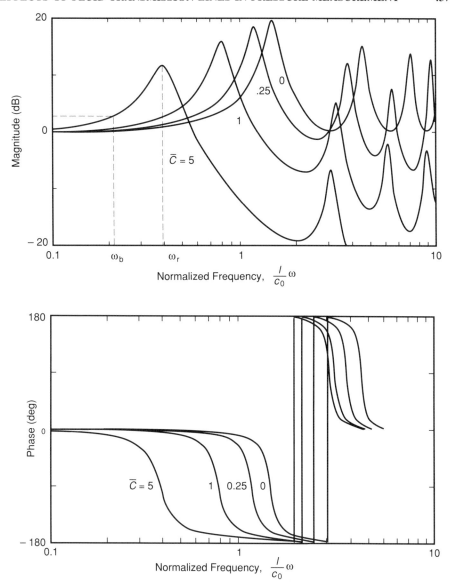

Fig. 10.45 Theoretical frequency response of a hydraulic transmission line with load capacitance and $D_n = 0.01$.

10^{-5} m^2/s(2.2×10^{-4} ft^2/s) and $c_0 = 1400$ m/s (4600 ft/s), $D_n = 1.59 \times 10^{-4}$. Using Eq. 10.13 with $\overline{C} = 0$, the first mode resonant frequency of the line is computed to be $\omega_r = 2.13 \times 10^4$ rad/s (3.4 kHz). The frequency, ω_b, at which the magnitude amplification has increased to $+3$ dB is approximately 1.07×10^4 rad/s (1.7 kHz). Thus, as long as the pressure signal being measured has frequencies less that 1.07×10^4 rad/s (1.7 kHz), there will be minimum distortion in the measurement.

With reference to the previous example, assume that an air bubble with 1 mm (0.04 in) diameter is trapped in the transducer's inlet. If the mean pressure is 2 atm and the fluid bulk modulus is 2 $\times 10^6$ kPa (290×10^3 psi), then from Eq. 10.12, considering only the effect of the bubble gives $\overline{C} = 0.65$. Using Eq. 10.13, $\omega_r = 1.45 \times 10^4$ rad/s (2.3 kHz) and $\omega_b \cong 7.25 \times 10^3$ rad/s (1.15 kz). Thus, the presence of the tiny air bubble has reduced the effective bandwidth of the transducer and line by a factor of 0.68.

In reference to the same example, assume that the compliance of the transducer is such that 10^{-6} m^3 of fluid intake is required for a pressure variation of 5000 kPa. Considering only the compliance term in Eq. 10.12 gives $\overline{C} = 141$. Using Eq. 10.15, $\omega_r = 1200$ rad/s (187 Hz) and $\omega_b = 600$ rad/s (94 Hz). Thus, the effect of the mechanical compliance of the transducer has reduced the effective bandwidth by a factor of 0.056.

10.6 CURRENT-LOOP SIGNAL TRANSMISSION

Anthony D. Wang

10.6.1 Background

Current-loop transmission had its first widespread use in the 1940s when it was necessary to drive various receiver mechanisms over long distances. DC current values as high as 200 mA were used to drive electrohydraulic devices. For indicators and recorders, 50 mA was adequate.

In the late 1950s, electronic process controls became commonplace and the 20 mA current loop standard was established for safety reasons: a dead short across the leads would not produce a spark. The decision to have a *live zero*, the 4 mA offset, was adopted so that an open in the line could be detected. There were other schemes as well. A popular alternative, the 10–50 mA range, was incorporated because it made the signals two and a half times easier to measure. Another old standard was 1–5 mA, which uses less power but also made resolution two and a half times harder to measure for a given load resistor.

By the late 1960s, the ISA (Instrument Society of America) assigned a committee to develop a "standard applicable to analogue dc signals used to transmit information between elements of systems for process control and monitoring." This committee was known as SP-50, and in 1975 it defined 4–20 mA as the standard. Most of the control industry today adheres to this standard.

10.6.2 Current-Loop Concept

Consider the situation illustrated in Fig. 10.46. The strain-gage bridge measurement is accomplished remotely from the control room. This distance could be anywhere from a hundred feet to several thousand feet away. The output of the bridge is amplified directly to provide a high-level output signal (0–10 V) for transmission. Unfortunately, the typical plant environment can be modeled as a high-impedance ac noise source that introduces noise levels on the order of several volts. This could be reduced by using shielded cable but at the expense of higher wiring costs. Even with shielded cable, there are other problems. The cable has finite resistance, which requires that the meter have a high-input impedance to keep the signal from being attenuated. But with higher meter impedance, more noise will be picked up!

In Fig. 10.47, the situation differs in that the amplifier's output drives a current source. The current loop (the path that the current traverses) eliminates losses due to the wire resistance because the voltage drops along the line do nothing to affect the current—its value stays constant. Noise pickup is all but eliminated by the high noise immunity of the current source. The meter can be terminated in a low impedance, which further resists noise pickup. This makes the shielded cable in Fig. 10.46 unnecessary. Relatively inexpensive, ordinary hookup wire can be used instead.

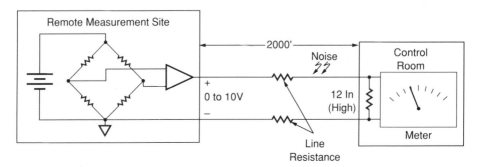

Fig. 10.46 Voltage signal transmission.

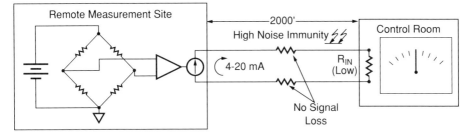

Fig. 10.47   Current signal transmission.

10.6.3   Advantages and Disadvantages

There are many reasons why the current-loop transmitter is still in widespread use today:

1. The standard exists.
2. Wire quality requirements are minimal. Concern for ohmic drops or shielding is not critical.
3. Many factories are already wired for current loop transmitters.
4. For remote measurements of one or two parameters, it still is unmatched for its low cost and ease of use.
5. The current required for the live zero can be used to power the remote transducer.
6. It is possible to connect the loads (receiving electronics) in a serial *daisy chain* in remote locations.
7. Electrically isolated transmitters can be interspersed all around the factory and yet be powered by one supply.

There are some disadvantages as well, especially considering the current state of the art:

1. It is slow. This usually has less to do with the transmitter than with the inductance and capacitance associated with long wire runs.
2. The receiving electronics needs to remove electrically the live zero from the actual signal.
3. Troubleshooting large loops or systems can be difficult.
4. The chance of wiring faults increases with the number of wires.
5. A large factory may require much more wire than necessary when using transmitters for several distributed smart satellite stations.

10.6.4   Types of Transmitters

See Fig. 10.48.

2-Wire: 2-wire transmitters draw power and provide signal over the same two wires.

3-Wire: 3-wire transmitters draw power from one wire and drive signal on the second, with a common return for both.

4-Wire: 4-wire transmitters draw power from two wires and provide signal over the other two wires.

Isolated: Isolated transmitters have electrical isolation (up to 1500 V) between the input, output, and/or supply leads.

10.6.5   Using Current Transmitters

Transmitters can be used to send information from transducers, or they can be used to drive actuators. Many systems use a resistor to convert the current to a voltage. A capacitor in parallel with the resistor filters the signal in order to minimize induced noise. However, the user should be careful where to place the cutoff frequency, since signal bandwidth also could be reduced.

Transmitter specifications cover the following:

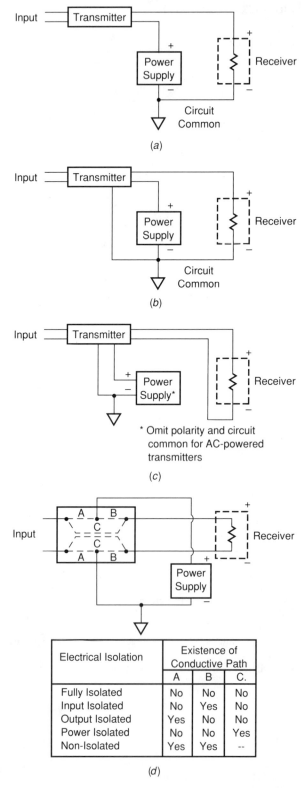

Fig. 10.48 Types of transmitters.

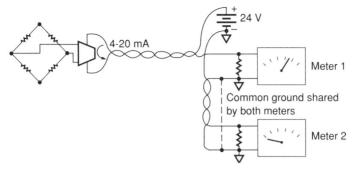

Fig. 10.49   Ground-loop error shorts meter 2.

1. Offset is the deviation of the live zero (4 mA, 10 mA) from ideal value. The offset can be expressed as units of current or as % of span.

2. Span is the change in output current for a full-scale change in input voltage. Span can be expressed as units of current or as % of span.

3. Nonlinearity is the deviation of the transfer function from a straight line drawn between the endpoints, usually expressed as % of span.

4. TC is the temperature coefficient expressed either as units of measurement versus temperature (e.g., mV/C, μA/C) or as unitless ratio versus temperature (e.g., %/C, ppm/C).

10.6.6   Electrical Considerations

The user should keep track of the grounds that are interspersed around the loop. If an earth ground is used at one receiver site to terminate the sense resistor, the second receiver in the series system should be isolated, or its sense resistor essentially will be shorted and no voltage will appear across it (Fig. 10.49). All the currents will have bypassed the receiver.

The user should not stack too many load resistors in series; this may run into a compliance limit on the transmitter. The compliance range is specified indirectly by the maximum load resistance versus supply voltage graphs (Fig. 10.50) that usually accompany the transmitters. The user should include the resistance of the interconnect wire in the calculations. One mile (1610 m) of 22 gage (AWG) twisted pair easily can introduce over 150 Ω in series with the load resistor.

10.6.7   Current Measurements

The most common way to measure a current signal is to place a resistor across the leads. This is then read by a digital voltmeter (DVM) or an analog-to-digital converter (ADC). The user should

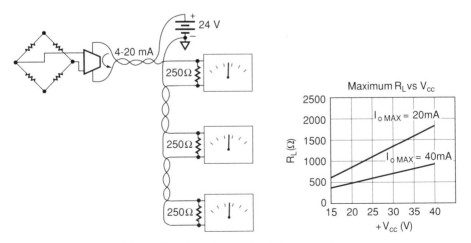

Fig. 10.50   Determining allowable line resistance.

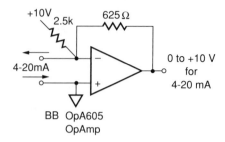

Fig. 10.51 Current-to-voltage converter.

carefully choose the resistor since it can contribute nonlinearity error by self-heating. This situation arises when the power dissipated by the resistor causes it to heat up and change value because of its own TCR (Temperature Coefficient of Resistance). This can be avoided by choosing a low TC resistor with an adequate power rating or, if unavailable, by paralleling higher-valued resistors to minimize the power dissipated by any one resistor.

Another way to read the current is by using an I-V (current-to-voltage) converter (Fig. 10.51). This circuit has the advantage of adding no voltage drop in series with the loop and of being able to subtract the live zero offset from the output. In general, DVMs in the ammeter mode should be avoided except for gross measurements. Most of these instruments are accurate to only 8 or 9 bits of resolution (0.4% or 0.2%, respectively).

One circuit that is custom designed for this purpose is the Burr-Brown RCV420, which can convert a 4–20 mA input to a 0–5 V output without any external components. This part presents a series 75 Ω resistance in the current path and can be daisy chained with very little degradation to the signal current (Fig. 10.52).

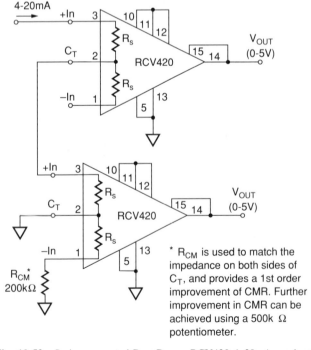

Fig. 10.52 Series-connected Burr-Brown RCV420 4–20mA receivers.

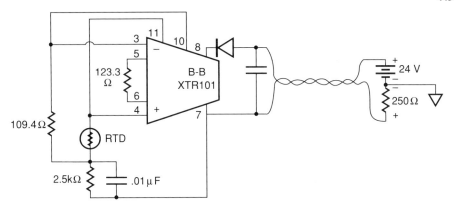

Fig. 10.53 RTD measurement with Burr-Brown XTR101.

10.6.8 Practical Example

The Burr-Brown XTR101 is a precision, low-drift, 4–20 mA, two-wire transmitter, integrated circuit. It has an instrumentation amplifier as a front end that can be programmed for various gains and has the availability of two matched, 1 mA current sources, which can be used to bias resistive transducers. Figure 10.53 shows this XTR101 as it is configured for RTD measurement. The platinum RTD measures 100 Ω at 0C and 200 Ω at +226C. The component values indicated allow the circuit to transmit 4 mA for +25C and 20 mA for +150C.

REFERENCES

1. E. E. Herceg, *Handbook of Measurement and Control*, Schaevitz Engineering, Pennsauken, NJ, 1976.

2. Bulletin No. 258B, Lebow Associates, Inc., Troy, Michigan.

3. W. H. Ko, B.X. Shao, C. D. Fung, W. J. Shen, and G. J. Yeh, "Capacitive Pressure Transducers with Integrated Circuits," *Sensors and Actuators*, Vol. 4, 1983, pp. 403–411.

4. G. F. Mauer, "A Transducer for the Measurement of Pulsatile Air Flow in High Vibration Environments," SAE Paper #840139, Society of Automobile Engineers, March 1984.

5. S. Y. Lee, "Variable Capacitance Signal Transduction and the Comparison with Other Transduction Schemes," Tutorial Proceedings of the Sixteenth International Aerospace Instrumentation Symposium, ISA Pittsburgh, Seattle, WA, May 1970.

6. James L. Sherard, "Piezometers in Earth Dam Impervious Section," Symposium on Instrumentation Reliability and Long Term Performance Monitoring of Embankment Dams, ASCE, New York, May 1981.

7. John B. McRae and J. Barrie Sellers, "Vibrating Wire Piezometer Manual," Geokon Inc., Lebanon, NH, 1982.

8. Ivor Hawkes and W. V. Bailey, "The Vibrating Wire Stressmeter Manual," Irad Gage Inc., Lebanon, NH, 1973.

9. Technical Bulletin 3001, Schaevitz Engineering, Pennsauken, NJ.

10. R. E. Goodson and R. G. Leonard, "A Survey of Modeling Techniques for Fluid Line Transients," *Journal of Basic Engineering*, ASME, June 1972.

11. C. Y. Hsue and D. A. Hullender, "Modal Approximations for the Fluid Dynamics of Hydraulic and Pneumatic Transmission Lines," *Fluid Transmission Line Dynamics 1983*, Special Publications, ASME WAM, Nov. 1983.

12. D. A. Hullender and A. J. Healey, "Rational Polynomial Approximations for Fluid Transmission Line Models," *Fluid Transmission Line Dynamics*, Special Publication for the ASME Winter Annual Meeting, Nov. 1981.

13. R. L. Woods, "A First-Order Square-Root Approximation for Fluid Transmission Lines," *Fluid Transmission Line Dynamics 1983*, Special Publication for the ASME Winter Annual Meeting, Nov. 1983.

14. R. L. Woods and C. H. Hsu, "Experimental Response of Noncircular Fluid Transmission Lines," Internal Report, Fluid Controls Center, The University of Texas at Arlington, Arlington, Jan. 1985.

15. R. L. Woods, C. H. Hsu, C. H. Chung, and D. R. Keyser, "Comparison of Theoretical and Experimental Fluid Line Responses with Source and Load Impedances," *Fluid Transmission Line Dynamics 1983*, Special Publication for the ASME Winter Annual Meeting, Nov. 1983.

BIBLIOGRAPHY

Benedict, R. P., *Principles of Temperature, Pressure, and Air Flow Measurement*, Omega Press, Stamford, 1985.

Bolton, W., *Engineering Instrumentation and Control*, Butterworths, London, 1980.

Burr-Brown Integrated Circuits Data Book, Vol. 33, Burr-Brown Corp., Tucson, 1989.

Considine, Douglas M., *Encyclopedia of Instrumentation and Control*, McGraw-Hill, New York, 1971.

Dally, J., W. Riley, and K. McConnell, *Instrumentation for Engineering Measurements*, Wiley, New York, 1984.

Doebelin, Ernest O., *Measurement Systems:* Application and Design, McGraw-Hill, New York, 1966.

Ewins, D. J., *Modal Testing: Theory and Practice*, Wiley, New York, 1984.

Foldvari, T. L., and K.S. Lion, "Capacitive Transducers," *Instruments and Control Systems*, Nov. 1964, pp. 77–85.

Graeme, Jerald G., *Designing with Operational Amplifiers,* McGraw-Hill, New York, 1977.

Graeme, Jerald G., Gene E. Tobey, and Lawrence P. Huelsman, Eds., *Operational Amplifiers: Design and Applications*, McGraw-Hill, New York, 1971.

Harvey, Glenn F., Ed., *Standards and Practices for Instrumentation*, Instrumentation Society of America, 4th ed., Pittsburgh, 1974.

Hordeski, Michael F., *Transducers for Automation*, Van Nostrand Rheinhold, New York, 1987.

Lion, K. S., "Capacitive Transducers," *Instruments and Control Systems*, June 1966, pp. 157–159.

Lion, K. S., "Nonlinear Twin-T Network for Capacitance Transducers," *Review of Scientific Instruments*, Vol. 15, No. 3, March 1964.

Lyons, Jerry L., *The Designer's Handbook of Pressure-Sensing Devices*, Van Nostrand Rheinhold, New York, 1980.

Mansfield, P.H., *Electrical Transducers for Industrial Measurement*, Butterworths, London, 1973.

Michelson, L., "Greater Precision for Noncontact Sensors," *Machine Design*, Dec. 6, 1979, pp. 117–121.

Norton, Harry N., *Handbook of Transducers for Electronic Measuring Systems*, Prentice-Hall, Englewood Cliffs, 1969.

Norton, Harry N., *Sensor and Analyzer Handbook*, Prentice-Hall, Englewood Cliffs, 1982.

Oliver, Frank J., *Practical Instrumentation Transducers*, Hayden Book Company, New York, 1971.

Seippel, Robert G., *Transducers, Sensors, and Detectors*, Reston Publishing Company, Inc., Reston, 1983.

Smith, John I., *Modern Operational Circuit Design*, Wiley, New York, 1971.

Woolvet, G. A., *Transducers in Digital Systems*, Peter Peregrinus Ltd., Stevenage, England, 1977.

CHAPTER 11
TEMPERATURE AND FLOW TRANSDUCERS

ROBERT J. MOFFAT

Department of Mechanical Engineering
Stanford University
Palo Alto, California

11.1 INTRODUCTION

There are hundreds of different transducers for temperature and flow measurements. The most common types will be discussed in this chapter; some others will be mentioned only briefly.

In flow measurements, only closed-channel flow measurement techniques are considered here, and only for "clean" fluids. Slurries and liquids carrying large objects are not treated.

In temperature measurements, current interests range from cryogenics (a few Kelvins) to plasmas (upwards of 10,000 K). Most applications, however, are in the range from room temperature to 2000 K, and that is where the bulk of this chapter will be concentrated.

The accuracy of a temperature or flow measurement depends not only on the sensor characteristics but on the interaction between the sensor and the system being instrumented. There are two primary classes of interactions: *system disturbance errors* (i.e., changes in the behavior of the system caused by the presence of the sensor) and *system/sensor interactions* (the sensor responding to more than one parameter of the system).

High-temperature measurements are subject to installation errors caused by heat transfer between the system and the transducer. The term *error* is defined as the difference between the observed value and the true value of the intended measurand. The output of a temperature sensor describes its own temperature, the *achieved temperature*, but the objective is usually to measure the temperature at a particular point in the solid, liquid, or gas into which the sensor is installed—the *available temperature*. There is often a significant difference between the available value and the achieved value because the sensor exchanges heat with its entire surroundings, not just with the immediate region around the sensor. It is not uncommon in high-temperature gas temperature measurements with unshielded sensors, for example, to have errors of several hundreds of degrees caused by radiation error, velocity error, or conduction error effects on the sensor. These errors cannot be accounted for by calibration of the sensor, nor can corrections be applied with any degree of certainty. The sensor must be protected by appropriate shielding. In most applications at high temperature, the installation errors are far larger than the calibration error of the sensor; hence, sensor accuracy does not mean the same as measurement accuracy.

Individual transducers are discussed in the following sections.

11.2   THERMOCOUPLES[*]

Thermocouples are the most commonly used electrical output transducers that measure temperature. They are inexpensive, small in size, and remarkably accurate when their peculiarities are understood.

11.2.1   Types and Ranges

Any pair of thermoelectrically dissimilar wires can be used as a thermocouple. The wires need only be joined together at one end (the measuring junction) and connected to a voltage-measuring instrument at the other end (the reference junction) to form a usable system. Whenever the measuring junction is at a different temperature than the reference junction, a voltage will be developed, which is related to the temperature difference between the two junctions. Several metallic materials are listed in Table 11.1 in order of thermoelectric polarity; each material in the listing is positive with respect to all beneath it. In an iron-palladium thermocouple, for example, the cold end of the iron wire will be positive with respect to the cold end of the palladium.

In some instances the operating temperatures of machinery elements have been measured using the machine structure as part of the thermoelectric circuit (cutting-tool tip temperatures, cam shaft/rocker arm contact temperatures, etc.). In such cases each material in the circuit must be calibrated, and all intermediate temperatures must be measured, in order to interpret the signal.

The alloys usually used for thermoelectric temperature measurement are listed in Table 11.2. These have been developed over the years for the linearity, stability, and reproducibility of their EMF vs. temperature characteristics, and for their high-temperature capability. Calibration data for thermocouples have been presented by the National Bureau of Standards, NBS Monograph 125, based on the International Practical Temperature Scale of 1968.

The noble metal and refractory metal thermocouples are used generally with extension wires of substitute materials, which are cheaper and easier to handle (more ductile). The extension wires used are described in Table 11.3. Except for the substitute alloys, thermocouple extension wire is of the same nominal composition as thermocouple wire, and differs from it mainly in the accuracy of its calibration and the type of insulation used. Extension wire is not calibrated as accurately as thermocouple-grade wire.

Thermocouple material can be purchased as individual bare wires, as flexible, insulated pairs of wires, or as mineral-insulated pairs swaged into stainless-steel tubes for high-temperature service. Prices range from a few cents to several dollars per foot, depending on the wire and the insulation. There are many suppliers; the San Francisco telephone directory, for example, lists 12 of them in the yellow pages!

11.2.2   Peripheral Equipment

Any instrument capable of reading low-DC voltages (on the order of millivolts) with 5–10 μV resolution will suffice for temperature measurements; the accuracy depends upon the voltmeter. The

TABLE 11.1   THERMOELECTRIC POLARITY ORDER OF METALLIC MATERIALS

100°C	500°C	900°C
antimony	Chromel	Chromel
Chromel	nichrome	nichrome
iron	copper	silver
nichrome	silver	gold
copper	gold	iron
silver	iron	$Pt_{90}Rh_{10}$
$Pt_{90}Rh_{10}$	$Pt_{90}Rh_{10}$	platinum
platinum	platinum	cobalt
palladium	cobalt	Alumel
cobalt	palladium	nickel
Alumel	Alumel	palladium
nickel	nickel	constantan
constantan	constantan	
copel	copel	
bismuth		

Source: Reference 4.

[*] Material in this section is substantially derived from Ref. 1 with permission, except where otherwise referenced.

TABLE 11.2 COMMON ALLOYS USED IN THERMOELECTRIC TEMPERATURE MEASUREMENT

Max. Temp.		Allowable Atmosphere (Hot)	Material Names	Type ANSI[2]	Color Code	Avg. Output mv/100°F	Accuracy[2]	
°F	°C						Std.	Spec.
5072	2800	Inert, H₂, Vac.	Tungsten/tungsten 26% rhenium	—	—	0.86	—	—
5000	2760	Inert, H₂, Vac.	Tungsten 5% rhenium/tungsten 26% rhenium	—	—	0.76	—	—
4000	2210	Inert, H₂, Vac.	Tungsten 3% rhenium/tungsten 25% rhenium	—	—	0.74	—	—
3270	1800	Oxidizing[1]	Platinum 30% rhodium/platinum 6% rhodium	B	—	0.43	1/2%	1/4%
2900	1600	Oxidizing[1]	Platinum 13% rhodium/platinum	R	—	0.64	1/4%	1/4%
2800	1540	Oxidizing[1]	Platinum 10% rhodium/platinum	S	—	0.57	1/4%	1/4%
2372	1300	Oxidizing[1]	Platinel II (5355)/platinel II (7674)[3]	—	—	2.20	±5/8%	—
2300	1260	Oxidizing	Chromel/Alumel[4], Tophel/Nial[5], Advance T1/T2[6], Therm. Kanthal P/N[7]	K	Yellow-red	2.20	4°F,3/4%	2°F,3/8%
1800	980	Reducing	Chromel[2]/constantan	E	Purple-red	4.20	1/2%	3/8%
1600	875	Reducing	Iron/constantan	J	White-red	3.00	4°F,3/4%	2°F,3/8%
750	400	Reducing	Copper/constantan	T	Blue-red	2.50	3/4%	3/8%

Source: Reference 3.

[1] Avoid contact with carbon, hydrogen, metallic vapors, silica, reducing atmosphere.
[2] Per ANSI C96.1 Standard
[3] © Englehard Corp.
[4] © Hoskins Mfg. Co.
[5] Wilber B. Driver Co.
[6] Driver-Harris Co.
[7] The Kanthal Corp.

TABLE 11.3   EXTENSION WIRES FOR THERMOCOUPLES

Thermocouple Material	Type	Extension Wire, Type[1]	Color[1]		
			(+)	(−)	Overall
Tungsten/tungsten 26% rhenium	—	Alloys 200/226[2]	—	—	—
Tungsten 5% rhenium/tungsten 26% rhenium	—	Alloys (405/426)[2]	White	Red	Red[2]
Tungsten 3% rhenium/tungsten 25% rhenium	—	Alloys (203/225)[2]	White/yellow	White/red	Yellow/red
Platinum/platinum rhodium	S,R	SX, SR	Black	Red	Green
Platinel II-5355/platinel II-7674	—	P2X[4]	Yellow	Red	Black[4]
Chromel/alumel, tophel/nial, Advance, Thermokanthal[3]	E	KK	Yellow	Red	Yellow
Chromel/constantan	K	EX	Purple	Red	Purple
Iron/constantan	J	JX	White	Red	Black
Copper/constantan	T	TX	Blue	Red	Blue

Source: Reference 3

[1] ANSI, except where noted otherwise.
[2] Designations affixed by Hoskins Mfg. Co.
[3] Registered trademark names, see Table 11.2 for identification of ownership.
[4] Englehardt Mfg. Co.

signal from a thermocouple depends upon the difference in temperature between the two ends of the loop; hence, the accuracy of the temperature measurement depends upon the accuracy with which the *reference junction* temperature is known as well as the accuracy with which the electrical signal is measured.

Galvanometric measuring instruments can be used, but since they draw current, the voltage available at the terminals of the instrument depends not only on the voltage output of the thermocouple loop but also on the resistances of the instrument and the loop. Such instruments are normally marked to indicate the external resistance for which they have been calibrated. Potentiometric instruments, either manually or automatically balanced, draw no current when in balance and therefore can be used with thermocouple loops of any resistance without error. High-input impedance voltmeters draw only minute currents and, except for very high-resistance circuits, pose no problems. When in doubt, check the input impedance of the instrument against the circuit resistance.

The input stages of many instruments have one side grounded. Ground loops can result from using a grounded-junction thermocouple with such an instrument. If the ground potential where the thermocouple is attached is different from the potential where the instrument is grounded, then a current may flow through the thermocouple wire. The voltage-drop in the wire due to the ground-loop current will mix with the thermoelectric signal and may cause an error.

11.2.3 Thermoelectric Theory

The EMF-temperature calibrations of the more common materials are shown qualitatively in Fig. 11.1, derived from NBS Monograph 125 and other sources. The EMF is the electromotive force, mv, that would be derived from thermocouples made of material X used with platinum when the cold end is at 0°C and the hot end is at T. Those elements commonly used as first names for thermocouple pairs [i.e., Chromel (-Alumel), iron (- constantan), copper (-constantan), etc.] have positive slopes in Fig. 11.1.

It can be shown from either the free-electron theory of metals or thermodynamic arguments alone that the output of a thermocouple can be rigorously described as the sum of the contributions from each length of material in the circuit—the junctions are merely electrical connections between the wires. Formally,

$$E_{\text{net}} = \int_0^L \epsilon_1 \frac{dT}{dx} \, dx + \int_L^0 \epsilon_2 \frac{dT}{dx} \, dx \tag{11.1}$$

where ϵ = the total thermoelectric power of the material; equal to the sum of the
           Thomson coefficient and the temperature derivative of the Peltier coefficient
       T = temperature

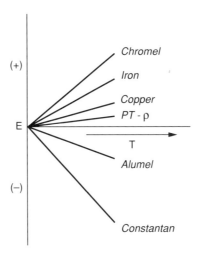

Fig. 11.1 EMF vs. temperature calibrations for several materials. (Reproduced from Ref. 1, with permission.)

x = distance along the wire
L = length of the wire

When the wire is uniform in composition, so that ϵ is not a function of position, then

$$E_{net} = \int_{T_0}^{T_L} \epsilon_1 \, d\theta + \int_{T_L}^{T_0} \epsilon_2 \, d\theta \tag{11.2}$$

Commercial wire is homogeneous within close limits, but used thermocouples may be far from uniform.

When only two wires are used in a circuit, it is customary to further simplify the problem. If both wires begin at one temperature (say T_0) and end at another (say T_L), the two integrals above can be collected:

$$E_{net} = \int_{T_0}^{T_L} (\epsilon_1 - \epsilon_2) \, d\theta \tag{11.3}$$

Three simplifications are built into this reduced equation:

1. ϵ is not a function of position (i.e., the wires are homogeneous).
2. There are only two wires.
3. Each wire begins at T_0 and ends at T_L.

These are the conditions for which the EMF-temperature tables are intended. If any of the three conditions is not met, the tables cannot be used to interpret the output.

11.2.4   Graphical Analysis of Circuits

It is possible to graphically analyze a thermocouple circuit and describe its output in terms of the calibration of its wires. The simplest practical circuit consists of two wires joined together at one end and connected directly to a measuring instrument, as shown in Fig. 11.2. This is referred to as

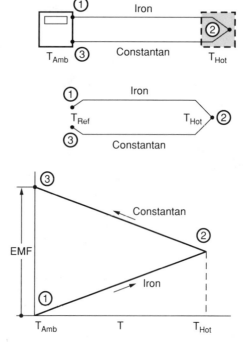

Fig. 11.2   Temperature measurement u͟s͟i͟n͟g the ambient temperature as the reference. (Reproduced from Ref. 1, with permission.)

the *pattern circuit*. The pattern circuit is thermoelectrically ideal (providing that the materials of the instrument do not affect the reading) since it contains no switches, connectors, or lead wires. The output of this system can be analyzed graphically using the calibration data shown in Fig. 11.1. The method used here is described in detail by Moffat[2].

The circuit in Fig. 11.2 requires frequent measurement of the temperature at the instrument terminals and is seldom used where accuracy of better than $\pm$ 2°F is required (chiefly because it is difficult to measure the temperatures of points 1 and 3 more accurately than $\pm$ 1°F).

A reference zone of controlled temperature eliminates the need for frequent measurements of the ambient temperature. The reference zone may be an ice point, a triple point, or an electrically controlled, high-temperature reference zone box. A circuit as in Fig. 11.3 assumes the reference temperature to be an ice-point bath.

The output of this circuit is the EMF between points 1 and 5 (those connected to the instrument terminals). The ideal circuit would have had the output given by EMF(2–4). The graphical construction shows that EMF(2–4) is equal to EMF(1–5) since the segments 1–2 and 4–5 each represent the same material (copper) over the same temperature interval $(T_{amb} - T_{ref})$, and the wires are connected so as to cancel these EMFs. Thus, the actual circuit is thermoelectrically equivalent to the ideal circuit. Note that the copper lead wires (1–2 and 4–5) play no role in determining the output of the circuit, provided that (1) the calibrations of the two pieces of copper wire are the same, and (2) the temperature intervals across the two copper wires are the same.

11.2.5 Zone-Box Circuits

When several thermocouples are used in a single test far from the measuring station, significant economies can sometimes be achieved by using a common zone box and substituting copper lead wires for much of the thermoelectric material. Such a circuit is shown in Fig. 11.4. The objective of the analysis is to determine the conditions under which the actual circuit is thermoelectrically equivalent to the pattern circuit.

The function of the zone box is to provide a region of uniform temperature within which connections can be made. The temperature of the zone box need not be constant, and it need not be known—it need only be uniform.

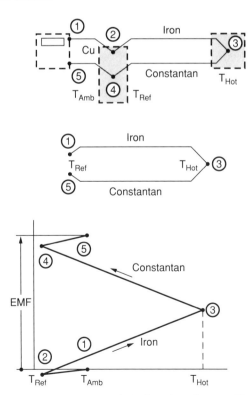

Fig. 11.3 Temperature measurement using an icebath as the reference. (Reproduced from Ref. 1, with permission.)

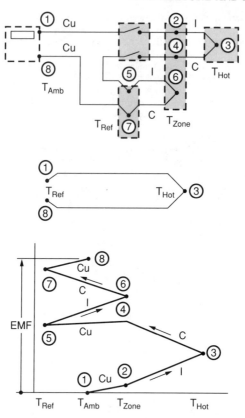

Fig. 11.4 Multiple measurements using a zone box and selector switch. (Reproduced from Ref. 1, with permission.)

The circuit consists of a reference bath, a selector switch, a read-out instrument, and a set of thermocouples extending from the zone box to their individual sensing points, connected together with copper wires. Providing the selector switch and the lead wires introduce no spurious EMFs, the behavior of any one thermocouple for this circuit should be the same as that of the pattern circuit shown. In the E-T diagram, the copper lead wires are shown passing through the selector switch with no acknowledgment of its existence—the switch is assumed to have a uniform temperature at the ambient temperature.

Commercially available zone-box and selector switch assemblies are sometimes made with the assumption that the junction 6 will be placed in the reference bath, and points 5 and 7 in the zone box. This requires reversing the polarity of the reference junction (i.e., the wire between points 5 and 6 must then be constantan in this example, and the wire between 6 and 7 must then be iron).

If a wiring diagram is not available, a test for reference junction polarity should be made. With the system connected, at any arbitrary temperature, the instrument reading should go up if the temperature of the reference junction goes down, and conversely.

11.2.6 The Laws of Thermoelectricity

Various authors have attempted to summarize the behavior of thermocouples through sets of laws ranging from three to six in number. One of the more detailed sets is given by Doebelin.[5] Each law can easily be proven by recourse to an EMF–temperature sketch. The first three from Doebelin's list are used as examples:

1. The thermal EMF of a thermocouple with junctions at T_{hot} and T_{ref} is totally unaffected by temperature elsewhere in the circuit if the two metals used are each homogeneous.

In Fig. 11.5, it is presumed that $T_{hot} < T_{candle}$. If the wire is uniform in calibration on both sides of the hot spot, the potential hysteresis loop closes and no net EMF is generated because of

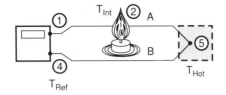

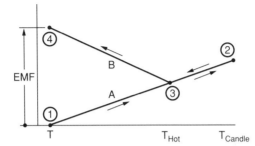

Fig. 11.5 Illustration of the law of interior temperatures. (Reproduced from Ref. 1, with permission.)

the hot spot. The principal importance of this law is that it establishes the conditions under which a thermocouple is a point sensor. If and only if wires A and B are uniform in composition, the thermocouple output is determined only by the temperatures T_{ref} and T_{hot} and is independent of the temperature distribution along the wire.

2. If a third homogeneous metal C is inserted into either A or B, as long as the two new thermojunctions are at like temperature, the net EMF of the circuit is unchanged irrespective of the temperature of C away from the junctions.

Figure 11.6 shows a third material inserted into the A leg and then heated locally. It is presumed that the temperatures at the two ends of C remain equal. If the material C is homogeneous, the EMF induced by the excursion in temperature from point 2 to point 3 is canceled by that from point 3 to point 4, and no net signal is produced.

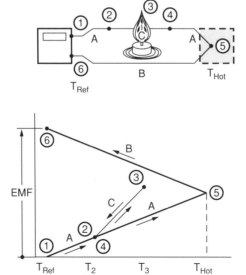

Fig. 11.6 Illustration of the law of inserted materials. (Reproduced from Ref. 1, with permission.)

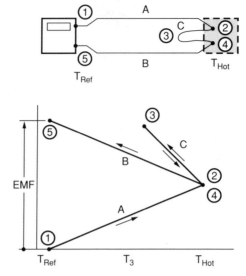

Fig. 11.7 Illustration of the law of intermediate materials. (Reproduced from Ref. 1, with permission.)

3. If metal C is inserted between A and B at one of the junctions, the temperature of C at any point away from the AC and BC junctions is immaterial. So long as the junctions AC and BC are both at the temperature T_1, the net EMF is the same as if C were not there.

Figure 11.7 illustrates this case. An intermediate material, C, is inserted between A and B at the measuring junction. The diagram once more shows no change in the net EMF if the inserted material is homogeneous and does not undergo a net temperature change.

The situation involving intermediate materials at the junction is of great practical importance because it addresses questions of manufacturing technique and how they affect thermocouple calibration. For example, the third material (C), used to connected the two materials A and B, might be the soft solder, silver solder, or braze material. The output of the thermocouple is independent of that third material, provided that the third material is homogeneous and begins and ends at the same temperature. In practice, these conditions are usually satisfied because the joining material is isothermal. If the third material is isothermal, it makes no contribution to the output of the thermocouple. This proof should not be taken as a blanket license to connect thermocouple wires together without due care—some installations may result in temperature gradients near the junction, and those are sensitive to the presence of a joining material. However, a well-designed probe assures an isothermal zone around the junction, and those probes are insensitive to the material used to join the thermoelements together.

11.2.7 Switches, Connectors, Zone Boxes, and Reference Baths

A thermocouple switch or connector must not produce EMFs that would contaminate the temperature signal. By their very nature, switches and connectors connect several materials together and are susceptible to generation of thermoelectric EMFs.

The principal defense against spurious EMFs is to ensure that the switch or connector is isothermal, not only on the whole but in detail. The mechanical energy dissipated as heat when the switch is moved appears first as a high-temperature spot on the oxide films of the two contacts. Substantial temperature gradients may persist for several milliseconds after a switch movement.

Connectors frequently are used to join a thermocouple to lead wires, often in a location near the test apparatus. Temperature gradients within connectors may generate spurious signals. It is important to insulate the outer shell and provide a good conduction path inside the connector.

Switches and connectors made of thermocouple grade alloys will minimize the troubles caused by poor thermal protection, but good thermal design is still necessary.

The errors that can be introduced by using a non-isothermal connector are illustrated in Fig. 11.8. An all-copper connector is presumed, with its connection points 2, 3, 5, and 6 as shown. Two cases are examined: one in which both the A and B wires enter at one temperature and leave at

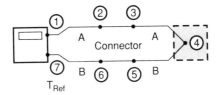

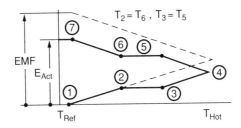

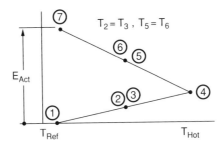

Fig. 11.8 The effects of connectors. (Reproduced from Ref. 1 with permission.)

another ($T_2 = T_6$ and $T_3 = T_5$, but $T_2 \neq T_3$), and one in which the A wire enters and leaves at one temperature while the B wire enters and leaves at some other temperature ($T_2 = T_3$ and $T_5 = T_6$, but $T_2 \neq T_5$). The $E-T$ diagram shows the latter situation causes no error regardless of the type of material in the connector, since there is never a temperature gradient along the connector material. In the first case, the resultant signal is in error, having "lost" the entire amount of the temperature interval across the connector.

Reference baths which provide stable and uniform zones for termination of thermocouple systems are available. These generally consist of a heated or cooled zone box thermostatically controlled to remain at some specified temperature. Laboratory users may well use an ice bath or triple-point cell. For most engineering purposes, a bath made with ice and water sufficiently pure for human consumption will be within a few hundredths of a degree of 0°C if certain simple precautions are followed.

A Dewar flask or vacuum-insulated bottle of at least 0.5 liter-capacity should be completely filled with ice crushed to particles about 1/4 to 1/2 cm in diameter. The flask is then flooded with water to fill the interstices between the ice particles. A glass tube resembling a small diameter (0.5 cm), thin-walled (0.5 mm) test tube should be inserted 6 or 8 cm deep into the ice pack and supported there with a cork or float. A small amount (0.5 cm deep) of silicone oil should be placed in the bottom of the tube to improve the thermal contact between the junction and the ice-water mixture. The reference thermocouple junction is then inserted into the tube until it touches the bottom of the tube and is secured at the top of the tube with a gas-tight seal. The object of the seal is to prevent atmospheric moisture from condensing inside the tube and causing corrosion of the thermocouples.

The assembly is shown in Fig. 11.9, along with a proper connection diagram. Note that the relative polarity of the connection is different from that shown in Fig. 11.4. If a connection like Fig 11.4 is desired, two glass tubes must be prepared, one each for the positive and negative elements. These tubes must be mounted farther apart than the size of the ice particles to assure adequate cooling. One thermoelement and one copper lead wire are put into each tube.

The principal requirement of the reference bath is that its temperature be known accurately. Any region of known temperature can serve as a reference bath.

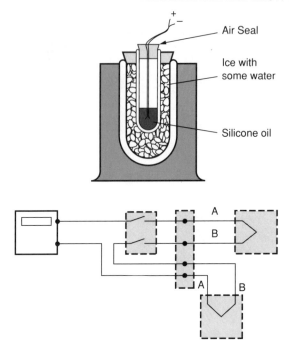

Fig. 11.9 Construction of a reference ice bath. (Reproduced from Ref. 1, with permission.)

Many instruments that provide their output in temperature units contain local reference regions and compensating circuits that augment the thermoelectric signal to account for the local reference temperature. Such instruments can be used only with the type of thermocouple for which they were intended, since the compensating network is specific to the calibration of the thermocouple being used.

11.2.8 Obtaining High Accuracy with Thermocouples

The temperature EMF tolerances quoted for thermocouples account for two types of deviations from the expectation values: (1) batch-to-batch differences in average calibration and, (2) point-to-point differences in local calibration along an individual thermocouple.

Calibration of individual thermocouples can account for the batch-to-batch differences but not the point-to-point variations along the wire.

For highest precision, three precautions should be taken:

1. Calibrate the individual thermocouples.
2. Minimize the working temperature difference (i.e., use a reference temperature near the working temperature and physically close by).
3. Install the thermocouple so the working temperature difference is stretched over as long a length of wire as possible.

11.2.9 Service-Induced Inhomogeneity Errors

When thermocouples are used in unfavorable environments or for very long times, the output voltage may drift with time due to development of inhomogeneities in regions of appreciable temperature gradient. Figure 11.10 illustrates the local effects of cold work on thermocouple calibration. It is difficult to identify this problem by subsequent recalibration, since the region of partially degraded material may be placed in a uniform temperature zone during recalibration and hence play no part in generating the signal under calibration conditions. If a thermocouple is suspected of being inhomogeneous, it should be tested for homogeneity along its entire length. If the test shows that the wire is homogeneous, then no calibration is required, since the used portion of the wire is the same as the unused portion. If the wire is not homogeneous, no recalibration can be of value because it will be impossible to place the temperature gradient in the same location for calibration as it was for service. This situation is described in Figs. 11.11 through 11.13.

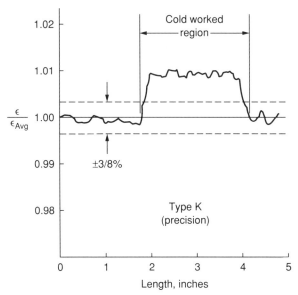

Fig. 11.10 Qualitative effects on calibration of type *K* materials, resulting from moderate cold work. (Reproduced from Ref. 1, with permission.)

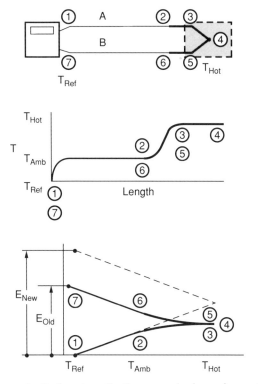

Fig. 11.11 Temperature distribution along the thermocouple, in service, and the resulting output after deterioration of the wires. (Reproduced from Ref. 1, with permission.)

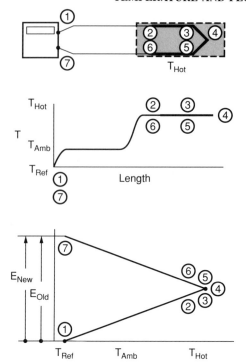

Fig. 11.12 Temperature distribution along the thermocouple, during recalibration, and the resulting output. (Reproduced from Ref. 1, with permission.)

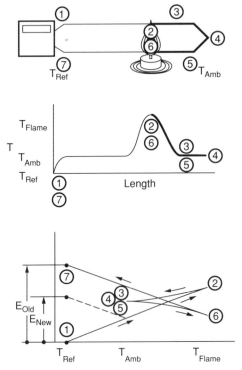

Fig. 11.13 Homogeneity testing with a local hot spot and the resulting output. (Reproduced from Ref. 1, with permission.)

Assume that the thermocouple is exposed to the unfavorable environment only near the hot end, as shown in Fig. 11.13, and that both wires become less active as a result of the reaction. The output will drop, as shown. If this defective thermocouple is placed in a usual calibration facility, all of the affected material will be in the region of uniform temperature. The EMF will be generated entirely by the material near the entrance of the furnace, which was never changed — the wires between points 1 and 2 and between 6 and 7 in Fig. 11.12. Under such conditions, a perfectly normal signal will be developed.

Figure 11.13 illustrates a test for homogeneity that can identify a defective thermocouple. To test a thermocouple for inhomogeneity, clamp the junction at a constant temperature and heat the wire in the suspected region. A homogeneous material will not produce any EMF, as shown earlier in Fig. 11.5. An inhomogeneous material will have different calibrations on the upslope and downslope sides of the temperature excursion and will produce a net EMF. This process is illustrated in Fig. 11.13.

The best technique for identifying a service-induced inhomogeneity is "side-by-side" running or comparison by replacement. Neither is very convenient, but on the other hand, there is no alternative. It is not within the present state of the art, regardless of how much effort is expended, to be able to interpret the readings from an inhomogeneous thermocouple in an arbitrary environment.

11.2.10 Thermoelectric Materials Connected in Parallel

Thermocouple materials are usually thought of as being connected in series with respect to the direction of the temperature gradient. There are situations, however, where two materials are in parallel electrical contact along their length. In such cases, the thermoelectric potential causes a distributed current to circulate in the materials. The net effect of such a configuration can be computed if the geometries and material properties are known.

There are three applications for which this effect is known to be of importance: (1) in the design of plated-junction thermopiles, (2) in attempts to precisely measure surface temperatures, using thermocouples attached with solder, and (3) in the case of distributed failure of thermocouple insulation, usually at high temperatures.

The essential features of this parallel circuitry can be illustrated by discussion of the thermopile design and the shunted thermocouple situation.

Copper may be plated onto a constantan wire to form a cylindrical thermocouple pair in which the copper and constantan contact throughout their entire length. When the two ends of this plated material are held at different temperatures, an EMF will be generated, which is a function of the relative electrical resistance of the two materials, as well as the usual thermoelectric parameters. The physical situation is shown in Fig. 11.14.

Gerashenko and Ionova[6] related the net EMF to that of a conventional copper–constantan thermocouple by the following:

$$\frac{\text{EMF}}{\text{EMF}_{\text{std}}} = \frac{R_{\text{const}}}{R_{\text{const}} + R_{\text{copper}}} \tag{11.4}$$

where R is the electrical resistance for the component named in the length direction.

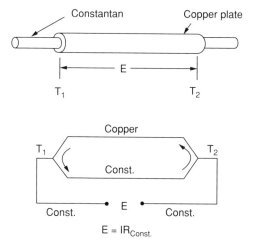

Fig. 11.14 The plated junction thermopile element and its equivalent circuit. (Reproduced from Ref. 1, with permission.)

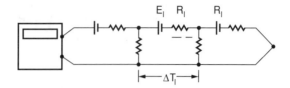

Fig. 11.15 Electrical network model of the distributed shunt effect. (Reproduced from Ref. 1, with permission.)

Thick copper platings will produce low resistance in the copper, and almost the entire output of the copper–constantan pair will be recovered.

There is another circumstance in which thermocouple elements become connected in parallel. When long thermocouples are used in hostile environments there can be significant effects on the output due to partial shunting of the thermocouple by breakdown of the insulation between the wires. This effect was discussed in 1964 by Tallman,[7] who presented a model for numerical calculations to evaluate the output which is useful in explaining the behavior. The circuit shown in Fig. 11.15 describes the behavior when the insulators serve only as resistive connections between the wires (i.e., do not participate by generating EMF). Only one node is shown in Fig. 11.15, although a real circuit would require many nodes for an accurate model.

Each $E_1 = \epsilon_{AB}(\Delta T_i)$

ϵ_{AB} = total thermoelectric power of the AB pair

$R_i = R_A + R_B$ for the ith interval

r_i = resistance between the A and B wires in the ith interval

ΔT_i = temperature difference across the ith interval

This network can be solved by several techniques and seems to accurately model the physical behavior of the distributed shunt effect.

11.2.11 Spurious EMFs Due to Corrosion and to Strain

Thermocouples produce DC signals in response to temperature differences. They can also become sensitive to AC electrical noise and ground loops, all of which are tractable and can be handled as noise problems would be handled in any system.

There are two problems peculiar to the thermocouple that bear mention: galvanic EMF generation and strain-induced EMFs.

Thermocouples necessarily involve pairs of dissimilar materials and hence are liable to galvanic EMF generation in the presence of electrolytes. The iron-constantan pair is the most vigorous, being capable of generating a 250 millivolt signal when immersed in an electrolyte.

Neither copper-constantan nor type K material shows significant galvanic effects.

The equivalent circuit is shown in Fig. 11.16. It is similar to the shunt problem, but the active EMF source in the shunt dominates it.

The output of the system will be the thermoelectric EMF corresponding to ($T_{\text{hot}} - T_{\text{ref}}$) plus or minus the IR drop in the circulating current loop.

Typical values for E_{shunt} and R_{shunt} are 250 mV and 1 MΩ/cm of wetted length (24-gauge, duplex-fiberglass-insulated material, wet with distilled water). The error induced by this signal depends on the resistance in the thermocouple loop and the location of the wet spot along the thermocouple. Wet iron-constantan thermocouples can produce large error signals.

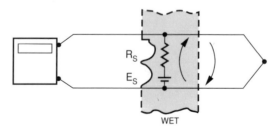

Fig. 11.16 Electrical network model of the galvanic EMF effect on a wet thermocouple. (Reproduced from Ref. 1, with permission.)

Although type K materials do not display appreciable galvanic EMF, they are strain sensitive (i.e., cold work causes a change in calibration), and during the act of straining, they generate appreciable EMFs. Temperatures measured on vibrating equipment appear to fluctuate, as a result of flexing of the thermocouple wires. Copper-constantan and iron-constantan are less active than type K, with copper-constantan least active. Spurious signals on the order of 50°C have been observed using type K materials on a vibrating apparatus (i.e., whole engine tests, etc.).

11.3 RESISTANCE TEMPERATURE DETECTORS[*]

The terms *resistance temperature detector* and *resistance thermometer* are used interchangeably to describe temperature sensors containing either a fine wire or a thin film metallic element, whose resistance increases with temperature. A small current (AC or DC) is passed through the element, and its resistance is measured. The temperature of the element is then deduced from the measured resistance using a calibration equation or a table.

11.3.1 Types and Ranges

Resistance temperature detectors can be designed for Standards and Calibration Laboratories or for field service, although the probe designs are vastly different. Field service probes are generally encased in stainless-steel protective tubes, with the wire or film elements bonded to sturdy support structures. They are made to handle considerable physical abuse. Laboratory standards probes are often enclosed in Quartz tubes, with the resistance wire mounted in a strain-free manner on a delicate mandrel. High-precision temperature measurement requires that the element be strain-free, so the electrical resistance is a function only of temperature, not of strain.

Resistance temperature detectors are well suited for single-point measurements in steady-state service at temperatures below 1000°C where long-time stability and traceable accuracy are required, and where reasonably good heat transfer conditions exist between the probe and its environment.

These detectors are not recommended for use in still air or in low conductivity environments unless the self-heating effect can be accounted for appropriately. They are not recommended for averaging service unless computer data acquisition and software averaging is contemplated. They are not recommended for transient service or dynamic temperature measurements unless specifically designed for such characteristics, since the usual probe is not amenable to simple time-constant compensation.

Resistance temperature detector probes tend to be larger than some of the alternative sensors and to require more peripheral equipment (i.e., bridges and linearizers).

11.3.2 Physical Characteristics of Typical Probes

Probes intended for field service concentrate on ruggedness and repeatability. Their calibrations may not agree with the standard expectation values of laboratory grade probes but should be repeatable. Drift tolerances on field service probes may be stated in terms of percent drift per 100 hours.

Probes for laboratory service are designed to ensure freedom from mechanical strain either due to fabrication or thermal expansion during service.

The bare sensing elements of resistance temperature detectors range in size from wafers .5 × 1.0·× 2.0 mm, with pig-tail leads, 0.25 mm in diameter and 2.5 cm in length to wire-wrapped mandrels, again with pig-tail leads, 4 mm in diameter and 2 cm in length. With protective tubes in place, typical probes range from 1.0 mm to 5.0 mm in diameter.

Some typical sensors and probes for field service are shown in Fig. 11.17. Stainless steel is used for the protection tubes on most such units. The sensing element can be either a wire wound on a mandrel or a thin film deposited on an insulating substrate. Wire resistance elements may also be bonded to their support structures and encapsulated in glass or ceramic, but strain-free, steel-jacketed probes are also available. Thin film elements are usually formed directly on the substrate by sputtering or vapor deposition.

Some, but not all, laboratory grade probes use Quartz for the protection tubes. For highest accuracy and best long-term stability, the laboratory grade probes use strain-free rigging, and a bifilar winding. This combination avoids the effects of mechanical strain on the resistance and also reduces pickup from electromagnetic or electrostatic fields. The windings are fragile, however, and the probes must be treated with great care. Figure 11.18 shows cut-away views of two laboratory grade probes.

[*] Substantial amounts of the material in this section are taken from Reference 8 with permission of the author.

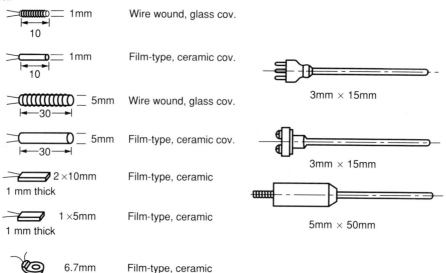

	1mm	Wire wound, glass cov.
⊢—10—⊣		
	1mm	Film-type, ceramic cov.
⊢—10—⊣		
	5mm	Wire wound, glass cov.
⊢—30—⊣		
	5mm	Film-type, ceramic cov.
⊢—30—⊣		
1 mm thick	2×10mm	Film-type, ceramic
1 mm thick	1×5mm	Film-type, ceramic
2.2 mm thick	6.7mm	Film-type, ceramic

3mm × 15mm

3mm × 15mm

5mm × 50mm

Fig. 11.17   Physical characteristics of some sensors and probes.[9]

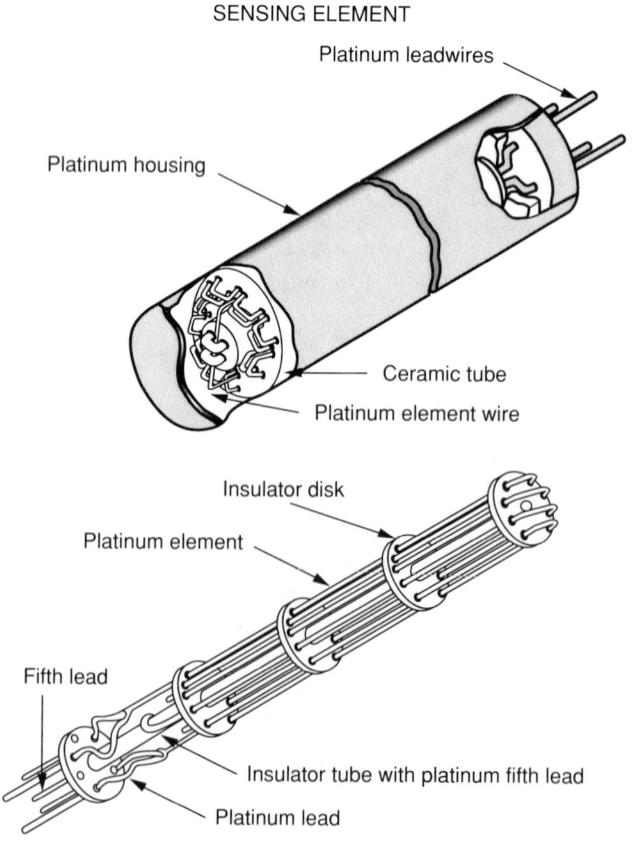

SENSING ELEMENT

Platinum leadwires

Platinum housing

Ceramic tube
Platinum element wire

Insulator disk
Platinum element

Fifth lead

Insulator tube with platinum fifth lead
Platinum lead

Fig. 11.18   Cutaway views of two laboratory grade probes.[10,11]

11.3.3 Electrical Characteristics of Typical Probes

Resistance temperature detectors are available commercially with resistances from 20 to 2000 Ω, 100 Ω being common. Platinum, nickel, copper, and manganin have been used in commercial resistance thermometers.

Typically, the resistance of a platinum resistance thermometer will change by about 0.39% per °C. The resistance of a 100 Ω probe, according to the European calibration curve, would change by about 0.385 Ω per °C in the range 0–100°C. If the probe was in still air at 0°C and was being driven at a constant current of 3.16 mA (which would, on an average probe, produce a self-heating effect of about 0.25°C in still air), the voltage drop across the probe would be 316.304 mV, which would increase by 1.2166 mV for each °C increase in air temperature. By contrast, this compares to about 0.050 mV per °C for a typical base metal thermocouple.

Twisted-pair lead wires are recommended. Care must be taken to avoid thermoelectric signal generation in resistance thermometer circuits using DC excitation (AC circuits convey the temperature information at carrier frequency, and they are not affected by thermoelectric signals).

11.3.4 Thermal Characteristics of Typical Probes

Interrogation of a resistance temperature detector dissipates power in the element, which goes off as heat transfer to the surroundings. This self-heating causes the sensing element to stabilize at a temperature higher than its surroundings. The amount of self-heating depends on three factors: (1) the power dissipated in the element, (2) the probe's internal thermal resistance, and (3) the external thermal resistance between the surface of the probe and the surrounding material whose temperature is to be measured. The self-heating temperature rise is given by

$$T_{sens} - T_{spec} = W(R_{int} + R_{ext}) \qquad (11.5)$$

A typical probe exposed to still air will display self-heating errors on the order of 0.1 to 1.0°C per mw (commercial probes of 1.5 to 5 mm in diameter). At 1 m/s air velocity, the self-heating error may only be 0.03 to 0.3°C, while in water (at 1 m/s velocity), the self-heating effect would be reduced by a factor of four or five, depending on the relative importance of the internal and the external thermal resistances, compared to the values in moving air.

Self-heating error can be kept small by using probes with low internal thermal resistance, by locating the probes in regions of high fluid velocity (or placing the probes in tight-fitting holes in structures), or by operating at very low power dissipation. Pulse interrogation can also be used; it will reduce self-heating regardless of the internal and external resistances. The temperature rise of an element subjected to a pulse of current depends on the duration of the pulse and the thermal capacitance of the sensor rather than on the resistances between the sensor and the specimen.

Resistance temperature detectors (RTDs) are subject to all of the installation errors and environmental errors of any immersion sensor and, because of the detectors' larger size, are usually affected more than thermocouples or thermistors. Since RTDs are usually selected by investigators who wish to claim high accuracy for their data, the higher susceptibility to environmental error may be a significant disadvantage.

The internal structure of most RTDs is sufficiently complex that their thermal response is not *first order*. As a consequence, it is very difficult to interpret transients. The term *time constant*, which applies only to first-order systems, should not be applied to RTDs in general. If a *time constant* is quoted for a resistance temperature detector, it may only mean ". . . the time required for 63.2% completion of the response to a step change," and it may not imply the other important consequences of first-order response. Quoted values, from one probe supplier, of "time to 90% completion" range from 9 to 140 seconds for probes of 1.0 to 4.5 mm in diameter exposed to air at 1 m/s.

11.3.5 Measuring Circuits

Resistance temperature detectors require a source of power and a means of measuring resistance or voltage.

Resistance can be deduced by voltage drop measurements across the resistor when the current is known or by comparison with a known resistor in a bridge circuit. Six circuits frequently used for *resistance thermometry* are shown in Fig. 11.19.

In the following paragraphs, these six circuits used for reading the resistance of the thermometer will be briefly discussed. The determination of probe temperature from probe resistance is discussed in a future section.

The 2–wire, constant current method is the simplest way to use a resistance thermometer. One simply provides it with a known current and measures the voltage drop across the probe and its lead wire (Fig. 11.19a). The resistance of the circuit is determined from Ohm's law in the cold and the hot condition, and the temperature is determined from the resistance ratio.

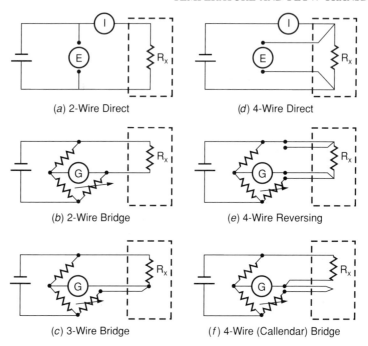

Fig. 11.19 Six circuits for the measurement of resistance. (Reproduced from Ref. 8 with permission.)

In this approach, the lead-wire resistance is ignored, which leads to underestimating the temperature change by approximately the ratio of the lead-wire resistance to the total circuit resistance:

$$\frac{R_{\text{TOT},H}}{R_{\text{TOT},C}} = 1 + \left(\frac{\Delta R_P}{R_{P,C} + R_L}\right) \tag{11.6}$$

Errors can be avoided by properly accounting for the lead-wire resistance, either by measuring it using a shorting plug to replace the probe or by calculating it from standard tabular data on wire resistance.

The 2–wire bridge circuit (Fig. 11.19b) suffers from the same lead-wire error as the 2–wire direct method but has one advantage: it can be used with an unstable power supply. The probe resistance can be measured in terms of the calibrated reference resistor, independent of the bridge voltage value.

The 3–wire bridge circuit (Fig. 11.19c) compensates, to a considerable extent, for the effect of lead-wire resistance. The circuit uses a matched pair of lead wires to connect the bridge to the probe. One member of the pair is inserted into each of the two arms of the bridge, so changes in the lead-wire resistance affect both arms. For a bridge with equal resistances in both arms, this provides an approximate compensation for lead-wire resistance.

Many commercial probes are supplied with 3–wire lead connection.

The most accurate technique for measuring the probe resistance is the 4–wire direct method (Fig. 11.19d).

The probe is driven from a known source of constant current through one pair of leads, while the voltage drop across the sensing resistor is measured using the other pair. Since there is no current flow in the voltage sensing leads, there is no error introduced by the lead-wire resistance. With current and voltage drop known, the resistance can be calculated directly using Ohm's law. Modern, regulated power supplies and high-impedance voltmeters make this an increasingly attractive option.

Two 4–wire bridges are shown: the 4–wire Callendar bridge and the reversing bridge. The Callendar bridge inserts equal lengths of lead wire in each of the two arms of the bridge, which provides good compensation for the effect of lead-wire resistance. The reversing bridge allows the operator to exchange lead wires, thus measuring the resistance in each arm with two different pairs of leads. The average of the two values is usually used as the best estimate, since the two measurements are equally believable.

Several manufacturers offer linearizing amplifiers as accessories for resistance thermometers. These devices produce signals that are linearly proportional to sensor temperature by correcting its nonlinear response.

The increasing use of computers in data interpretation has reduced the need for linearizing circuitry in laboratory and research work, but the commercial market still prefers linear systems for control and monitoring.

In some cases, the linearizing amplifier is provided as an integral part of the probe, which may also contain its own power supply. The output of such a probe is linearly proportional to the probe temperature.

11.3.6 The Standard Relationships for Temperature vs. Resistance

Platinum resistance thermometers are sold in the United States under two different calibrations (U.S. and European) and are subject to tolerances, which can also be described by either of two standards.

The two calibrations differ in their expected values of *alpha*: the average temperature coefficient of resistivity (Ω/Ω–C) over the interval 0–100°C. The European standard (DIN 43670) specifies alpha to be 0.003850 while the U.S. standard specifies 0.003925 per °C. The IPTS-68 specifies 0.003925 for probes acceptable as standards. The value of alpha increases with increasing purity, and values as high as 0.003927 have been observed, according to Norton.[12] Norton also mentions that thin films of platinum do not follow either of the two standard calibration curves just mentioned for bulk platinum but tend to approach the European standard value.

Table 11.4 illustrates the differences in resistance of U.S. and European calibrations.

From the user's standpoint, the important issue is, "How much different are the temperatures deduced from a measured resistance?" Figure 11.20 shows the temperature difference (European-U.S.) as a function of temperature level.

The equations proposed for data interpretation (i.e., the Callendar equation or the Callendar-Van Dusen equation) are attempts at fitting the tabular data—they are not the sources of the tables.

Just as there are different standards for the resistance-temperature relationship, so there are different standards for interchangeability among commercial probes. The hot resistance of a probe at a particular temperature depends on its cold resistance and its average value of alpha up to the operating temperature. Standards for interchangeability must acknowledge both sources of difference. Interchangeability is typically discussed in terms of "percent of reading" (in °C), with values from 0.1% to 0.6% being available in commercial probes for field service.

The resistance-temperature characteristic of a probe may drift (i.e., change with time) while the probe is in service. Manufacturers of laboratory grade probes will specify the expected drift rate, usually in terms of the expected temperature error over an interval of time. Two sample specifications are "0.01°C per 100 hours" for a low-resistance, high-temperature probe (0.22 Ω at 0°C, 1100°C maximum service temperature) and "0.01°C per year" for a moderate-resistance, moderate-temperature probe (25.5 Ω at 0°C, 250°C maximum service temperature). Drift of the resistance-temperature relationship appears related to changes in the grain structure of the platinum. Such changes take place more rapidly at high temperatures. Probes are annealed at high temperatures after fabrication in order to reduce the effects of manufacturing strains.

11.3.7 Interpreting Temperature from Resistance: Common Practice

The final step in determining the measured temperature is to interpret the probe resistance through a calibration relationship. There are several options depending on the accuracy required.

TABLE 11.4   RESISTANCES OF A PLATINUM RESISTANCE THERMOMETER

Temperature °C	U.S. Curve Alpha = 0.003925	European Curve Alpha = 0.003850
− 200	17.14	18.53
− 100	59.57	60.20
0	100.00	100.00
+ 100	139.16	138.50
+ 200	177.13	175.84
+ 300	213.93	212.03
+ 400	249.56	247.08
+ 500	284.02	280.93
+ 600	317.28	313.65

Source: Reference 14.

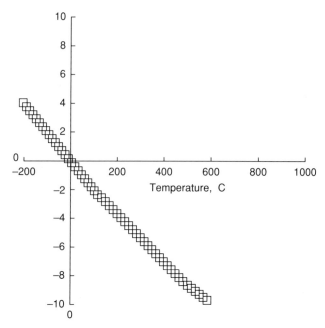

Fig. 11.20   Comparison of the European and U.S. resistance curves. (Reproduced from Ref. 8 with permission.)

For rough work near the ambient, it is often good enough to simply take the nominal value of alpha as a constant and divide the observed change in resistance by the value of alpha to find the temperature change.

For work between 0°C and 100°C, the *constant alpha* method is quite accurate, since alpha was determined in this interval.

In general, manufacturers will provide resistance vs. temperature tables for their probes. These tables should be used only with the probes for which they were intended because they may not have any generality. Manufacturers' tables are slightly more accurate than the constant alpha method.

For high quality, laboratory-grade platinum resistance thermometers (used between 0 and 630°C), the Callendar equation describes the IPTS-68 temperature within ±0.045°C, according to Benedict.[13] The difference between IPTS-68 and the Callendar equation varies across the range of temperatures, as shown in Fig. 11.21.

The Callendar equation is:

$$t = 100\left(\frac{R_t - R_o}{R_{100} - R_0}\right) + \delta\left(\frac{t}{100} - 1\right)\left(\frac{t}{100}\right) \tag{11.7}$$

where R_t = Resistance at temperature t

R_o = Resistance at the ice point

R_{100} = Resistance at 100°C

δ = Constant evaluated at Zinc point (419.58)

As shown in Fig. 11.21, the tolerance of ±0.045°C represents an agreed-upon difference between the Callendar equation and the IPTS-68 equation.

For most measurement purposes, the Callendar equation can be considered accurate within ±0.045°C over the range 0 to 630°C.

Below 0°C, temperature according to the Callendar equation deviates significantly from the IPTS-68 temperature. An additional term converts the Callendar equation into the Callendar-Van Dusen equation, which is much closer to the IPTS-68 temperature and is recommended for field service work between −190°C and 0°C.

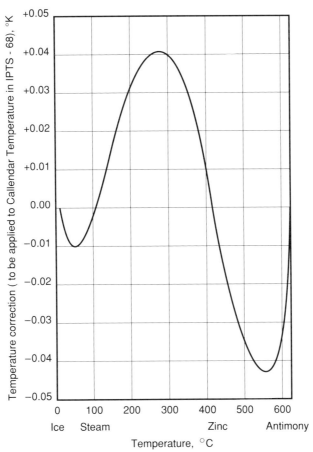

Fig. 11.21 The difference between the Callendar equation and EPTS-68. (From Benedict,[13] with permission.)

The Callendar-Van Dusen equation is

$$t = 100\left(\frac{R_t - R_o}{R_{100} - R_o}\right) + \delta\left(\frac{t}{100} - 1\right)\left(\frac{t}{100}\right) + \beta\left(\frac{t}{100} - 1\right)\left(\frac{t}{100}\right)^3 \quad (11.8)$$

where

R_t = Resistance at temperature t

R_o = Resistance at the ice point

R_{100} = Resistance at 100°C

δ = Constant determined at Zinc point (419.58)

If $\qquad t > 0, \beta = 0$

If $\qquad t < 0, \beta$ = Constant determined at Oxygen point (-182.962°C)

According to Benedict,[13] the disagreement between the Callendar-Van Dusen equation and IPTS-68 does not exceed ±0.03°C from −190 to 0°C. The difference varies as shown in Fig. 11.22.

11.4 THERMISTORS*

Thermistors are temperature-sensitive resistors whose resistance varies inversely with temperature. The resistance of a 5000 Ω thermistor temperature sensor may go down by 20 Ω for each degree

*Substantial portions of the material in this section are taken from Ref. 8, with permission.

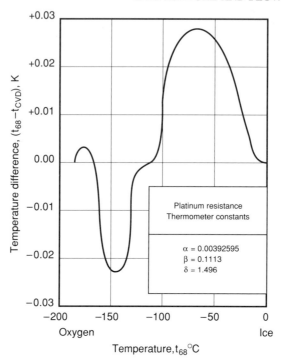

Fig. 11.22 The difference between the Callendar-Van Dusen equation and IPTS-68. (From Benedict,[13] with permission).

Celsius increase in temperature (in the vicinity of the initial temperature). Driven by a 1.0 mA current source, this yields a signal of 200 mV/°C.

Thermistors are used frequently in systems where high sensitivity is required. It is not uncommon to find thermistor data logged to the nearest 0.001°C. This does not mean the data are accurate to 0.001°C, but the data are readable to that precision.

A thermistor probe is sensitive to the same environmental errors that afflict any immersion sensor: its accuracy depends on the care with which it was designed for that particular environment.

11.4.1 Types and Ranges

Thermistor probes can be used between −183°C (the Oxygen point) and +327°C (the Lead point).[15] But most applications are between −80°C and +150°C.[16] The sensitivity of a thermistor (i.e., the % change in resistance per °C change in thermistor temperature) varies markedly with temperature, being highest at the lowest temperatures.

The long-time stability of thermistor probes is open to some question, although several months of accurate usage between calibrations seem attainable. The evidence on drift and its causes is not clear. It would be prudent, as with any temperature measuring system, to make provision for periodic recertification of thermistor probes on a time scale established by experience within the system itself.

If accurate measurements are required, calibration facilities are needed, and this need poses some problems. Few instruments are capable of providing a transfer calibration for thermistors to their limit of readability, since few are that sensitive. Melting point baths or precision-grade resistance thermometry are needed to capitalize on the available precision.

Thermistors have strongly nonlinear output. Linearizing bridges are available, but these add to cost. Nonlinearity is not a significant issue when the data will be interpreted by computer.

11.4.2 Physical Characteristics of Typical Probes

Thermistor probes range in size from 0.25 mm spherical beads (glass-covered) to 6 mm diameter steel-jacketed cylinders. Lead wires are proportionately sized. Disks and pad-mounted sensors are available in a wide range of shapes, usually representing a custom design gone commercial. Aside from the unmounted spherical probes and the cylindrical probes, there is nothing standard about the probe shapes.

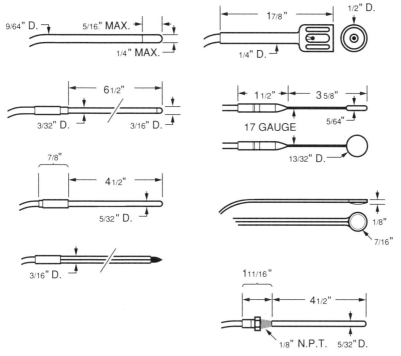

Fig. 11.23 Representative thermistor probes for temperature measurement. (Courtesy of Omega Engineering, F4 of 1985 catalog.)

Figure 11.23 shows some representative shapes of commercially available probes.

For medical applications, thermistor probes are often encapsulated in sterilizable, flexible vinyl material. Such probes are frequently taped to a patient's skin and used as the control sensor for the temperature regulating system.

The thermistor element itself is fabricated using the techniques of powder metallurgy. At one company, a mixture of metallic oxides is compressed into a disk and sintered. The mixture's composition, the sintering temperature, and the atmosphere in the furnace determine the resistance and the resistance-temperature coefficient of the thermistor. The faces of the disk are plated with silver, and the resistance of the thermistor is adjusted by removing material from the edges until the desired value has been obtained. The lead wires are attached, and the assembled thermistor is then potted in epoxy.

The process is shown schematically in Fig. 11.24.

High-temperature thermistors (1000°C) have been made from doped ceramic materials but with only moderate success. There has not been much activity in this area as yet.

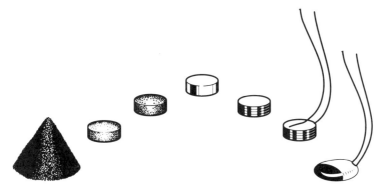

Fig. 11.24 The thermistor manufacturing cycle.

**TABLE 11.5   VARIATIONS OF THERMISTOR TEMPERATURE
COEFFICIENT WITH TEMPERATURE**

Temp (°C)	Condition	%/°C
−183	Liquid Oxygen	−61.8
−80	Dry Ice	−13.4
−40	Frozen Mercury	−9.2
0	Ice Point	−6.7
25	Room Temperature	−5.2
100	Boiling Water	−3.6
327	Melting Lead	−1.4

Source: Reference 15.

11.4.3   Electrical Characteristics of Typical Probes

Thermistor probes vary in resistance from a few hundred ohms to megohms. Resistance is frequently quoted at 25°C with no power dissipation in the thermistor. The commercial range is from about 2000 Ω to 30,000 Ω.

The resistance-temperature characteristic of a thermistor depends on the material of which it is made. Representative values of the sensitivity coefficient (% change in resistance per °C) are given in Table 11.5.

To illustrate the range of resistances encountered in practice, Table 11.6 shows the resistance as a function of temperature for a typical probe whose resistance is 2252 Ω at 25°C.

Thermistor resistance data are frequently shown as: logarithm of resistance ratio vs. $1/T$. Such a presentation emphasizes the "almost linearity" of a single-term exponential description of the resistance-temperature characteristic. The data listed in Table 11.6 are shown in Fig. 11.25.

Proprietary probes are available that *linearize* thermistors by placing them in combination with other resistors to form a circuit whose overall resistance varies linearly with temperature over some range. These compound probes can be summed, differenced, and averaged like any linear sensor.

Modern manufacturing practices allow matched sets of thermistors to be made interchangeable within ±0.1°C.

11.4.4   Thermal Characteristics of Typical Probes

Thermistor probes are generally interrogated using a current of 1 to 10 μA, either AC or DC. With a probe resistance of 10 KΩ, this results in power dissipation of 0.01W, that must be transferred from the probe into its surrounding material. At the first instant of application, this current has no significant effect on the measured resistance of the probe, but in steady state, it results in the probe running slightly above the temperature of the medium into which it is installed. This is referred to as the *self-heating* effect. Since thermistors are often used where very small changes in temperature are important, even small amounts of self-heating may be important.

The self-heating response is frequently discussed in terms of the *dissipation constant* of the probe, in mW per °C. Dissipation constants can range from 0.005 mW/°C to several W/°C and can be estimated with acceptable accuracy, given the geometry and thermal properties of the probe and its installation. The dissipation constant should not be used to correct a reading for the self-heating effect, the uncertainty in the correction is too high. If a calculation shows the self-heating effect to be significant, pulse interrogation should be used.

Dissipation constants for two representative probes are given in Table 11.7.

**TABLE 11.6   VARIATION OF THERMISTOR RESISTANCE
WITH TEMPERATURE**

Temp (°C)	Res. (Ω)	Temp (°C)	Res. (Ω)
−80	1.66 M	0	7355.0
−40	75.79 K	25	2252.0
−30	39.86 K	100	152.8
−20	21.87 K	120	87.7
−10	12.46 K	150	41.9

Source: Reference 17.

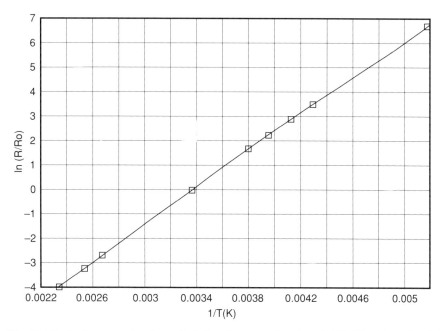

Fig. 11.25 Resistance ratio of a typical thermistor as a function of $1/T$ (K). (Reproduced from Ref. 8, with permission.)

The self-heating effect must be considered in calibration as well as in use. Calibration baths frequently use stirred oil as the medium. Table 11.7 shows the self-heating effect in the oil bath to be very low compared to that in air at 5 m/s. A probe calibrated in an oil bath and used in air would be subject to different self-heating effects, and the user should be aware   that.

The transient response of thermistors is more complex than that of thermocouples, and they are not as well suited to transient measurements. The difference arises from the thermal response of the thermistor material itself. Thermistor materials typically have lower thermal diffusivity than metals and hence do not equilibrate their internal temperatures as rapidly, all other factors being the same. Thus the highest frequency at which a thermistor will yield a first-order response is far lower than that for a thermocouple. During rapid changes, the temperature is not uniform inside the thermistor, and the average resistance no longer describes the average temperature. Since the thermistor resistance is not a linear function of temperature, the transient resistance cannot easily be converted to temperature.

Thermistor probes are susceptible to environmental errors (radiation, velocity, and conduction errors) as are all other immersion temperature sensors, and the same guidelines and design rules apply.

Thermistor sensors can be built into complex probe designs, as can any temperature sensor, and the structure of the probe will determine its steady state and transient behavior. The thermistor has relatively low thermal conductivity and is frequently encapsulated in epoxy, glass, or vinyl, which also have low thermal conductivity. On this account, the transient response of a thermistor probe will be slower and more complex than that of a metallic sensor.

TABLE 11.7   REPRESENTATIVE THERMAL DISSIPATION CONSTANTS FOR TWO DIFFERENT THERMISTOR PROBE DESIGNS

Environment	1.0 cm Disk	5.0 cm Glass Cylinder
Still Air	8(mW/°C)	1(mW/°C)
Air at 5 m/s	35	—
Still Oil	55	3.5
Still Water	—	5
Oil at 1 m/s	250	—

Source: Reference 15.

11.4.5   Measuring Circuits and Peripheral Equipment

Electrical resistance is easily measured within ordinary levels of accuracy by a number of techniques, and any method can be used with thermistors. The logarithmic variation of resistance with temperature means that over small ranges a constant percent error in resistance translates into a fixed error in temperature level. For example, with a thermistor whose sensitivity coefficient (see Table 11.5) is 5% per °C, a resistance measurement accurate to ±0.1% of reading will yield a temperature measurement accurate within ±0.02°C everywhere in the range. Accuracy of this order can be achieved by 2–wire and 4–wire systems using off-the-shelf instruments and thermistors.

When thermistors are used to resolve very small differences in temperature (for example, into the mK resolution range), special precautions must be taken to measure the resistance with the required accuracy. For the sensitivity mentioned above, fifty parts per million corresponds to about 1 mK. Measurements in the 1–100 mK range are more typical of standards room practice than of field measurements, yet some applications require this precision.

11.4.6   Determining Temperature from Resistance

Experimental evidence shows that the resistance of a thermistor varies inversely, and nearly exponentially, with temperature. This suggests expressing the resistance-temperature relationship as a polynomial in $\ln(R)$:

$$1/T = A_0 + A_1\ln(R) + A_2\ln(R)^2$$

$$+ A_3\ln(R)^3 \cdots + A_N\ln(R)^N \qquad (11.9)$$

Note the R is dimensional in this equation, hence the values of the constants would depend on the units used for R.

A form retaining terms through the cubic, sufficiently accurate for small ranges in temperature, was used by Steinhart and Hart.[18] Their results indicated that the second-order terms had no effect, but later work has shown some advantage in retaining these terms. However, the Steinhart-Hart equation is still widely used:

$$1/T = A_0 + A_1\ln(R) + A_3[\ln(R)]^3 \qquad (11.10)$$

Arguments concerning the need for dimensional homogeneity led Bennet[19] to propose a dimensionless form, which is independent of the units used:

$$1/T - 1/T_0 = A_1\ln(R/R^0) + A_2\ln(R/R_0)^2$$

$$+ \cdots + A_N \ln(R/R_0)^N \qquad (11.11)$$

Justification for this form can be seen in the near-linearity of Fig. 11.25.

This general form has given rise to several simplified forms. For a first approximation, only one term is needed:

$$R = R_0\exp[A_1(1/T_0 - 1/T)] \qquad (11.12)$$

This form reveals the exponential nature of the thermistor resistance response to changes in temperature.

Seren et al.[20] compared four different forms of the calibration equation for goodness of fit to a data set covering a range of 200 K, published by Bosson, Guttmann, and Simmons,[21] and accepted the data set as reliable. Their results are shown in Table 11.8. The following calculations were used to arrive at those results.

SRE = Standard Relative Error, the root mean square of the relative errors at the test points.

$$\text{SRE} = \left[\sum_{i=1}^{n} \frac{\left(T_{c,i} - T_{oi}\right)^2/T_{oi}^2}{n}\right]^{1/2} \qquad (11.13)$$

RMS error = Root Mean Square error.

$$\text{RMS} = \left[\frac{1}{n}\sum_{i=1}^{n}\left(T_{ci} - T_{oi}\right)^2\right]^{1/2} \qquad (11.14)$$

TABLE 11.8 THE ACCURACY OF LEAST-SQUARES FITS OF FOUR DIFFERENT POLYNOMIAL EXPRESSIONS FOR THERMISTOR CALIBRATION

Form of the Equation for $1/T$	SRE	RMS Error	Mean Error
1. $A_0 + A_1 \ln(R)$	0.00601	1.999	0.1925
2. $A_0 + A_1 \ln(R)$ $\quad + A_2[\ln(R)]^2$	0.000428	0.1353	−0.0026
3. $A_0 + A_1 \ln(R)$ $\quad + A_2[\ln(R)]^2$ $\quad + A_3[\ln(R)]^3$	0.000139	0.0392	0.00008
4. $A_0 + A_1 \ln(R)$ $\quad + A_3[\ln(R)]^3$	0.000138	0.0380	−0.00001

Source: Reference 21.

Mean error = Algebraic mean error.

$$\text{Mean} = \frac{1}{n} \sum_{i=1}^{n} \left(T_{c,i} - T_{o,i} \right) \tag{11.15}$$

where T_c = Calculated temperature (K) at measured resistance R_i.

T_i = Observed temperature at measured resistance R_i.

The fourth form is the Steinhart-Hart equation. It seems, from this comparison, that there is little to choose from between the two cubic fits as far as goodness of fit is concerned, and since the Steinhart-Hart form requires only three constants, it would be preferred.

A calibration equation such as the Steinhart-Hart equation would be used with a commercial probe only when the highest accuracy is required. In most cases, the manufacturer's calibration is sufficiently accurate.

Thermistor probes are sold with resistance vs. temperature calibration tables accurate within ±0.1 or 0.2 K, depending on the probe grade purchased. These tables are typically in 1 K increments. For

TABLE 11.9 RESISTANCE AND RELATIVE RESISTANCE OF TYPICAL PROBES

Temp °C	$1/T$ (K)	Probe 1		Probe 2	
		Resistance	Relative Resistance	Resistance	Relative Resistance
−80	.00518	1660000	225.6968	2211000	225.7044
−70	.00493	702300	95.4861	935600	95.5084
−60	.00469	316500	43.0320	421500	43.0278
−50	.00448	151050	20.5370	201100	20.5288
−40	.00429	75790	10.3046	101000	10.3103
−30	.00412	39860	5.4194	53100	5.4206
−20	.00395	21870	2.9735	29130	2.9737
−10	.00380	12460	1.6941	16600	1.6946
0	.00366	7355	1.0000	9796	1.0000
10	.00353	4482	.6094	5971	.6095
20	.00341	2814	.3826	3748	.3826
30	.00330	1815	.2468	2417	.2467
40	.00319	1200	.1632	1598	.1631
50	.00310	811.30	.1103	1081	.1104
60	.00300	560.30	.0762	746.30	.0762
70	.00292	394.50	.0536	525.40	.0536
80	.00283	282.70	.0384	376.90	.0385
90	.00275	206.10	.0280	274.90	.0281
100	.00268	152.80	.0208	203.80	.0208
110	.00261	115	.0156	153.20	.0156
120	.00254	87.70	.0119	116.80	.0119
130	.00248	67.80	.0092	90.20	.0092
140	.00242	53	.0072	70.40	.0072
150	.00236	41.90	.0057	55.60	.0057

Source: Reference 3.

computer interpretation, they should be fit to the Steinhart-Hart form and the coefficients determined for least error.

For the highest precision in measurement of temperature level, the purchased thermistors must be calibrated against laboratory-grade temperature standards at enough points to determine all of the constants in the model equation selected. Good practice would suggest one redundant calibration point to provide a closure check.

When an application requires more precision than provided by the manufacturer's tables but less than the highest achievable, a transfer standard instrument can be used. A strain-free platinum resistance thermometer or a Quartz crystal thermometer can be used as such a transfer standard. Typically, the comparison is made in a well-stirred, temperature-regulated oil bath or in a comparison block of nickel or copper in an electrically heated oven.

Values of resistance and resistance divided by resistance at 0°C are listed in Table 11.9 for two typical commercial probes as functions of temperature (°C) and $1/T$ (K).

11.5 FIBER-OPTIC BLACK-BODY RADIATION THERMOMETRY[*]

A fiber-optic black-body radiation sensor measures the intensity of the radiation emitted from a black-body cavity at the end of an optical fiber and deduces temperature from the intensity, using well-established laws of radiant emission.

The system has four main elements: the cavity, the high-temperature fiber, the low-temperature fiber, and the processor. Two of these elements are exposed to the high-temperature environment. A typical high-temperature fiber is a single-crystal sapphire fiber, about 1.25 mm in diameter and usually between 100 and 250 mm long. The cavity can be formed by sputtering a thin metallic film onto the end of the sapphire fiber, typically about 5 μm. The metal layer can be covered by a film of alumina (also about 5 μm) to protect it from oxidation or erosion.

11.5.1 Physical Characteristics

The general arrangement and the principal features of the data interpretation are shown in Fig. 11.26.

The cavity radiates as a "dark-gray body" (almost a black-body) and the high-temperature fiber conducts the radiant energy from the cavity to the low-temperature fiber, which in turn delivers it to the signal processor. The low-temperature fiber can be of great length, without degrading the signal, since loss coefficients for the low-temperature fiber material are quoted in percent per kilometer.

11.5.2 Electrical Characteristics

The processing of the radiant signal can be discussed in terms of the system pattern function: a set of equations that show how the system is supposed to work. The pattern function for the fiber-optic system is presented below. A final calibration adjusts the output to bring it into conformance with the calibration results.

$$I'(T_i) = \int_0^\infty \frac{C_1 C_d(\lambda, T_i)}{\lambda^5 [\exp(C_2, \lambda T) - 1]} F(\lambda) BS(\lambda, T_i) LTF(T_i) d\lambda \qquad (11.16)$$

and

$$I = I' K_{\text{sensor}} K_{\text{cable}} K_{\text{detector}} F(T_i) \qquad (11.17)$$

In these equations,

$$I'(T_i) = \text{Detector Current, pattern function}$$

$$I(T_i) = \text{Actual Detector Current, corrected for } T_i$$

$$\lambda = \text{Wavelength}$$

$$T = \text{Temperature (K)}$$

$$T_i = \text{Temperature of the signal processor box}$$

$$C_1 = \text{A radiation constant: } 3.743 \times 10^8 \ (\text{W} \cdot \mu\text{m}^4/\text{m}^2)$$

$$C_2 = \text{A radiation constant: } 1.4387 \times 10^4 \ (\mu\text{m} \cdot \text{k})$$

[*] Substantial portions of the material in this section are taken from Ref. 8, with permission.

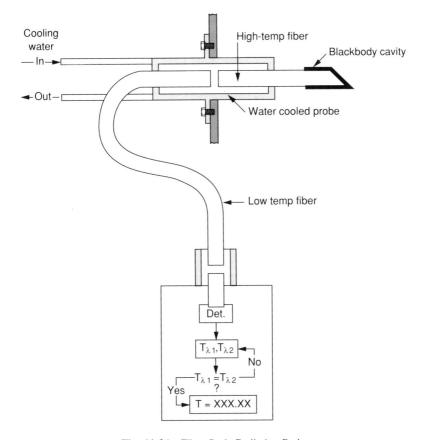

Fig. 11.26 Fiber-Optic Radiation Probe.

$C_d(\lambda, T_i)$ = A calibration constant for the photodiode

$F(\lambda)$ = Narrow bandpass filter function

$BS(\lambda, T_i)$ = Beam Splitter function

$LTF(T_i)$ = Low-temperature fiber transmission function

The dependence on T_i reflects changes in the optical properties of the mirrors and prisms used in steering the beams to their respective sensors.

The temperature range is 600 to 2000°C. Accuracy begins to fail at lower temperatures, while the Sapphire fiber cannot tolerate higher temperatures.

11.5.3 Precision and Accuracy

The calibration precision is limited only by the precision with which the electrical signals can be read, assuming the equation governing the radiant emission to be absolutely accurate. The measurement accuracy depends on how well the sensor is brought into equilibrium with the specimen temperature.

The fiber-optic probe is sensitive to all the usual installation errors, which could be calculated from its physical and thermal properties. But it introduces one new concern: radiation loss along the transparent fiber. This is not an error mode that has been recorded in the literature, but it amounts to only a small fraction of the sensor's total radiation error under usual conditions. In addition, there is a small (3 to 10 K) error introduced because of radiation lost through the sides of the hot fiber near the cavity, the Brewster loss. These new error modes are not viewed as deterrents to the use of the probe as a reference system for high temperatures, rather they are part of the correction which must be applied at final calibration.

11.5.4 Spatial Resolution

The sensing capsule is 1.25 mm in diameter and can be from 1 to 15 mm long. The optimum length is calculated on the basis of minimizing the radiation error of the cavity.

11.5.5 Temporal Resolution

Standard thin-film units will respond to 10 kHz fluctuations in temperature. Response is flatter than that of a thermocouple, rolling off at only 3 db per octave instead of 6.

11.6 ELECTRON NOISE THERMOMETERS[*]

The electron noise method of temperature measurement uses for its signal the voltage developed by thermal agitation of the electrons in a resistor. The voltage is small (on the order of microvolts) and at high frequencies (up to 1 GHz). But, of greatest importance, the signal is linearly related to the absolute temperature (for frequencies less than kT/h, defined later) by a known physical law. Furthermore, the voltage is independent of the resistor material. The signal using multiple measurements can be made independent of the resistor value and insensitive to background noise (electrical noise). The signal is broadband with a zero mean and a high bandwidth. These characteristics allow rejection, by filtering, of environmental noise without loss of measurement accuracy.

The theory is described by Decreton, et al.[22]

Electric noise due to thermal agitation was first described by Nyquist (1928)[23] and Johnson (1928).[24] An unloaded passive network always presents at its ends a voltage, Vn, fluctuating statistically around zero. This mean-squared noise voltage is given by

$$\overline{Vn^2} = 4kTR\frac{(hf)/(kT)}{\exp[(hf)/(kT)] - 1} \qquad (11.18)$$

where h and k are Planck's and Boltzmann's constants respectively, T the absolute temperature, f the frequency, and R the resistance. For temperatures above 100 K and frequencies below 1 GHz, Eq. 11.18 can be accurately approximated by

$$\overline{Vn^2} = 4kTR \qquad (11.19)$$

Equation 11.19 represents a frequency-independent white-noise signal. In a practical measurement, a given frequency bandwidth (df) is imposed, and the true rms voltage (Vn) is then given by

$$(Vn) = (4kTRdf)^{(1/2)} \qquad (11.20a)$$

When the current (I_n) is measured instead of the voltage, the equivalent model of the noise signal gives the rms current:

$$(I_n) = (4kTdf/R)^{(1/2)} \qquad (11.20b)$$

On multiplying Eqs. 11.20a and 11.20b, the noise power (P_n) is obtained:

$$(P_n) = 4kTdf \qquad (11.21)$$

which is independent of the resistance value and thus is strictly linearly related to the absolute temperature.

Sophisticated signal extraction techniques are required. This technique is useful mainly at the temperature extremes (very low and very high) when other sensors fail. The technique is promising. Results have been reported up to 1000°C with errors less than 0.5% of reading over a period of several months.[25]

11.7 ACOUSTIC VELOCITY PROBES[*]

The velocity of stress-wave propagation through a gas, liquid, or solid can be used to measure temperature. In gases and liquids, the acoustic velocity is used (normal stress or pressure-wave propagation), whereas both normal and shear stress waves have been used in solids. Two signal

---

[*] Substantial portions of the material in this section are taken from Ref. 8, with permission.

processing techniques are available: detection of resonance and measurement of the time-of-travel. These techniques have been used in both gases and solids, as described by Tasman and Richter.[26]

The acoustic velocity in a perfect gas is given by

$$a = [(C_p/C_v)RT]^{1/2} \tag{11.22}$$

Where C_p and C_v are the specific heats at constant pressure and at constant volume, respectively, R is the gas constant, and T is the temperature. Shear stress waves are not present in gases.

The normal stress propagation velocity in a solid (the extensional wave velocity) is

$$V_{ex} = [E/D]^{1/2} \tag{11.23}$$

where E and D are the modulus of elasticity and the density, respectively.

The shear stress propagation velocity in a solid is

$$V_{sh} = K[G/D]^{1/2} \tag{11.24}$$

where K is a shape factor, and G and D are the shear modulus and density respectively. Torsional waves sent down a cylindrical rod are one embodiment of shear waves. A typical Quartz crystal thermometer uses shear-wave propagation in the resonance mode to achieve its very high precision (on the order of .0001 K).

Tuning-fork resonance probes of pure sapphire have been run at 2000°C in a laboratory environment with an apparent stability of 1% or better by Bell et al.[27]

While capable of high precision under favorable conditions, the resonance mode is very sensitive to deposits or accidental contact between the resonator and its support. It is also inherently a line-averaging technique since it is the temperature of the oscillating stem, not that of the mass at the end that determines the resonant frequency. For those reasons, most acoustic velocity systems proposed for high-temperature work use the time-of-flight technique known as the *pulse-echo* technique.

Tasman and Richter[26] offer the following description of the pulse-echo system:

> The pulse-echo technique measures the transit time of a single sound pulse through a thin wire. A magnetostrictive transducer transforms an electrical pulse into a sound pulse, which is injected into a transmission line. Discontinuities in the line, and the end of the line, reflect (part of) the sound pulse, which is then converted back into electrical signals by the transmitting transducer. The actual sensor is the wire joining the discontinuities. Almost any kind of discontinuity will do to produce an echo, even a kink in the wire. One sensor may contain several discontinuities, apart from its end, and thus measure the longitudinal temperature profile over several consecutive sections.

For good resolution, the sensor material should have a large change in propagation velocity over the temperature range to be encountered. Thoriated tungsten is one candidate material. Tasman and Richter quote the round trip time difference between echos as increasing from about 4.5 to about 5.5 μs/cm when the temperature changes from 20 to 2700°C. Time differences can be measured to about ±1 ns, which means that a 50 mm long sensor could resolve temperature to ±0.2 K at 2000°C (in theory). Practical considerations seem to limit the resolution to about ±3 K for the 50 mm sensor. Tungsten is very sensitive to oxidation at high temperatures, and pure tungsten cannot be used as an acoustic velocity sensor because it continues to recrystallize at high temperatures, causing a continuous shift in calibration. Thoria blocks the grain growth, however, and the calibration of thoriated tungsten is stable (in the absence of oxygen).

11.8   TEMPERATURE-SENSITIVE COATINGS

Paints and crayons are available that are designed as temperature indicators,[29] up to 2500°F (1371°C) with a quoted accuracy of ±1%. Two types are available: phase change and color change. When the phase change materials melt, they yield easily discernible evidence that their event temperature has been exceeded. Color change materials are subtler and less easy to interpret. There have been some complaints of calibration shifts of these paints when used on heavily oxidized materials, which are believed due to alloying of the oxide with the paint.

These paints are nonmetallic and, therefore, have different radiation properties than metals. They should be used only over small areas of metallic surfaces (small compared with the metal thickness), or else their different emissivities will lead to a shift in the operating temperature of the parts. Since nonmetals tend to have higher emissivities than metals, the painted regions may have different radiation properties than the substrate material. Trial specimens should be checked if precise data are needed. If the specimens are to be heated by radiation from a high-temperature source, the different radiation properties can have a significant effect (2 to 3% in temperature).

Paints are useful in examining the distribution of temperature over a surface. Discrete spots should be used, rather than lines, since contact with molten material facilitates the melting of neighboring material even if it has not yet reached its own melting point.

The paints are cheap and easy to apply. They provide a good "first look" capability.

The principal disadvantages of the paints and crayons are that they require visual interpretation and they are one-shot, irreversible indicators. They read peak temperature during the test cycle, yet they cannot record whether the peak was reached during steady state or during soak-back.

11.9   FLOW RATE

The objective of flow rate measurement is to determine the quantity of matter flowing. In some instances a flowmeter returns this information directly, but in most cases the flowmeter signal is derived from some property of the flow: volume, heat transfer rate, momentum flux, etc. In most cases the flowmeter signal requires correction for pressure, temperature, or viscosity before the flow rate is known.

Interest in mass flow rate stems from a basic principle of engineering: The creation rate of mass is zero. If a flow is measured in mass flow units, then unless the pipe leaks or more fluid is added, the mass flow rate is the same everywhere along the pipe, regardless of changes in the density of the fluid flowing. This is not so for a volume-based measurement. The volume flow rate represented by a fixed mass flow rate of gas depends upon the molecular weight, the temperature, and the pressure of the gas. Hence, as either T or P change, the volume flow measure would change, even though the mass flow measure remained constant.

The term *SCFM (Standard Cubic Feet per Minute)* is frequently used in gas flow metering. The number of SCFM corresponding to a particular flow rate is a measure of the volume flow rate that would have been observed had the same mass flow rate of the same gas occurred at standard conditions of temperature and pressure. In other words, it is the volume flow rate that the actual mass flow rate would have produced had it been delivered at standard temperature and pressure.

11.9.1   Nomenclature

Mass: Quantity of matter. When weighed under standard gravitational conditions (i.e., on the average surface of the earth), one pound mass weighs one pound (exerts a force of one pound) and 1 kg exerts a force of 9.8 N. Symbol: lb_m or kg.

Mass Flow Rate: The rate at which matter passes the measuring location. Pounds mass (or kg) per unit time. Symbol: W or $\dot{m}$ depending on the author.

Volume Flow Rate: The volume occupied by the mass passing the measuring location. Cubic feet (or cubic meters) per unit time. Symbol: Q.

Standard Conditions: An arbitrarily chosen pair of values (different organizations may use different values) for temperature and pressure used to describe a standard state for measurement of the density of a gas. (e.g., 70°F and 14.7 psia, or 20°C and 760 mm Hg).

Standard Density (. . .of a gas): The density (lb_m per cubic foot or kg/m^3) of the gas in question when its temperature and pressure are those of Standard Conditions.

The following symbols and terms are commonly used in flow metering situations. In the rate statements that follow, the time base is taken as minutes or seconds.

$$W = \text{Mass flow rate, } lb_m/\text{min or kg/s}$$

$$Q = \text{Volume flow rate, } ft^3/\text{min or } m^3/s$$

also frequently used:

$$GPM = \text{Gallons Per Minute}$$

$$CFM = \text{Cubic Feet per Minute}$$

$$ACFM = \text{Actual Cubic Feet per Minute, } ft^3/\text{min}$$

$$SCFM = \text{Standard Cubic Feet per Minute, } ft^3/\text{min}$$

Consider a flow meter operating at 2 atm and 70°F. The density of a gas at those conditions is twice its density at Standard Conditions. Consider a flow through the meter such that 100 ft³ of gas at 70°F and 2 atm passes through the meter each minute. This 100 ft³ of gas would occupy 200 ft³ if it were at standard conditions. In this case, the flow rate could be described either as 100 ACFM (at 70°F and 2 atm) or 200 SCFM. If ACFM is quoted, then the temperature and pressure must also be specified. The term *SCFM* describes the mass flow, but expresses it in terms of the volume that that mass flow would occupy if it were at Standard Conditions.

11.9.2 Basic Principles Used in Flow Measurement

Measurement of flow rate can be accomplished using many different physical principles. The basic flow rate equation is

$$W = \rho A V \qquad (11.25)$$

where

$$W = \text{Flow rate, lb}_m/\text{sec (kg/s)}$$

$$A = \text{Area, ft}^2(\text{m}^2)$$

$$V = \text{Velocity, ft/s (m/s)}$$

$$\rho = \text{Density, lb}_m/\text{ft}^3(\text{kg/m}^3)$$

To measure flow, any metering system must provide enough information to evaluate all three terms. Usually, two of the terms are fixed, and flow is evaluated by observing the change in the remaining term. An orifice used on water (presumed to be constant density) fixes A and ρ, leaving W proportional to V. A variable area meter fixes ρ and V, leaving W proportional to A. Any combination of physical laws that permits evaluation of ρ, A, and V can form the basis for a flow metering system.

Flow meters can be divided into three generic groups depending on their sensing principle: conservation-based, rate-based, and dynamic.

The first group consists of flow meters which depend upon a conservation principle for their output. There are three conservation laws which can be related to mass flow: conservation of mass, conservation of momentum, and conservation of energy. Conservation of momentum provides the basis for a large class of meters: orifices, nozzles, venturi meters, drag disks, and obstruction meters. Conservation of energy (usually thermal energy) has been used in the construction of flowmeters for small flow rate. The conservation of mass law is used implicitly in all systems.

The second large group of flowmeters depend upon *rate equations*. There are many natural phenomena whose rate depends upon fluid velocity: viscous drag, heat transfer, mass transfer, displacement of a tracer particle, etc. Any rate process sensitive to flow rate can be used as the basis for a flowmeter.

The third class of flowmeters, the dynamic meters, contains those whose signal depends on some dynamic aspect of a flow field. This class of meters depends upon the repeatability of certain unstable behaviors of flow fields; for example, vortex precession in a swirling flow in an adverse pressure gradient and vortex shedding from a bluff body.

In the remaining sections the more common types of flowmeters will be discussed in terms of the following items: physical appearance, output data and peripheral equipment, equations computed for off-design operation, and sources for more information.

11.9.3 Orifice, Nozzle, and Venturi Meters

Orifice, nozzle, and venturi meters are momentum-based, fixed-area meters. The flow-related signal reveals the pressure difference between two points on the meter body as shown in Fig. 11.27. These meters accelerate the fluid stream by imposing a contraction on the flow area and then decelerate the flow by expanding back to the initial pipe diameter. As the fluid accelerates, its static pressure goes down. As it decelerates, the pressure rises again but not without loss. Losses are related to irreversibilities in the flow field, such as eddy structures and turbulence. Mechanical energy dissipated by these mechanisms comes from flow work done by the main stream. As a result, there is a loss in total pressure.

A principal determinant of "Which meter to choose?" is the irrecoverable loss in pressure associated with the meter. As shown in Fig. 11.27, orifice meters have the largest loss for a given signal among these three candidates and venturi meters have the smallest. As might be expected, venturi meters are more expensive than orifice meters. *Fluid Meters—Their Theory and Application*[31] provides extensive data on the losses for different styles of orifice, nozzle, and venturi meters, and data on the precautions that must be observed in installing them.

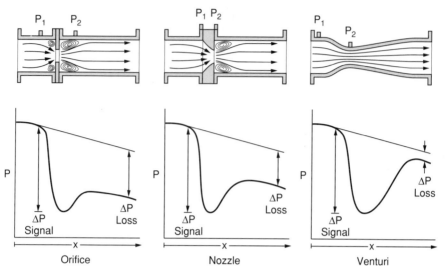

Fig. 11.27 Orifice, nozzle, and venturi meter pressure distributions. (From Moffat,[30] with permission.)

Measurement of flow rate with these meters requires two pressure sensors and one temperature sensor. The accuracy of the flow rate measurement will depend upon the accuracy of these instruments.

The equation given below takes its nomenclature from the previous reference.

$$W_h = 359 \ C F d^2 F_a Y \sqrt{h_w \gamma} \tag{11.26}$$

where

W_h = Flow rate, lb_m/hr

C = Coefficient of discharge, dimensionless

F = Coefficient of approach, dimensionless

d = Minimum diameter of the orifice, nozzle, or venturi, inches (or m)

F_a = Coefficient of expansion, dimensionless

Y = Compressibility factor, dimensionless

h_w = Pressure difference, $P_1 - P_2$, inches of water

γ = Density of the fluid flowing, at the location of measurement of P_1, lb_m/ft^3

Of the seven required terms, two are data or derived from data (h_w and γ), while five are calibration coefficients tabulated within the reference. The product of C and F is tabulated directly (K).

The coefficient K is a function of the flow rate, hence Eq. 11.26 cannot be solved explicitly— an iterative procedure is required. A first estimate of $K = C \times F = 0.60$ should be used for a first try. Using this K value, the calculated flow will generally be within a few percent and will serve to fix the exact value of K (within fractions of a percent).

It is apparent from Eq. 11.26 that the signal (ΔP) is proportional to W_n^2, the square of the flow rate. If the flow is fluctuating and an average ΔP is observed, then this corresponds to the time-average of the square of the instantaneous flow, which is not equal to the square of the time-average flow. As a result, orifice, nozzle, and venturi meters are not recommended for pulsating flows.

The most authoritative source of data regarding installation precautions for this class of meters is found in Reference 31. A good general reference is Benedict.[32]

In general, orifice flowmeters are accurate within ±1/2% of flow rate even without individual calibrations, when used properly. To achieve this accuracy, the orifice meter must be installed in strict compliance with the prescribed conditions for the approach and exit flow paths. The specifications in Reference 31 cannot be abridged without loss of generality and should be treated as the final arbiter of an installation.

11.9.4   Variable Area Meters

A variable area meter consists of a tapered tube (usually of glass) with a float inside as shown in Fig. 11.28. The clearance area between the outer diameter of the float and the inner diameter of the tube is a function of float position. When flow passes through the meter it lifts the float under constant pressure drop conditions until the clearance area is sufficient to pass the existing flow. The output of the meter is *float position*. For gas flow measurements, pressure and temperature must be measured upstream of the float to establish the density.

Variable area meters are usually purchased for use in a specific application and delivered calibrated for those conditions. The scale is calibrated directly in flow rate units. If the pressure and temperature of the flow remain at their design values, the float position gives flow rate directly.

A typical calibration legend for such a meter would read:

Full scale = 1.44 SCFM

Gas specific gravity of 1.38

When metered at 25 psig and 70°F

The float position then directly reads in percentage of full scale flow provided that the inlet pressure is 25 psig, the inlet temperature is 70°F, and the specific gravity of the gas flowing is 1.38.

For off-design operation of the variable area meter it is necessary to correct for changes in pressure, temperature, or specific gravity.

A sufficiently accurate equation for gas flow service can be derived from a force balance on the float in combination with Eq. 11.25, thus resulting in the fundamental equation for flow measurement.

$$W_{USE} = W_{calib}\sqrt{\frac{\rho_{USE}}{\rho_{calib}}} \qquad (11.27)$$

Where: W_{USE} = Mass flow rate under actual conditions of use, lb_m/sec (or kg/sec)

   W_{calib} = Mass flow rate under calibration conditions, for the same float position; lb_m/sec or (kg/sec)

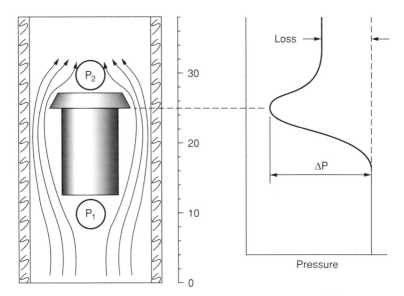

Fig. 11.28   Variable area flowmeter pressure distribution. (From Moffat,[30] with permission.)

ρ_{USE} = Density of the fluid flowing, lb_m/ft^3 or (kg/m^3)

ρ_{calib} = Density of the calibration fluid, lb_m/ft^3 or (kg/m^3)

Example.   Assume that the meter previously described is to be used in a new application where the fluid is air (specific gravity = 1.00), the pressure is 100 psig, and the temperature is 70°F. What would be the full scale flow rate in SCFM (AIR)?

$$W_{USE} = W_{calib}\sqrt{\frac{\rho_{USE}}{\rho_{calib}}}$$

$$\rho_{USE} = \frac{P_{USE}}{R_{AIR}T_{USE}} = \frac{(100 + 14.7)(144)}{53.3 \times (460 + 70)}$$

$$= 0.585\frac{lb_m}{ft^3}$$

$$\rho_{calib} = 1.38\left(\frac{P_{calib}}{R_{AIR}T_{calib}}\right)$$

$$= 1.38\frac{(25 + 14.7)(144)}{(53.3)(460 + 70°F)} = .279$$

$$W_{calib} = (1.44)(1.38)(.075) = 0.149\ lb_m/sec$$

$$W_{USE} = (.149)\sqrt{\frac{.585}{.279}} = 0.216\ lb_m/sec$$

$$SCFM_{(USE)} = \frac{W_{USE}}{\rho_{STD,AIR}} = \frac{W_{USE}}{0.075} = 2.877\ SCFM$$

This correction equation was derived under the assumption that the float was held up only by the pressure force (i.e., buoyancy and viscous drag were both negligible). When metering liquids of high density, the buoyant force must be accounted for, and if the viscosity of the fluid exceeds the *viscosity immunity ceiling* for a given meter, then the viscous drag must be taken into account. Precautions for dealing with these two cases are covered by the manufacturer's instructions. A meter purchased for gas service cannot be converted to liquid service by the preceding equation, or vice versa; a new calibration is required.

Variable area meters are generally limited in accuracy to ±1% or ±2% of full-scale reading.

It is important to measure the actual density of the fluid flowing in the meter to ensure that the calibration conditions are properly met. Pressure and temperature must be measured at the meter, just upstream of the float.

Irreversibilities involved in the mixing of the annular jet introduce losses in pressure that are nearly independent of flow rate and roughly equal to the pressure required to hold up the float. More accurate meters generally have higher losses (up to 3 or 4 psi used with air).

11.9.5   Laminar Flowmeters

Commercial laminar flowmeters consist of a *matrix* or *core* of small diameter passages arranged so that the pressure drop across this core can be measured. The meter must be sized properly so the flow in these passages will remain laminar, even at the highest rated flow. If the flow remains laminar, then the pressure drop is linearly related to the volume flow rate through the core. General commercial practice is to provide flow straightening sections upstream and downstream of the core and measure the pressures in the space between the flow straighteners and the core, as shown in Fig. 11.29.

Laminar flowmeters produce a pressure difference, usually 0–8 inches of water, which must be measured using an appropriate auxiliary instrument. The meter responds to volume flow, hence it is also necessary to measure the density of the fluid flowing. If the composition is known, density can be calculated knowing temperature and pressure at the inlet.

The pressure drop across a well-designed laminar flowmeter is linearly proportional to the volume flow rate (ACFM) divided by the viscosity of the fluid flowing. The pressure drop is independent of density.

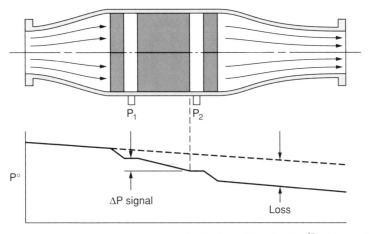

Fig. 11.29   The laminar flowmeter pressure distribution. (From Moffat,[30] with permission.)

Each meter is accompanied by a calibration curve, which typically reads:

Air Flow in Cubic Feet per Minute at 70°F and 29.92 Inches of Mercury Absolute vs. Pressure
   Drop, Inches of Water.

The SCFM flow rate can be taken directly from the calibration curve if the meter is being used
at 29.92 inches of Mercury and 70°F. If the pressure is 2 atm, however, then the density, the mass
flow, and the number of SCFM corresponding to the same pressure drop signal would be doubled.
Tables of pressure corrections, often provided with each meter, simplify this calculation.

If the temperature of the metered flow is higher than 70°F, the density is less than standard
density and the viscosity of the flow is higher. Temperature correction factors, often provided in
tabular form, account for both effects. The temperature correction factor for air flow is roughly
proportional to absolute temperature raised to the 1.7 power, and it accounts for both the decrease
in density and the increase in viscosity.

The equation for a laminar flow meter is:

$$SCFM = (CHART\ CFM)(CP)(CT)$$

or

$$SCFM = (CHART\ CFM)\left(\frac{P_{act}}{29.92}\right)\left(\frac{530}{460 + T_{act}}\right)^{1.7} \qquad (11.28)$$

Laminar flow meters respond linearly to flow, hence they can be used to directly measure the
time-average of a fluctuating flow. Commercial meters will usually average properly up to about
100 Hz.

11.9.6   Instability Meters

Two modern flow metering systems derive their signal from the frequency of occurrence of an
unstable fluid-dynamic phenomenon. One uses a swirling flow in a divergent passage to generate
a precessing stall region. The other introduces a prismatic bluff body into a pipe and generates a
periodic vortex trail. In each case, the frequency of the unstable event is related to the volume flow
rate through the meter. Each claims 100:1 useable range, and each produces a pulse train whose
repetition rate is linearly related to flow.

The Swirlmeter (Registered Trademark of the Fischer and Porter Co.) is shown schematically
in Fig. 11.30. The meter has no moving parts. Stationary blades impart a swirling motion to the
flow, which then enters a diverging passage. The rotating flow field develops a stall (a region of
relatively low axial velocity), which rotates in the diverging passage. An anemometer button in the
wall senses the low-speed region each time it passes, generating a pulse train. The frequency of this
pulse train is linearly related to the volume flow rate.

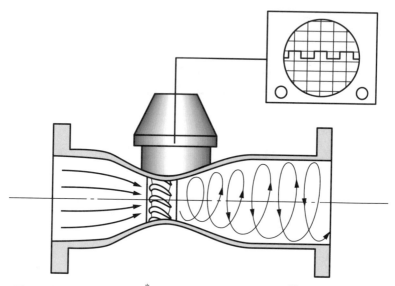

Fig. 11.30 The swirlmeter,* schematically. (From Moffat,[30] with permission.)

Mass flow rate calculation requires knowledge of the volume flow and the density. The equation is

$$\text{ACFM} = \frac{\text{total counts per minute}}{\text{calibration constant}} \tag{11.29}$$

$$W = \text{ACFM} \times \rho_{\text{act}} \tag{11.30}$$

$$\text{SCFM} = \text{ACFM} \times \frac{\rho_{\text{act}}}{\rho_{\text{STD}}} \tag{11.31}$$

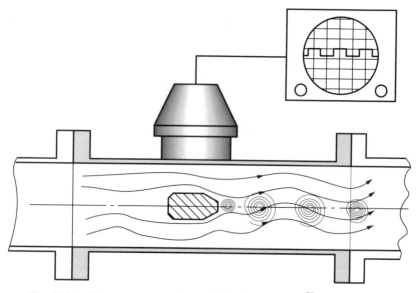

Fig. 11.31 The vortex meter, schematically. (From Moffat,[30] with permission)

* Registered trade mark, Fischer and Porter Co.

Vortex-shedding flowmeters use a prismatic bluff body to generate the instability and one of several means to sense the oscillations of the stagnation streamline, which occur when the bluff body is shedding a vortex trail. A plan view schematic is shown in Fig. 11.31.

Again, these meters can operate over a turn-down ratio of 100/1 or better and claim $\pm 0.25\%$ accuracy.

The advantages of the instability meters are due to their pulse train signal, which is not easily obscured by random noise and can be transmitted over long distances.

REFERENCES

1. R. J. Moffat, *Experimental Methods in the Thermosciences*, Department of Mechanical Engineering, Stanford University, 1978, pp. 1–26.

2. R. J. Moffat, "The Gradient Approach to Thermocouple Circuitry," *Temperature, Its Measurement and Control in Science and Industry*, Vol. 3, Part 2, pp. 33–38, Reinhold Publishing Co., New York, 1962.

3. R. J. Moffat, *Experimental Methods in the Thermosciences*, Department of Mechanical Engineering, Stanford University, 1980, pp. 2–3.

4. Robert P. Benedict, *Fundamentals of Temperature, Pressure, and Flow Measurement*, 2nd ed., John Wiley & Sons, New York, 1977.

5. E. O. Doebelin, *Measurement Systems: Application and Design*, McGraw-Hill, New York, 1966.

6. O. A. Geraschenko and N. N. Ionova, "Thermal EMF of Plated Thermocouples" UDC 536, 532, Translated from *Ismeritel'naya Tekhnika*, No. 1, Jan. 1966, pp. 65–66.

7. C. Tallman, Personal Communication, 1964.

8. R. J. Moffat, *High Temperature Meaurements in Gases—The State of the Art, 1985*, Moffat Thermosciences, Los Altos, CA, 1985.

9. Omega Engineering, Inc., *Temperature Measurement Handbook and Encyclopedia*, Stamford, CT., 1984, pp. E1–E18.

10. Rosemount Engineering Product Bulletin 2086, Rev. 10/78.

11. Rosemount Engineering Product Bulletin 2108, Rev. 6/74.

12. Harry N. Norton, *Sensor and Analyser Handbook*, Prentice-Hall, Englewood Cliff, NJ, 1982, p. 337.

13. Robert P. Benedict, *Fundamentals of Temperature, Pressure, and Flow Measurement*, 2nd ed., John Wiley & Sons, 1977, pp. 38–9.

14. Omega Engineering, Inc., *Temperature Measurement Handbook*, Stanford, CT, 1983, pp. T45–T48.

15. Victory Engineering Corporation Technical Bulletin MCT-181, Springfield, NJ, 1968.

16. Yellow Springs Instrument Co., "YSI Precision Thermistor Products," Yellow Springs, OH, 1980.

17. Omega Engineering, Inc.,*Temperature Measurement Handbook and Encyclopedia*, Stamford, CT, 1985, pp. T75–T76.

18. J. S. Steinhart and S. R. Hart, "Calibration Curves for Thermistors," *Deep Sea Research*, Vol. 15, 1968, p. 497.

19. A. S. Bennet, "The Calibration of Thermistors over the Temperature Range of 0–30°C," *Deep Sea Research*, Vol. 19, No. 157, 1972.

20. L. Seren, C. B. Panchal, and D. M. Rote, "Temperature Sensors for OTEC Applications," *ANL/OTEC-PS-12*, Argonne National Laboratory, U.S. Department of Energy, May 1984.

21. G. Bosson, F. Guttmann, and L. M. Simmons, "A Relationship between Resistance and Temperature of Thermistors," *Journal of Applied Physics*, Vol. 21, 1950, pp. 1267–68.

22. M. Decreton, L. Binard, C. Delrez, W. Hebel, and W. Schubert, "High-Temperature Measurements by Noise Thermometer," *High Temperature–High Pressure*, Vol. 12, 1980, pp. 395–402.

23. H. Nyquist, *Physics*, Rev. ed., Vol. 32, 1928, pp. 110–13.

24. J. B. Johnson, *Physics*, Rev. ed., Vol. 32, 1928, pp. 97–110.

25. H. Brixy, R. Hecker, K. F. Rittinghaus, and H. Howener, "Applications of Noise Thermometry in Industry under Plant Conditions," *Temperature, Its Measurement and Control in Science and Industry*, Reinhold Publishing Co., NY, 1982, pp. 1225–37.

26. H. A. Tasman and J. Richter, "Unconventional Methods of Temperature Measurement," *High Temperatures–High Pressure*, Vol. 11, 1979, pp. 87–101.

27. J. F. W. Bell, A. A. Fathamani, and T. N. Seth, 1975, cited in Von der Hardt et al., 1975, vol. 2, pp. 649–80.

28. S. S. Fam, L. C. Lynworth, and E. H. Carnevale, *Instrument Control Systems*, Vol. 42, 1969, pp. 107–10.

29. Omega Engineering, Inc., *The Temperature Measurement Handbook and Encyclopedia*, Stamford, CT, 1985, p. Q4.

30. R. J. Moffat, "Introduction to Flow Rate Measurement," *Experimental Methods in the Thermosciences*, Mechanical Engineering Department, Stanford University, 1978.

31. *Fluid Meters—Their Theory and Application*, 6th ed. (ASME, 1971).

32. R. P. Benedict, *Fundamentals of Temperature, Pressure, and Flow Measurement*, 2nd ed., John Wiley & Sons, New York, 1977.

CHAPTER 12
SIGNAL PROCESSING AND TRANSMISSION

WILLIAM J. KERWIN

Department of Electrical and Computer Engineering
University of Arizona
Tucson, Arizona

JOSEPH L. GARBINI
RICHARD A. DOWNS

Department of Mechanical Engineering,
University of Washington
Seattle, Washington

A. G. BOLTON

South Australian Institute of Technology

KENNETH C. FISCHER

Corporate Technology Department
Allen-Bradley Company
Milwaukee, Wisconsin

12.1 PASSIVE SIGNAL PROCESSING

12.1.1 Introduction

We will be concerned here with systems having an electrical input and output. Very powerful filtering techniques are available for this case, and frequently the electrical system is the lowest in cost.

We can specify either a phase characteristic or a magnitude characteristic, but only rarely can both be specified. The greater the filtering requirement—that is, the steeper the slope in the magnitude beyond the cut-off frequency—the greater will be the cost. If we need to eliminate specific finite frequencies, then we must use a more sophisticated and more costly filter having response zeros at finite frequencies. In addition, as we obtain a steeper slope beyond cut-off, we worsen the time domain response; that is, the overshoot to a step input is greater and the ringing continues for a longer time, leading to a very long settling time. It may therefore be necessary to temper the steady-state filtering requirements in order to achieve acceptable time domain response. We will start with some details of notation and some examples.

LaPlace Transform

We will use the Laplace operator, $s = \sigma + j\omega$. Steady-state impedance is thus Ls and $1/Cs$, respectively, for an inductor (L) and a capacitor (C), and admittance is $1/Ls$ and Cs.

Transfer Functions

We will consider only lumped, linear, constant, bilateral elements and we will define the transfer function $T(s)$ as response over excitation.

$$T(s) = \frac{\text{signal output}}{\text{signal input}} = \frac{N(s)}{D(s)}$$

The roots of the numerator polynomial $N(s)$ are the zeros of the system and the roots of the denominator $D(s)$ are the poles of the system (the points of infinite response). If we substitute $s = j\omega$ into $T(s)$ and separate the result into real and imaginary parts (numerator and denominator) we obtain

$$T(j\omega) = \frac{A_1 + jB_1}{A_2 + jB_2} \tag{12.1}$$

Then the magnitude of the function, $|T(j\omega)|$, is

$$|T(j\omega)| = \left(\frac{A_1^2 + B_1^2}{A_2^2 + B_2^2}\right)^{\frac{1}{2}} \tag{12.2}$$

and the phase $\overline{T(j\omega)}$ is

$$\overline{T(j\omega)} = \tan^{-1}\frac{B_1}{A_1} - \tan^{-1}\frac{B_2}{A_2} \tag{12.3}$$

Analysis. Although mesh or nodal analysis can always be used, since we will consider only ladder networks we will use a method commonly called "linearity," or "working your way through." The method starts at the output and assumes either one volt or one ampere as appropriate and uses Ohm's law and Kirchoff's current law only.

Example 12.1. Determine the transfer function of the circuit of Fig. 12.1.
   Let $v_o = 1$ V; then

$$i_5 = 1, \; i_4 = s, \; i_3 = 1 + s$$
$$v_1 = v_o + i_3(2s) = 1 + 2s(1 + s)$$
$$i_2 = sv_1 = 2s^3 + 2s^2 + s$$
$$i_1 = i_2 + i_3 = 2s^3 + 2s^2 + 2s + 1$$
$$v_i = v_1 + i_1 = 2s^3 + 4s^2 + 4s + 2$$

$$\frac{v_o}{v_i} = T(s) = \frac{1}{2s^3 + 4s^2 + 4s + 2}$$

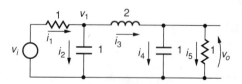

Fig. 12.1   Doubly-terminated third-order low-pass filter (in Ω, h, f).

Example 12.2.   Determine the steady-state magnitude and phase of the transfer function derived in Example 12.1.

$$\frac{v_o}{v_i} = T(s) = \frac{1}{2s^3 + 4s^2 + 4s + 2} = \frac{1}{2}\left(\frac{1}{s^3 + 2s^2 + 2s + 1}\right)$$

Substituting $s = j\omega$ (steady state)

$$T(j\omega) = \frac{1}{2}\left(\frac{1}{1 - 2\omega^2 + j(2\omega - \omega^3)}\right)$$

and from Eq. 12.2

$$|T(j\omega)| = \frac{1}{2}\left(\frac{1}{\sqrt{[(1 - 2\omega^2)^2 + (2\omega - \omega^3)^2]}}\right) = \frac{1}{2}\left(\frac{1}{\sqrt{(\omega^6 + 1)}}\right)$$

and from Eq. 12.3

$$\underline{/T(j\omega)} = 0° - \tan^{-1}\frac{2\omega - \omega^3}{1 - 2\omega^2}$$

The values used for the circuit of Fig. 12.1 were normalized; that is, they are all near unity in ohms, henrys, and farads. These values simplify computation and, as we will see later, can easily be scaled to any desired set of actual element values. In addition, this circuit is low-pass due to the shunt capacitors and the series inductor. By low-pass we mean a circuit that passes the lower frequencies and attenuates higher frequencies. The cut-off frequency is the point at which the magnitude is 0.707 (-3 dB) of the dc level and is the dividing line between the pass band and the stop band. In the above example we see that the magnitude of v_o/v_i at $\omega = 0$ (DC) is 0.50 and that at $\omega = 1$ rad/s we have

$$|T(j\omega)| = \frac{1}{2}\frac{1}{\sqrt{(\omega^6 + 1)}} = 0.3535 \tag{12.4}$$

Thus we see that the normalized element values used here give us a cut-off frequency of 1 rad/s.

12.1.2   Low-Pass Filter Functions

The most common function in signal processing is the Butterworth.[1] It is a function that has only poles (i.e., no finite zeros) and has the flattest magnitude possible in the pass band. This function is also called MFM (maximally flat magnitude). The derivation of this function is illustrated by taking a general all-pole function of third-order with a dc gain of 1 as below:

$$T(s) = \frac{1}{as^3 + bs^2 + cs + 1} \tag{12.5}$$

The squared magnitude is

$$|T(j\omega)|^2 = \frac{1}{(1 - b\omega^2)^2 + (c\omega - a\omega^3)^2} \tag{12.6}$$

or

$$|T(j\omega)|^2 = \frac{1}{a^2\omega^6 + (b^2 - 2ac)\omega^4 + (c^2 - 2b)\omega^2 + 1} \tag{12.7}$$

MFM requires that the coefficients of the numerator and the denominator match term by term (or be in the same ratio) except for the highest power. Since the numerator and denominator DC values are equal (both unity) we have the matching condition for this case.

Therefore

$$c^2 - 2b = 0; \; b^2 - 2ac = 0 \tag{12.8}$$

We will also impose a normalized cut-off (-3 dB) at $\omega = 1$ rad/s; that is,

$$|T(j\omega)|_{\omega=1} = \frac{1}{\sqrt{(a^2 + 1)}} = 0.707 \tag{12.9}$$

Thus we find $a = 1$, then $b = 2$, $c = 2$ are solutions to the flat magnitude conditions of Eq. 12.8 and our third-order Butterworth function is

$$T(s) = \frac{1}{s^3 + 2s^2 + 2s + 1} \tag{12.10}$$

Table 12.1 gives the Butterworth denominator polynomials up to $n = 5$.
    In general, for all Butterworth functions the normalized magnitude is

$$|T(j\omega)| = \frac{1}{\sqrt{(\omega^{2n} + 1)}} \tag{12.11}$$

Note that this is down 3 dB at $\omega = 1$ rad/s for all n.
    This may, of course, be multiplied by any constant less than one for circuits whose DC gain is deliberately set to be less than one. The circuit of Fig. 12.1 has a DC gain of 0.5.

Example 12.3.   A low-pass Butterworth filter is required whose cut-off frequency (-3 dB) is 4 kHz and in which the response must be down 45 dB at 10 kHz. Normalizing to a cut-off frequency of 1 rad/s, the -45 dB frequency is

$$\frac{10k}{4k} = 2.5 \text{ rad/s}$$

thus

$$-45 = 20 \log \frac{1}{\sqrt{2.5^{2n} + 1}}$$

therefore $n = 5.65$. Since n must be an integer, a sixth-order filter is required for this specification.

    There is an extremely important difference between the singly-terminated (DC gain $= 1$) and the doubly-terminated filters (DC gain $= 0.5$). As was shown by John Orchard,[2] the sensitivity in the pass band (ideally at maximum output) to all L, C components in an L, C filter with *equal* terminations is *zero*. This is true regardless of the circuit.
    This, of course, means component tolerances and temperature coefficients are of much less importance in the doubly-terminated (equal values) filter. For this type of Butterworth low-pass filter (normalized to equal $1 - \Omega$ terminations), Takahasi[3] has shown that the normalized element values are exactly given by

$$L, C = 2 \sin \left(\frac{(2k - 1)\pi}{2n} \right) \tag{12.12}$$

for any order, n, where k is the L or C element from 1 to n.

TABLE 12.1   BUTTERWORTH DENOMINATOR POLYNOMIALS

$$s + 1$$
$$s^2 + \sqrt{2}s + 1$$
$$s^3 + 2s^2 + 2s + 1$$
$$s^4 + 2.6131s^3 + 3.4142s^2 + 2.6131s + 1$$
$$s^5 + 3.2361s^4 + 5.2361s^3 + 5.2361s^2 + 3.2361s + 1$$

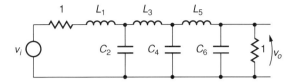

Fig. 12.2 Doubly-terminated sixth-order low-pass filter (in Ω, h, f).

Example 12.4. Design a normalized ($\omega_{-3dB} = 1$ rad/s) doubly-terminated (i.e., source and load = 1 Ω) Butterworth low-pass filter of order 6; that is, $n = 6$. The form of the filter is shown in Fig. 12.2.

The element values from Eq. 12.12 are

$$L_1 = 2 \sin \frac{(2 - 1)\pi}{12} = 0.5176 \text{ H}$$

$$C_2 = 2 \sin \frac{(4 - 1)\pi}{12} = 1.4141 \text{ F}$$

$$L_3 = 2 \sin \frac{(6 - 1)\pi}{12} = 1.9319 \text{ H}$$

The values repeat for C_4, L_5, C_6 so that

$$C_4 = L_3, \; L_5 = C_2, \; C_6 = L_1$$

Thomson Functions

The Thomson function is one in which the time delay of the network is made maximally flat.[4] This implies a linear phase characteristic since the steady-state time delay is the negative of the derivative of the phase. This function has excellent time-domain characteristics and is used wherever excellent step response is required. These functions have very little overshoot to a step input and have far superior settling times compared to the Butterworth functions. The slope near cut-off is more gradual than the Butterworth. Table 12.2 gives the Thomson denominator polynomials.[5] The numerator is a constant equal to the dc gain of the circuit multiplied by the denominator constant. The cut-off frequencies are *not* all 1 rad/s. They are given in Table 12.2.

Chebyshev Functions

A second function defined in terms of magnitude, the Chebyshev has an equal ripple character within the pass band.[6] The ripple is determined by ϵ below.

$$\epsilon = \sqrt{(10^{A/10} - 1)} \tag{12.13}$$

where $A = $ dB of ripple; then for a given order n, we define v.

$$v = \frac{1}{n} \sinh^{-1}\left(\frac{1}{\epsilon}\right) \tag{12.14}$$

Table 12.3 gives denominator polynomials for the Chebyshev functions. In all cases, the cut-off frequency (defined as the end of the ripple) is 1 rad/s. The -3 dB frequency for the Chebyshev function is

TABLE 12.2   THOMSON DENOMINATOR POLYNOMIALS

	ω_{-3dB} (in rad/s)
$s + 1$	1.0000
$s^2 + 3s + 3$	1.3617
$s^3 + 6s^2 + 15s + 15$	1.7557
$s^4 + 10s^3 + 45s^2 + 105s + 105$	2.1139
$s^5 + 15s^4 + 105s^3 + 420s^2 + 945s + 945$	2.4274

TABLE 12.3   CHEBYSHEV DENOMINATOR POLYNOMIALS

$$s + \sinh v$$

$$s^2 + (\sqrt{2}\,\sinh v)s + \sinh^2 v + \tfrac{1}{2}$$

$$(s + \sinh v)[s^2 + (\sinh v)s + \sinh^2 v + \tfrac{3}{4}]$$

$$[s^2 + (0.75637\,\sinh v)s + \sinh^2 v + 0.85355] \times [s^2 + (1.84776\,\sinh v)s + \sinh^2 v + 0.14645]$$

$$(s + \sinh v)[s^2 + (0.61803\,\sinh v)s + \sinh^2 v + 0.90451] \times [s^2 + (1.61803\,\sinh v)s + \sinh^2 v + 0.34549]$$

$$\omega_{-3\mathrm{dB}} = \cosh\left[\frac{\cosh^{-1}(1/\epsilon)}{n}\right] \tag{12.15}$$

The magnitude in the *stop band* ($\omega > 1$ rad/s) for the normalized filter is

$$|T(j\omega)|^2 = \frac{1}{1 + \epsilon^2\cosh^2(n\cosh^{-1}\omega)} \tag{12.16}$$

for the singly-terminated filter. For equal terminations the above magnitude is multiplied by one-half ($\tfrac{1}{4}$ in Eq. 12.16).

Example 12.5. What order of singly-terminated Chebyshev filter having 0.25 dB ripple (A) is required if the magnitude must be -60 dB at 15 kHz and the cut-off frequency (-0.25 dB) is to be 3 kHz? The normalized frequency for a magnitude of -60 dB is

$$\frac{15\mathrm{k}}{3\mathrm{k}} = 5 \text{ rad/s}$$

Thus for a ripple of $A = 0.25$ dB, we have from Eq. 12.13

$$\epsilon = \sqrt{(10^{A/10} - 1)} = 0.2434$$

and solving Eq. 12.16 for n with $\omega = 5$ rad/s and $|T(j\omega)| = -60$ dB, we obtain $n = 3.93$. Therefore we must use $n = 4$ to meet these specifications.

Inverse Chebyshev Function

The inverse Chebyshev function has an MFM pass band and an equal ripple stop band. This requires finite $j\omega$-axis zeros. We will consider only the two-zero function since it is adequate for most requirements. This function has two advantages over the Butterworth function: (1) a steeper cut-off slope for a given order and (2) the zero can be placed at any specific frequency in the stop band where, for example, a particular signal must be highly attenuated.

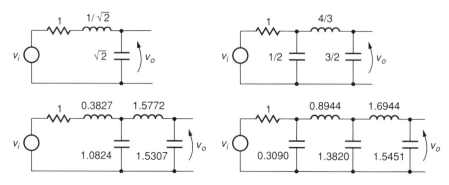

Fig. 12.3   Singly-terminated Butterworth filter element values (in Ω, h, f).

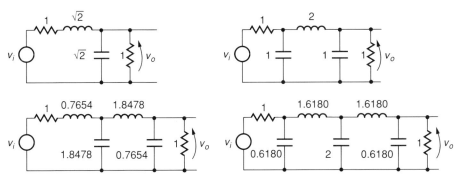

Fig. 12.4 Doubly-terminated Butterworth filter element values (in Ω, h, f).

12.1.3 Low-Pass Filters

Introduction

Normalized element values are given here for both singly- and doubly-terminated filters. The source and load resistors are normalized to 1 Ω. Scaling rules will be given in Section 12.3 that will allow these values to be modified to any specified impedance value and to any cut-off frequency desired. In addition, we will cover the transformation of these low-pass filters to high-pass or band-pass filters.

Butterworth Filters

For $n = 2$, 3, 4, or 5, Fig. 12.3 gives the element values for the singly-terminated filters and Fig. 12.4 gives the element values for the doubly-terminated filters. All cut-off frequencies (-3 dB) are 1 rad/s.

Thomson Filters

Singly- and doubly-terminated Thomson filters of order $n = 2, 3, 4, 5$ are shown in Figs. 12.5 and 12.6. All time delays are one second. The cut-off frequencies are given in Table 12.2.

Chebyshev Filters

The amount of ripple can be specified as desired, so that only a selective sample can be given here. We will use 0.1 dB, 0.25 dB and 0.5 dB. All cut-off frequencies (end of ripple for the Chebyshev function) are at 1 rad/s. Since the maximum power transfer condition precludes the existence of an equally terminated even-order filter, only the odd orders are given for the doubly-terminated case. Figure 12.7 gives the singly-terminated Chebyshev filters for $n = 2$, 3, 4, and 5 and Fig. 12.8 gives the doubly-terminated Chebyshev filters for $n = 3$ and $n = 5$.

Inverse Chebyshev Filters

For the two-zero, three-pole case shown in Fig. 12.9, the element values are given in Table 12.4 for various zero positions. In addition, the stop band peak return magnitude is also given. Note that R_L is either 1 Ω or ∞, so that both singly- and doubly-terminated filters are included.

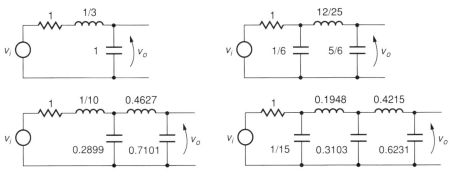

Fig. 12.5 Singly-terminated Thomson filter element values (in Ω, h, f).

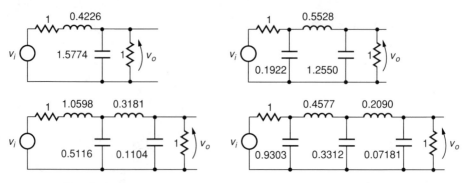

Fig. 12.6 Doubly-terminated Thomson filter element values (in Ω, h, f).

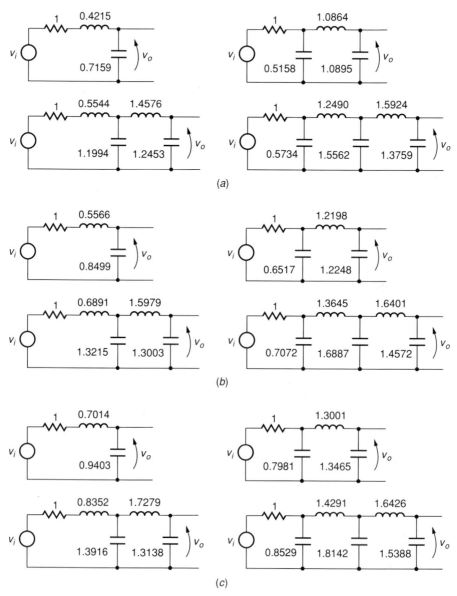

Fig. 12.7 Singly-terminated Chebyshev filter element values (in Ω, h, f): (*a*) 0.1 dB ripple; (*b*) 0.25 dB ripple; (*c*) 0.50 dB ripple.

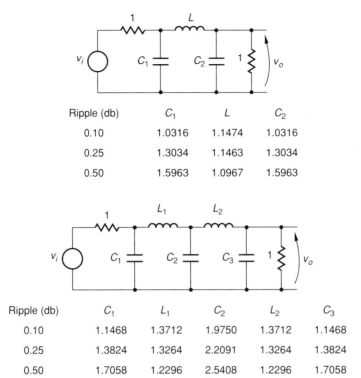

Fig. 12.8 Doubly-terminated Chebyshev filter element values (in Ω, h, f).

Ripple (db)	C_1	L	C_2
0.10	1.0316	1.1474	1.0316
0.25	1.3034	1.1463	1.3034
0.50	1.5963	1.0967	1.5963

Ripple (db)	C_1	L_1	C_2	L_2	C_3
0.10	1.1468	1.3712	1.9750	1.3712	1.1468
0.25	1.3824	1.3264	2.2091	1.3264	1.3824
0.50	1.7058	1.2296	2.5408	1.2296	1.7058

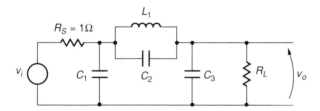

Fig. 12.9 Two-zero, three-pole inverse Chebyshev filter.

TABLE 12.4 TWO-ZERO, THREE-POLE INVERSE CHEBYSHEV FILTER ELEMENT VALUES (IN Ω, H, f)

$\omega_{-\infty}$	ω_{max}	$|T(j\omega)|_{\omega_{max}}$(dB)	R_L	C_1	L_1	C_2	C_3
2	3.4641	-29.87	1	0.8172	1.6344	0.1530	0.8172
2	3.4641	-23.87	∞	0.2556	0.9687	0.2581	1.3787
3	5.1962	-41.90	1	0.9230	1.8460	0.06019	0.9230
3	5.1962	-35.90	∞	0.4013	1.1794	0.09421	1.4447
4	6.9282	-49.86	1	0.9574	1.9149	0.03264	0.9574
4	6.9282	-43.86	∞	0.4461	1.2482	0.05007	1.4688
5	8.6602	-55.88	1	0.9730	1.9459	0.02056	0.9730
5	8.6602	-49.88	∞	0.4659	1.2793	0.03127	1.4800

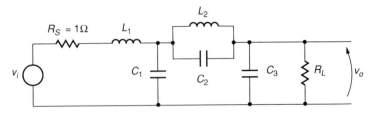

Fig. 12.10 Two-zero, four-pole inverse Chebyshev filter.

TABLE 12.5 TWO-ZERO, FOUR-POLE INVERSE CHEBYSHEV FILTER ELEMENT VALUES (IN Ω, H, f)

$\omega_{-\infty}$	ω_{max}	$|T(j\omega)|_{\omega_{max}}$(dB)	R_L	L_1	C_1	L_2	C_2	C_3
2	2.8284	−39.63	1	0.7350	1.6723	1.5189	0.16460	0.5816
2	2.8284	−33.63	∞	0.3666	0.8649	1.2338	0.20260	1.3890
3	4.2426	−55.19	1	0.7500	1.7703	1.7121	0.06490	0.6918
3	4.2426	−49.19	∞	0.3761	0.9928	1.4327	0.07756	1.4693
4	5.6568	−65.65	1	0.7570	1.8050	1.7726	0.03526	0.7247
4	5.6568	−59.65	∞	0.3791	1.0332	1.4973	0.04174	1.4965
5	7.0711	−73.60	1	0.7600	1.8205	1.8001	0.02222	0.7396
5	7.0711	−67.60	∞	0.3804	1.0512	1.5265	0.02620	1.5089

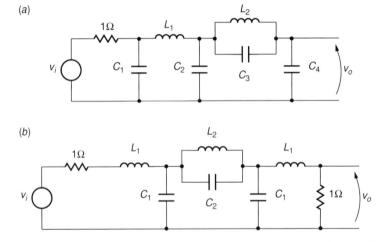

Fig. 12.11 Two-zero, five-pole inverse Chebyshev filter; (a) singly-terminated; (b) doubly-terminated.

TABLE 12.6 TWO-ZERO, FIVE-POLE INVERSE CHEBYSHEV FILTER ELEMENT VALUES (IN Ω, H, f)

$\omega_{-\infty}$	ω_{max}	$|T(j\omega)|_{\omega_{max}}$(dB)	C_1	L_1	C_2	L_2	C_3	C_4
(a) Singly-terminated, Fig. 12.11a								
2	2.5820	−42.22	0.2981	0.8649	1.18240	1.3678	0.18280	1.3996
3	3.8730	−61.30	0.3045	0.8824	1.29910	1.5568	0.07137	1.4828
4	5.1640	−74.26	0.3066	0.8878	1.33640	1.6184	0.03862	1.5105
5	6.4550	−84.16	0.3075	0.8902	1.35300	1.6461	0.02430	1.5231
(b) Doubly-terminated, Fig. 12.11b								
2	2.5820	−48.24	1.4401	0.5962	0.14812	1.6878		
3	3.8730	−67.32	1.5432	0.6091	0.05947	1.8683		
4	5.1640	−80.28	1.5767	0.6131	0.03243	1.9272		
5	6.4550	−90.18	1.5918	0.6149	0.02047	1.9537		

The two-zero, four-pole filter is shown in Fig. 12.10 and the element values for the various zero positions are shown in Table 12.5.

The two-zero, five-pole filter is shown in Fig. 12.11 and the element values for various zero positions are shown in Table 12.6.

Interpolation to find element values for zero frequencies between those given in Tables 12.4, 12.5, and 12.6 is quite adequate, particularly for the doubly-terminated (equal values) case.

12.2 ACTIVE SIGNAL PROCESSING

12.2.1 Introduction

The addition of active elements can be done in two ways: (1) as a buffer or gain element only; or (2) to provide feedback so as to eliminate the need for inductance and still be able to realize the functions we previously discussed (passive RC circuits cannot realize any of those functions above first order). This second category of networks is called active RC.

12.2.2 RLC Synthesis by Buffer Isolation

By factoring a transfer function into quadratic terms and first-degree terms as needed and isolating the sections by buffer amplifiers, the synthesis problem becomes trivial as is shown by the next example. It is, in fact, synthesis by inspection. This is useful when we wish to realize functions not covered in the preceding sections.

Example 12.6. Design a network realizing

$$T(s) = \left(\frac{s + 3}{s^2 + s + 4}\right)\left(\frac{s/2}{s^2 + (s/2) + 2}\right)$$

This can be split into two sections; first,

$$T_1(s) = \frac{s + 3}{s^2 + s + 4} = \frac{1 + (3/s)}{s + 1 + (4/s)}$$

$$T_2(s) = \frac{s/2}{s^2 + (s/2) + 2} = \frac{1/2}{s + 1/2 + (2/s)}$$

Note that all quadratics when divided by s have terms that are inductive, resistive, and capacitive, in that order, and we can realize each of the above factors with a simple L-section voltage divider by inspection as in Fig. 12.12. By separating the two sections with a buffer amplifier, we prevent interaction between the sections which would otherwise change the transfer function.

Note that the numerator of the second factor is a resistor of 0.5 Ω, and the corresponding denominator term was also 0.5 Ω. If any numerator term were larger than the corresponding

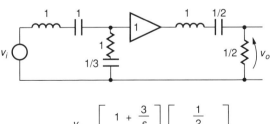

$$T(s) = \frac{v_o}{v_i} = \left[\frac{1 + \dfrac{3}{s}}{s + 1 + \dfrac{4}{s}}\right]\left[\frac{\dfrac{1}{2}}{s + \dfrac{1}{2} + \dfrac{2}{s}}\right]$$

Fig. 12.12   RLC synthesis using buffer amplifiers (in Ω, h, f).

denominator term, we would obtain a negative result when the numerator was subtracted from the denominator to obtain the series branch of the L-section. But since we are using buffer amplifiers, we can divide the numerator by any arbitrary number so that negative elements do not occur and then use a corresponding buffer amplifier gain to achieve exactly the desired transfer function. Another highly useful result using this method of synthesis is that different impedance scaling factors can be applied to the separate sections without affecting the result. Frequency scaling factors must, of course, be the same for all sections.

This method is not the most economical of elements, but this is of little consequence in the early stages of the design of an instrumentation system. After the design is complete and it has been determined that the particular filter is the one required, more sophisticated design methods can then be used to save elements if desired.

12.2.3   Active Feedback

Introduction

A very large number of active feedback synthesis methods have been developed in order to eliminate the need for inductance in filter networks. Each method has applicability to certain frequency ranges or to a particular pole, Q. For a quadratic factor, $s^2 + \beta s + \gamma$, the pole, Q, is $Q = \sqrt{\gamma}/\beta$. For many networks the Q-sensitivity to amplifier gain, K, defined as

$$S_K^Q = \frac{K}{Q} \frac{\partial Q}{\partial K} \tag{12.17}$$

is a strong function of Q, thereby limiting the application of those networks to low-Q systems only. In addition, some networks have a passive element sensitivity problem which also limits their applicability to low-Q designs. Passive element sensitivities are

$$S_R^Q = \frac{R}{Q} \frac{\partial Q}{\partial R} \tag{12.18}$$

$$S_C^Q = \frac{C}{Q} \frac{\partial Q}{\partial C} \tag{12.19}$$

To simplify this very large field,[7] we will consider only three active RC configurations. First, we will discuss networks suited to low-Q applications (the Q of the poles of most low-pass filters is quite low) that use low-gain voltage amplifiers, so that operation is possible over a very wide frequency range. Second, we will present a network suited to high-Q applications. Third will be a single-amplifier network that provides $j\omega$-axis zeros.

Low-Gain, Second-Order, Low-Pass, Active RC Filters

An early class of active RC filters[8,9] used individual quadratic RC sections with feedback from a voltage amplifier as shown in Fig. 12.13. The transfer function is given in Eq. 12.20.

$$T(s) = \frac{K/C}{s^2 + [(1.1/C) + 1 - K]s + 1/C} \tag{12.20}$$

As can be seen, the amplifier gain reduces the s coefficient by direct subtraction and when this coefficient becomes less than two (for $C = 1$ f), we have complex poles and therefore can duplicate an RLC network of the same order. These structures have high sensitivity to amplifier gain change. From the transfer function of Eq. 12.20, we can determine that

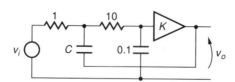

Fig. 12.13   Second-order active RC low-pass filter (in Ω, f).

$$S_K^Q = \frac{K}{(1.1/C) + 1 - K} \qquad (12.21)$$

and thus for $C = 1$ F, $Q = 10$, $K = 2.0$ we have $S_K^Q = 20$; that is, a 1% change in K produces a 20% change in Q! We must therefore restrict use of this network to low-Q systems.

Example 12.7. Using the positive-gain active RC structure of Fig. 12.13, design a fourth-order low-pass Chebyshev filter of 0.03 dB ripple, and determine the DC gain and the Q-sensitivities to amplifier gain change.

The fourth-order Chebyshev polynomial of 0.03 dB ripple (obtained from Table 12.3 using Eqs. 12.13 and 12.14) is

$$(s^2 + 0.6748s + 1.6309)(s^2 + 1.6291s + 0.9238)$$

Thus we find for the first factor

$$\frac{1.1}{C_1} + 1 - K_1 = 0.6748$$

$$\frac{1}{C_1} = 1.6309$$

$$C_1 = 0.6132 \text{ F}$$

$$K_1 = 2.1191$$

and for the second factor

$$\frac{1.1}{C_2} + 1 - K_2 = 1.6291, \quad \frac{1}{C_2} = 0.9238$$

$$C_2 = 1,0825 \text{ F}, \quad K_2 = 0.3871$$

Cascading these two networks, we obtain the complete fourth-order Chebyshev filter shown in Fig. 12.14.

This normalized fourth-order Chebyshev low-pass filter can now be scaled in impedance and frequency as desired (Section 12.3). The normalized cut-off frequency is $\omega_{-0.03\text{dB}} = 1$ rad/s. The DC gain is $(2.1191)(0.3871) = 0.8203$, and the Q-sensitivity to changes in gain of the first operational amplifier is

$$S_{K_1}^Q = \frac{K_1}{(1.1/C_1) + 1 - K_1} = 3.14$$

and for the second operational amplifier

$$S_{K_2}^Q = \frac{K_2}{(1.1/C_2) + 1 - K_2} = 0.24$$

Higher-Order, Single-Amplifier, Low-Pass Filters

Higher-order transfer functions can be obtained with a single amplifier.[7,10] The most useful are the third- and fourth-order as shown in Figs. 12.15 and 12.16.

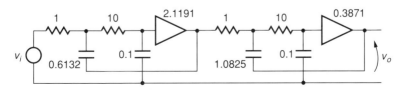

Fig. 12.14 Fourth-order active RC Chebyshev filter (0.03 dB ripple) (in Ω, f).

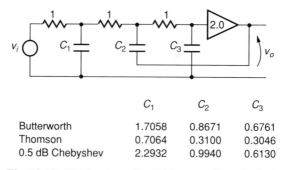

	C_1	C_2	C_3
Butterworth	1.7058	0.8671	0.6761
Thomson	0.7064	0.3100	0.3046
0.5 dB Chebyshev	2.2932	0.9940	0.6130

Fig. 12.15 Third-order active RC low-pass filter (in Ω, f).

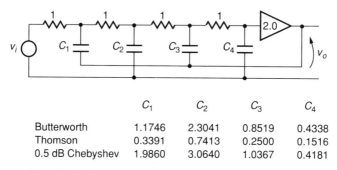

	C_1	C_2	C_3	C_4
Butterworth	1.1746	2.3041	0.8519	0.4338
Thomson	0.3391	0.7413	0.2500	0.1516
0.5 dB Chebyshev	1.9860	3.0640	1.0367	0.4181

Fig. 12.16 Fourth-order active RC low-pass filter (in Ω, f).

State-Variable Second-Order

The state-variable active RC synthesis method was developed in 1967.[11] It is a more complex structure requiring three amplifiers for a low-pass, band-pass or high-pass second-order structure and four amplifiers for a complete biquadratic function. The reason for considering such a complex structure is that the sensitivity to both active and passive elements is very low (all < 1 *and* independent of the Q). In addition, the minimum number of capacitors is used (two) even when a complete biquadratic is required. A normalized schematic is shown in Fig. 12.17 for the three-amplifier structure.

Even though all three types of filters are obtained simultaneously, the primary use of this structure is for band-pass filters since that is when we need low Q-sensitivity because of the high-Q nature of these circuits. The band-pass output transfer function (assuming ideal operational amplifiers), the Q and the center frequency, ω_o, are

$$T(s) = -\left(\frac{R_2(1 + R)}{1 + R_2}\right)\left(\frac{s}{s^2 + [(1 + R)/(1 + R_2)]s + R}\right) \tag{12.22}$$

$$Q = \frac{(1 + R_2)\sqrt{R}}{1 + R} \tag{12.23}$$

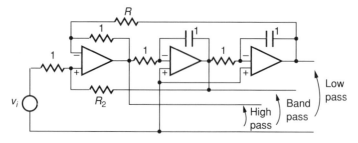

Fig. 12.17 Second-order state-variable low-pass filter (in Ω, f).

$$\omega_o = \sqrt{R} \qquad (12.24)$$

As can be seen, no subtractions exist, which was the cause of the high sensitivity in the previous Sallen and Key circuit shown in Fig. 12.13. These state-variable structures can easily be cascaded to produce higher-order transfer functions, and, of course, they can be used for high-Q circuits because of their very low Q-sensitivity.

Example 12.8. Design a second-order band-pass filter having a $Q = 50$ at a center frequency of 10 kHz using the state-variable network. Use a minimum resistor value of 10 kΩ. What is the center frequency gain?

From Eq. 12.22 we normalize to $\omega_o = 1$ rad/s.

$$\omega_o = \sqrt{R} = 1 \qquad R = 1\,\Omega$$

and from Eq. 12.23, for $R = 1\,\Omega$

$$Q = \frac{1 + R_2}{2} = 50 \qquad R_2 = 99\,\Omega$$

The center-frequency gain is

$$-\frac{R_2(1 + R)}{1 + R_2}\left(\frac{1 + R_2}{1 + R}\right) = -99$$

The amplifier gain must be at least twice the Q required at the operating frequency, and must be much greater than twice the Q if the value of R_2 calculated is to be correct. We can see that this method is restricted primarily to the audio frequencies, unless we use high performance operational amplifiers.

$j\omega$-Axis Zeros

To design filters such as the inverse Chebyshev, we must have $j\omega$-axis zero capability. The most commonly used RC method for $j\omega$-axis zeros is the twin-T. By providing appropriate feedback and network loading, a pair of $j\omega$-axis zeros and a pair of *independent* complex poles can be obtained.[12] The schematic is shown in Fig. 12.18 for the case where the zeros are beyond the poles (greater radial distance from the origin). The transfer function is also shown in Fig. 12.18.

Example 12.9. Design an inverse Chebyshev two-zero, four-pole active RC low-pass filter with a cut-off frequency of 5 kHz and zero output at 10 kHz.

Using the MFM conditions for a two-zero, four-pole function normalized to cut-off at 1 rad/s and a zero at 2 rad/s we find

$$T(s) = \frac{s^2 + 4}{3\,s^4 + 8.1828s^3 + 11.1599s^2 + 9.0154s + 4}$$

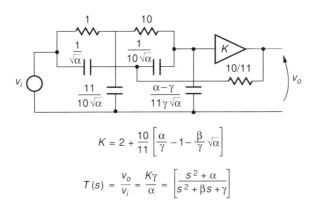

$$K = 2 + \frac{10}{11}\left[\frac{\alpha}{\gamma} - 1 - \frac{\beta}{\gamma}\sqrt{\alpha}\right]$$

$$T(s) = \frac{V_o}{V_i} = \frac{K\gamma}{\alpha} = \left[\frac{s^2 + \alpha}{s^2 + \beta s + \gamma}\right]$$

Fig. 12.18 Two-$j\omega$-axis-zero, two-pole active RC network (in Ω, f).

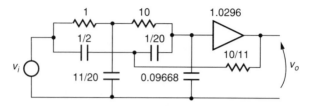

Fig. 12.19 Two-$j\omega$-axis-zero, two-pole active RC network (in Ω, f).

Factoring, we obtain

$$T(s) = \frac{s^2 + 4}{(3s^2 + 6.1293s + 3.8375)(s^2 + 0.6845s + 1.0423)}$$

When a frequency scaling factor of $2\pi(5000)$ is applied to this $T(s)$, we will have a cut-off frequency of 5 kHz and a zero at 10 kHz as required (see Section 12.3 for detailed frequency scaling rules).
   We will realize the two-zero, two-pole network first. Factoring out 1/3, we have

$$T_1(s) = \frac{1}{3}\left(\frac{s^2 + 4}{s^2 + 2.0431s + 1.2792}\right) = \frac{1}{3}\left(\frac{s^2 + \alpha}{s^2 + \beta s + \gamma}\right)$$

With these values of α, β, γ and from Fig. 12.18, we obtain the network shown in Fig. 12.19.
   There is a multiplier of (γ/α) $K = 0.3293$, but this has no effect on the pole and zero positions. Now we must realize the second factor, $T_2(s)$ where

$$T_2(s) = \left(\frac{1}{s^2 + 0.6845s + 1.0423}\right)$$

and cascade it with $T_1(s)$ to realize $T(s)$. Since we are using building blocks in which the amplifier is at the output, it doesn't matter which network is placed first. Using the network of Fig. 12.13 we find

$$\frac{1}{C} = 1.0423$$

$$C = 0.9594$$

$$\frac{1.1}{C} + 1 - K = 0.6845$$

$$K = 1.4620$$

Cascading these two networks as shown in Fig. 12.20, we obtain the complete $T(s)$. Note that the DC gain achieved is $(1.0296)(1.4620) = 1.5053$.

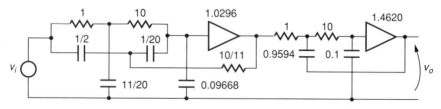

Fig. 12.20 Two-zero, four-pole inverse Chebyshev active RC low-pass filter (in Ω, f).

12.3   FILTER DESIGN

12.3.1   Introduction

We now consider the steps necessary to convert normalized filters into actual filters by scaling both in frequency and in impedance. In addition, we will cover the transformation laws that convert low-pass filters to high-pass filters and low-pass filters to band-pass filters.

12.3.2   Scaling Laws and a Design Example

Since all data previously given is for normalized filters, it is necessary to use the scaling rules to design a low-pass filter for a specific signal processing application.

   Rule 1. All impedances may be multiplied by any constant without affecting the transfer voltage ratio.
   Rule 2. To modify the cut-off frequency, divide all inductors and capacitors by the ratio of the desired frequency to the normalized frequency.

Example 12.10.   Design a low-pass filter of MFM type (Butterworth) to operate from a 600 Ω source into a 600 Ω load, with a cut-off frequency of 500 Hz. The filter must be at least 36 dB below the DC level at 2 kHz, that is, -42 dB (DC level is -6 dB).
   Since 2 kHz is four times 500 Hz, it corresponds to $\omega = 4$ rad/s in the normalized filter. Thus at $\omega = 4$ rad/s we have

$$-42 \text{ db} = 20 \log \frac{1}{2} \left| \frac{1}{\sqrt{4^{2n} + 1}} \right|$$

therefore, $n = 2.99$, so $n = 3$ must be chosen. The 1/2 is present because this is a doubly-terminated (equal values) filter so that the DC gain is 1/2.
   Thus a third-order, doubly-terminated Butterworth filter is required. From Fig. 12.3 we obtain the normalized network shown in Fig. 12.21a.
   The impedance scaling factor is $600/1 = 600$ and the frequency scaling factor is $2\pi 500/1 = 2\pi 500$; that is, the ratio of the desired radian cut-off frequency to the normalized cut-off frequency (1 rad/s). Note that the impedance scaling factor increases the size of the resistors and inductors, but reduces the size of the capacitors. The result is shown in Fig. 12.21b.

12.3.3   Transformation Rules, Passive Circuits

All information given so far applies only to low-pass filters; yet we frequently need high-pass or band-pass filters in signal processing.[13]

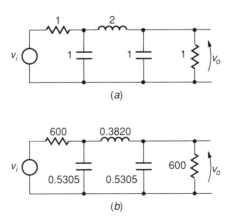

Fig. 12.21   Third-order Butterworth low-pass filter: (a) normalized (in Ω, h, f); (b) scaled (in Ω, h, f).

Low-Pass to High-Pass Transformation

To transform a low-pass filter to high-pass, we first scale it to a cut-off frequency of 1 rad/s if it is not already at 1 rad/s. This allows a simple frequency rotation about 1 rad/s of $s \rightarrow 1/s$. All L's become C's, all C's become L's, and all values reciprocate. The cut-off frequency does not change.

Example 12.11. Design a third-order, high-pass Butterworth filter to operate from a 600 Ω source to a 600 Ω load with a cut-off frequency of 500 Hz.

Starting with the normalized third-order low-pass filter of Fig. 12.3 for which $\omega_{-3} = 1$ rad/s, we reciprocate all elements and all values to obtain the filter shown in Fig. 12.22a for which $\omega_{-3} = 1$ rad/s.

Now we apply the scaling rules to raise all impedances to 600 Ω and the radian cut-off frequency to $2\pi 500$ rad/s as shown in Fig. 12.22b.

Low-Pass to Band-Pass Transformation

To transform a low-pass filter to a band-pass filter we must first scale the low-pass filter so that the cut-off frequency is equal to the bandwidth of the normalized band-pass filter. The normalized center frequency of the band-pass filter is $\omega_0 = 1$ rad/s. Then we apply the transformation $s \rightarrow s + 1/s$. For an inductor

$$Z = Ls \text{ transforms to } Z = L\left(s + \frac{1}{s}\right)$$

For a capacitor

$$Y = Cs \text{ transforms to } Y = C\left(s + \frac{1}{s}\right)$$

The first step is then to determine the Q of the band-pass filter where

$$Q = \frac{f_0}{B} = \frac{\omega_0}{B_r}$$

(f_0 is the center frequency in Hz and B is the 3 dB bandwidth in Hz). Now we scale the low-pass filter to a cut-off frequency of $1/Q$ rad/s, then series tune every inductor, L, with a capacitor of value $1/L$ and parallel tune every capacitor, C, with an inductor of value $1/C$.

Example 12.12. Design a band-pass filter centered at 100 kHz having a 3 dB bandwidth of 10 kHz starting with a third-order Butterworth low-pass filter. The source and load resistors are each to be 600 Ω.

The Q required is

$$Q = \frac{100 \text{ kHz}}{10 \text{ kHz}} = 10, \text{ or } \frac{1}{Q} = 0.1$$

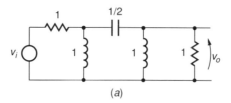

(a)

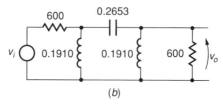

(b)

Fig. 12.22 Third-order Butterworth high-pass filter: (a) normalized (in Ω, h, f); (b) scaled (in Ω, h, f).

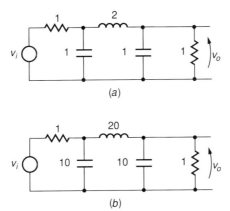

Fig. 12.23   Third-order Butterworth low-pass filter: (a) normalized (in Ω, h, f); (b) scaled (in Ω, h, f).

Scaling the normalized third-order low-pass filter of Fig. 12.23a to $\omega_{-3\,\text{dB}} = 1/Q = 0.1$ rad/s, we obtain the filter of Fig. 12.23b.

Now, converting to band-pass with $\omega_0 = 1$ rad/s, we obtain the normalized band-pass filter of Fig. 12.24a.

Next, scaling to an impedance of 600 Ω and to a center frequency of $f_0 = 100$ kHz ($\omega_0 = 2\pi 100$ k rad/s), we obtain the filter of Fig. 12.24b.

12.3.4   Transformation Rules, Active Circuits

In order to obtain active RC circuits when we transform an active RC low-pass circuit, we need different transformation rules.

Low-Pass to High-Pass Transformation

We cannot use the low-pass to high-pass transformation used for passive circuits because we would obtain an RL network, not an RC network. If, however, we multiply all impedances by s (the impedance scaling rule tells us that we can multiply by any constant, real or complex), we will obtain an RL low-pass network which, when transformed to high-pass, will give us the RC high-pass we want. We start with the normalized low-pass filter (Butterworth, $\omega_{-3\,\text{dB}} = 1$ rad/s) and obtain the normalized high-pass Butterworth filter ($\omega_{-3\,\text{dB}} = 1$ rad/s) as shown in Fig. 12.25.

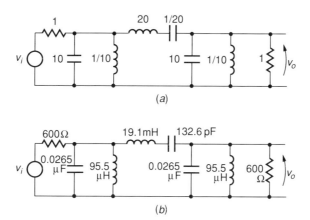

Fig. 12.24   Sixth-order Butterworth band-pass filter ($Q = 10$): (a) normalized, $\omega_0 = 1$ rad/s (in Ω, h, f); (b) scaled.

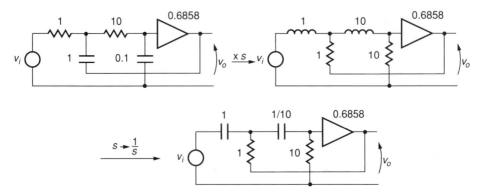

Fig. 12.25 Low-pass to high-pass transformation of an active RC second-order Butterworth filter (in Ω, h, f).

Low-Pass to Band-Pass Transformation

We have no general rule here; however, earlier in this section we designed a band-pass filter using the state-variable method. This is a very practical band-pass synthesis method and doesn't require starting with a low-pass prototype. Higher-order band-pass filters only require cascading several second-order sections which are individually designed by using each of the quadratic factors of the higher-order function. These can be obtained from a low-pass function by making the substitution $s = Q\lambda$, where Q is the band-pass filter Q desired. Then substitute $\lambda = s + 1/s$ to convert the function to band-pass, factor into quadratic sections and realize each one with the state-variable network. All of the individual second-order sections are then connected in cascade. If there are any first-order factors these can easily be added in cascade.

Example 12.13. Transform the normalized second-order Butterworth low-pass function to a $Q = 10$ band-pass function ($\omega_0 = 1$ rad/s), and factor it into two band-pass quadratics.

Given

$$T(s) = \frac{1}{s^2 + \sqrt{2}s + 1}$$

substitute

$$s = Q\lambda = 10\lambda$$

to yield

$$T(\lambda) = \frac{1}{100\lambda^2 + 10\sqrt{2}\lambda + 1}$$

Let

$$\lambda = s + \frac{1}{s}$$

then

$$T(s) = \frac{1}{100}\left[\frac{s^2}{s^4 + 0.1414s^3 + 2.01s^2 + 0.1414s + 1}\right]$$

Factoring gives

$$T(s) = \frac{1}{100}\left[\frac{s}{s^2 + 0.0682s + 0.9317}\right]\left[\frac{s}{s^2 + 0.0732s + 1.0733}\right]$$

12.4 INTRODUCTION TO DIGITAL FILTERING

Digital filtering (or digital signal processing) is concerned with the processing of signals. The purpose of such processing may be to estimate characteristic parameters of a signal (i.e., speech recognition) or to transform a signal into a form which is more desirable (i.e., remove noise). Signal processing is important in such diverse fields as biomedical engineering, acoustics, sonar, seismology, speech communication, and data communication. Digital filtering information is contained within general digital signal processing texts[14] or, if presented under the specific title, will usually include a wide range of basic signal processing knowledge. Suitable introductory texts,[15,16] and a comprehensive summary of the terms used in digital signal processing[17] are available.

Digital signal processing emerged in the 1950s with the need for sophisticated signal processing of geophysical data[18] which analog filters could not provide. Some of the advantages of digital filters over analog filters are that they have sharper rolloffs, better stability, and are less susceptible to power supply and temperature fluctuations. Digital filters can also be used for adaptive filtering, which allows real-time changes in a filter's characteristics[19]. A digital filter may be realized in either "hardware" or "software" form. As hardware, a suitable set of digital electronic circuits are interconnected to provide the essential building blocks of a digital filtering operation—storage, delay, addition, subtraction, and multiplication by constants. A major advantage of such hardware is speed. The alternative "software" approach is to program a computer as a digital filter. While this method is slower than with the use of hardware, it is sufficiently fast for many applications and is often more versatile than a hardware implementation.

Since the design of analog filters is a well-defined process, one traditional approach to the design of digital filters involves the transformation of an analog filter into a digital filter meeting prescribed specifications. For this reason, digital filtering is explained here in terms of systems analogous to analog filters.

12.5 IMPLEMENTING ANALOG TECHNIQUES IN DIGITAL FORMATS[*]

12.5.1 Introduction

The performance of a digitally implemented filter can often resemble that of an analog equivalent so closely that for all practical purposes they are identical. This chapter introduces some properties of digital systems by outlining how analog techniques can be implemented digitally.

12.5.2 Review of Analog Filters

Analog filters are composed of electrical components: resistors, inductors, capacitors, and amplifiers. Filters are often classified according to the frequency ranges over which they ideally transmit or reject signal components. An *ideal* "low-pass" filter would permit low frequency components to pass without distortion, while completely removing high frequency components. Similarly, a "high-pass" filter transmits only high frequency components. And "band-pass" and "band-reject" filters pass or reject signal components only within specified frequency bands.

Figure 12.26 shows the overshoot and ripple properties for various second-order filters. Higher-order filters improve the frequency discrimination with a little increase in the overshoot. These are implemented in both active analog and digital filters by adding cascaded second-order stages. Stages with the highest damping factors should be earliest in the sequence, otherwise the signal levels within the filter can be quite large.

Table 12.7 gives values for the pole locations for several types of low-pass filters. The Bessel and Chebyshev filters are normalized so that the 3 dB point is at unity frequency. Various normalizing relationships are in use and should be checked before using any tables. Note that the ripple in the even-order Chebyshev filters is above the steady-state gain while it is below for filters of odd order.

Using the filter design tables the analog filter can be described by

$$T(s) = \frac{\omega^2}{s^2 + 2\beta\omega s + \omega^2} \tag{12.25}$$

An implementation of this equation using ideal integrators is given in Fig. 12.27. It has a pass-band gain of unity, and the signal levels at the outputs of the integrators are nearly equal. Also, the design relationships are very convenient to implement.

---

[*] Section 12.5 is reproduced by permission of John Wiley & Sons Ltd., from P. H. Sydenham, ed., *Handbook of Measurement Science*, Vol. I, *Theoretical Fundamentals*, Wiley, 1982, Section 10.1.

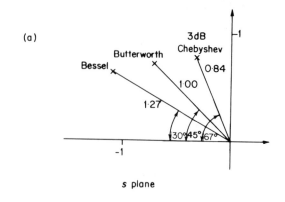

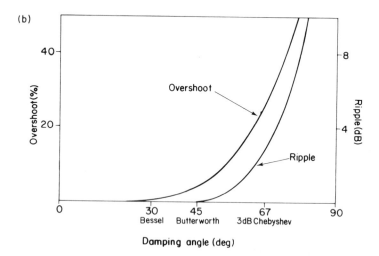

Fig. 12.26 (*a*) Pole locations and (*b*) corresponding overshoot and ripple.

TABLE 12.7   SOME SELECTED POLE LOCATIONS FOR LOW-PASS FILTERS

Type	Order	ω	β	a	b
Bessel	1	1.0000		1.0000	
	2	1.2720	0.8660	1.1016	0.6360
	3	1.3227		1.3227	
		1.4476	0.7235	1.0474	0.9993
	4	1.4302	0.9580	1.3701	0.4102
		1.6034	0.6207	0.9952	1.2571
Butterworth	1	1.0000		1.0000	
	2	1.0000	0.7071	0.7071	0.7071
	3	1.0000		1.0000	
		1.0000	0.5000	0.5000	0.8660
	4	1.0000	0.9239	0.9239	0.3827
		1.0000	0.3827	0.3827	0.9239
3 dB Chebyshev	1	1.0024		1.0024	
	2	0.8414	0.3882	0.3224	0.7772
	3	0.2986		0.2986	
		0.9161	0.1630	0.1493	0.9038
	4	0.4427	0.4645	0.2056	0.3920
		0.9503	0.0896	0.0852	0.9465

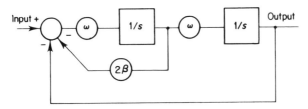

Fig. 12.27 Analog filter implementation of (Eq. 12.25) using integration.

Digital integration approximates analog integration for large sampling rates. In its simplest form it is implemented using the addition operation. This allows a useful range of digital filters to be implemented quite conveniently. Peculiarities of digital filters will be examined after basic design techniques have been described.

12.5.3 Basic Design Techniques

In Fig. 12.28 the analog signal to be filtered is converted to digital form using an analog-to-digital (A/D) converter, processed using a computing element such as a microprocessor, and again converted to analog form. Various other sources of digital signal data, including a transmission system, data storage, or a previous digital filter, could arise. Similarly, the output could be used in a digital form in a variety of ways. The simple analog input and output system serves here as a useful introduction.

The analog quantity is represented within the digital system in *binary, weighted* form. It is usually most convenient to represent negative quantities using two's complement notation. The two's complement of a binary number is derived by replacing 0's digits with 1's, and 1's with 0's, then adding 1. This allows the same arithmetic program instructions to be used with both positive and negative quantities. Replication of the analog system by the digital system relies on various assumptions. These assumptions will be examined in detail after the basic design procedure has been outlined. They are, with respect to time, that

1. The *sampling rate* is larger than any frequency component present at the analog input.
2. The sampling interval is smaller than any time constant within the filter.

and, with respect to magnitudes, that

3. The binary data represent the analog signal with an arbitrary degree of precision.
4. The analog quantity is never too large to be represented within the digital system.

and, with respect to coefficients, that

5. The coefficient values can be implemented precisely.

Digital integration can be performed by adding successive input values. The program flowchart of Fig. 12.29 will provide digital integration which nearly approximates analog integration given the previous assumptions. The sampling interval, T, determines the constant of integration. The output from the digital integrator will be $1/T$ times greater than that from an analog integrator. In a microprocessor, the time delay, T, can be obtained using a fixed number of instructions with a known execution time or an external clock which periodically interrupts another program to perform the filter algorithm. The filter time constants are proportional to the sampling interval, T. It is, therefore, important to fix its value. This gives an advantage to digital filters. The interval can be varied to tune filters of arbitrary complexity with precision. In the digital filter design it is convenient to normalize frequencies and time constants to the sampling interval.

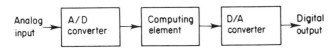

Fig. 12.28 One basic digital filtering arrangement.

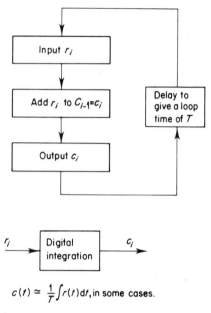

$$c(t) \simeq \frac{1}{T}\int r(t)\,dt, \text{in some cases.}$$

Fig. 12.29   Program flowchart for digital integration.

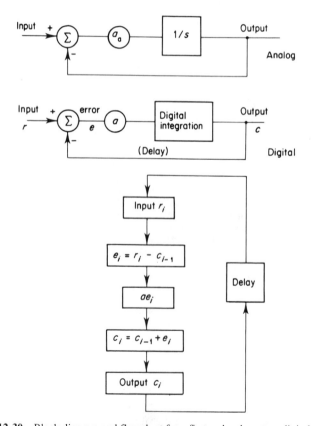

Fig. 12.30   Block diagram and flowchart for a first-order, low-pass digital filter.

The useful first-order filter given in Fig. 12.30 uses the integration algorithm. It approximates an analog filter having the transfer function

$$T(s) = \frac{1}{\tau s + 1} \quad \text{where } \tau = 1/a_a \quad (12.26)$$

The subscript a stands for analog. In the corresponding digital filter the coefficient a must be adjusted to allow for the digital integration because it has an output $1/T$ times greater than that of the analog integrator. It is convenient to normalize the time constant and cut-off frequency of the filter to the sampling interval

$$\tau_n = \tau/T \text{ and } \omega_n = \omega T \quad (12.27)$$

The coefficient of the digital filter is then $a = 1/\tau_n$.

Example 12.14.   A first-order, low-pass digital filter with a cut-off frequency of 100 Hz ($\omega_c = 628$ rad/s, $\tau = 1.6$ ms) is to be implemented.
The steps necessary to design such a filter are

1. Choose the sampling interval. Use about $\frac{1}{10}$ times the time constant

$$T = 0.1 \text{ ms}$$

2. Calculate the normalized time constant

$$\tau_n = \tau/T = 16$$

3. Evaluate the coefficient

$$a = 1/\tau_n = 2^{-4}$$

4. Use the flowchart of Fig. 12.30 with $a = 2^{-4}$ and $T = 0.1$ ms to implement a digital processor.

Extension of this design procedure to second-order systems uses the block diagram and corresponding program flowchart of Figs. 12.31 and 12.32. In this case it is necessary to use tables to find the pole locations. Again the cut-off frequency can be normalized to the sampling interval as shown in the following example.

Example 12.15.   A second-order, 3 dB, Chebyshev low-pass filter with a cut-off frequency of 40 Hz ($\omega_c = 251$ rad/s) is required.

1. Choose a sampling interval less than one tenth the time constant.

$$T = 4 \text{ ms}$$

2. Calculate the cut-off frequency normalized to the sampling interval.

$$\omega_{cn} = \omega_c T = 0.1004$$

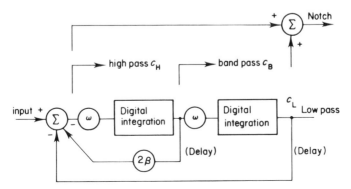

Fig. 12.31   Block diagram of a second-order digital filter.

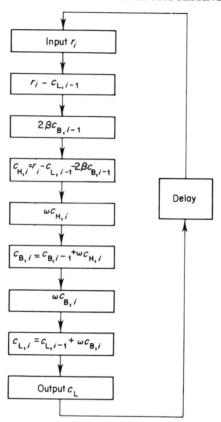

Fig. 12.32 Program flowchart for a second-order digital resonator.

3. Select the required pole location using the tables for analog filters.

$$\omega = 0.8414 \qquad \beta = 0.3882$$

4. Evaluate the filter coefficients.

$$\omega_f = \omega_c T_{\text{tables}} = 0.1004 \times 0.8414 = 0.08448$$

$$\beta_f = \beta_{\text{tables}}$$

Note that the damping factor is not altered by the scaling for frequency.

5. Implement the filter using the program flowchart of Fig. 12.32.

The step response for this filter is given in Fig. 12.33.

12.5.4 Review of Design Assumptions

It is necessary to examine the assumptions which made the digital filter closely resemble the analog equivalent. This allows the selection of important design aspects such as register lengths, sampling rates, and any necessary preparation of the analog signal before conversion to the digital form.

The first assumption was that the sampling rate was larger than any frequency component present at the input to the A/D converter. If higher frequencies were present, *aliasing* would occur. This means that the high-frequency components appear as low-frequency components within the passband after sampling. Time and frequency domain representations of aliasing are given in Ref. 14.

The frequency domain interpretation is based on the fact that the product of two sinusoids gives both the sum and the difference frequency components. Sampling at intervals of T can be regarded as multiplication by all sinusoids of the form $\cos(2\pi n/t)$ where $n = 0, 1, \ldots, \infty$. In this way

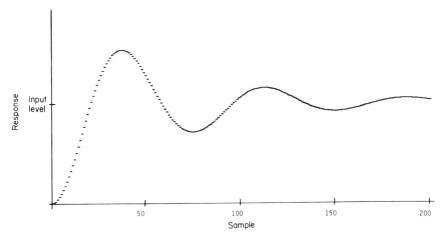

Fig. 12.33   Step response of filter design for Example 12.15.

the frequency components above half the sampling frequency (the Nyquist frequency) are folded back as unwanted lower frequency components. These high-frequency components must be removed using anti-aliasing low-pass filters before the A/D conversion as shown in Fig. 12.34. The frequency domain representation of aliasing gives an estimate of the amount of filtering required; the frequency components remaining above the Nyquist frequency will appear within the passband with the same magnitude.

It is of interest that use can be made of this scheme to provide selective bandpass filters. The A/D converter can act as a frequency translator, the anti-aliasing filter becoming a band-pass filter. Quadrature components are needed to complete the frequency translation.

A second assumption was that the sampling interval is smaller than any time constant in the filter. Longer sampling intervals mean that the digital integration no longer approximates the analog integration and the filter's output will differ from the designed response. Generally, sampling rates ten times any time constant are adequate to give an accurate approximation. This can be improved using compensation. Bolton[21] has derived relationships that maintain the magnitude frequency response. For the sampling interval, $T = 1$,

$$a_y = (1 + a)\cos(b) - 1 \quad b_y = (1 + a)\sin(b) \tag{12.28}$$

then

$$\omega = [(a_y^2 + b_y^2)/(a_y^2 + b_y^2 + 2a_y + 1)]^{1/2} \tag{12.29}$$

$$2\beta = \omega[2a_y/(a_y^2 + b_y^2) + 1] \tag{12.30}$$

For lower sampling rates the bilinear transform, described in the following section, provides a more useful filter.

A third assumption was that the analog data were represented with an arbitrary degree of precision. Finite digital register lengths mean there must be some *magnitude quantization*.[14]

The input and output (A/D) conversion must have a sufficiently small quantization level to represent the signal with the desired accuracy. Eight-bit conversions will give a level of 2^{-8} or 0.4%

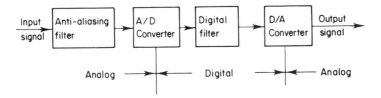

Fig. 12.34   Anti-aliasing (low-pass) filter before A/D conversion.

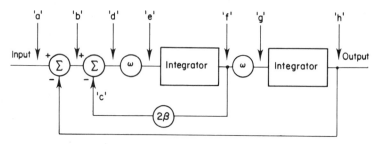

Fig. 12.35 Signal points where underflow and overflow can occur.

of the full analog range. However, the signal levels within the filter are also quantized; usually these effects are more severe than those input and output conversions. Various signals are labeled in Fig. 12.35.

The dominant effect of quantization within the digital filter occurs at point 'e' of Fig. 12.35. The difference signal, $r - e$, is multiplied by the coefficient, a, to give a component at 'e.' The coefficient is much less than unity, particularly for large sampling rates. Underflow at 'e' will occur when the original difference $r - e$ is $1/a$ times larger than the quantization level. This can cause a *deadband* in the output which is $1/a$ times the quantization level. An original quantization level of 0.4% can be expected to give a deadband of 4% even if the sampling rate is only ten times the time constant of the filter. This must be avoided.

One solution is to use longer registers within the filter, leaving the input and output data conversions unchanged. In practice, use of 16-bit registers and time constants about ten times greater than the sampling interval provides workable operating levels. Larger sampling rates make the filter more ideal and reduce the aliasing filter requirements. However, *double precision* arithmetic may be needed to reduce the deadband effect. An alternative scheme is to cascade digital filters, the first having a faster sampling rate and smaller time constants. In effect this is a digital aliasing filter. Another approach is to use an instruction to detect whenever the signal at 'b' is greater than zero. If in these cases a least significant bit is added at 'g,' the deadband will be removed and there will be only a minor distortion of the filter characteristic. This also removes all small-signal limit cycles.[21]

A fourth assumption was that *magnitude overload* does not occur. With digital systems magnitude overload causes a discontinuity in the output to the other extreme value. A slight overload in the positive direction will cause a large negative value. This contrasts with analog systems which usually saturate.

The effect on the output of the filter is much greater in digital systems where overloads can cause sustained oscillations. These can be avoided by providing a saturating characteristic at points 'd,' 'f,' and 'h' of Fig. 12.35 using programming techniques. Alternatively the signal levels and input conditions can be confined so that overloads can never occur. With the Butterworth, Thomson, and Bessel filters the signal levels within the filter should never exceed twice the maximum input signal level. In higher-order filters the stages must be in order of ascending Q, with the signal passing through the more damped stages first. The 3 dB Chebyshev filters require an additional factor of 2.

Finally, it was assumed that the coefficients could be implemented precisely. With the type of digital filter described here the response never critically depends on the values of the coefficients.[21,22] Having rounded the value of a coefficient for implementation, it may be helpful to assess its effect on the filter's response by evaluating the pole locations for these new values. With some types of digital filters the response depends critically on proportionate changes in the coefficients to the point where the available transfer functions are severely limited. However, with the filters described so far, the ability to realize the coefficients with precision provides stability advantages over corresponding analog types.

12.5.5 Extension of Low-Pass Designs

Extension of the low-pass design to give a high-pass response uses the transform $s \rightarrow 1/s$, which means poles at $r/ \pm 0$ are relocated at $1/r/ \pm 0$. The output from the digital filter is then taken from the point shown in Fig. 12.31.

The transform to give a band-pass output is $s \rightarrow (s^2 + \omega^2_{center})/s$. The transfer function at the band-pass output of Fig. 12.31 approximates

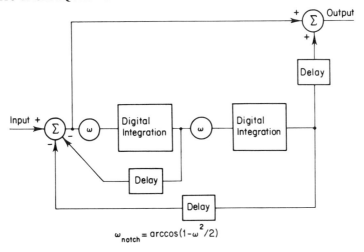

$$\omega_{notch} = arccos(1-\omega^2/2)$$

Fig. 12.36 A notch filter structure.

$$T(s) = \frac{\omega s}{s^2 + 2\beta \omega s + \omega^2}$$

and this can be used to implement band-pass designs.

The notch output is quite deep at high sampling rates. At low sampling rates the special notch structure of Fig. 12.36 is required to provide a deep notch.[21] Three delay terms are shown. Each is obtained by using the value from the previous loop. This means the notch addition is performed after the high-pass output is computed but before the low-pass output.

12.6 ANALYTIC TECHNIQUES FOR SAMPLED SYSTEMS[*]

12.6.1 Introductory Remarks

At low sampling rates the comparison between analog and digital filters becomes inadequate for analysis or design. The transform that represents digital systems without approximation is the Z-transform, in the same way as the Laplace operator, s, is used with continuous systems. The Z-transform is now introduced and a filter design example given to serve as an introduction to specialist texts.

12.6.2 Sampling

When an A/D converter samples a value of a signal, subsequent changes in the signal are not registered by the computer until the next sample is taken. It is as if the value of the signal is held constant during the time interval. However, holding the value of the sampled function constant is mathematically more complicated than representing the sampled output by a series of delta (δ) functions. The delta or impulse functions, $\delta(t - T)$, is a function of time which is zero everywhere except at time $t = T$. At $t = T$ its value goes to infinity for an infinitesimally small duration. The area of a unit impulse is one unit. The function $2\delta(t - T)$ has an area, or integral, of two units, and so on. A sampling stage is described mathematically as

$$f^*(t) = \sum_{n=0}^{\infty} \delta(t - nT)f(t) \qquad (12.31)$$

Sampling is represented in Fig. 12.37.

[*] Section 12.6 is reproduced by permission of John Wiley & Sons Ltd., from P. H. Sydenham, ed., *Handbook of Measurement Science*, Vol. 1, *Theoretical Fundamentals*, Wiley, 1982, Section 10.2.

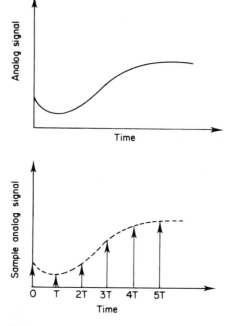

Fig. 12.37   Sampling of an analog signal at intervals of T.

12.6.3   Convolution

Sampled systems can be used to show the significance of convolution when determining the output of linear systems to known input functions. Consider a system which has the response to a unit impulse given in Fig. 12.38 (dotted curve). The output of this system decays by 50% at each sampling interval which corresponds to a time constant $\tau = T/\ln 2$. The impulse response, $g^*(t)$, is a series of functions

$$g^*(t) = \delta(t) + 0.5\delta(t - T) + 0.25\delta(t - 2T) + \cdots \tag{12.32}$$

If the input function to this system is a sampled ramp, the output at any time is the sum of the components corresponding to each of the input samples. This is because the system is linear, for which the principle of superposition applies.

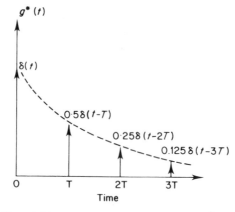

Fig. 12.38   Impulse response of a system, $g^*(nT)$.

When evaluating the output of $g^*(t)$ to $r^*(t)$ at, say, $t = 4T$ it is necessary to sum the four components present. These are 4, 1.5, 0.5, and 0.125, which add to 6.125. This summation can be represented as

$$c^*(4T) = \sum_{n=1}^{4} r^*(nT)g^*((4-n)T) \tag{12.33}$$

where the function g has been turned about $t = 4T$. The output at any time $t = mT$ can be found using the more general convolution summation

$$c^*(mT) = \sum_{n=0}^{m} r^*(nT)g^*((m-n)T) \tag{12.34}$$

The convolution summation lends itself to a numerical evaluation using a digital computer. This is particularly useful when it is not convenient to describe the functions in analytic form.

12.6.4 The Z-Transform

All the functions used so far have been functions of time. The use of the Laplace and the Z-transforms involve functions of other variables which transform to functions of time. The transforms are devised so that the convolution operation in the time domain corresponds to multiplication in the Laplace, or z, domains.

The Z-transform is defined so that a sampled function of time, $f^*(t)$, is transformed to a polynomial function of z^{-1}, $F(z)$. The power of z^{-1} is given by the time of the impulse and the coefficient of the term is given by its magnitude. In general, if

$$f^*(t) = \sum_{n=0}^{\infty} a_n \delta(t - nT) \tag{12.35}$$

then

$$F(z) = \sum_{n=0}^{\infty} a_n z^{-n} \tag{12.36}$$

The example previously evaluated in the time domain can be used to illustrate the use of the z domain. The ramp input becomes

$$R(z) = 0z^0 + 1z^{-1} + 2z^{-2} + 3z^{-3} + 4z^{-4} + \cdots \tag{12.37}$$

and

$$G(z) = 1z^0 + 0.5z^{-1} + 0.25z^{-2} + 0.125z^{-3} + \cdots \tag{12.38}$$

The output function in the z domain, $C(z)$, is given by the product of these two functions. It can be seen that the coefficient of z^{-4} in this product is

$$\text{term in } z^{-4} = 1z^{-1} \times 0.125z^{-3} + 2z^{-2} \times 0.25z^{-2}$$

$$+ 3z^{-3} \times 0.5z^{-1} + 4z^{-4} \times z^0$$

$$= 6.125z^{-4}$$

The convolution summation in the time domain has been obtained using the product in the z domain. All of the output impulses are represented by the function $C(z)$; the output at nT being given by the coefficient of z^{-n}.

Most useful functions can be expressed as the ratio of finite polynomials in the z domain. This allows simple and convenient multiplication operations in the z domain. Remembering that

$$\frac{1}{1-x} = 1 + x + x^2 + x^3 + x^4 + \cdots \tag{12.39}$$

$$G(z) = \frac{1}{(2z)^0} + \frac{1}{(2z)^1} + \frac{1}{(2z)^2} + \frac{1}{(2z)^3} + \cdots \tag{12.40}$$

$$= \frac{1}{1 - 1/(2z)} = \frac{z}{z - \frac{1}{2}}$$

Transform tables are useful to find expressions such as those for $R(z)$, where

$$R(z) = \frac{z}{(z-1)^2} \tag{12.41}$$

when the sampling interval $T = 1$ unit.

Functions in the z domain can be transformed into the time domain by obtaining the z domain function as a sum of terms in z^{-n}. Usually the function in the z domain contains a polynomial in the denominator, so synthetic division is used to obtain the series of terms. An alternative technique for obtaining the inverse Z-transform is the method of residues.[14] This is very similar to the corresponding Laplace method. If $a/\underline{\alpha}$ is the residue at a singular complex pole at $z_p/\underline{\beta}$, the component from this pole at $t = nT$ is

$$2a|z_p|^{n-1} \cos[(n-1)\beta + \alpha] \tag{12.42}$$

It can be seen that if $|z_p| > 1$ and the pole lies outside the unit circle, the system is unstable. The unit circle gives the boundary of stability and corresponds to the $j\omega$ axis of the s-plane.

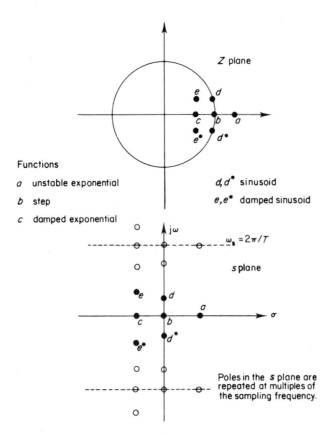

Functions

a   unstable exponential

b   step

c   damped exponential

d, d^* sinusoid

e, e^* damped sinusoid

Fig. 12.39   Corresponding impulse invariant responses in the z- and s-planes.

The z domain can be related to the Laplace domain using the fact that the operator z^{-1} in the z domain corresponds to the operator e^{-sT} in the Laplace domain. Often the equality $z = e^{sT}$ is inferred to relate the Laplace and the z domains. In fact the equality is not strictly correct since, for example, $e^{sT/2}$ gives a delay of $T/2$ while $z^{1/2}$ is not defined. The relationship between the Laplace and z domains can be inferred by comparing the responses in the time domain for given pole locations. When the impulse response from poles in the Laplace domain at $a \pm jb$ is sampled, the poles in the z domain for the corresponding response are at $e^a\!\!\underline{/\pm b}$.

In this sense $z = e^{sT}$. This is called *impulse invariance* and can be used as a design basis for filters. It should be remembered that the impulses do in fact differ because one is sampled, so the step, frequency, and phase responses will differ also. Figure 12.39 shows some examples of corresponding impulse invariant responses in the Laplace and z domains.

12.7 FILTER DESIGN USING THE BILINEAR TRANSFORM[*]

If a transfer function, $T(s)$, is given in the Laplace domain, it can be implemented in the z domain using the *frequency-warped bilinear* transform. This is

$$s = \omega \cot(\omega/2)\frac{z-1}{z+1} \qquad (12.43)$$

where ω is normalized to the sampling interval. The implementation is approximate; however, when the warping term is included, the sinusoidal magnitude and phase response at ω are identical to the original Laplace domain function. In filter design ω is set equal to the cut-off frequency so that the salient properties of the filter are preserved.

Example 12.16. A second-order Butterworth digital filter is to be implemented having a cut-off frequency at 1 kHz (6.28 krad/s) using a sampling frequency of 10 kHz and the delay implementation. To design such a filter we use the following steps:

1. From filter tables the poles are at $1\underline{/\pm 135°}$ which gives a transfer function

$$T(s) = \frac{1/2}{(s + 1/\sqrt{2})^2 + 1/2}$$

This provides a gain of unity for arbitrarily small frequencies.
2. The normalized frequency

$$\omega = \omega_c T = 6.28 \times 10^3 \times 10^{-4} = 0.628$$

3. The frequency transformation required for the analog design can be combined with the frequency warping relationship using the substitution

$$s = \omega^2 \cot(\omega/2)\frac{z-1}{z+1} = 0.628 \times 1.934 \frac{z-1}{z+1} = 1.21\frac{z-1}{z+1} \qquad (12.44)$$

4. Find $T(z)$ from

$$T(z) = \frac{0.137z^2 + 0.274z + 1}{z^2 + 0.333z + 0.262} \qquad (12.45)$$

This implementation is given in block diagram form in Fig. 12.40. It is the preferred implementation for very low sampling rate systems because it provides zeros in the stop band which give better attenuation in this region. However, for large sampling rates its coefficient values become critical and the signal magnitudes vary considerably throughout the filter. Also, at large sampling rates the provision of zeros at $z = -1$ is not necessary and they require additional computation. Analog filters with carefully placed zeros in the stop band can be implemented digitally using either design technique and the corresponding advantages realized.

[*] Section 12.7 is reproduced by permission of John Wiley & Sons Ltd., from P. H. Sydenham, ed., *Handbook of Measurement Science*, Vol. 1, *Theoretical Fundamentals*, Wiley, 1982, Section 10.3.

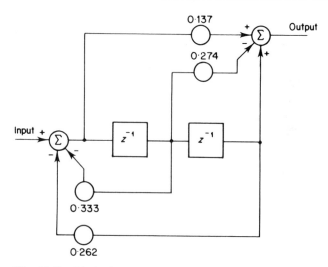

Fig. 12.40   Block diagram of a delay implementation digital filter.

12.8   OTHER TYPES OF DIGITAL FILTERS

The introduction to digital filtering techniques presented here concentrates on the implementation of approximations to analog filters. This approach to filtering postulates that the useful signal lies in one frequency band and unwanted signals lie in another band. The filtering techniques presented here can be used to remove these unwanted disturbances from a signal. However, if the signal is random, then more sophisticated filters must be used to extract the pertinent information. Normally, the filtering of random signals is referred to as estimation, because most estimation filters are statistical and estimation is a well-defined statistical technique. Statistical approaches to filtering postulate that certain statistical properties are possessed by the useful signal and the unwanted noise. Measurements are available of the sum of the signal and the noise, and the task is to eliminate by some means as much of the noise as possible through processing of the measurements by a filter. One of the most widely used techniques for reducing the effects of noise in measurements is the Kalman filter. A Kalman filter is a method based on the least-squares approach of estimating the true value of a set of variables from a set of noisy measurements. Dean[23] and Brown[24] provide a good introduction to Kalman filters. For a more detailed description of the mathematics behind Kalman filtering, Anderson and Moore[25] provide a good reference source.

The Kalman filter is a type of filter known as a recursive filter. Recursive filters use not only data values to compute the output values, but also past values of the output. Recursive filters tend to be used in systems where there are very long runs of data. Some of the different recursive filters are Box-Jenkins filters,[26] Bayesian filters,[27] infinite impulse response (IIR) filters, autoregressive moving average (ARMA) filters, and autoregressive integrated moving average (ARIMA) filters. Another type of filter is a nonrecursive, which uses only data values to compute the output. They are the simplest to understand, design, and use. Some of the different nonrecursive filters are finite impulse response (FIR) filters, tapped delay line filters, and moving average filters.

Recent advances in the computer industry have made it possible to use these more complex filters in many applications. Because of the success that has been obtained with the use of these filters, digital signal processing techniques are taking over a wide variety of instrumentation, numerical processing, and sensing and control functions that were formerly filled by purely analog systems. These techniques offer better precision and accuracy at a lower cost than do analog components. In addition, digital signal processing elements are less susceptible to drift, and are relatively insensitive to supply voltage variations. Digital signal processing chips are a special class of microcomputer chips designed to support high speed, numeric-intensive applications. DSP microcomputer chips require the same degree of hardware/software support as a general purpose microcomputer does for any application. Information on hardware for digital signal processing can be obtained from the following makers of digital signal processing chips: Analog Devices, ATT Technology Systems, Motorola, National Semiconductor Corp., NEC Electronics, and Texas Instruments.

References are a somewhat difficult problem in an introductory article that covers a rapidly developing field of research. The references *Digital Signal Processing* and *Digital Signal Processing II* by the IEEE Press are collections of published papers that were regarded as being of special importance.[28,29] The books by Oppenheim and Schafer[14] and by Rabiner and Gold[30] are excellent ones that start from the assumption that the analog filter theory is already known. The book by Hamming[16] is a good introductory text for those with no knowledge of analog filters. Texts to provide an understanding of the relationship between control systems and signal processing are those by Astrom and Wittenmark[31] and Franklin and Powell.[32] A list of digital signal processing programs is available in the IEEE book *Programs for Digital Signal Processing*.[33] To keep up to date on the latest research in signal processing, the journals *Signal Processing* by North-Holland and the *IEEE Transactions on Acoustics, Speech and Signal Processing* should be consulted.

12.9   DIGITAL SIGNAL TRANSMISSION

12.9.1   Characteristics

With digital signal transmission, the information is represented by a small number (most often two) of discrete levels. This signal is sent over a transmission medium, where it can be regenerated several times before reaching its final destination without degradation.

Parallel/Serial Transmission

Digital signals are transmitted either in parallel or serially. Parallel transmission uses a separate wire for each bit of the digital signal. Along with the signal wires there is also a strobe wire to notify the receiver when the signal wires all contain valid information (Fig. 12.41a).

Parallel transmission is most often used when the distance between the transmitter and the receiver is small and the signaling rate is high. As distances increase, the cost of parallel transmission increases rapidly because of the large number of wires needed and the number of wire drivers and receivers that need to be upgraded to operate with longer wires.

Serial transmission is most often used when there is a large separation (greater than 10 m) between the transmitter and receiver. With serial transmission, the bits are sent down a single wire one after the other (Fig. 12.41b).

The trade-off between parallel and serial transmission is performance versus cost. Serial transmission offers low cost as it has only a single driver, a single receiver, and a single pair of wires. Parallel transmission is more costly, but it offers higher performance by virtue of its multiple signal lines. Finally, serial transmission is more error prone because there is not a separate timing signal to indicate when the data signal is valid.

Asynchronous/Synchronous Transmission[34]

Digital signals which are transmitted serially are either asynchronous or synchronous. Asynchronous transmission is favored for low speed signals (19.2 kbit/s and below) because it offers the lowest cost. Synchronous signal transmission is preferred for higher speed transmission (19.2 kbit/s and above) because it transfers the data more efficiently.

With asynchronous signal transmission the data may arrive at irregular intervals. Start and stop bits are used to frame each character. This gives the receiver time to synchronize so it can properly receive the data between the start and stop bits (Fig. 12.42). These two nonsignal bits decrease the efficiency of asynchronous signal transmission.

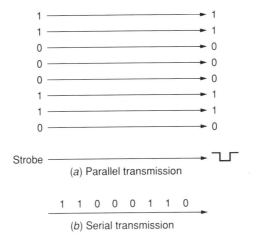

(a) Parallel transmission

1 1 0 0 0 1 1 0

(b) Serial transmission

Fig. 12.41   (a) Parallel and (b) serial signal transmission.

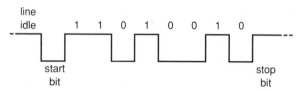

Fig. 12.42   Asynchronous data character format.

With synchronous signal transmission, the data arrives at regular intervals along with a clock pulse. The transmitted clock pulse keeps the receiver synchronized (it knows when to look at the data) with the transmitter (Fig. 12.43). Synchronous signal transmission is more efficient than asynchronous because it does not require a start or stop bit. It is more complex and costly, however, since a clock pulse must be sent with the data.

Bit Error Rate

Bit error rate (BER) is analogous to signal-to-noise ratio used with analog signal transmission. A bit error occurs when noise corrupts the signal to the point that the receiver detects the wrong digital level. Bit error rate equals the number of wrongly detected bits divided by the total number of bits sent. Bit errors can be detected, using methods outlined in Section 12.9.6.

12.9.2   Advantages Over Analog Signal Transmission

Media Independence

The reliable and accurate transmission of analog signals requires well-characterized media. Digital signal transmission does not require the same degree of characterization. Also, digital signal transmission is more adaptable from media to media since only two levels must be correlated.

Noise Immunity

With an analog signal any noise corrupts the signal. With digital transmission, the noise must corrupt the signal to the point that the receiver detects the wrong digital level. Thus the small number of discrete levels with a digital signal gives it increased noise immunity over an equivalent analog signal. In particular, over noisy communication media a digitized signal will have a better signal-to-noise ratio at the output of the receiver than its analog counterpart.

Compatibility with Signal Processing Equipment

Most signal processing equipment uses digital signals for internal operations. With digital signal transmission the incoming digital signal can be regenerated, stored and manipulated a large number of times without degeneration of signal quality.

12.9.3   Digital Transmission of Analog Signals[35]

Digital transmission of analog signals is accomplished by sampling the analog signal and representing the samples as discrete values. If done correctly the analog signal can be accurately reconstructed, if necessary, at the receiver.

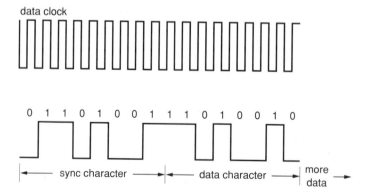

Fig. 12.43   Synchronous data character format.

Sampling Rate

The first step in digitizing an analog signal is determining the correct sampling rate. A band-limited signal of bandwidth, W, must be sampled at or above a rate of $2W$. This minimum sampling rate is the Nyquist rate.

In practical systems, sampling must take place moderately faster than the Nyquist rate. Nonideal filtering, nonideal sampling, and the fact that signals are not truly time-continuous all contribute to the need for an increased sampling rate.

Sampling the analog signal produces a series of pulses, each having an analog amplitude, occurring at the sampling rate. To represent these as digital values they are quantized using an analog-to-digital converter.

Signal Quantization

Quantization refers to the approximation of a continuous value by a finite number of predetermined levels. Sampling converts a continuous time signal to a discrete time signal. Quantization converts a continuous amplitude signal to a discrete amplitude signal. Figure 12.44 illustrates how these two operate on a continuous signal.

The degree to which a quantized signal disagrees with the continuous amplitude signal is known as quantization error. This error is determined from the minimum and maximum allowed digital levels, the number of intermediate levels between them, and the method of quantization used. For binary signal representation the number of bits used to represent the signal, N, is related to the number of levels, L, by Eq. 12.46. Thus for an 8-bit representation, 256 discrete levels are possible.

$$L = 2^N \tag{12.46}$$

If the difference between intermediate levels is the same (linear quantization), the signal-to-noise ratio for the quantized signal is given in Eq. 12.47.

$$\frac{S_q}{N_q} - 20 \log_{10} L \tag{12.47}$$

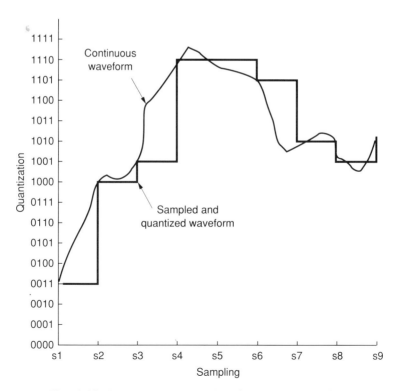

Fig. 12.44 Sampling and quantization of a continuous waveform.

Quantization can also be accomplished in a nonlinear fashion. This allows the waveform to be represented by fewer bits. Nonlinear quantization is accomplished by first compressing the waveform logarithmically and then using a linear quantization. In this way the difference between the minimum and maximum values of the waveform is reduced.

Another quantization technique, known as differential quantization, uses the previous sample as the relative zero point of the next sample. If sampled rapidly enough, the change from sample to sample is small and can be represented by a smaller number of bits (or represented more accurately by the same number of bits).

12.9.4   Digital Signal Representation

Communications between various types of equipment require an agreed upon method of representing the digital information being passed between them. The methods outlined below are the most commonly used. These methods differ primarily in the number of bits used to represent the information. For binary digital signals, the number of discrete information entities (characters) is 2^n where n is the number of bits used for each character.

Binary

Binary representation of digital signals is accomplished by using the bits as is without interpretation by the communication equipment at either end. This is highly efficient for transmitting data from one intelligent device to another, but does not work well when human interface is required.

TABLE 12.8   ASCII CHARACTER SET

b_4 b_3 b_2 b_1	ROW	$\begin{smallmatrix}0&&\\&0&\\&&0\end{smallmatrix}$ 0	$\begin{smallmatrix}0&&\\&0&\\&&1\end{smallmatrix}$ 1	$\begin{smallmatrix}0&&\\&1&\\&&0\end{smallmatrix}$ 2	$\begin{smallmatrix}0&&\\&1&\\&&1\end{smallmatrix}$ 3	$\begin{smallmatrix}1&&\\&0&\\&&0\end{smallmatrix}$ 4	$\begin{smallmatrix}1&&\\&0&\\&&1\end{smallmatrix}$ 5	$\begin{smallmatrix}1&&\\&1&\\&&0\end{smallmatrix}$ 6	$\begin{smallmatrix}1&&\\&1&\\&&1\end{smallmatrix}$ 7
0 0 0 0	0	NUL	DLE	SP	0	@	P	`	p
0 0 0 1	1	SOH	DC1	!	1	A	Q	a	q
0 0 1 0	2	STX	DC2	"	2	B	R	b	r
0 0 1 1	3	ETX	DC3	#	3	C	S	c	s
0 1 0 0	4	EOT	DC4	$	4	D	T	d	t
0 1 0 1	5	ENQ	NAK	%	5	E	U	e	u
0 1 1 0	6	ACK	SYN	&	6	F	V	f	v
0 1 1 1	7	BEL	ETB	'	7	G	W	g	w
1 0 0 0	8	BS	CAN	(	8	H	X	h	x
1 0 0 1	9	HT	EM	)	9	I	Y	i	y
1 0 1 0	10	LF	SUB	*	:	J	Z	j	z
1 0 1 1	11	VT	ESC	+	;	K	[	k	{
1 1 0 0	12	FF	FS	,	‹	L	\	l	|
1 1 0 1	13	CR	GS	-	=	M	]	m	}
1 1 1 0	14	SO	RS	.	›	N	^	n	~
1 1 1 1	15	SI	US	/	?	O	_	o	DEL

This material is reproduced with permission from American National Standard Code for Information Interchange, ANSI X3.4, copyright 1977 by the American National Standards Institute. Copies of this standard may be purchased from the American National Standards Institute at 1430 Broadway, New York, NY 10018.

ASCII[36]

American Standard Code for Information Interchange (ASCII) is widely used. It is an 8-bit code where 7 bits are used to represent the data (128 characters) and one bit is used for parity (even for asynchronous, odd for synchronous). ASCII supports alphanumeric, graphics, and control characters. A chart outlining ASCII is given in Table 12.8.

12.9.5 Digital Signal Encoding

Several methods are used to change the characteristics of the raw digital data stream before it is sent over the transmission medium. These encoding methods change the frequency characteristics of the digital signal to better suit the media and, often, provide a clock signal embedded in with the data for receiver synchronization. Several of the most common schemes are outlined below and illustrated in Fig. 12.45.

NRZ

Non-return-to-zero encoding transmits the bits as is. Since the data is left as is, there are periods where identical bits are sent and no transitions occur. This results in a large low-frequency component which may cause problems on certain media. This also makes clock recovery difficult and unpredictable. With asynchronous transmission a minimum transition density is guaranteed by the start and stop bits, which is enough to keep the transmit and receive clocks synchronized at lower data rates.

NRZI

Non-return-to-zero-inverting encoding is an attempt to force a minimum transition density to occur. It does this by making a transition whenever there is a ONE present (NRZI-MARK) or whenever there is a ZERO present (NRZI-SPACE). There is still a possibility with either of these that no transitions will occur for long periods, but it is much less than with NRZ.

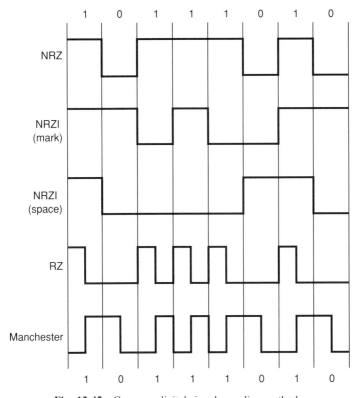

Fig. 12.45 Common digital signal encoding methods.

RZ

Return-to-zero encoding returns the signal level to zero half way through each bit period. This provides transitions during a long string of transmitted ONEs occurring at twice the bit rate. It does not provide transitions during the transmission of a long string of ZEROs. RZ encoding is most effective with bipolar signals. In this case, a ONE is represented by a positive value, a ZERO by a negative value, and an idle state by a zero value. Now, both ZEROs and ONEs have transitions associated with them. Not all media perform well with bipolar signals, however.

Manchester

Manchester encoding uses a signal transition in the center of every bit period. If successive bits of the same value are transmitted, a transition occurs at each bit boundary as well. A negative mid-bit transition represents a ZERO and a positive mid-bit transition represents a ONE. Manchester encoding has a guaranteed transition density, with at least one transition every bit period. Manchester encoding allows effective clock recovery, and operates equally well over most media.

12.9.6   Digital Signal Error Detection

VRC

Vertical Redundancy check follows each character with a parity bit. It is used with short data transfers over a relatively noise-free channel. An even number of errors within a character are not detected by a VRC (see Fig. 12.46).

LRC

Longitudinal Redundancy check follows each message block with a parity character. The parity character's bits each correspond to a parity bit for that bit position in the characters in the preceding message block. LRCs are used with VRCs for short data transfers over noisy channels (see Fig. 12.46).

CRC

A Cyclic Redundancy check is used for large data transfers where the combination VRC-LRC would lead to a great amount of overhead. With the CRC, the message is viewed as a polynomial. A remainder is generated by dividing the message polynomial by a designated CRC polynomial. The message is then sent with the remainder (8 to 32 bits) appended.

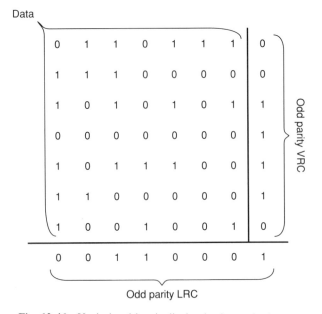

Fig. 12.46   Vertical and longitudinal redundancy checksums.

TABLE 12.9   COMMON CRC POLYNOMIALS

CRC-16 $= X^{16} + X^{15} + X^2 + 1$

CRC-16 Reverse $= X^{16} + X^{14} + X + 1$

SDLC (IBM, CCITT) $= X^{16} + X^{12} + X^5 + 1$

SDLC Reverse $= X^{16} + X^{11} + X^4 + 1$

CRC-12 $= X^{12} + X^{11} + X^3 + X^2 + X + 1$

LRCC-16 $= X^{16} + 1$

LRCC-8 $= X^8 + 1$

The receiver performs an identical division on the message and appended remainder. If the result is zero, the message is assumed valid. Some of the various standard CRC polynomials are shown in Table 12.9. Hardware to implement these standard polynomials is readily available.

12.9.7   Digital Signal Media

Data Rate–Bandwidth Relationship

According to Nyquist, the maximum data rate on a noise-free channel of bandwidth W is $2W(\log_2(L))$ where L is the number of levels used to represent the data. For binary digital signals (where $L = 2$), the maximum bit rate on a noiseless channel is $2W$. While this data rate cannot be achieved (there are no noise-free channels), the relationship points out two things. First, the channel bandwidth and data rate are dependent on one another. Second, the more noise-free the channel the closer we can come to this upper limit. The media outlined below are those typically used for in-plant communication. They each have unique bandwidth and noise characteristics.

Twisted Pair

A twisted pair (Fig. 12.47a) uses two conductors twisted together at a fixed rate and sometimes shielded by a wrapping of metal foil. This type of cable is the most widely used because of its low cost. As data rates and distance requirements increase, a twisted pair becomes less favorable because of its relatively high cable capacitance (which affects bandwidth) and limited noise immunity. Upper limits for data rates are in the 1 to 5 Mbit/s range. Distances at these data rates are specified in Section 12.9.8 as part of the RS-422A standard.

Twinax

A twinax cable (Fig. 12.47b) uses a twisted pair for signal transmission, but surrounds the twisted pair with a well-characterized dielectric and a braided metal shield. This yields a balanced transmission medium with a well-specified impedance. Twisting the signal wires provides protection from low-frequency magnetic field pickup where the braided shield has little effect. Neither of the signal wires are used as shield with twinax, so the protection against ground loops and common mode is very good. The upper limit of signaling rates with twinax cable is in the 1 to 5 Mbit/s range.

Coax

Coax cable (Fig. 12.47c) consists of a center conductor surrounded by a dielectric layer and a concentric cylindrical outer conductor. The dielectric is chosen to give the desired electrical characteristics. The outer conductor is braided or solid depending on the noise immunity and mechanical characteristics desired. Coax cable offers high bandwidth and good noise immunity (if properly installed). Coax cable used to transmit baseband signals is typically small diameter and uses a braided outer conductor. Broadband CATV coax has a larger diameter—1/2 in. (13 mm) or larger—and uses a solid outer conductor.

The outer conductor of a coax cable acts as a shield and as a signal path, which may allow common mode noise and ground loops to degrade the bit error rate on the cable. Careful grounding practices must be used to reduce this susceptibility. This type of noise generally occurs below the 5 MHz cutoff of broadband coax systems.

Fiber Optics

Fiber optic cable (Fig. 12.47d) consists of one or more ultrapure glass or plastic strands surrounded by a buffer, strength member, and outer jacket. With fiber optics, digital signals are transmitted as pulses of light. Since fiber optics is an all-dielectric medium, it is immune to electrical noise and does not emit electrical radiation. Fiber optics can support very high data rates over long distances. Typical data-grade fiber optic cable has an attenuation of 5 dB/km and a bandwidth of 300 MHz-km.

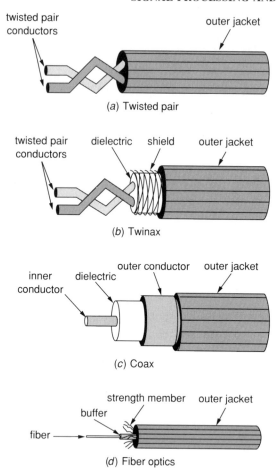

Fig. 12.47 Typical in-plant communication media.

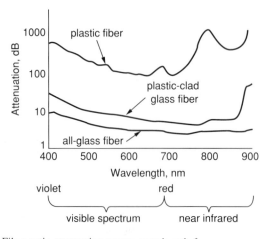

Fig. 12.48 Fiber optic attenuation versus wavelength for some common fiber types.

TABLE 12.10   RS-232C PIN ASSIGNMENTS

Pin Number	Description
1	Protective ground
2	Transmitted data
3	Received data
4	Request-to-send
5	Clear-to-send
6	Data set ready
7	Signal ground
8	Received line signal detector
9	(Reserved for data set testing)
10	(Reserved for data set testing)
11	Unassigned
12	Secondary RX line signal detector
13	Secondary clear-to-send
14	Secondary transmitted data
15	TX signal element timing (DCE source)
16	Secondary received data
17	RX signal element timing (DCE source)
18	Unassigned
19	Secondary request-to-send
20	Data terminal ready
21	Signal quality detector
22	Ring indicator
23	Data signal rate selector (DCE/DTE source)
24	TX signal element timing (DTE source)
25	Unassigned

Reprinted by permission of Electronic Industries Association (EIA), 2001 Eye Street, N.W., Washington, DC 20006.

The attenuation with fiber optic cable is dependent on the wavelength of the light being transmitted. This relationship is shown for several fiber types in Fig. 12.48.

12.9.8   Equipment Interfaces for Digital Signal Transmission

RS-232C[36]

RS-232C is an EIA Standard addressing the serial asynchronous or synchronous connection between computers, modems, and terminals over distances of under 50 ft (16 m) and data rates below 20,000 bit/s. RS-232C is a physical interface standard, outlining the connector type and pin assignments (Table 12.10), as well as an electrical interface standard, outlining the signal characteristics (Fig. 12.49). While it is a common interface, it is intended that RS-232C be gradually replaced by RS-449, RS-423A and RS-422A.

RS-449[36]

RS-449 is intended as a replacement for the functional/physical portion of RS-232C. RS-449 provides more functions than RS-232C. As a result it specifies a 37-pin connector and an ancillary 9-pin connector for secondary channel interchange. Signaling rates of up to 2 Mbit/s are allowed under

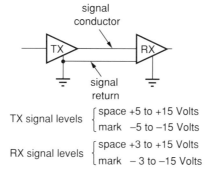

Fig. 12.49   RS-232C functional circuit diagram.

TABLE 12.11   RS-449 PIN ASSIGNMENTS

Pin Number	Description
Primary 37-Position Connector	
1	Shield
2	Signaling rate indicator
3	Spare
4	Send data
5	Send timing
6	Receive data
7	Request-to-send
8	Receive timing
9	Clear-to-send
10	Local loopback
11	Data mode
12	Terminal ready
13	Receiver ready
14	Remote loopback
15	Incoming call
16	Select freq/Signal rate select
17	Terminal timing
18	Test mode
19	Signal ground
20	Receive common
21	Spare
22	Send data
23	Send timing
24	Receive data
25	Request-to-send
26	Receive timing
27	Clear-to-send
28	Terminal in service
29	Data mode
30	Terminal ready
31	Receiver ready
32	Select standby
33	Signal quality
34	New signal
35	Terminal timing
36	Standby indicator
37	Send common
Secondary 9-Pin Connector	
1	Shield
2	Secondary receiver ready
3	Secondary send data
4	Secondary receive data
5	Signal ground
6	Receive common
7	Secondary request-to-send
8	Secondary clear-to-send
9	Send common

RS-449. As a replacement for RS-232C, RS-449 has not caught on as fast as its electrical/signal counterparts RS-423A and RS-422A. Table 12.11 outlines the pin assignments for RS-449.

RS-423A[36]

RS-423A specifies electrical characteristics for unbalanced voltage digital interface circuits (summarized in Fig. 12.50). It is an electrical superset of RS-232C and is intended to provide interconnection with existing RS-232C interfaces. RS-423A is specified to operate with data signaling rates of under 100 kbit/s. The distances supported by RS-423A are dependent on the data signaling rate and media used. Figure 12.51 shows the signaling rate versus distance relationship for a typical medium using RS-423A.

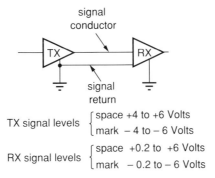

TX signal levels $\begin{cases} \text{space} +4 \text{ to } +6 \text{ Volts} \\ \text{mark} \ -4 \text{ to } -6 \text{ Volts} \end{cases}$

RX signal levels $\begin{cases} \text{space} +0.2 \text{ to } +6 \text{ Volts} \\ \text{mark} \ -0.2 \text{ to } -6 \text{ Volts} \end{cases}$

Fig. 12.50   RS-423C functional circuit diagram.

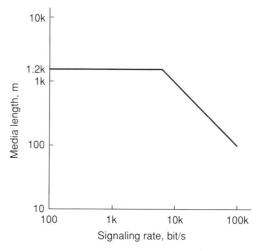

Fig. 12.51   RS-423A signaling rate versus distance for 24 AWG twisted pair wire (adapted from EIA Std. RS-423A, by permission. Electronic Industries Association, 2001 Eye Street, N.W., Washington, D.C. 20006).

RS-422A[36]

RS-422A specifies electrical characteristics for balanced-voltage digital interface circuits (Fig. 12.52). The balanced nature of RS-422A provides higher data rates and better noise immunity than RS-232 or RS-423. RS-422A provides data signaling rates of up to 10 Mbit/s. The distances supported by

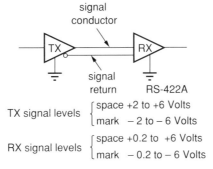

TX signal levels $\begin{cases} \text{space} +2 \text{ to } +6 \text{ Volts} \\ \text{mark} \ -2 \text{ to } -6 \text{ Volts} \end{cases}$

RX signal levels $\begin{cases} \text{space} +0.2 \text{ to } +6 \text{ Volts} \\ \text{mark} \ -0.2 \text{ to } -6 \text{ Volts} \end{cases}$

Fig. 12.52   RS-422A functional circuit diagram.

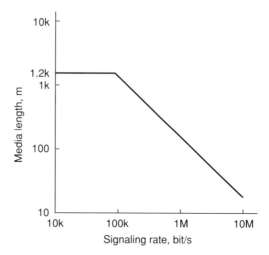

Fig. 12.53 RS-422A signaling rate versus distance for 24 AWG twisted pair wire (published with permission of Electronic Industries Association, 2001 Eye Street, N.W., Washington, D.C. 20006).

RS-422A are dependent on the data signaling rate and media used. Figure 12.53 shows the signaling rate versus distance relationship for a typical media using RS-422A.

RS-485

RS-485 is a modification of the RS-422A standard to provide multidrop operation. With RS-485, the same data rates as RS-422A are supported. Unlike RS-422A, which allows only one transmitter on the media, up to 32 transceivers (transmitter/receiver pairs) are allowed to share the media. This configuration is becoming more popular with the increase of intelligent devices.

IEEE-488

The IEEE-488 interface bus standard is targeted at remotely programmable and unprogrammable test instruments, displays, and terminals. It has found wide applicability, however, as a peripheral communications interface.

IEEE-488 specifies a parallel-bus structure 24 lines wide, with 8 lines used for data transfer, 8 for control, and the remainder for signal grounds. The bus length can be up to 20m and supports data rates of up to 1 Mbyte/s (1 byte = 8 bits). From 2 to 15 devices can share an IEEE-488 bus.

CAMAC

Computer Automated Monitor And Control (CAMAC) is an interface standard which was originally designed for high-energy physics data acquisition. It is based on "crates" with each crate containing 25 slots. Within each crate are data acquisition cards and a crate controller. An 86-pin backplane (dataway) is used to interconnect the cards and the controller within the crate. Crates can be interconnected, via each crate controller, either over a high-speed parallel "highway" (IEEE Std. 596) consisting of 66 twisted pairs, or a lower speed serial highway (IEEE Std. 595) based on RS-422A.

12.9.9 Higher Level Protocols

Protocols provide data framing, error control, message sequencing, data transparency, and media control. To accomplish these ends, protocols use one of three methods: (1) special characters (character-oriented); (2) a byte count to determine the size of each field (byte-count-oriented); (3) special bit patterns to delineate messages (bit-oriented).

Protocols at a still higher level than the ones outlined below exist today or are in the process of being specified. While these protocols are as necessary as the ones outlined for successful communication, they are not directly involved with signal transmission and will not be considered in the brief overview which follows.

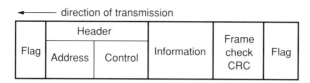

Fig. 12.54 General packet format for SDLC, ADCCP, HDLC.

SDLC, ADCCP, HDLC

IBM's Synchronous Data Link Control (SDLC), ANSI's Advanced Data Communications Control Procedures (ADCCP), and ISO's High-Level Data Link Control (HDLC) are very similar bit-oriented protocols. These protocols often serve as the next level above the serial equipment interfaces for the transfer of data and control information between intelligent devices. To accomplish the transfer of data and control information, these protocols use a packet format for communications. Each packet contains addressing, control, data, and error checking information. This allows the shared use of a common media by several stations. The general packet format used by these protocols is shown in Fig. 12.54.

IEEE-802.3/Ethernet[37]

The IEEE-802.3 standard specifies a physical and low-level logical interface for Local Area Networks. In particular, 802.3 specifies a media (baseband coax cable), a range of signaling rates (1–10 Mbit/s), a packet format, and an access method (Carrier Sense Multiple Access with Collision Detect—(CSMACD)). This standard is closely related to Ethernet, a de facto standard proposed by Xerox, Intel, and Digital Equipment Corporation.

CSMACD is a probabilistic access method, which allows many stations to share a common medium. With CSMACD a station monitors the cable for activity; if it does not hear anything and it has something to send, it does so. If two stations do this simultaneously, they will detect that their signals are colliding a short time later and each will stop transmitting, wait a random time, and start the process all over. Because the time it takes to secure the use of the media is probabilistic, CSMACD is not popular in manufacturing environments. *IEEE-802.4* is more popular for those environments.

IEEE-802.4[38]

The IEEE-802.4 token bus local area network standard specifies the use of several media (baseband and broadband coax), several signaling rates (1–10 Mbit/s), a packet format similar to 802.3, and an access method (token passing). Token passing offers deterministic access to the shared media by the attached stations.

The token (a specially formatted packet) is passed from station to station in a logical ring. When a station obtains the token it gains permission to exclusively use the shared medium. The station passes the token along to the next station when it is done sending its data, or passes the token along immediately if it has nothing to send at the time it receives the token. The 802.4 standard takes into account the error conditions that might be encountered when using token passing (e.g., lost token, multiple tokens).

GLOSSARY OF TERMS

baud The number of signal transitions per second. May be less than, equal to, or greater than the bit rate, depending on the encoding method used.

bit Contraction of binary digit. The fundamental unit of digital information.

bit rate The number of bits transferred across the communication link per second (see **baud**).

CSMACD An access method where each station monitors the shared medium for activity. If activity is detected, the station defers transmission; if it detects no activity, it transmits. If two or more stations begin transmission at the same time, their signals collide, each one stops, and waits some random time before starting the bus monitoring process over again.

DCE Data Communication Equipment

DTE Data Terminal Equipment

full-duplex Concurrent, two-way data transfer.

half-duplex Sequential, two-way data transfer

mark The idle state in digital transmission, designated as a binary 1.

parity A bit added such that there is an even number of ones in the checked sequence (*even parity*) or an odd number of ones (*odd parity*).

polling An access method whereby one station (master) queries each of the other stations (slaves) one at a time in a predetermined logical sequence. Each slave station responds to the master when it is polled.

protocol Procedures and conventions that govern the transfer of data.

simplex Permanent, one-way data transfer.

space The signal state in digital transmission, designated as a binary 0.

token passing An access method whereby each station gains permission to use the shared media by obtaining a *token*. Typically, a token is a specially formatted message.

REFERENCES

1. S. Butterworth, "On the Theory of Filter Amplifiers," *Wireless Engineering*, Vol. 7, 536–541, 1930.

2. H. J. Orchard, "Inductorless Filters," *Electronics Letters*, Vol. 2, 224–225, 1966.

3. L. Weinberg and P. Slepian, "Takahasi's Results on Tchebycheff and Butterworth Ladder Networks," *IRE Transactions, Professional Group on Circuit Theory*, CT-7, No. 2, 88–101, 1960.

4. W. E. Thomson, "Maximally Flat Delay Networks," *IRE Transactions*, CT-6, 235, 1959.

5. L. Storch, "Synthesis of Constant Delay Ladder Networks Using Bessel Polynomials," *Proceedings of the IRE*, Vol. 42, 1666–1675, 1954.

6. A. Papoulis, "On the Approximation Problem in Filter Design," *IRE National Convention Record*, Vol. 5, Pt. 2, 175–185, 1957.

7. W. J. Kerwin et al., in L. P. Huelsman, Ed., *Active Filters: Lumped, Distributed, Integrated, Digital and Parametric*, McGraw-Hill, New York, 1970.

8. R. P. Sallen and E. L. Key, "A Practical Method of Designing RC Active Filters," *IRE Transactions on Circuit Theory*, CT-2, 74–85, 1955.

9. W. J. Kerwin, R. S. Aikens, and W. H. Gross, "A Sensitivity Comparison of Single and Multiple Amplifier Equal Capacitor Active RC Structures," *Proceedings 6th Asilomar Conference on Circuits and Systems*, Asilomar, 1972.

10. R. S. Aikens and W. J. Kerwin, "Single Amplifier, Minimal RC, Butterworth, Thomson and Chebyshev filters to sixth order," *Proceedings of the International Filter Symposium*, Santa Monica, 1972.

11. W. J. Kerwin, L. P. Huelsman, and R. W. Newcomb, "State Variable Synthesis for Insensitive Integrated Circuit Transfer Functions," *IEEE Journal on Solid State Circuits*, SC-2, 87–92, 1967.

12. W. J. Kerwin and L. P. Huelsman, "Design of High Performance Band-Pass Filters," *Proceedings of the IEEE 1966 International Convention*, IEEE, New York, 1966.

13. L. Weinberg, *Network Analysis and Synthesis*, McGraw-Hill, New York, 1962.

14. Alan V. Oppenheim and Ronald W. Schafer, *Digital Signal Processing*, Prentice-Hall, Englewood Cliffs, NJ, 1975.

15. R. E. Bogner and A. G. Constantinides, *Introduction to Digital Filtering*, Wiley, Chichester, 1975.

16. R. W. Hamming, *Digital Filters*, Prentice-Hall, Englewood Cliffs, NJ, 1983.

17. Lawrence R. Rabiner et al., "Terminology in Digital Signal Processing," *IEEE Trans. Audio Electroacoustics*, Vol. AU-20, 322–337, December 1972.

18. E. A. Robinson and S. Treital, "Principles of Digital Filtering," *Geophysics*, Vol. 29, 395–404, 1964.

19. Bernard Widrow and Samuel D. Stearns, *Adaptive Signal Processing*, Prentice-Hall, Englewood Cliffs, NJ, 1985.

20. F. J. Hill and G. R. Peterson, *Introduction to Switching Theory and Logical Design*, Wiley, New York, 1974.

21. A. G. Bolton, "Design and Implementation of Digital Filters for Microprocessor Implementation Using Fixed Point Arithmetic," *Proc. IEE*, Vol. 128, Part G, No. 5, 245–50, 1981.

22. A. G. Bolton and B. R. Davis, "Evaluation of Coefficient Sensitivities for Second Order Digital Resonators," *Proc. IEE*, Vol. 128, Part G, No. 3, 127–8, 1981.

23. G. C. Dean, "An Introduction to Kalman Filters," *Measurement and Control*, Vol. 19, pp. 69–73, March 1986.

24. Robert G. Brown, *Introduction to Random Signal Analysis and Kalman Filtering*, John Wiley and Sons, New York, 1983.

25. Brian D. Anderson and John B. Moore, *Optimal Filtering*, Prentice-Hall, Englewood Cliffs, NJ, 1979.

26. G. E. Box and G. M. Jenkins, *Time Series Analysis: Forecasting and Control*, Holden Day, Oakland, 1976.

27. A. Jazwinski, *Stochastic Processes and Filtering Theory*, Academic Press, New York, 1970.

28. Lawrence R. Rabiner, Ed., *Digital Signal Processing*, IEEE Press, New York, 1977.

29. Digital Signal Processing Committee, Ed., *Digital Signal Processing II*, IEEE Press, New York, 1975.

30. Lawrence R. Rabiner and B. Gold, *Theory and Application of Digital Signal Processing*, Prentice-Hall, Englewood Cliffs, NJ, 1975.

31. Karl J. Astrom and Bjorn Wittenmark, *Computer Controlled Systems, Theory and Design*, Prentice-Hall, Englewood Cliffs, NJ, 1984.

32. Gene F. Franklin and J. David Powell, *Digital Control of Dynamic Systems*, Addison-Wesley, Reading, MA, 1980.

33. Digital Signal Processing Committee, Ed., *Programs for Digital Signal Processing*, IEEE Acoustics, Speech and Signal Processing Society, IEEE Press, New York, 1979.

34. John E. McNamara, *Technical Aspects of Data Communications*, Digital Press.

35. K. Sam Shanmugan, *Digital and Analog Communication Systems*, Wiley, New York, 1979.

36. Harold C. Folts, Ed., *Data Communication Standards*, McGraw-Hill, New York, 1982.

37. *IEEE-802.3 Draft D*, IEEE, 1982.

38. *IEEE-802.4 Draft D*, IEEE, 1982.

CHAPTER **13**

DATA ACQUISITION AND DISPLAY SYSTEMS

DANIEL T. MIKLOVIC

Weyerhaeuser Company
Tacoma, Washington

PHILIP C. MILLIMAN

Weyerhaeuser Company
Tacoma, Washington

13.1 OVERVIEW

In order to control any process or to understand what occurs during a process reaction, the controller (a human or machine) must have information about what is occurring. In the simplest of control loops, the measured variable must be converted, then some response must occur. A Data Acquisition System (DAS) collects and displays that information. In its simplest terms, a DAS may be an analog meter or recorder. In its most complex terms, it crosses the boundaries that define a control system. We will briefly discuss chart recorders and meters, and we will continue through more complex systems such as process monitors, which may perform rather complex data acquisition but may have only simple alarm and control activity.

13.2 DATA ACQUISITION

Data acquisition includes the following: (1) acquiring raw data from the process being measured, (2) converting it to usable units, and (3) putting it into a form that can be displayed.

13.2.1 Raw Data—Its Acquisition and Conversion

For the most part, the data, such as pressure, temperature, and flow rate, that we are interested in acquiring tends to be analog in nature. Usually the only digital status we are interested in is the binary status of some piece of equipment, for example, whether or not a motor is on or off, an alarm signal is activated, or a threshold is exceeded. With the exception of analog meters, recorders, and analog oscilloscopes, the data is handled digitally. The microprocessors in data acquisition systems are digital devices and deal only with digital data. A digital-to-analog (D/A) and analog-to-digital (A/D) converter is required to bring analog information from or to the process. The analog to digital conversion process begins with the conversion and scaling of the analog signal from a transducer to a level compatible with the circuitry doing the conversion. Before discussing scaling and linearization we shall reexamine some of the basics of A/D and D/A conversion.

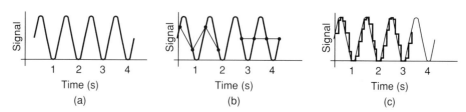

Fig. 13.1   (*a*) True wave form; (*b*) aliasing of data; (*c*) sampled every tenth second.

One of the problems with A/D conversion includes the proper choice of sampling interval. A sine wave with a period of 1 s (Fig. 13.1*a*) is measured with several sampling intervals. One-half s (Fig. 13.1*b*) and 0.1 s (Fig. 13.1*c*) intervals both provide different impressions of what is actually happening. The 1 s sampling rate being in phase with the sine wave, yields the impression that we are measuring a nonvarying level. The 0.5 s sampling rate yields several different results depending on what phase shift is encountered. This is known as aliasing (Ref. 1, p. 153; Ref. 2, pp. 122–125). If a 0.1 s period is used we finally begin to obtain a realistic idea of what the waveform truly looks like.

Accuracy and precision are dependent on the sampling interval as well as the resolution of the system (Ref. 3, Glossary and p. 39). When dealing with the A/D conversion process the step size or number of bits used is critical when determining the system precision and accuracy (Ref. 2, pp. 78–81). Figure 13.2 illustrates the difference between accuracy and precision. Table 13.1 illustrates the effect the number of bits has on the precision.

From the previous information it should be clear that the number of digital bits involved is a critical issue. Eight–, 10–, and 12-bit D/A and A/D converters are quite common. Sixteen-bit conversion is rarer and is included to provide a comparison for discussion and preparation for systems currently beginning to be manufactured. Also to be considered when selecting converters is the settling time the converter requires. This is the time from receiving the encode/decode instruction that the device takes to produce the true output. Typical settling times range from 50 ns to 20 μs, with 200 to 300 ns being typical. After a brief review of conversion techniques typical requirements will be discussed.

There are three major methods of A/D conversion that will be briefly reviewed here. For additional information see Chapter 6 of this handbook.

Ramp Generator

A microprocessor sends digital values in a ramping sequence to a D/A chip whose output is fed to a comparator. When the unknown equals the ramped signal, the comparator triggers the microprocessor which then retains the appropriate value. The advantages of this scheme include its straightforward and simple design. The disadvantages include the fact that it is slow and that most D/A output is based on current, so that conversion to voltage is required and scaling can be complex. This type of converter also has a tendency to miss peak signals. See Fig. 13.3 for an illustration of the problem.

Integrator

The integrator is used somewhat more frequently than the ramp converter and comes in single–, dual–, or triple-slope form, of which the dual-slope form is the most popular. The device operates by integrating the input analog signal by means of charging a capacitor for a fixed time, applying a negative known voltage, timing how long it takes to get to the zero crossing, and finally scaling

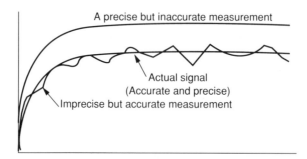

Fig. 13.2   Difference between accuracy and precision.

TABLE 13.1   RELATIONSHIP BETWEEN NUMBER OF BITS AND PRECISION

No. of Bits	Steps	Resolution on 5 Volt Measurement (V)	% Resolution
8	256	0.01950	0.3900
10	1024	0.00488	0.0980
12	4096	0.00122	0.0240
16	65,536	0.00008	0.0015

the time to a voltage level. It has the advantage of being very accurate and relatively immune to noise, such as that from power lines. For instance, one could reject power line noise by selecting a conversion period equal to the period of the power line signal, typically 16.7 ms (corresponding to 60 Hz). Its major disadvantage is that typical conversion times are slow (above 10 μs).

Successive Approximation

The successive approximator is the most common converter (Ref. 2, p. 91). It is somewhat similar to the ramp converter but uses the sign of the comparator difference to adjust the D/A instead of a fixed ramp. See Fig. 13.4 for an illustration of this technique. The major advantages of this scheme are its speed and accuracy. Its major disadvantages are that it requires a D/A and the conversion circuitry. As large-scale integrated circuit technology develops, this will become less of a problem. Another problem is that the converter is sensitive to instantaneous noise. Techniques such as averaging and filtering of data are discussed in Section 13.2.2. Trade-offs are discussed in Ref. 4 (p. 216).

The actual signals which need to be converted are either voltages (typically from 0 to 2.5, −2.5, 10, or −10 V) or a current signal (typically 4–20 mA). As indicated, for a converter to work and still have reasonable resolution, some form of scaling is required as well as some form of sign handling. Section 13.2.2 covers this and linearization in greater detail.

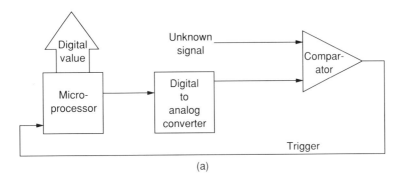

(a)

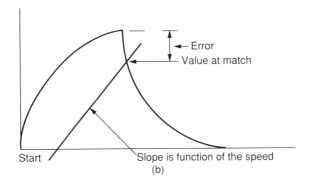

(b)

Fig. 13.3 (a) Ramp generator block diagram; (b) measurement error of ramp generator convertor.

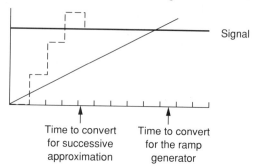

Fig. 13.4 Comparison of conversion times between the ramp generator and the successive approximator.

Collection Interval

When collecting data the sampling interval is critical. The need to collect sufficient data to perform the desired analysis must be balanced against the cost of collecting too much data. The maximum speed at which the data can be collected is basically a function of two constraints:

1. Computing power and speed of the system collecting the data as described in Section 13.3.2
2. Conversion technique

The conversion technique refers to the method of conversion as described above and the method by which multiple signals are converted. Commonly, instruments or cards which do data acquisition handle two, four, eight, or 16 single or differential inputs (signals). In all but the most expensive systems these signals are multiplexed in analog form and then converted by a single A/D converter. This can result in the skewing of data if the signals are not sampled and "held" all at the same instant in time. With 16 signals the shift between the first and the middle signal is 50% of the scan interval which can affect the analysis of the results. In the sample and hold arrangement all signals are latched into holding capacitors at the same instant in time and after the conversion the capacitors are discharged. Thus, no time skewing takes place. The drawback to such a scheme is that unless very small capacitances are used or special switching is used to drain the capacitors, the response speed may suffer. This also adds complexity and cost to the converter.

Time-Based versus Event-Driven Collection

Data acquisition and collection is either time-based or event-driven. In a time-based system all the data is gathered at predetermined times. For simplicity there is usually a single fixed interval when all signals are acquired and recorded. This simplifies the circuitry, costs, and the later analysis of the data. Where multiple signals are acquired and displayed at different intervals a time stamp must be associated with each value to facilitate later analysis. Analog recorders are a classic example of devices that record data against a continuous analog time scale.

Event-based systems collect data values based on some event other than time. They must have a time stamp associated with data values since one cannot assume the time at which a data value was captured. While correlation and analysis of the data is more difficult, there are many systems which have data recorded in this fashion. Batch processes, such as mixing a tankful of chemicals, often have data collected only at the start and end of the process. The data during the actual reaction may be of little use. In a system such as this the sampling interval may be several minutes, hours, or even days in length.

The presentation of data can take several forms. It is important that the data be collected in a format that facilitates further analysis. Data that is never analyzed need never have been collected in the first place. The form of presentation affects the ease with which the analysis may be performed. The display methods will be discussed later, but the organization of the data will be briefly dealt with here. Based on the collection frequency and the amount of data, the data is best organized to facilitate the analysis method. If the data are to be analyzed by a person, numerical printout or graphical presentation is the most effective. In the event the data are to be further analyzed by a computer, then numerical or encoded values are best used.

13.2.2 Engineering Units Conversion

Often the data obtained from a process are not in the form or units desired. This section describes several methods of transforming data to produce proper units, reduce storage quantity, and reduce noise.

Transformation of Data

There are many reasons why process measurements might need to be transformed in order to be useful. Usually the signals obtained will be values whose units (e.g., voltage, current, etc.) are other than the desired units (e.g., temperature, pressure, etc.). For example, the measurement from a pressure transducer may be in the range from 4–20 mA. To use this as a pressure measurement in psi, one would need to convert it using some equation. As environmental conditions change, the performance characteristics of many sensors change. A parametric model (equation) can be used to convert between types of units or to correct for changes in the parameters of the model. The parameters for this equation may be derived through a process known as calibration. This involves determining the parameters of some equation by placing the sensor in known environmental conditions (such as freezing or boiling water) and recording the voltage or other measurable quantity it produces. Some (normally simple) calculations will then produce the parameters desired. See the following discussion of simple linear fit for the procedure for a simple, two-parameter equation. The complexity of the model increases when the measured value is not directly proportional to the desired units (nonlinear).

Simple Linear Fit

The simplest formula for converting a measured value to the desired units is a simple linear equation (p. 126).[2] The form of the equation is $y = ax + b$, where x is the measured value, y represents the value in the units desired, and a and b are parameters to adjust the slope and offset, respectively. The procedure for finding a and b is as follows:

1. Create a known state for the sensor in the low range. An example would be to put an RTD temperature sensor in ice water.
2. Determine the value obtained from the sensor.
3. Create a known state for the sensor in the high range. An example would be to immerse the sensor in boiling water.
4. Determine the value obtained from the sensor.
5. Calculate the values of a and b from these values using the equations in Table 13.2.

Figure 13.5 demonstrates example relationships between measured values and engineering units.

Nonlinear Relationships

Generally, there is not a simple linear relationship between the engineering units and the measured units. Instead, for a constantly rising pressure or temperature, the measured value would form some curve. If possible, we use a portion of the sensor's range where it is linear, and we can use the equation described above. When this is not possible, we have to characterize the sensor by a different equation, which could be a polynomial, a transcendental, or a combination of a series of functions (Fig. 13.6). This variety of formulas should make one point clear: without an understanding of the basic model of the sensor, one cannot know what type of conversion to use. Many sensors have known distributions of data. Be aware of the effect environmental conditions have on the sensor readings. If the characteristics of a sensor are unknown, then the sensor must be measured under a variety of conditions to determine the basic relationship between the measured values and engineering units. Some knowledge of the theory of the sensor's mechanism will help to give an idea of which model to use. The development and evaluation of a model is beyond the scope of this chapter.

Filtering

Even after data is converted to the appropriate units it may have characteristics that hide the key information one is attempting to get. For instance, the data may have occasional fluctuations caused by factors other than the process, or the process may have short-term perturbations, which are not really an indication of the major process factors. Filtering is a technique that allows one to retain the essence of the data while minimizing the effects of fluctuations. The data may then appear to be "smoothed." In fact, the terms "filtering" and "smoothing" are often interchanged. Filtering can

TABLE 13.2   CALCULATION OF STRAIGHT LINE COEFFICIENTS

$$a = \frac{\text{Actual High Value} - \text{Actual Low Value}}{\text{Measured High Value} - \text{Measured Low Value}}$$

$$b = \text{Actual Low Value} - (a \times \text{Measured Low Value})$$

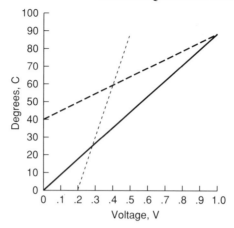

			Legend			
	Voltage for 0°C	Voltage for 100°C	Equation for a	Equation for a	Value for b	Equation for b
————	0	1	$\dfrac{100-0}{1-0}$	$0-100$	0	$C=100*V$
- - - - - - - -	−1	1	$\dfrac{100-0}{1-(-1)}$	$0-50(-1)$	50	$C=50*V+50$
· — — — —	.2	.6	$\dfrac{100-0}{.6-.2}$	$0-250(.2)$	−50	$C=250*V-50$

Fig. 13.5   Conversion to engineering units.

occur when the data is still in an analog state (Ref. 2, p. 54) or can occur after the data is converted to digital data (digital filtering). Measurement variability comes from a variety of sources. The process itself may undergo fluctuations that result in variation in measurement, but that are only temporary and should be ignored. For instance, if the level of an open tank of water were to be measured, but waves caused fluctuations in the height of a float, then the exact value at any given time would not be an accurate reflection of the level of the tank. The sensor itself may have fluctuations due to variability in its method for acquiring data. For instance, the proximity of a 60 Hz line may induce a 60 Hz sinusoidal variation in the signal (measured value). One form of filtering involves combining the currently measured value with previously measured values to give a composite value. This composite value, since it contains more than one measured value, will tend to throw away transitory information and retain information that has existed over more than one scan. Examples of filtering approaches are shown in Table 13.3.

The simple average is useful when repeated samples are taken at approximately the same point in time. The more samples, the more random noise is removed. The formula for an average is shown in Table 13.3a. However, if the noise appeared for all the samples (as when all the samples are taken at just the time that a wave ripples through a tank) then this average would still have the noise value. A moving average can be taken over time (Table 13.3b) with the same formula, but each value would be from the same sensor, only displaced in time.[5,6] One of the disadvantages of this approach is that one has to keep a list of previous values at least as long as the time span one wishes to average. A simpler approach is the first-order digital filter, where a portion of the current sample is combined with a portion of previous samples. The formula for a first-order digital filter is described in Table 13.3c. Alpha in the table is a factor selected by the user. The more one wants the data filtered, the smaller the choice of alpha; the less one wants filtered, the larger the choice of alpha. Alpha must be between 0 and 1, inclusive. The first-order digital filter suffers from tending to move actual events forward in time. The moving average can tend to move data forward in time also, unless data retrieved *after* the point to be smoothed is included in the data to be averaged (often called windowing or boxcar, Refs. 5, 6). These filters can be cascaded, that is, the output of a filter can be used as the input to another filter. A danger with any filter is that valuable information might be lost. Some laboratory instruments such as a gas chromatograph may have profiles that correspond to

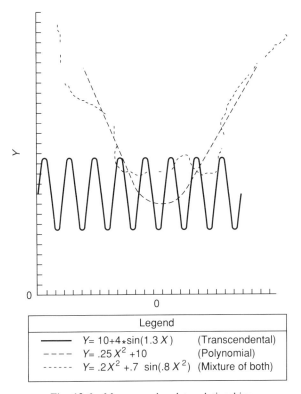

Legend

——	$Y= 10+4*\sin(1.3\,X)$	(Transcendental)
----	$Y= .25\,X^2 +10$	(Polynomial)
------	$Y= .2\,X^2 +.7\ \sin(.8\,X^2)$	(Mixture of both)

Fig. 13.6 More complex data relationships

certain types of data (a peak may correspond to the existence of an element). The data acquisition system can be trained to look for these profiles through a preexisting set of instructions, or the human operator could indicate which profiles correspond to an element and the data acquisition system would build a set of rules. Voice recognition systems often operate on a similar set of procedures. The operator speaks some words on request into the computer and it builds an internal profile of how the operator speaks to use later on new words. One area where pattern matching is used is in error-correcting serial data transmission. When serial data is being transmitted, a common practice to reduce errors is to insert known patterns into the data stream before and after the data. What if there is noise on the line? One can then look for a start-of-message pattern and an end-of-message pattern. Any data coming over the line which is not bracketed by these characters would be ignored or flagged as an extraneous transmission.

TABLE 13.3 FILTERING APPROACHES

a. Repeated sample average
Take a number (N) of samples at once, and average them
$$\frac{1}{N}\sum \text{Value}(i)$$

b. Finite length average (moving)
Take the average of the last N measurements, averaging them to obtain a current calculated value

c. Digital filters
$$y_i = (1 - \alpha)y_{i-1} + \alpha x_{i-1}$$

Reference 9, p. 538.

Compression Techniques

For high-speed or long-duration data collection sessions there may be massive amounts of data collected. A variety of techniques are available to reduce the amount of data storage required. Occasionally, only summary information is required anyway. Statistical information can be retained and the actual data collected can be discarded. While certain statistical information (such as the mode of the data) requires the raw data to be stored, the average, total, standard deviation, and correlation can be obtained purely from summary information (Table 13.4).

When one needs to record at least a portion of the original data for historical reasons there are other techniques which allow the data to be compressed while retaining much of the original data. One of the easiest techniques to implement is to sample the data, saving every so many readings or saving data at fixed time intervals. If data is sampled at fixed time intervals then it would be the same as only collecting data at a slower rate. Another approach would be to save only data that have changed (a special case of event-triggered data logging) and saving a time stamp with the data. This leads to the concept of compression of the data. Rather than just sample the data, why not save all the unique values of the data, discarding values which are the same or within some limits of the previous reading? This really applies best to continuous processes. Quite significant space reduction can be maintained in processes that are slowly changing and have only occasional large upsets. Variations of this technique can provide additional improvements. For instance, rather than just checking to see if the current value is the same as or within some limits from the previous reading, see if it is on the same line or curve as the previous value. This can result in a great reduction of storage requirements at the loss of a slight amount of accuracy in reconstruction. The more flexible the compression technique, the more work must be done to reconstruct the data later for examination. For instance, if the user of the data acquisition system wants to retrieve a datum point which lies on a line that has been reduced to a point on a line segment, the user must determine which line segment is wanted using the time stamp for the beginning of the line segment interval and then must recalculate the point from the equation.

13.2.3 Data Display

Once the data is acquired from a sensor, there are a variety of ways it can be handled. The current value can be inspected, values can be stored for inspection later, values can be trended, alarm conditions can be detected and reported, or some output back to the process can be performed.

TABLE 13.4 WAYS TO STORE AVERAGE, TOTAL, STANDARD DEVIATION, AND CORRELATION FROM ONLY SUMMARY INFORMATION

Averages

Keep a running sum of data, and count of readings
average = sum of values / count of values

Totals

Keep a running sum of data

Standard Deviation

$$s = \sqrt{\frac{n \sum_{i=1}^{n} x_i^2 - \left(\sum_{i=1}^{n} x_i \right)^2}{n(n-1)}}$$

p. 473
Reprinted with permission from CRC Standard Mathematical Tables, 24th Edition. Copyright CRC Press, Boca Raton

Correlation

$$r = \frac{n \sum x_i y_i - \left(\sum x_i \right)\left(\sum y_i \right)}{\sqrt{\left[n \sum x_i^2 - \left(\sum x_i \right)^2 \right]\left[n \sum y_i^2 - \left(\sum y_i \right)^2 \right]}}$$

p. 477
Reprinted with permission from CRC Standard Mathematical Tables, 24th Edition. Copyright CRC Press, Boca Raton

Range

Save largest and smallest values

TABLE 13.5   DATA COLLECTION RATES: EXAMPLES

Discrete Manufacturing Operations:	
Assembly Line Manufacturing/Assembly	0.01 to several s
Video Image Processing	0.001 s
Parts Machining	0.002 to 0.02 s
Continuous Processes:	
Paper Machine	1 to 60 s
Boiler	few to several min
Refinery	seconds to min
Dissolving Operations	1 to 20 min

Current Value Inspection

Often, one wants to see the data as it is being collected. This can be of critical importance in experiments which are hard or costly to repeat, allowing the researcher to react to situations as they occur. As it is collected, each data item will be called the current value for that sensor. Current data are usually stored in high-speed storage (the computer main memory). As new values are obtained, they replace the value from the last reading. The collection rate can vary widely (Table 13.5). For instance, detecting the profile at 10 mm intervals for a log moving at 100 m/min requires values to be obtained for each sensor 167 times/s. In continuous processes, data may only need to be acquired once/min, as in monitoring the level of a large vat.

Display Types

Current data must be put into some display form for inspection. Representations can include a simple digital readout, coded values, or an analog representation (Fig. 13.7). There are a wide variety of digital readouts available. A digital watch is an example. The hour display is a number whose value is the same as the value of the quantity being measured. The fifth hour of the day would be represented by a 5. Coded values are symbols which represent ranges of states of the quantity being measured. For instance, +5 V may be represented on a display by a light, by the digit 1, by the word ON, or by the word TRUE. It is difficult to separate the numerical readout from some form of coding. For instance, in the previous digital watch example, there could be two hours numbered 5 in one day. Therefore, rather than a direct correspondence between the measured quantity and the displayed representation, there may be some grouping, such as A.M. and P.M.

Often, one can better understand the data being obtained by an analog representation (Ref. 7, pp. 243–254). This involves representing the measured quantity by some other continuously variable quantity such as position, intensity, or rotation. A common example is the traditional wristwatch. The hours and minutes are determined by the position of a line indicator on a circle. A common analog representation for data acquisition is the faceplate. This is a bar graph where the height of the bar corresponds to the value being measured. Often, lines or symbols may be overlaid on the bar to indicate high or low ranges. A frequent indicator is the meter. A needle rotates in a circle with the degrees of movement corresponding to the value obtained from a sensor. Many voltmeters use this technique.

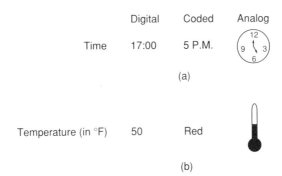

	Digital	Coded	Analog
Time	17:00	5 P.M.	

(a)

	Digital	Coded	Analog
Temperature (in °F)	50	Red	

(b)

Fig. 13.7 Comparison of digital, coded, and analog data representation.

Inspection of Stored Data

Stored data can be inspected in much the same manner as current data. The primary difference is that a new copy of data is made for each new collection interval. As discussed before, much of these data may be only summarized or compressed. Occasionally more data is required for short periods of time. Collecting data forever would overflow the memory capability. Two approaches for collecting and later reporting of larger amounts of stored data are the snapshot and the trigger. Both are based on collecting data only during a given period of time but each uses a slightly different method. The snapshot method involves predetermining the start and finish times of the collection interval. This limits the chance of catching events that occur unexpectedly and of saving data prior to some event one may be waiting for. The trigger method involves continually acquiring data and discarding it as it gets too old. When an event occurs that is of interest, the old data is retained and new data is recorded throughout a given time interval. The trigger method is particularly useful for discovering the causes of unusual events.

Report Formats

The list is the simplest report format—just list the values of data items on a screen or page. The trend plot is an alternative report format that graphically shows the history of the value of a datum point (Fig. 13.8).

Section 13.3.3 describes display formats as used in data acquisition systems.

13.2.4   System Capability

The relationships that a data acquisition system has with other systems are important to understand. They can be characterized by three terms: stand-alone, host-based, and distributed.

Stand-Alone Systems

Stand-alone systems do not interact with any other data acquisition system. A scientist may want to have an inexpensive box to collect some data for his local analysis. These systems have an advantage over others because they don't have a complex communications system and tend to be inexpensive. As the needs of the observer grow and analysis becomes more important, these systems may become a hindrance because of their inflexibility and lack of integration with other systems. As the scope of data acquisition projects increases, the power of computation required often increases beyond the capabilities of stand-alone systems. A conversion effort is then required at potentially great cost to move the data acquisition function, or at least the data, to another machine.

Host-Based Systems

Host-based systems are based on a large computer system, usually a general purpose computer. They are powerful options for data acquisition because they have great room for growth, powerful capabilities for communications with other systems, and have a variety of data analysis software. Host-based systems have a wide variety of developed software such as data acquisition systems. They also have disadvantages in that they are more expensive than the other options, often give slower video display terminal update speed, and the observer is often contending with other users of the system for use of its resources.

Distributed Systems

Distributed systems are a powerful approach to data acquisition systems because they combine some of the best of both stand-alone and host-based systems. The data acquisition portion is located on a small processor that has communication capability to a host computer system. The small system

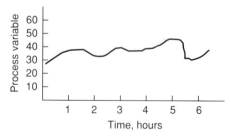

Fig. 13.8   Trend plot.

collects the data, possibly reducing some of it to a more compact form, and then sends it to the host system for analysis. The host system can analyze the data when it has the available time to do so. Only the data acquisition portion needs to be very responsive to the process. If the data acquisition task gets too big for the small system, the cost of expansion is limited to moving the data acquisition software to a new computer and doesn't require any changes to the host computer portion. The major disadvantage of distributed systems is that they suffer from a more complex overall architecture even though the individual parts are simple. This leads to problems with understanding error sources and increases the potential errors because of more parts. Unless the communications are designed carefully, the messages sent between the small system and the host system may be inflexible, causing problems when one wants to change the type of data being collected. Distributed systems may be expensive because of the number of components, but often fit well with environments where one already has a host computer. For instance, some companies have several VAX computers as host computers. They may use personal computers to collect data and then send the data over serial lines to the VAX hosts for analysis.

System Error Analysis

The errors that can occur at different stages in the data acquisition process must be analyzed, as they can add up to make the data meaningless. For instance, one may have very accurate sensors, but by the time the data reaches the host computer it might have been converted to integer data or from real to integer and back to real again. This can lead to dangerous assumptions about the accuracy of the received numbers, because each conversion can cause rounding or other errors. It is the responsibility of the person setting up the acquisition system and the analyst to examine each source of potential error, discover how much it is, and reduce it to the point where it will not have a significant impact on the conclusions to be derived from the data.

13.3   DATA ACQUISITION SYSTEMS

Factors to be considered when discussing data acquisition systems include understanding the differences between classes of systems, their storage capacities, their operator interfaces, their ability to link to other systems, and the software tools to carry out data acquisition.[4]

13.3.1   Classes of Systems

The components interconnected to build a data acquisition system range from individual functional components to devices that can stand alone and often perform the full task. Table 13.6 gives useful classes of data acquisition systems.

Data Converters

At the lowest end of the scale are the data translation, signal conditioning, and input/output devices. A name for this class could be data converter. A limited amount of intelligence is built-in, usually a fixed conversion function of some type. Typical devices in the lowest class are digital panel meters, filters, and signal processors that linearize signals. Some transducers even have these features built-in. At the high end of the scale in this class are devices used for user input. Examples of these are keyboards, bar code readers, optical character readers, and input tablets.

Data Recorders

The classic data acquisition devices begin at their lowest end with simple data recorders. The characteristics of these devices are: (1) visual storage of information, usually but not always on paper; and (2) ability to order or display data as a function of time. An example of a device not using paper is a digital storage oscilloscope; it does, however, retain the ability to repeatedly display the acquired data. Paper-based storage devices are much more common and these devices usually display their data either in analog or digital form. Analog display is the most common. Strip chart recorders are the oldest and most prevalent of these devices. The measured variable is displayed on one axis while time passage (the other axis) is indicated by the paper movement under an ink pen or other marking device. The pen may use ink on plain paper. Felt-tipped cartridges are seeing increasing use in slow-response, low-cost units. High-response speed units still rely on nib-tipped pens and an ink reservoir. While messier and not as convenient as the felt-tipped units, the low mass of the hollow point nib is necessary for high response speed. Mid-range units use thermal heads which require special paper. Here a trade-off between material cost, unit speed, and overall cost has been made. The special paper often has a cost two or three times that of plain paper.

TABLE 13.6   CLASSES OF DATA ACQUISITION SYSTEMS

Data Converters
Action:
   Data translation
   Signal conditioning
   Input/output devices
Examples:
   Digital panel meters
   Filters
   Keyboards
   Bar code readers
   Optical character readers

Data Recorders
Action:
   Visual storage of information
   Ability to order data as a function of time
Examples:
   Strip chart recorders
   Digital storage oscilloscope

Data Loggers
Action:
   Printing numbers
   Storage of data values
   Replay at later time
   Signal processing
   Triggering of actions
Examples:
   Data loggers
   Serial line analyzers (special case)
   Data acquisition systems

Process Monitors
Action:
   Computer-based
   Data manipulation
   Data storage
   Data analysis
   Often multitasking
Examples:
   Personal computers
   Process control computers
   Minicomputers

The fastest type of recorder is the oscillographic recorder; it uses a light or electron beam scanned across a sensitized paper. As digital storage oscilloscopes expand in capability and decrease in cost, the demand for oscillographic recorders is decreasing. Their complexity and high supplies cost make them warranted in special circumstances only.

A unique type of recorder is the X-Y recorder (or plotter), where two signals are plotted against each other and time is not indicated. In the digital area the signals are first converted to a human interpretable format, then printed.

Data Loggers

The next class of device, the data logger, is different from the data recorder in the way that it reports, stores, and manipulates the data. The simplest data logger is merely a printer that prints data from an A/D converter after some simple form of conversion, say, to a hexadecimal number. The conversion process may be carried further and the values converted to decimal numbers, with a time stamp or interval mark associated with the data. Often multiple channels of data may be accumulated. Careful selection of the data logger is warranted because of the limited value of a strip of paper with a series of numbers on it. Printing speed is also important; five lines/s is a common rate. Where samples are taken every minute or so and the information is considered historical in nature, this basic data logger may be a cost-effective solution.

Expanding the capability of the data logger by adding signal processing circuitry and more complex triggering mechanisms creates the more common versions of the data loggers. These offer excellent power-to-cost-performance ratios. Data storage is a common and valuable feature. Associated with this storage capability is the ability to replay this information to a computer at a later time. The ability to scale the data and the ability to alarm when thresholds are exceeded are valuable features. Some unique devices in this class are serial line analyzers, which are used to monitor and analyze data transmitted serially between computers.

Process Monitors

At the highest end of the spectrum are process monitors. These are computer-based systems that provide significant data processing capability as well as the basic data acquisition and storage functions. These functions may include statistical analysis, complex filtering, or even limited decision capability. The ancillary functions occur while the data is being collected and the computer/process monitor often relies on a multi-tasking operating system to facilitate this. The Foxboro Fox 1/A is an example of a product in the high end of this class.

13.3.2 Data Storage Capability

The data storage capability of a device is an important consideration, particularly when defining a system that will gather historical data to be used for trend analysis, product quality determinations, or other functions that impact production. In an intelligent unit, the data storage capability is primarily a function of the architecture used. That is, it is a function of the computer hardware used.

The microprocessors used in data acquisition systems directly affect both the speed at which the device stores data as well as the amount of data that it can store. Several issues impact the ability of systems to sample and store data. These include the speed of the clock driving the microprocessor, the schema for accessing memory, the data width (as handled internally), and the usable versus maximum memory.

The clock speed is a function of the hardware (microprocessor technology) and varies not only from family to family, but within families of microprocessors. It is usually a published specification of the device, at least for general-purpose computers.

As mentioned previously, a speed-related issue is the memory and storage. If the system collects a variable infrequently relative to the mass storage access time of the system, then it may be able to store the data on a diskette or tape. If it does not have this time, it must store all data in memory for the duration of the test period. This may drastically reduce the maximum length of tests.

Another issue is the number of variables collected versus time. While 60 variables collected once/s and one variable collected 60 times/s would appear to place the same load on the system, the overhead associated with the former may be significantly higher because of the need to keep track of different sensors. It may require three or more times the effort and a corresponding amount of computational time to accomplish. Also, ordered lists with headers require more memory as the number of variables increases.

Usable versus maximum memory is another key issue when considering microcomputer-based systems. The size (number of bits) of the architecture limits the maximum addressable amount of memory. The operating system or monitor occupies some of the memory and the "user" program (in our case, that which does the data collection) requires yet more. A very basic system which advertises 64K of memory may have only 30K (about 7000 real values) for data storage.

Eight-Bit Machines

(8080/85, Z80, 6502, 6800: Radio Shack TRS-80, Apple IIe, Osborne, Kaypro, Commodore 64)

These machines usually support 48 or 64 kbytes of RAM although some have the capability of utilizing 128 kbytes in a sometimes awkward switching arrangement. These systems are best at low-speed, low-volume tasks.

16-Bit Machines

(8086/88, 68000, PDP-11: Apple/Macintosh, IBM PC, DEC PDP-11)

These machines can handle memory sizes from 64 kbytes to four or 16 Mbytes depending on microprocessor type. Many have powerful multi-user operating systems and most have a real-time multi-tasking operating system available as an option. Best applications are for moderate size tasks with medium to medium-fast speed requirements. They are often used as stand-alone workstations or front-ends for dedicated systems.

32-Bit Machines

(68020/68030, 80286/80386, Z80000, RISC, DEC VAX, IBM PC/AT/PS2, Apple/Macintosh II)

Generally, these begin at memory sizes of 1 Mbyte up to several Gbytes. They often support many users, with a corresponding decrease in apparent speed as they approach saturation. The IBM PC's using 80286 chips are really crossover devices, handling internal data as 32-bit words but interfacing to the world with 16-bit words. As multitasking operating systems are more available for these machines, their use will become more prevalent. OS/2 is such an operating system. DEC VAX computers cover a range from desktop workstations to large minicomputers. Mill host or process management computers are increasingly using this technology.

Array Processors

These machines usually feature 64-bit internal word length and contain millions of bytes of data. They are ultra-fast number crunchers with many applications such as pattern recognition, image processing, and matrix calculations. The cheapest start at about $5,000 and can cost over $1,000,000.

13.3.3   Displays

There are several types of displays encountered when examining data displayed in an analog representation. For single-variable instantaneous representation, the formats in Fig. 13.9 are most common. Often the indicators may be color-coded or use symbols to represent such things as alarm limits exceeded, valve open, or motor on. A multivariable display often has the variables plotted against each other, the sum of the variables or the difference of the variables plotted against time, as well as the individual variables displayed against the same axis.

If any permanent record of the data acquired is furnished, it is often in the form of a plot. Analog recorders provide this form of record as do oscillographic recorders. Data loggers may provide an analog presentation but usually provide a digital representation, increasingly often supplying both. Data loggers without an analog display usually provide time-stamp information. This is needed to synchronize data with real-world events. The benefits of printed data are that it is permanent in nature and not easily altered. There are numerous indicators which may or may not be part of a display subsystem. These range from analog meter movements to LED/LCD simulations of the meter movement. They may also have digital readouts of the peak or instantaneous values. The main limitation of these display devices is that they can only display a single variable at a time, although through switching arrangements several variables may be displayed from a single indicator (not at the same time).

Video displays provide the most information and can present the same data in several different forms. These devices are usually Cathode Ray Tube (CRT)-based, although LCD, electroluminescent, and plasma displays are becoming available.[10] LCD color shutters are also seeing wider applications, particularly in oscilloscopes. Most CRTs use raster scan technology similar to the standard television receiver. The raster scan is a useful low-cost technique that supports color. The vector scan technique is similar to oscilloscopes and some video games. It provides higher resolution and faster speed for simple displays, but is often more expensive and thus limited to CAD applications. CRT issues include glare, readability, and fragility (Ref. 7, p. 238).

13.3.4   Data Communications

Data communications are involved in many aspects of data acquisition systems. The communications between the sensing and control elements and data acquisition devices, as well as the communications between the data acquisition system and other computer systems, can be carried out in many ways. This section will cover some aspects of communications, especially as they pertain to computer systems. For additional information, see Section 12.9.

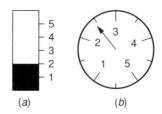

(a)   (b)

Fig. 13.9   Single-variable instantaneous representation. (*a*) Bar graph (*b*) meter.

TABLE 13.7 TIME SEQUENCE OF BITS SENT OVER A SERIAL COMMUNICATIONS LINE

Character "A" is (in bit form)

Bit Number	6	5	4	3	2	1	0
Bit Value	1	0	0	0	0	0	1

The communications are using the RS232C communications standard and sending the ASCII character A.

Bit Value	Time
Start bit	0 s after start
1 (bit 0 of A)	1/9600 s after start
0 (bit 1 of A)	2/9600 s after start
0 (bit 2 of A)	3/9600 s after start
0 (bit 3 of A)	4/9600 s after start
0 (bit 4 of A)	5/9600 s after start
0 (bit 5 of A)	6/9600 s after start
1 (bit 6 of A)	7/9600 s after start
parity bit	8/9600 s after start
stop bit	9/9600 s after start

When discussing communications, several terms need to be defined. The following paragraphs provide some basis for the rest of this section.

Serial Communications

A serial communication link means that data sent over a communications line is spread out over time on one physical data path. For instance, if a character is sent from a sensor to a computer, each bit making up the character (normally eight bits) will be sent one after the other (Table 13.7). This is often useful for low-cost, low-speed (usually less than 10,000 cps) rates of data transfer.

Parallel Communications

A serial communication link may not require very many wires, but the time spent to transfer data can add up. A way to improve the speed of communications is to use parallel communication links. This is done by having a number of wires to carry data. For instance, sending the same 'A' over a nine-wire bus would only require one transfer (Table 13.8).

Point-to-Point Communications

Point-to-point communications refers to sending a message between two computers that are physically connected by a data path; that is, the wires go directly from the sender to the receiver. This often allows very simple communications when only two computers are involved. In situations

TABLE 13.8 TIME SEQUENCE OF BITS SENT OVER A PARALLEL COMMUNICATIONS INTERFACE

Character "A" is (in bit form)

Bit Number	6	5	4	3	2	1	0
Bit Value	1	0	0	0	0	0	1

The communications are using a hypothetical 9-line parallel communications bus sending the ASCII character A.

Bit Value	Time
synch bit	0 s after start
1 (bit 0 of A)	0 s after start
0 (bit 1 of A)	0 s after start
0 (bit 2 of A)	0 s after start
0 (bit 3 of A)	0 s after start
0 (bit 4 of A)	0 s after start
0 (bit 5 of A)	0 s after start
1 (bit 6 of A)	0 s after start
parity bit	0 s after start

If the bus could handle the same rate-of-change of bits as the serial interface, then the next character could be sent 1/9600 s after the first character (the A).

where several computers need to communicate, point-to-point communications can involve a lot of overhead. A message can potentially be routed through computers to other computers; each computer taking information from one point-to-point pathway can pass this information on through to another computer over another point-to-point pathway. This can involve much overhead, since computers must keep track of routes from each computer to the others.

Networks

A network may be point-to-point or may be bus-like in that several computers may be connected to the same wires. The term "network" refers to a system where two or more devices are able to communicate in a standard fashion. Figure 13.10 gives several examples of network configurations. Of the approaches diagrammed in Fig. 13.10, the bus and ring are becoming the most common. They provide the least tendency toward bottlenecks. If a system is host-based and all communications are meant to go to the host, then the star or hierarchy may allow the most control over the system. The most common network trend for manufacturing operations and building-wide office systems appears to be the bus approach.

OSI Standard Review

As discussed in Section 12.4, the International Standards Organization has developed a set of standards for discussing communications between cooperating systems called the Open Systems Interconnect (OSI) Model.[11] This defines communications protocols in terms of seven layers (Table 13.9). We will look at it from a slightly different viewpoint in this chapter. The OSI model is the most significant development in communications. It does not define the interfaces for communications, it just standardizes terminology and provides a way of distinguishing between the different sub-types of communications activities.

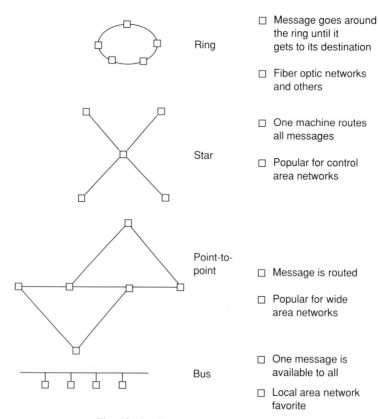

Fig. 13.10   Common network topologies.

TABLE 13.9 OPEN SYSTEMS INTERCONNECT MODEL

Layer	Principle	Example
7. Application	Application	Millwide reporting
6. Presentation	Display, Format, Edit	Convert ASCII to EBCDIC
5. Session	Establish Communications	Log onto remote computer
4. Transport	Virtual Circuits	Make sure all message parts got there in order
3. Network	Route to other networks	Talk to Arpanet
2. Data Link	Correct Errors Synchronize Communications	Send character down-line Send ACK-NAK
1. Physical	Electrical Interface	Wire and voltages

OSI Standard—Layer 7 (Application)

Most of the programming for data acquisition systems would be classed as Application Layer (Layer 7). This would cover analysis and interpretation of data. Much work needs to be done to standardize activities in this layer. Sections 13.3.1 and 13.3.5 give a hint toward some ways to approximate the use of standards by using systems that give general-purpose facilities for data acquisition and control, and using general-purpose, standard languages.

OSI Standard—Layer 6 (Presentation)

When communicating between sensors and data acquisition computers, and between data acquisition computers and other computers, one of the most common problems is the conversion of data from the normal format of one system to that of another. For instance, a common problem when dealing with an IBM system is having to convert data from ASCII (American National Standard Code for Information Interchange) to EBCDIC (Extended Binary Coded Decimal Interchange Code). The customary way to handle this in the past was to write a program to convert the data. With adequate facilities at layer 6, the communications facility would automatically convert data to the proper format. It is possible that there may come a standardized sensor input format such as channel number, value, and units. It is also possible that communications systems would have the capability of converting this from internal formats to the standard format or vice versa, thus allowing even greater flexibility in choosing components to collect or process data.

OSI Standard—Layers 5 through 1

Layers 5 through 1 have great impact on the ability to share information between computer systems. The ideal goal would be for all communications systems to support standard communications protocols for all the layers of the OSI model. In reality, the cost of a communications system could be prohibitive, particularly when all that is desired is a simple, inflexible exchange of prearranged data between two machines, such as collecting data from a data logger where only a specific instrument will be used, with little probability of change. One would be paying for flexibility and not using it. On the other hand, when real flexibility is desired, one approach could be to use remote smart sensor systems and communicate from them to a central data acquisition system over a standard communications network. New sensors could be easily added. Wiring costs might be reduced because a number of sensors could tap into the same communications lines. Almost all of the current standard large computer networks only have facilities that correspond to layers 1 through 3 (network layer). The common tools for small computer communication include even less of the capability of the OSI model. The common RS232C standard only applies to the lowest layer (layer 1). An often-used wide area network standard, X.25, only applies up to layer 3.

Standards Based on OSI Model

There are several standards efforts based on the OSI model that should be watched, as they may have special significance for data acquisition systems. An emerging standard which will certainly have a very large impact on factory communications is the Manufacturing Automation Protocol (MAP).[12]

This is based on a token-bus network especially designed for the factory environment. MAP is aimed at implementing facilities for the entire Open Systems Interconnect Model, even including facilities at layer 7, such as file transfer and and commands for robot controllers. A closely related standard is the Technical/Office Protocol, which is very similar to MAP, but based on alternative lower layers.[13]

Computer Bus Interfaces

A computer bus is a high-speed pathway between the parts of a computer system. In some ways it is different from communications networks because it is specially designed to take advantage of the particular computer architecture. A bus is usually only able to operate over short distances and normally has a number of parallel wires, allowing data to be transferred at a higher rate. The degree of standardization of bus communications varies greatly. Most of the functions of a bus correspond to layers 1 and 2 of the OSI model.

Most computer manufacturers have bus interfaces specific to their computer systems. Many computers have more than one bus. For instance, the Digital Equipment Company VAX series of computers has several bus interfaces: the MASSBUS, the UNIBUS, the CIBUS, and an internal memory bus.

Several computer bus interfaces have standards that describe their interface. These include the S100, STD, VME, MULTIBUS, and QBUS.[14] These bus interfaces are fairly common and have a wide variety of sensors available for them. One caution is that these are often specific to certain types of computers. The connections need to be planned in advance, determining available bus interfaces for the data acquisition computer being bought and finding available sensors for that bus. The S100 and STD bus interfaces are used quite frequently for mix-and-match systems, where one buys a card frame, a central processor circuit board, one or more interface cards, and the sensors required for the application.[15] The VME bus was originally advocated by Motorola and the MULTIBUS was introduced by Intel. These and other bus interfaces developed by chip manufacturers tend to be supported by custom interface card manufacturers because of the wide use of the chips produced by Motorola and Intel. The QBUS is the microcomputer bus for the Digital Equipment Company and is one of the more popular computer company-sponsored bus interfaces.

Instrumentation Bus Interfaces

Standard instrument bus interfaces are more likely to be specified by the engineer performing data acquisition than are computer bus interfaces. These are not necessarily as fast as the computer bus interfaces, but are less tied to a specific computer company and are often less expensive than computer bus interfaces. The most widely used instrumentation bus is the IEEE 488 bus, also known as the General Purpose Instrument Bus or the Hewlett-Packard Instrument Bus (GPIB or HP-IB).[16] This is a popular bus for data acquisition, since many laboratory instruments interface with it. It is important to know the maximum number of sensors and the longest distance data must travel when selecting an instrumentation bus, as bus interfaces are usually limited in the number of interfaces allowed and are often limited to short distances.

Field Bus

The Instrument Society of America and the International Electrotechnical Commission have established standards committees to develop what is known as Fieldbus. This standard, when complete, will define digital instrumentation communications for industrial purposes and replace the present 4–20 mA current loop standard.

Benefits of Standard Communications

When implementing data acquisition and display systems, the ability to communicate in a standard fashion can play a large role in the cost of the system. This is realized in a variety of ways.

1. Different sensors can be connected to a system without having to buy a whole new system.
2. Data can be sent to other systems as needed for further processing.
3. As technology or need changes, portions of a system can be mixed and matched.
4. Increased competition from vendors will tend to bring prices down.
5. A standard will have many people using products based on the standard, resulting in more vendors and greater availability of parts with a greater variety of options.

Problems with Standard Communications

One difficulty with communications protocols is that they often require sophisticated programs to communicate properly. This can sometimes force one to use a larger computer. The necessity for

TABLE 13.10 NUMBER OF PLACES ERRORS CAN OCCUR

	Smart Sensor	In-Computer Sensor
Interface to Process	X	X
Format to ASCII	X	
Transmit to Computer	X	
Convert to Binary	X	
Analyze Data	X	X

error checking and routing will tend to decrease the maximum speed at which one can send a message. Some smart sensors send their values to data acquisition computers in ASCII form over RS232C interfaces. For high-speed data acquisition needs, this may not be fast enough. For instance, collecting 10 samples at 960 cps can take 0.1 s just for the communications portion. The sensor must acquire the data and convert it into ASCII. The data acquisition computer must convert that to internal binary representation to use it in calculations. This can all add to the time to acquire data in the form desired. It also increases the number of places where errors can occur. Table 13.10 compares the number of situations in which errors can occur for the case of a smart remote sensor versus the case of collecting data directly on a data acquisition computer.

13.3.5 Data Manipulation Languages

Overview

Some instructions must be given to any data acquisition system to acquire appropriate data. These range from simply answering questions from a program to actually writing the program oneself.

Menu-Driven Interfaces

The simplest way to enter data acquisition instructions is through a question/answer or menu selection process. The data acquisition system prompts the user for parameters to control the data acquisition process. This is often termed a menu-driven system, especially where the observer selects from options presented on a screen. Many data loggers and large-scale process control systems use this approach. A limitation with this approach is that the creators of these systems had to think of the appropriate questions and responses. It is impossible to conceive of all cases, so only a subset of possible situations are provided. These systems lack flexibility to handle unusual situations. Another problem is that many menu-driven data acquisition systems often require that a great many questions be answered. An experienced user might like to save time and only answer necessary questions, or to answer them in some shorthand or high-speed form. For instance, an experienced observer could request the environment to be the same as the last run, but to collect data 5 times/s rather than 2 times/s. Some systems require the observer to go through the entire setup sequence to change a single parameter.

Common Programming Languages

When greater flexibility is desired than the question/answer approach of menu-driven systems, the observer must use some sort of computer language. A computer language is a set of instructions that allow one to instruct the computer how to carry out its operations. Common languages include BASIC and Fortran.[17] BASIC is a simple computer language that comes standard on a large portion of the personal computers sold today. It is suitable for small programs (sets of instructions), is usually interpreted (does not require compilation, so one can try out an idea and find out if it works almost immediately), and is familiar to a large number of people. A problem with interpreted BASIC is that it may run too slowly for the application. The trend is now toward vendors providing both interpreted and compiled versions of BASIC. Simple programs that acquire data from a few sensors and store it are appropriate for BASIC. Most BASIC implementations use line numbers, do not allow parameter-passing to subroutines, and do not allow structured data types. This is changing, but BASIC is not evolving equally across computer systems. This means that for systems over a few hundred lines BASIC is not recommended, as the management of data and organization of the program become very difficult. Fortran is the most popular scientific programming language. It overcomes many of the problems of BASIC for medium-to-large systems, allowing subroutines to have names and parameters to be passed to subroutines. It also allows some structured programming to be carried out using the IF-THEN-ELSE statement. Fortran subroutines can be developed separately from each other, allowing them to be easily used by different programs and allowing different people to work on different parts of the same system simultaneously.

Modern Engineering Languages

C is a popular language often used on systems that are based on the UNIX operating system. It allows the programmer to do high-level programming while allowing manipulation of machine-level structures such as the bits in memory locations. Its use as a portable language has increased its popularity. It is quickly becoming the most popular programming language on a wide variety of computer architectures.

Pascal was defined by Niklaus Wirth as a teaching tool. It is very similar to C, allowing complex data structures to be defined, while providing stronger controls over the operations on data.

A significant problem with all of the languages discussed so far is that they do not have built-in capability to perform real-time functions such as synchronization and communication with other programs. Languages similar to Pascal, but which include facilities to allow real-time programming, include Modula-2 and Ada. Ada is not yet well established outside the military-industrial sector, but both reflect a trend in programming where language developers have realized that great gains in productivity can be made by standardizing the coordination of cooperating programs.

Assembly Language

There are many times when none of the programming languages described above will solve a problem because it requires too much speed, space, or special interfaces to devices such as sensors. Assembly languages are languages designed specifically for the computer they are implemented on. They are closely tied to the architecture and are usually very low level, requiring many instructions to do the same function as the languages discussed above. The real power of assembly language is that the developer can write a program that can do anything the machine can do. Data acquisition systems often require assembly language routines because of the wide variety of sensors the computers must deal with. This has changed somewhat recently because of the increasing use of standard serial interfaces to acquire data from sensors. One device driver (a special set of instructions to control the input and output of data from a device) can be written to handle a serial interface and then can be connected to any sensor that uses serial interfaces of the same standard.

Process-Oriented Languages

As one moves toward the more complex data acquisition and control systems, there are process-oriented languages. These range from enhancements to normal high-level languages to keyword-based systems. The Instrument Society of America (ISA) extensions to Fortran allow it to be used in situations where multiple tasks and activities must run in a coordinated fashion. Other vendors have developed their own languages or extensions to existing languages; for example, Hewlett-Packard BASIC, FOXBORO BLOCK Process language, and Analog Devices BASIC.

Special process languages have some of the best and worst of the other languages described above. They will solve the problem they were designed for quite well, but if asked to handle unusual situations or devices, they can quickly become cumbersome or unable to handle the situation.

The recent developments in object-oriented environments and improvements in user interfaces have resulted in easier ways to set up instruments and control their operation. These allow the user of a system to quickly and easily customize its appearance to meet his or her needs. This field is just emerging and may result in a third class of ways to interact with instruments.

Standardization is also a problem. Some data acquisition extensions are unique to the vendor, resulting in problems when one has to transfer a program from one system to another.

Discussion

There are several conclusions that can be drawn from the above discussion. If, for example, speed is an issue, then an interpreted language may be too slow to perform the functions needed. A compiled language will normally run several times faster than an interpreted version of the same program. On the other hand, an interpreted language often allows one to try out simple ideas, make changes, and get immediate feedback. This allows the developer to avoid running several programs entailing a long wait for a simple program change. Recent products create an integrated environment where compilation time is very fast and is thus not an issue.

There is a wide variation across vendors for nonstandardized tools. For instance, BASIC is only recently becoming standardized. Most manufacturers have implemented BASIC interpreters that are supersets of the standard BASIC. They perform simple commands in much the same way across vendors, but if one wishes to perform graphics, catch errors, generate sounds, communicate with devices, coordinate activities with other programs, or access the system clock, the commands may be quite different. The same applies for Fortran, Pascal, C, and process languages. The Ada language has many of these problems, but represents an improvement in its definition of ways to coordinate activities with other programs and with its facilities to isolate enhancements from the language.

As a general rule of thumb, for simple applications which do not require great flexibility in program functions such as generating alarms, unusual graphics types, control of the process, or integration into larger systems, it is appropriate to use the simple question/answer or menu-type systems. When the system must be very flexible or customized, then it may be more appropriate to write a program in one of the languages mentioned above. Evaluate this very carefully, for the cost of implementing a program is often much higher than expected. For instance, if one wanted to perform simple data acquisition and storage from a sensor, the cost to write a program would probably cost more than buying a data logger and entering the parameters for data collection. Writing a program involves analysis, design, writing and debugging of the program, and testing of results. The cost of documenting a program is often a very large unplanned cost. If the results are intended to be used to make economic or process-related decisions, then the program must be tested carefully.

13.4   EXAMPLE SYSTEMS

The best way to understand the concepts described in preceding sections is to give concrete examples from the industry. We do not mean to endorse any products, but rather have chosen products that we are most familiar with or which help to differentiate one from the other and which have achieved a fairly broad application base. Also, we have chosen specific applications to demonstrate the use of a particular system. This is in no way intended to be the only use of that system as it may have a wide range of potential uses.

13.4.1   Hewlett-Packard Model 680

An example of a data recorder is the Hewlett-Packard Model 680. This is a strip chart recorder. It is used in situations where a hard-copy plot is desired of some measurements of a process. Table 13.11 describes key features of the unit. The HP680 has a few operator controls for paper movement, pen control, and paper speed. The only output is the pen mark on the paper. Input is analog, in several ranges. A full-scale change of input can be plotted in 0.5 s. There is no output from the instrument to actuators or capability to send the data to another computer. The system has a simple operator interface, as mentioned above. This allows ease of use, but limits the flexibility.

TABLE 13.11   HEWLETT-PACKARD MODEL 680 STRIP CHART RECORDER

Interface
   Process Inputs
      Number and type—1 analog input, several ranges
      Scan speed—pen response is 0.5 s full scale
      Throughput speed—same as scan speed
   Process Output—none
   Operator
      Display—single pen plot
      Entry—dials for paper position and speed, pen control
   Data Communications—none

Storage—none

Portability
   Weight—5 kg
   Size—165 × 197 × 219 mm

Flexibility
   Hardware Modularity
      Options—small number, mostly chart control
         and input range
      Bus—none
   Programming—none

Other Factors
   Environment—not known
   Price—around $2000

Courtesy of Hewlett-Packard

13.4.2 Micronic M203

An example of a data acquisition system at the low end of the range is the Micronic M203. This is a hand-held unit. It is more than a self-contained data recorder because it can relay data to another computer, but it is a very low-level data acquisition system because it does no processing of the data. It is often used in situations where the process is not instrumented. A technician may call out numbers to another, who will enter them on the keyboard. The data may later be transferred to another computer for analysis. Table 13.12 describes key features of the unit. The Micronic has a small keypad and readout, reflecting the limited flexibility of the system. Input is either through the keypad or through an optional bar code wand (some models support a caliper). Data can be taken at the speed at which the operator can enter data at the keypad or move the wand across a bar code. There is no provision for analog or digital input, nor for output to actuators. The transfer to another computer is over an RS232C communications line. This can occur at rates up to 120 cps. The system is quite capable of storing all the information in RAM memory that an operator can enter over a period of several hours, up to 16,000 numbers. The system can be set up to give prompts on the readout such as 'Diameter:'. Still, the setup is through simple keyword or parameter entry, and programming is not possible.

13.4.3 Acurex Autodata Ten/30

The Acurex Autodata Ten/30 is a flexible data logger. It is used to acquire data, do some simple conversion, send outputs to the process, and send data to another computer. As such, the system can be used for everything from a simple data recorder to a process control system. It does not have the flexibility of a computer system, being programmable in a menu fashion. The system is desktop sized. It is configured to have a large number of analog and digital inputs. It operates based on a scan cycle, meaning that data is taken at periodic time intervals. Table 13.13 describes key features of the unit. The Autodata Ten/30 has a small display, membrane keypad, and printer on its front panel. The keypad has all the letters of the alphabet, but they are arranged in alphabetical order, indicating their use to be for occasional entry rather than for typing. Process input can come from 120 channels

---

TABLE 13.12   MICRONIC M203

Interface
  Process Input
    Number and type—keypad, UPC bar code wand
    Scan speed—not known, human manual entry
    Throughput speed
  Process Ouput—none
  Operator
    Display—16 characters, 7 segment LCD
    Keyboard—18-key pad
  Data Communications
    30 to 120 cps
    Acoustic coupler or V24 modem interface

Storage
  Capacity—16,000 numbers
  Mechanism—RAM memory

Portability
  Weight—150 g
  Size—138 × 65 × 17 mm

Flexibility
  Hardware Modularity
    Options—UPC wand, some models have a caliper
    Bus—none
  Programming
    Language—keyword based, very limited
    Operating System—none
    Routines Available—none

Other Factors
  Environment—−5°C to +50°C noncondensing
  Price—low cost

TABLE 13.13   ACUREX AUTODATA TEN/30

Interface
Process Input
  Number and type
    120 analog or digital channels in main box
    880 additional in expander chassis
  Scan speed
    continuous
    triggered
    programmed: 1 s to 99 hours
  Throughput speed
    68 scans/s low resolution mode
    37 scans/s standard resolution mode
    10 scans/s high resolution mode
Process Output—analog or digital
Operator
  Display—5 inch diagonal monitor CRT
  Keyboard—full alphanumeric ASCII, alphabetical order
  Printer—6 lines/s, 22 columns, 64-character ASCII
Data Communications
  30 to 960 cps
  RS232C

Storage
Capacity—1 scan (1000 points)
Mechanism—RAM memory
Optional—up to 750K additional RAM (150,000 readings)

Capacity—approx 2.7 Mb (1680 in. at 1600 bpi)
Mechanism—magnetic tape unit

Portability
Weight—29.5 kg
Size—238 mm high × 584 mm deep × 451 mm wide

Flexibility
Hardware Modularity
  Options—wide variety of input and output cards
  Bus—RS422
Programming
  Language—keyword based, fairly broad range
  Operating System—none
  Routines Available—MATHPAC

Other Factors
Environment—0°C to 50°C, noncondensing humidity
Price—mid to high cost (above $5000 without I/O)
Common Mode Voltage Tolerance—250 V

in the main system box, and an additional 880 inputs can be obtained with an optional expander chassis. The system has several communications options, including magnetic tape and RS232C (up to 960 cps). Data can be taken at a variety of speeds, up to 68 channels/s. The system is capable of storing a series of scans in memory or on tape and then later playing it back. The operator can watch selected channels on the display and can even request notification if certain conditions (alarms) are met. The system is programmed through forms/parameter entry, and programming is not as flexible as with a computer-based system. This is also not as flexible as a large-scale process control system, such as the FOXBORO 1A, which allows a variety of process blocks to be defined and combined, and has graphics to support display of data.

13.4.4   Hewlett-Packard 3045DL with 85F Scientific Computer

The Hewlett-Packard 3045DL with the 85F Scientific Computer is an example of a system that can be used to perform the entire range of activities from data recording through process control. A common use is for data acquisition. Data acquisition can be performed by programming the system in the BASIC language to acquire data and to do some manipulation on it. This makes this system

TABLE 13.14 HELWETT-PACKARD 3054DL AND 85F

Interface
 Process Input
  Number and type
   approximately 100 analog and/or digital channels
  Scan speed—from 48 to 5000 readings per second
  Throughput speed—not known
 Process Output—digital
 Operator
  Display—CRT diameter 16 lines × 32 characters
  Keyboard—full alphanumeric ASCII, 20-key pad
 Data Communications
  HPIB
  Serial

Storage
 Capacity—32,000 bytes
 Mechanism—RAM memory

 Capacity—544,000 bytes
 Mechanism—electronic disk (RAM memory)

 Capacity—210,000 bytes
 Mechanism—magnetic tape unit

Portability
 Weight—unknown
 Size—150 × 419 × 45.2 mm (HP 85)

Flexibility
 Hardware Modularity
  Options—HP 85–wide variety of computer peripherals
  Bus—HPIB
 Programming
  Language—three levels–menu, keyword, BASIC language
   programming
  Operating System—none
  Routines Available—data acquisition subroutines

Other Factors
 Environment—unknown
 Price—approximately $10,000

Courtesy of Hewlett-Packard

a very flexible option for the scientist. In comparison to the Autodata Ten/30, this system is much more flexible, allowing the engineer to perform special operations on the data to match his needs. Table 13.14 describes key features of the system. The HP3045DL in conjunction with the HP85F has a full typewriter keyboard, a 20-key pad, a printer, a graphics display, and a tape drive. Process input can come from approximately 100 channels of analog and/or digital input. The system has several communications options, including magnetic tape, and HPIB. The system is quite capable of storing a series of scans in memory and then later playing them back; the trade-off is between the amount of the 16K memory reserved for the program and the amount reserved for the data. The operator can watch selected channels on the display and can even request notification if certain conditions (alarms) are met. The system can be programmed through forms/parameter entry, through keywords, or through BASIC language programs.

13.4.5 Digital Equipment Company MicroVAX

The Digital Equipment Company MicroVAX represents a computer with maximum flexibility. It is a fully programmable computer system with sophisticated tools for creating highly customized systems. It has several operating systems, suited for combination real-time/general purpose use (VAX/VMS), highly optimized real-time programming (VAX/ELN), and industry standard operating systems (VAX/ULTRIX—a UNIX implementation). This system can be used to perform the range of activities from data recording to highly sophisticated and customized process monitoring and control. A common use is for process monitor systems. Programs can be developed in a variety of languages for a wide range of data acquisition and control instruments. There are programs available that can

TABLE 13.15   DIGITAL EQUIPMENT COMPANY MICROVAX

Interface
   Process Input
      Number and type—wide variety of channels
      Scan speed—very fast
      Throughput speed—very fast
   Process Output—wide variety
   Operator
      Display—CRT 24 lines × 80 characters
      Keyboard—full alphanumeric ASCII, 48 additional function keys
      Printers, tape drives, and other computer peripheral devices
   Data Communications
      QBUS
      DECNET
      HPIB
      Serial

Storage
   Capacity—4,000,000 bytes (minimum)
   Mechanism—RAM memory

   Capacity—800,000 bytes dual floppy drives
   Mechanism—dual floppy diskette drives

   Capacity—70,000,000 bytes (minimum)
   Mechanism—fixed disk

Portability
   Weight—desktop or rack mountable

Flexibility
   Hardware Modularity
      Options—very wide variety
      Bus—QBUS, IEEE 488
   Programming
      Language—wide variety of languages
      Operating System—VAX/VMS, VAX/ELN, UNIX
      Routines Available—wide variety of programs, including
         data base, statistical, and data acquisition

Other Factors
   Environment—unknown
   Price—approximately $17,000

be used as is. This makes this system an extremely flexible option for the scientist, although it may also require the services of a programmer to fully use these capabilities. Table 13.15 describes key features of the system. There are a wide variety of products in the Digital Equipment line which are compatible with the MicroVAX. In addition, there is a large number of vendors supplying programs and computer support equipment. Key options are the DECNET network interface, allowing this computer to be a highly integrated part of a network of all the computers in the Digital Equipment line. The QBUS interface is an industry de facto standard, with a number of vendors making peripherals to connect to it. Programs to support large data bases, very large programs, many instruments or users, and large amounts of storage are available.

13.4.6   Other Approaches

Researchers often have needs that don't fit into the mold of standard data acquisition systems. Other approaches include acquiring special-purpose sensors, developing data acquisition systems from microcomputers, or purchasing large-scale process monitoring and control systems for highly customized processes.

Special-Purpose Sensors

New sensors are becoming available every day. One of the most flexible is the analog-to-digital converter that has its own serial interface so that data can be sent directly to any computer. This frees the computer from dealing with a lot of hardware interfacing details and simplifies the computer programmer's job.

Microcomputers

Microcomputers are ideal candidates for data acquisition.[18,19] They are inexpensive, run a wide variety of programs, and are small enough to move from site to site. The IBM PC and its various families of related computers are examples of personal computers that have been used in a variety of data acquisition situations.[20,21] Programs are available, such as ONSPEC and LABTECH, to perform many of the tasks required for data acquisition and analysis, thus freeing the researcher for other tasks. Special purpose cards are available from companies such as Metrabyte and Data Translation. Most of these cards come with data acquisition software to facilitate interfacing to the process. The IBM PC allows some simple programming capability with the BASIC language. Personal computer data acquisition systems are most appropriate for collecting laboratory data, since there is usually little environmental hazard. The Macintosh computer has been emerging as a useful laboratory computer. They may be used in plants and mills, but this requires careful consideration of the operating environment. There is a series of ruggedized personal computers becoming available now.

Other microcomputer systems are based on the S100 and Standard Bus. There are many data acquisition products intended for these buses. Their advantages are that the researcher needs only to acquire those portions required for a specific data collection effort, and the parts are fairly inexpensive. What is lost is the system integration service offered by packaged systems. For instance, the researcher must often order the central processing unit, operating system, memory, interface cards, and peripherals separately. This can be a substantial task, and the quality of software is often not as great as that for highly packaged personal computers such as the IBM PC. Standard bus systems are often useful when one requires an inexpensive, yet fairly custom hardware solution to a data acquisition problem.

If very high data rates or unusual sensors are required, researchers can make their own microcomputers through acquiring chips and other technology. This is the most flexible option for the researcher, allowing the assembly of systems that perform at the highest speed with custom parts. The researcher has a heavy responsibility in this case. The parts must not only be assembled for standard bus systems, but the proper power supplies and chips must be selected. Programming these systems may also be expensive since assembly language is often necessary. Microsecond response is feasible with this approach.

Large-Scale Special-Purpose Systems

On the other end of the scale are the large process control manufacturing systems. For instance, the Honeywell Distributed Manufacturing Control System 3000 (DMCS 3000) is a system organized around time-based processes, where data is usually examined based on some time interval. A user of this system can create a control application from some simple forms, yet DMCS 3000 has the flexibility to allow special-purpose programs to be written, large amounts of data to be stored, data to be passed back and forth between different computers and operator stations, and highly customized graphic displays to be created. Systems of this type are useful in the petrochemical and energy management area.

A similar approach for the more discrete-oriented systems (where actions are the direct result of some event occurring) could be characterized by the Allen Bradley PLC 3. This is a process controller, a computer specifically designed to replace the ladder logic of older relay-based systems. The system is not organized toward data as much as the DMCS 3000, but does allow communication between computers and closely matches the terminology of plant engineers.

13.5   SUMMARY

There is a wide variety of options for data acquisition and display. This broad area can be overwhelming. We have attempted to reduce the confusion by focusing on several aspects of data acquisition and display systems.

1. Data Acquisition and Manipulation
    Understanding the data required and what is to be done with it is the first step in developing a data acquisition system.

2. Characteristics of Systems
    An understanding of the capabilities of data acquisition systems makes up the other half of the requirements, allowing the engineer/scientist to choose between competing options.

We gave examples of systems which cover various aspects of the range of data acquisition and display systems. Remember that these are only representative examples; there are many other systems that also fit these ranges.

Acknowledgments

The authors are grateful to the guidance and help of Chet Nachtigal, to the patience of our wives, and to Virgil L. Laing for his assistance in reviewing this work.

REFERENCES

1. R. C. Hallgren, "Putting the Apple II to Work: Part 1: The Hardware," *BYTE*, Vol. 9, No. 4, 1984.

2. C. D. Johnson, *Microprocessor-Based Process Control*, Prentice-Hall, Englewood Cliffs, New Jersey, 1984.

3. P. W. Murrill, *Fundamentals of Process Control Theory*, ISA, 1981.

4. M. A. Needler and D. E. Baker, *Digital and Analog Controls*, Prentice-Hall, Englewood Cliffs, New Jersey, 1985.

5. J. G. Liscouski, "Connecting Computer and Experiments: Noise Rejection Through Software," *CAL—Computer Applications in the Laboratory (Intelligent Instruments and Computers—Applications in the Laboratory)*, Vol. 2, No. 4, April, 1984.

6. S. C. Gates, "Laboratory Data Collection with an IBM PC," *BYTE*, Vol. 9, No. 5, 1984.

7. R. W. Bailey, *Human Performance Engineering: A Guide for Systems Designers*, Prentice-Hall, Englewood Cliffs, New Jersey, 1982.

8. W. H. Beyer, Ed., *CRC Standard Mathematical Tables*, 24th ed., CRC Press, Boca Raton, 1976.

9. J. D. Wright and T. F. Edgar, "Digital Computer Control and Signal Processing Algorithms," in D. A. Mellichamp, ed., *Real-Time Computing*, Van Nostrand Reinhold, New York, 1983.

10. D. Allen, "Flat Panels: Beyond the CRT," *Computer Graphics World*, Vol. 9, No. 2, 1986.

11. "Open Systems Interconnection—Basic Reference Model," Draft Proposal 7498, 97/16 N719, American National Standards Institute, New York, 1981.

12. M. A. Kaminski, Jr., "Protocols for Communicating in the Factory," *IEEE Spectrum*, Vol. 23, No. 4, 1986.

13. S. A. Farowich, "Communicating in the Technical Office," *IEEE Spectrum*, Vol. 23, No. 4, 1986.

14. R. E. Floyd and R. C. Stanley, "Data Buses Simplify Microprocessor System Integration for Industrial Control," *Control Engineering*, Vol. 32, No. 13, 1985.

15. R. M. Genet, L. J. Boyd, and D. J. Sauer, "Interfacing for Real-Time Control," *BYTE*, Vol. 9, No. 4, 1984.

16. B. Milne, "Computers and Instruments: United They Stand!" *Electronic Design*, Vol. 32, No. 3, 1984.

17. T. J. Harrison, "Process Control Programming Language Standards: Accomplishments and Challenges," *InTech*, Vol. 28, No. 11, 1981.

18. A. Krigman, "Data Acquisition Systems: The Microprocessor Revolution," *InTech*, Vol. 30, No. 7, 1983.

19. "Analog I/O Boards Make Micro Data-Acquisition System," *Systems and Software*, Vol. 3, No. 11, 1984.

20. S. J. Bailey, "Desktop to Plant Floor—Personal Computers Can Help the Control Engineer," *Control Engineering*, Vol. 30, No, 13, 1983.

21. S. J. Bailey, "Personal Computers: Frugal Path to Specialized Control Systems," *Control Engineering*, Vol. 31, No. 7, 1984.

BIBLIOGRAPHY

Bailey, R. W., *Human Performance Engineering: A Guide for Systems Designers*, Prentice-Hall, Englewood Cliffs, New Jersey, 1982.

Beyer, W. H., ed., *CRC Standard Mathematical Tables*, 24th ed., CRC Press, Boca Raton, 1976.

Helms, H. L., *The McGraw-Hill Computer Handbook*, McGraw-Hill, New York, 1983.

Johnson, C. D., *Microprocessor-Based Process Control*, Prentice-Hall, Englewood Cliffs, New Jersey, 1984.

Mellichamp, D. A., ed., *Real-Time Computing*, Van Nostrand Reinhold, New York, 1983.

Murrill, P. W., *Fundamentals of Process Control Theory*, Instrument Society of America, 67 Alexander Drive, Research Triangle Park, NC, 27709, 1981.

Needler, M. A., and D. E. Baker, *Digital and Analog Controls*, Prentice-Hall, Englewood Cliffs, New Jersey, 1985.

Terman, F. E., and J. M. Pettit, *Electronic Measurements*, McGraw-Hill, New York, 1952.

MAGAZINES THAT CARRY RELEVANT INFORMATION

American Laboratory
BYTE
Computer Design
Computer Graphics World
Control Engineering
IEEE Spectrum
Electronic Design
Electronic Products
InTech
Intelligent Instruments in the Laboratory (formerly *CAL—Computer Applications in the Laboratory*)
Robotics Age
Systems and Software

OTHER SOURCES

Product vendors often will include good advice along with product catalogs or give free or inexpensive seminars. Examples include

Action Instruments computer products handbook
Analog Devices applications notes
Data Translation Microcomputer I/O products catalog
Cyber Research IBM PC Enhancement Handbook for scientists and engineers
HP Introduction to Data Acquisition seminar
DataMyte Handbook, DataMyte Corporation, 1984

CHAPTER 14
CLOSED-LOOP CONTROL SYSTEM ANALYSIS

SUHADA JAYASURYA

Department of Mechanical Engineering
Texas A & M University
College Station, Texas

14.1 INTRODUCTION

The field of control has a rich heritage of intellectual depth and practical achievement. From the water clock of Ctesibius in ancient Alexandria, where feedback control was used to regulate the flow of water, to the space exploration and the automated manufacturing plants of today, control systems have played a very significant role in technological and scientific development. James Watt's flyball governor (1769) was essential for the operation of the steam engine, which was, in turn, a technology fueling the Industrial Revolution. The fundamental study of feedback begins with James Clerk Maxwell's analysis of system stability of steam engines with governors (1868). Giant strides in our understanding of feedback and its use in design resulted from the pioneering work of Black, Nyquist, and Bode at Bell Labs in the 1920s. Minorsky's work on ship steering was of exceptional practical and theoretical importance. Tremendous advances occurred during World War II in response to the pressing problems of that period. The technology developed during the war years led, over the next 20 years, to practical applications in many fields.

Since the 1960s, there have been many challenges and spectacular achievements in space. The guidance of the Apollo spacecraft on an optimized trajectory from the earth to the moon and the soft landing on the moon depended heavily on control engineering. Today, the shuttle relies on automatic control in all phases of its flight. In aeronautics, the aircraft autopilot, the control of high-performance jet engines, and ascent/descent trajectory optimization to conserve fuel are typical examples of control applications. Currently, feedback control makes it possible to design aircraft that are aerodynamically unstable (such as the X-29) so as to achieve high performance. The National Aerospace Plane will rely on advanced control algorithms to fly its demanding missions.

Control systems are providing dramatic new opportunities in the automotive industry. Feedback controls for engines permit federal emission levels to be met, while antiskid braking control systems provide enhanced levels of passenger safety. In consumer products, control systems are often a critical factor in performance and thus economic success. From simple thermostats that regulate temperature in buildings to the control of the optics for compact disk systems, from garage door openers to the head servos for computer hard disk drives, and from artificial hearts to remote manipulators, control applications have permeated every aspect of life in industrialized societies.

In process control, where systems may contain hundreds of control loops, adaptive controllers have been available commercially since 1983. Typically, even a small improvement in yield can be quite significant economically. Multivariable control algorithms are now being implemented by several large companies. Moreover, improved control algorithms also permit inventories to be

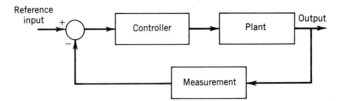

Fig. 14.1 Closed-loop system configuration.

reduced, a particularly important consideration in processing dangerous material. In nuclear reactor control, improved control algorithms can have significant safety and economic consequences. In power systems, coordinated computer control of a large number of variables is becoming common. Over 30,000 computer control systems have been installed in the United States alone. Again, the economic impact of control is vast.

Accomplishments in the defense area are legion. The accomplishments range from the antiaircraft gunsights and the bombsights of World War II to the missile autopilots of today and to the identification and estimation techniques used to track and designate multiple targets.

A large body of knowledge has come into existence as a result of these developments and continues to grow at a very rapid rate. A cross section of this body of knowledge is collected in this chapter. It includes analysis tools and design methodologies based on classical techniques. In presenting this material, some familiarity with general control system design principles is assumed. As a result, detailed derivations have been kept to a minimum with appropriate references.

14.1.1 Closed-Loop versus Open-Loop Control

In a closed-loop control system the output is measured and used to alter the control inputs applied to the plant under control. Figure 14.1 shows such a closed-loop system.

If output measurements are not utilized in determining the plant input, then the plant is said to be under open-loop control. An open-loop control system is shown in Fig. 14.2.

An open-loop system is very sensitive to any plant parameter perturbations and any disturbances that enter the plant. So an open-loop control system is effective only when the plant and disturbances are exactly known. In real applications, however, neither plants nor disturbances are exactly known *a priori*. In such situations open-loop control does not provide satisfactory performance. When the loop is closed as in Fig. 14.1 any plant perturbations or disturbances can be sensed by measuring the outputs, thus allowing plant input to be altered appropriately.

The main reason for closed-loop control is the need for systems to perform well in the presence of uncertainties. It can reduce the sensitivity of the system to plant parameter variations and help reject or mitigate external disturbances. Among other attributes of closed-loop control is the ability to alter the overall dynamics to provide adequate stability and good tracking characteristics. Consequently, a closed-loop system is more complex than an open-loop system due to the required sensing of the output. Sensor noise, however, tends to degrade the performance of a feedback system thus making it necessary to have accurate sensing. Therefore sensors are usually the most expensive devices in a feedback control system.

14.1.2 Supervisory Control Computers[1]

With the advent of digital computer technology computers have found their way into control applications. The earliest applications of digital computers in industrial process control were in a so-called supervisory control mode, and the computers were generally large-scale mainframes (minis and micros had not yet been developed). In supervisory control the individual feedback loops are locally controlled by typical analog devices used prior to the installation of the computer. The main function of the computer is to gather information on how the entire process is operating, then feed this into an overall process model located in the computer memory, and then periodically send signals to the set points of individual local analog controllers so that the overall performance of the system is enhanced. A conceptual block diagram of such supervisory control is shown in Fig. 14.3.

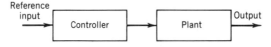

Fig. 14.2 Open-loop system configuration.

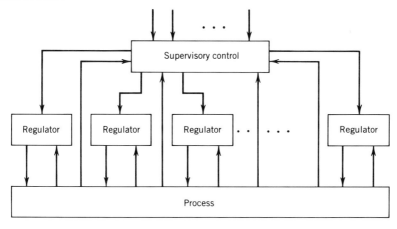

Fig. 14.3 Supervisory control configuration.

14.1.3 Hierarchical Control Computers[1]

In order to continuously achieve optimum overall performance the function of the supervisory control computer can be broken up into several levels and operated in a hierarchical manner. Such a multilevel approach becomes useful in controlling very complex processes, with large numbers of control loops with many inputs and controlled variables. In the hierarchical approach the system is subdivided into a hierarchy of simpler control design problems rather than attempting direct synthesis of a single comprehensive controller for the entire process. Thus the controller on a given "level of control" can be less complex due to the existence of lower-level controllers that remove frequently occurring disturbances. At each higher level in the hierarchy, the complexity of the control algorithm increases, but the required frequency of execution decreases. Such a system is shown in Fig. 14.4.

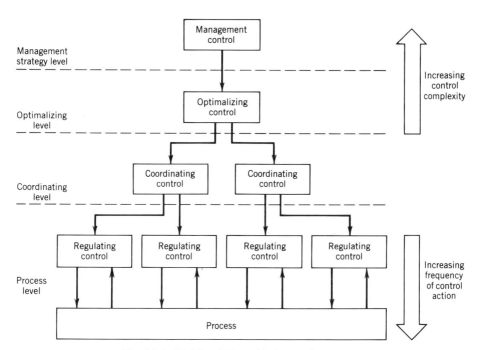

Fig. 14.4 Multilevel hierarchical control structure.

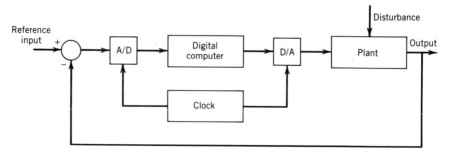

Fig. 14.5   Direct digital control (DDC) configuration.

14.1.4   Direct Digital Control (DDC)

In direct digital control all the analog controllers at the lowest level or intermediate levels are replaced by a central computer serving as a single time-shared controller for all the individual feedback loops. Conventional control modes such as PI or PID (proportional integral differential) are still used for each loop, but the digital versions of the control laws for each loop reside in software in the central computer. In DDC the computer input and output are multiplexed sequentially through the list of control loops, updating each loop's control action in turn and holding this value constant until the next cycle. A typical DDC configuration is shown in Fig. 14.5.

14.1.5   Hybrid Control

Combinations of analog and digital methods based on individual loop controllers is known as hybrid control. It is clear that any real control system would have both analog and digital features. For example, in Fig. 14.5, the plant outputs and inputs are continuous (or analog) quantities. The inputs and outputs of the digital computer are nevertheless digital. The computer processes only numbers. All control algorithms, however, need not be implemented digitally. As an example, for most feedback control loops, the PID algorithm implemented with analog circuitry is superior to the PID algorithm implemented in digital software. The derivative mode for instance requires very high resolution to approach the quality of analog performance. Thus it is sometimes advantageous to use hybrid systems for controller implementation.

14.1.6   Real Systems with Digital Control

Real systems are typically mixtures of analog and digital signals. Typical plant inputs and outputs are analog signals, that is, actuators and sensors are analog systems. If a digital computer is included in the control loop, then digitized quantities are needed for their processing. These are usually accomplished by using analog-to-digital (A/D) converters. In order to use real-world actuators, computations done as numbers need to be converted to analog signals by employing digital to analog (D/A) converters. One of the main advantages of digital control is the ease with which an algorithm can be changed on line by altering the software rather than hardware.

14.2   LAPLACE TRANSFORMS

Often in designing control systems, linear time-invariant (LTI) differential equation representations of physical systems to be controlled are sought. These are typically arrived at by appropriately linearizing the nonlinear equations about some operating point. The general form of such an LTI ordinary differential equation representation is

$$a_n \frac{d^n y(t)}{dt^n} + a_{n-1} \frac{d^{n-1} y(t)}{dt^{n-1}} + \cdots + a_1 \frac{dy(t)}{dt} + a_0 y(t)$$

$$= b_m \frac{d^m}{dt^m} u(t) + b_{m-1} \frac{d^{m-1}}{dt^{m-1}} u(t) + \cdots + b_1 \frac{du(t)}{dt} + b_0 u(t) \qquad (14.1)$$

where $y(t) =$ output of the system
          $u(t) =$ input to the system
             $t =$ time
       $a_j, b_j =$ physical parameters of the system

and $n \geq m$ for physical systems.

The ability to transform systems of the form given by Eq. 14.1 to algebraic equations relating the input to the output is the primary reason for employing Laplace transform techniques.

14.2.1   Single-Sided Laplace Transform

The Laplace transform $\mathcal{L}[f(t)]$ of the time function $f(t)$ defined as

$$f(t) = \begin{cases} 0 & t < 0 \\ f(t) & t \geq 0 \end{cases}$$

is given by

$$\mathcal{L}[f(t)] = F(s) = \int_0^\infty e^{-st} f(t) \, dt \qquad t > 0 \tag{14.2}$$

where s is a complex variable $= \sigma + j\omega$. The integral of Eq. 14.2 cannot be evaluated in a closed form for all $f(t)$ but when it can, it establishes a unique pair of functions, $f(t)$, in the time domain and its companion $F(s)$ in the s domain. It is conventional to use uppercase letters for s functions and lowercase for t functions.

Example 14.1.   Determine the Laplace transform of the unit step function $u_s(t)$:

$$u_s(t) = \begin{cases} 0 & t < 0 \\ 1 & t \geq 0 \end{cases}$$

By definition

$$\mathcal{L}[u_s(t)] = U_s = \int_0^\infty e^{-ts} 1 dt = \frac{1}{s}$$

Example 14.2.   Determine the Laplace transform of $f(t)$:

$$f(t) = \begin{cases} 0 & t < 0 \\ e^{-\alpha t} & t \geq 0 \end{cases}$$

$$\mathcal{L}[f(t)] = F(s) = \int_0^\infty e^{-ts} e^{-\alpha t} dt = \frac{1}{s + \alpha}$$

Example 14.3.   Determine the Laplace transform of the function $f(t)$ given by

$$f(t) = \begin{cases} 0 & t < 0 \\ t & 0 \leq t \leq T \\ T & T \leq t \end{cases}$$

By definition

$$F(s) = \int_0^\infty e^{-ts} f(t) dt$$

$$= \int_0^T e^{-ts} t \, dt + \int_T^\infty e^{-ts} T dt$$

$$= \frac{1}{s^2} - \frac{e^{-Ts}}{s^2} = \frac{1}{s^2}(1 - e^{-Ts})$$

In transforming differential equations, entire equations need to be transformed. Several theorems useful in such transformations are given next without proof.

T1. Linearity Theorem

$$\mathcal{L}[\alpha f(t) + \beta g(t)] = \alpha \mathcal{L}[f(t)] + \beta \mathcal{L}[g(t)] \tag{14.3}$$

T2. Differentiation Theorem

$$\mathcal{L}\left[\frac{df}{dt}\right] = sF(s) - f(0) \tag{14.4}$$

$$\mathcal{L}\left[\frac{d^2f}{dt^2}\right] = s^2F(s) - sf(0) - \frac{df}{dt}(0) \tag{14.5}$$

$$\mathcal{L}\left[\frac{d^nf}{dt^n}\right] = s^nF(s) - s^{n-1}f(0) - s^{n-2}\frac{df}{dt}(0) - \cdots - \frac{d^{n-1}}{dt^{n-1}}(0) \tag{14.6}$$

T3. Translated Function (Fig. 14.6)

$$\mathcal{L}[f(t - \alpha)u_s(t - \alpha)] = e^{-\alpha s}F(s + \alpha) \tag{14.7}$$

T4. Multiplication of $f(t)$ by $e^{-\alpha t}$

$$\mathcal{L}[e^{-\alpha t}f(t)] = F(s + \alpha) \tag{14.8}$$

T5. Integration Theorem

$$\mathcal{L}[\int f(t)dt] = \frac{F(s)}{s} + \frac{f^{-1}(0)}{s} \tag{14.9}$$

where $f^{-1}(0) = \int f(t)dt$ evaluated at $t = 0$.

T6. Final-Value Theorem. If $f(t)$ and $df(t)/dt$ are Laplace transformable, if $\lim_{t \to \infty} f(t)$ exists, and if $F(s)$ is analytic in the right-half s plane including the $j\omega$ axis, except for a single pole at the origin, then

$$\lim_{t \to \infty} f(t) = \lim_{s \to 0} sF(s) \tag{14.10}$$

T7. Initial Value Theorem. If $f(t)$ and $df(t)/dt$ are both Laplace transformable, and if $\lim_{s \to \infty} sF(s)$ exists, then

$$f(0) = \lim_{s \to \infty} sF(s) \tag{14.11}$$

Example 14.4 The time function of Example 14.3 can be written as

$$f(t) = tu_s(t) - (t - T)u_s(t - T) \tag{14.12}$$

and

$$\mathcal{L}[f(t)] = \mathcal{L}[tu_s(t)] - \mathcal{L}[(t - T)u_s(t - T)] \tag{14.13}$$

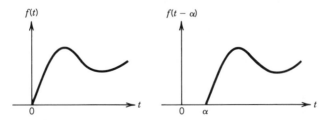

Fig. 14.6   Plots of $f(t)$ and $f(t - \alpha)u_s(t - \alpha)$.

But

$$\mathcal{L}[t u_s(t)] = \frac{1}{s^2} \qquad (14.14)$$

By using Eqs. 14.14 and 14.7 in Eq. 14.13, we get

$$F(s) = \frac{1}{s^2} - e^{-Ts}\frac{1}{s^2} = \frac{1}{s^2}(1 - e^{-Ts})$$

14.2.2   Transforming LTI Ordinary Differential Equations

The Laplace transform method yields the complete solution (the particular solution plus the complementary solution) of linear differential equations. Classical methods of finding the complete solution of a differential equation require the evaluation of the integration constants by use of the initial conditions. In the case of the Laplace transform method, initial conditions are automatically included in the Laplace transform of the differential equation. If all initial conditions are zero, then the Laplace transform of the differential equation is obtained simply by replacing d/dt with s, d^2/dt^2 with s^2, and so on.

Consider the differential equation

$$5\frac{d^2 y}{dt^2} + 3y = f(t) \qquad (14.15)$$

with $\dot{y}(0) = 1, y(0) = 0$. By taking the Laplace transform of Eq. 14.15, we get

$$\mathcal{L}\left[5\frac{d^2 y}{dt^2} + 3y\right] = \mathcal{L}[f(t)]$$

Now using T1 and T2 yield

$$5[s^2 Y(s) - sy(0) - y(0)] + 3Y(s) = F(s)$$

Thus

$$Y(s) = \frac{F(s) + 5}{5s^2 + 3} \qquad (14.16)$$

For a given $f(t)$, say $f(t) = u_s(t)$, the unit step function

$$F(s) = \frac{1}{s} \qquad (14.17)$$

Substituting Eq. 14.17 in Eq. 14.16 gives

$$Y(s) = \frac{5s + 1}{s(5s^2 + 3)}$$

$y(t)$ can then be found by computing the inverse Laplace transform of $Y(s)$. That is

$$y(t) = \mathcal{L}^{-1}[Y(s)] = \mathcal{L}^{-1}\left\{\frac{5s + 1}{s(5s^2 + 3)}\right\}$$

This will be discussed in Section 14.2.4.

14.2.3   Transfer Function

The transfer function of a linear time-invariant system is defined to be the ratio of the Laplace transform of the output to the Laplace transform of the input under the assumption that all initial conditions are zero.

For the system described by the LTI differential equation (14.1),

$$\frac{\mathcal{L}[\text{output}]}{\mathcal{L}[\text{input}]} = \frac{\mathcal{L}[y(t)]}{\mathcal{L}[u(t)]} = \frac{Y(s)}{U(s)}$$

$$= \frac{b_m s^m + b_{m-1} s^{m-1} + \cdots + b_1 s + b_0}{a_n s^n + a_{n-1} s^{n-1} + \cdots + a_1 s + a_0} \tag{14.18}$$

It should be noted that the transfer function depends only on the system parameters characterized by a_i's and b_i's in Eq. 14.18. The highest power of s in the denominator of the transfer function defines the order n of the system. The order n of a system is typically greater than or equal to the order of the numerator polynomial m.

Equation 14.18 can be further written in the form

$$\frac{Y(s)}{U(s)} = G(s) = \frac{b_m}{a_n} \frac{\prod\limits_{j=1}^{m} (s + z_i)}{\prod\limits_{j=1}^{n} (s + p_j)} \tag{14.19}$$

The values of s making Eq. 14.19 equal to zero are called the system zeros, and the values of s making Eq. 14.19 go to ∞ are called poles of the system. Hence $s = -z_i, i = 1, \ldots, m$ are the system zeros and $s = -p_j, j = 1, \ldots, n$ are the system poles.

14.2.4   Partial Fraction Expansion and Inverse Transform

The mathematical process of passing from the complex variable expression $F(s)$ to the time domain expression $f(t)$ is called an inverse transformation. The notation for the inverse Laplace transformation is $\mathcal{L}^{-1}$, so that

$$\mathcal{L}^{-1}[F(s)] = f(t)$$

Example 14.5   The inverse Laplace transform of $1/s$ is the unit step function:

$$\mathcal{L}[u_s(t)] = \frac{1}{s}$$

Hence

$$\mathcal{L}^{-1}\left(\frac{1}{s}\right) = u_s(t)$$

Time functions for which Laplace transforms are found in a closed form can be readily inverse Laplace transformed by writing them in pairs. Such pairs are listed in Table 14.1. When transform pairs are not found in tables, other methods have to be used to find the inverse Laplace transform. One such method is the partial fraction expansion of a given Laplace transform.

If $F(s)$, the Laplace transform of $f(t)$, is broken up into components

$$F(s) = F_1(s) + F_2(s) + \cdots + F_n(s)$$

and if the inverse Laplace transforms of $F_1(s), F_2(s), \ldots, F_n(s)$ are readily available, then

$$\mathcal{L}^{-1}[F(s)] = \mathcal{L}^{-1}[F_1(s)] + \mathcal{L}^{-1}[F_2(s)] + \cdots + \mathcal{L}^{-1}[F_n(s)]$$

$$= f_1(t) + f_2(t) + \cdots + f_n(t) \tag{14.20}$$

where $f_1(t), f_2(t), \ldots, f_n(t)$ are the inverse Laplace transforms of $F_1(s), F_2(s), \ldots, F_n(s)$, respectively.

For problems in control systems, $F(s)$ is frequently in the following form:

$$F(s) = \frac{B(s)}{A(s)}$$

where $A(s)$ and $B(s)$ are polynomials in s, and the degree of $B(s)$ is equal to or higher than that of $A(s)$.

If $F(s)$ is written as in Eq. 14.19 and if the poles of $F(s)$ are distinct, then $F(s)$ can always be expressed in terms of simple partial fractions as follows:

TABLE 14.1   LAPLACE TRANSFORM PAIRS

	$f(t)$	$F(s)$
1	Unit impulse $\delta(t)$	1
2	Unit step $u_s(t)$	$\dfrac{1}{s}$
3	t	$\dfrac{1}{s^2}$
4	e^{-at}	$\dfrac{1}{s+a}$
5	te^{-at}	$\dfrac{1}{(s+a)^2}$
6	$\sin \omega t$	$\dfrac{\omega}{s^2+\omega^2}$
7	$\cos \omega t$	$\dfrac{s}{s^2+\omega^2}$
8	$t^n\,(n=1,2,3...)$	$\dfrac{n!}{s^{n+1}}$
9	$t^n e^{-at}\,(n=1,2,3,...)$	$\dfrac{n!}{(s+a)^{n+1}}$
10	$\dfrac{1}{b-a}(e^{-at}-e^{-bt})$	$\dfrac{1}{(s+a)(s+b)}$
11	$\dfrac{1}{b-a}(be^{-bt}-ae^{-at})$	$\dfrac{s}{(s+a)(s+b)}$
12	$\dfrac{1}{ab}[1+\dfrac{1}{a-b}(be^{-at}-ae^{-bt})]$	$\dfrac{1}{s(s+a)(s+b)}$
13	$e^{-at}\sin \omega t$	$\dfrac{\omega}{(s+a)^2+\omega^2}$
14	$e^{-at}\cos \omega t$	$\dfrac{s+a}{(s+a)^2+\omega^2}$
15	$\dfrac{1}{a^2}(at-1+e^{-at})$	$\dfrac{1}{s^2(s+a)}$
16	$\dfrac{\omega_n}{\sqrt{1-\zeta^2}}e^{-\zeta\omega_n t}\sin \omega_n\sqrt{1-\zeta^2}\,t$	$\dfrac{\omega_n^2}{s^2+2\zeta\omega_n s+\omega_n^2}$
17	$\dfrac{-1}{\sqrt{1-\zeta^2}}e^{-\zeta\omega_n t}\sin(\omega_n\sqrt{1-\zeta^2}\,t-\phi)$ $\phi=\tan^{-1}\dfrac{\sqrt{1-\zeta^2}}{\zeta}$	$\dfrac{s}{s^2+2\zeta\omega_n s+\omega_n^2}$
18	$1-\dfrac{1}{\sqrt{1-\zeta^2}}e^{-\zeta\omega_n t}\sin(\omega_n\sqrt{1-\zeta^2}\,t+\phi)$ $\phi=\tan^{-1}\dfrac{\sqrt{1-\zeta^2}}{\zeta}$	$\dfrac{\omega_n^2}{s(s^2+2\zeta\omega_n s+\omega_n^2)}$

$$F(s)=\frac{B(s)}{A(s)}=\frac{\alpha_1}{s+p_1}+\frac{\alpha_2}{s+p_2}+\cdots+\frac{\alpha_n}{s+p_n} \qquad (14.21)$$

where the α_j's are constant. Here α_j is called the residue at the pole $s=-p_j$. The value of α_j is found by multiplying both sides of Eq. 14.21 by $(s+p_j)$ and setting $s=-p_j$, giving

$$\alpha_j=\left[(s+p_j)\frac{B(s)}{A(s)}\right]_{s=p_j} \qquad (14.22)$$

Noting that

$$\mathscr{L}^{-1}\left[\frac{1}{s+p_j}\right]=e^{-p_j t}$$

the inverse Laplace transform of Eq. 14.21 can be written as

$$f(t) = \mathcal{L}^{-1}[F(s)] = \sum_{j=1}^{n} \alpha_j e^{-p_j t} \tag{14.23}$$

Example 14.6 Find the inverse Laplace transform of

$$F(s) = \frac{s+1}{(s+2)(s+3)}$$

The partial fraction expansion of $F(s)$ is

$$F(s) = \frac{s+1}{(s+2)(s+3)} = \frac{\alpha_1}{s+2} + \frac{\alpha_2}{s+3}$$

where α_1 and α_2 are found by using Eq. 14.22 as follows:

$$\alpha_1 = \left[(s+2)\frac{s+1}{(s+2)(s+3)}\right]_{s=-2} = -1$$

$$\alpha_2 = \left[(s+3)\frac{s+1}{(s+2)(s+3)}\right]_{s=-3} = 2$$

Thus

$$f(t) = \mathcal{L}^{-1}\left[\frac{-1}{s+2}\right] + \mathcal{L}^{-1}\left[\frac{2}{s+3}\right]$$

$$= -e^{-2t} + 2e^{-3t} \qquad t \geq 0$$

Partial fraction expansion when $F(s)$ involves multiple poles: Consider

$$F(s) = \frac{K \prod_{i=1}^{m}(s+z_i)}{(s+p_1)^r \prod_{j=r+1}^{n}(s+p_j)} = \frac{B(s)}{A(s)}$$

where the pole at $s = -p_1$ has multiplicity r. The partial fraction expansion of $F(s)$ may then be written as

$$F(s) = \frac{B(s)}{A(s)} = \frac{\beta_1}{s+p_1} + \frac{\beta_2}{(s+p_1)^2} + \cdots + \frac{\beta_r}{(s+p_1)^r}$$

$$+ \sum_{j=r+1}^{n} \frac{\alpha_j}{s+p_j} \tag{14.24}$$

The α_j's can be evaluated as before using Eq. 14.22. The β_i's are given by

$$\beta_r = \left[\frac{B(s)}{A(s)}(s+p_1)^r\right]_{s=-p_1}$$

$$\beta_{r-1} = \left\{\frac{d}{ds}\left[\frac{B(s)}{A(s)}(s+p_1)^r\right]\right\}_{s=-p_1}$$

$$\vdots$$

$$\beta_{r-j} = \frac{1}{j!}\left\{\frac{d^j}{ds^j}\left[\frac{B(s)}{A(s)}(s+p_1)^r\right]\right\}_{s=-p_1}$$

$$\vdots$$

$$\beta_1 = \frac{1}{(r-1)!}\left\{\frac{d^{r-1}}{ds^{r-1}}\left[\frac{B(s)}{A(s)}(s+p_1)^r\right]\right\}_{s=-p_1}$$

The inverse Laplace transform of Eq. 14.24 can then be obtained as follows:

$$f(t) = \left[\frac{\beta_r}{(r-1)!}t^{r-1} + \frac{\beta_{r-1}}{(r-2)!}t^{r-2} + \cdots + \beta_2 t + \beta_1 \right]e^{-p_1 t}$$

$$+ \sum_{j=r+1}^{n} \alpha_j e^{-p_1 t} \qquad t \geq 0 \qquad (14.25)$$

Example 14.7. Find the inverse Laplace transform of the function

$$F(s) = \frac{s^2 + 2s + 3}{(s+1)^3}$$

Expanding $F(s)$ into partial fractions, we obtain

$$F(s) = \frac{B(s)}{A(s)} = \frac{\beta_1}{s+1} = \frac{\beta_2}{(s+1)^2} + \frac{\beta_3}{(s+1)^3}$$

where β_1, β_2, and β_3 are determined as already shown.

$$\beta_3 = \left[\frac{B(s)}{A(s)}(s+1)^3 \right]_{s=-1} = 2$$

similarly $\beta_2 = 0$ and $\beta_1 = 1$. Thus we get

$$f(t) = \mathcal{L}^{-1}\left[\frac{1}{s+1} \right] + \mathcal{L}^{-1}\left[\frac{2}{(s+1)^3} \right]$$

$$= e^{-t} + t^2 e^{-t} \qquad t \geq 0$$

14.2.5 Inverse Transform by a General Formula

Given the Laplace transform $F(s)$ of a time function $f(t)$, the following expression holds true:

$$f(t) = \frac{1}{2\pi j} \int_{c-j\infty}^{c+j\infty} F(s)e^{ts}\,ds \qquad (14.26)$$

where c, the abscissa of convergence, is a real constant and is chosen larger than the real parts of all singular points of $F(s)$. Thus the path of integration is parallel to the $j\omega$ axis and is displaced by the amount c from it (see Fig. 14.7). This path of integration is to the right of all singular points. Equation 14.26 need be used only when other simpler methods cannot provide the inverse Laplace transformation.

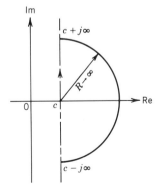

Fig. 14.7 Path of integration.

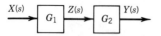

Fig. 14.8 Two blocks in cascade.

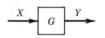

Fig. 14.10 Multiplication $Y = GX$.

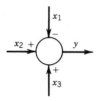

Fig. 14.9 Summing point.

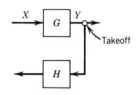

Fig. 14.11 Takeoff point.

14.3 BLOCK DIAGRAMS

A control system consists of a number of components. In order to show the functions performed by each component, we commonly use a diagram called a **block diagram**.

In a block diagram, all system variables are linked to each other through blocks. The block is a symbol for the mathematical operation on the input signal to the block that produces the output. The transfer functions of the components are usually entered in the blocks, with blocks connected by arrows to indicate the direction of the flow of signals. A basic assumption in block diagrams is that there is no loading between blocks. Figure 14.8 shows a simple block diagram with two blocks in cascade. The arrowheads pointing toward blocks indicate inputs and those pointing away indicate outputs. These arrows are referred to as signals. The following basic components allow the generation of many complex block diagrams.

Addition. Signal addition is represented by a summing point as shown in Fig. 14.9.

It should be noted that there can be only one signal leaving a summing point though any number of signals can enter a summing point. For the summing point shown in Fig. 14.9, we have

$$-x_1 + x_2 + x_3 = y$$

The sign placed near an arrow indicates whether the signal is to be added or subtracted.

Multiplication. Multiplication is denoted by a symbol as shown in Fig. 14.10. Here X the input and Y the output are related by the expression

$$Y = GX$$

Takeoff Point. If a signal becomes an input to more than one element, then a takeoff point as shown in Fig. 14.11 is employed.

A typical block diagram using these elements is shown in Fig. 14.12.

In the block diagram of Fig. 14.12 it is assumed that G_2 will have no back reaction (or loading) on G_1; G_1 and G_2 usually represent two physical devices. If there are any loading effects between the devices, it is necessary to combine these components into a single block. Such a situation is given in Section 14.3.2.

14.3.1 Block Diagram Reduction

Any number of cascaded blocks representing nonloading components can be replaced by a single block, the transfer function of which is simply the product of the individual transfer functions. For

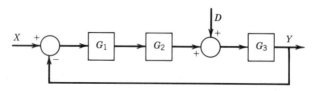

Fig. 14.12 Typical block diagram.

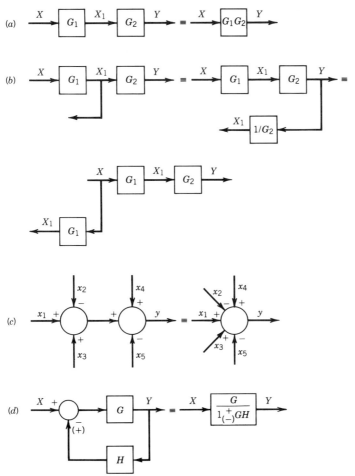

Fig. 14.13 Basic block diagram reduction rules.

example, the two elements in cascade shown in Fig. 14.12 can be replaced by a single element $G_0 = G_1G_2$. Some fundamental block diagram reduction rules are shown in Fig. 14.13.

In simplifying a block diagram a general rule is to first move takeoff points and summing points to form internal feedback loops as shown in Fig. 14.13d. Then remove the internal feedback loops and proceed in the same manner.

Example 14.8. Reduce the block diagram shown in Fig. 14.14.

First, in order to eliminate the loop $G_3G_4H_1$, we move H_2 behind block G_4 and therefore obtain Fig. 14.15a. Eliminating the loop $G_3G_4H_1$, we obtain Fig. 14.15b. Then eliminating the inner loop containing H_2/G_4, we obtain Fig. 14.15c. Finally, by reducing the loop containing H_3, we obtain the closed-loop system transfer function as shown in Fig. 14.15d.

14.3.2 Transfer Functions of Cascaded Elements[3]

Many feedback systems have components that load each other. Consider the system of Fig. 14.16. Assume that e_i is the input and e_o is the output. In this system the second stage of the circuit (R_2C_2 portion) produces a loading effect on the first stage (R_1C_1 portion). The equations for this system are

$$\frac{1}{C_1}\int (i_1 - i_2)dt + R_1i_1 = e_i \tag{14.27}$$

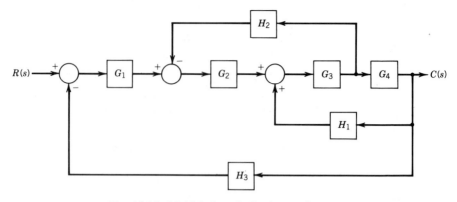

Fig. 14.14 Multiple-loop feedback control system.

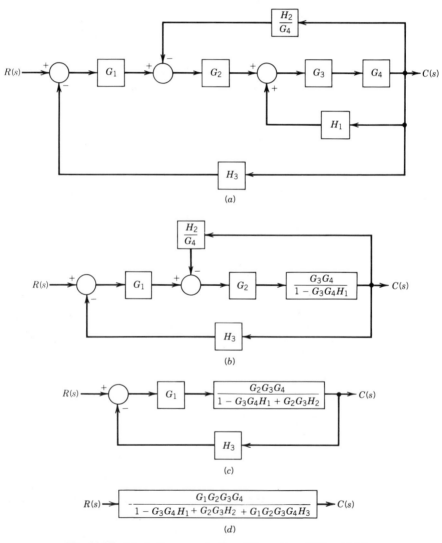

Fig. 14.15 Block diagram reduction of the system of Fig. 14.14.

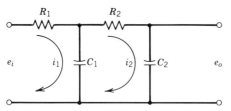

Fig. 14.16 Electrical system.

and

$$\frac{1}{C_1}\int(i_1 - i_2)\,dt + R_2 i_2 = -\frac{1}{C_2}\int i_2\,dt = -e_0 \tag{14.28}$$

Taking the Laplace transform of Eqs. 14.27 and 14.28, assuming zero initial conditions, and simplifying yield

$$\frac{E_o(s)}{E_i(s)} = \frac{1}{R_1 C_1 R_2 C_2 s^2 + (R_1 C_1 + R_2 C_2 + R_1 C_2)s + 1} \tag{14.29}$$

The term $R_1 C_2 s$ in the denominator of the transfer function represents the interaction of two simple RC circuits.

This analysis shows that if two RC circuits are connected in cascade so that the output from the first circuit is the input to the second, the overall transfer function is not the product of $1/(R_1 C_1 s + 1)$ and $1/(R_2 C_2 s + 1)$. The reason for this is that when we derive the transfer function for an isolated circuit, we implicitly assume that the output is unloaded. In other words, the load impedance is assumed to be infinite, which means that no power is being withdrawn at the output. When the second circuit is connected to the output of the first, however, a certain amount of power is withdrawn and then the assumption of no loading is violated. Therefore if the transfer function of this system is obtained under the assumption of no loading, then it is not valid. Chapter 5 deals with this type of problem in greater detail.

14.4   z TRANSFORMS

One of the mathematical tools commonly used to deal with discrete-time systems is the z transform. The role of the z transform in discrete-time systems is similar to that of the Laplace transform in continuous-time systems. Laplace transforms allow the conversion of linear ordinary differential equations with constant coefficients into algebraic equations in s. The z transformation transforms linear difference equations with constant coefficients into algebraic equations in z.

14.4.1   Single-Sided z Transform

If a signal has discrete values $f_0, f_1, \ldots, f_k, \ldots$, we define the z transform of the signal as the function

$$F(z) = \mathscr{Z}\{f(k)\}$$

$$= \sum_{k=0}^{\infty} f(k)z^{-k} \qquad r_0 \le |z| \le R_0 \tag{14.30}$$

and is assumed that one can find a range of values of the magnitude of the complex variable z for which the series of Eq. 14.30 converges. This z transform is referred to as the one-sided z transform. The symbol $\mathscr{Z}$ denotes "the z transform of." In the one-sided z transform it is assumed $f(k) = 0$ for $k < 0$, which in the continuous-time case corresponds to $f(t) = 0$ for $t < 0$.

Expansion of the right-hand side of Eq. 14.30 gives

$$F(z) = f(0) + f(1)z^{-1} + f(2)z^{-2} + \cdots + f(k)z^{-k} + \cdots \tag{14.31}$$

The last equation implies that the z transform of any continuous-time function $f(t)$ may be written in the series form by inspection. The z^{-k} in this series indicates the instant in time at which the amplitude $f(k)$ occurs. Conversely, if $F(z)$ is given in the series form of Eq. 14.31, the inverse z transform can be obtained by inspection as a sequence of the function $f(k)$ that corresponds to the values of $f(t)$ at the respective instants of time. If the signal is sampled at a fixed sampling period T, then the sampled version of the signal $f(t)$ given by $f(0), f(1), \ldots, f(k)$ correspond to the signal values at the instants $0, T, 2T, \ldots, kT$.

14.4.2 Poles and Zeroes in the z Plane

If the z transform of a function takes the form

$$F(z) = \frac{b_0 z^m + b_1 z^{m-1} + \cdots + b_m}{z^n + a_1 z^{n-1} + \cdots + a_n} \tag{14.32}$$

or

$$F(z) = \frac{b_0(z - z_1)(z - z_2) \cdots (z - z_m)}{(z - p_1)(z - p_2) \cdots (z - p_n)}$$

then p_i's are the poles of $F(z)$ and z_i's are the zeroes of $F(z)$.

In control studies, $F(z)$ is frequently expressed as a ratio of polynomials in z^{-1} as follows:

$$F(z) = \frac{b_0 z^{-(n-m)} + b_1 z^{-(n-m+1)} + \cdots + b_m z^{-n}}{1 + a_1 z^{-1} + a_2 z^{-2} + \cdots + a_n z^{-n}} \tag{14.33}$$

where z^{-1} is interpreted as the unit delay operator.

14.4.3 z Transforms of Some Elementary Functions

Unit Step Function. Consider the unit step function

$$u_s(t) = \begin{cases} 1 & t \geq 0 \\ 0 & t < 0 \end{cases}$$

whose discrete representation is

$$u_s(k) = \begin{cases} 1 & k \geq 0 \\ 0 & k < 0 \end{cases}$$

From Eq. 14.30

$$\mathcal{Z}\{u_s(k)\} = \sum_{k=0}^{\infty} 1 \cdot z^{-k} = \sum_{k=0}^{\infty} z^{-k} = 1 + z^{-1} + z^{-2} + \cdots$$

$$= \frac{1}{1 - z^{-1}} = \frac{z}{z - 1} \tag{14.34}$$

Note that the series converges for $|z| > 1$. In finding the z transform, the variable z acts as a dummy operator. It is not necessary to specify the region of z over which $\mathcal{Z}\{u_s(k)\}$ is convergent. The z transform of a function obtained in this manner is valid throughout the z plane except at the poles of the transformed function. The $u_s(k)$ is usually referred to as the unit step sequence.

Exponential Function. Let

$$f(t) = \begin{cases} e^{-at} & t \geq 0 \\ 0 & t < 0 \end{cases}$$

The sampled form of the function with sampling period T is

$$f(kT) = e^{-akT} \qquad k = 0, 1, 2, \ldots$$

By definition

$$F(z) = \mathcal{Z}\{f(k)\}$$

$$= \sum_{k=0}^{\infty} f(k)z^{-k}$$

$$= \sum_{k=0}^{\infty} e^{-akt}z^{-k}$$

$$= 1 + e^{-at}z^{-1} + e^{-2at}z^{-2} + \cdots$$

$$= \frac{1}{1 - e^{-at}z^{-1}}$$

$$= \frac{z}{z - e^{-at}}$$

14.4.4 Some Important Properties and Theorems of the z Transform

In this section some useful properties and theorems are stated without proof. It is assumed that the time function $f(t)$ is z transformable and that $f(t)$ is zero for $t < 0$.

P1. If

$$F(z) = \mathcal{Z}\{f(k)\}$$

then

$$\mathcal{Z}\{af(k)\} = aF(z) \tag{14.35}$$

P2. If $f_1(k)$ and $g_1(k)$ are z transformable and α and β are scalars, then $f(k) = \alpha f_1(k) + \beta g_1(k)$ has the z transform:

$$F(z) = \alpha F_1(z) + \beta G_1(z) \tag{14.36}$$

where $F_1(z)$ and $G_1(z)$ are the z transforms of $f_1(k)$ and $g_1(k)$, respectively.

P3. If

$$F(z) = \mathcal{Z}\{f(k)\}$$

then

$$\mathcal{Z}\{a^k f(k)\} = F\left(\frac{z}{a}\right) \tag{14.37}$$

T1. Shifting Theorem. If $f(t) \equiv 0$ for $t < 0$ and $f(t)$ has the z transform $F(z)$, then

$$\mathcal{Z}\{f(t + nT)\} = z^{-n}F(z)$$

and

$$\mathcal{Z}\{f(t + nT)\} = z^{n}\left[F(z) - \sum_{k=0}^{n-1} f(kT)z^{-k}\right] \tag{14.38}$$

T2. Complex Translation Theorem. If

$$F(z) = \mathcal{Z}\{f(t)\}$$

then

$$\mathcal{Z}\{e^{-at}f(t)\} = F(ze^{at}) \tag{14.39}$$

T3. Initial-Value Theorem. If $F(z) = \mathcal{Z}\{f(t)\}$ and if $\lim_{z \to \infty} F(z)$ exists, then the initial value $f(0)$ of $f(t)$ or $f(k)$ is given by

$$f(0) = \lim_{z \to \infty} F(z) \tag{14.40}$$

T4. Final-Value Theorem. Suppose that $f(k)$, where $f(k) \equiv 0$ for $k < 0$, has the z transform $F(z)$ and that all the poles of $F(z)$ lie inside the unit circle, with the possible exception of a simple pole at $z = 1$. Then the final value of $f(k)$ is given by

$$\lim_{k \to \infty} f(k) = \lim_{z \to 1} \left[(1 - z^{-1}) F(z) \right] \tag{14.41}$$

14.4.5 Pulse Transfer Function

Consider the linear time-invariant discrete-time system characterized by the following linear difference equation:

$$y(k) + a_1 y(k-1) + \cdots + a_n y(k-n) = b_0 u(k) + b_1 u(k-1) + \cdots + b_m u(k-m) \tag{14.42}$$

where $u(k)$ and $y(k)$ are the system's input and output, respectively, at the kth sampling or at the real time kT; T is the sampling period. To convert the difference Eq. 14.42 to an algebraic equation, take the z transform of both sides of Eq. 14.42 by definition:

$$\mathcal{Z}\{y(k)\} = Y(z) \tag{14.43a}$$

$$\mathcal{Z}\{u(k)\} = U(z) \tag{14.43b}$$

By referring to Table 14.2, the z transform of Eq. (14.42) becomes

$$Y(z) + a_1 z^{-1} Y(z) + \cdots + a_n z^{-n} y(z) = b_0 U(z) + b_1 z^{-1} U(z) + \cdots + b_m z^{-m} U(z)$$

or

$$\left[1 + a_1 z^{-1} + \cdots + a_n z^{-1} \right] Y(z) = \left[b_o + b_1 z^{-1} + \cdots + b_m z^{-m} \right] U(z)$$

which can be written as

$$\frac{Y(z)}{U(z)} = \frac{b_o + b_1 z^{-1} + \cdots + b_m z^{-m}}{1 + a_1 z^{-1} + \cdots + a_n z^{-n}} \tag{14.44}$$

Consider the response of the linear discrete-time system given by Eq. 14.44, initially at rest when the input $u(t)$ is the delta "function" $\delta(kT)$

$$\delta(kT) = \begin{cases} 1 & k = 0 \\ 0 & k \neq 0 \end{cases}$$

since

$$\mathcal{Z}\{\delta(kT)\} = \sum_{k=0}^{\infty} \delta(kT) z^{-k} = 1$$

$$U(z) = \mathcal{L}\{\delta(kT)\} = 1$$

and

$$Y(z) = \frac{b_0 + b_1 z^{-1} + \cdots + b_m z^{-m}}{1 + a_1 z^{-1} + \cdots + a_n z^{-n}} = G(z) \tag{14.45}$$

Thus $G(z)$ is the response of the system to the delta input (or unit impulse) and plays the same role as the transfer function in linear continuous-time systems. The function $G(z)$ is called the pulse transfer function.

14.4.6 Zero- and First-Order Hold

Discrete-time control systems may operate partly in discrete time and partly in continuous time. Replacing a continuous-time controller with a digital controller necessitates the conversion of numbers to continuous-time signals to be used as true actuating signals. The process by which a discrete-time sequence is converted to a continuous-time signal is called data hold.

TABLE 14.2   TABLE OF z TRANSFORMS[a]

No.	$\mathcal{F}(s)$	$f(nT)$	$F(z)$
1	—	$1, n = 0; 0, n \neq 0$	1
2	—	$1, n = k; 0, n \neq k$	z^{-k}
3	$\dfrac{1}{s}$	$1(nT)$	$\dfrac{z}{z-1}$
4	$\dfrac{1}{s^2}$	nT	$\dfrac{Tz}{(z-1)^2}$
5	$\dfrac{1}{s^3}$	$\dfrac{1}{2!}(nT)^2$	$\dfrac{T^2}{2}\dfrac{z(z+1)}{(z-1)^3}$
6	$\dfrac{1}{s^4}$	$\dfrac{1}{3!}(nT)^3$	$\dfrac{T^3}{6}\dfrac{z(z^2+4z+1)}{(z-1)^4}$
7	$\dfrac{1}{s^m}$	$\displaystyle\lim_{a\to 0}\dfrac{(-1)^{m-1}}{(m-1)!}\dfrac{\delta^{m-1}}{\delta a^{m-1}}e^{-unT}$	$\displaystyle\lim_{a\to 0}\dfrac{(-1)^{m-1}}{(m-1)!}\dfrac{\delta^{m-1}}{\delta a^{m-1}}\dfrac{z}{z-e^{-aT}}$
8	$\dfrac{1}{s+a}$	e^{-anT}	$\dfrac{z}{z-e^{-aT}}$
9	$\dfrac{1}{(s+a)^2}$	nTe^{-anT}	$\dfrac{Tze^{-aT}}{(z-e^{-aT})^2}$
10	$\dfrac{1}{(s+a)^3}$	$\dfrac{1}{2}(nT)^2e^{-anT}$	$\dfrac{T^2}{2}(e^{-aT})\dfrac{z(z+e^{-aT})}{(z-e^{-aT})^3}$
11	$\dfrac{1}{(s+a)^m}$	$\dfrac{(-1)^{m-1}}{(m-1)!}\dfrac{\delta^{m-1}}{\delta a^{m-1}}(e^{-anT})$	$\dfrac{(-1)^{m-1}}{(m-1)!}\dfrac{\delta^{m-1}}{\delta a^{m-1}}\dfrac{z}{z-e^{-aT}}$
12	$\dfrac{a}{s(s+a)}$	$1-e^{-anT}$	$\dfrac{z(1-e^{-aT})}{(z-1)(z-e^{-aT})}$
13	$\dfrac{a}{s^2(s+a)}$	$\dfrac{1}{a}(anT-1+e^{-anT})$	$\dfrac{z[(aT-1+e^{-aT})z+(1-e^{-aT}-aTe^{-aT})]}{a(z-1)^2(z-e^{-aT})}$
14	$\dfrac{b-a}{(s+a)(s+b)}$	$(e^{-anT}-e^{-bnT})$	$\dfrac{(e^{-aT}-e^{-bT})z}{(x-e^{-aT})(z-e^{-bT})}$
15	$\dfrac{s}{(s+a)^2}$	$(1-anT)e^{-anT}$	$\dfrac{z[z-e^{-aT}(1+aT)]}{(z-e^{-aT})^2}$
16	$\dfrac{a^2}{s(s+a)^2}$	$1-e^{-anT}(1+anT)$	$\dfrac{z[z(1-e^{-aT}-aTe^{-aT})+e^{-2aT}-e^{-aT}+aTe^{-aT}]}{(z-1)(z-e^{-aT})^2}$
17	$\dfrac{(b-a)s}{(s+a)(s+b)}$	$be^{-bnT}-ae^{-anT}$	$\dfrac{z[z(b-a)-(be^{-aT}-ae^{-bT})]}{(z-e^{-aT})(z-e^{-bT})}$
18	$\dfrac{a}{s^2+a^2}$	$\sin anT$	$\dfrac{z\sin aT}{z^2-(2\cos aT)z+1}$
19	$\dfrac{s}{s^2+a^2}$	$\cos anT$	$\dfrac{z(z-\cos aT)}{z^2-(2\cos aT)z+1}$
20	$\dfrac{s+a}{(s+a)^2+b^2}$	$e^{-anT}\cos bnT$	$\dfrac{z(z-e^{-aT}\cos bT)}{z^2-2e^{-aT}(\cos bT)z+e^{-2aT}}$
21	$\dfrac{b}{(s+a)^2+b^2}$	$e^{-anT}\sin bnT$	$\dfrac{ze^{-aT}\sin bT}{z^2-2e^{-aT}(\cos bT)z+e^{-2aT}}$
22	$\dfrac{a^2+b^2}{s((s+a)^2+b^2)}$	$1-e^{-anT}\left(\cos bnT+\dfrac{a}{b}\sin bnT\right)$	$\dfrac{z(Az+B)}{(z-1)(z^2-2e^{-aT}(\cos bT)z+e^{-2aT})}$
			$A = 1-e^{-aT}\cos bT-\dfrac{a}{b}e^{-aT}\sin bT$
			$B = e^{-2aT}+\dfrac{a}{b}e^{-aT}\sin bT-e^{-aT}\cos bT$

[a] $\mathcal{F}(s)$ is the Laplace transform of $f(t)$ and $F(z)$ is the transform of $f(nT)$. Unless otherwise noted, $f(t) = 0, t < 0$, and the region of convergence of $F(z)$ is outside a circle $r < |z|$ such that all poles of $F(z)$ are inside r.

In a conventional sampler, a switch closes to admit an input signal every sample period T. In practice, the sampling duration is very small compared with the most significant time constant of the plant. Suppose the discrete-time sequence is $f(kT)$, then the function of the data hold is to specify the values for a continuous equivalent $h(t)$ where $kT \le t < (k+1)T$. In general, the signal $h(t)$ during the time interval $kT < t < (k+1)T$ may be approximated by a polynomial in τ as follows:

$$h(kT + \tau) = a_n \tau^n + a_{n-1}\tau^{n-1} + \cdots + a_1\tau + a_0 \tag{14.46}$$

where $0 \le \tau < T$. Since the value of the continuous equivalent must match at the sampling instants, one requires

$$h(kT) = f(kT)$$

Hence Eq. 14.46 can be written as

$$h(kT + \tau) = a_n \tau^n + a_{n-1}\tau^{n-1} + \cdots + a_1\tau + f(kT) \tag{14.47}$$

If an nth-order polynomial as in Eq. 14.47 is used to extrapolate the data, then the hold circuit is called an nth-order hold. If $n = 1$, it is called a first-order hold (the nth-order hold uses the past $n + 1$ discrete data $f[(k-n)T]$). The simplest data hold is obtained when $n = 0$ in Eq. 14.47, that is, when

$$h(kT + \tau) = f(kT) \tag{14.48}$$

where $0 \le \tau < T$ and $k = 0, 1, 2 \ldots$. Equation 14.48 implies that the circuit holds the amplitude of the sample from one sampling instant to the next. Such a data hold is called a zero-order hold. The output of the zero-order hold is shown in Fig. 14.17.

Zero-Order Hold

Assuming that the sampled signal is 0 for $k < 0$, the output $h(t)$ can be related to $f(t)$ as follows:

$$h(t) = f(0)\big[u_s(t) - u_s(t-T)\big] + f(T)\big[u_s(t-T) - u_s(t-2T)\big]$$
$$+ f(2T)\big[u_s(t-2T) - u_s(t-3T)\big] + \cdots$$

$$= \sum_{k=0}^{\infty} f(kT)\{u_s(t-kT) - u_s[t-(k+1)T]\} \tag{14.49}$$

Since the $\mathscr{L} | u_s(t-kT) | = e^{-kTs}/s$, the Laplace transform of Eq. 14.49 becomes

$$\mathscr{L}[h(t)] = H(s) = \sum_{k=0}^{\infty} f(kT)\frac{e^{-kTs} - e^{-(k+1)Ts}}{s}$$

$$= \frac{1 - e^{-Ts}}{s}\sum_{k=0}^{\infty} f(kT)e^{-kTs} \tag{14.50}$$

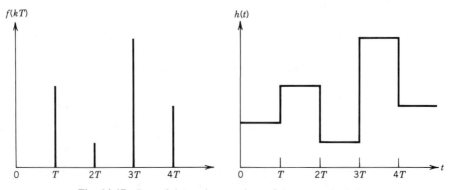

Fig. 14.17 Input $f(kt)$ and output $h(t)$ of the zero-order hold.

The right-hand side of Eq. 14.50 may be written as the product of two terms:

$$H(s) = G_{h0}(s)F^*(s) \tag{14.51}$$

where

$$G_{h0}(s) = \frac{1 - e^{-Ts}}{s}$$

$$F^*(s) = \sum_{k=0}^{\infty} f(kT)e^{-kTs} \tag{14.52}$$

In Eq. 14.51, $G_{h0}(s)$ may be considered the transfer function between the output $H(s)$ and the input $F^*(s)$ (see Fig. 14.18). Thus the transfer function of the zero-order hold device is

$$G_{h0}(s) = \frac{1 - e^{-Ts}}{s} \tag{14.53}$$

First-Order Hold

The transfer function of a first-order hold is given by

$$G_{h1}(s) = \frac{1 - e^{-Ts}}{s} \frac{Ts + 1}{T} \tag{14.54}$$

From Eq. 14.47 for $n = 1$,

$$h(kT + \tau) = a_1\tau + f(kT) \tag{14.55}$$

where $0 \leq \tau < T$ and $k = 0, 1, 2, \ldots$. By using the condition $h[(k - 1)T] = f[(k - 1)T]$ the constant a_1 can be determined as follows:

$$h[(k - 1)T] = -a_1T + f(kT) = f[(k - 1)T]$$

or

$$a_1 = \frac{f(kT)_f[(k - 1)T]}{T}$$

Hence Eq. 14.55 becomes

$$h(kT + \tau) = f(kT) + \frac{f(kT) - f[(k - 1)T]}{T}\tau \tag{14.56}$$

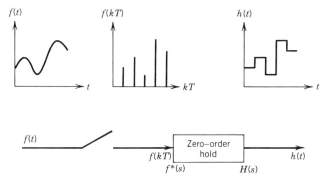

Fig. 14.18   Sampler and zero-order hold.

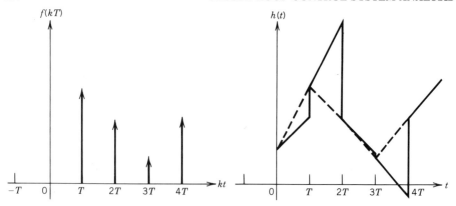

Fig. 14.19 Input $f(kT)$ and output $h(t)$ of a first-order hold.

where $0 \leq \tau < T$. The extrapolation process of the first-order hold is based on Eq. 14.56 and is a piecewise-linear signal as shown in Fig. 14.19.

14.5 CLOSED-LOOP REPRESENTATION

The typical feedback control system has the feature that some output quantity is measured and then compared with a desired value, and the resulting error is used to correct the system output. A block diagram representation of a closed-loop or feedback system is shown in Fig. 14.20.

In this figure, r is the reference input, w is a disturbance, and y is the output. Transfer functions G_p, H and G_c denote, respectively, the plant dynamics, sensor dynamics, and the controller. The influence of r and w on the output y can be determined using elementary block diagram algebra as

$$Y(s) = \frac{G_c G_p}{1 + G_c G_p H} R(s) + \frac{G_p}{1 + G_c G_p H} W(s) \tag{14.57}$$

where $R(s) = \mathcal{L}[r(t)]$, $W(s) = \mathcal{L}[w(t)]$, and $Y(s) = \mathcal{L}[y(t)]$.

14.5.1 Closed-Loop Transfer Function

In Eq. 14.57 if $W(s) = 0$, then

$$Y(s) = \frac{G_c G_p}{1 + G_c G_p H} R(s)$$

or alternatively a transfer function called the closed-loop transfer function between the reference input and the output is defined:

$$\frac{Y(s)}{R(s)} = G_{cl}(s) = \frac{G_c G_p}{1 + G_c G_p H} \tag{14.58}$$

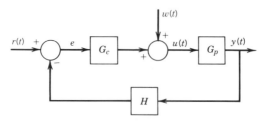

Fig. 14.20 Closed-loop system with disturbance.

14.5.2 Open-Loop Transfer Function

The product of transfer functions within the loop, namely G_cG_pH, is referred to as the open-loop transfer function or simply the loop transfer function:

$$G_{ol} = G_cG_pH \tag{14.59}$$

14.5.3 Characteristic Equation

The overall system dynamics given by Eq. 14.57 is primarily governed by the poles of the closed-loop system or the roots of the closed-loop characteristic equation (CLCE):

$$1 + G_cG_pH = 0 \tag{14.60}$$

It is important to note that the closed-loop characteristic equation is simply

$$1 + G_{ol} = 0 \tag{14.61}$$

This latter form is the basis of root locus and frequency domain design techniques discussed in Chapter 15.

The roots of the characteristic equation are referred to as poles. Specifically, the roots of the open-loop characteristic equation are referred to as open-loop poles and those of the closed loop are called closed-loop poles.

Example 14.9. Consider the block diagram shown in Fig. 14.21. The open-loop transfer function

$$G_{ol} = \frac{K_1(s + 1)(s + 4)}{s(s + 2)(s + 3)}$$

The closed-loop transfer function of the system is

$$G_{cl} = \frac{K_1(s + 1)(s + 4)}{s(s + 2)(s + 3) + K_1(s + 1)(s + 4)}$$

The open-loop characteristic equation (OLCE) is the denominator polynomial of G_{ol} set equal to zero. Hence

$$OLCE \equiv s(s + 2)(s + 3) = 0$$

and the open-loop poles are 0, -2, and -3.

The closed-loop characteristic equation is

$$s(s + 2)(s + 3) + K_1(s + 1)(s + 4) = 0$$

and its roots are the closed-loop poles appropriate for the specific gain value K_1.

If the transfer functions are rational (i.e., are ratios of polynomials), then they can be written as

$$G(s) = \frac{K\prod\limits_{i-1}^{m}(s + z_i)}{\prod\limits_{j-1}^{n}(s + p_j)} \tag{14.62}$$

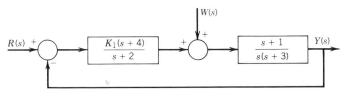

Fig. 14.21 Closed-loop system.

When the poles of the transfer function $G(s)$ are distinct, $G(s)$ may be written in partial fraction form as

$$G(s) = \sum_{j=1}^{n} \frac{A_j}{s + p_j} \tag{14.63}$$

Hence

$$g(t) = \mathcal{L}^{-1}[G(s)] = \mathcal{L}^{-1}\left[\sum_{j=1}^{n} \frac{A_j}{s + p_j}\right]$$

$$= \sum_{j=1}^{n} A_j e^{-p_j t} \tag{14.64}$$

Since a transfer function $G(s) = \mathcal{L}[\text{output}]/\mathcal{L}[\text{input}]$, $g(t)$ of of Eq. 14.64 is the response of the system depicted by $G(s)$ for a unit impulse $\delta(t)$, since $\mathcal{L}[\delta(t)] = 1$.

The impulse response of a given system is key to its internal stability. The term *system* here is applicable to any part of (or the whole) closed-loop system.

It should be noted that the zeros of the transfer function only affect the residues A_j. In other words, the contribution from the corresponding transient term $e^{-p_j t}$ may or may not be significant depending on the relative size of A_j. If, for instance, a zero $-z_k$ is very close to a pole $-p_l$ then the transient term $A e^{-p_j t}$ would have a value close to zero for its residue A_l. As an example, consider the unit impulse response of the two systems:

$$G_1(s) = \frac{1}{(s + 1)(s + 2)} = \frac{1}{s + 1}\frac{1}{s + 2} \tag{14.65}$$

$$G_2(s) = \frac{(s + 1.05)}{(s + 1)(s + 2)} = \frac{0.05}{s + 1} + \frac{0.95}{s + 2} \tag{14.66}$$

From Eq. 14.65, $g_1(t) = e^{-t} - e^{-2t}$, and from Eq. 14.66, $g_2(t) = 0.05e^{-t} + 0.95e^{-2t}$.

Note that the effect of the term e^{-t} has been modified from a residue of 1 in G_1 to a residue of 0.05 in G_2. This observation helps to reduce the order of a system when there are poles and zeros close together. In $G_2(s)$, for example, little error would be introduced if the zero at -1.05 and the pole at -1 are neglected and the transfer function approximated by

$$G_2(s) = \frac{1}{s + 2}$$

From Eq. 14.64 it can be observed that the shape of the impulse response is determined primarily by the pole locations. A sketch of several pole locations and corresponding impulse responses is given in Fig. 14.22.

The fundamental responses that can be obtained are of the form $e^{-\alpha t}$ and $e^{\lambda t}\sin\beta t$ with $\beta > 0$. It should be noted that for a real pole its location completely characterizes the resulting impulse response. When a transfer function has complex conjugate poles the impulse response is more complicated.

14.5.4   Standard Second-Order Transfer Function

A standard second-order transfer function takes the form

$$G(s) = \frac{\omega_n^2}{s^2 + 2\zeta\omega_n s + \omega_n^2} = \frac{1}{\frac{s^2}{\omega_n^2} + 2\zeta\frac{s}{\omega_n} + 1} \tag{14.67}$$

Parameter ζ is called the damping ratio, and ω_n is called the undamped natural frequency. The poles of the transfer function given by Eq. 14.67 can be determined by solving its characteristic equation:

$$s^2 + 2\zeta\omega_n s + \omega_n^2 = 0 \tag{14.68}$$

giving the poles

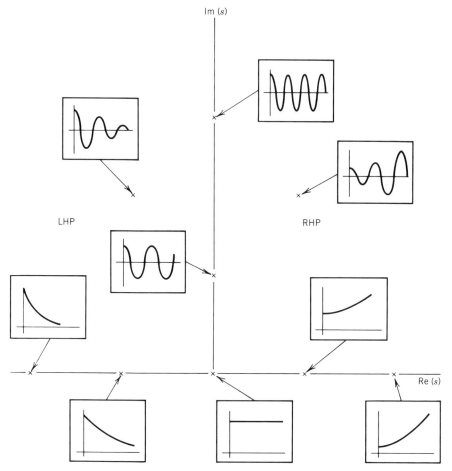

Fig. 14.22 Impulse responses associated with pole locations.

$$s_{1,2} = -\zeta\omega_n \pm j\sqrt{1 - \zeta^2}\,\omega_n = -\sigma \pm j\omega_d \qquad (14.69)$$

where $\sigma = \zeta\omega_n$ and $\omega_d = \omega_n\sqrt{1 - \zeta^2}$, $\zeta < 1$.

The two complex conjugate poles given by Eq. 14.69 are located in the complex s plane as shown in Fig. 14.23.

It can be easily seen that

$$|s_{1,2}| = \sqrt{\zeta^2\omega_n^2 + \omega_d^2} = \omega_n \qquad (14.70)$$

and

$$\cos\beta = \zeta \qquad 0 \le \beta \le \pi/2 \qquad (14.71)$$

When the system has no damping, $\zeta = 0$, the impulse response is a pure sinusoidal oscillation. In this case the undamped natural frequency ω_n is equal to the damped natural frequency ω_d.

14.5.5   Step Input Response of a Standard Second-Order System

When the transfer function has complex conjugate poles, the step input response rather than the impulse response is used to characterize its transients. Moreover, these transients are almost always used as time domain design specifications. The unit step input response of the second-order system given by Eq. 14.67 can be easily shown to be

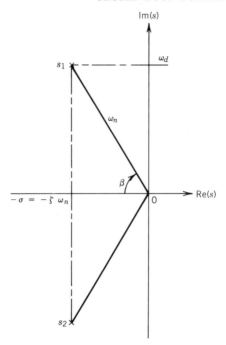

Fig. 14.23 Complex conjugate poles in the s plane.

$$y(t) = 1 - e^{\sigma T}\left(\cos \omega_d t + \frac{\sigma}{\omega_d} \sin \omega_d t\right) \tag{14.72}$$

where $y(t)$ is the output.

A typical unit step input response is shown in Fig. 14.24.

From the time response of Eq. (14.72) several key parameters characterizing the shape of the curve are usually used as time domain design specifications. They are the rise time t_r, the settling time t_s, the peak overshoot M_p, and the peak time t_p.

Rise Time, t_r. The time taken for the output to change from 10% of the steady-state value to 90% of the steady-state value is usually called the rise time. There are other definitions of rise time.[3] The basic idea, however, is that t_r is a characterization of how rapid the system responds to an input.

A rule of thumb for t_r is

$$t_r \simeq 1.8/\omega_n \tag{14.73}$$

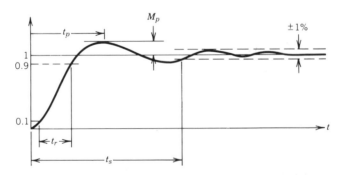

Fig. 14.24 Definition of the rise time t_r, settling time t_s, and overshoot time M_p.

Settling Time, t_s. This is the time required for the transients to decay to a small value so that $y(t)$ is almost at the steady-state level. Various measures of "smallness" are possible: 1, 2, and 5% are typical.

$$\text{For 1\% settling,} \qquad t_s \approx \frac{4.6}{\zeta \omega_n}$$

$$\text{For 2\% settling,} \qquad t_s \approx \frac{4}{\zeta \omega_n} \qquad (14.74)$$

$$\text{For 5\% settling,} \qquad t_s \approx \frac{3}{\zeta \omega_n}$$

Peak Overshoot, M_p. The peak overshoot is the maximum amount by which the output exceeds its steady-state value during the transients. It can be easily shown that

$$M_p = e^{-\pi \zeta / \sqrt{1 - \zeta^2}} \qquad (14.75)$$

Peak Times, t_p. This is the time at which the peak occurs. It can be readily shown that

$$t_p = \frac{\pi}{\omega_d} \qquad (14.76)$$

It is important to note that given a set of time domain design specifications, they can be converted to an equivalent location of complex conjugate poles. Figure 14.25 shows the allowable regions for complex conjugate poles for different time domain specifications.

For design purposes the following synthesis forms are useful:

$$\omega_n \geq \frac{1.8}{t_r} \qquad (14.77a)$$

$$\zeta \geq 0.6(1 - M_p) \text{ for } 0 \leq \zeta \leq 0.6 \qquad (14.77b)$$

$$\sigma \geq \frac{4.6}{t_s} \qquad (14.77c)$$

14.5.6 Effects of an Additional Zero and an Additional Pole[3]

If the standard second-order transfer function is modified due to a zero as

$$G_1(s) = \frac{(s/\alpha\zeta\omega_n + 1)\omega_n^2}{s^2 + 2\zeta\omega_n + \omega_n^2} \qquad (14.78)$$

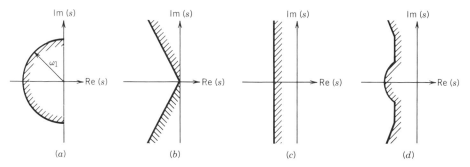

Fig. 14.25 Graphs of regions in the s plane for certain transient requirements to be met: (a) rise-time requirements, (ω_n); (b) oveshoot requirements, (ζ); (c) settling time requirements, (ζ); (d) composite of (a), (b), and (c).

or due to an additional pole as

$$G_2(s) = \frac{\omega_n^2}{(s/\alpha\zeta\omega_n + 1)(s^2 + 2\zeta\omega_n s + \omega_n^2)} \tag{14.79}$$

it is important to know how close the resulting step input response is to the standard second-order step response. Following are several features of these additions:

1. For a second-order system with no finite zeros, the transient parameters can be approximated by Eq. 14.77.
2. An additional zero as in Eq. 14.78 in the left-half-plane will increase the overshoot if the zero is within a factor of 4 of the real part of the complex poles. A plot is given in Fig. 14.26.
3. An additional zero in the right half plane will depress the overshoot (and may cause the step response to undershoot). This is referred to as a nonminimum phase system.
4. An additional pole in the left-half-plane will increase the rise time significantly if the extra pole is within a factor of 4 of the real part of the complex poles. A plot is given in Fig. 14.27.

14.6   STABILITY

We shall distinguish between two types of stabilities: external and internal stability. The notion of external stability is concerned with whether or not a bounded input gives a bounded output. In this type of stability we notice that no reference is made to the internal variables of the system. The implication here is that it is possible for an internal variable to grow without bound while the output remains bounded. Whether or not the internal variables are well behaved is typically addressed by the notion of internal stability. Internal stability requires that in the absence of an external input the internal variables stay bounded for any perturbations of these variables. In other words internal stability is concerned with the response of the system due to nonzero initial conditions. It is reasonable to expect that a well-designed system should be both externally and internally stable.

The notion of asymptotic stability is usually discussed within the context of internal stability. Specifically, if the response due to nonzero initial conditions decays to zero asymptotically, then the system is said to be asymptotically stable. A linear time-invariant (LTI) system is asymptotically stable if and only if all the system poles lie in the open left-half-plane (i.e., the left half s plane excluding the imaginary axis). This condition also guarantees external stability for LTI systems. So in the case of LTI the notions of internal and external stability may be considered equivalent.

For LTI systems, knowing the locations of the poles or the roots of the characteristic equation would suffice to predict stability. The Routh–Hurwitz stability criterion is frequently used to obtain stability information without explicitly computing the poles for LTI. This criterion will be discussed in Section 14.6.1.

For nonlinear systems, stability cannot be characterized that easily. As a matter of fact there are many definitions and theorems for assessing stability of such systems. A discussion of these topics is beyond the scope of this handbook. Interested reader however may refer to Ref. 9.

14.6.1   Routh–Hurwitz Stability Criterion

This criterion allows one to predict the status of stability of a system by knowing the coefficients of its characteristic polynomial. Consider the characteristic polynomial of an nth-order system:

$$P(s) = a_n s^n + a_{n-1}s^{n-1} + a_{n-2}s^{n-2} + \cdots + a_1 s + a_0$$

A necessary condition for asymptotic stability is that all the coefficients $\{a_i\}$'s be positive. If any of the coefficients are missing (i.e., are zero) or negative, then the system will have poles located in the closed right half plane. If the necessary condition is satisfied, then the so-called Routh array needs to be formed to make conclusions about the stability. A necessary and sufficient condition for stability is that all the elements in the first column of the Routh array be positive.

To determine the Routh array, the coefficients of the characteristic polynomial are arranged in two rows, each beginning with the first and second coefficients and followed by even-numbered and odd-numbered coefficients as follows:

s^n	a_n	a_{n-2}	a_{n-4} $\cdots$
s^{n-1}	a_{n-1}	a_{n-3}	a_{n-5} $\cdots$

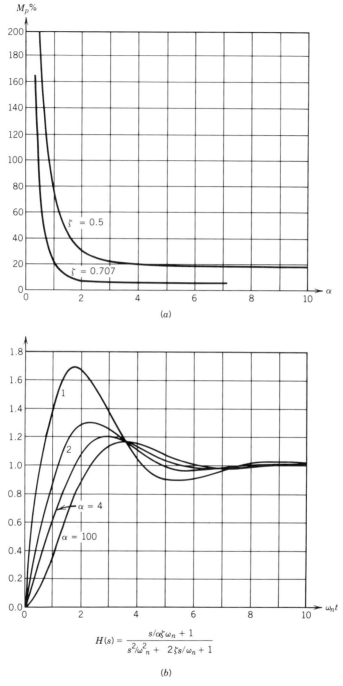

Fig. 14.26 Effect of an extra zero on a standard second-order system; (*a*) normalized rise time versus α; (*b*) step response versus α, for $\zeta = 0.5$.

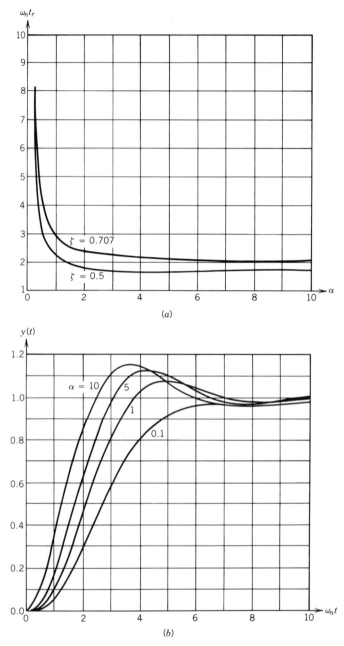

Fig. 14.27 Effect of an additional pole on a standard second-order system; (a) normalized rise time versus α; (b) step response versus α, for $\zeta = 0.5$.

The following rows are subsequently added to complete the Routh array:

$$
\begin{array}{cccccc}
s^n & a_n & a_{n-2} & a_{n-4} & \cdots \\
s^{n-1} & a_{n-1} & a_{n-3} & a_{n-5} & \cdots \\
s^{n-2} & b_1 & b_2 & b_3 & \cdots \\
s^{n-3} & c_1 & c_2 & c_3 & \cdots \\
\cdot & \cdot & \cdot & \cdot & \cdots \\
s^2 & * & * \\
s & * \\
s^0 &
\end{array}
$$

where the elements from the third row on are computed as follows:

$$
b_1 = \frac{a_{n-1}a_{n-2} - a_n a_{n-3}}{a_{n-1}} \qquad c_1 = \frac{b_1 a_{n-3} - a_{n-1} b_2}{b_1}
$$

$$
b_2 = \frac{a_{n-1}a_{n-4} - a_n a_{n-5}}{a_{n-1}} \qquad c_2 = \frac{b_1 a_{n-5} - a_{n-1} b_3}{b_1}
$$

$$
b_3 = \frac{a_{n-1}a_{n-6} - a_n a_{n-7}}{a_{n-1}} \qquad c_3 = \frac{b_1 a_{n-7} - a_{n-1} b_4}{b_1}
$$

Note that the elements of the third row and of rows thereafter are formed from the two previous rows using the two elements in the first column and other elements for successive columns. Normally there will be $n + 1$ elements in the first column when the array is completed.

The Routh-Hurwitz criterion states that the number of roots of $P(s)$ with positive real parts is equal to the number of changes of sign in the first column of the array. This criterion requires that there be no changes of sign in the first column for a stable system. A pattern of $+$, $-$, $+$ is counted as two sign changes, one going from $+$ to $-$ and another from $-$ to $+$.

If the first term in one of the rows is zero or if an entire row is zero, then the standard Routh array cannot be formed and the use of special techniques, described in the following discussion become necessary.

Special Cases

1. Zero in the first column, while some other elements of the row containing a zero in the first column are nonzero.
2. Zero in the first column, and the other elements of the row containing the zero are also zero.

CASE 1: In this case the zero is replaced with a small positive constant $\epsilon > 0$, and the array is completed as before. The stability criterion is then applied by taking the limits of entries of the first column as $\epsilon \to 0$. For example, consider the following characteristic equation:

$$
s^5 + 2s^4 + 2s^3 + 4s^2 + 11s + 10 = 0
$$

The Routh array is then

1	2	11	
2	4	10	
0	6	0	
ϵ	6	0	first column zero replaced by ϵ
c_1	10		
d_1	0		

where

$$c_1 = \frac{4\epsilon - 12}{\epsilon} \quad \text{and} \quad d_1 = \frac{6c_1 - 10\epsilon}{c_1}$$

As $\epsilon \to 0$, we get $c_1 \simeq -12/\epsilon$, and $d_1 \simeq 6$. There are two sign changes due to the large negative number in the first column. Therefore the system is unstable, and two roots lie in the right half plane. As a final example consider the characteristic polynomial

$$P(s) = s^4 + 5s^3 + 7s^2 + 5s + 6$$

The Routh array is

1	7	6
5	5	
6	6	
ϵ	$\leftarrow$	zero replaced by $\epsilon > 0$
6		

If $\epsilon > 0$, there are no sign changes. If $\epsilon < 0$, there are two sign changes. Thus if $\epsilon = 0$, it indicates that there are two roots on the imaginary axis, and a slight perturbation would drive the roots into the right half plane or the left half plane. An alternative procedure is to define the auxiliary variable,

$$z = \frac{1}{s}$$

and convert the characteristic polynomial so that it is in terms of z. This usually produces a Routh array with nonzero elements in the first column. The stability properties can then be deduced from this array.

CASE 2: This case corresponds to a situation where the characteristic equation has equal and opposite roots. In this case if the ith row is the vanishing row, an auxiliary equation is formed from the previous i_{-1} row as follows:

$$P_1(s) = \beta_1 s^{i+1} + \beta_2 s^{i-1} + \beta_3 s^{i-3} + \cdots$$

where $\{\beta_i\}$'s are the coefficients of the $(i - 1)$th row of the array. The ith row is then replaced by the coefficients of the derivative of the auxiliary polynomial, and the array is completed. Moreover, the roots of the auxiliary polynomial are also roots of the characteristic equation. As an example, consider

$$s^5 + 2s^4 + 5s^3 + 10s^2 + 4s + 8 = 0$$

for which the Routh array is

s^5	1	5	4
s^4	2	10	8
s^3	0	0	Auxiliary equation: $2s^4 + 10s^2 + 8 = 0$
s^3	8	20	
s^2	5	8	
s^1	7.2		
s^0	8		

There are no sign changes in the first column. Hence all the roots have nonpositive real parts with two pairs of roots on the imaginary axis, which are the roots of

$$2s^4 + 10s^2 + 8 = 0$$

$$= 2(s^2 + 4)(s^2 + 1)$$

Thus equal and opposite roots indicated by the vanishing row are $\pm j$ and $\pm 2j$.

The Routh-Hurwitz criterion may also be used in determining the range of parameters for which a feedback system remains stable. Consider, for example, the system described by the closed-loop characteristic equation

$$s^4 + 3s^3 + 3s^2 + 2s + K = 0$$

The corresponding Routh array is

$$
\begin{array}{llll}
s^4 & 1 & 3 & K \\
s^3 & 3 & 2 & \\
s^2 & \frac{7}{3} & K & \\
s^1 & s - (9/7)K & & \\
s^0 & K & &
\end{array}
$$

If the system is to remain asymptotically stable, we must have

$$0 < K < 14/9$$

The Routh-Hurwitz stability criterion can also be used to obtain additional insights. For instance information regarding the speed of response may be obtained by a coordinate transformation of the form $s + a$ where $-a$ characterizes rate. For this additional detail the reader is referred to Ref. 10.

14.6.2 Polar Plots

The frequency response of a system described by a transfer function $G(s)$ is given by

$$y_{ss}(t) = X|G(j\omega)|\sin\left[\omega t + \underline{/G(j\omega_1)}\right]$$

where $X \sin \omega t$ is the forcing input and $y_{ss}(t)$ is the steady-state output. A great deal of information about the dynamic response of the system can be obtained from a knowledge of $G(j\omega)$. The complex quantity $G(j\omega)$ is typically characterized by its frequency-dependent magnitude $|G(j\omega)|$ and the phase $\underline{/G(j\omega)}$. There are many different ways in which this information is handled. The manner in which the phasor $G(j\omega)$ traverses the $G(j\omega)$ plane as a function of ω is represented by a polar plot. This is represented in Fig. 14.28. The phasor OA corresponds to a magnitude $|G(j\omega_1)|$ and the phase $\underline{/G(j\omega_1)}$. The locus of point A as a function of the frequency is the polar plot. This representation is useful in stating the Nyquist stability criterion.

14.6.3 Nyquist Stability Criterion

This is a method by which the closed-loop stability of a linear time-invariant system can be predicted by knowing the frequency response of the loop transfer function. A typical closed-loop characteristic equation may be written as

$$1 + GH = 0$$

where GH is referred to as the loop transfer function.

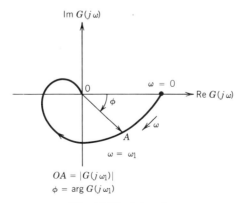

$$OA = |G(j\omega_1)|$$
$$\phi = \arg G(j\omega_1)$$

Fig. 14.28 Polar plot.

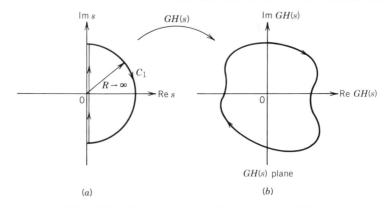

Fig. 14.29 Nyquist contour and its mapping by $GH(s)$.

The Nyquist stability criterion[3] is based on a mapping of the so-called Nyquist contour by the loop transfer function. Figure 14.29a shows the Nyquist contour, and Fig. 14.29b shows its typical mapped form in the GH plane. Let C_1 be the contour in the s plane to be mapped. Utilizing this information and the poles of GH, the Nyquist stability criterion can be stated as follows:

A closed-loop system is stable if and only if the mapping of C_1 in the GH plane encircles the $(-1,0)$ point N number of times in the counterclockwise direction, where N is the number of poles of GH with positive real parts.

Clearly if GH does not have any poles in the right half-plane, then C_1 should not encircle the $(-1,0)$ point if the closed-loop system is to be stable.

If there are poles of GH on the imaginary axis, then the Nyquist contour should be modified with appropriate indentations as shown in Fig. 14.30. An example is given to map such an indented contour with the corresponding loop transfer function. Consider

$$GH(s) = \frac{1}{s(s + 1)}$$

The contour Γ in the s plane is shown in Fig. 14.31a, where an indentation is affected by a small semicircle of radius ϵ, where $\epsilon > 0$. When mapped by $GH(s)$, the contour of Fig. 14.31b is obtained. In order to effect the mapping, the points on the semicircles are represented as $s = \epsilon e^{j\phi}$ on the small indentation with $\phi \epsilon[-\pi/2, \pi/2]$ and $\epsilon \to 0$ and $s = R e^{j\theta}$ with $\theta \epsilon[-\pi/2, \pi/2]$ and $R \to \infty$ on the infinite semicircle. We observe that the $(-1,0)$ point is not encircled. Since $GH(s)$ does not have any poles in the right half s plane from the Nyquist stability criterion, it follows that the closed-loop system is stable. In Table 14.3 several loop transfer functions, with the appropriate Nyquist contour and mapped contours are given.

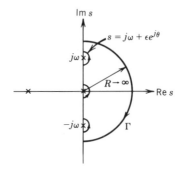

Fig. 14.30 Nyquist contour with indentations.

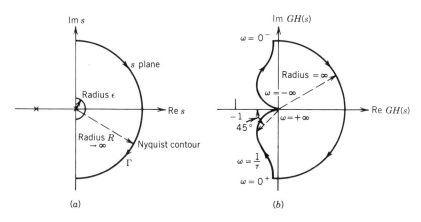

Fig. 14.31 Nyquist contour and mapping for $GH(s) = K/s(\tau s + 1)$.

TABLE 14.3 LOOP TRANSFER FUNCTIONS

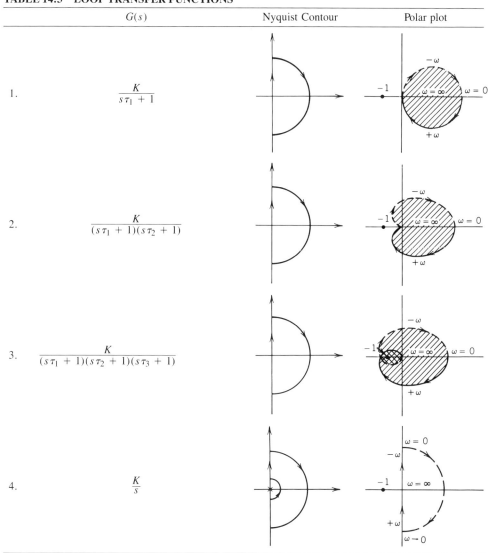

	$G(s)$	Nyquist Contour	Polar plot
1.	$\dfrac{K}{s\tau_1 + 1}$		
2.	$\dfrac{K}{(s\tau_1 + 1)(s\tau_2 + 1)}$		
3.	$\dfrac{K}{(s\tau_1 + 1)(s\tau_2 + 1)(s\tau_3 + 1)}$		
4.	$\dfrac{K}{s}$		

(Continued)

TABLE 14.3   LOOP TRANSFER FUNCTIONS (*Continued*)

	$G(s)$	Nyquist Contour	Polar plot
5.	$\dfrac{K}{s(s\tau_1 + 1)}$		
6.	$\dfrac{K}{s(s\tau_1 + 1)(s\tau_2 + 1)}$		
7.	$\dfrac{K(s\tau_a + 1)}{s(s\tau_1 + 1)(s\tau_2 + 1)}$ $\tau_a < \dfrac{\tau_1 \tau_2}{\tau_1 + \tau_2}$ $\tau_a < \dfrac{\tau_1 \tau_2}{\tau_1 + \tau_2}$		
8.	$\dfrac{K}{s^2}$		
9.	$\dfrac{K}{s^2(s\tau_1 + 1)}$		
10.	$\dfrac{K(s\tau_1 + 1)}{s^2(s\tau_1 + 1)}$ $\tau_2 > \tau_1$ $\tau_a > \tau_1$		

TABLE 14.3 LOOP TRANSFER FUNCTIONS (*Continued*)

	$G(s)$	Nyquist Contour	Polar plot
11.	$\dfrac{K}{s^3}$		
12.	$\dfrac{K(s\tau_a + 1)}{s^3}$		
13.	$\dfrac{K(s\tau_a + 1)(s\tau_b + 1)}{s^3}$		
14.	$\dfrac{K(s\tau_a + 1)(s\tau_b + 1)}{s(s\tau_1 + 1)(s\tau_2 + 1)(s\tau_3 + 1)(s\tau_4 + 1)}$		
15.	$\dfrac{K(s\tau_a + 1)}{s^2(s\tau_1 + 1)(s\tau_2 + 1)}$		

14.7   STEADY-STATE PERFORMANCE AND SYSTEM TYPE

In the design of feedback control systems the steady-state performance is also of importance in many instances. This is in addition to the stability and transient performance requirements. Consider the closed-loop system shown in Fig. 14.32.

Here $G_p(s)$ is the plant and $G_c(s)$ is the controller. When steady-state performance is important, it is advantageous to consider the system error due to an input. From the block diagram

$$E(s) = R(s) - Y(s) \tag{14.80}$$

$$Y(s) = G_c(s)G_p(s)E(s) \tag{14.81}$$

Substituting Eq. (14.81) in (14.80) yields

$$\frac{E(s)}{R(s)} = \frac{1}{1 + G_c(s)G_p(s)} \tag{14.82}$$

called the error transfer function. It is important to note that the error dynamics are described by the same poles as those of the closed-loop transfer function. Namely, the roots of the characteristic equation $1 + G_c(s)G_p(s) = 0$.

Given any input $r(t)$ with $\mathcal{L}[r(t)] = R(s)$, the error can be analyzed by considering the inverse Laplace transform of $E(s)$ given by

$$e(t) = \mathcal{L}^{-1}\left[\frac{1}{1 + G_c(s)G_p(s)}R(s)\right] \tag{14.83}$$

For the system's error to be bounded, it is important to first assure that the closed-loop system is asymptotically stable. Once the closed-loop stability is assured, the steady-state error can be computed by using the final-value theorem. Hence

$$\lim_{t \to \infty} e(t) = e_{ss} = \lim_{s \to 0} sE(s) \tag{14.84}$$

By substituting for $E(s)$ in Eq. 14.84

$$e_{ss} = \lim_{s \to 0} s\frac{1}{1 + G_c(s)G_p(s)}R(s) \tag{14.85}$$

14.7.1   Step Input

If the reference input is a step of magnitude c, then from Eq. (14.85)

$$e_{ss} = \lim_{s \to 0} s\frac{1}{1 + G_c(s)G_p(s)}\frac{c}{s} = \frac{c}{1 + G_cG_p(0)} \tag{14.86}$$

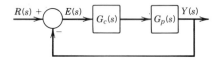

Fig. 14.32   Closed-loop configuration.

Equation 14.86 suggests that in order to have small steady-state error, the low-frequency gain of the open-loop transfer function $G_c G_p(0)$ must be very large. It is typical to define

$$K_p = G_c G_p(0) \tag{14.87}$$

as the position error constant. With this definition the steady-state error due to a step input of magnitude c can be written as

$$e_{ss} = \frac{c}{1 + K_p} \tag{14.88}$$

Thus a high value of K_p corresponds to a low steady-state error. If the steady-state error is to be zero, then $K_p = \infty$. The only way that $K_p = \infty$ is when the open-loop transfer function has at least one pole at the origin, that is, $G_c G_p(s)$ must be of the form

$$G_c G_p(s) = \frac{1}{s^N} \frac{\displaystyle\prod_{i=1}^{m}(s + z_i)}{\displaystyle\prod_{j=1}^{n}(s + p_j)} \tag{14.89}$$

where $N \geq 1$. When $N \geq 1$,

$$G_c G_p(0) = \frac{1}{0} \frac{\displaystyle\prod_{i=1}^{m} z_i}{\displaystyle\prod_{j=1}^{n} p_j} \to \infty \tag{14.89}$$

Hence, it can be concluded that for the steady-state error due to a step input to be zero, the open-loop transfer function must have at least one free integrator. The value of N specifies the type of the system. If $N = 1$ it is called a type I, when $N = 2$ it is called a type II system, and so on. So to get zero steady-state error for a step input, the system loop transfer function must be at least type I.

14.7.2 Ramp Input

If the reference input is a ramp $ct u_s(\pm)$ where $u_s(t)$ is the unit step, then from Eq. (14.85)

$$e_{ss} = \lim_{s \to 0} s \frac{1}{1 + G_c G_p(s)} \frac{c}{s^2} = \lim_{s \to 0} \frac{c}{s G_c G_p(s)} \tag{14.90}$$

From Eq. 14.90 for small steady-state errors $\lim_{s \to 0} s G_c G_p(s) = K_v$ must be large. K_v is called the velocity error constant and

$$e_{ss} = \frac{c}{K_v} \tag{14.91}$$

As in the case with the step input for e_{ss} to be small, K_v must be very large. For zero steady-state error with a ramp input, $K_v = \infty$. From Eq. 14.90 it is clear that for $K_v = \infty$, $G_c G_p(s)$ must be at least type II. Thus

$$K_v = \lim_{s \to 0} s G_c G_p(s) = \lim_{s \to 0} s \frac{1}{s^2} = \frac{\displaystyle\prod_{i=1}^{m}(s + z_i)}{\displaystyle\prod_{j=1}^{n}(s + p_j)} = \infty$$

14.7.3 Parabolic Input

If the reference input is a parabolic input of the form

$$r(t) = c \frac{t^2}{2} u_s(t)$$

TABLE 14.4   STEADY-STATE ERROR IN TERMS OF ERROR CONSTANTS

Type	$cu_s(t)$	$ctu_s(t)$	$\frac{t^2}{2}u_s(t)$
0	$\dfrac{c}{1 + K_p}$	∞	∞
1	0	$\dfrac{c}{K_v}$	∞
2	0	0	$\dfrac{c}{K_a}$
3	0	0	0

then the steady-state error becomes

$$e_{ss} = \lim_{s \to 0} \frac{c}{s^2 G_c G_p(s)} \tag{14.92}$$

From Eq. 14.92 for small steady-state errors

$$K_a = \lim_{s \to 0} s^2 G_c G_p(s) \tag{14.93}$$

must be very large. K_a is called the parabolic error constant or the acceleration error constant. For zero steady-state error due to a parabolic input, $K_a = \infty$. Therefore the system open-loop transfer function must be at least type III.

Table 14.4 shows the steady-state errors in terms of the error constants K_p, K_v, and K_a.

In steady-state performance considerations it is important to guarantee both closed-loop stability and the system type. It should also be noticed that having a very high loop gain as given by Eq. 14.89 is very similar to having a free integrator in the open-loop transfer function. This notion is useful in regulation type problems. In tracking systems the problem is compounded by the fact that the system type must be increased. Increasing the system type is usually accompanied by instabilities. To illustrate this consider the following example.

Example 14.10.   Synthesize a controller for the system shown in Fig. 14.33a so that the steady-state error due to a ramp input is zero.

The open-loop system is unstable. For tracking purposes transform the system to a closed-loop one as shown in Fig. 14.33b. To have zero steady-state error for a ramp, the system's open-loop transfer function must be at least type II and must be of a form that guarantees closed-loop stability. Therefore let

$$G_c(s) = \frac{G_1(s)}{s^2} \tag{14.94}$$

The closed-loop characteristic equation (CLCE) is

$$1 + \frac{G_1(s)}{s^2} \frac{1}{s - 2} = 0 \tag{14.95}$$

That is, $s^3 - 2s^2 + G_1(s) = 0$.

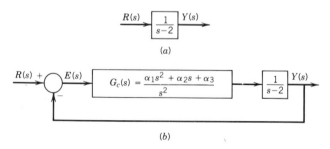

(a)

(b)

Fig. 14.33   Tracking system; (a) open-loop system; (b) closed-loop system.

Now for closed-loop stability, $G_1(s)$ must be of the form

$$G_1(s) = \alpha_1 s^2 + \alpha_2 s + \alpha_3$$

yielding a CLCE:

$$s^3 + (\alpha_1 - 2)s^2 + \alpha_2 s + \alpha_3 = 0 \tag{14.96}$$

From the Routh-Hurwitz stability criterion it can be shown that for asymptotic stability $\alpha_1 > 2$, $\alpha_3 > 0$, and $(\alpha_1 - 2)\alpha_2 - \alpha_3 > 0$. So pick $\alpha_1 = 4$, $\alpha_2 = 1$, $\alpha_3 = 1$ yielding the controller

$$G_c(s) = \frac{4s^2 + s + 1}{s^2} \tag{14.97}$$

In an actual design situation the choice of α_1, α_2, and α_3 will be dictated by the transient design specifications.

14.7.4 Indices of Performance

Sometimes the performance of control systems is given in terms of a performance index. A performance index is a number that indicates the "goodness" of system performance. Such performance indices are usually employed to get an optimal design in the sense of minimizing or maximizing the index of performance. The optimal parameter values depend directly upon the performance index chosen. Typical performance indices are $\int_0^\infty e^2(t)dt$, $\int_0^\infty te^2(t)dt$, $\int_0^\infty |e(t)|dt$, $\int_0^\infty t|e(t)|dt$. In control system design the task is to select the controller parameters to minimize the chosen index.

14.7.5 Integral-Square-Error (ISE) Criterion

According to the ISE, the quality of system performance is evaluated by minimizing the integral

$$J = \int_0^\infty e^2(t)dt \tag{14.98}$$

A system designed by this criterion tends to show a rapid decrease in a large initial error. Hence the response is fast and oscillatory. Thus the system has poor relative stability. ISE is of practical significance because $\int e^2(t)dt$ resembles power consumption for some systems.

14.7.6 Integral of Time-Multiplied Absolute-Error (ITAE) Criterion

According to this criterion, the optimum system is the one that minimizes the performance index:

$$J = \int_0^\infty t|e(t)|dt \tag{14.99}$$

This criterion weighs large initial errors lightly, and errors occurring late in the transient response are penalized heavily. A system designed by use of ITAE has a characteristic that the overshoot in the transient response is small and oscillations are well damped.

14.7.7 Comparison of Various Error Criteria[3]

Figure 14.34 shows several error performance curves. The system considered is

$$\frac{C(s)}{R(s)} = \frac{1}{s^2 + 2\zeta s + 1} \tag{14.100}$$

The curves of Fig. 14.34 indicate that $\zeta = 0.7$ corresponds to near optimal value with respect to each of the performance indices used. At $\zeta = 0.7$ the system given by Eq. 14.100 results in rapid response to a step input with approximately 5% overshoot.

Table 14.5 summarizes the coefficients that will minimize the ITAE performance criterion for a step input to the closed-loop transfer function[5]

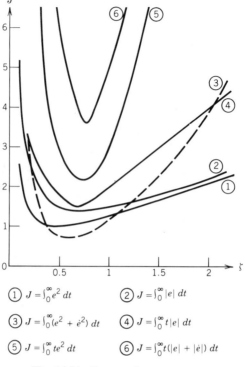

$$\text{①} \quad J = \int_0^\infty e^2 \, dt \qquad\qquad \text{②} \quad J = \int_0^\infty |e| \, dt$$

$$\text{③} \quad J = \int_0^\infty (e^2 + \dot{e}^2) \, dt \qquad \text{④} \quad J = \int_0^\infty t|e| \, dt$$

$$\text{⑤} \quad J = \int_0^\infty te^2 \, dt \qquad\qquad \text{⑥} \quad J = \int_0^\infty t(|e| + |\dot{e}|) \, dt$$

Fig. 14.34 Error performance curves.

$$\frac{C(s)}{R(s)} = \frac{a_0}{s^n + a_{n-1}s^{n-1} + \cdots + a_1 s + a_0} \tag{14.101}$$

Table 14.6[2] summarizes the coefficients that will minimize the ITAE performance criterion for a ramp input applied to the closed-loop transfer function:

$$\frac{C(s)}{R(s)} = \frac{b_1 s + b_0}{s^n + b_{n-1}s^{n-1} + \cdots + b_1 s + b_0} \tag{14.102}$$

Figure 14.35 shows the response resulting from optimum coefficients for a step input applied to the normalized closed-loop transfer function given in Eq. 14.101, for ISE, IAE (integral absolute-error), and ITAE.

TABLE 14.5   OPTIMAL FORM OF THE CLOSED-LOOP TRANSFER FUNCTION BASED ON THE ITAE CRITERION FOR STEP INPUTS

$$\frac{C(s)}{R(s)} = \frac{a_0}{s^n + a_{n-1}s^{n-1} + \cdots + a_1 s + a_0}, \; a_0 = \omega_n^n$$

$$s + \omega_n$$

$$s^2 + 1.4\omega_n s + \omega_n^2$$

$$s^3 + 1.75\omega_n s^2 + 2.15\omega_n^2 s + \omega_n^3$$

$$s^4 + 2.1\omega_n s^3 + 3.4\omega_n^2 s^2 + 2.7\omega_n^3 s + \omega_n^4$$

$$s^5 + 2.8\omega_n s^4 + 5.0\omega_n^2 s^3 + 5.5\omega_n^3 s^2 + 3.4\omega_n^4 s + \omega_n^5$$

$$s^6 + 3.25\omega_n s^5 + 6.60\omega_n^2 s^4 + 8.60\omega_n^3 s^3 + 7.45\omega_n^4 s^2 + 3.95\omega_n^5 s + \omega_n^6$$

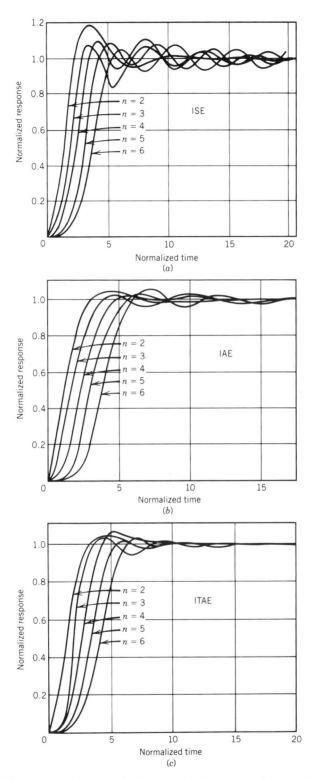

Fig. 14.35 (*a*) Step response of a normalized transfer function using optimum coefficients for ISE, (*b*) IAE, and (*c*) ITAE.

TABLE 14.6   OPTIMUM COEFFICIENTS OF $T(s)$ BASED ON THE ITAE CRITERION FOR A RAMP INPUT

$$s^2 + 3.2\omega_n s + \omega_n^2$$

$$s^3 + 1.75\omega_n s^2 + 3.25\omega_n^2 s + \omega_n^3$$

$$s^4 + 2.41\omega_n s^3 + 4.93\omega_n^2 s^2 + 5.14\omega_n^3 s + \omega_n^4$$

$$s^5 + 2.19\omega_n s^4 + 6.50\omega_n^2 s^3 + 6.30\omega_n^3 s^2 + 5.24\omega_n^4 s + \omega_n^5$$

14.8   SIMULATION FOR CONTROL SYSTEM ANALYSIS

Often in control system design the governing differential equations are reduced to a form that facilitates the shaping of a controller. The model order reductions involve things such as linearization, neglecting fast dynamics, parasitics, time delays, and so on. Once a controller is synthesized for a system, after making simplifying assumptions, it becomes necessary to try out the controller on the actual system. This involves experimentation. Computer simulation may also be viewed as experimentation where the synthesized controller is inserted into a very sophisticated model of the actual physical system and the system response computed. Extensive simulation studies can reduce the cost of experimentation and may even at times serve as the final controller to be implemented on the actual physical plant. So in simulation the burden of experiment is transformed into generating accurate models for all the components in a system without over simplifying assumptions. Simulations can be performed on analog, digital, or hybrid computers[6].

14.8.1   Analog Computation[7]

An analog computer is a machine in which several physical components can be selected and interconnected in such a way that the equations describing the operation of the computer are analogous to that of the actual physical system to be studied. It is a continuous-time device operating in a real-time parallel mode, making it particularly suitable for the solution of differential equations and hence for the simulation of dynamic systems. The most commonly used analog computer is the electronic analog computer in which voltages at various places within the computer are proportional to the variable terms in the actual system. The ease of use, and the direct interactive control over the running of such a computer, allows full scope for engineering intuition and makes it a valuable tool for the analysis of dynamic systems and the synthesis of any associated controllers. A facility frequently useful is the ability to slow down or speed up the problem solution. The accuracy of solution, since it is dependent on analog devices, is generally of the order of a few percent but, for the purposes of system analysis and design, higher accuracy is seldom necessary; also, this accuracy often matches the quality of the available input data.

The basic building block of the analog computer is the high-gain DC amplifier, represented schematically by Fig. 14.36. When the input voltage is $e_i(t)$, the output voltage is given by

$$e_0(t) = -Ae_i(t) \tag{14.103}$$

where A, the amplifier voltage gain, is a large constant value.

If the voltage to the input of an amplifier, commonly referred to as the summing junction, exceeds a few microvolts, then the amplifier becomes saturated or overloaded because the power supply cannot force the output voltage high enough to give the correct gain. Therefore, if an amplifier is to be operated correctly, its summing junction must be very close to ground potential, and is usually treated as such in programming.

When this is used in conjunction with a resistance network as shown in Fig. 14.37, then the resulting circuit can be used to add a number of voltages.

If

$$R_1 = R_2 = R_3$$

then

$$V_0 = -(V_1 + V_2 + V_3) \tag{14.104}$$

If R_1, R_2, R_3 are arbitrary, then

$$V_0 = -\left(\frac{R_f}{R_1}V_1 + \frac{R_f}{R_2}V_2 + \frac{R_f}{R_3}V_3\right) \tag{14.105}$$

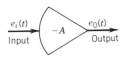

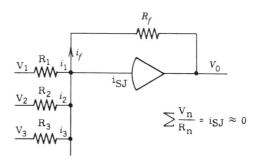

Fig. 14.36 High-gain DC amplifier.

$$\sum \frac{V_n}{R_n} = i_{SJ} \approx 0$$

Fig. 14.37 Current balance.

If there is only one voltage input, then

$$V_0 = \frac{R_f}{R_1} V_1 \cdots \text{multiplication by a constant} \tag{14.106}$$

It should be noted that in all cases there is a sign inversion. Usually the available ratios R_f/R_1 and so on are standardized to 1 and 10, the appropriate gain being selectable as required. The complete circuit comprising high-gain amplifier, input resistors, and feedback element is termed an **operational amplifier**. It is given symbolically as shown in Fig. 14.38.

In order to multiply a voltage by a constant other than 10, use is made of a grounded potentiometer (usually a 10-turn helical potentiometer), as shown in Fig. 14.39. This permits multiplication by a constant in the range 0 to 1.

Since electrical circuits are very nearly linear, the generation of even the simplest nonlinearity, like a product, requires special consideration. One way this can be done is to use the feedback control concept as shown in Fig. 14.40. Two or more potentiometers are mounted rigidly with a

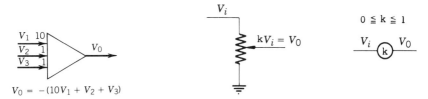

$$V_0 = -(10V_1 + V_2 + V_3)$$

Fig. 14.38 Symbol for summing amplifier.

$kV_i = V_0$

$0 \leq k \leq 1$

Fig. 14.39 Potentiometer.

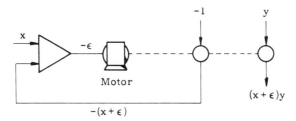

Fig. 14.40 Servomultiplier.

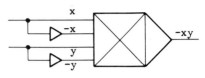

Fig. 14.41 Diode.   **Fig. 14.43** Quarter-square multiplier.

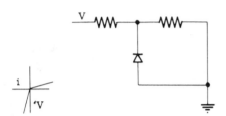

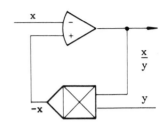

Fig. 14.42 Diode network.   **Fig. 14.44** Division.

common shaft turning the sliders exactly together. A negative unit voltage, which commonly is 10-100 V, is applied across one of the potentiometers. The slider voltage is added to the voltage representing one of the factors. If they differ, then the motor turns the shaft in the direction that eliminates the error. The second factor voltage is applied across the other potentiometer. Its slider then reads the product. This device is called a servomultiplier. It is quite slow in operation because of the motor. Faster multiplication requires the use of special nonlinear networks using diodes. A diode allows current to flow in one direction only if polarities are in the indicated directions in the symbolic representations of Fig. 14.41. Combinations of diodes and resistors can be made such that the current-voltage curve is a series of straight-line segments. The circuit of Fig. 14.42 gives the current indicated there. Almost any shape of curve can be approximated by straight-line segments in this way, in particular a network can be made in which the current into the summing junction of an amplifier is approximately proportional to the square of the voltage applied.

The quarter-square multiplier uses two of these circuits, based on the identity

$$xy = \frac{1}{4}[(x + y)^2 - (x - y)^2]$$ (14.107)

Its symbol is shown in Fig. 14.43. In most modern computers both polarities of input voltage must be provided. This has been indicated by two amplifiers on the inputs.

Division of voltages is accomplished by using a multiplier in the feedback path of an amplifier, as indicated in Fig. 14.44.

With the quarter-square multiplier, multiplication and division may be performed accurately at high frequencies. These complete the description of how the ordinary algebraic operations are performed on an analog computer.

Next, we turn to the operations of the calculus, which enable differential equations to be solved. Consider the circuit of Fig. 14.45. The charge on a capacitor is the integral of the current, therefore the derivative of the charge is the current. The charge is the potential times the capacitance, which is fixed, thus current balance around the summing junction gives

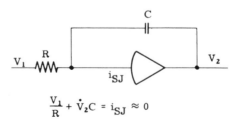

$$\frac{V_1}{R} + \dot{V_2}C = i_{SJ} \approx 0$$

Fig. 14.45 Current balance.

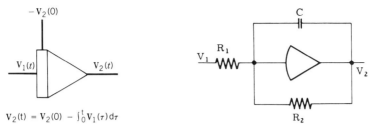

$$V_2(t) = V_2(0) - \int_0^t V_1(\tau)\,d\tau$$

Fig. 14.46 Integration.   **Fig. 14.47** Time lag.

$$\frac{V_1}{R} + C\dot{V}_2 = 0 \tag{14.108}$$

or

$$V_2(t) = V_2(0) - \frac{1}{RC}\int_0^t V_1(\tau)\,d\tau \tag{14.109}$$

Thus this circuit is an accurate integrator with a proportionality factor or time constant of RC. A common symbol is shown in Fig. 14.46. The initial condition can be set in with a special network furnished with most computers.

Differentiation can be performed with an input capacitor and a feedback resistor; however, it is not generally recommended because input signals are often contaminated by high-frequency noise, and the differentiator circuit amplifies this noise in direct proportion to its frequency. Since differential equations can always be written in terms of integrals, this causes no inconvenience.

The circuit of Fig. 14.47 generates a simple time lag, as can be derived from its current balance:

$$\frac{V_1}{R_1} + \dot{V}_2 C + \frac{V_2}{R_2} = 0 \tag{14.110}$$

or

$$\dot{V}_2 + \frac{1}{R_2 C}V_2 = -\frac{V_1}{R_1 C} \tag{14.111}$$

The transfer function of this circuit is

$$G(s) = \frac{V_2}{V_1}(s) = \frac{-R_2/R_1}{R_2 Cs + 1} = \frac{-R_2/R_1}{\tau s + 1}$$

Its time response to a pulse input is a simple exponential decay, with its time constant equal to $R_2 C$.

The circuit corresponding to a damped sinusoid is shown in Fig. 14.48. Its differential equation is

$$\ddot{y} + 2\zeta\omega_n\dot{y} + \omega_n^2 y = \omega_n^2 f(t) \tag{14.112}$$

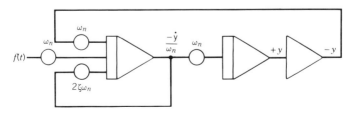

Fig. 14.48 Sinusoid.

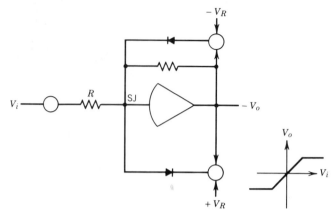

Fig. 14.49 Saturation.

The transfer function is

$$G(s) = \frac{\omega_n^2}{s^2 + 2\omega_n \zeta s + \omega_n^2} = \frac{1}{\dfrac{s^2}{\omega_n^2} + \dfrac{2\zeta}{\omega_n}s + 1} \tag{14.113}$$

A particular precaution that must be taken with analog computers is in scaling voltages to represent other physical variables. This must be done consistently for all variables and must be done in such a way that no amplifier is ever overloaded. Since amplifier outputs can never be fully predicted, some problems may require rescaling because of a bad initial guess about the range of a dependent variable. This is particularly troublesome if nonlinear equipment is involved in the simulation. The necessity for proper prior scaling is one of the major disadvantages of analog computation.

With these basic circuit elements, the control engineer can proceed directly from a block diagram of a physical system to a wired program board without writing down differential equations because the circuits corresponding to the common transfer functions are combinations of those shown in Figs. 14.46–14.48. Special nonlinearities can be simulated by special networks. Simple ones such as multiplication, squaring, exponentiation, and generation of logarithms are often prewired directly to the program board.

Therefore, by suitable interconnection of components, even the most complicated system can be simulated in full dynamic character. Frequency response, stability limits, sensitivity to parameter changes, and effects of noise can all be tested and studied. Adjustments of the simulated controls for optimum rejection of noise can be done. Different types of control can be tried experimentally.

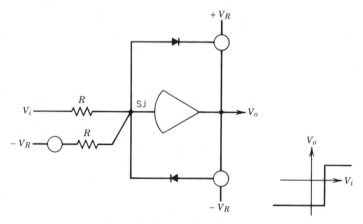

Fig. 14.50 On-off controller.

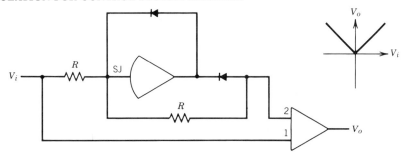

Fig. 14.51 Rectifier.

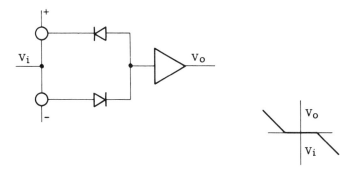

Fig. 14.52 Deadzone.

If care has been taken in formulating the system dynamics, then the analog computer can be made to behave precisely like the device it simulates.

Figures 14.49–14.53 show some special diode networks for simulating some of the special discontinuous nonlinearities that have been discussed previously.

Modern analog computers are generally equipped with additional types of special equipment, some of which are now listed and described below:

1. Comparator: equivalent to the on-off control of Fig. 14.50.
2. Resolver: a circuit for rapidly and continuously converting from polar to Cartesian coordinates or the reverse.
3. Relays and switches: operated by voltages generated in the simulation.
4. Memory amplifiers: An integrator into which an initial condition is set, but to which no integrating input is patched, will "remember" the initial condition.

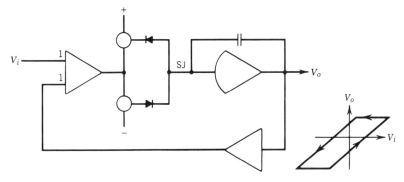

Fig. 14.53 Hysteresis or backlash.

5. Repetitive operation at high speed: solutions can be generated to repeat hundreds of times per second.

6. Outputs can be displayed on an oscilloscope, plotted on coordinate paper, or on continuous recording equipment.

All these features make an analog computer a highly useful tool for design of control systems.

14.8.2 Digital Computation

An analog computer simultaneously solves all of the differential equations that model a physical system, and continuous voltage signals represent the variables of interest. This enables the machine to operate in real time. Significant disadvantages of analog simulation are the high cost of the computer due to the multiplicity of elements with demanding performance specifications, difficulties with scaling to avoid overloading of amplifiers, and relatively limited accuracy and repeatability due in part to amplifier drift. As a consequence of the very rapid development of digital computer hardware and software giving ever greater capability and flexibility at reducing cost, system simulation is inevitably being carried out more and more on the digital computer. There is effectively no problem of overloading, enabling wide ranges of parameter variation to be accommodated.

The solution of a differential equation involves the process of integration, and for the digital computer analytical integration must be replaced by some numerical method that yields an approximation to the true solution. A number of special programming languages referred to as continuous system simulation languages (CSSL) or simply as simulation languages are available as analytical tools to study the dynamic behavior of a wide range of systems without the need for a detailed knowledge of computing procedures. The languages are designed to be simple to understand and use, and they minimize programming difficulty by allowing the program to be written as a sequence of relatively self-descriptive statements. Numerous different languages with acronyms such as ACSL, CSMP, CSSL, DYNAMO, DARE, MIMIC, TELSIM, ENPORT, SCEPTRE, and SIMNON have been developed by computer manufacturers, software companies, universities, and others, some for specific families of machines and others for wider applications. Due to standardization a number of the languages tend to be broadly similar. Symbolic names are used for the system variables, and the main body of the program is written as a series of simple statements based on the system state equations, block diagrams, or bond graphs. To these are added statements specifying initial parameter values and values of system constants, and simple command statements controlling the running of the program and specifying the form in which the output is required. The short user-written program is then automatically translated into a FORTRAN (or other high-level language) program that is then compiled, loaded, and executed to produce a time history of the variables of interest in print or plot form. System constants and initial conditions can be altered and the program rerun without the need to retranslate and recompile. Many of the languages are designed to be run interactively from a graphics terminal and have the facility of displaying on the screen whichever output is of interest and, if the solution is not developing as desired, the capability of interrupting the program, changing parameters, and rerunning immediately.

The typical steps to be taken by a CSSL are shown in Fig. 14.54. The major process blocks include:

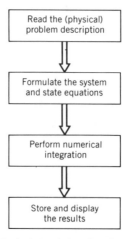

Fig. 14.54 Typical simulation flowchart of the CSSL.

1. Reading the initial problem description, in which the circuit, bond graph, schematic, or block diagram information is communicated to the computer.
2. Formulating the system and state equations.
3. Performing the numerical integration of the state equations by a suitable method.
4. Storing and displaying results.

Most simulation languages differ with regard to use by the engineer in steps 1 and 2; for example, SCEPTRE for circuit descriptions, ENPORT for bond graph descriptions, CSMP for block diagrams, and DARE for state equations. The best guide to a program's characteristics is its user's manual.

In summary, the advantages of digital simulation are: (i) simple program writing, (ii) accuracy and reproducibility, (iii) cost-effectiveness, (iv) where interactive facilities exist, keyboard entry of program and running, (v) inspection of plots on graphics display screen, and (vi) on-line modification and rerunning.

14.8.3 Hybrid Computation

Hybrid computation is a combination of analog and digital computations. In hybrid computations analog and digital machines are used in such a way that a problem can be programmed by exploiting the most efficient features of each. Much of the cost of such a machine arises from the rather complicated interface equipment required to make the necessary conversions between analog and digital signals and vice versa. The cost of an analog or hybrid computer is now justified only for certain large simulations where fast computing speed attained by parallel operation is important. Even this advantage may disappear as parallel computation is introduced into digital systems.

REFERENCES

1. E. O. Doebelin, *Control System Principles and Design*, New York, Wiley, 1985.
2. R. C. Dorf, *Modern Control Systems*, Reading, Mass., Addison-Wesley, 1986.
3. K. Ogata, *Modern Control Engineering*, Englewood Cliffs, N.J., Prentice-Hall, 1970.
4. G. F. Franklin, J. D. Powell, and A. Emami-Naeini, *Feedback Control of Dynamic Systems*, Reading, Mass., Addison-Wesley, 1986.
5. D. Graham and R. C. Lathrop, "The Synthesis of Optimum Response: Criteria and Standard Forms,"*AIEE Transactions*, Part II, 72, pp. 273–288, 1953.
6. G. A. Korn and J. V. Wait, *Digital Continuous System Simulation*, Englewood Cliffs, N.J., Prentice-Hall, 1978.
7. O. W. Eshbach and M. Souders, (Eds.)., *Handbook of Engineering Fundamentals*, 3rd ed., New York, Wiley, 1975.
8. R. C. Rosenberg and D. C. Karnopp, *Introduction to Physical System Dynamics*, New York, McGraw-Hill, 1983.
9. M. Vidyasagar, *Nonlinear Systems Analysis*, Englewood Cliffs, N.J., Prentice Hall, 1978.
10. Y. Takahashi, M. J. Rabins, and D. M. Auslander, *Control and Dynamic Systems*, Reading, Mass., Addison-Wesley, 1972.
11. G. F. Franklin and J. D. Powell, *Digital Control of Dynamic Systems*, Reading, Mass., Addison-Wesley, 1980.
12. B. C. Kuo, *Automatic Control Systems*, Englewood Cliffs, N.J., Prentice-Hall, 1982.

BIBLIOGRAPHY

Bode, H. W., *Network Analysis and Feedback Amplifier Design*, Van Nostrand, 1945.

Chestnut, H., and Mayer, R. W., *Servomechanisms and Regulating Systems Design*, 2nd ed., Vol. 1, New York, Wiley, 1959.

D'Azzo, J. J., and Houpis, C. H., *Linear Control System Analysis and Design*, New York, McGraw-Hill, 1988.

Distefano, J. J. III, Stubberud, A. R., and Williams, I. J., *Feedback and Control Systems* (Schaum's Outline Series), New York, Shaum Publishing, 1967.

Dransfield, P., *Engineering Systems and Automatic Control*, Englewood Cliffs, N.J., Prentice-Hall, 1968.

Elgerd, O. I., *Control Systems Theory*, New York, McGraw-Hill, 1967.

Evans, W. R., *Control-System Dynamics*, New York, McGraw-Hill, 1954.

Eveleigh, V. W., *Introduction to Control Systems Design*, New York, McGraw-Hill, 1972.

Horowitz, I. M., *Synthesis of Feedback Systems*, New York, Academic Press, 1963.

Houpis, C. H., and Lamont, G. B., *Digital Control Systems Theory, Hardware, Software*, New York, McGraw-Hill, 1985.

Kuo, B. C., *Digital Control Systems*, New York, Holt, Rinehart and Winston, 1980.

Melsa, J. L., and Schultz, D. G., *Linear Control Systems*, New York, McGraw-Hill, 1969.

Nyquist, H., "Regeneration Theory,"*Bell System Tech. J.*, Vol. II, pp. 126–147, 1932.

Palm, N. J. III, *Modeling, Analysis and Control of Dynamic Systems*, New York, Wiley, 1983.

Phillips, C. L., and Nagle, H. T., Jr., *Digital Control System Analysis and Design*, Englewood Cliffs, N.J., Prentice-Hall, 1984.

Ragazzini, J. R., and Franklin, G. F., *Sampled Data Control Systems*, New York, McGraw-Hill, 1958.

Raven, F. H., *Automatic Control Engineering*, 4th ed., New York, McGraw-Hill, 1987.

Shinners, S. M., *Modern Control Systems Theory and Application*, Reading, Mass., Addison Wesley, 1972.

Truxal, J. G., *Automatic Feedback Control Synthesis*, New York, McGraw-Hill, 1955.

CHAPTER **15**

CONTROL SYSTEM PERFORMANCE MODIFICATION

Suhada Jayasuriya

Department of Mechanical Engineering
Texas A&M University
College Station, Texas

15.1 INTRODUCTION

Chapter 14 presents a wide variety of tools for analyzing closed-loop control systems. This chapter focuses on the development of additional analytical tools for closed-loop control systems, tools aimed at performance modification and improvement. Each successive tool presented in this chapter is useful in its ability to predict system performance and to pinpoint the appropriate modifications so that the closed-loop system will meet its required performance objectives.

Chapter 16 also addresses the problem of system modifications, but there the emphasis is to work with the adjustable parameters of a device called the servocontroller to achieve the necessary result.

15.2 GAIN AND PHASE MARGIN

The Nyquist stability criterion may be conveniently used to define certain measures of relative stability or robustness. We note that $(-1,0)$ point in the GH plane plays a crucial role in determining the closed-loop stability of a system. If a system's stability status is known, one might be interested in knowing how stable the system is due to changes in parameters. For example, if the system remains stable despite large changes in parameters, then it is said to possess a high degree of relative stability or robustness. Gain margin and phase margin are two measures typically employed to characterize this robustness. They characterize how close the Nyquist plot is to encircling the $(-1,0)$ point in the GH plane.

15.2.1 Gain Margin

This is a measure of how much the loop gain can be raised before closed-loop instability results. The basic definition of gain margin (GM) is apparent from Fig. 15.1.

15.2.2 Phase Margin

The phase margin (PM) is the difference between the phase of $GH(j\omega)$ and 180° when the $GH(j\omega)$ crosses the circle with unit magnitude. A positive phase margin corresponds to a case where the Nyquist locus does not encircle the $(-1,0)$ point. This is shown in Fig. 15.2.

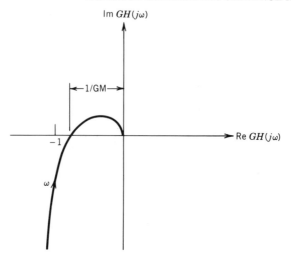

Fig. 15.1 Gain margin.

A stable system corresponds to gain and phase margins that are positive. In some cases, however, the PM and GM notions break down. For first- and second-order systems, the phase never crosses the 180° line; hence the gain margin is always ∞. For higher-order systems, it is possible to have more than one crossing of the unit amplitude circle and more than one crossing of the 180° line. In such situations the GM and PM are somewhat misleading. Furthermore, nonminimum phase systems exhibit stability criteria that are opposite to those previously defined.

15.2.3   Gain-Phase Plots

The graphical representation of the frequency response of the system $G(s)$ using either

$$G(j\omega) = G(s)|_{s=j\omega} = \operatorname{Re} G(j\omega) + j\operatorname{Im} G(j\omega)$$

or

$$G(j\omega) = |G(j\omega)|e^{j\phi(\omega)}$$

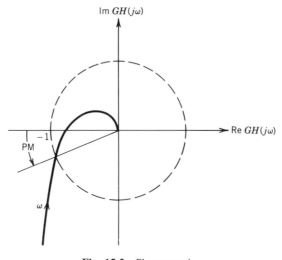

Fig. 15.2   Phase margin.

where

$$\phi(\omega) = \underline{/G(j\omega)}$$

is known as the polar plot. The coordinates of the polar plot are the real and imaginary parts of $G(j\omega)$ as shown in Fig. 15.3.

Example 15.1.   Obtain the polar plot of the transfer function

$$G(s) = \frac{K}{s(\tau s + 1)}$$

The frequency response is given by

$$G(j\omega) = \frac{K}{j\omega(j\tau\omega + 1)}$$

Then the magnitude and the phase can be written as

$$|G(j\omega)| = \frac{K}{\omega\sqrt{\omega^2\tau^2 + 1}}$$

and

$$G(j\omega) = -\frac{\pi}{2} - \tan^{-1}\omega\tau$$

If $|G(j\omega)|$ and $\underline{/G(j\omega)}$ are computed for different frequencies an accurate plot can be obtained. A quick idea can, however, be gained by simply doing a limiting analysis at $\omega = 0$, $\omega = \infty$, and the corner frequency $\omega = 1/\tau$.
We note that for

$$\omega = 0 \qquad |G(j\omega)| \to \infty \qquad \underline{/G(j\omega)} = \frac{-\pi}{2}$$

$$\omega = \infty \qquad |G(j\omega)| \to 0 \qquad \underline{/G(j\omega)} = -\pi$$

$$\omega = \frac{1}{\tau} \qquad |G(j\omega)| = \frac{K\tau}{\sqrt{2}} \qquad \underline{/G(j\omega)} = -\frac{3\pi}{4}$$

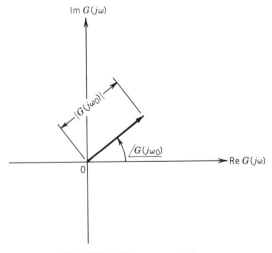

Fig. 15.3   Polar representation.

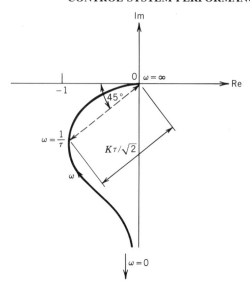

Fig. 15.4   Polar plot for $G(s) = K/s(\tau s + 1)$.

The polar plot is shown in Fig. 15.4.
   A gain-phase plot is where the frequency response information is given with respect to a Cartesian frame with vertical axis for gain and the horizontal for phase.

15.2.4   Polar Plot as a Design Tool in the Frequency Domain

As a design tool its best use is in determining relative stability with respect to gain and phase margin. If the uncertainty in the transfer function can be characterized by bounds on the gain-phase plot, then it allows one to determine what type of compensation needs to be provided for the system to perform in the presence of such uncertainties. As an example consider the closed-loop system shown in Fig. 15.5.

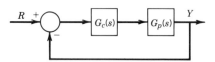

Fig. 15.5   Closed-loop system.

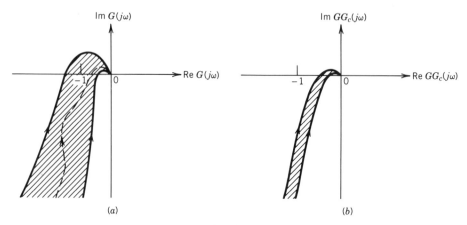

Fig. 15.6   Polar plots for an uncertain system.

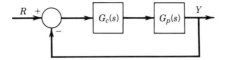

Fig. 15.7 Closed-loop system with $G(s) = K/[s(s + 1)(0.0125s + 1)]$.

Suppose the gain-phase plot is known to lie in the shaded region in the $G(j\omega)$ plane as shown in Fig. 15.6.

Since the shaded region includes the $(-1,0)$ point, and the true gain-phase plot for the plant can lie anywhere in the shaded region, the system can potentially be unstable. If stabilization in the presence of uncertainty is the primary design issue, then we would require a $G_c(s)$ so that it would reshape the high-frequency part of the polar plot with a reduced band of uncertainty. The reduction in the band of uncertainty is a required feature of any sound feedback system design. Qualitatively one would expect to reshape the polar plot to something that looks like what is shown in Fig. 15.6b. Moreover, knowing the important frequency ranges will allow one to be more concerned with relevant portions of the gain-phase plot for reshaping. To further illustrate the basic philosophy of a typical design in the frequency domain, consider the plant transfer function

$$G_p(s) = \frac{K}{s(1 + s)(1 + 0.0125s)}$$

in the feedback configuration shown in Fig. 15.7.

It is required that when a ramp input is applied to the closed-loop system, the steady-state error of the system does not exceed 1% of the amplitude of the input ramp. Using steady-state error computations we find that the minimum K should be such that

$$\text{Steady-state error} = e_{ss} = \lim_{s \to 0} \frac{1}{sG_p(s)} = \frac{1}{K} \leq 0.01$$

that is, $K \geq 100$.

It can be easily verified that with $G_c(s) = 1$ the system is unstable for $K > 81$, implying that a controller $G_c(s)$ must be designed to satisfy the steady-state performance and relative stability requirements. Putting it another way, the controller must be able to keep the zero-frequency gain of $sG_pG_c(s)$ effectively at 100 while maintaining a prescribed degree of relative ability. The principal of the design in the frequency domain is best illustrated by the polar plot of $G_p(s)$ shown in Fig. 15.8. In practice, the Bode diagram is preferred for design purposes because it is simpler to construct. The polar plot is used mainly for analysis and added insight.

As shown in Fig. 15.8, when $K = 100$, the polar plot of $G_p(s)$ encloses the $(-1,0)$ point, and the closed-loop system is unstable. Let us assume that we wish to realize a PM $= 30°$. This means

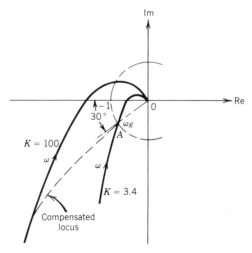

Fig. 15.8 Polar plot for open-loop system transfer function of Fig. 15.7.

that the polar plot must pass through point A (with magnitude 1 and phase $-150°$). If K is the only adjustable parameter to achieve this PM, the desired value of $K \approx 3.4$, as shown in Fig. 15.8. But, K cannot be set to 3.4 since the ramp error constant would only be 3.4 sec^{-1}, and the steady-state error requirement will not be satisfied.

Since the steady-state performance of the system is governed by the characteristics of the transfer function at low frequency, and the damping or the transient behavior of the system is governed by the relatively high-frequency characteristics, as Fig. 15.8 shows, to simultaneously satisfy the transient and the steady-state requirements, the frequency locus of $G_p(s)$ has to be reshaped so that the high-frequency portion of the plot follows the $K = 3.4$ trajectory and the low-frequency portion follows the $K = 100$ trajectory. The significance of this reshaping of the frequency locus is that the compensated locus shown in Fig. 15.8 will be coincident with the high-frequency portion yielding PM $= 30°$, while the zero-frequency gain is maintained at 100 to satisfy the steady-state requirement.

When we inspect the loci of Fig. 15.8, we see that there are at least two alternatives in arriving at the compensated locus:

1. Starting from the $K = 100$ locus and reshaping the locus in the region near the gain cross-over frequency ω_g, while keeping the low-frequency region of $G_p(s)$ relatively unaltered.
2. Starting from the $K = 3.4$ locus and reshaping the low-frequency portion of $G_p(s)$ to obtain an error constant $= 100$ while keeping the locus near $\omega = \omega_g$ relatively unchanged.

In the first approach, the high-frequency portion of $G_p(s)$ is pushed in the counterclockwise (ccw) direction, which means that more phase is added to the system in the positive direction in the proper frequency range. This scheme is basically phase-lead compensation and controllers used for this purpose are often of the high-pass filter type. The second approach apparently involves the shifting of the low-frequency part of the $K = 3.4$ trajectory in the clockwise (cw) direction, or alternatively, reducing the magnitude of $G_p(s)$ with $K = 100$ at the high-frequency range. This scheme is often referred to as phase-lag compensation since more phase lag is introduced to the system in the low-frequency range. The controllers used for this purpose are often referred to as low-pass filters.

15.3 HALL CHART

In typical frequency response design only the open-loop transfer function is plotted. Therefore it is useful to know how the closed-loop performance is related to the open loop. Hall charts provide a convenient way of carrying out a frequency response design with closed-loop performance specifications. One important consideration is the maximum closed-loop gain. Another is the closed-loop phase. A Hall chart primarily consists of constant closed-loop gain loci and constant closed-loop phase loci. A design would then proceed by drawing the open-loop polar plot on the Hall chart.

For a unity negative feedback system as shown in Fig. 15.9 the closed-loop transfer function is

$$\frac{C(s)}{R(s)} = \frac{G(s)}{1 + G(s)} \tag{15.1}$$

In the following discussion we assume that the polar plot of $G(j\omega)$ is known.

15.3.1 Constant-Magnitude Circles

The loci on which the closed-loop magnitude

$$\left| \frac{C(s)}{R(s)} \right| = \left| \frac{G(s)}{1 + G(s)} \right| = M = \text{constant}$$

are referred to as constant-magnitude loci. In fact these loci are circles in the $G(j\omega)$ plane. This can be established by noting a typical point on the $G(j\omega)$ plot as $X + jY$.

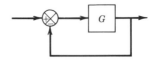

Fig. 15.9 Unity negative feedback system.

Then

$$M = \frac{|X + jY|}{|1 + X + jY|}$$

and

$$M^2 = \frac{X^2 + Y^2}{(1 + X)^2 + Y^2}$$

Hence

$$X^2 + \frac{2M^2}{M^2 - 1}X + \frac{M^2}{M^2 - 1} + Y^2 = 0$$

which can be written as

$$\left(X + \frac{M^2}{M^2 - 1}\right)^2 + Y^2 = \frac{M^2}{(M^2 - 1)^2} \tag{15.2}$$

Equation 15.2 is the equation of a circle with center at $X = -M^2/(M^2 - 1)$, $Y = 0$ and with radius $|M/(M^2 - 1)|$. A family of constant M circles is shown in Fig. 15.10. Given a point $P \equiv (X_1, Y_1)$ on an open-loop polar plot $G(j\omega)$, the corresponding closed-loop magnitude can be determined by locating the M circle passing through that point.

Graphically the intersection of the $G(j\omega)$ plot and the constant M locus gives the value of M at the frequency denoted on the $G(j\omega)$ curve. If it is desired to keep the value of the maximum closed-loop gain M_r less than a certain value, the $G(j\omega)$ curve must not intersect the corresponding M circle at any point, and at the same time must not enclose the $(-1, j0)$ point. The constant M circle with the smallest radius that is tangent to the $G(j\omega)$ curve gives the value of M_r, and the resonant frequency ω_r is read off at the tangent point on the $G(j\omega)$ curve.

15.3.2 Constant-Phase Circles

The loci of constant phase of the closed-loop system can also be determined in the $G(j\omega)$ plane by a method similar to that used for constant M loci. With reference to Eq. 15.1 the phase of the closed-loop system corresponding to the point $P = X + jY$ is written as

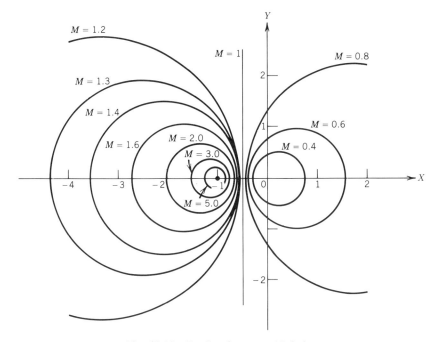

Fig. 15.10 Family of constant M circles.

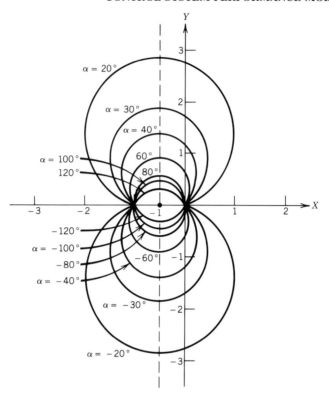

Fig. 15.11 Family of constant N circles.

$$\phi = \tan^{1}\left(\frac{Y}{X}\right) - \tan^{1}\left(\frac{Y}{1+X}\right) \tag{15.3}$$

Taking the tangent on both sides of Eq. 15.3 and rearranging yields

$$\left(X+\frac{1}{2}\right)^2 + \left(Y-\frac{1}{2N}\right)^2 = \frac{1}{4} + \left(\frac{1}{2N}\right)^2 \tag{15.4}$$

where $N = \tan\phi$.

Equation 15.4 represents a family of circles with center at $(-1/2, 1/2N)$ and with radius $\sqrt{1/4 + 1/(2N)^2}$. The constant-phase loci are shown in Fig. 15.11.

The use of constant magnitude and phase circles enables one to find the entire closed-loop frequency response from the open-loop frequency response $G(j\omega)$ without calculating the magnitude and phase of the closed-loop transfer function at each frequency. The intersections of the $G(j\omega)$ locus and the M circles and N circles give the values of M and N at frequency points on the $G(j\omega)$ locus.

15.3.3 Closed-Loop Frequency Response for Nonunity Feedback Systems

The constant M and N circles are limited to closed-loop systems with unity negative feedback, whose transfer function is given by Eq. 15.1. When a system has nonunity feedback, the closed-loop transfer function is

$$\frac{C(s)}{R(s)} = \frac{G(s)}{1 + G(s)H(s)} \tag{15.5}$$

and constant M loci derived earlier cannot be directly applied. However, with a slight modification constant M and N loci can still be applied to systems with nonunity feedback. We modify Eq. 15.5 as

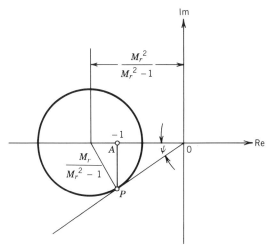

Fig. 15.12 M circle.

$$\frac{C(s)}{R(s)} = \frac{1}{H(s)} \frac{G(s)H(s)}{1 + G(s)H(s)}$$

The magnitude and phase angle of $G_1(s)/[1 + G_1(s)]$ where $G_1(s) = G(s)H(s)$, may be obtained easily by plotting the $G_1(j\omega)$ locus and reading the values of M and N at various frequency points. The closed-loop frequency response $C(j\omega)/R(j\omega)$ may then be obtained by multiplying $G_1(j\omega)/[1 + G_1(j\omega)]$ by $1/H(j\omega)$.

15.3.4 Closed-Loop Amplitude Ratio

In obtaining suitable performance, the adjustment of gain is usually the first consideration. The adjustment of gain is usually based on the maximum closed-loop gain or the resonant peak. That is the gain K which must be chosen so that over the entire frequency range the closed-loop amplitude ratio M_r is not exceeded.

Consider first isolating the circle corresponding to M_r as shown in Fig. 15.12. Then a tangent line to the M_r circle is drawn from the origin, which makes an angle ψ with the real line.

If $M_r > 1$, then

$$\sin \psi = \left| \frac{M_r/(M_r^2 - 1)}{M_r^2/(M_r^2 - 1)} \right| = \frac{1}{M_r}$$

It can be shown that the line drawn from P, perpendicular to the negative real axis, intersects this axis at the $(-1,0)$ point. These two facts, namely $\sin \psi = 1/M_r$ and that the normal from P passes through $(-1,0)$, can be used to determine the appropriate gain K.

Example 15.2. Consider the system shown in Fig. 15.13a: determine K so that $M_r = 1.4$. First sketch the polar plot of

$$\frac{G(j\omega)}{K} = \frac{1}{j\omega(1 + j\omega)}$$

as shown in Fig. 15.13b. The value of ψ corresponding to $M_r = 1.4$ is obtained from

$$\psi = \sin^{-1} \frac{1}{M_r} = \sin^{-1} \frac{1}{1.4} = 45.6°$$

The next step is to draw a line OP that makes an angle $\psi = 45.6°$ with the negative real axis. Then draw the circle that is tangent to both the $G(j\omega)/K$ locus and the line OP. The perpendicular line

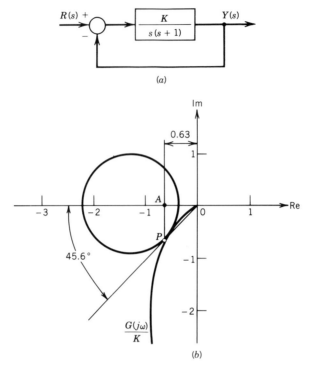

Fig. 15.13   (a) Closed-loop system; (b) determination of the gain K using an M circle.

drawn from the point P intersects the negative real axis at $(-0.63,0)$. Then the gain K of the system is determined as follows:

$$K = \frac{1}{0.63} = 1.58$$

15.4   NICHOLS CHART

Both the gain and phase plots are generally required to analyze the performance of a closed-loop system. A major disadvantage in working with polar plots is that the curve no longer retains its original shape when a simple modification such as the change of the loop gain is made to the system. In design, however, not only the loop gain must be altered but often series or feedback controllers are to be added to the original system that require the complete reconstruction of the resulting open-loop transfer function. For design purposes it is more convenient to work with Bode diagrams or gain-versus-phase plots. The latter representation with corresponding M and N circles superimposed on it is referred to as the **Nichols chart.** In a gain-versus-phase plot the entire $G(j\omega)$ is shifted up or down vertically when the gain is altered. A Nichols chart is shown in Fig. 15.14.

This chart is symmetric about the $-180°$ axis. The M and N loci repeat for every $360°$, and there is symmetry at every $180°$ interval. The M loci are centered about the critical point (0 db, $-180°$).

15.4.1   Closed-Loop Frequency Response from That of Open Loop

It is quite easy to determine the closed-loop frequency response from that of the open loop by using the Nichols chart. If the open-loop frequency response curve is superimposed on the Nichols chart, the intersections of the open-loop frequency response curve $G(j\omega)$ and the M and N loci give the magnitude M and phase angle ϕ of the closed-loop frequency response at each frequency point. If the $G(j\omega)$ locus does not intersect the $M = M_r$ locus but is tangent to it, then the resonant peak value of the closed-loop frequency response is given by M_r. The resonant frequency is given by the frequency at the point of tangency.

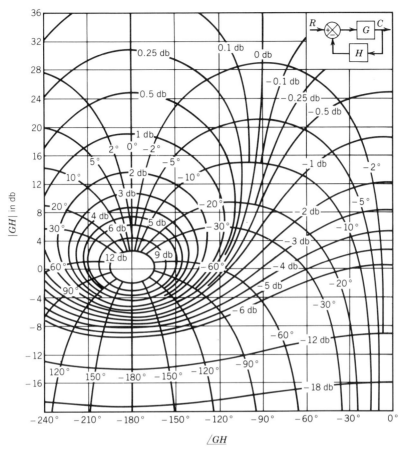

Fig. 15.14 Nichols chart.

As an example consider the unity negative feedback system with the following open-loop transfer function:

$$G(s) = \frac{K}{s(s + 1)(0.5s + 1)}; \qquad K = 1$$

In order to find the closed-loop frequency response by use of the Nichols chart, the $G(j\omega)$ locus is first constructed. (It is easy to first construct the Bode diagram and then to transfer values to the Nichols chart.) The closed-loop frequency response curves (gain and phase) may be constructed by reading the magnitude and phase angles at various frequency points on the $G(j\omega)$ locus from the M and N loci as shown in Fig. 15.15. Since the $G(j\omega)$ locus is tangent to $M = 5$ db locus, the peak value of the closed-loop frequency response is $M_r = 5$ db, and the resonant frequency is 0.8 rad/sec.

The bandwidth of the closed-loop system can easily be found from the $G(j\omega)$ locus in the Nichols chart. The frequency at the intersection of the $G(j\omega)$ locus and the $M = -3$ db locus gives the bandwidth. The gain and phase margins can be read directly from the Nichols chart.

If the open-loop gain K is varied, the shape of the $G(j\omega)$ locus in the Nichols chart remains the same, but is shifted up (for increasing K) or down (for decreasing K) along the vertical axis. Therefore the modified $G(j\omega)$ locus intersects the M and N loci differently, resulting in a different closed-loop frequency response curve.

15.4.2 Sensitivity Analysis Using the Nichols Chart[12]

Consider a unity feedback system with the transfer function

$$\frac{C(s)}{R(s)} = \frac{G(s)}{1 + G(s)} = G_{cl}(s)$$

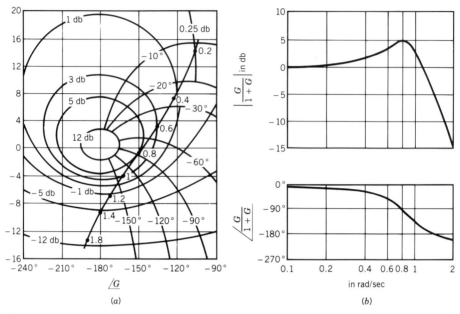

Fig. 15.15 (a) Plot of $G(j\omega)$ superimposed on Nichols chart; (b) closed-loop frequency response curves.

The sensitivity of $G_{cl}(s)$ with respect to $G(s)$ is defined as

$$S_G^{G_{cl}}(s) = \frac{dG_{cl}(s)/G_{cl}(s)}{dG(s)/G(s)}$$

which yields

$$S_G^{G_{cl}}(s) = \frac{1}{1 + G(s)} \tag{15.6}$$

Clearly the sensitivity function is a function of the complex variable s.

To design a system with a prescribed sensitivity the Nichols chart is quite convenient. Equation 15.6 is written as

$$S_G^{G_{cl}}(j\omega) = \frac{G^{-1}(j\omega)}{1 + G^{-1}(j\omega)}$$

which clearly indicates that the magnitude and phase of $S_G^{G_{cl}}(j\omega)$ can be obtained by plotting $G^{-1}(j\omega)$ on the Nichols chart and making use of the constant M loci for constant sensitivity function. Since the vertical coordinate of the Nichols chart is in decibels, the $G^{-1}(j\omega)$ curve on the Nichols chart can be easily obtained if $G(j\omega)$ is already available since

$$|G^{-1}(j\omega)|_{db} = -|G(j\omega)|_{db}$$

$$\underline{/G^{-1}(j\omega)} = -\underline{/G(j\omega)}$$

As an example consider the unity feedback system with the open-loop transfer function

$$G(s) = \frac{400,000K}{s(s + 49)(s + 991)}$$

the function $G^{-1}(j\omega)$ is plotted on the Nichols chart as shown in Fig. 15.16, for $K = 2.94$. The intersections of $G^{-1}(j\omega)$ curve with M loci give the magnitude of $S_G^{G_{cl}}(j\omega)$ at the corresponding frequencies. Figure 15.16 indicates several interesting points with regard to the sensitivity of the feedback system. The sensitivity function approaches 0 db or unity as $\omega \to \infty$: $S_G^{G_{cl}} \to 0$

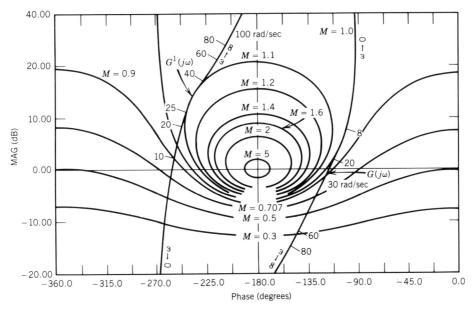

Fig. 15.16 Determination of the sensitivity function S_G^M in the Nichols chart.

as $\omega \to 0$. A peak value of 1.1 db is reached at $\omega = 25$ rad/sec. This means that the closed-loop system is most sensitive to a change of $G(j\omega)$ at this frequency and more generally in this frequency range.

15.5 ROOT LOCUS

Poles and zero locations of a dynamic system characterize the system performance in a significant way. The root locus method allows one to investigate the closed-loop pole patterns of a dynamic system with respect to a single parameter.

A typical closed-loop characteristic equation of a feedback system can be written as

$$1 + G(s)H(s) = 0 \tag{15.7}$$

where $G(s)H(s)$ is the open-loop transfer function.

If $G(s)H(s)$ has a single parameter K as a variable, then by rewriting as

$$1 + KG(s)H(s) = 0 \tag{15.8}$$

a standard procedure for obtaining the closed-loop poles corresponding to any K is the Evans root locus method.

15.5.1 Angle and Magnitude Conditions

The closed-loop characteristic equation (Eq. 15.8) can be written as

$$KG(s)H(s) = -1 = e^{j(2\pi/\pm\pi)} \tag{15.9}$$

Thus, any point s_0 satisfying the condition

$$\underline{/KG(s_0)H(s_0)} = (2l \pm 1)\pi \tag{15.10}$$

satisfies Eq. 15.8. If $K > 0$, then Eq. 15.10 reduces to

$$\underline{/G(s_0)H(s_0)} = (2l \pm 1)\pi \tag{15.11}$$

and is commonly called the angle condition. All points s_0 in the complex plane satisfying this angle condition satisfy the closed-loop characteristic equation and hence are said to lie on the root locus.

If s_0 is a point on the root loci, then the corresponding value of K may be computed by noting that

$$|K||G(s_0)||H(s_0)| = 1 \tag{15.12}$$

which is called the amplitude condition.

By studying the angle condition in detail of the closed-loop characteristic equation,

$$\underline{/1 + KG(s)H(s)} = 1 + K\frac{\prod_{i=1}^{m}(s + z_i)}{\prod_{j=1}^{n}(s + p_j)}$$

a set of rules can be developed for constructing the root locus easily. These rules are given next without proof.[3]

Rule 1. The system root loci have n branches originating at the n open-loop poles $-p_j, j = 1, \ldots, n$ with the value of $K = 0$.

Rule 2. Out of the n branches m number of branches will terminate on m finite zeros $-z_i, i = 1, 2, \ldots, m$ of the open-loop transfer function at $K = \infty$.

Rule 3. The remaining $n - m$ branches will go to ∞ along asymptotes as $K \to \infty$. The asymptotes are straight lines meeting at a point on the real line called the hub with specific orientation as given in Rule 4.

Rule 4. a. The asymptotes meet at the hub

$$\sigma = \frac{\sum_{j=1}^{n} \text{poles} - \sum_{i=1}^{m} \text{zeros}}{n - m}$$

$$= \frac{\sum_{j=1}^{n}(-p_j) - \sum_{i=1}^{m}(-z_i)}{n - m}$$

   b. The $n - m$ asymptote angles are given by

$$\theta_N = \pm \frac{180° N}{n - m}$$

where N takes on values $1, 3, 5, 7, \ldots$. For each N, two angles are computed and the procedure repeated until $n - m$ angles are obtained.

Rule 5. If to the right of a point on the real axis there lies an odd number of open-loop poles and zeros, then it is a point on the root loci.

Rule 6. If two open-loop poles or two open-loop zeros are connected, then there must be a break point between the two (Fig. 15.17).

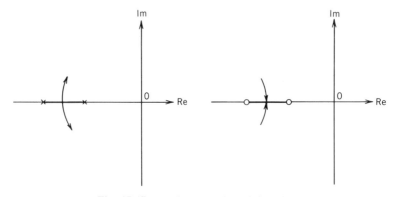

Fig. 15.17   Breakaway and break-in points.

If an open-loop pole $-p_l$ and an open-loop zero $-z_q$ are connected, in most cases it may be considered as a full branch of the root loci, that is, that the closed-loop pole corresponding to the open-loop pole $-p_l$ starts at $-p_l$ for $K = 0$ and reaches the closed-loop pole signified by the open-loop zero $-z_q$ as $K = \infty$.

Note. Exceptions to this rule exist. Some typical situations are depicted in Fig. 15.18a.

In order to determine the occurrence of such multiple break points, the next rule may be used.

Rule 7. The break points may be computed by determining points for which $dK/ds = 0$. Since

$$1 + KGH = 0$$

$$-K = \frac{1}{GH} = \frac{B(s)}{A(s)}$$

$$\frac{dK}{ds} = B(s)\frac{dA(s)}{ds} - A(s)\frac{dB(s)}{ds} = 0$$

Break points coupled with information from rule 6 makes it rather easy to pin down the branches.

Rule 8. The points at which the branches cross the imaginary axis can be determined by letting $s = j\omega$ in the characteristic equation.

Rule 9. The angle condition is made use of to determine the angle by which a branch would depart from a pole or would arrive at a zero as $K \to \infty$.
   A point s_0 is considered very near the pole (zero) and the angle $G(s_0)H(s_0)$ is computed. The fact that s_0 is very near the pole (zero) makes all but one angle fixed. Thus by employing the angle condition, the unknown angle of departure (arrival) can be computed.
   An example is given next to illustrate the various rules for constructing a root locus.

Example 15.3. Consider the closed-loop characteristic equation (CLCE)

$$1 + KG(s)H(s) = 1 + \frac{K(s + 6)}{s(s + 1)(s + 4)}$$

R1: $n = 3 \Rightarrow$ 3 branches originating at $0, -1, -4$ at $K = 0$.
R2: $m = 1 \Rightarrow$ 1 branch terminates at -6, at $K = \infty$.
R3: $n - m = 2$ branches approach ∞ along asymptotes.
R4: Hub $\sigma = (0 - 1 - 4 - (-6))/2 = 0.5$.

$$\text{Asymptote angles} \quad \pm\frac{180° \ N}{n - m} = \pm\frac{180° \ N}{2}$$

$$\text{Set } N = 1 \Rightarrow \pm 90°$$

R1–R4: Yield the sketch of Fig. 15.18b.
R5: Sections on the real line are 0 to -1 and -4 to -6.
R6: There must be a breakaway point between 0 and -1.
R7: Break points $dK/ds = 0$.

$$\Rightarrow (s + 6)(3s^2 + 10s + 4) - (s^3 + 5s^2 + 4s) = 0$$

$$s^3 + 11.5s^2 + 30s + 12 = 0$$

$$(s + 0.49)(s + 7.89)(s + 3.12) = 0$$

$s = -7.89$ and -3.12 are unacceptable from R5. Therefore the only breakaway point is at -0.49.
   A sketch of the root loci is given in Fig. 15.18c.

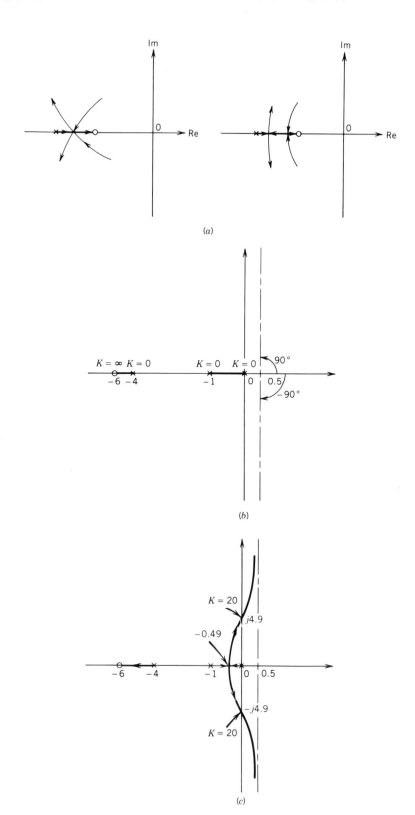

Fig. 15.18 (*a*) Breakaway and break-in possibilities; (*b*) sketch of root loci of Example 15.3 resulting from Rules R1-R4; (*c*) root loci for Example 15.3 where $G(s) = K(s+6)/s(s+1)(s+4)$.

R8: Imaginary axis crossings:

$$\text{CLCE} \qquad s^3 + 5s^2 + (4 + K)s + 6K = 0$$

Now let $s = j\omega$:

$$(j\omega)^3 + 5(j\omega)^2 + (4 + K)j\omega + 6K = 0$$

$$(6k - 5\omega^2) + j\omega\big[(4 + K) - \omega^2\big] = 0$$

Therefore

$$6K = 5\omega^2$$

$$\omega[4 + K - \omega^2] = 0$$

yielding

$$K = 0 \qquad \omega = 0$$

and

$$K = 20 \qquad \omega = \pm4.9$$

Example 15.4. Consider the unity negative feedback system shown in Fig. 15.19a. Obtain the loci of the closed-loop poles as α is varied from 0 to ∞, that is, obtain the root loci for $0 \le \alpha \le \infty$.

Solution. The root loci is the points s satisfying the closed-loop characteristic equation:

$$1 + \frac{750}{(s + 5)(s + 10)(s + \alpha)} = 0$$

In order to utilize the rules for constructing the root loci the CLCE is rearranged in the form of Eq. 15.8.
By expanding and rearranging we get

$$CLCE = s(s + 5)(s + 10) + \alpha(s + 5)(s + 10) + 750 = 0$$

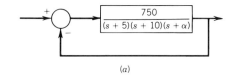

(a)

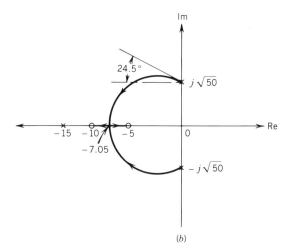

(b)

Fig. 15.19 (a) Closed-loop system of Example 15.4; (b) root loci of system of Example 15.4 where $G(s) = \alpha(s + 5)(s + 10)/(s + 15)(s + 50)$.

or

$$1 + \frac{\alpha(s + 5)(s + 10)}{(s + 15)(s^2 + 50)} = 0 \qquad (15.13)$$

Equation 15.13 has three poles at -15, $+j\sqrt{50}$, $-j\sqrt{50}$ and two finite zeros at -5, -10.

R1: $n = 3$ implies there are three branches originating at -15, $-j\sqrt{50}$, $+j\sqrt{50}$ at $K = 0$.
R2: $m = 2$ implies two of the three branches terminate on -5, -10 at $K = \infty$.
R3: $n - m = 1$ implies that there is one asymptote.
R4: For a single asymptote a hub does not exist

$$\text{asymptote angle} = \pm\frac{180° N}{n - m} = \pm\frac{180° N}{1}$$

that is, $\theta_1 = 180°$

R5: There is a section of the root loci between $-\infty$ and -15 and between -10 and -5.
R6: Since the two zeros -5 and -10 are connected, there must be a break-in point between -5 and -10. The section -15 to $-\infty$ forms a full branch.
R7: Break points: Since

$$G(s)H(s) = \frac{A(s)}{B(s)} = \frac{s^2 + 15s + 50}{s^3 + 15s^2 + 50s + 750}$$

$$\frac{d\alpha}{ds} = (s^2 + 15s + 50)(3s^2 + 30s + 50) - (s^3 + 15s^2 + 50s + 750)(2s + 15) = 0$$

that is,

$$s^4 + 30s^3 - 85s^2 - 8750 = 0 \qquad (15.14)$$

$s = -7.05$ is a break point.

Remark. To obtain the break points the fourth-order polynomial in s given in (15.14) must be factored. However, knowing that there must be a break point in the range -5 and -10 (Rule 6), it is quite easy to find the breakaway point. If all roots of (15.14) are found anyway, then only those points yielding $\alpha \leq 0$ are admissible as break points.

For this example R1–R6 give all the essential information to sketch the root loci of Fig. 15.19b. If the angles of departure are needed we can employ rule 9.

Consider the point s_0 closer to $+j\sqrt{50}$ and write down the angle condition:

$$\underline{/(s_0 + 5)} + \underline{/(s_0 + 10)} - \underline{/(s_0 + 15)} - \underline{/(s_0 + j\sqrt{50})} - \underline{/(s_0 - j\sqrt{50})} = \pi$$

Now let $s_0 \to j\sqrt{50}$ to yield

$$\underline{/(5 + j\sqrt{50})} + \underline{/(10 + j\sqrt{50})} - \underline{/(15 + j\sqrt{50} - j2\sqrt{50})} - \theta = \pi$$

$$\tan^{-1}\frac{\sqrt{50}}{5} + \tan^{-1}\frac{\sqrt{50}}{10} - \tan^{-1}\frac{\sqrt{50}}{15} - \frac{\pi}{2} - \theta = \pi$$

$$\theta = -3.57 \text{ rad} = -204.5°$$

Some typical root loci plots are shown in Table 15.1.

15.5.2 Time-Domain Design Using the Root Locus

Time domain performance specifications can often be related in an approximate sense to closed-loop pole locations. If suitable pole locations for a certain time domain performance can be effectively identified, then the root loci can be used to locate the closed-loop poles at those locations by appropriate compensation. Compensation can be provided by introducing additional dynamics into

TABLE 15.1 TYPICAL ROOT LOCI PLOTS

No.	$G_0(s)$	Root loci	No.	$G_0(s)$	Root loci
1	$\dfrac{1}{s - p_1}$	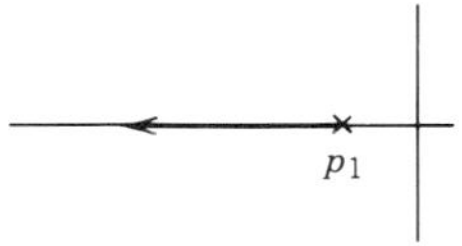	6	Same as 5 $p_2 < z_1 < p_1$	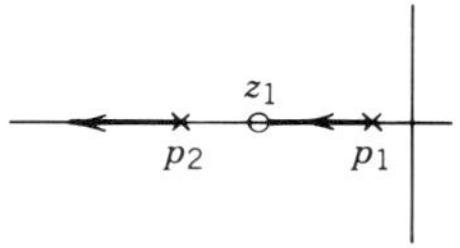
2	$\dfrac{s - z_1}{s - p_1}$	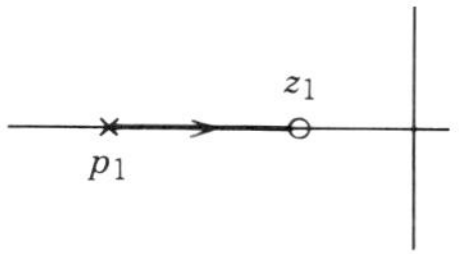	7	Same as 5 $p_1, p_2 < z_1$	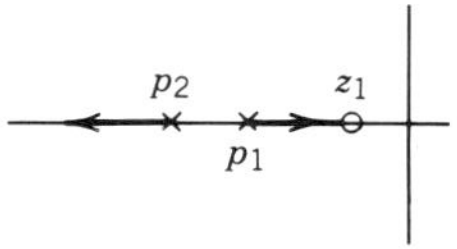
3	$\dfrac{1}{(s - p_1)(s - p_2)}$	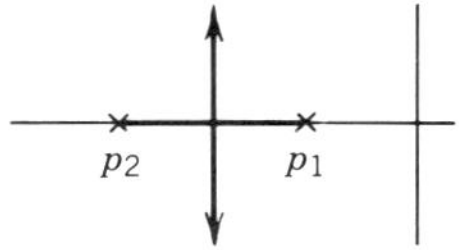	8	Same as 5 p_1, p_2 complex	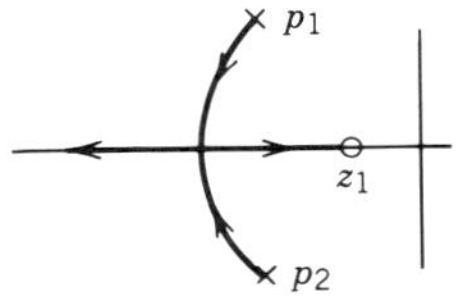
4	$\dfrac{1}{(s - p_1)(s - p_2)}$ p_1, p_2 complex	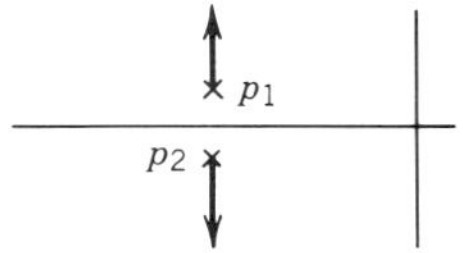	9	$\dfrac{1}{(s - p_1)(s - p_2)(s - p_3)}$	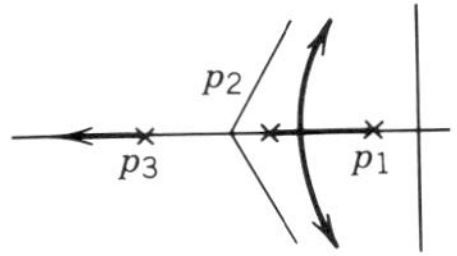
5	$\dfrac{(s - z_1)}{(s - p_1)(s - p_2)}$ $z_1 < p_1, p_2$	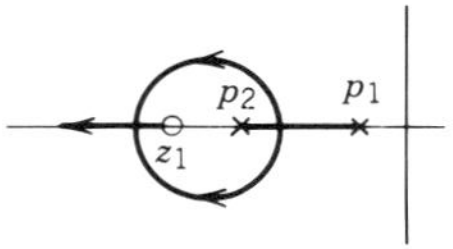	10	$\dfrac{1}{s(s - p_2)(s - p_3)}$	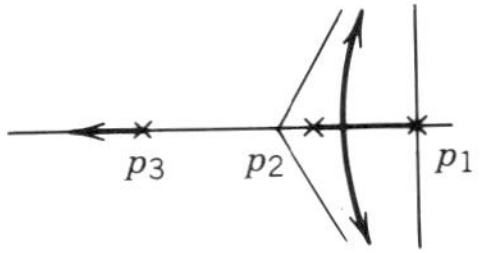

(Continued)

635

TABLE 15.1 TYPICAL ROOT LOCI PLOTS (*Continued*)

No.	$G_0(s)$	Root loci	No.	$G_0(s)$	Root loci
11	$\dfrac{1}{(s-p)^3}$		16	Same as 15	
12	$\dfrac{1}{(s-p_1)(s-p_2)(s-p_3)}$ p_2, p_3 complex		17	Same as 15	
13	Same as 12		18	Same as 15	
14	Same as 12		19	$\dfrac{(s-z_1)(s-z_2)}{(s-p_1)(s-p_2)(s-p_3)}$	
15	$\dfrac{(s-z_1)}{(s-p_1)(s-p_2)(s-p_3)}$		20	Same as 19	

21	$\dfrac{(s-z_1)(s-z_2)}{(s-p)^3}$ $z_2 < z_1 < p$	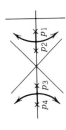	25	$\dfrac{1}{(s-p_1)(s-p_2)(s-p_3)(s-p_4)}$ All poles real	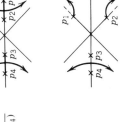
22	$\dfrac{(s-z_1)(s-z_2)}{s(s-p_2)(s-p_3)}$ $z_2 < p_3 < z_1 < p_2 < 0$	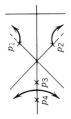	26	Same as 25 2 poles real 2 poles complex	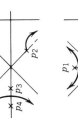
23	Same as 22	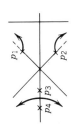	27	Same as 26	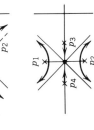
24	Same as 22	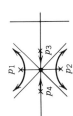	28	Same as 25 All poles complex	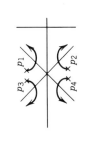

the feedback system in the form of increased poles and zeros (PID control, lead, lag, lead-lag, etc.). We shall now consider some examples to illustrate this time domain design philosophy.

Example 15.5. Obtain the root loci for a system with an open-loop transfer function

$$G(s) = \frac{K}{s(s + 2)}$$

  a. Indicate the location of closed-loop poles when $K = 4$ and determine the damping ratio ζ and the natural frequency ω_n corresponding to $K = 4$.

  b. It is now required to double the natural frequency while keeping the same damping ratio. Design a compensator for satisfying the new design specifications.

Solution. a. Suppose the closed-loop system is as shown in Fig. 15.20*a*. Then its root loci are as shown in Fig. 15.20*b*. To find the poles at $K = 4$, solve the closed-loop characteristic equation (CLCE)

$$1 + \frac{4}{s(s + 2)} = 0 \quad \text{or} \quad s^2 + 2s + 4 = 0$$

the roots are $s_{1,2} = -1 \pm j\sqrt{3}$:

$$\zeta = \cos\beta = 0.5$$

  b. Since the natural frequency is to be doubled keeping the same damping ratio, we need to move the closed-loop poles at A and A' in Fig. 15.21 so that $\overline{OB} = 2\overline{OA}$.

  In order to satisfy the design specifications, the modified root loci must be made to pass through B and B'. To reshape the original root loci, additional poles and zeros are required. We give below a simple way to appropriately modify the root loci. Conceptually, the modification takes place as shown in Fig 15.22.

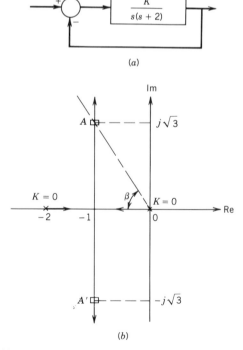

(a)

(b)

Fig. 15.20   (*a*) Time-domain design example; (*b*) sketch of root loci.

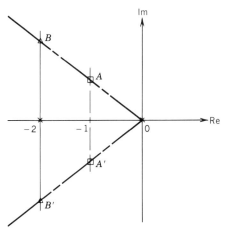

Fig. 15.21   Desired pole locations.

If $G_c(s)$ is chosen to cancel the pole at -2 by selecting

$$G_c(s) = \frac{s + 2}{s + p}$$

then we only need to find p so that the modified root locus passes through B. This is quite easy to do by noting that the pole-zero cancellation at -2 leaves us with a second-order system with the two open-loop poles at 0 and $-p$.

By selecting $p = +\ 4$ it is easy to verify that the modified root loci are as shown in Fig. 15.22. So the compensator

$$G_c(s) = \frac{s + 2}{s + 4}$$

will work.

Remark.   Pole-zero cancellation as was done here must be avoided if it lies in the right half plane. Since any real system model has parameter uncertainty, exact cancellation is almost impossible to achieve. When this is the case such an attempted cancellation will leave an uncompensated unstable

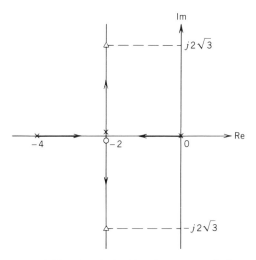

Fig. 15.22   Root loci with pole-zero cancellation.

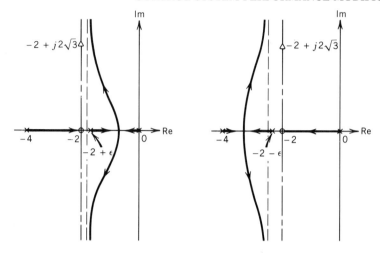

Fig. 15.23   Effect of nonexact pole-zero cancellation on the root loci.

mode in the closed-loop system. Even in the case of a stable approximate cancellation the dynamics can change. In order to see this consider in Example 15.5 the pole at -2 to be uncertain (say $-2 \pm \epsilon$, $\epsilon > 0$) and that a zero is exactly located at -2. Let us consider the two cases with the pole at $-2 + \epsilon$ and $-2 - \epsilon$ (Fig. 15.23).

We note that the modified root loci do not pass through B, B' when $\epsilon \neq 0$, implying that the time domain performance will be affected.

Example 15.6.   Consider the system shown in Fig. 15.24a where $K \geq 0$ and α and β are unknown constants. In order to identify K, α, and β, the following information about the system is provided:

1. All the poles of the open-loop transfer function are in the closed left half s plane.
2. When the closed-loop system is excited by the input $r(t) = t u_s(t)$, the trace of Fig. 15.24b is obtained ($\infty > e_1 > 0$).
3. When the gain K is doubled, the impulse response of Fig. 15.24c is observed.

Determine K, α, and β.

Solution.   Since the closed-loop system has a finite steady-state error $e_1 < \infty$ for a ramp input, the system should be type I. Thus we require either α or β to be zero. Let $\alpha = 0$. So it only remains to determine β and K.

Now the root loci for the system can be sketched in the following manner.

From the root locus it is clear that at a certain gain value the system goes unstable. From Fig. 15.24c we know that when the gain is doubled, the system has two closed-loop eigenvalues at A and A'. From the impulse response trace the frequency of oscillation is

$$\omega = \frac{2\pi}{\pi/10}$$

or

$$\omega = 20 \text{ rad/sec}$$

Thus we know that when the gain is doubled, there are two closed-loop poles at $\pm j20$. The corresponding CLCE therefore is

$$P(s) = (s + a)(s + j20)(s - j20) = 0$$

or

$$s^3 + as^2 + 400s + 400a = 0$$

We also know that

$$P(s) = 1 + \frac{2K}{s(s + \beta)(s + 40)} = 0$$

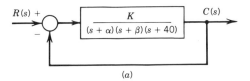

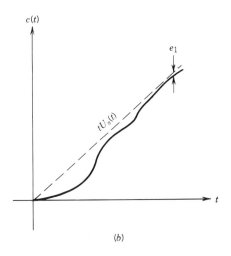

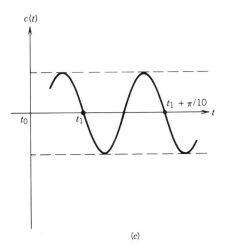

Fig. 15.24 (*a*) System of Example 15.5; (*b*) response due to a ramp; (*c*) response due to an impulse.

or

$$s^3 + (40 + \beta)s^2 + 40\beta s + 2K = 0$$

By matching coefficients

$$40 + \beta = a$$

$$40\beta = 400$$

$$2K = 400a$$

Therefore $\beta = 10$, $a = 50$, and $K = 10,000$. Hence the open-loop transfer function is

$$G(s) = \frac{10,000}{s(s + 10)(s + 40)}$$

15.5.3   Time-Domain Response versus s Domain Pole Locations

Given a transfer function $G(s)$ the pole locations can be found. These pole locations essentially describe the type of time response to be expected. The basic response can be effectively characterized by the impulse response $g(t)$ given by

$$g(t) = \mathcal{L}^{-1}[G(s)]$$

If

$$G(s) = \frac{K\prod_{i=1}^{m}(s + z_i)}{\prod_{j=1}^{n}(s + p_j)} \qquad n \geq m$$

then

$$g(t) = \sum_{j=1}^{n} a_j e^{-p_j t}$$

We note that any real pole contributes an exponential behavior into the time response and a complex conjugate pair contributes an exponential oscillation. A pure imaginary pair of poles leads to a sustained oscillation. Various components to be expected are shown in Fig. 15.25.

The role of zeros of the transfer function is to affect the relative weights a_j in the impulse response. For example, if a pole and a zero are close together, the net contribution to the overall response from such a pair will be negligible. If they cancel each other (say $- p_k$ by $- z_j$), then the coefficient associated with the term $e^{-p_k t}$, is zero. This idea can often be used to reduce the order of a dynamic system, that is, remove all pole-zero pairs close to one another. However, care should be exercised not to remove right half plane poles and zeros. (See Example 15.5 of Section 15.5.2.)

To note the effect of zero locations on the time response consider a second-order oscillatory system with a single zero, that is, consider the transfer function written in the normalized form

$$G(s) = \frac{(s/\alpha\zeta\omega_n) + 1}{(s/\omega_n)^2 + 2\zeta s/\omega_n + 1} = (s/\alpha\zeta\omega_n + 1)G_0(s)$$

The zero is located at $s = -\alpha\zeta\omega_n$, so if α is large, the zero is far removed from the poles and will have little effect on the response of $G_0(s)$. If $\alpha = 1$, the zero is at the value of the real part of the poles and could be expected to have a substantial influence on the response of $G_0(s)$. The step response curves for $\zeta = 0.5$ and for several values of α are plotted in Fig. 15.26. We see that the major effect of the zeros is to increase the overshoot M_p with very little influence on the settling time. A plot of M_p versus α is given in Fig. 15.27. If α is negative, then the zero is in the right half s plane. In this case an undershooting phenomenon as shown in Fig. 15.28 occurs.

In addition, it is useful to know the effect of an extra pole on the standard second-order response $G_0(s)$. In this case consider the transfer function

$$G(s) = \frac{1}{(s/\alpha\zeta\omega_n + 1)}G_0(s)$$

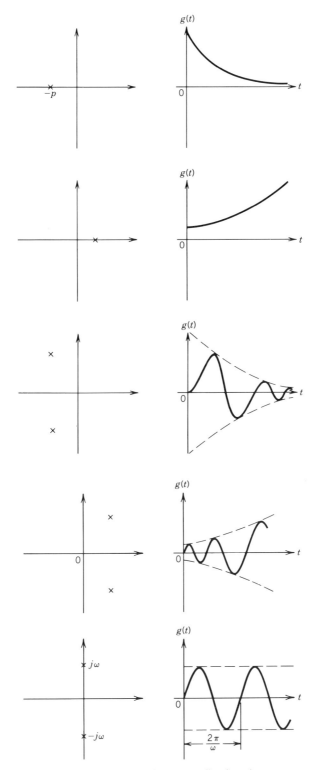

Fig. 15.25 Pole locations and corresponding impulse responses.

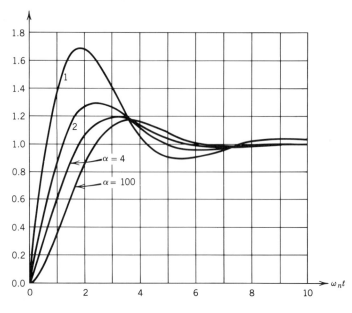

Fig. 15.26 Plots of the step response of a second-order system with an extra zero ($\zeta = 0.5$).

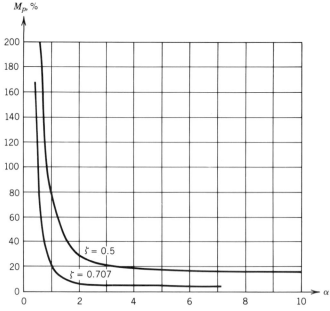

Fig. 15.27 Plot of overshoot M_p as a function of normalized zero location α. At $\alpha = 1$, the real part of the zero equals the real part of the pole.

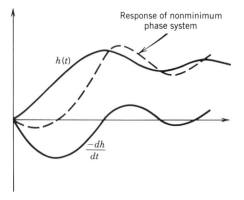

Fig. 15.28 Plot of the response of a second-order system with a right half-plane zero: a nonminimum phase system.

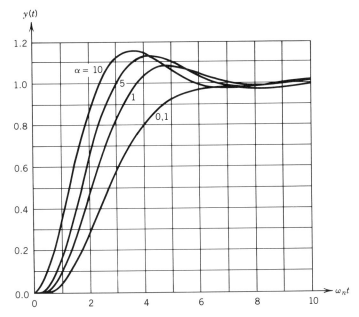

Fig. 15.29 Plot of step response for several third-order systems with $\zeta = 0.5$.

Plots of the step response for this case are shown in Fig. 15.29 for $\zeta = 0.5$ and for several values of α. In this case the major effect is to increase the rise time, shown in Fig. 15.30. For a detailed discussion of the effect of a zero and a pole location on a standard second-order response the reader may refer to Ref. 4.

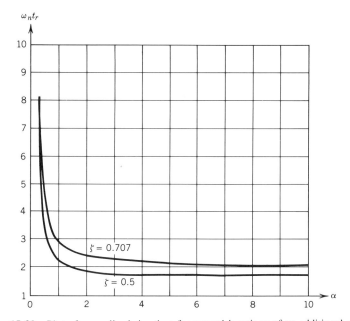

Fig. 15.30 Plot of normalized rise time for several locations of an additional pole.

15.6   POLE LOCATIONS IN THE z DOMAIN

For discrete-time systems the input-output relation is given by the pulse transfer function. A typical pulse transfer function $G(z)$ is of the form

$$G(z) = \frac{K\prod_{i=1}^{m}(z - \beta_i)}{\prod_{j=1}^{n}(z - p_j)} \qquad n \geq m$$

The poles of the pulse transfer function are p_j, $j = 1, \ldots, n$. As in the case of continuous time, the pole locations determine the stability properties of the system represented by its pulse transfer function. In the z domain poles have to lie inside the unit circle $|z| = 1$ for asymptotic stability. Thus the open left half s plane is equivalent to the interior of the unit circle in the z domain. The exterior of the unit circle (i.e., $|z| > 1$) represents the unstable region in the z domain.

15.6.1   Stability Analysis of Closed-Loop Systems in the z Domain

Consider a unity negative feedback system with the closed-loop pulse transfer function

$$\frac{C(z)}{R(z)} = \frac{G(z)}{1 + G(z)} \tag{15.15}$$

The stability of the system defined by Eq. 15.15, as well as of other types of discrete-time control system, may be determined from the locations of the closed-loop poles in the z plane or the roots of the closed-loop characteristic equation

$$P(z) = 1 + G(z) = 0$$

as follows:

1. For the system to be stable, the closed-loop poles or the roots of the characteristic equation must lie within the unit circle in the z domain. Any closed-loop pole outside the unit circle makes the system unstable.
2. If a simple pole lies at $z = 1$ or $z = -1$, then the system becomes marginally stable. Also, the system becomes marginally stable if a single pair of complex conjugate poles lie on the unit circle in the z domain. Any multiple closed-loop pole on the unit circle makes the system unstable.
3. Closed-loop zeros do not affect the absolute stability and therefore may be located anywhere in the z plane. Thus, a linear time-invariant single-input—single-output discrete-time closed-loop system becomes unstable if any of the closed-loop poles lies outside the unit circle or any multiple closed-loop pole lies on the unit circle in the z domain.

15.6.2   Performance Related to Proximity of Closed-Loop Poles to the Unit Circle

In the continuous-time case or in the s domain the transient performance of a system can be characterized by the s plane pole locations. Recall that the overshoot is related to the damping ratio ζ.

Damping Ratio ζ.   In the s plane a constant damping ratio may be represented by a radial line from the origin. A constant damping ratio locus (for $0 \leq \zeta \leq 1$) in the z plane is a logarithmic spiral. Figure 15.31 shows constant ζ loci in both the s plane and the z plane.

If all the poles in the s plane are specified as having a damping ratio not less than a specified value ζ_1, then the poles must lie to the left of the constant damping ratio line in the s plane (shaded region). In the z plane, the poles must lie in the region bounded by logarithmic spirals corresponding to $\zeta = \zeta_1$ (shaded region).

Damped Natural Frequency ω_d.   The rise time or the speed of response depends on the damped natural frequency ω_d and the damping ratio ζ of the dominant complex conjugate closed-loop poles. In the s plane the constant ω_d loci are horizontal lines, while in the z plane they are radial lines emanating from the origin.

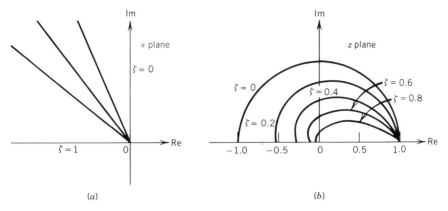

Fig. 15.31 (a) Constant ζ loci in the s plane; (b) constant ζ loci in the z plane.

Settling Time t_s. The settling time is determined by the value of attenuation σ of the dominant closed-loop poles $-\sigma \pm j\omega_d$. If the settling time is specified, it is possible to draw a line, $\sigma = -\sigma_1$, in the s plane corresponding to a given settling time. The region to the left of the line, $\sigma = -\sigma_1$, in the s plane corresponds to the interior of a circle with radius $e^{-\sigma_1 T}$ in the z plane as shown in Fig. 15.32.

Remark. To transform s-plane pole locations to the z domain, the transformation $z = e^{sT}$ where T is the sampling time is employed.

15.6.3 Root Locus in the z Domain

The root locus method for continuous-time systems can be extended to discrete-time systems without modifications, except that the stability boundary is changed from the $j\omega$ axis in the s plane to the unit circle in the z plane. The reason for being able to extend the root locus method is that the characteristic equation for the discrete-time system is of the same form as that for the root loci in the s plane. For the discrete-time case the closed-loop characteristic equation is

$$1 + G(z) = 0$$

Exactly the same rules as used for the continuous-time case in the s plane can be used for the discrete case too. (See Section 15.5 for root loci construction in the s plane.)

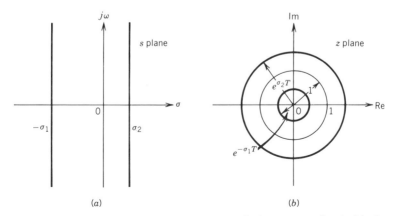

Fig. 15.32 (a) Constant attenuation lines in the s plane; (b) the corresponding loci in the z plane.

Example 15.7. Consider the closed-loop characteristic equation

$$1 + K \frac{(z + 1)(z - 0.5)}{(z - 1)(z - 0.9)(z + 0.6)} = 0$$

R1: The system has $n = 3$ poles indicating three branches. They start at 1, 0.9, and -0.6 with $K = 0$.

R2: There are two finite zeros ($m = 2$). Hence two of the branches terminate on the zeros -0.9 and 0.5 at $K = \infty$.

R3: One branch ($n - m = 1$) will go to ∞ along an asymptote.

R4: Sections of the root loci on the real line are between 0.9 and 1.0, -1.0 and $-\infty$, and -0.6 and 0.5; with this information the sketch shown in Fig. 15.33 can be easily obtained. If additional features are needed, the rest of the root loci construction rules can be applied without change.

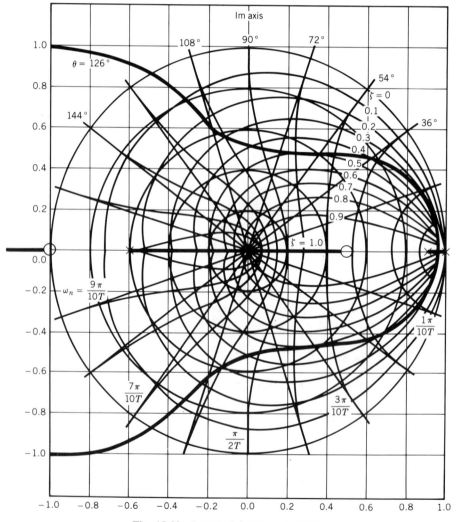

Fig. 15.33 Root loci for Example 15.7.

15.7   CONTROLLER DESIGN

In the previous Sections 14.1–15.6 some useful tools for designing single-input–single-output systems were given. Once the open-loop system is described, either by its set of poles and zeros or by its frequency response, the root locus method or frequency response method will indicate whether the feedback system can be given an acceptable transient response by adjustment of the loop gain. The steady-state accuracy can then be determined.

Often good transient performance and good steady-state performance cannot both be achieved simply by adjusting a single parameter such as a loop gain. When this is the case, it is necessary to modify the system dynamics. Either the dynamic properties of some components in the loop need to be altered or additional components need to be inserted into the loop. The process of modifying the system dynamics so as to allow the performance specifications to be met by subsequent loop gain adjustment is known as compensation.

Transient specifications are typically based on a step input response. By specifying the rise time, overshoot, and settling time, the response is confined to within the shaded region of Fig. 15.34. It is then assumed that a system whose step response satisfies these constraints will have an acceptable transient response to any kind of input.

In the frequency domain the bandwidth and the resonant peak of the closed-loop frequency response are measures roughly corresponding to rise time and overshoot, respectively. Specification of these parameters constrains the magnitude of the closed-loop frequency response to the region shown in Fig. 15.35. An alternative way of constraining the transient response by frequency domain criteria is to stipulate the smallest acceptable gain and phase margins.

Often used compensators are the so-called three-term controllers (PID), lag, and lead compensators. These controller or compensator designs are discussed later. They are often done on a trial-and-error basis and can be designed in the s domain or the z domain depending on the type of application. Continuous-time or s-domain compensators can often be converted to equivalent z-domain compensators by techniques such as pole-zero maps, hold equivalence, and Butterworth pole configurations.[11]

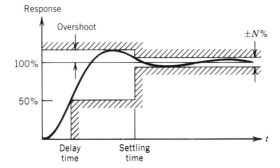

Fig. 15.34   Unit step response specifications.

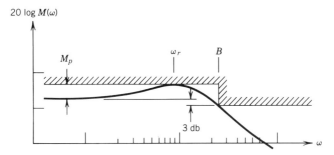

Fig. 15.35   Closed-loop frequency response specifications.

REFERENCES

1. E. O. Doebelin, *Control System Principles and Design*, Wiley, New York, 1985.
2. R. C. Dorf, *Modern Control Systems*, Addison-Wesley, Reading, MA, 1986.
3. K. Ogata, *Modern Control Engineering*, Prentice-Hall, Englewood Cliffs, NJ, 1970.
4. G. F. Franklin, J. D. Powell, and A. Emami-Naeini, *Feedback Control of Dynamic Systems*, Addison-Wesley, Reading, MA, 1986.
5. D. Graham and R. C. Lathrop, "The Synthesis of Optimum Response: Criteria and Standard Forms," *AIEE Transactions*, Part II, 72, pp. 273-288, 1953.
6. G. A. Korn and J. V. Wait, *Digital Continuous System Simulation*, Prentice-Hall, Englewood Cliffs, NJ, 1978.
7. O. W. Eshbach and M. Souders (Eds.), *Handbook of Engineering Fundamentals*, 3rd ed., Wiley, New York, 1975.
8. R. C. Rosenberg and D. C. Karnopp, *Introduction to Physical System Dynamics*, McGraw-Hill, New York, 1983.
9. M. Vidyasagar, *Nonlinear Systems Analysis*, Prentice Hall, Englewood Cliffs, NJ, 1978.
10. Y. Takahashi, M. J. Rabins, and D. M. Auslander, *Control and Dynamic Systems*, Addison-Wesley, Reading, MA, 1972.
11. G. F. Franklin and J. D. Powell, *Digital Control of Dynamic Systems*, Addison-Wesley, Reading, MA, 1980.
12. B. C. Kuo, *Automatic Control Systems*, Prentice-Hall, Englewood Cliffs, NJ, 1982.

BIBLIOGRAPHY

Bode, H. W., *Network Analysis and Feedback Amplifier Design*, Van Nostrand, New York, 1945.

Chestnut, H., and R. W. Mayer, *Servomechanisms and Regulating Systems Design*, 2nd ed., Vol. 1, Wiley, New York, 1959.

D'Azzo, J. J., and C. H. Houpis, *Linear Control System Analysis and Design*, McGraw-Hill, New York, 1988.

Distefano, J. J., III, A. R. Stubberud, and I. J. Williams, *Feedback and Control Systems* (Schaum's Outline Series), Schaum Publishing, New York, 1967.

Dransfield, P., *Engineering Systems and Automatic Control*, Prentice Hall, Englewood Cliffs, NJ, 1968.

Elgerd, O. I., *Control Systems Theory*, McGraw-Hill, New York, 1967.

Evans, W. R., *Control-System Dynamics*, McGraw-Hill, New York, 1954.

Eveleigh, V. W., *Introduction to Control Systems Design*, McGraw-Hill, New York, 1972.

Horowitz, I. M., *Synthesis of Feedback Systems*, Academic Press, New York, 1963.

Houpis, C. H., and G. B. Lamont, *Digital Control Systems: Theory, Hardware, Software*, McGraw-Hill, New York, 1985.

Kuo, B. C., *Digital Control Systems*, Holt, Rinehart and Winston, New York, 1980.

Melsa, J. L., and D. G. Schultz, *Linear Control Systems*, McGraw-Hill, New York, 1969.

Nyquist, H., "Regeneration Theory," *Bell System Tech. J.*, Vol. II, pp. 126–147, 1932.

Palm, N. J., III, *Modeling, Analysis and Control of Dynamic Systems*, Wiley, New York, 1983.

Phillips, C. L., and H. T. Nagle, Jr., *Digital Control System Analysis and Design*, Prentice-Hall, Englewood Cliffs, NJ, 1984.

Ragazzini, J. R., and G. F. Franklin, *Sampled Data Control Systems*, McGraw-Hill, New York, 1958.

Raven, F. H., *Automatic Control Engineering*, 4th ed., McGraw Hill, New York, 1987.

Shinners, S. M., *Modern Control Systems Theory and Application*, Addison Wesley, Reading, MA, 1972.

Truxal, J. G., *Automatic Feedback Control Synthesis*, McGraw-Hill, New York, 1955.

CHAPTER **16**

SERVOACTUATORS FOR CLOSED-LOOP CONTROL

KARL N. REID

Oklahoma State University
Stillwater, Oklahoma

SYED HAMID

Halliburton Services
Duncan, Oklahoma

16.1   INTRODUCTION

16.1.1   Definitions

A servoactuator is an open-loop system that controls the linear or rotary motion of a load in response to an input command (Fig. 16.1*a*). Feedback may be used with a servoactuator to produce a closed-loop system referred to as a servosystem (Fig. 16.1*b*). Servoactuators are normally "rate-type" systems, in that an input command results in an output velocity for steady-state operation. Position feedback must be used with the rate-type system to produce a servosystem for position control. If high-accuracy velocity control is required, velocity feedback may be used with the servoactuator. Or, if high-accuracy force (or torque) control is required, force (or torque) feedback may be used.

The term "servomotor" designates the various types of higher-level energy converters such as electrical and hydraulic motors. The servomotor provides the muscle function of the servoactuator. The "modulator" provides a conversion of the low-power input command (for the servoactuator) or the error signal (for the servosystem) to a high-power output that operates the servomotor. The "transducer" provides the feedback in the case of the servosystem. The input to the servoactuator or servosystem can be electronic, mechanical, hydraulic, or pneumatic. And depending on the energy conversion medium, servoactuators can be of the electromechanical, electrohydraulic, electropneumatic, or hydromechanical types.

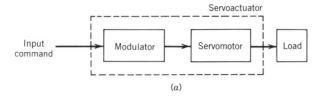

(a)

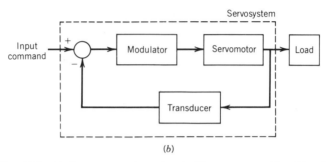

(b)

Fig. 16.1   (*a*) Servoactuator (open-loop); (*b*) servosystem (closed-loop).

16.1.2   Applications

Early development of servoactuators and servosystems was predominantly in electropneumatics (in the process control industry) (Ref. 1). With the advent of microprocessors and the development of high-coercive strength magnetic materials (such as Samarium Cobalt and Neodium), electromechanical servosystems find the largest applications in modern industry (Ref. 2). Table 16.1 describes the servo components (modulator, servomotor, and transducer) for the various implementations. Applications range from fairly simple open-loop systems such as the hydraulic controls on a backhoe, to complex feedback systems in robotics and aerospace vehicles. Figure 16.2 shows a typical linear output servosystem designed for use in a wide variety of motion control applications. A three-phase brushless motor is modulated by a pulse-width-modulated (PWM) controller (not shown in Fig. 16.2; see Ref. 3). A ballscrew is used to convert rotary motion to linear motion. Feedback is provided by a tachometer.

TABLE 16.1   SERVOSYSTEM COMPONENTS

System Type	Modulator	Servomotor	Transducer
Electromechanical	Amplifier	Servomotor AC, DC Linear/ Rotary	Position, velocity, torque
	Driver	Brushless servomotor	
	Translator/driver	Stepper motor	
Electrohydraulic	Servovalve	Hydraulic motor cylinder	Position, velocity force, pressure torque
Electropneumatic	Servovalve	Airmotor, cylinder	Position, velocity, force, pressure torque,

(a)

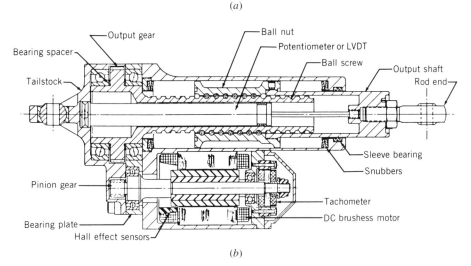

(b)

Fig. 16.2 (a) Electromechanical servosystem; (b) cross-section of an electromechanical servosystem. (Courtesy of Moog, Inc., East Aurora, NY)

16.1.3 Mathematical Models

Mathematical models of the various components of a servoactuator are needed for component selection to meet a given set of performance specifications. These specifications may consist of moving a given load through a given displacement or velocity profile in a specified time, or equivalently, following displacement or velocity commands generated by other subsystems or by an operator. The mathematical model of a component describes the steady-state and/or dynamic performance characteristics of that component. Mathematical models are presented in this chapter for the components typically used in high-performance servoactuators and servosystems. Examples are presented in this chapter to illustrate the use of mathematical models in the prediction of steady-state and dynamic performance of servoactuators and servosystems.

16.2   ELECTRICAL SERVOMOTORS

Electrical servomotors may be classified by the following characteristics:

1. Type of power (DC or AC)
2. Type of motion executed (continuous or discrete, rotary or translatory)
3. Type of commutation (mechanical or electronic)
4. Method of magnetic field generation (permanent magnet or electromagnetic)

Accordingly there are DC and AC servomotors of both the permanent magnet and field-wound types. Stepper motors belong to the discrete motion type. The rather uncommon linear motor executes translatory motion. Brushless DC motors are of the electronic commutation type. For the sake of simplicity, electrical servomotors are broadly classified here into four categories: DC and AC servomotors, stepper motors, and linear servomotors.

Electrical servomotors offer several advantages over their hydraulic and pneumatic counterparts. These advantages include: (a) compactness (facilitated by availability of high coercive strength magnetic materials such as Samarium Cobalt or Neodium), (b) low cost, (c) high reliability, (d) cleanliness, (e) ease of control function implementation, (f) portability due to operation at low DC voltage levels, and (g) large bandwidth due to high torque/inertia ratios.

16.3   DIRECT-CURRENT SERVOMOTORS

Direct-current (DC) servomotors offer certain advantages over alternating-current (AC) servomotors. These advantages are higher reliability, smaller size, and lower cost. Use of epoxy resins and improved brush designs combined with superior magnetic materials contribute to these advantages. Direct-current servomotors are compatible with thyristors (SCR) and transistor amplifiers, which facilitates control implementation. Typical DC servomotors range in power from fractional horse-power to several thousand horsepower. Conventional brushed DC motors theoretically can be used as servomotors. However, in lower horsepower levels (10 hp or less) they are not preferred.

16.3.1   Brushed DC Servomotors

In the DC servomotor, the interaction of two magnetic fields (either one or both generated electrically) results in mechanical motion of an armature. A typical permanent-magnet DC motor is illustrated in Fig. 16.3. The permanent magnet is sometimes replaced by a field winding to generate the

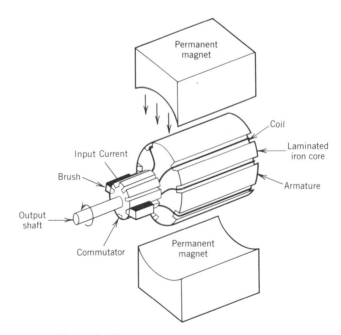

Fig. 16.3   Conventional permanent-magnet motor.

TABLE 16.2 DC SERVOMOTOR CLASSIFICATION

Motor Type	Configuration	Typical Steady-State Characteristics	Salient Features
Permanent magnet			No power required for field generation Runs cooler Torque-speed characteristics is linear Compactness
Straight series			Large starting torque
Split series			Allows quick reversing
Shunt			Low starting torque Finite speed at zero torque
Compound			High starting torque Complex circuitry required for reversing

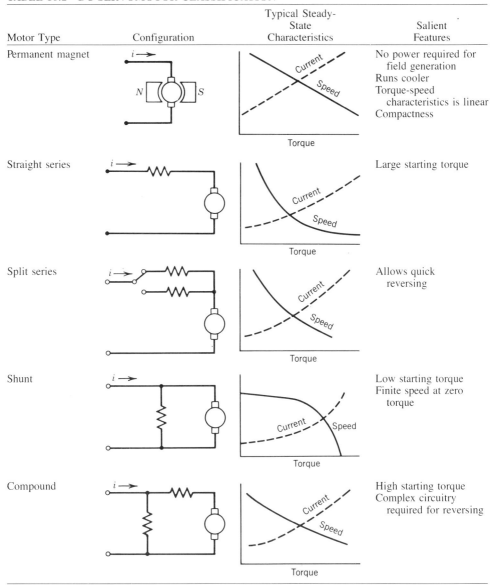

magnetic field. The field winding may be connected in three different ways to the armature winding: series, shunt, or compound. Table 16.2 summarizes the basic features of the various configurations along with the resultant performance characteristics. Table 16.3 shows typical upper limits of DC servomotor performance (Ref. 4).

The split-series field-wound motor has two windings, one for each direction of rotation. A manual switch is usually employed to activate the appropriate winding. The two windings of the compound motor are always excited and result in a high starting torque with good linearity. All of the field-wound motors are self-excited with the residual magnetism.

Permanent-Magnet Motors

Permanent-magnet (PM) motors are the most extensively used for servomotors because they generate less heat and have higher efficiency and more compactness than field-excited motors. There are three types of PM motors with mechanical commutation: (1) iron core, (2) surface wound, and (3)

TABLE 16.3   UPPER LIMITS OF DC SERVOMOTOR PERFORMANCE

Motor Type	Maximum Power (hp)	Maximum Speed (rpm)	Torque/Inertia Ratio (rad/s^2)	Maximum Bandwidth (rad/s)
Moving coil	0.5–1	4500–5500	200–250	1500
Printed circuit	7	3000–4000	130–220	1000
Permanent magnet	10–15	850–3000	15–30	100

Reproduced from Ref. 4.

moving coil. Figure 16.4 shows the construction of the three types. Details of the advantages and disadvantages of each type may be found in Ref. 5.

Mathematical Model of a Permanent-Magnet Servomotor

Comprehensive presentations on mathematical modeling of DC servomotors are given in Refs. 6, 7, 8, and 9. A simplified dynamic model is presented here.

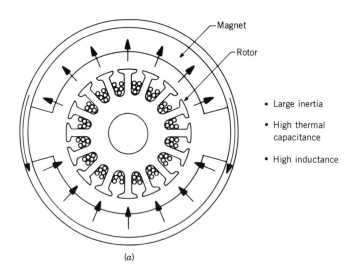

Magnet

Rotor

- Large inertia
- High thermal capacitance
- High inductance

(a)

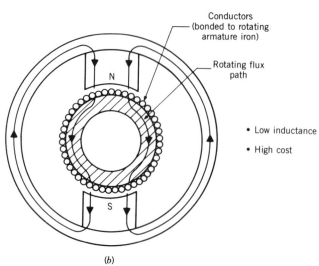

Conductors
(bonded to rotating armature iron)

Rotating flux path

- Low inductance
- High cost

(b)

Fig. 16.4 (a) Iron core; (b) surface wound.

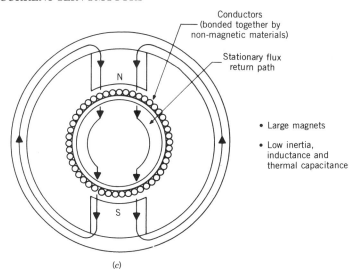

Conductors
(bonded together by
non-magnetic materials)

Stationary flux
return path

- Large magnets

- Low inertia,
  inductance and
  thermal capacitance

(c)

Fig. 16.4 (c) Moving coil (Taken from Reference 5).

The mathematical model of a permanent-magnet DC motor is obtained by lumping the inductance and resistance of the armature winding as shown in Fig. 16.5. The resulting equations are given.

Voltage Equations

$$v_a = L_a \frac{di}{dt} + R_a i + e_b \tag{16.1}$$

$$e_b = K_E \Omega_m \tag{16.2}$$

Torque Balance Equation

$$K_T i = J_m \frac{d\Omega_m}{dt} + B_m \Omega_m + T_{fm} + T_L \tag{16.3}$$

Taking Laplace transforms of Eqs. 16.1, 16.2, and 16.3 gives after algebraic manipulation.

$$\Omega_m(s) = G_1(s) V_a(s) - G_2(s) [T_{fm}(s) + T_L(s)] \tag{16.4}$$

where the transfer functions G_1 and G_2 are given by

$$G_1(s) = \frac{K_T}{R_a B_m (\tau_e s + 1)(\tau_m s + 1) + K_T K_E} \tag{16.5}$$

$$G_2(s) = \frac{R_a(\tau_e s + 1)}{R_a B_m (\tau_e s + 1)(\tau_m s + 1) + K_T K_E} \tag{16.6}$$

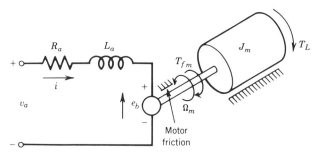

Fig. 16.5 Lumped parameter model of a permanent motor.

and the parameters are defined as follows:

B_m = Viscous damping in motor (N·m·s/rad)
e_b = Back EMF (V)
i = Current through armature (A)
J_m = Polar moment of inertia of armature (N·m·s^2/rad)
K_E = $z'\phi P/60$ = motor voltage constant or back EMF constant (V·s/rad)
K_T = $z'\phi P/2\pi$ = motor torque constant (N·m/A)
L_a = Armature inductance (H)
P = Number of poles
R_a = Armature resistance (ohms)
s = Laplace operator
t = Time (s)
T_{fm} = Coulomb friction torque in motor (N·m)
T_L = External load torque
v_a = Voltage applied to armature (V)
$V_a(s)$ = Laplace transform of armature voltage $v_a(t)$
z' = Number of conductors per parallel path in armature
θ_m = Angular position of motor shaft (rad)
ϕ = Magnetic flux per pole (Wb)
τ_e = L_a/R_a = electrical time constant (s)
τ_m = J_m/B_m = mechanical time constant (s)*
Ω_m = Angular velocity of motor (rad/s)
$\Omega_m(s)$ = Laplace transform of motor angular velocity

Equation 16.4 can be simplified if the armature inductance is small (making the electrical time constant τ_e negligible) and the Coulomb friction and load torque are assumed zero. The result is

$$\frac{\Omega_m(s)}{V_a(s)} = \frac{K_m}{\tau s + 1} \tag{16.7}$$

where

$$K_m = \frac{K_T}{R_a B_m + K_T K_E} \quad \text{(motor constant)} \tag{16.8}$$

and

$$\tau = \frac{R_a J_m}{R_a B_m + K_T K_E} \tag{16.9}$$

Reference 9 discusses cases where the electrical time constant cannot be neglected.

The preceding mathematical models assume a voltage input. For applications where a current amplifier is used, the following approximate model should be used:

$$\frac{\Omega_m(s)}{I(s)} = \frac{K'_m}{\tau_m s + 1} \tag{16.10}$$

$I(s)$ is the Laplace transform of the current input i. The motor constant in this case is

$$K'_m = \frac{K_T}{B_m} \tag{16.11}$$

---

* Some servomotor manufacturers define τ_m differently. For example, Electro-Craft (Ref. 9) defines the mechanical time constant as $\tau_m = (R_a J_m)/(K_T K_E)$.

In principle, the models developed can be applied to all of the DC motors of the various types with the appropriate input conditions. These models describe the open-loop response. For closed-loop systems with velocity or position feedback, an appropriate closed-loop transfer function can be derived easily by making use of the motor dynamic model. An example of a closed-loop system is given in Section 16.10.2.

Numerical Example

For a Motomatic PM servomotor model number E350-MG (Ref. 10), the following specifications are given:

$$
\begin{aligned}
K_T &= 3.4 \text{ in·oz/A } (0.024 \text{ N·m/A}) \\
K_E &= 2.5 \text{ V/krpm } (0.024 \text{ V·s/rad}) \\
R_a &= 12.4 \text{ ohms} \\
J_m &= 2.5 \times 10^{-4} \text{ in·oz·s}^2/\text{rad } (1.8 \times 10^{-6} \text{ N·m·s}^2/\text{rad}) \\
B_m &= 0.015 \text{ in·oz/krpm } (1.01 \times 10^{-6} \text{ N·m·s/rad}) \\
T_{fm} &= 0.5 \text{ in·oz } (3.5 \times 10^{-3} \text{ N·m}) \\
T_{\max} &= 2.5 \text{ in·oz } (1.8 \times 10^{-2} \text{ N·m}) \\
I_{\max} &= 0.75 \text{ A} \\
\Omega_{\max} &= 10{,}500 \text{ rpm at no load } (1{,}099 \text{ rad/s}) \\
L_a &= 3.1 \text{ mH} \\
\tau_e &= 0.25 \times 10^{-3} \text{ s} \\
R_{th} &= \text{thermal resistance} = 13°\text{C/W}
\end{aligned}
$$

The mechanical time constant can be computed as

$$\tau_m = \frac{J_m}{B_m} = 1.75 \text{ s} \tag{16.12}$$

Since $\tau_e \ll \tau_m$, Eq. 16.7 can be used to determine the dynamic response if the Coulomb friction and load torque are neglected. In this case, the time constant is

$$\tau = \frac{R_a J_m}{R_a B_m + K_T K_E} = 0.037 \text{ s} \tag{16.13}$$

and the motor constant is

$$K_m = \frac{K_T}{R_a B_m + K_T K_E} = 0.39 \text{ krpm/V} \quad (40.8 \text{ rad/V·s}) \tag{16.14}$$

The transfer function of Eq. 16.7 becomes

$$\frac{\Omega_m(s)}{V_a(s)} = \frac{40.8}{0.037s + 1} \tag{16.15}$$

For a step input of one volt, the motor speed is given by

$$\Omega_m(s) = \frac{1}{s}\left(\frac{40.8}{0.037s + 1}\right) \tag{16.16}$$

The inverse Laplace transform gives the step response as follows:

$$\Omega_m(t) = 40.8(1 - e^{-t/0.037}) \tag{16.17}$$

16.3.2 Brushless DC Servomotors

The development of brushless DC servomotors was an outgrowth of semiconductor devices even though the first patent was obtained with vacuum tube technology (Ref. 11). The basic construction of a brushless DC motor eliminates mechanical commutation. Instead, the commutation process

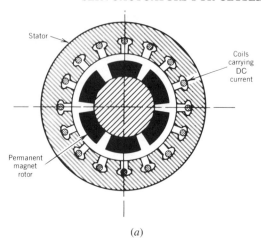

(a)

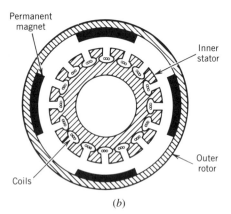

(b)

Fig. 16.6 Cross section of typical brushless DC motors; (a) inner rotor - outer stator type.; (b) inner stator - outer rotor type. (Taken from Reference 5.)

is accomplished electronically with no moving contacts. Hence, the problems associated with mechanical commutation such as brush wear particles, EMI, or arcing are eliminated. Elimination of arcing makes DC servomotors excellent candidates for applications requiring explosion-proof safety classification.

Construction

Typically, brushless motors have an inner rotor and outer stator and a configuration such as the one shown in Fig. 16.6a. However, the other configuration (i.e., inner stator and outer rotor) is also possible (see Fig. 16.6b). The former configuration with the outer stator carrying electrical windings provides excellent thermal dissipation characteristics, since both the iron and copper losses occur in the stator and the stator is better exposed to the ambient for convective heat transfer. This feature allows brushless motors to be operated at higher speeds and hence provides higher power-to-weight ratio.

Brushless DC motors range from 1 to 40 in. (0.025 to 1.02 m) in diameter with 6 in·oz (4.24 $\times$ 10^{-2} N·m) to 1,650 ft·lb (2237 N·m) of torque capability (see Table 16.4). Typical applications include memory disk drives, videotape recorders, and use as position servos in cryogenic compressors and fuel pumps.

TABLE 16.4   BRUSHLESS DC MOTOR PERFORMANCE DATA

Magnet Type	Power (W)	Peak Torque (in·oz)	Electrical Time Constant (s)	Mechanical Time Constant (s)	Torque/Inertia ratio (rad/s²)
Ceramic	25–900	10–600	0.0002–0.0016	0.0221–0.7400	413–11,400
Alnico	20–280	6–5,000	0.0001–0.0030	0.0065–0.1330	465–57,500
Rare earth	25–6000	10–316,000	0.0001–0.0140	0.0024–0.0291	137–100,000

The rotors are permanent magnets made from one of three primary materials: ceramic, ALNICO, and rare earth (such as Samarium Cobalt). Ceramic rotors are used in applications where cost consideration is important. Rare-earth magnets are the most expensive but provide exceptional performance. ALNICO magnets are of medium cost and provide medium magnetic strengths.

Operation

The brushless motor is operated by generating a rotating magnetic field that is 90° (electrical) out of phase with the rotor. Position sensors are used to determine the rotor position. These position sensors are of three types: photo transistor, electromagnetic, and Hall effect generators.

Commutation

An electronic module consisting of logic circuits and power amplification circuits is used to drive the motor (Refs. 3, 4, 12, and 13). This module receives rotor position information from the position sensors. The angle through which the rotor turns during the firing of a winding is called the "conduction angle." Figure 16.7 shows schematically a two-phase brushless motor with the driver electronics.

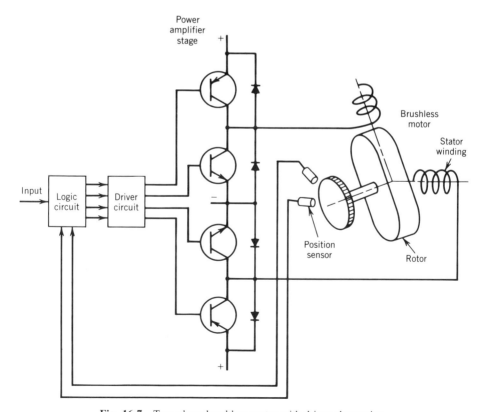

Fig. 16.7   Two-phase brushless motor with driver electronics.

Three-phase, three-step, half-wave motor controller

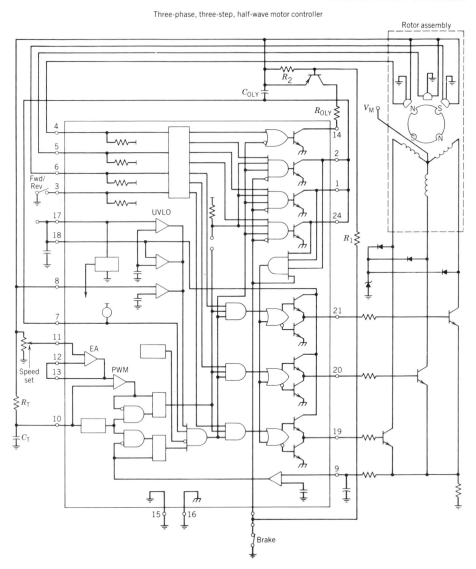

Fig. 16.8 Three-phase brushless motor controller circuit (taken from *Machine Design*, June 9, 1988, p. 140).

Figure 16.8 shows the controller circuit for a three-phase brushless motor. Each phase requires a pair of switches for commutation. Since the cost of the motor is dependent on the number of switches, there is a tendency to keep the number of phases to a minimum. Typically, three-phase motors with six switches are used. The current through the windings may be varied in a sinusoidal or a square-wave manner. The latter excitation results in a small torque ripple (17% average to peak for a two-phase motor, and 7% for a three-phase motor).

Ideally, a sinusoidal torque function results in a constant torque. But sinusoidal torque function generation is technically difficult and uneconomical. An alternate approach is to design the spatial variation of the magnetic field (possible by means of high coercive strength magnets) to obtain a trapezoidal torque function while the input current has a square waveform (easily generated by simple transistor control circuitry such as shown in Fig. 16.8). The motor torque is then approximately constant and is proportional to the maximum value of current during each cycle. The trapezoidal torque generation scheme also results in higher efficiency.

The locations of the position sensors relative to the rotor are aligned to result in appropriate timing for proper commutation. When properly commutated, a brushless motor duplicates the torque-speed characteristics of a brush-type DC motor.

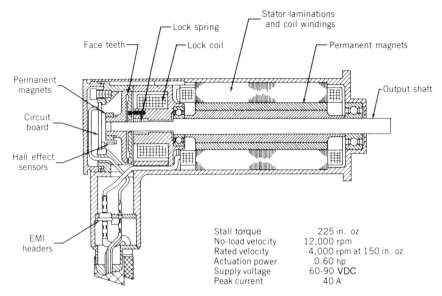

Stall torque   225 in. oz
No-load velocity   12,000 rpm
Rated velocity   4,000 rpm at 150 in. oz
Actuation power   0.60 hp
Supply voltage   60-90 VDC
Peak current   40 A

Fig. 16.9 Cross section of a brushless DC motor (courtesy of Moog, Inc., East Aurora, NY).

The power output of the brushless motor is effectively controlled by pulse-width modulation (PWM) or pulse-frequency modulation (PFM) methods. A linear (i.e., class A) power amplifier can also be used for power control. However, use of this type of amplifier produces back-EMF conduction during the zero-voltage portions of the voltage modulation, and thereby increases viscous damping. This effect can be eliminated by using a current amplifier rather than a voltage amplifier (see Ref. 9).

Figure 16.9 shows a cross section of a brushless motor developed for use as a fin actuator. Hall effect sensors are used for position measurement.

Mathematical Model

The mathematical model required to represent a brushless DC motor is identical to that of a brush-type DC motor. Therefore the equations given in Section 16.3.1 are applicable.

Numerical Example

Table 16.5 shows the specifications for a ceramic magnet, inside rotation-type DC brushless motor manufactured by Magnetic Technology (Refs. 14 and 15).

TABLE 16.5 PERFORMANCE DATA FOR MAGNETIC TECHNOLOGY MODEL 2800-153-084 BRUSHLESS DC MOTOR

Peak torque $=$ 40 in·oz
Power at peak torque $=$ 175 W
Electrical time constant $=$ 0.0005 s
Mechanical time constant $=$ 0.054 s
Damping factor $=$ 0.064 in·oz·s/rad
Moment of inertia $=$ 0.0035 in·oz·s^2/rad
Total breakaway torque $=$ 1.5 in·oz
Temperature rise $=$ 2°C/Watt
Maximum allowable winding temperature $=$ 155°C
Weight $=$ 19 oz
Number of poles $=$ 8
Number of phases $=$ 2
Resistance $=$ 8.4 ohms
Inductance $=$ 4.2 mH
Voltage at peak torque $=$ 38.2 V
Current at peak torque $=$ 4.57 A
Torque constant $=$ 8.7 in·oz/A
Voltage constant $=$ 0.0617 V·s/rad

If this motor is operating at peak torque, the steady-state motor speed is given by (from Eq. 16.4, dropping the dynamic terms):

$$\Omega_m = \frac{K_T v_a}{R_a B_m + K_T K_E} = 309 \text{ rad/s} = 2,953 \text{ rpm} \tag{16.18}$$

and the temperature rise is

$$\Delta T = (2°\text{C/W})\,(175 \text{ W}) = 350°\text{C} \tag{16.19}$$

This temperature rise is greater than the 155°C maximum allowable winding temperature. Thus the motor cannot be operated at peak torque indefinitely. The ambient temperature should be added to the temperature rise to arrive at the operating temperature of the winding.

16.4 ALTERNATING-CURRENT SERVOMOTORS

Alternating-current servomotors are used in applications requiring smooth speed control. These motors find widespread use in stationary industrial applications due to the ready availability of AC power (Refs. 16 and 17). In a majority of these applications, two-phase AC servomotors are used because of simplicity of the associated controls. Figure 16.10 shows a schematic of a two-phase AC servomotor. The operation of the two-phase AC servomotor is similar to an induction motor except the voltages applied to the two windings (fixed phase and control phase) are generally unequal and out of phase. The AC voltage applied to the fixed phase is held constant, and the one applied to the control phase is varied to control the motor speed. The two phases are generally 90 degrees out of phase. Changing the phase angle from +90 to -90 degrees reverses the direction of rotation of the motor. The appropriate phase angle is achieved through capacitors or other phase-shift circuits.

16.4.1 Types of AC Servomotors

Depending on the rotor construction, AC servomotors are classified into three types: (1) squirrel cage, (2) solid iron, and (3) drag cup (see Fig. 16.11). The squirrel-cage construction of the rotor is exactly the same as that of a standard induction motor. The inherent disadvantage of the squirrel-cage construction is cogging, or non-uniform armature rotation, which is minimized by skewing the bars of the cage relative to the rotor axis. The solid-iron rotor eliminates cogging. However, this configuration has a low torque-to-inertia ratio. The torque-to-inertia ratio is improved by the use of drag-cup construction (see Fig. 16.11c). The efficiency of AC servomotors is fairly low (5–20%), which necessitates external cooling.

16.4.2 Mathematical Model

Steady-State Model

Assuming linearity (i.e., operation without magnetic saturation) and using the method of symmetrical components (Ref. 18), mathematical models can be developed for the AC servomotor.

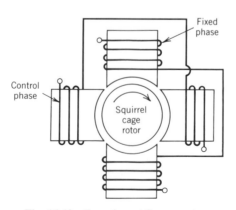

Fig. 16.10 Two-phase AC servomotor.

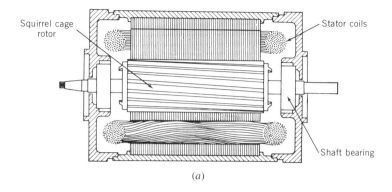

(a)

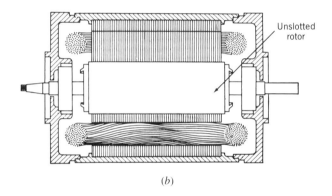

(b)

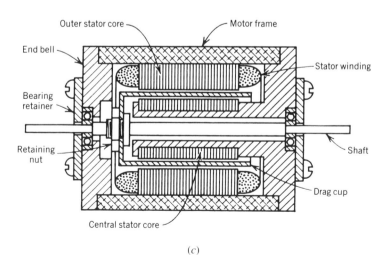

(c)

Fig. 16.11 Types of AC servomotors (taken from John E. Gibson and Franz B. Tuteur, *Control System Components*, McGraw-Hill, pp. 279-280); (a) squirrel-cage rotor; (b) solid-iron rotor; (c) drag-cup rotor.

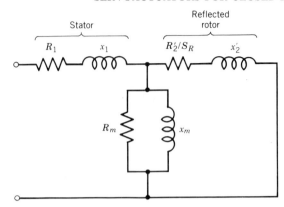

Fig. 16.12   Equivalent circuit diagram of one phase of an AC servomotor.

Figure 16.12 shows an equivalent circuit of one phase of the AC servomotor. The torque developed by a two-phase servomotor in which the control phase lags the reference phase by an angle θ is given by

$$T_m = \frac{1}{4}\left[F_1(1 + 2k\sin\theta + k^2) - F_2(1 - 2k\sin\theta + k^2)\right]\left\|v_m\right\|^2 \qquad (16.20)$$

The functions F_1 and F_2 are defined as follows:

$$F_1 = \frac{2}{\Omega_s}\left|\frac{Z_m}{Z_1(Z_2 + Z_m) + Z_2 Z_m}\right|^2 \frac{R_2'}{S_R} \qquad (16.21)$$

$$F_2 = \frac{2}{\Omega_s}\left|\frac{Z_m}{Z_1(Z_2' + Z_m) + Z_2' Z_m}\right|^2 \frac{R_2'}{2 - S_R} \qquad (16.22)$$

where

f = Frequency of AC voltages (Hz)

k = v_c/v_m

n_1 = Number of turns in winding of one pole of the stator

n_2 = Number of turns of rotor winding ($n_2 = 1$ for squirrel-cage type rotor)

R_1 = Resistance of stator (ohms)

R_2 = Resistance of rotor (ohms)

R_2' = $R_2(n_1/n_2)^2$

R_m = Resistance due to magnetic field (ohms)

S_R = Slip ratio = $(\Omega_s - \Omega_m)/\Omega_s$

v_c = Voltage applied to control phase (V)

v_m = Voltage applied to fixed phase (V)

X_1 = Inductive reactance of stator

X_2 = Inductive reactance of rotor

X_2' = $X_2(n_1/n_2)^2$

X_m = Inductive reactance due to magnetic field

Z_1 = Stator impedance = $R_1 + jX_1$

Z_2 = Reflected impedance of rotor at S_R = $(R_2'/S_R) + jX_2'$

Z_2' = Reflected impedance of rotor at $(2 - S_R)$ = $R_2'/(2 - S_R) + jX_2'$

Z_m = Impedance due to magnetic field generation in the stator = $\left[(1/R_m) + (1/jX_m)\right]^{-1}$

θ = Phase angle between fixed and control voltages

Ω_m = Motor speed (rad/s)

Ω_s = Synchronous speed (rad/s)

Figure 16.13 shows typical torque-speed characteristics for a two-phase AC motor. The characteristics are clearly nonlinear. However, about the origin they may be treated as linear.

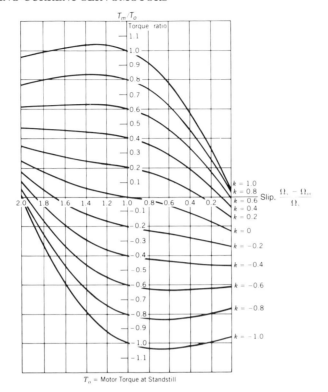

Fig. 16.13 Torque-speed characteristics of a two-phase AC servomotor (taken from John E. Gibson and Franz B. Tueter, *Control System Components*, McGraw-Hill, p. 288).

Dynamic Model

Combining the equation of motion, the electrical lag (due to stator inductance and resistance), and the torque-speed characteristic of Eq. 16.20, yields a nonlinear dynamic model. For an approximate static analysis, Eq. 16.20 may be linearized about an operating point. For many servosystem applications the most interesting and useful operating point is $k = 0$ and $S_R = 1.0$. This linearization yields

$$\Delta\Omega_m = A\Delta v_c - \frac{1}{B}\Delta T_m \tag{16.23}$$

where

$$A = \frac{\partial\Omega_m}{\partial v_c} \tag{16.24}$$

$$\frac{1}{B} = -\frac{\partial\Omega_m}{\partial T_m} \tag{16.25}$$

The symbol Δ indicates small variations from the steady-state operating point. Including the electrical and mechanical dynamics gives the transfer function

$$\Delta\Omega_m(s) = \frac{A\Delta V_c(s) - (1/B)\Delta T_m(s)}{(\tau_e s + 1)(\tau_m s + 1)} \tag{16.26}$$

where
B_m = Damping coefficient of motor (N·m·s)
J_m = Polar moment of inertia of rotor (N·m·s²/rad)
L_1 = Inductance of stator (H)
R_1 = Resistance of stator (ohms)
s = Laplace variable

ΔT_m = Change in load torque reflected to motor shaft
$\Delta T_m(s)$ = Laplace transform of change in load torque reflected to motor shaft
$\tau_e = L_1/R_1$ = Electrical time constant (s)
$\tau_m = J_m / B_m$ = Mechanical time constant (s)
$\Delta\Omega_m(s)$ = Laplace transform of change in motor speed

Computation of the constants A and B of Eq. 16.26 from the torque expression of Eq. 16.20 is rather tedious and requires measurements of the rotor impedances over a speed range of $-\Omega_s$ to $+\Omega_s$. As an alternative, an approximate expression for the torque has been developed (see Ref. 18) as follows:

$$T_m = \frac{S_R}{C_1 + C_2 S_R^2} \tag{16.27}$$

where C_1 and C_2 are constants which are determined from two points on an experimentally measured torque-speed characteristic. The constants A and B in Eq. 16.23 are related to C_1 and C_2 as follows:

$$A = \frac{2\Omega_s}{v_m} \frac{(C_1 + C_2)}{(C_1 - C_2)} \sin\theta \tag{16.28}$$

$$B = \frac{1}{2\Omega_s} \frac{(C_1 - C_2)}{(C_1 + C_2)^2} \tag{16.29}$$

Numerical Example

Specifications for a typical 2-phase AC servomotor are given as follows:

Number of poles $= P = 4$
Stator resistance $= R_1 = 10$ ohms
Stator inductance $= L_1 = 3$ mH
Moment of inertia of rotor $= J_m = 5.4 \times 10^{-5}$ in·oz·s²/rad (3.8×10^{-7} N·m·s²/rad)
Locked torque $= 9.5$ in·oz (6.71×10^{-2} N·m)
Torque at 1000 rpm $= 5.4$ in·oz (3.8×10^{-2} N·m)
Voltages: $v_c = 110$ V at 90°; $v_m = 110$ V at 0°
Frequency $= 60$ Hz
Synchronous speed $\Omega_s = \dfrac{120f}{P} = 1800$ rpm (188.3 rad/s)
Slip ratio at standstill $= S_{R1} = 1$
Slip ratio at 1000 rpm $= S_{R2} = \dfrac{\Omega_s - \Omega_m}{\Omega_s} = 0.44$

Substituting $T = 9.5$ in·oz at $S_R = 1$ in Eq. 16.27 gives

$$\frac{1}{C_1 + C_2} = 9.5 \tag{16.30}$$

Similarly, substituting $T = 5.4$ in·oz at $S_R = 0.44$ gives

$$\frac{0.44}{C_1 + C_2 (0.44^2)} = 5.4 \tag{16.31}$$

Solving Eqs. 16.30 and 16.31 gives

$$C_1 = 0.076 \ 1/\text{in·oz} \ (11 \ 1/\text{N·m})$$

$$C_2 = 0.029 \ 1/\text{in·oz} \ (4.1 \ 1/\text{N·m})$$

Substituting numerical values in Eqs. 16.28 and 16.29 gives

$$A = 7.79 \ \text{rad/V·s}$$

$$B = 1.59 \times 10^{-2} \ \text{in·oz·s/rad} (1.12 \times 10^{-4} \ \text{N·m·s/rad})$$

From the definitions for the motor time constants

$$\tau_e = 3 \times 10^{-4} \text{ s}$$

$$\tau_m = 4.6 \times 10^{-2} \text{ s}$$

From Eq. 16.26, the motor transfer function is given as

$$\Delta\Omega_m(s) = \frac{7.79\,\Delta V_c(s) - 62.9\,\Delta T_m(s)}{(3 \times 10^{-4}\,s + 1)\,(4.6 \times 10^{-2}\,s + 1)} \qquad (16.32)$$

Generally the mechanical time constant is much greater than the electrical time constant, as is the case for this example. The term $(\tau_e s + 1)$ in the transfer function of Eq. 16.26 often may be neglected without introducing significant error.

16.5 STEPPER MOTORS

Stepper motors (also called step motors) represent a significant breakthrough in the area of electromechanical actuation. These are incremental motors that by their very nature are compatible with digital systems. A stepper motor converts an electrical pulse into an equivalent rotary displacement. Since their introduction in the early 1930s, stepper motors have evolved into sophisticated designs (Refs. 19, 20, 21, and 22). The stepper motor possesses some inherent advantages over a conventional servomotor: (1) it is compatible with digital processors; (2) open-loop control is possible, which eliminates stability problems associated with closed-loop servos; (3) the step error is noncumulative; and (4) the brushless design provides easy maintenance and ruggedness. As a result of these advantages, stepper motors find widespread usage in industrial applications such as drives for TV antennas, NC machines, computer drives, hydraulic valve positioning, and other high-performance feedback control systems. The primary disadvantage of stepper motors is their low efficiency, which restricts their use to fractional horsepower applications.

16.5.1 Operation

The principle of operation of a stepper motor is illustrated by the single-stack, variable-reluctance, three-phase motor shown in Fig. 16.14. The stator has six poles that are wound in a three-phase configuration. The soft-iron rotor has four poles. When phase 1 is powered by a DC voltage, one pair of rotor poles will line up with the phase 1 stator poles. When phase 1 is switched off and phase 2 is turned on, the rotor will turn clockwise through 30° until one pair of teeth align with the phase 2 poles at C. Similarly, if phase 2 is switched off and phase 3 is turned on, the rotor will rotate another 30° in the clockwise direction. Thus, with each switching the rotor advances through one step of 30°. So the "step angle" is 30°. (The most commonly used step angle is 1.8°.) The direction of rotation may be reversed by switching in a 1.3.2.1.3.2 . . . sequence.

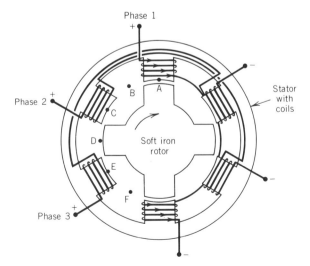

Fig. 16.14   Single-stack variable reluctance three-phase stepper motor.

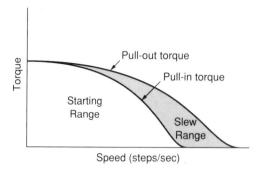

Fig. 16.15 Typical relation between pull-in torque and pull-out torque of a stepper motor (taken from B. C. Kuo, *Incremental Motion Control-Step Motors and Control Systems*, Vol. II, Champaign, IL, SRL Publishing, p 101).

In the preceding scheme only one phase winding is switched on at a time. This scheme is termed "one-phase-on" switching. An alternate switching scheme is called "two-phase-on" switching. In two-phase-on switching, two windings are turned on simultaneously. Referring to Fig. 16.14, if the switching sequence is 12.23.31.12 . . . , the rotor will rotate in a clockwise direction one step of 30° at a time. So, the two-phase-on scheme does not alter the step size for a three-phase stepper motor such as shown in Fig. 16.14. However, for a four-phase variable-reluctance (VR) motor, a switching sequence such as 1.12.2.23.3.34.4.41.1 . . . results in reducing the step size in half. This sequence is called "half stepping." Half stepping results in about 41% more torque (for a four-phase motor) compared to the single-stepping scheme. By varying the relative magnitude of the voltages applied to the two windings, the rotor can be made to rotate in fractional increments of a step. This scheme is termed "microstepping."

When the motor is energized, the holding torque of a stepper motor theoretically varies with the rotor position as a sinusoidal. Figure 16.15 shows a typical torque-speed characteristic for a stepper motor. The torque above which the stepper motor definitely loses steps is termed the "pull-out torque." The torque below which a stepper does not lose any steps (in an open-loop mode) is termed the "pull-in torque." The speed range between the pull-in and pull-out torques is called the "slew range." The slew range represents an unstable region of operation.

16.5.2 Types of Stepper Motors

In the early years of stepper motor development, stepper design included mechanical detenting and solenoid controls (Ref. 19). However, these designs have been replaced by more rugged and efficient designs. The latter designs may be broadly classified as (1) permanent-magnet steppers, (2) variable-reluctance steppers, (3) hybrid steppers, (4) electromechanical steppers, and (5) electrohydraulic steppers. Figure 16.16 shows configurations of each of these various types of stepper motors. The most commonly used are the permanent-magnet and the three-phase and four-phase variable-reluctance types of stepper motors.

The stator windings of variable-reluctance stepper motors sometimes are wound with two windings of opposite polarity per pole. This approach is termed a "bifilar winding." Figure 16.17 shows a bifilar-wound stepper motor. This arrangement provides more torque and improves damping, but the disadvantage is more complex circuitry.

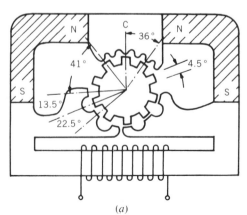

(*a*)

Fig. 16.16 Stepper motor configurations (taken from B. C. Kuo, *Incremental Motion Control-Step Motors and Control Systems*, Vol. II, SRL Publishing Co., Champaign, IL, pp. 12-14); (*a*) permanent stator stepper motor.

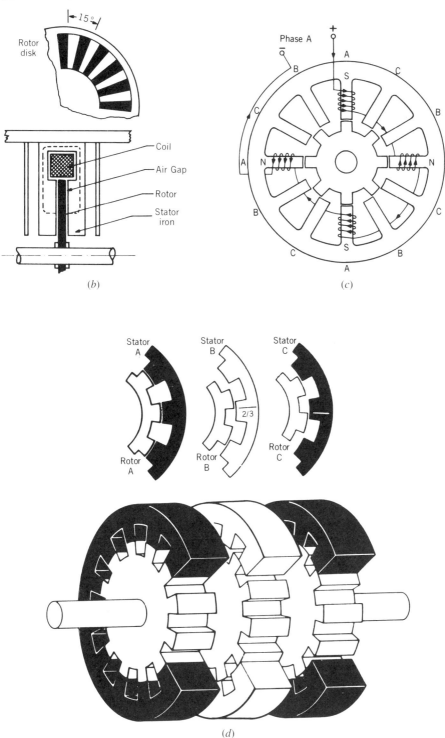

Fig. 16.16 (b) Single-stack, variable reluctance, axial-gap stepper motor; (c) single-stack, variable reluctance, radial-gap stepper motor; (d) multi-stack, variable reluctance, radial-gap stepper motor.

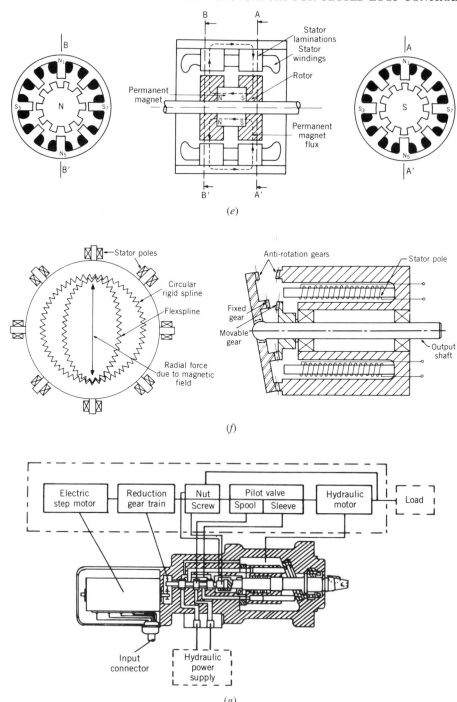

Fig. 16.16 (*e*) Hybrid stepper motor; (*f*) electromechanical stepper motor; (*g*) electrohydraulic stepper motor.

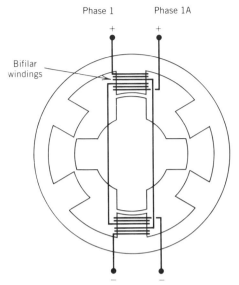

Fig. 16.17 Bipolar-wound stepper motor.

16.5.3 Mathematical Model of a Permanent-Magnet Stepper Motor

The mathematical model of a stepper is generally much more complex than a conventional DC motor since the voltages applied to the various phases change in a discontinuous fashion. These discontinuities in the applied voltages result directly in corresponding discontinuities in the phase currents. This effect is further complicated by the spatial variation of the magnetic reluctance. Reference 20 gives detailed mathematical models for permanent-magnet and variable-reluctance stepper motors. Computer codes (in FORTRAN IV) are available in Ref. 20 for these stepper motors.

The mathematical models of stepper motors are inherently nonlinear due to discontinuities in input voltages and due to the transcendental spatial variation of the self and mutual inductances. Hence these models do not lend themselves to a frequency-domain analysis.

16.5.4 Numerical Example

Table 16.6 gives the specifications of a Crouzet Model No. 82 940.0 Stepper Motor. The motor is to be used to drive a rotary viscometer that has a rotary inertia of 3.88×10^{-3} in·oz·s^2/rad (2.74×10^{-5} N·m·s^2/rad), a constant frictional torque of 1.3 in·oz (9.18×10^{-3} N·m), and a viscous damping coefficient of 0.96 in·oz·s/rad (6.8×10^{-3} N·m·s/rad). The motor is required to accelerate the viscometer from 5.2 to 13.1 rad/s in a maximum of 0.1 s.

The maximum torque developed by the motor may be estimated as follows:

$$T_m = (J_m + J_L) \frac{d\Omega_m}{dt} + B_L \Omega_m + T_f \tag{16.33}$$

where

B_L = Rotary damping coefficient of viscometer cup
J_L = Polar moment of inertia of viscometer cup
J_m = Polar moment of inertia of motor
T_f = Friction torque on viscometer cup
Ω_m = Motor shaft speed

From the specifications and Eq. 16.33: $d\Omega_m/dt = 78.5$ rad/s^2 and $T_m = 14.2$ in·oz (0.10 N·m). From the torque-speed characteristics of Table 16.6 it can be seen that the characteristic labeled b can be used. So a series resistance of 9 ohms should be used to meet the torque requirements.

The electrical and mechanical time constants may be estimated as follows:

$$\tau_e \approx \frac{L}{R} = 2.7 \times 10^{-3} \text{ s}$$

$$\tau_m \approx \frac{J_m + J_L}{B_L} = 5.3 \times 10^{-3} \text{ s}$$

TABLE 16.6   STEPPER MOTOR SPECIFICATIONS (CROUZET MODEL 82940.0)

Step Angle = 7.5°
Number of phases = 2 (bipolar)
Resistance per phase = 9 ohms
Inductance per phase = 24 mH
Current per phase = 0.55 A
Maximum input power = 10 W
Maximum voltage = 6.7 V (continuous duty)
Holding torque at 6V = 21.9 in·oz
Detent torque = 1.67 in·oz
Rotor inertia = 1.19×10^{-3} in·oz·s^2/rad
Maximum coil temperature = 248°F

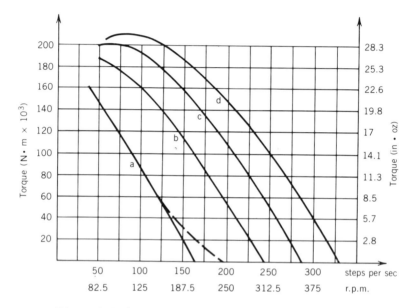

a = Using constant voltage drive with Rs (series resistance) = 0 (L/R)
b = Using constant voltage drive with Rs (series resistance) = R *Motor* ($L/2R$)
c = Using constant voltage drive with Rs (series resistance) = $2R$ Motor ($L/3R$)
d = Using constant voltage drive with Rs (series resistance) = $3R$ Motor ($L/4R$)

Since the time allowed for the acceleration of the load is 0.1 s, which is more than an order of magnitude greater than either time constant, the motor should have adequate dynamic response.

For a more exact dynamic response determination, the motor parameters may be determined experimentally (as described in Chapter 6 of Ref. 20) and used in the dynamic models.

16.6   ELECTRICAL MODULATORS

This section describes electrical modulators used for the various types of servomotors described in Sections 16.3 through 16.5. The term "modulator," as defined earlier, designates components employed for conversion of the command signal to appropriate means to modulate the power flow to the servomotor (see Fig. 16.1). Modulators for the various types of electrical servomotors differ significantly.

16.6.1   Direct-Current Motor Modulators

Modulators used in servoactuator applications with DC motors usually contain two stages of amplification: a first-stage voltage amplifier followed by a second-stage power amplifier. Voltage amplifiers

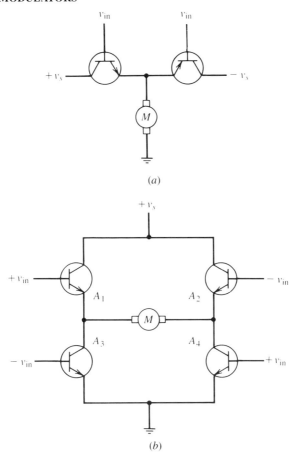

Fig. 16.18 Electrical modulator configurations; (*a*) type T; (*b*) type H.

are generally quite linear in performance. Power amplifiers are used in two different configurations, namely Type T and Type H as shown in Fig. 16.18. The Type T configuration employs two power sources and only two power transistors. The Type H configuration uses only one power source but four transistors. The Type T configuration lends itself readily to current-feedback schemes. However, the Type H is more commonly used because of a single power source requirement. Type H configurations can be operated in two different modes, resulting in bipolar and unipolar drives.

Linear Amplifiers

Linear amplifiers that are DC amplifiers used in the output stage (as H or T configuration) provide gains typically in the range of 2 to 10. At power levels above 200 Watts, these amplifiers require external cooling (e.g., fans) to overcome excessive heat generation. This problem is particularly severe in high-performance servo applications that typically have low-impedance rotors operating at low speed and high torque conditions. Sometimes this problem is overcome by incorporating a dual-mode current-limiting device which permits high currents for short durations (for overcoming inertia) and then imposes a lower limit for longer durations.

Linear amplifiers are generally configured as voltage amplifiers, but in some applications a current-source configuration is employed. Reference 5 discusses the details of voltage- and current-source configurations, the associated mathematical models, and the influence of the two configurations on the dynamic response of the motor.

Switching amplifiers overcome heat generation problems by switching the voltage on and off at high frequencies. This switching limits the maximum allowable inductance and results in shorter time constants and increased bandwidth. One drawback is that the RFI noise level generated is much higher than with linear amplifiers. There are predominantly two types of switching used in servoactuator applications: (1) pulse-width modulation (PWM) and (2) frequency modulation (FM). PWM amplifiers are most commonly used.

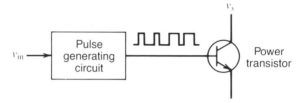

Fig. 16.19   PWM amplifier configuration.

PWM Amplifiers

In pulse-width-modulation amplifiers, the voltage applied to the servomotor is varied by changing the pulse width of a high, constant-frequency (typically 10 kHz) train of pulses. Figure 16.19 shows a schematic of a PWM amplifier. Figure 16.20a shows a photograph of a typical PWM amplifier and Fig. 16.20b shows the associated circuit diagram. Figure 16.21 shows the firing sequence for unipolar and bipolar configurations.

(a)

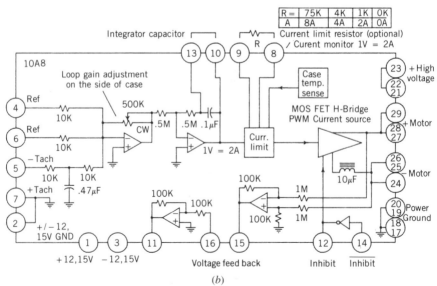

(b)

Fig. 16.20   PWM servo amplifier IC (courtesy of Advanced Motion Controls, Van Nuys, CA); (a) PWM servo amplifier photograph; (b) PWM servo amplifier circuit diagram.

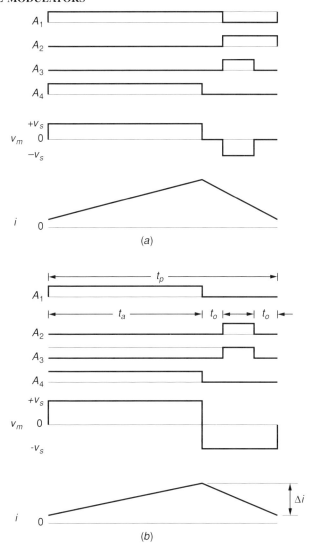

Fig. 16.21 Phases, motor voltage and current PWM amplifier; (*a*) Unipolar drive; (*b*) bipolar drive.

Mathematical Model—Bipolar Drive PWM Amplifier

The average voltage v_{ma} applied over one cycle of the PWM signal is given by (see Ref. 9, Section 4.5.3)

$$v_{ma} = \frac{2v_s t_a}{t_p} - v_s \qquad (16.34)$$

and the change in current over one cycle is given by

$$\Delta i = \frac{2v_s t_a}{L}\left(1 - \frac{t_a}{t_p}\right) \qquad (16.35)$$

where

i = Current
L = Inductance of armature winding

t_p = Time period of PWM signal (see Fig. 16.21)

t_a = Actuation time $= (t_p/2) - t_o - t_p v_{in}/v_c$ (see Fig. 16.21)

t_o = Time delay (see Fig. 16.21)

v_c = Peak-to-peak amplitude of a triangular wave voltage signal of time period t_p

v_{in} = Input voltage signal applied to amplifier

v_s = Supply voltage

Equations 16.34 and 16.35 are valid for operating conditions where the instantaneous motor current i remains positive throughout a cycle. The actuation time t_a is determined from the pulse generation circuit characteristics in conjunction with the input signal (see Ref. 9).

Mathematical Model—Unipolar Drive PWM Amplifier

The average voltage and current amplitudes for a PWM amplifier in a unipolar drive mode are given by

$$v_{ma} = \frac{t_a v_s}{t_p} \tag{16.36}$$

and

$$\Delta i = \frac{v_s t_a}{L}\left(1 - \frac{t_s}{t_p}\right) \tag{16.37}$$

For operating conditions where the instantaneous motor current becomes zero for a part of the cycle, the mathematical models can be found in Refs. 5, 20, 23, and 24.

16.6.2 Stepper Motor Modulators

Modulators for stepper motors also are called "drives." There are three types: oscillator-translators, indexers, and microstepping controls. Oscillators provide a variable-frequency pulse train to the translator. The function of the translator is to convert the pulse train into appropriate signals to operate the power transistors, thereby directing power to the stepper motor. The variable-frequency (3,000 to 15,000 pulses per second) capability of the oscillators provides for accurate manual control of speed. Similarly, a pulse generator in conjunction with a translator provides manual position control. Figure 16.22 shows a typical oscillator-translator package. Most commercial units also provide for half-stepping and electronic damping.

Fig. 16.22 Oscillator-translator modulator for stepper motor control (Courtesy of Superior Electric Co., Bristol, CT); (*a*) photograph of modulator.

Fig. 16.22 (*b*) Photograph of circuit card.

Indexers are generally programmable microprocessors that perform the functions of the oscillator-translator package, in addition to providing features such as numerical processing, programming, and communications (RS232 and RS274) with computers. The software controls provide such features as setting upper and lower limits of stepper motion and controlling slewing rate. Figure 16.23 shows a typical indexer. Indexers generally operate on 20 to 100 VDC power.

Stepper motors suffer from mechanical resonances (generally in the range of 50 to 250 steps per minute) due to the excitation resulting from the square shape of the current pulses. Microstepping eliminates this problem by energizing multiple windings simultaneously and by controlling the currents. Hence, modulators for microstepping essentially control the simultaneous currents to the various phases achieving as high as 50,000 microsteps per revolution. Switching frequencies range from 20 to 200 kHz.

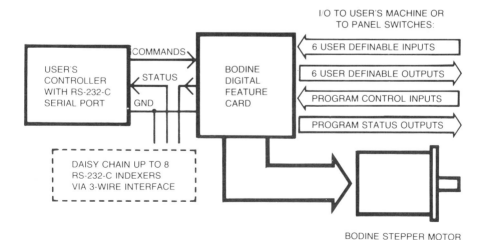

Fig. 16.23 Linear-motion hydraulic servomotors (actuators); Indexer-type modulator for stepper motor control (courtesy of Bodine Electric Co., Chicago, IL).

16.7 HYDRAULIC SERVOMOTORS

Hydraulic servomotors or "actuators" convert hydraulic power into mechanical motion. This mechanical motion can be either linear or rotary. Because hydraulic servomotors provide large actuating force or torque capabilities, they are commonly used in heavy equipment industries. Also, due to their high mechanical stiffness, fast dynamic response, and high power-to-weight ratios, hydraulic servomotors find widespread use in aircraft, missiles, and space vehicles, as well as critical industrial systems where high performance and high power control are needed (Ref. 25).

16.7.1 Linear-Motion Servomotors

Linear-motion servomotors or actuators provide a translatory motion of a load along a straight line. The motion of the load can be controlled by modulating the flow of hydraulic fluid into or out of the servomotor. There are two basic types of linear-motion servomotor designs: unbalanced and balanced designs as shown schematically in Fig. 16.24. The unbalanced design is shorter, while the balanced design has equal extension and return rates when the servomotor is driven by a symmetrical modulator (control valve). High-performance hydraulic and electrohydraulic servosystems require servomotors with low leakage and low friction. Correspondingly, seals are critical elements in servomotor design. Construction details and mounting arrangements are discussed in Refs. 26, 27, and 28.

Since servomotors must move heavy loads quickly and accurately, they should act as very stiff structural members. In a well-designed servomotor, the fluid columns on either side of the piston are the most compliant portions of the structure. The spring rate of one fluid column is given by $K = A\beta/L$, where A is the net column area, L is the column length, and β is the fluid bulk modulus of elasticity. Servomotors used for high dynamic performance systems are designed to have the minimum stroke and the shortest permissible connecting passages in order to minimize the volume of fluid under compression. Modulators often are integrated into the servomotor body to minimize lengths of connecting passages (see Section 16.9.2).

16.7.2 Rotary-Motion Servomotors

Rotary-motion servomotors are functionally more versatile than their linear counterparts. They are commonly employed even in linear-motion applications through a rack-and-pinion type kinematic conversion. However, linear-motion servomotors also are employed for rotary applications through corresponding kinematic conversion. Rotary actuators can be either simple reversible hydraulic motors of continuous-rotation type or high-performance type with limited angular displacement. Most rotary servomotors used in medium- and high-performance hydraulic and electrohydraulic servosystems are vane (continuous or limited rotation) or piston type (continuous rotation) servomotors.

Rotary-vane servomotors are commonly used in industrial applications where medium to high

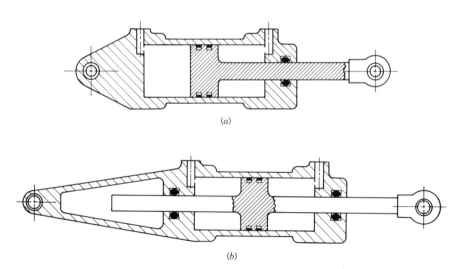

(a)

(b)

Fig. 16.24   Linear-motion hydraulic servomotors (actuators); (a) double-acting unbalanced actuator; (b) double-acting balanced actuator.

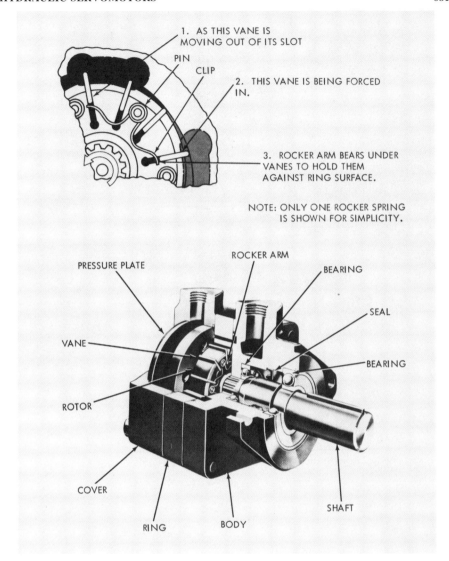

Fig. 16.25   Balanced vane-type hydraulic servomotor (courtesy of Vickers, Inc., Troy, MI).

performance is required and size and weight are not at a high premium. Such applications include earth moving equipment, agricultural machinery, and materials processing and handling equipment. Figure 16.25 shows a typical continuous-rotation rotary-vane servomotor capable of operating at speeds up to 4000 rpm (419 rad/s) and pressures up to 2500 psi (1.72×10^7 N·m^2).

In aircraft, missile, and spacecraft applications where high performance and small size and weight are required, piston-type servomotors are commonly used. Figure 16.26 shows a typical "in-line" piston-type servomotor and Fig. 16.27 shows a typical "bent-axis" piston-type servomotor. Such servomotors are capable of operating at speeds up to 8000 rpm (838 rad/s) and pressures up to 5000 psi (3.44×10^7 N/m^2).

16.7.3   Mathematical Models

An hydraulic servomotor (either linear or rotary-motion type) is a rate-type device. That is, a given flow rate into the servomotor results in a certain velocity (or speed for a rotary servomotor).

Figure 16.28 shows schematics of linear and rotary servomotors with definitions of the variables.

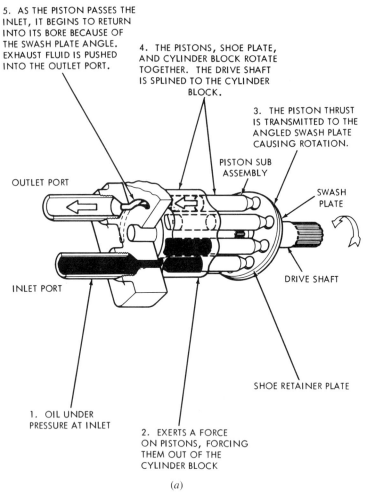

5. AS THE PISTON PASSES THE INLET, IT BEGINS TO RETURN INTO ITS BORE BECAUSE OF THE SWASH PLATE ANGLE. EXHAUST FLUID IS PUSHED INTO THE OUTLET PORT.

4. THE PISTONS, SHOE PLATE, AND CYLINDER BLOCK ROTATE TOGETHER. THE DRIVE SHAFT IS SPLINED TO THE CYLINDER BLOCK.

3. THE PISTON THRUST IS TRANSMITTED TO THE ANGLED SWASH PLATE CAUSING ROTATION.

PISTON SUB ASSEMBLY

OUTLET PORT

SWASH PLATE

DRIVE SHAFT

INLET PORT

SHOE RETAINER PLATE

1. OIL UNDER PRESSURE AT INLET

2. EXERTS A FORCE ON PISTONS, FORCING THEM OUT OF THE CYLINDER BLOCK

(a)

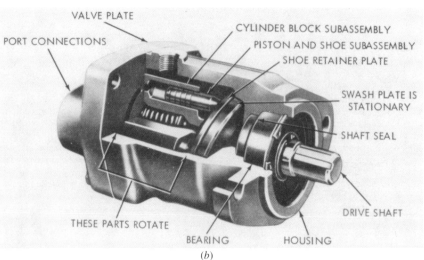

VALVE PLATE

PORT CONNECTIONS

CYLINDER BLOCK SUBASSEMBLY

PISTON AND SHOE SUBASSEMBLY

SHOE RETAINER PLATE

SWASH PLATE IS STATIONARY

SHAFT SEAL

DRIVE SHAFT

THESE PARTS ROTATE

BEARING   HOUSING

(b)

Fig. 16.26 In-line piston servomotor (courtesy of Vickers, Inc., Troy, MI); (a) operation of servomotor; (b) cutaway of servomotor.

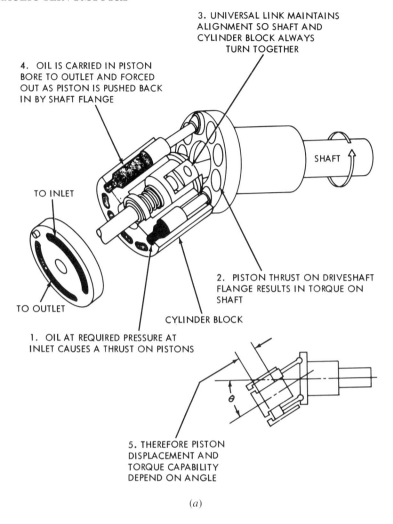

3. UNIVERSAL LINK MAINTAINS
ALIGNMENT SO SHAFT AND
CYLINDER BLOCK ALWAYS
TURN TOGETHER

4. OIL IS CARRIED IN PISTON
BORE TO OUTLET AND FORCED
OUT AS PISTON IS PUSHED BACK
IN BY SHAFT FLANGE

SHAFT

TO INLET

TO OUTLET

2. PISTON THRUST ON DRIVESHAFT
FLANGE RESULTS IN TORQUE ON
SHAFT

CYLINDER BLOCK

1. OIL AT REQUIRED PRESSURE AT
INLET CAUSES A THRUST ON PISTONS

θ

5. THEREFORE PISTON
DISPLACEMENT AND
TORQUE CAPABILITY
DEPEND ON ANGLE

(*a*)

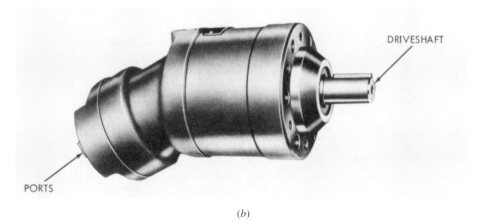

DRIVESHAFT

PORTS

(*b*)

Fig. 16.27 Bent-axis piston servomotor (courtesy of Vickers, Inc., Troy, MI); (*a*) operation of servomotor; (*b*) photograph of servomotor.

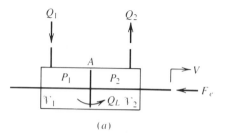

(a)

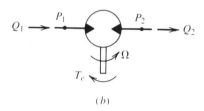

(b)

Fig. 16.28 Nomenclature of hydraulic servomotors.

Dynamic models are given in the following descriptions. It is assumed that all structural components are rigid and that external leakage to the environment is negligible. Models that include external leakage effects are presented in Ref. 29.

Linear Servomotor (Balanced Design)

Force balance

$$\frac{dV}{dt} = \frac{1}{M_a} (AP_m - B_m V - F_f - F_c - F_e) \tag{16.38}$$

Chamber continuity

$$\rho_1 Q_1 - \frac{(\rho_1 + \rho_2)}{2} Q_L = \frac{d}{dt} (\rho_1 \mathcal{V}_1) \tag{16.39}$$

$$\frac{\rho_1 + \rho_2}{2} Q_L - \rho_2 Q_2 = \frac{d}{dt} (\rho_2 \mathcal{V}_2) \tag{16.40}$$

Equation of state

$$\frac{d\rho_1}{dt} = \frac{\rho_1}{\beta} \frac{dP_1}{dt} \tag{16.41}$$

$$\frac{d\rho_2}{dt} = \frac{\rho_2}{\beta} \frac{dP_2}{dt} \tag{16.42}$$

Chamber volume

$$\mathcal{V}_1 = \mathcal{V}_{1i} + AY \tag{16.43}$$

or

$$\frac{d\mathcal{V}_1}{dt} = AV$$

$$\mathcal{V}_2 = \mathcal{V}_{2i} - AY \tag{16.44}$$

or

$$\frac{d\mathcal{V}_2}{dt} = -AV$$

where

A = Effective area of servomotor piston
B_m = Viscous damping coefficient in servomotor
C_2 = Leakage coefficient
F_c = Force due to Coulomb friction in servomotor
F_e = External load force on servomotor piston
F_f = Force due to stiction in servomotor
M_a = Mass of moving parts in servomotor
P_1 = Pressure at inlet
P_2 = Pressure at outlet
P_m = $P_1 - P_2$ = Pressure drop across servomotor
Q_1 = Flow rate into servomotor
Q_2 = Flow rate out of servomotor
Q_L = $C_2 P_m$ = Internal leakage flow rate
V = Velocity of servomotor piston
$\mathcal{V}_1$ = Fluid volume under compression at the inlet
$\mathcal{V}_{1i}$ = Fluid volume under compression at inlet for $Y = 0$
$\mathcal{V}_2$ = Fluid volume under compression at the outlet
$\mathcal{V}_{2i}$ = Fluid volume under compression at outlet for $Y = 0$
Y = Linear displacement of servomotor piston from neutral position
β = Fluid bulk modulus of elasticity
ρ_1 = Fluid mass density at inlet
ρ_2 = Fluid mass density at exit

Rotary Servomotor

Torque balance

$$\frac{d\Omega}{dt} = \frac{1}{J_a} (D_m P_m - B_r \Omega - T_f - T_c - T_e) \tag{16.45}$$

Chamber continuity

$$\rho_1 Q_1 - \frac{(\rho_1 + \rho_2)}{2} Q_L = \frac{d}{dt} (\rho_1 \mathcal{V}_1) \tag{16.46}$$

$$\frac{(\rho_1 + \rho_2)}{2} Q_L - \rho_2 Q_2 = \frac{d}{dt} (\rho_2 \mathcal{V}_2) \tag{16.47}$$

Equation of state

$$\frac{d\rho_1}{dt} = \frac{\rho_1}{\beta} \frac{dP_1}{dt} \tag{16.48}$$

$$\frac{d\rho_2}{dt} = \frac{\rho_2}{\beta} \frac{dP_2}{dt} \tag{16.49}$$

Chamber volume

$$\mathcal{V}_1 = \mathcal{V}_{1i} + D_m \theta \tag{16.50}$$

or

$$\frac{d\mathcal{V}_1}{dt} = D_m \Omega$$

$$\mathcal{V}_2 = \mathcal{V}_{2i} - D_m \theta \tag{16.51}$$

TABLE 16.7   HYDRAULIC SERVOMOTOR PARAMETERS

Z	V	Ω
K	A/B_m	D_m/B_r
τ	M_a/B_m	J_a/B_r

or

$$\frac{d\mathcal{V}_2}{dt} = -D_m\Omega$$

where

B_r = Viscous damping coefficient in servomotor
C_2 = Leakage coefficient
D_m = Displacement of servomotor
J_a = Polar moment of inertia of servomotor moving parts
P_1 = Pressure at inlet
P_2 = Pressure at outlet
P_m = $P_1 - P_2$ = Pressure drop across servomotor
Q_1 = Flow rate into servomotor
Q_2 = Flow rate out of servomotor
Q_L = C_2P_m = Internal leakage flow rate
T_c = Torque due to Coulomb friction in servomotor
T_e = External load torque on servomotor
T_f = Torque due to stiction in servomotor
$\mathcal{V}_1$ = Fluid volume under compression at inlet
$\mathcal{V}_{1i}$ = Fluid volume under compression at inlet for $\theta = 0$
$\mathcal{V}_2$ = Fluid volume under compression at outlet
$\mathcal{V}_{2i}$ = Fluid volume under compression at outlet for $\theta = 0$
β = Fluid bulk modulus of elasticity
ρ_1 = Fluid mass density at inlet
ρ_2 = Fluid mass density at outlet
θ = Angular displacement of servomotor
Ω = Angular velocity of servomotor

Simplified Mathematical Model

A simplified mathematical model for hydraulic servomotors can be obtained by assuming that (1) stiction and Coulomb friction effects are negligible, (2) internal leakage is negligible, (3) external load force (or torque) is zero, and (4) the fluid is incompressible. The result is

$$\frac{Z(s)}{P_m(s)} = \frac{K}{\tau s + 1} \tag{16.52}$$

where the parameters are as defined in Table 16.7.

16.8   HYDRAULIC MODULATORS

Two basic approaches are used to modulate the flow of a high-pressure fluid to a work producing device (servomotor). First, modulation may be accomplished by varying the displacement of a rotary pump which supplies fluid directly to a rotary servomotor. A variation on this approach is to use a variable-displacement servomotor and a fixed-displacement pump. Such systems, referred to as pump-displacement controlled (or motor-displacement controlled) servoactuators, are not as commonly used in high-performance applications as are valve-controlled servoactuators. The reader is referred to Refs. 29, 30, 31, 32, and 33 for more detailed discussion of pump-displacement controlled servoactuators. The discussion here is limited to the second modulation approach (i.e., servoactuators that employ servovalves as modulators).

16.8.1 Servovalve Design and Operation

Modern servovalves employ one or more of several types of metering devices: flapper-nozzle, poppet, spool, sliding plate, rotary "plate," and jet pipe. Servovalves are typically made in one-, two-, or three-stage configurations. Single- and two-stage configurations are the most common. Generally, the flapper-nozzle valve or the jet-pipe valve is used in the first stage of a two-stage servovalve. The spool-type valve is the most commonly used for single-stage servovalves and for the second stage of two-stage servovalves. Servovalves may be of the two-way, three-way, or four-way type (Ref. 30). Four-way types are used in most servosystems where bidirectional motion is required. Three-way types can be used where only unidirectional motion is required; bidirectional control can be achieved if a load biasing scheme is used. Depending on the internal design and application, servovalves may provide flow-rate control or pressure control. Special designs are available which employ flow rate, pressure, or dynamic pressure feedback within the valve (Refs. 34 and 35).

Servovalves may have a mechanical, hydraulic, pneumatic, or electrical input. Most servosystems used in high-performance applications today use electrohydraulic servovalves where an electrical input signal is converted to a mechanical motion through a torque motor. In single-stage valves, the torque motor actuates the control valve, which in turn modulates the flow of hydraulic fluid under pressure from a high-pressure source to a linear- or rotary-motion servomotor. In two-stage valves, the torque motor actuates the first-stage (or pilot) valve which is typically a flapper-nozzle or jet-pipe valve. The hydraulic output from the first stage drives the second stage (typically a spool-type valve), which in turn modulates the flow from the source to the servomotor. Reference 36 is a detailed history of electrohydraulic servomechanisms with special emphasis on electrohydraulic servovalves.

Figures 16.29 and 16.30 show the design features of a modern two-stage electrohydraulic servovalve. The double-coil, double-air-gap torque motor is "dry" (i.e., it is in an environmentally sealed compartment isolated from hydraulic fluid by the flexure tube). The first stage or pilot valve is a symmetrical, double-nozzle (four-way) flapper-nozzle valve. The flapper is attached to the upper (free) end of the flexure tube. The second stage is a spool valve that slides in a bushing with mating rectangular slots formed by electric discharge machining. The spool-bushing tolerance is held to 0.5 micrometers. Mechanical force feedback from the second-stage spool to the torque motor is provided by a cantilever spring attached to the flapper at the upper end, and to the spool through a ball joint.

Figure 16.31 is a cross section of a two-stage electrohydraulic servovalve that employs a jet-pipe valve as the first stage. Otherwise the valve is virtually the same in design and operation as the valve shown in Figs. 16.29 and 16.30. The jet-pipe valve can pass contamination particles as large as 200 microns, whereas the flapper-nozzle valve can only pass 50-micron particles. Good fluid filtration can negate the importance of these differences.

A cutaway diagram of a single-stage swing-plate servovalve is shown in Fig. 16.32a; a cross section of the valve is shown in Fig. 16.32b. This is an industrial valve that has a high dynamic response (natural frequency about four times that of the two-stage servovalves mentioned earlier). However, the swing-plate valve is considerably heavier than the two-stage aerospace-type valves in Figs. 16.30 and 16.31.

Table 16.8 shows typical performance capabilities of various types of servovalves.

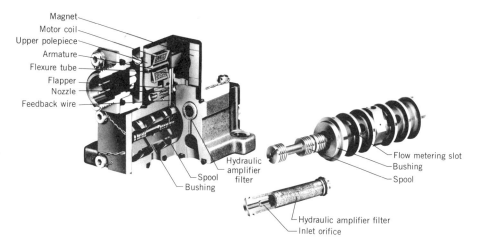

Fig. 16.29 Design features of a two-stage electrohydraulic servovalve (courtesy of Moog, Inc., East Aurora, NY).

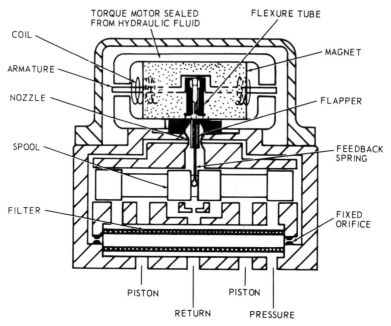

Fig. 16.30 Cross section of two-stage electrohydraulic servovalve with a flapper-nozzle first stage and spool second stage (courtesy of Moog, Inc., East Aurora, NY).

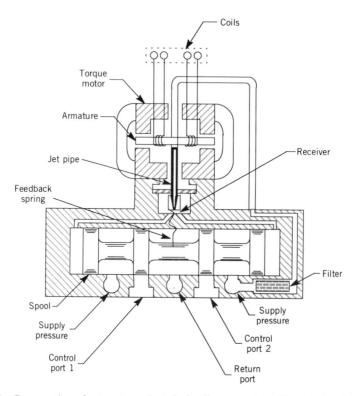

Fig. 16.31 Cross section of a two-stage electrohydraulic servovalve with a jet-pipe first stage and spool second state (courtesy of Abex Corporation, Aerospace Division, Oxnard, CA).

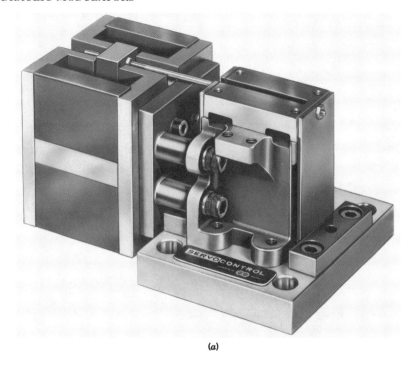

(a)

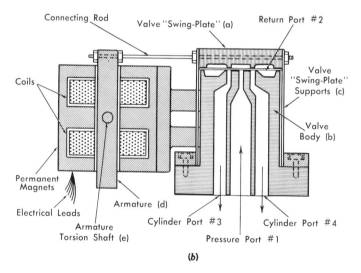

(b)

Fig. 16.32 Cross section of a single-stage, 'swing-plate' electrohydraulic servovalve (courtesy of The Oilgear Company, Milwaukee, WI).

TABLE 16.8   PERFORMANCE SPECIFICATIONS OF SERVOVALVES

Valve Type	Maximum Working Pressure (psi)	Maximum Flow at 1000 psi Pressure Drop (gpm)	Frequency at 90° Lag (Hz)	Hysteresis (%)	Resolution (%)
Spool					
One stage	5000	3500	200	0.1	0.01
Two stage	7000	1000	200	1	0.01
Three stage	4500	300	200	1	0.01
Flapper-nozzle/spool					
One and two stage	5000	1000	500	0.2	0.01
Three stage	5000	1000	500	1	0.01
Jet pipe/spool					
One and two stage	4500	300	500	2	0.1
Three stage	4500	300	200	2	0.1
Sliding plate					
One and two stage	300	40	150	3	0.1

Source: From Ref. 25.

16.8.2   Mathematical Model of a Spool-Type Valve

Spool-type valves are the most popular due to their ease of construction. They are also easier to analyze than other types of valves. Figure 16.33 shows a typical spool-valve configuration and defines the important variables and parameters. The valve has three "energy ports" where energy or power flows from or to the environment of the valve. Correspondingly, the valve can be modeled using the three mathematical equations given in functional form as follows:

$$Q_m = f(x, P_m, P_s) \tag{16.53}$$

$$Q_s = f(x, P_m, P_s) \tag{16.54}$$

$$F_v = f(x, P_m, P_s) \tag{16.55}$$

Equation 16.53 gives the pressure-flow-displacement characteristics of the valve. These characteristics are needed in the dynamic analysis of a servoactuator which employs the valve. Equation 16.54 is used to compute the required flow rate from the source and will not be considered further here. Equation 16.55 is used to calculate the force required to move the spool (e.g., force output requirement of the torque motor in the case of an electrohydraulic servovalve).

The steady-state pressure-flow-displacement characteristics of the spool valve are characterized by the nonlinear orifice equation (Refs. 30 and 37):

$$Q_m = C_d w(x + U) \sqrt{\frac{P_s - P_m}{\rho}} \tag{16.56}$$

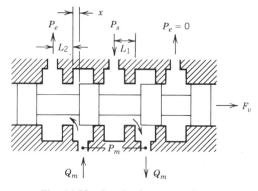

Fig. 16.33   Spool valve nomenclature.

where

C_d = Effective coefficient of discharge
P_s = Supply pressure
P_m = Pressure drop across the servomotor (see Fig. 16.33)
Q_m = Flow rate to the servomotor (see Fig. 16.33)
U = Underlap of spool with respect to sleeve (see Ref. 30); $U = 0$ for an "idealized" valve
w = Circumferential width of metering ports in the valve
x = Displacement of the spool from its neutral position
ρ = Mass density of fluid

This model assumes that the flow rates through the metering orifices are steady, the fluid is incompressible, and the valve exhaust pressure $P_e = 0$. A linearized form of this model facilitates the dynamic analysis of a servoactuator containing the valve. The nonlinear model may be linearized by considering small changes of all variables about an initial steady-state operating point, with the result:

$$\Delta Q_m = K_1 \Delta x - C_1 \Delta P_m \qquad (16.57)$$

where

$$K_1 = \left. \frac{\partial Q_m}{\partial x} \right|_{P_{m0}, x_0} = C_d w \sqrt{\frac{P_s - P_{m0}}{\rho}} \qquad (16.58)$$

$$C_1 = - \left. \frac{\partial Q_m}{\partial P_m} \right|_{x_0, P_{m0}} = \frac{C_d w x_0}{2 \sqrt{\rho(P_s - P_{m0})}} \qquad (16.59)$$

The terms ΔQ_m, Δx, and ΔP_m represent small changes of the corresponding variables about the steady-state operating point x_0, P_{m0}. The constants K_1 and C_1 are evaluated at the operating point. These expressions assume that the valve port shape does not vary with displacement.

The static and dynamic behavior of the valve spool (Eq. 16.55) can be modeled by considering the forces which act on the spool (Ref. 30). These forces include the externally imposed force (input) as well as steady and unsteady flow forces resulting from flow through the orifices. Additional forces that may be present include the viscous damping between the spool and the sleeve and any mechanical spring forces acting on the spool (not shown in Fig. 16.33). The force balance equation for the spool shown in Fig. 16.33 is

$$F_v = m_s \ddot{x} + \frac{\mu A_s}{h} \dot{x} + \left[C_d w \sqrt{2\rho\left(\frac{P_s - P_m}{2}\right)} (L_1 - L_2) \right] \dot{x} + \left[2C_d C_v w \left(\frac{P_s - P_m}{2}\right) \cos \theta_j \right] x \qquad (16.60)$$

where

A_s = Net shear area of spool lands
C_d = Metering orifice discharge coefficient
C_v = Metering orifice velocity coefficient (see Ref. 30)
F_v = External force on spool (e.g., imposed by torque motor)
h = Radial clearance between spool and sleeve (valve body)
L_1 = Length of fluid column to be accelerated at inlet (see Fig. 16.33)
L_2 = Length of fluid column to be accelerated at outlet (see Fig. 16.33)
m_s = Mass of spool
P_m = Pressure drop across servomotor
P_s = Supply pressure
w = Circumferential width of metering ports in valve
x = Displacement of spool
$\dot{x}$ = Velocity of spool
$\ddot{x}$ = Acceleration of spool
ρ = Mass density of fluid
μ = Absolute viscosity of fluid
θ_j = Effective angle of fluid jet (see Ref. 30)

The fourth term on the right-hand side of Eq. 16.60 is the steady flow-induced force and the third term on the right-hand side is the unsteady flow-induced force. The steady flow force is a "spring-like" force that always opposes the motion of the spool, and hence is a stabilizing force. The unsteady flow force is a "damping-like" force that changes its direction of action depending on the flow direction, and hence it can be a stabilizing or destabilizing force. The valve is dynamically stable if $(L_1 - L_2) > 0$. A more complete discussion of the dynamic modeling of the valve spool is given in Ref. 30.

16.8.3 Mathematical Models for an Electrohydraulic Servovalve

The steady-state pressure-flow characteristics for an electrohydraulic servovalve of the type shown in Fig. 16.30 are identical to those of the spool-type valve in the previous section except the input x is replaced by the current I. That is, in the steady state, the motion of the spool in the electrohydraulic servovalve is directly proportional to the current input to the valve. The steady-state pressure-flow-current characteristics for the "idealized" electrohydraulic servovalve (e.g., Fig. 16.30; spool matched perfectly with sleeve such that effective underlap $U = 0$) are given by the equation

$$Q_m = K_v I \sqrt{\frac{P_s - P_m}{\rho}}$$

(16.61)

where

$\quad K_v$ = A size factor
$\quad I$ = Current input to servovalve
$\quad P_s$ = Supply pressure
$\quad P_m$ = Pressure drop across the servomotor
$\quad Q_m$ = Control flow rate to the servomotor
$\quad \rho$ = Mass density of fluid

This model assumes that the exhaust pressure $P_e = 0$. Equation 16.61 may be linearized for operation about an initial steady-state operating point with the result:

$$\Delta Q_m = K_1 \Delta I - C_1 \Delta P_m$$

(16.62)

where

$$K_1 = \left.\frac{\partial Q_m}{\partial I}\right|_{P_{m0}, I_0} \qquad \text{(Flow sensitivity)}$$

(16.63)

$$C_1 = -\left.\frac{\partial Q_m}{\partial P_m}\right|_{I_0, P_{m0}} \qquad \text{(Flow-pressure sensitivity)}$$

(16.64)

Equation 16.62 is valid for cases when $U \neq 0$ as well. The terms ΔQ_m, ΔI, and ΔP_m represent small changes of the corresponding variables about the initial steady-state operating point I_0, P_{m0}. The constants K_1 and C_1 are evaluated at the operating point.

Typical steady-state pressure-flow-current characteristics for an "idealized" electrohydraulic servovalve that employs a spool-valve second stage (governed by Eq. 16.61) are shown in Fig. 16.34. Characteristics for other types of electrohydraulic servovalves are given in Refs. 34 and 35.

Another important characteristic of an electrohydraulic servovalve is its hysteresis due to the characteristics of the permanent magnets in the torque motor. The hysteresis characteristic is determined from a measurement of the output flow rate as a function of the input current for a constant (usually zero) pressure drop across the valve (load pressure drop). A typical hysteresis characteristic is shown in Fig. 16.35. The slope of the flow-current curve is the "flow sensitivity" of the valve (i.e., K_1 in Eq. 16.63).

It is often convenient in the dynamic analysis of servoactuators to have an approximate dynamic model for the servovalve. Experience has shown that linearized transfer functions based on empirical approximations from measured servovalve responses are adequate for most system designs. Reference 39 outlines considerations underlying the determination of approximate transfer function models for electrohydraulic servovalves. Figure 16.36 shows typical frequency-response plots for an electrohydraulic flow-control servovalve, along with approximate transfer functions. For a frequency range of 0–50 Hz, the following first-order expression has been found to be adequate for two-stage electrohydraulic servovalves:

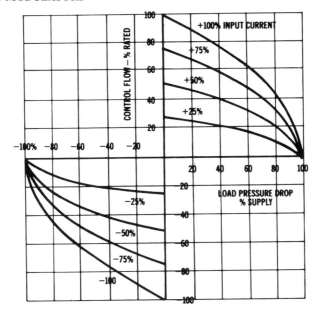

Fig. 16.34 Typical steady-state pressure-flow characteristics of an electrohydraulic servovalve. (Courtesy of Moog Inc., East Aurora, NY)

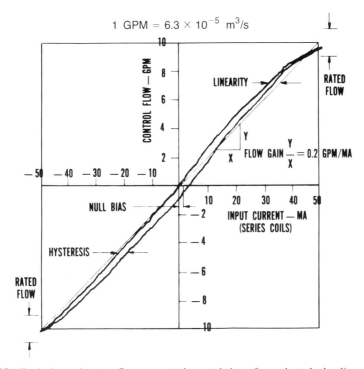

Fig. 16.35 Typical steady-state flow-current characteristics of an electrohydraulic servovalve. (Courtesy of Moog, Inc., East Aurora, NY)

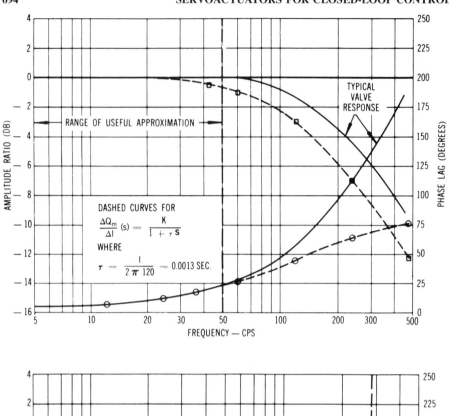

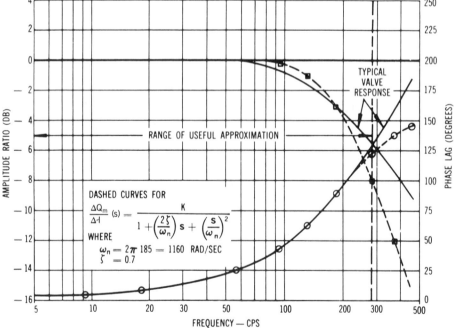

Fig. 16.36   Typical dynamic behavior of electrohydraulic servovalves (taken from Reference 39).

TABLE 16.9   TYPICAL DYNAMIC CHARACTERISTICS OF TWO-STAGE ELECTROHYDRAULIC FLOW-CONTROL SERVOVALVES

| Flow-control servovalve | Max. flow capacity at 3000 psi gpm | Approximate dynamics 3000 psi 100°F Peak-to-Peak Input at 50% rated current | | |
| | | 1st order | 2nd order | |
		τ sec	f_n cps	ζ
A	2	0.0013	240	0.5
B	6	0.0015	200	0.5
C	12	0.0020	160	0.55
D	18	0.0023	140	0.6
E	30	0.0029	110	0.65

Source: From Reference 39.

$$\frac{\Delta Q_m(s)}{\Delta I(s)} = \frac{K_1}{\tau s + 1} \tag{16.65}$$

where

$Q_m(s)$ = Laplace transform of control flow rate to the servomotor

$I(s)$ = Laplace transform of the current input to the servovalve

K_1 = Servovalve flow sensitivity at $P_m = 0$, $I = 0$

τ = Apparent servovalve time constant (s)

Typical time constants for electrohydraulic flow-control servovalves are given in Table 16.9.

If a good approximation is desired over a wider frequency range, the following second-order model may be preferred:

$$\frac{\Delta Q_m(s)}{\Delta I(s)} = \frac{K_1}{(s/\omega_n)^2 + (2\zeta/\omega_n)s + 1} \tag{16.66}$$

where

$\omega_n = 2\pi f_n$ = Apparent natural frequency (rad/s)

ζ = Apparent damping ratio (dimensionless)

Typical values of f_n and ζ for two-stage electrohydraulic flow control servovalves are given in Table 16.9.

16.9   ELECTROMECHANICAL AND ELECTROHYDRAULIC SERVOSYSTEMS

16.9.1   Typical Configurations of Electromechanical Servosystems

An electrical servomotor may be combined with an electrical or electronic modulator to form an electromechanical servoactuator. The addition of a feedback transducer forms a servosystem. Figure 16.2 shows an electromechanical linear-motion servosystem which incorporates a rotary brushless DC servomotor and tachometer feedback; the electronic modulator is not shown.

16.9.2   Typical Configurations of Electrohydraulic Servosystems

A servoactuator comprising an electrohydraulic servovalve and a servomotor may be combined with an electronic servoamplifier (or modulator) and an appropriate feedback transducer to form a high-performance servosystem. Schematic diagrams of three typical electrohydraulic servosystems are shown in Fig. 16.37.

Figure 16.38 shows a photograph and a cross section of a high-performance servosystem which incorporates an electrohydraulic servovalve, an axial-piston servomotor, and a tachometer. The servo-

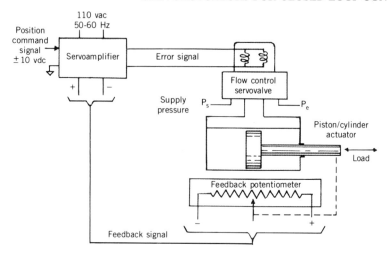

(a) Linear position servosystem

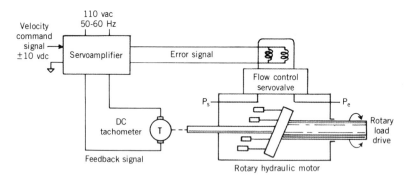

(b) Rotary velocity servosystem

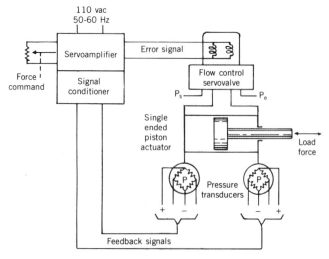

(c) Force servosystem

Fig. 16.37 Typical electrohydraulic servoststems (courtesy of Moog, Inc., East Aurora, NY); (a) linear position servosystem; (b) rotary velocity servosystem; (c) force servosystem.

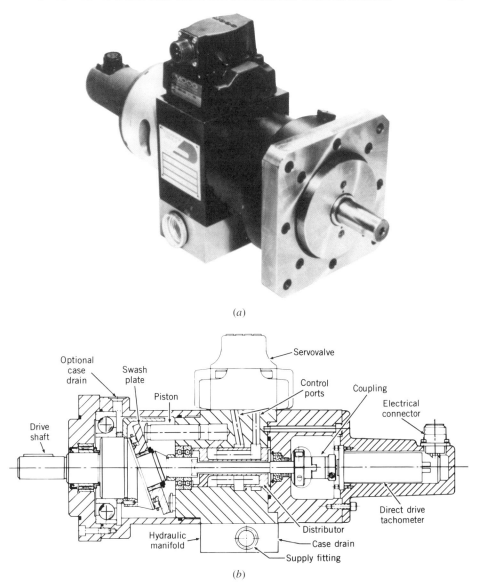

Fig. 16.38   Servosystem with rotary actuator, tachometer, and electrohydraulic servovalve (courtesy Moog Inc., East Aurora, NY); (*a*) photograph of Moog-Donzelli servosystem; (*b*) cross section of Moog-Donzelli servosystem.

amplifier is not shown. Direct manifold mounting of the servovalve results in a small compressed oil volume, and therefore high torsional stiffness and fast dynamic response.

Digital control is becoming an important technique for producing near optimum performance from electrohydraulic servosystems. Figure 16.39 shows a block diagram and a cutaway of a fully integrated digital electrohydraulic position control system (Ref. 40). A two-stage electrohydraulic servovalve drives a linear-motion servomotor. A microcomputer and other electronics integrated within the servovalve housing drive the valve and provide signal conditioning for the position feedback transducer. A ferromagnetic digital position-measurement transducer is mounted within the servomotor housing. The feedback management technique uses the microcomputer to model the system and provide digital velocity and acceleration signals without the use of separate sensors (i.e., a digital observer is used). Eight gains are required for accurate and smooth control, and these are calculated within the digital controller to generate the control signal to the torque motor.

(*a*)

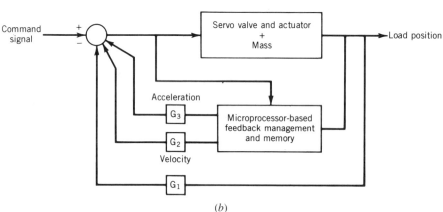

(*b*)

Fig. 16.39 Servosystem with linear actuator, position transducer, and microprocess-based electro-hydraulic servovalve (courtesy of Vickers, Inc., Troy, MI); (*a*) cutaway of servosystem; (*b*) block diagram of servosystem.

16.9.3 Comparison of Electromechanical and Electrohydraulic Servosystems

Precision motion and force control can be achieved using either electromechanical or electrohydraulic servosystems. The actual choice between electromechanical and electrohydraulic must be based on a number of factors and trade offs. Studies by Moog, Inc. have produced the following conclusions (Ref. 3):

1. Brushless motors with high-energy Samarium-Cobalt magnets make possible lightweight electromechanical actuators having good response and high efficiency.

2. Electromechanical actuation is currently an alternative to electrohydraulic actuation for applications requiring up to approximately three to four horsepower. Higher power electromechanical actuation systems are limited at the present time by the lack of reliable, compact, lightweight electronics.

TABLE 16.10   COMPARISON OF ELECTROMECHANICAL AND ELECTROHYDRAULIC SERVOSYSTEMS

Electromechanical	Electrohydraulic
Advantages	
lower cost than electrohydraulic	mature technology
momentary overdrive capability	very high reliability
low quiescent power	highest actuation performance
low system weight in low HP range	smaller and lighter weight in high HP range
packaging flexibility	continuous power output capability
(conventional or pancake motors)	continuous stall torque capability
(different types of gear reduction)	wide temperature capability
easy check-out	high vibration and acceleration capability
single responsibility for servoelectronics and	proven long-term storability
actuators	nuclear hardenable
	no EMI generation
	simple low-power servoelectronics
Disadvantages	
more complex electronics	usually higher cost
(communication logic for brushless motors)	generally requires more complex power
(high power drive with current limiting)	conversion equipment
motor inertia-into-stops problems	requires clean hydraulic fluid
overheating with high static loads	quiescent power loss
requires motion reduction/conversion	
generates EMI	
more difficulty nuclear hardening	
high power electromechanical actuation not	
yet proven	

3. Electrohydraulic actuation has a proven record in a variety of aerospace and industrial applications requiring high power levels.

Table 16.10 (Ref. 3) compares advantages and disadvantages of electromechanical and electrohydraulic servosystems.

16.10   STEADY-STATE AND DYNAMIC BEHAVIOR OF SERVOACTUATORS AND SERVOSYSTEMS

Mathematical models of the components presented in previous sections of this chapter may be combined with a load model to describe the behavior of the servosystem. The system model may be used to study the steady-state and dynamic behavior of the system for various values of system parameters and operating conditions. A number of commercially available digital simulation codes are available for determining the performance in the time or frequency domain.

The combination of a modulator and a servomotor with a load (with or without gearing) forms an open-loop system. Position or velocity feedback may be used to provide a special performance feature (e.g., use of position feedback to convert an open-loop velocity control system to a position control system) or to improve performance.

16.10.1   Electromechanical Servoactuators

Figure 16.40 shows schematically a servoactuator comprising a permanent-magnet DC servomotor and an electronic amplifier used to control the velocity of a rotary inertia load. Gearing is used to match the motor torque capability with the load requirements. Chapter 5 discusses the concept of impedance matching to achieve an optimum gear ratio in a system such as this.

A simplified dynamic model may be derived based on the following assumptions:

1. The amplifier bandwidth is considerably greater than that of the servomotor-load portion of the system.
2. The gears have zero backlash and infinite stiffness.
3. The connecting shafts all have infinite stiffness.

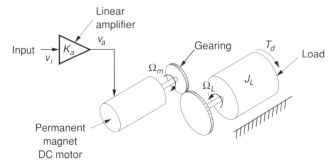

Fig. 16.40 Schematic diagram of an electromechanical servoactuator.

Assumptions 2 and 3 eliminate some important dynamic effects that may need to be included in some cases (see Ref. 41).

Definitions of the critical parameters and variables are as follows:

$$B_L \ = \ \text{Viscous damping in the load (N·m·s/rad)}$$

$$J_L \ = \ \text{Polar moment of inertia of the load (N·m·s}^2\text{/rad)}$$

$$K_a \ = \ \text{Voltage gain of amplifier (V/V)}$$

$$K_{ac} \ = \ \text{Current gain of amplifier (A/V)}$$

$$n \ = \ \text{Gear ratio, } \Omega_m/\Omega_L, \theta_m/\theta_L$$

$$T_e \ = \ \text{External load torque (N·m)}$$

$$T_d(s) \ = \ \text{Laplace transform of } T_d$$

$$T_{fL} \ = \ \text{Coulomb friction torque in load (N·m)}$$

$$T_m \ = \ \text{Total load torque reflected to the motor shaft (N·m)}$$

$$T_L \ = \ \text{Total load torque (N·m)}$$

$$v_a \ = \ \text{Voltage output of amplifier (V)}$$

$$v_i \ = \ \text{Voltage input to amplifier (V)}$$

$$V_i(s) \ = \ \text{Laplace transform of } v_i$$

$$\theta_L \ = \ \text{Angular displacement of the load (rad)}$$

$$\Omega_L \ = \ \text{Angular velocity of the load (rad/s)}$$

Other parameters are as defined in Section 16.3.

Mathematical Model

The basic equations which describe the dynamic behavior of the servoactuator in Fig. 16.40 are as follows:

Amplifier equation:

$$v_a = K_a v_i \tag{16.67}$$

Motor equations:

Electrical:

$$v_a = K_E \Omega_m + L_a \frac{di}{dt} + R_a i \tag{16.68}$$

Torque balance:

$$K_T i = J_m \frac{d\Omega_m}{dt} + B_m \Omega_m + T_{fm} + T_m \tag{16.69}$$

Gear equations:

$$n = \Omega_m/\Omega_L = \theta_m/\theta_L \tag{16.70}$$

Loss-less power transfer:

$$T_m\Omega_m = T_L\Omega_L \tag{16.71}$$

Load torque balance:

$$T_L = J_L\frac{d\Omega_L}{dt} + B_L\Omega_L + T_{fL} + T_e \tag{16.72}$$

If the gearing has backlash, the relationship between θ_m and θ_L is as shown in Fig. 16.41. Similarly, if there is a Coulomb friction load torque, the relationship between the torque and the load velocity is as shown in Fig. 16.42. These nonlinear effects can be included in the model, but the resulting set of algebraic and differential equations can be studied only through digital simulation. Then the solution must be obtained for a specified input command; a general solution is not possible. It is often sufficient in analysis underlying preliminary design to simplify the model by linearizing the equations and eliminating nonlinear effects such as backlash and Coulomb friction. Such a linearized model can be very useful in gaining an understanding of the general behavior of a system and the sensitivity of the performance to parameter changes.

Equations 16.67 through 16.72 can be linearized by assuming small variations of all variables about an initial steady-state operating point. For the system considered here, the linearized dynamic model in the Laplace domain is given by

$$\Delta\Omega_m(s) = G_1(s)\Delta V_i(s) - G_2(s)\Delta T_e(s) \tag{16.73}$$

where the transfer functions $G_1(s)$ and $G_2(s)$ are

$$G_1(s) = \frac{K_aK_T}{R_aB_t(\tau_e s + 1)(\tau_m s + 1) + K_TK_E} \tag{16.74}$$

$$G_2(s) = \frac{(R_a/n)(\tau_e s + 1)}{R_aB_t(\tau_e s + 1)(\tau_m s + 1) + K_TK_E} \tag{16.75}$$

and where

$$\tau_e = L_a/R_a$$

$$\tau_m = J_t/B_t$$

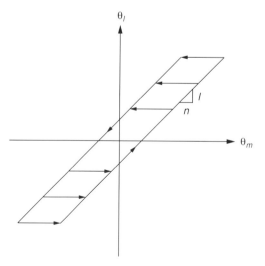

Fig. 16.41 Relationship between θ_m and θ_L due to backlash.

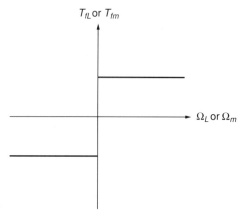

Fig. 16.42 Coulomb friction characteristic.

$$J_t = J_m + J_L/n^2$$

$$B_t = B_m + B_L/n^2$$

A block diagram of Eq. 16.73 is given in Figure 16.43.

In most cases the mechanical time constant (τ_m) is much greater than the electrical time constant (τ_e). In this case, the transfer functions of Eqs. 16.74 and 16.75 simplify to first-order forms as follows:

$$G_1'(s) = \frac{K_1}{\tau_1 s + 1} \tag{16.76}$$

$$G_2'(s) = \frac{K_2}{\tau_2 s + 1} \tag{16.77}$$

where

$$K_1 = \frac{K_a K_T}{R_a B_t + K_T K_E} \tag{16.78}$$

$$K_2 = \frac{R_a}{n(R_a B_t + K_T K_E)} \tag{16.79}$$

$$\tau_1 = \tau_2 = \frac{R_a B_t \tau_m}{R_a B_t + K_T K_E}$$

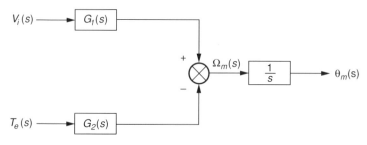

Fig. 16.43 Block diagram of a servoactuator with voltage input.

If the mechanical and electrical time constants are comparable, the transfer functions can be expressed in the canonical form as

$$G_1(s) = \frac{K_1}{(s^2/\omega_n^2) + (2\zeta/\omega_n)s + 1} \tag{16.80}$$

$$G_2(s) = \frac{K_2(\tau_e s + 1)}{(s^2/\omega_n^2) + (2\zeta/\omega_n)s + 1} \tag{16.81}$$

where K_1 and K_2 are as defined in Eqs. 16.78 and 16.79, respectively, and

$$\omega_n = \sqrt{\frac{1}{\tau_e \tau_m}\left(1 + \frac{K_T K_E}{R_a B_t}\right)} \tag{16.82}$$

$$\zeta = \frac{1}{2}\left[\frac{\tau_e + \tau_m}{\sqrt{\tau_e \tau_m \left[1 + (K_T K_E / R_a B_t)\right]}}\right] \tag{16.83}$$

If a current amplifier is used instead of a voltage amplifier, the resulting linearized dynamic model of the system is given by

$$\Delta\Omega m(s) = G_3'(s)\Delta I(s) - G_4'(s)\Delta T_e(s) \tag{16.84}$$

where

$$G_3'(s) = \frac{K_T K_{ac}}{B_t(\tau_m s + 1)} \tag{16.85}$$

$$G_4'(s) = \frac{1}{B_t n(\tau_m s + 1)} \tag{16.86}$$

That is, the electrical time constant is not a factor if the current amplifier is used.

Numerical Example

A Magnetic Technology Inc. DC servomotor (Model 3069-237/045; see Ref. 14) is used to rotate an inertia load. System parameters are as follows:

Motor:

$$B_m = 1.9 \text{ in·oz·s/rad } (1.34 \times 10^{-2} \text{ N·m·s/rad})$$

$$J_m = 0.016 \text{ in·oz·s}^2/\text{rad } (1.13 \times 10^{-4} \text{ N·m·s}^2/\text{rad})$$

$$K_E = 0.236 \text{ V·s/rad}$$

$$K_T = 33.4 \text{ in·oz/A } (0.24 \text{ N·m/A})$$

$$L_a = 4.8 \text{ mH}$$

$$R_a = 4.5 \text{ ohms}$$

Gearing:

$$n = 3$$

Load:

$$B_L = 35 \text{ in·oz·s/rad } (0.25 \text{ N·m·s/rad})$$

$$J_L = 0.2 \text{ in·oz·s}^2/\text{rad } (1.41 \times 10^{-3} \text{ N·m·s}^2/\text{rad})$$

Amplifier:

$$K_a = 8 \text{ V/V}$$

Substituting these parameters into Eqs. 16.78, 16.79, 16.82, and 16.83 gives the following gains and performance factors:

$$K_1 = 7.87 \text{ rad/s·V}$$

$$K_2 = 0.044 \text{ rad/in·oz·s (6.23 rad/N·m·s)}$$

$$\omega_n = 68.5 \text{ Hz}$$

$$\zeta = 1.27$$

and from the definition for τ_e,

$$\tau_e = L_a/R_a = 1.07 \times 10^{-3} \text{ s}$$

Since the damping ratio (ζ) is greater than one, the system will have an overdamped dynamic response. The transfer functions are as follows:

$$G_1(s) = \frac{7.87}{5.40 \times 10^{-6}s^2 + 5.88 \times 10^{-3}s + 1} \tag{16.87}$$

$$G_2(s) = \frac{0.13(1.07 \times 10^{-3}s + 1)}{5.40 \times 10^{-6}s^2 + 5.88 \times 10^{-3}s + 1} \tag{16.88}$$

Equations 16.87 and 16.88 can be used to determine the transient response or the frequency response.

16.10.2 Electromechanical Servosystems

Only servosystems including DC servomotors are considered here. The reader should consult Ref. 41 for a more comprehensive treatment of other types of electromechanical servosystems.

Figure 16.44 shows a closed-loop servosystem that utilizes tachometer feedback around the servoactuator discussed in the previous section. Through the use of velocity feedback, greater accuracy can be achieved than with the open-loop system.

Mathematical Models

An approximate linearized model may be derived for the servosystem by combining the open-loop system model (Eq. 16.73) with the following equations:

Summation:

$$v_i = v_{\text{ref}} - v_f \tag{16.89}$$

Tachometer:

$$v_f = K_t \Omega_m \tag{16.90}$$

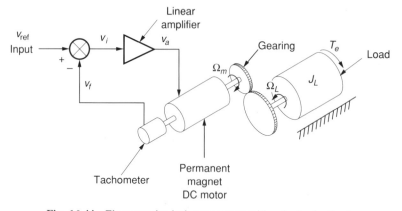

Fig. 16.44   Electromechanical servosystem with velocity feedback

Amplifier:

$$v_a = K_a v_i \tag{16.91}$$

Equation 16.90 assumes that the tachometer is mounted directly on the motor shaft, which is normally the case. Reference 18 discusses cases where the tachometer is mounted in other configurations. Also, Ref. 18 presents more accurate tachometer models to account for the magnetic coupling of the tachometer to the motor field. The basic form of the tachometer model in that case is that of a high-pass filter with a phase shift that is dependent on the angular orientation of the tachometer to the motor.

Equations 16.73, 16.89, 16.90, and 16.91 may be combined to obtain the following closed-loop servosystem model:

$$\Delta\Omega_m(s) = G_3(s)\Delta V_i(s) - G_4(s)\Delta T_e(s) \tag{16.92}$$

where

$$G_3(s) = \frac{K_a K_T}{R_a B_t(\tau_e s + 1)(\tau_m s + 1) + K_T K_E + K_f K_a K_T} \tag{16.93}$$

$$G_4(s) = \frac{(R_a/n)(\tau_e s + 1)}{R_a B_t(\tau_e s + 1)(\tau_m s + 1) + K_T K_E + K_f K_a K_T} \tag{16.94}$$

All parameters and variables are defined in the previous section and in Fig. 16.44. The second-order transfer functions $G_3(s)$ and $G_4(s)$ can be expressed in canonical form as follows:

$$G_3 = \frac{K_3}{(s^2/\omega_{n1}^2) + (2\zeta_1/\omega_{n1})s + 1} \tag{16.95}$$

$$G_4 = \frac{K_4(\tau_e s + 1)}{(s^2/\omega_{n1}^2) + (2\zeta_1/\omega_{n1})s + 1} \tag{16.96}$$

where

$$K_3 = \frac{K_a K_T}{R_a B_t + K_T K_E + K_f K_a K_T} \tag{16.97}$$

$$K_4 = \frac{R_a}{n(R_a B_t + K_T K_E + K_f K_a K_T)} \tag{16.98}$$

$$\omega_{n1} = \sqrt{\frac{1}{\tau_e \tau_m}\left(1 + \frac{K_T K_E + K_f K_a K_T}{R_a B_t}\right)} \tag{16.99}$$

$$\zeta_1 = \frac{1}{2}\frac{\tau_e + \tau_m}{\left[\sqrt{\tau_e \tau_m\left(1 + \frac{K_T K_E + K_f K_a K_T}{R_a B_t}\right)}\right]} \tag{16.100}$$

From the expressions for the closed-loop natural frequency (ω_{n1}) and damping ratio (ζ_1), it is apparent that increasing the feedback gain (K_f) increases the speed of response but decreases the degree of stability.

As in the case with the open-loop system, if the electrical time constant is much smaller than the mechanical time constant, the second-order transfer functions of Eqs. 16.95 and 16.96 reduce to the first-order forms given below:

$$G_5(s) = \frac{K_5}{\tau_5 s + 1} \tag{16.101}$$

$$G_6(s) = \frac{K_6}{\tau_6 s + 1} \tag{16.102}$$

where

$$K_5 = K_3 \quad \text{(see Eq. 16.97)}$$

$$K_6 = K_4 \quad \text{(see Eq. 16.98)}$$

and

$$\tau_5 = \tau_6 = \frac{R_a B_t \tau_m}{R_a B_t + K_T K_E + K_f K_a K_T}$$

Similar expressions may be derived for the case with a current amplifier instead of a voltage amplifier. It is preferable to use a current amplifier where cost is not prohibitive.

Numerical Example

Consider a closed-loop electromechanical servosystem comprising the open-loop system of Section 16.10.1 with velocity feedback added. For the same numerical values as in the numerical example of Sec. 16.10.1, and with $K_f = 3$ V/krpm, the following results are obtained.

$$K_3 = 6.42 \text{ rad/s} \cdot \text{V}$$

$$K_4 = 0.036 \text{ rad/in} \cdot \text{oz} \cdot \text{s} \ (5.10 \text{ rad/N} \cdot \text{m} \cdot \text{s})$$

$$\omega_{n1} = 75.8 \text{ Hz}$$

$$\zeta_1 = 1.143$$

16.10.3   Electrohydraulic Servoactuators

Figure 16.45 is a physical representation of an electrohydraulic servoactuator used to position an inertia load. A two-stage electrohydraulic servovalve modulates the flow of hydraulic fluid to a double acting hydraulic motor (actuator). The current input to the torque motor of the servovalve, I, is provided by an electronic servoamplifier of gain K_a.

A simplified dynamic model may be derived to relate the voltage input to the servoamplifier, E_s, to the velocity output of the load, Y. The following basic assumptions simplify the analysis.

1. The amplifier bandwidth is considerably greater than that of the servovalve and the servomotor-load portions of the system.

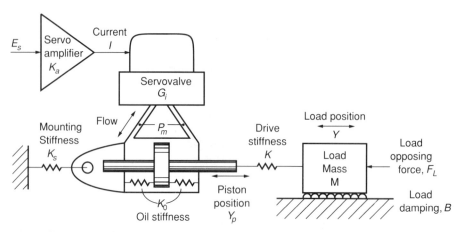

Fig. 16.45   Physical representation of an electrohydraulic servoactuator (adapted from Reference 35).

2. The supply pressure, P_s, and the exhaust pressure, P_e, are constant.

3. The bulk modulus of elasticity and viscosity of the hydraulic fluid are constant.

4. The leakage flow rate past the actuator piston is linearly proportional to the pressure drop across the piston; that is, $Q_L = C_2 P_m$ (see Fig. 16.28 and Eqs. 16.39 and 16.40).

5. All connecting passages are rigid and sufficiently short in length and large in diameter to eliminate any resistance or transmission line effects.

6. The mass of the moving parts in the servomotor (M_a) and the viscous dampling in the servomotor (B_m) are small compared to the mass (M) and the viscous dampling (B) associated with the load.

7. The Coulomb friction and stiction forces in the servomotor and load are negligible.

Mathematical Model

The typical servoamplifier is governed by the equation

$$I = K_a E_s \tag{16.103}$$

Equation 16.61 describes the steady-state behavior of the electrohydraulic servovalve. This equation can be used in the dynamic analysis of the servoactuator if the servovalve dynamics are negligible compared to the servomotor-load dynamics. Otherwise, either Eq. 16.65 or 16.66 should be used. Equations 16.38 through 16.44 describe the dynamic behavior of the servomotor. The external load force in Eq. 16.38 may be expressed as follows:

$$F_e - F_L = M\frac{d^2Y}{dt^2} + B\frac{dY}{dt} + K(Y_p - Y) \tag{16.104}$$

The set of algebraic equations outlined above cannot be combined into a single dynamic equation or transfer function because of nonlinearities. Also, since the set of equations is rather complex, conclusions about the performance of the system cannot be drawn without actually solving the equations for a variety of conditions. A computer-based analysis is the only practical method of determining the steady-state and dynamic performance of the system unless the equations are linearized in some fashion.

A linearized model, which assumes small perturbations of all variables about an initial steady-state operating condition, can be very useful in quickly assessing system dynamic behavior. Such a model is particularly useful for preliminary design and as a reference when an analysis is made using the set of nonlinear describing equations.

It can be shown (Ref. 45) that for the case when $\mathcal{V}_{1i} = \mathcal{V}_{2i} = \mathcal{V}_i$ (see Fig. 16.28), and under the assumptions listed above, the linearized model may be expressed as follows:

$$\left[\left(\frac{K_3 M}{K_2 B + A^2}\right)s^2 + \left(\frac{K_2 M + BK_3}{K_2 B + A^2}\right)s + 1\right]s(\Delta Y)$$

$$= \left(\frac{K_a A}{K_2 B + A^2}\right)G_1(s)(\Delta E_s) - \left(\frac{K_2 + K_3 s}{K_2 B + A^2}\right)(\Delta F_L) \tag{16.105}$$

where

$K_2 = C_1 + C_2$

$C_1 = $ Servovalve pressure-flow sensitivity (see Eq. 16.64)

$C_2 = $ Internal leakage flow rate coefficient

$K_3 = \dfrac{A^2}{K_t} = A\left(\dfrac{1}{K_0} + \dfrac{1}{K_s} + \dfrac{1}{K}\right)$

$K_0 = \dfrac{2\beta A^2}{\mathcal{V}_i} = $ Stiffness of the sealed chamber

$K = $ Stiffness of the load drive

$K_s = $ Stiffness of the structural mounting

and $G_i(s)$ is the transfer function for the servovalve. When the servovalve dynamics are negligible compared to the servomotor-load dynamics, $G_i(s) = K_1$ (see Eq. 16.63). When the servovalve dynamics are of the same order as the servomotor-load dynamics, $G_i(s)$ is given by the right-hand side of either Eq. 16.65 or Eq. 16.66.

Four important measures of performance can be observed from Eq. 16.105 without actually solving the equation. The steady-state gain or sensitivity of the servoactuator is

$$\frac{\Delta \dot{Y}}{\Delta E_s}\bigg|_{\text{steady state; } \Delta F_L = 0} = \frac{K_a K_1 A}{K_2 B + A^2} \tag{16.106}$$

The steady-state load sensitivity of the servoactuator is

$$\frac{\Delta \dot{Y}}{\Delta F_L}\bigg|_{\text{steady state; } \Delta E_s = 0} = \frac{-K_2}{K_2 B + A^2} \tag{16.107}$$

Also observable from Eq. 16.105 are the natural frequency and the damping ratio associated with the servomotor-load portion of the system. The natural frequency is

$$\omega_{ns} = \sqrt{\frac{K_2 B + A^2}{K_3 M}} \tag{16.108}$$

and the damping ratio is

$$\zeta_s = \frac{K_2 M + K_3 B}{2\sqrt{K_3 M (K_2 B + A^2)}} \tag{16.109}$$

The valve dynamics are often negligible compared to the servomotor-load dynamics. In this case, Eq. 16.105 reduces to a second-order linear differential equation of the generalized form:

$$\left[\left(\frac{1}{\omega_{ns}^2}\right)s^2 + \left(\frac{2\zeta_s}{\omega_{ns}}\right)s + 1\right]s(\Delta Y)$$

$$= \frac{K_a K_1 A}{(K_2 B + A^2)} \times \Delta E_s - \frac{(K_2 + K_3 s)}{(K_2 B + A^2)}\Delta F_L \tag{16.110}$$

In most cases, the term $K_2 B$ in Eqs. 16.105 through 16.110 is small compared to the term A^2. Simplified forms of these equations result.

The dynamic behavior of the open-loop system may be viewed in terms of the speed of response and the degree of damping. A fast system dynamic response requires a small value of τ and a large value of ω_{ns}. In order for ω_{ns} to be large, $K_3 M$ must be small compared to $(K_2 B + A^2)$. The value of M is usually fixed, but the designer often has some latitude in varying K_3. The value of K_3 can be minimized by making the volumes within the servomotor chambers small. Increasing the ram area also provides for an increase in ω_{ns}, but the effect is not as great as it first appears since $K_3 = f(\mathcal{V}_i)$ and $\mathcal{V}_i = f(A)$. Also, the actuator area is often set by such practical considerations as maximum load or acceleration requirements.

The degree of damping in the open-loop system is governed by Eq. 16.109. In most practical systems, $\zeta_s < 1$, and K_3, M, and A are fixed by other considerations. Then the damping may be increased by increasing the load damping, B, or the value of the effective leakage coefficient, K_L. The value of K_L may be increased by increasing either C_1 or C_2; that is, increasing the valve underlap or the leakage across the actuator piston. Since normally $K_2 B \ll A^2$, Eq. 16.107 shows that an increase in K_2 results in an increase in sensitivity to load disturbances. Likewise, an increase in the valve underlap (or the use of cross-port leakage) results in an increase in quiescent power dissipation. Clearly, there is a trade-off between degree of damping, steady-state load sensitivity, and quiescent power dissipation.

16.10.4 Electrohydraulic Servosystems

Figure 16.46 is a physical representation of a typical electrohydraulic servosystem intended to accurately position an inertia load. The addition of position feedback converts the "rate-type" open-loop system (servoactuator) into a position control system. The linearized model given by Eq. 16.110 describes the open-loop portion of this position control system, as shown in the block diagram of Fig. 16.47.

When the change in the external load force is zero (i.e., $\Delta F_L = 0$), the dynamic model for the closed-loop system may be written as

$$\frac{\Delta Y}{\Delta E_c} = \frac{1}{K_p}\left[\frac{1}{K_{Lp}(1/\omega_{ns}^2)s^2 + (2\zeta_s/\omega_{ns})s + 1)(\tau s + 1)(s) + 1}\right] \tag{16.111}$$

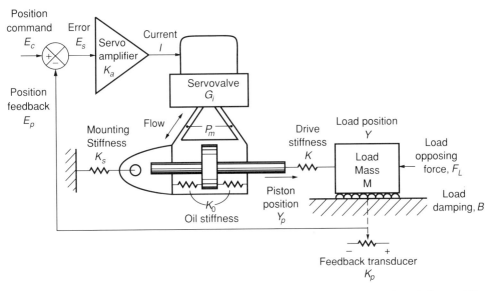

Fig. 16.46 Physical representation of an electrohydraulic servomechanism (taken from Reference 35).

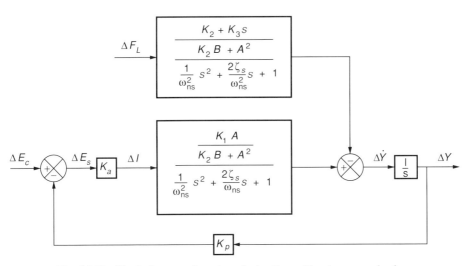

Fig. 16.47 Block diagram of an electrohydraulic positional servomechanism.

where

$$K_{Lp} = \frac{K_p K_a K_1 A}{K_2 B + A^2} = \text{Loop gain}$$

For the case where the servovalve dynamics are negligible compared to the servomotor-load dynamics, Eq. 16.111 reduces to the following third-order differential equation:

$$\frac{\Delta Y}{\Delta E_c} = \frac{1}{K_p}\left[\frac{1}{(1/K_{Lp})(1/\omega_{ns}^2)s^2 + ((2\zeta_s/\omega_{ns})s + 1s + 1)}\right] \qquad (16.112)$$

In the steady state, Eqs. 16.111 and 16.112 both reduce to

$$\frac{\Delta Y}{\Delta E_c}\bigg|_{\text{steady state}} = \frac{1}{K_p} \tag{16.113}$$

The integration in the forward loop results in a system with a zero steady-state error for a constant input and a steady-state output that is dependent only on the system input and the position feedback gain. (See Section 14.7 for a discussion of the following errors for systems with non-constant inputs.) That is, the accuracy with which ΔY follows the input ΔE_c depends only on the accuracy of the feedback measurement device, and not on the accuracy of the forward loop elements.

When the change in the control input is zero (i.e., $\Delta E_c = 0$), the steady-state load sensitivity is

$$\frac{\Delta Y}{\Delta F_L}\bigg|_{\text{steady state}} = \frac{K_2}{K_p K_a K_1 A} \tag{16.114}$$

That is, an increase in the feedback gain results in a decrease in the load sensitivity or an increase in the system stiffness. Likewise, an increase in the cross-port leakage in the servomotor or servovalve (i.e., increase in K_2) results in a decrease in the system stiffness.

In most practical cases the roots of the quadratic in Eqs. 16.111 and 16.112 are conjugate complex (i.e., $\zeta_s < 1$), or in physical terms, this portion of the system is "underdamped." Consequently, when position feedback is employed, limitations exist in the maximum value of the position feedback gain that can be used while still ensuring system stability. Limitations also exist in the input-output sensitivity and the system stiffness to load disturbances, since these characteristics are dependent on the position feedback gain.

System relative stability can be viewed conveniently by employing the root locus technique. Figure 16.48 illustrates root-locus plots for three important cases: (a) valve dynamics modeled by a second-order differential equation, (b) valve dynamics modeled by a first-order differential equation, and (c) negligible valve dynamics. These plots illustrate that the loop gain must be set below some value in order to ensure stability.

When the dynamics of the servovalve are negligible compared to the dynamics of the servomotor-load portion of the system, the system model is third-order. Considerable study has been made of third-order dynamic systems. The Routh absolute stability criterion can be employed to determine the maximum value of the feedback gain that can be used and still maintain stability. For the simplified model given by Eq. 16.112, the maximum value of the feedback gain is

$$K_p = \frac{2\zeta_s \omega_{ns}(K_2 B + A^2)}{K_a K_1 A} \tag{16.115}$$

and the corresponding maximum value of the loop gain is

$$L_{Lp} \equiv \frac{K_p K_a K_1 A}{K_2 B + A^2} = 2\zeta_s \omega_{ns} \tag{16.116}$$

Equation 16.116 shows that the maximum value of loop gain depends only on the values of the natural frequency and damping ratio of the open-loop system.

In general, the loop gain is a critical parameter. An increase in loop gain results in a decrease in load sensitivity (or increase in stiffness), an increase in the speed of response, and a decrease in the degree of stability.

For a given electrohydraulic position control system designed to meet certain steady-state performance requirements (e.g., load sensitivity), an optimum or best value of loop gain (and therefore, the best feedback gain) exists. This optimum value represents the best compromise between speed of response and degree of stability. Figure 16.49 illustrates responses of the system output (ΔY) to step changes in the system input (ΔE_c) for four different values of the loop gain (Eq. 16.112).

A comprehensive study of Eq. 16.112 by Meyfarth (Ref. 46) has shown that the optimum response characteristics to a step input signal are obtained when

$$\zeta_s = 0.5 \tag{16.117}$$

$$K_{Lp} = 0.34\omega_{ns} \tag{16.118}$$

In many practical casess, the open-loop system damping ratio (ζ_s) is well below 0.5. The resulting load resonance places severe limitations on the maximum level of loop gain that can be used, and therefore limitations on the quality of steady-state and dynamic performance that can be achieved with the closed-loop system. That is, it may not be possible to simultaneously satisfy the steady-state load sensitivity (or stiffness) and dynamic performance requirements without special enhancements to the system.

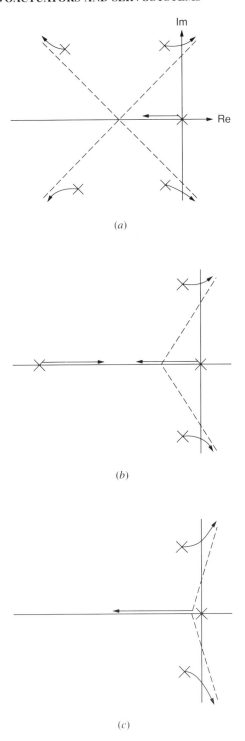

Fig. 16.48 Typical root locus plots for an electrohydraulic positional servomechanism with and without servovalve dynamics; (a) With second-order servovalve model (Eq. 16.66); (b) With first-order servovalve model (Eq. 16.65); (c) With static servovalve model (Eq. 16.62).

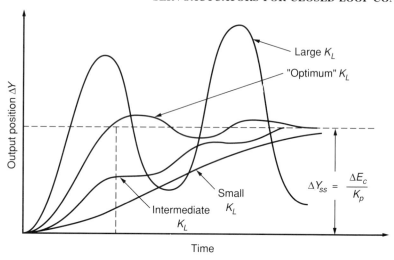

Fig. 16.49 Typical step responses for a third-order linear system.

16.10.5 Hydraulic Compensation

One of the features of electrohydraulic servosystems is the ease with which electronic feedback and forward-loop compensation networks can be employed to produce improved dynamic performance. Such techniques are discussed in Chapter 17. The following discussion is limited to techniques for improving the damping within the hydraulic portion of the system; these techniques minimize the need to consider other electronic compensation techniques.

Techniques for Improving Damping

Three well-known techniques may be employed to introduce additional damping into the open-loop system. First, *underlap* may be introduced into the servovalve, thereby increasing the valve flow-pressure sensitivity (see Eq. 16.64). Second, a leakage path may be provided across the servomotor (i.e., increased value of C_2 in the equation for leakage flow rate across the piston, $Q_L = C_2P_m$). Finally, load force (or load pressure) feedback may be provided around the servovalve-servomotor. The first and second techniques are simple and flexible, but are often undesirable because they result in decreased steady-state stiffness and increased steady-state power dissipation. The third technique also results in decreased steady-state stiffness, but avoids the problem of increased steady-state power dissipation. All three techniques result in an effective modification of the pressure-flow-current characteristics of the servovalve.

Load force (or pressure) feedback is generally preferred in high-performance systems. This feature may be implemented electrically, that is through feedback of the measured force directly to the servoamplifier. This electrical feedback approach results in a significant increase in system complexity and often a reduction in reliability. These problems can be avoided by direct use of the load pressure itself to reposition the servovalve spool. Figure 16.50 shows a servovalve in which load pressure is fed back to stub shafts located at the ends of the valve spool.

Experimentally determined steady-state flow-pressure-current characteristics for this "pressure-feedback servovalve" are shown in Fig. 16.51. Clearly, pressure feedback results in a reshaping of the characteristics from those of the conventional servovalve (see Fig. 16.34). In particular, the flow-pressure sensitivity (see Eq. 16.64) or slope of the characteristic curves (C_1) is increased in the vicinity of $I = 0$, $P_m = 0$ when pressure feedback is used. But an increase in this slope results in an increase in K_2 and therefore the load sensitivity (see Eq. 16.114).

The *dynamic pressure feedback* servovalve combines the best features of the conventional valve (see Fig. 16.30) and the pressure-feedback servovalve (Fig. 16.50). The DPF servovalve shown in Fig. 16.52 is similar to the conventional servovalve except for the addition of a high-frequency-pass network (hydraulic resistance and capacitance circuit) to achieve dynamic pressure feedback. Static load pressure feedback is eliminated in this design. The design and application of the DPF servovalve is discussed in detail in Ref. 35.

Examples

A comparative study of the performance of a position control system (see Fig. 16.53) with different servovalves was conducted by Moog, Inc. (Ref. 35). The system considered had a load resonant

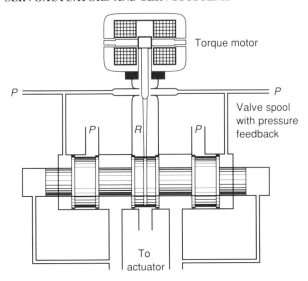

Fig. 16.50 Electrohydraulic servovalve with pressure feedback (taken from Reference 35).

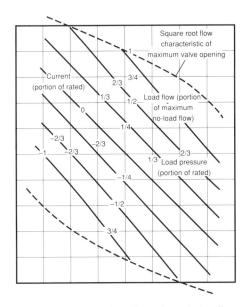

Fig. 16.51 Pressure-flow-current characteristics of an electrohydraulic servovalve with pressure feedback (taken from Reference 35).

frequency of 10.3 Hz and a damping ratio of 0.02. Tests were conducted with four different servovalves: (a) a conventional flow-control servovalve (Fig. 16.30), (b) a flow-control servovalve with a bypass orifice, (c) a flow-control servovalve with load pressure feedback (Fig. 16.50), and (d) a flow-control servovalve with dynamic pressure feedback (Fig. 16.52). In each test the amplifier gain was adjusted such that the peak amplitude ratio was 1.25 (or +2 db). In cases (b), (c), and (d) the damping was controlled to give an equivalent load damping ratio of 0.6. Measured performance results are given in Fig. 16.54 and Table 16.11. When a conventional flow-control servovalve is used as in case (a), the low value of loop gain required to produce stability results in significantly poorer dynamic performance. But only case (d), with a dynamic pressure feedback servovalve, combines good dynamic performance with a significant improvement in static stiffness to external load disturbances.

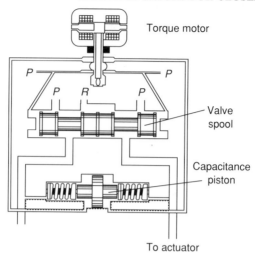

Fig. 16.52 Electrohydraulic servovalve with dynamic pressure feedback (taken from Reference 35).

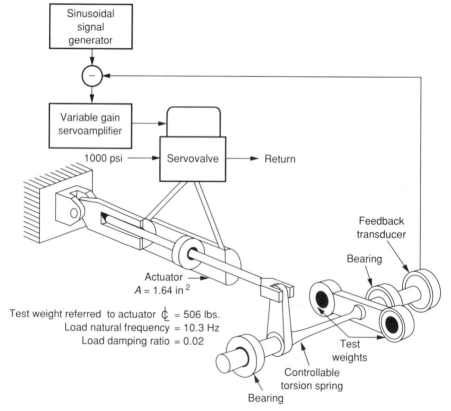

Fig. 16.53 Pictorial diagram of experimental test set up (taken from Reference 35).

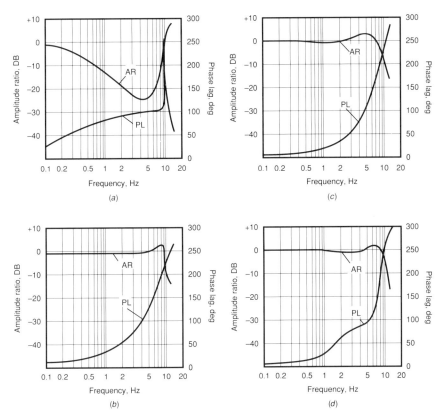

Fig. 16.54 Measured frequency responses for electrohydraulic servosystem with different servo-valves; (*a*) Measured system response with flow control servovalve; (*b*) Measured system response with flow control servovalve and bypass orifice; (*c*) Measured system response with pressure feedback servovalve; (*d*) Measured system response with servovalve (taken from Reference 35).

TABLE 16.11   COMPARITIVE PERFORMANCE OF VARIOUS ELECTROHYDRAULIC POSITION SERVOMECHANISMS

Servo Configuration	Bandwidth (± 2 db) Hz	90° Phase lag Hz	Static Load Stiffness lbs/in
Flow Control Servovalve	0.15	0.37	9,000
Flow Control Servovalve with Bypass Orifice	8.8	5	5,100
Flow Control Servovalve with Pressure feedback	8.8	5	2,500
DPF Servovalve	9.2	5	60,000

Source: From Reference 35.

16.10.6   Range of Control for Electrohydraulic and Electromechanical Servosystems

Two principal parameters characterize the range of control for most electrohydraulic servomechanisms (Ref. 36): power level and dynamic response. Figure 16.55 is a graph of control power versus control dynamics showing three dominant ranges of control for electrohydraulic servomechanisms. The middle region of the graph represents the range where electrohydraulic servomechanisms traditionally have dominated applications. Yet developments of high-effeciency and high-speed-of-response rare earth servomotors have led to increased numbers of electromechanical servomechanism applications in parts of this region. For applications requiring low power levels and low dynamic response, electromechanical solutions are generally preferred. There remains a region involving high power level and high dynamic response where neither electrohydraulic nor electromechanical solutions are available.

Areas of typical applications of electrohydraulic servosystems are plotted on the control power versus control dynamics graphs of Figs. 16.56 and 16.57. Future needs for both electromechanical and electrohydraulic servomechanisms are lower cost, higher energy efficiency, and higher dynamic response.

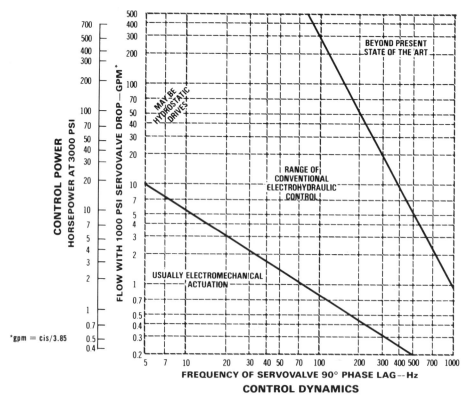

Fig. 16.55   Range of Control for Electrohydraulic Servomechanisms (taken from Reference 36). Courtesy Moog, Inc., East Aurora, NY.

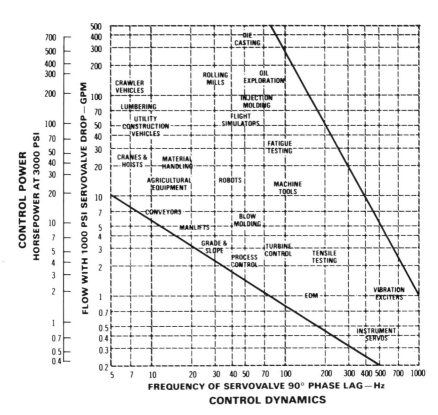

Fig. 16.56 Spectrum of industrial applications of electrohydraulic servomechanisms. (Courtesy of Moog, Inc., East Aurora, NY.)

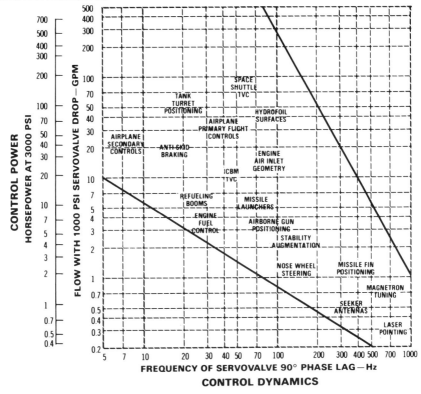

Fig. 16.57 Spectrum of Aerospace Applications of electrohydraulic servomechanisms. (Courtesy of Moog, Inc., East Aurora, NY.)

REFERENCES

1. G. S. Brown and D. P. Campbell, *Principles of Servomechanisms*, New York, Wiley, 1948.
2. M. F., Marx, and T. D. Lewis, "Electromagnetic Force Motor Design Using Rare Earth-Cobalt Permanent Magnets," NAECON 1977 RECORD, pp. 1119–1126.
3. M. A. Davis, "High Performance Electromechanical Servoactuation Using Brushless DC Motors," Technical Bulletin 150, Moog Inc., East Aurora, NY, April 1984.
4. "Electrical and Electronics 1986," *Machine Design*, Vol. 58, No. 12, May 15, 1986, pp. 6–48.
5. B. C. Kuo and J. Tal, *Incremental Motion Control—DC Motors and Control Systems*, Vol. I, SRL Publishing, Champaign, IL, 1978.
6. J. B. Leonard, "Electromechanical Primary Flight Control Activation Systems for Fighter/Attack Aircraft," Paper No. 821435, Society of Automotive Engineers, 1982.
7. B. Sawyer and J. T. Edge, "Design of a Somarium Cobalt Brushless DC Motor for Electromechanical Automator Applications," NAECON 1977 RECORD, pp. 1108–1112.
8. M. J. Cronin, "Design Aspects of Systems in All-Electric Aircraft," Paper No. 821436, Society of Automotive Engineers, 1982.
9. Electro-Craft Corp., *DC Motors, Speed Controls, Servo Systems*, 5th ed., Hopkins, MN, 1980.
10. *Speed and Position Control Systems Distributor Catalog*, Robbins Myers, Inc., Hopkins, MN, Form No. MM-7400-00, January 1988.
11. E. Aha, "Brushless DC Motors—A Tutorial Study," *Motion*, March/April 1987, pp. 20–26.
12. S. Meshkat, "Servo System Design—A Tutorial Study. Vectorial Control of Brushless DC Motors and AC Induction Motors," *Motion*, September/October 1986, pp. 19–23.
13. R. Benzer, "Single-Chip Brushless Motor Controller," *Machine Design*, June 9, 1988, pp. 140–144.
14. *Motion Control Engineering Handbook—DC Servo/Tachometers/Brushless DC*, Magnetic Technology, Inc., Canoga Park, CA, 1985.
15. *Direct Drive Engineering Handbook—DC Torque Motors/Tachometers/Brushless DC*, Magnetic Technology, Inc., Canoga Park, CA, 1985.
16. B. K. Bose, "Adjustable Speed AC Drives—A Technology Status Review," *Proceedings of the IEEE*, Vol. 70, No. 2, February 1982.
17. A. Kusko and D. G. Galler, "Survey of Microprocessors in Industrial Motor Drive Systems," *IEEE, IAS*, pp. 435–438, 1982.
18. J. E. Gibson and F. B. Tuteur, *Control System Components*, New York, McGraw–Hill, 1958, pp. 276–304.
19. B. C. Kuo, *Theory and Applications of Step Motors*, New York, West Publishing, 1974.
20. B. C. Kuo, *Incremental Motion Control—Step Motors and Control Systems*, Vol. II, SRL Publishing, Champaign, IL, 1979.
21. "1.8 deg. PM Hybrid Stepper Motors and Controls," Catalog ST-1, Bodine Electric, Chicago, IL, 1981.
22. B. H. Carlisle, "Stepping Motors: Edging into Servomotor Territory," *Machine Design*, Vol. 58, No. 26, November 6, 1986, pp. 88–100.
23. M. A. Lewis, "Design Strategies for High-Performance Incremental Servos," *Proceedings of the Sixth Annual Symposium on Incremental Motion Control Systems and Devices*, edited by B. C. Kuo, University of Illinois, 1977, pp. 141–151.
24. "Motor Selector, DC Motor Selection and Servo System—Design Software for the IBM PC and Compatibles," Par Tech Engineering, Windham, ME.
25. *Machine Design*, Fluid Power Reference Issue, September 17, 1987.
26. R. P. Lambeck, *Hydraulic Pumps and Motors—Selection and Application for Hydraulic Power Control Systems*, Marcel Dekker, New York, 1983.
27. *Fluid Power Design Engineers Handbook*, Parker-Hannifin Corp., Cleveland, OH, 1973.
28. G. B. Keller, *Hydraulic System Analysis*, Industrial Publishing, Cleveland, OH, 1969.
29. H. E. Merritt, *Hydraulic Control Systems*, Wiley, New York, 1967.
30. J. F. Blackburn, G. Reethof, and J. L. Shearer, *Fluid Power Control*, The MIT Press, Cambridge, MA, 1960.
31. E. Lewis and H. Stern, *Design of Hydraulic Control Systems*, McGraw-Hill, New York, 1962.
32. M. Guillon, *Hydraulic Servo Systems*, Plenum Press, New York, 1969.
33. J. Watton, *Fluid Power Systems*, Prentice-Hall International, Hertfordshire, 1989.

34. Moog Inc., *Electrohydraulic Servomechanisms in Industry,* East Aurora, NY (undated).

35. L. H. Geyer, "Controlled Damping Through Dynamic Pressure Feedback," Technical Bulletin 101, Moog, Inc., East Aurora, NY, April 1972.

36. R. H. Maskrey and W. J. Thayer, "A Brief History of Electrohydraulic Servomechanisms," *ASME Transactions, Journal of Dynamic Systems, Measurement and Control,* Vol. 100, June 1978.

37. K. N. Reid, "Fluid Power Control I," Course Notes, School of Mechanical and Aerospace Engineering, Oklahoma State University, Stillwater, OK, 1987.

38. *Industrial Hydraulics Manual,* No. 935100-A, Vickers, Inc., Troy, MI, Copyright by Sperry Rand Corp., First Edition 1970.

39. W. J. Thayer, "Transfer Functions for Moog Servovalves," Technical Bulletin 103, Moog Inc., East Aurora, NY, January 1965.

40. G. J. Blickley, "Servo Valve Becomes Digital Actuator," *Control Engineering,* June 1986, pp. 76–77.

41. E. B. Canfield, *Electromechanical Control Systems and Devices,* Wiley, New York, 1965, pp. 143–197.

42. T. P. Neal, "Performance Estimation for Electrohydraulic Control Systems," Technical Bulletin 126, Moog Inc., East Aurora, NY, November 1974.

43. *Stepper Motors: Permanent Magnet 7.5° Step Angle,* Catalog No. STM-REM 987-5M-RL, Crouzet Control Inc., Schaumburg, IL, 1987.

44. W. E. Wilson, "Performance Criteria for Positive-Displacement Pumps and Fluid Motors," *Transaction of the ASME,* May 1986.

45. J. L. Shearer, "Dynamic Characteristics of Valve-Controlled Hydraulic Servomotors," *ASME Transactions,* Vol. 76, August 1954, pp. 895–903.

46. P. F. Meyfarth, "Analytical Comparison of the Linear and Bang-Bang Control of Pneumatic Servomechanisms," Thesis, Department of Mechanical Engineering, Massachusetts Institute of Technology, Cambridge, MA, 1958.

47. P. F. Meyfarth, "Dynamic Response Plots and Design Charts for Third-Order Linear Systems," Massachusetts Institute of Technology, Dynamic Anaysis and Control Laboratory, Research Memorandum 7401-3, September 1958.

BIBLIOGRAPHY

Chitayat, A., "Brushless DC Linear Motors," *Motion,* pp. 22–23, September/October 1987.

Fitzgerald, A., and C. Kingley, *Electric Machinery,* McGraw-Hill, New York, 1961.

Humphrey, W. M., *Introduction to Servomechanical System Design,* Prentice Hall, Englewood Cliffs, NJ, 1973.

Koopman, "Operating Characteristics of 2-phase Servomotors," *Transactions of the AIEE,* Vol. 68, pp. 319–329, 1949.

Kusko, A., *Solid-State DC Motor Drives,* MIT Press, Cambridge, Mass., 1969.

Marinko, J. A., "PWM Servo-Amplifiers: A Tutorial Study," *Motion,* July/August 1987, pp. 3–8.

Mazurkiewicz, J., "Brushless Motors and Brushless Motor Controllers," *Motion,* pp. 14–19, November/December 1986.

Meshkat, S., "Vectorial Control of Brushless DC Motors and AC Induction Motors," *Motion,* pp. 19–23, September/October 1986.

Persson, E. K., "Brushless DC Motors in High-Performance Servo Systems," *Proceedings, Fourth Annual Symposium on Incremental Motion Control Systems and Devices,* edited by B. C. Kuo, University of Illinois, 1975, pp. T1–T16.

Proceedings, Conference on Small Electrical Machines, Institution of Electrical Engineers, London, March 30–31, 1976.

Puchstein, L., and Conrad, *AC Machines,* 3rd ed., Wiley, New York, 1954.

Schept, B., "Servo System Design—A Tutorial Study," Part 3 of 8, *Motion,* 4th Quarter, pp. 22–30, 1985.

Small and Special Electrical Machines, Publication No. 202, Institution of Electrical Engineers, London, September 1981.

Tal, J., "Quantization Errors in Digital Servo Systems," *Motion,* pp. 10–13, September/October 1986.

Yeaple, F., *Fluid Power Design Handbook,* Marcel Dekker, New York, 1984.

CHAPTER 17
CONTROLLER DESIGN

T. PETER NEAL

Aircraft Controls Division
Moog, Inc.
East Aurora, New York

17.1 INTRODUCTION

The purpose of this chapter is to provide a basis for the specification and functional design of electronic servocontrollers. No attempt is made to treat the subject of electronic circuit design. Instead, the goal is to aid the engineer in selecting and applying a suitable off-the-shelf controller or in specifying the controller requirements to a circuit designer. The emphasis is on position, velocity, or force control of mechanical loads, although many of the techniques are applicable to controller design in general. Specialized subjects, such as multiaxis control and adaptive control are beyond the scope of this chapter.

As a starting point, it is presumed that a servoactuator has been selected and mounted, together with a suitable power supply, drive amplifier, and mechanical drive mechanism. In addition, the primary feedback transducer has been chosen, and a simple loop closure has been analyzed to determine whether the specified closed-loop performance can be obtained. The process of accomplishing these tasks is treated in Chapter 16. If a simple loop closure provides adequate performance, the controller design problem primarily consists of making some basic decisions concerning electronic implementation.

In many applications, a simple loop closure is inadequate; and more elaborate controller functions are required. These latter cases are the primary subject of this chapter. The thrust of the discussion is synthesis of the controller function, rather than analysis of an existing design. For this reason, classical frequency-domain techniques will be used extensively, all starting from block diagrams and transfer functions based on the Laplace operator.[1,2,3] These techniques, many of which are graphical, are particularly useful in the early design stages. Since most people intuitively relate to time-domain responses, relationships between the frequency-domain results and time histories will be discussed as appropriate.

If the control system is to be implemented in digital hardware or software, it is certainly possible to handle the entire design task using the mathematics of digital control systems.[4,5] This approach has not been used here because the so-called classical techniques based on the Laplace transform are more illustrative and generally more familiar. For the control system applications being considered here, it is generally necessary to keep the resolution high and the sampling interval small. In this case, controller characteristics described as transfer functions in Laplace form can be accurately transformed into various mathematical forms appropriate for digital implementation, for example the Z-transform.

17.2 FUNDAMENTALS OF CLOSED-LOOP PERFORMANCE

To properly design a servocontroller, it is necessary to maintain a clear picture of the desired end result, namely, the achievement of some predetermined performance goals. These performance specifications should be established early in the design. As a minimum, they should define the desired static and dynamic accuracy, bandwidth (response time), and stability. The following sections offer a brief review of these important factors and how they relate to basic loop parameters.

17.2.1   Accuracy and Loop Gain

The most basic requirement of a servomechanism is probably static accuracy; that is, the controlled variable must accurately hold the command set point. Referring to the generalized block diagram of Fig. 17.1, several sources of inaccuracy can be described. An external disturbance can cause the load to move without any change in the command signal. The load will continue to move until the resulting error signal causes the actuator to balance the disturbance. Anomalies in the actuator and load must also be offset by a finite error signal. Examples are temperature-induced null shifts, hysteresis, threshold, friction, and lost motion. The magnitude of these error signals is minimized if the amplifier gain is high.

Ideally, the amplifier gain would be set high enough that the accuracy of the servo becomes dependent only upon the accuracy of the transducer itself. In practice, however, the amplifier gain is limited by stability considerations. Therefore, it is desirable to provide high gains at low frequencies for accuracy and low gains at high frequencies to minimize stability problems. Since the rate of amplitude roll-off with frequency is directly related to phase lag, excessive roll-off can create more stability problems than it solves. A good compromise is to make the entire forward path look like an integrator over the frequency range of interest (a type I system). This technique is very commonly used to give nearly infinite static gain and a linear gain roll-off with frequency, at the cost of 90° phase lag.

It is important to note that some servoloops contain an inherent integrator, which complicates the accuracy-versus-stability problem. For example, many actuators are inherently rate devices when operated open loop, so that a steady input results in a proportional velocity output. A velocity servo using such an actuator will inherently have a proportional forward loop, and an integrating servoamplifier can be used (Fig. 17.2). However, the corresponding position servo will inherently

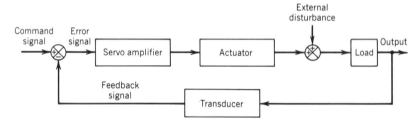

Fig. 17.1   Generalized servomechanism.

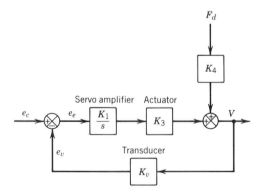

Fig. 17.2   Simplified velocity servo.

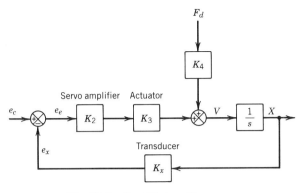

Fig. 17.3 Simplified position servo.

have an integration in the forward loop, as shown in Fig. 17.3. In this latter case, the use of an integrating servoamplifier can cause severe stability problems. From these figures, transfer functions can be written for the closed-loop responses to command and disturbance inputs. Note that the dynamic response characteristics of all elements have been neglected, and it is assumed that the load includes no spring to ground:

$$\frac{V}{e_c} = \frac{1}{K_v}\frac{1}{s/K_{vv} + 1} \tag{17.1}$$

$$\frac{V}{F_d} = K_4\frac{s}{s + K_{vv}} \tag{17.2}$$

$$\frac{X}{e_c} = \frac{1}{K_x}\frac{1}{s/K_{vx} + 1} \tag{17.3}$$

$$\frac{X}{F_d} = \frac{K_4}{K_{vx}}\frac{1}{s/K_{vx} + 1} \tag{17.4}$$

where e_c = command signal, volts

e_e = error signal, volts

e_v = velocity feedback signal, volts

e_x = position feedback signal, volts

F_d = disturbance force applied to the load, N

K_1 = integrating servoamplifier gain, (volts/sec)/volts

K_2 = proportional servoamplifier gain, volts/volts

K_3 = actuator gain, (mm/sec)/volts

K_4 = actuator velocity droop due to force disturbance, (mm/sec)/N

K_v = velocity transducer gain, volts/(mm/sec)

K_x = position transducer gain, volts/mm

K_{vv} = open-loop gain of velocity servo = $K_1 K_3 K_v$, sec^{-1}

K_{vx} = open-loop gain of position servo = $K_2 K_3 K_x$, sec^{-1}

X = load position, mm

$V = \dot{X}$, mm/sec

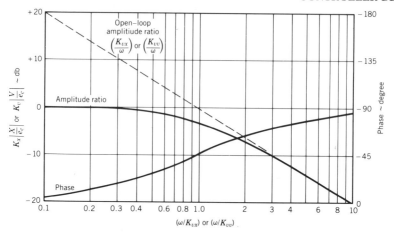

Fig. 17.4 Simplified response to commands.

As shown in Fig. 17.4, the velocity and position responses to commands are both characterized by a first-order lag having a break frequency equal to the open-loop gain. However, the responses to disturbance forces are quite different in the two cases (Fig. 17.5). When the disturbance is downstream of the integrator (velocity servo), the servo error is $(K_4 F_d)$ at high frequencies, but rolls off at frequencies (in radians per second) below K_{vv}, and is zero statically. When the disturbance is upstream of the integrator (position servo), there is a static error inversely proportional to the open-loop gain, which rolls off at frequencies above K_{vx}. Note that Eqs. 17.1–17.4 remain reasonably valid when the dynamic response characteristics of the various open-loop elements are considered, for those cases in which K_{vv} (or K_{vx}) is well below the lowest break frequencies of those elements. At higher loop gains, the closed-loop dynamics can change considerably, as shown in the following discussion.

The conclusions regarding the effects of disturbance forces on servo accuracy can be generalized to any forward-loop offset or uncertainty. Referring again to Fig. 17.2, it can be seen that the integrating amplifier will compensate for any forward-loop offset downstream of the integrator, so that the static errors are zero. From Fig. 17.3, it is apparent that static errors due to offsets upstream of the integration can be quantified as

$$\frac{X_e}{V_0} = \frac{1}{K_{vx}} \qquad\qquad (17.5)$$

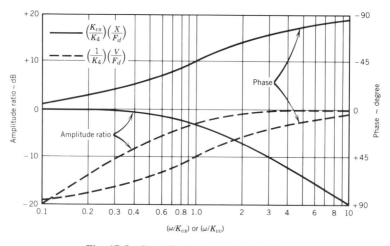

Fig. 17.5 Simplified response to disturbances.

where X_e = output error = $X_c - X$, mm

V_0 = forward-loop offset, converted to an equivalent open-loop offset in V, mm/sec

X_c = position command = e_c/K_x, mm

Even when no forward-loop offsets or disturbances are present, servos exhibit following errors, sometimes called tracking errors. In servos having forward-loop integrations, these quasi-static errors result whenever the command signal changes at a constant rate, as in a position-tracking servo. For the position servo of Fig. 17.3, the following error is

$$\frac{e_e}{\dot{e}_c} = \frac{X_e}{\dot{X}_c} = \frac{1}{K_{vx}} \tag{17.6}$$

For the velocity servo of Fig. 17.2, the following error is

$$\frac{e_e}{\dot{e}_c} = \frac{V_e}{\dot{V}_c} = \frac{1}{K_{vv}} \tag{17.7}$$

where V_e = output error = $V_c - V$, mm/sec

V_c = velocity command = e_c/K_v, mm/sec

Note that the following error for the position servo is the position error resulting from a steady rate of change of position command. For the velocity servo, the following error is the velocity error resulting from a steady rate of change of velocity command.

The servo errors discussed thus far have been those that can be minimized by a tight servoloop (high loop gain). To these must be added errors in the transducer mechanism. Even if infinite loop gain were achievable, the servo can be no more accurate than the transducer itself. The most important types of transducer inaccuracies are repeatability, resolution, and linearity. Errors due to transducer location and mounting geometry must also be taken into account.

Many of the foregoing concepts can be applied to servos in general. For example, a force or pressure servo working against a spring load is similar to a position servo in the sense that output force is proportional to actuator position. Also, temperature control servos tend to behave like position servos since the controlling device tends to provide heat flow proportional to temperature error, and thermal loads tend to produce temperature rate of change proportional to heat flow.

17.2.2 Dynamic Response and Stability

As discussed in Section 17.2.1, open-loop gain has a strong influence on servo accuracy. High loop gains also provide fast dynamic response in most cases. However, stability considerations will limit the maximum useful loop gain. The dynamic response and stability of a servo are determined by the dynamic characteristics of the various loop components. In many situations, the forward-loop dynamics are dominated by a relatively small number of low-frequency lag elements, and the transducer dynamics are negligible. In these cases, it is often possible to obtain an adequate estimate of servo performance by approximating the combined forward-loop characteristics with an integrator plus a first-order or second-order lag. The adequacy of this approximation can be determined by the match of the frequency response gain and phase for the frequency range in which the phase lag is less than 180°. Using these two rather basic dynamic forms, the relationships among loop gain, stability, and dynamic response are easily seen.

A block diagram using the basic dynamic forms is shown in Fig. 17.6,

where U = generalized controlled variable

U_c = generalized command input

D = generalized disturbance input

G_d = open-loop response of U to D

The first-order lag is typical of simple temperature control systems and dc servos having short electrical time constants. The second-order lag is often representative of electrohydraulic servos and dc servos having long electrical time constants. The closed-loop responses to command inputs are

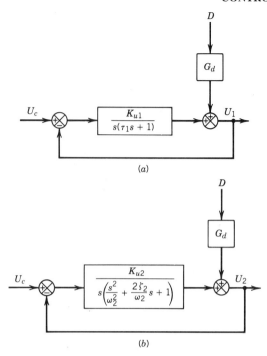

Fig. 17.6 Basic dynamic configurations; (*a*) first-order lag; (*b*) second-order lag.

$$\frac{U_1}{U_c} = \frac{1}{(\tau_1/K_{u1})s^2 + (1/K_{u1})s + 1} \tag{17.8}$$

and

$$\frac{U_2}{U_c} = \frac{1}{s^3/K_{u2}\omega_2^2 + (2\zeta_2/K_{u2}\omega_2)s^2 + (1/K_{u2})s + 1} \tag{17.9}$$

Representative root loci are shown in Fig. 17.7 for both forms. In Fig. 17.7*a*, the lag combines with the integrator to produce second-order closed-loop poles. The closed-loop natural frequency increases with loop gain, while the damping ratio decreases. In the case of Fig. 17.7*b*, the closed-

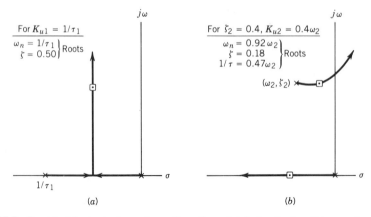

Fig. 17.7 Root loci for basic dynamic configurations; (*a*) first-order lag; (*b*) second-order lag.

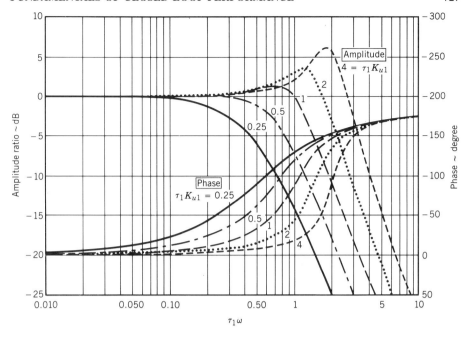

Fig. 17.8 Closed-loop frequency responses for U_1/U_c.

loop transfer function consists of a first-order and a second-order lag. The break frequency of the first-order lag increases with loop gain, while the second-order damping ratio rapidly decreases. In both loop closures, there are clearly trade-offs between closed-loop bandwidth and stability.

There are numerous methods for quantifying the relationships between bandwidth and stability. Closed-loop frequency responses to command inputs are shown for both basic forms in Figs. 17.8 and 17.9, while Figs. 17.10 and 17.11 present the corresponding step responses. Useful numerical measures of stability are phase margin, gain margin, and damping ratio of the closed-loop complex pair. These are given in Figs. 17.12 and 17.13. Note that the gain margin for Fig. 17.12 is infinite.

Referring to Fig. 17.6, closed-loop responses to disturbance inputs can be written as

$$\frac{U_1}{D} = \frac{G_d}{K_{u1}} \frac{s(\tau_1 s + 1)}{(\tau_1/K_{u1})s^2 + (1/K_{u1})s + 1} \tag{17.10}$$

and

$$\frac{U_2}{D} = \frac{G_d}{K_{u2}} \frac{s[s^2/\omega_2^2 + (2\zeta_2/\omega_2)s + 1]}{(s^3/K_{u2}\omega_2^2) + (2\zeta_2/K_{u2}\omega_2)s^2 + (1/K_{u2})s + 1} \tag{17.11}$$

To determine the final dynamic form of these responses, it is necessary to have a transfer function for G_d. This transfer function can be obtained from the physical model of the system, by deriving the response of the controlled variable to the disturbance input with the controller output equal to zero. As an example, consider a dc motor driving an inertial load and having a short electrical time constant. For an integrating velocity loop, the disturbance transfer function has the form

$$G_d = \frac{K_d}{\tau_1 s + 1} \tag{17.12}$$

and

$$\frac{U_1}{D} = \frac{K_d}{K_{u1}} \frac{s}{(\tau_1/K_{u1})s^2 + (1/K_{u1})s + 1} \tag{17.13}$$

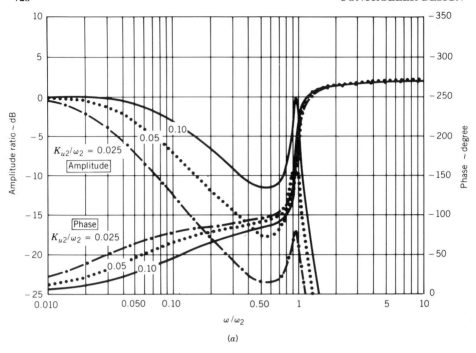

(a)

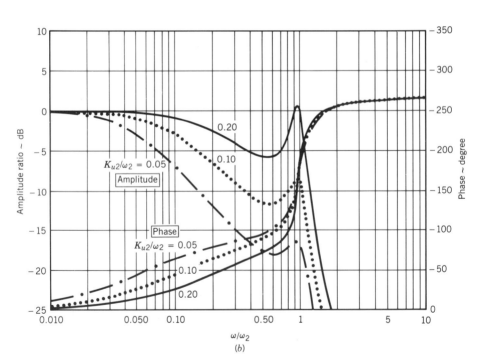

(b)

Fig. 17.9 Closed-loop frequency responses for U_2/U_c; (a) $\zeta_2 = 0.1$; (b) $\zeta_2 = 0.2$.

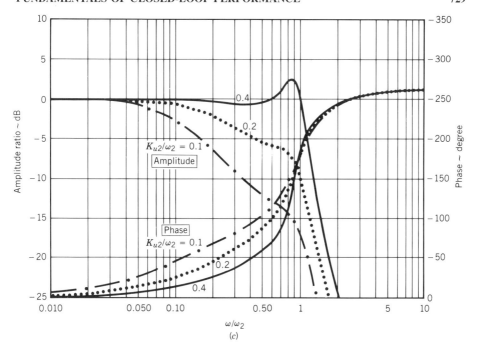

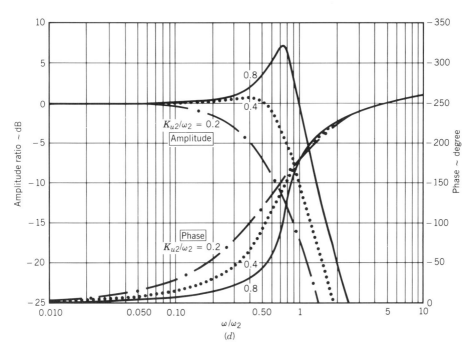

Fig. 17.9 (c) $\zeta_2 = 0.4$; (d) $\zeta_2 = 0.8$.

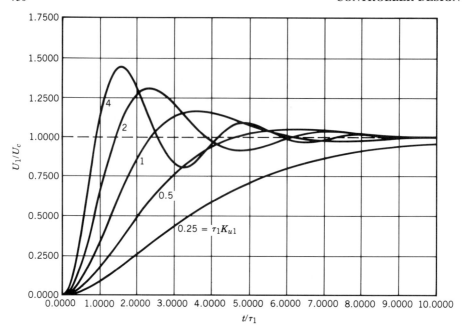

Fig. 17.10   Closed-loop step responses for U_1/U_c.

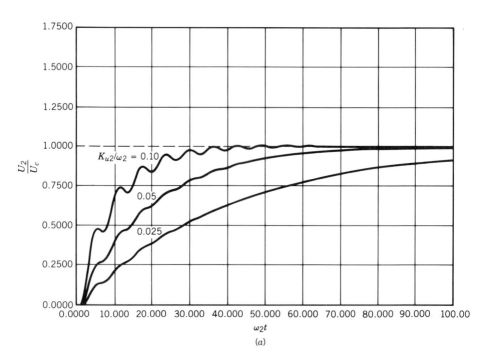

Fig. 17.11   Closed-loop step responses for U_2/U_c; (a) $\zeta_2 = 0.1$.

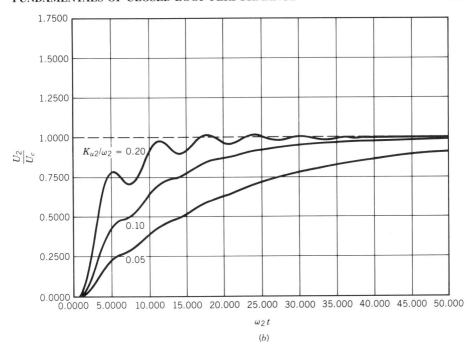

(b)

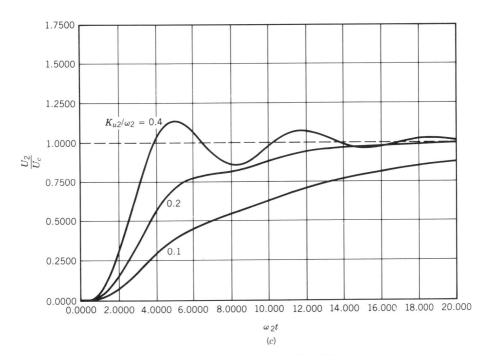

(c)

Fig. 17.11 (b) $\zeta_2 = 0.2$; (c) $\zeta_2 = 0.4$.

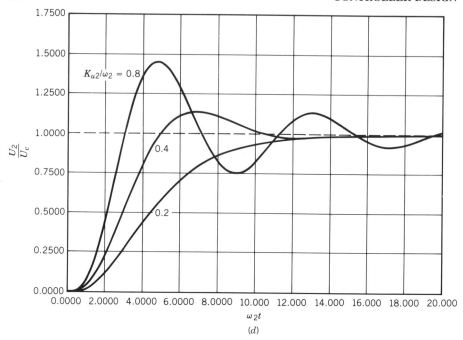

Fig. 17.11 (d) $\zeta_2 = 0.8$.

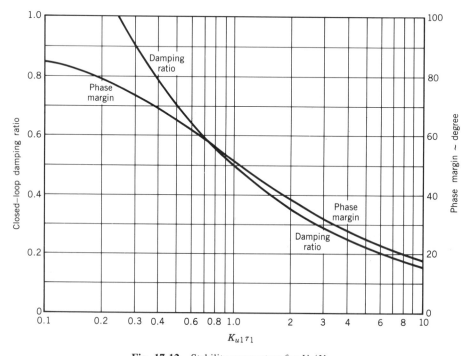

Fig. 17.12 Stability parameters for U_1/U_c.

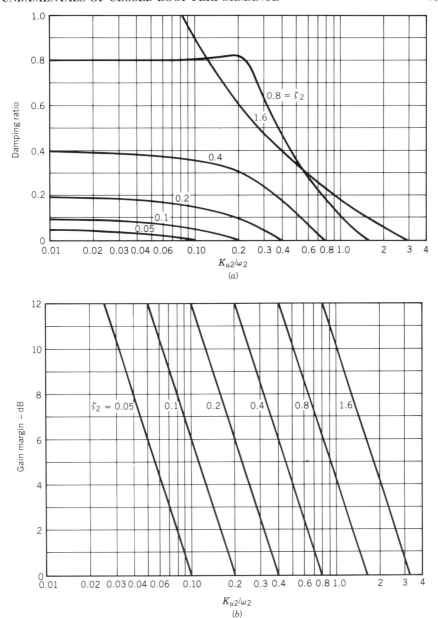

Fig. 17.13 Stability parameters for U_2/U_c; (a) closed-loop damping ratio; (b) gain margin.

For a position loop

$$G_d = \frac{K_d}{s(\tau_1 s + 1)} \tag{17.14}$$

and

$$\frac{U_1}{D} = \frac{K_d}{K_{u1}} \frac{1}{(\tau_1/K_{u1})s^2 + (1/K_{u1})s + 1} \tag{17.15}$$

It is instructive to compare Eqs. 17.13 and 17.15 with Eqs. 17.2 and 17.4, respectively.

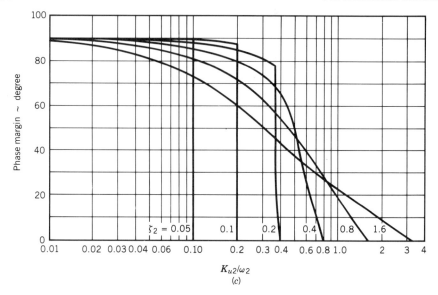

Fig. 17.13 (c) phase margin.

17.3 FREQUENCY COMPENSATION TO IMPROVE OVERALL PERFORMANCE

In Section 17.2, it is clear that open-loop dynamic characteristics impose profound limitations upon closed-loop performance. However, it is often possible to extend these limitations by modifying the inherent open-loop dynamics with frequency compensation (shaping). There are many techniques for designing compensators. The best technique to use is a function of the particular open-loop dynamics under consideration, as well as closed-loop performance goals. The following sections describe some techniques that are useful in various commonly encountered situations.

17.3.1 Well-Damped Systems

As mentioned in Section 17.2.1, an ideal form for the combined forward-loop transfer function is an integrator. This ensures very high gains at low frequencies, a linear gain roll-off with frequency, and only 90° of phase lag. For systems in which the dominant open-loop poles are reasonably well damped, loop gain is usually limited by phase lag. This is clearly illustrated by Figs. 17.12 and 17.13, in which phase margins deteriorate faster than gain margins (except when ζ_2 is low).

An obvious way to improve phase margins is to make the open-loop transfer function look like an integrator out to higher frequencies. This can be accomplished by using a lead compensator, whose zeros are identical to the dominant forward-loop poles. To make the compensator physically realizable, it must have at least as many poles as zeros, but these poles can be placed at higher frequencies. The net effect of such a lead compensator is to move the break frequencies of the forward-loop poles to higher frequencies. For the example of Fig. 17.6a, the form of the compensator would be

$$G_{c1} = \frac{\tau_{cz}s + 1}{\tau_{cp}s + 1} \tag{17.16}$$

where $\tau_{cz} = \tau_1$. For the example of Fig. 17.6b, the form would be

$$G_{c2} = \frac{s^2/\omega_{cz}^2 + (2\zeta_{cz}/\omega_{cz})s + 1}{s^2/\omega_{cp}^2 + (2\zeta_{cp}/\omega_{cp})s + 1} \tag{17.17}$$

where $\omega_{cz} = \omega_2$ and $\zeta_{cz} = \zeta_2$. In both cases, the closed-loop performance is now determined by the compensator poles since the original poles are canceled by the compensator zeros. This augmented performance can be quantified by using Figs. 17.8–17.13, with τ_{cp}, ω_{cp}, ζ_{cp} substituted for τ_1, ω_2, ζ_2, respectively.

There are a number of practical limitations on the use of lead compensation, primarily related to the large high-frequency gain of the compensator itself. For the examples of Eqs. 17.16 and 17.17, the high-frequency gains are τ_{cp}/τ_{cz} and $(\omega_{cp}/\omega_{cz})^2$, respectively. As a minimum, this characteristic will amplify any high-frequency electrical noise in the system. With reasonable care in the electrical design, high-frequency compensator gains of 10 or more are often practical. For a first-order compensator, this means that the forward-loop break frequencies can be boosted by a factor of 10, while the boost is only the square root of this factor for a second-order compensator. Another problem associated with the gain boost of a lead compensator is that poorly damped high-frequency modes can be excited or even destabilized. This latter effect will be further discussed in Section 17.3.3. The practicality of lead compensation in any given application can be best determined experimentally (additional high-frequency lags are sometimes needed).

The high-frequency noise situation is improved considerably for systems in which the integrator is electronic (e.g., the velocity servo described in Section 17.2.1). In this case, the integrator can replace one of the compensator poles so that Eqs. 17.16 and 17.17 are replaced by

$$G_{c3} = \frac{\tau_{z3}s + 1}{s} \qquad (17.18)$$

and

$$G_{c4} = \frac{s^2/\omega_{z4}^2 + (2\zeta_{z4}/\omega_{z4})s + 1}{s(\tau_{p4}s + 1)} \qquad (17.19)$$

Because of the noise-attenuating effect of the electronic integrator, $1/\tau_{p4}$ can often be set a factor of 10 greater than ω_{z4}. As previously discussed, the ratio of ω_{cp}/ω_{cz} in Eq. 17.17 is often limited to $\sqrt{10}$. As shown in the frequency responses of Figs. 17.14 and 17.15, the open-loop characteristics of a lead-compensated system will exhibit substantially improved phase characteristics when the integration is electronic rather than inherent. This makes it possible to greatly improve the closed-loop bandwidth of the system for a given level of stability.

In some cases, the open-loop dynamics may be dominated by a low-frequency lead term rather than a lag. When this occurs, a canceling lag compensator can often prevent instabilities due to poorly damped high-frequency modes. Lag compensation is normally well-behaved and suffers none of the noise problems that limit the use of lead compensation.

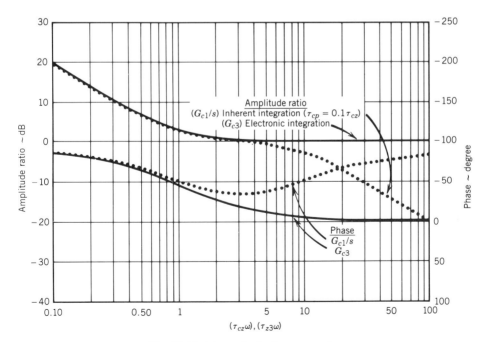

Fig. 17.14 First-order lead compensation.

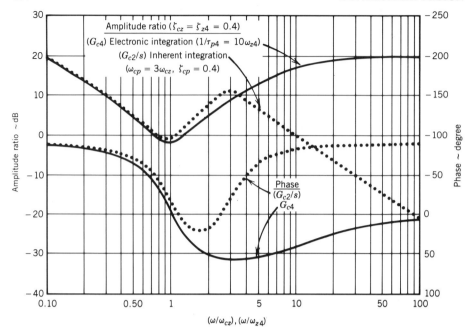

Fig. 17.15 Second-order lead compensation.

Transfer functions describing the system open-loop characteristics are not always available. Sometimes the only available system description is an experimental frequency response. When this is the case, a compensator can often be designed by graphically subtracting the experimental amplitude ratio (in decibels) and phase from those of an integrator having the same low-frequency gain. This "ideal" compensator can then be approximated with an appropriate transfer function.

17.3.2 Poorly Damped Systems

Section 17.3.1 discussed the benefits of lead compensation whose zeros are identical to the dominant forward-loop poles of the system. Theoretically, this technique can be used for any forward-loop transfer function. However, when the dominant forward-loop poles are second order and poorly damped, practical considerations render the technique highly risky in many cases. The basic problem is that the amplitude and phase of a poorly damped pair of poles change very rapidly with frequency in the vicinity of the resonant peak. If the compensator zeros are not precisely matched to the poles, the combined forward-loop transfer function can easily exhibit 180° of phase lag in the vicinity of substantial local peaking.

To illustrate the potential stability problem, consider the system of Fig. 17.6b, with $\zeta_2 = 0.10$. Suppose that the lead compensator of Eq. 17.7 is added with $\omega_{cz} = \omega_2$, $\zeta_{cz} = 0.10$, $\omega_{cp} = 3\omega_{cz}$, and $\zeta_{cp} = 0.80$. Theoretically, the forward loop is now dominated by the integrator and the compensator poles. Referring to Fig. 17.9d, it can be seen that a well-behaved response can be obtained with a loop gain $K_{u2} = 1.2\omega_{cz}$. However, suppose that ω_2 shifts to a lower value. For example, with an electrohydraulic servoactuator driving an inertial load, the "hydraulic resonance" can change 50% or more over the stroke range of the cylinder. Figure 17.16 shows how the closed-loop roots and the open-loop frequency response change with variations in ω_2.

Note that a reduction of ω_2 to $0.89\omega_{cz}$ will cause the closed-loop system to become unstable (0 db at $-180°$ phase). Even if the natural frequency of the forward-loop poles does not change at all, the poles remain poorly damped in the closed-loop transfer function. Although they are masked by the compensator zeros with regard to command inputs, they may be excited by disturbance inputs to the system. The same general comments apply to the use of notch filters. The notch is intended to attenuate the resonant peak, without the bandwidth boost of the lead compensator described in the present example ($\omega_{cp} = \omega_{cz}$, instead of $3\omega_{cz}$).

However, it is possible to design a compensator that will improve the damping of these poles. To accomplish this, it is useful to recall that for any system in which the number of poles exceeds the number of zeros by two or more, the sum of the real parts of all the poles is not changed when

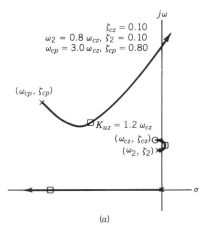

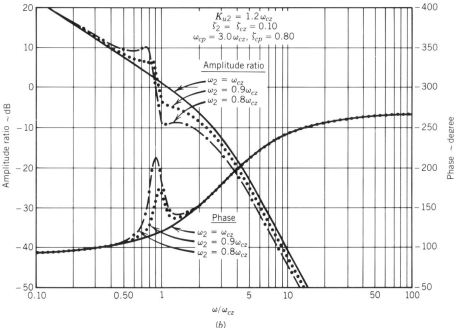

Fig. 17.16 Effects of variations in forward-loop poles on stability of a lead compensated loop; (a) root locus; (b) open-loop frequency response.

the loop is closed.[1] In this case, closing the loop will cause some roots to become more stable and others to become less stable (usually those that were poorly damped to begin with). On the other hand, if the number of poles exceeds the number of zeros by one or less, it is possible for the loop closure to improve the stability of all the roots. Therefore, to be effective in damping a poorly damped dynamic mode, lead compensation of sufficiently high order is required. To illustrate this concept, again consider the system of Fig. 17.6b, with $\zeta_2 = 0.1$. Even if we consider ideal lead compensators having no poles, Fig. 17.17 makes it clear that damping cannot be improved unless the order of the lead is 2 or more.

As mentioned in Section 17.3.1, there are important practical constraints on the use of second-order lead compensation. If the forward-loop integrator is electronic in nature, making it practical to achieve a ratio of compensator pole-zero break frequencies on the order of 10, the improvements in damping indicated by Fig. 17.17c are indeed possible. Such a compensator is defined by Eq. 17.19, with $\omega_{z4} = \omega_2$, $\zeta_{z4} = 0.4$, and $1/\tau_{p4} = 10\omega_{z4}$. The resulting root locus, shown in

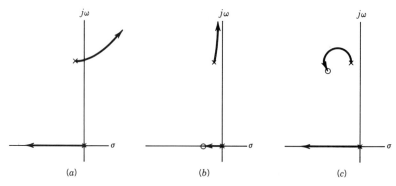

Fig. 17.17 Effect of lead compensation order on closed-loop roots; (a) no compensation; (b) first-order lead; (c) second-order lead.

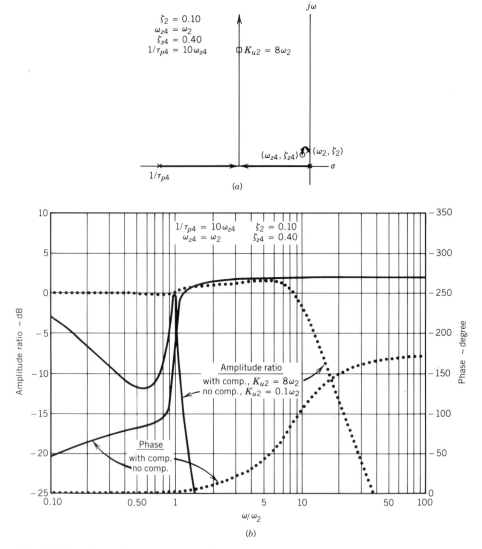

Fig. 17.18 Effects of second-order lead compensation on a poorly damped system (electronic integration); (a) root locus; (b) closed-loop frequency response.

Fig. 17.18a, indicates that open-loop gain on the order of $8\omega_2$, is possible with good stability (gain margin = 6 db and phase margin = 50°). To achieve the same gain margin without compensation, the open-loop gain would be limited to $0.1\omega_2$. Closed-loop frequency responses to commands for both cases are given in Fig. 17.18b.

If the forward-loop integration is inherent rather than electronic, the ratio of compensator pole-zero break frequencies may be limited to values as low as 3 (Section 17.3.1). Using the compensator of Eq. 17.17, a good compromise is $\omega_{cz} = \omega_2$, $\zeta_{cz} = 0.4$, $\omega_{cp} = 3\omega_{cz}$, and $\zeta_{cp} = 0.4$. The resulting impact on performance is shown in Fig. 17.19. To maintain a 6-db gain margin, the open-loop gain must be reduced to $1.0\omega_2$. Even with this reduced gain, the phase margin is only 30°. If the compensator is located in the forward loop, the closed-loop response to commands exhibits considerable peaking as shown in Fig. 17.19b. This figure also illustrates that if the compensator is located in the feedback loop, the compensator zeros do not appear in the closed-loop response to commands, and the response exhibits less peaking with more phase lag.

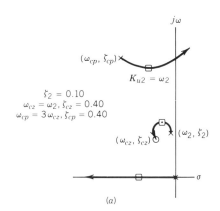

(a)

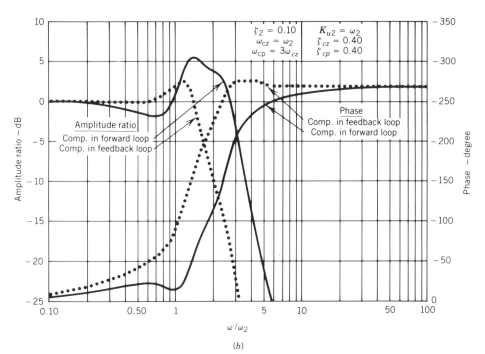

(b)

Fig. 17.19 Effects of second-order lead compensation on a poorly damped system (inherent integration); (a) root locus; (b) closed-loop frequency response.

A more straightforward method for improving the performance of poorly damped systems is by the use of lag compensation. A first-order low-pass filter in the forward loop will slow the degradation of closed-loop damping ratio as loop gain is increased. In addition, the lag compensator will attenuate abrupt command or disturbance inputs to the system, thereby reducing their ability to excite the poorly damped mode. Figure 17.20 shows the effects of a lag compensator optimized for the system of Fig. 17.6 with $\zeta_2 = 0.10$. The effects of an optimized lag compensator on various stability parameters are illustrated in Fig. 17.21 for several values of ζ_2. Comparisons with Fig. 17.13 show that lag compensation can provide a substantial improvement in loop gain for low ζ_2 but is of little use for $\zeta_2 > 0.3$. For $\zeta_2 = 0.10$, lag compensation allows K_{u2} to be increased from 0.10 to 0.27 sec^{-1} with comparable levels of stability (gain margin = 6 db). It should be noted that very little advantage is provided by the use of a higher-order low-pass filter instead of the first-order type.

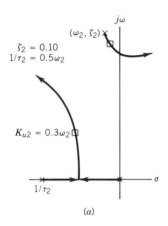

(a)

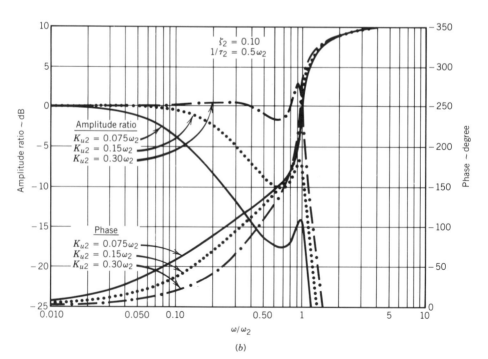

(b)

Fig. 17.20 Effects of a simple lag compensator on the closed-loop response of a poorly damped system; (a) root locus; (b) frequency response.

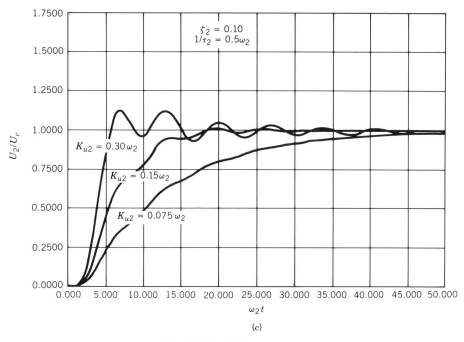

Fig. 17.20 (c) Step response.

In summary, the use of lead compensation to improve the damping of a poorly damped system is not straightforward, and the end result may be a marginal improvement in closed-loop performance. Lag compensation is more straightforward but offers only a modest improvement in performance. If substantial improvement in system damping and performance is required, the use of inner feedback loops is generally more effective, as explained in Section 17.4.

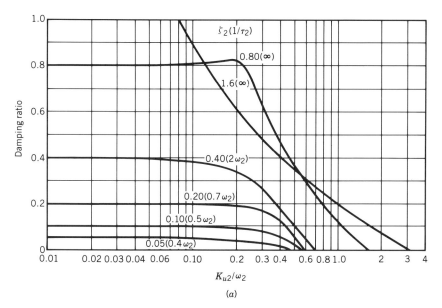

Fig. 17.21 Effects of a simple lag compensator on the stability parameters of a poorly damped system; (a) closed-loop damping ratio.

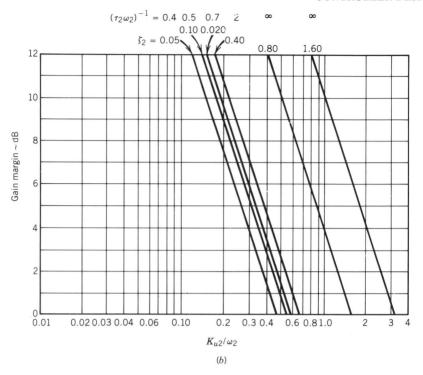

(b)

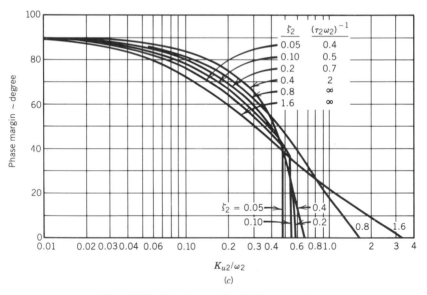

(c)

Fig. 17.21 (b) gain margin; (c) phase margin.

17.3.3 Higher-Order Effects

In the foregoing sections, the discussion has centered around systems whose forward-loop dynamics can be approximated by relatively simple lag elements. While this approach is often entirely adequate for controller design and performance estimation, the designer should be aware of several potential limitations. These limitations usually involve the higher-order, higher-frequency dynamic modes that were unknown or neglected in the early stages of the design. Examples are structural modes, transducer dynamics, and drive amplifier dynamics.

Higher-order modes associated with actuator mounting structure, actuator-load mechanical connections, and transducer mounting are usually second order and poorly damped. Often these modes are characterized as poorly damped pole-zero combinations. In either case, loop gains selected to obtain adequate performance and stability from the dominant low-frequency modes could be high enough to drive the poorly damped high-frequency modes unstable. This is particularly likely if lead compensation is used to improve the dynamics of the dominant modes. High-frequency stability is probably best assessed by a root locus plot or Bode plot that includes estimates of all the system dynamic modes. Often, accurate description of the higher-order modes is difficult at the design stage. For this reason, it is worthwhile investigating the effects of various high-frequency compensation techniques that can be added when the system is first evaluated experimentally. For example, notch filters are often useful in reducing the effects of high-frequency structural modes at the cost of some low-frequency phase lag.

Many higher-order modes are reasonably well damped. The dynamics of drive amplifiers, transducer signal conditioners, and notch filters are typical examples. The primary influence of such modes is that they introduce phase lags that can destabilize the low-frequency dominant modes or limit the effectiveness of lead compensation used to improve low-frequency behavior. A useful approximation of these effects can be made if there is reasonable frequency separation of the higher-order modes from the dominant lower-frequency modes (a factor of 5 or more). In this case, the system's low-frequency dynamic behavior can usually be assessed by replacing the high-frequency modes with a single first-order lag. The time constant of this first-order approximation, τ_3, can be determined as follows:

$$\tau_3 = \frac{\phi_3}{57.3\omega_3} \tag{17.20}$$

where ϕ_3 is the net phase lag (degrees) of all the combined high-frequency modes, measured at ω_3 (radians per second). The frequency ω_3 should be approximately equal to the highest natural frequency of the lower-frequency dominant modes. It is probably best not to use this simplification if ϕ_3 approaches 30°.

If frequency-response data are available, they can be used to estimate ϕ_3. However, at the design stage, it is likely that rough estimates of higher-order natural frequencies and damping ratios are the only information available. In this case, ϕ_3 can be determined by adding the phase lag contributions of the individual high-frequency modes:

$$\phi_3 = \sum \phi_4 + \sum \phi_5 \tag{17.21}$$

$$\phi_4 = 57.3\tau_4\omega_3 \tag{17.22}$$

$$\phi_5 = 115\zeta_5\frac{\omega_3}{\omega_5} \tag{17.23}$$

where τ_4 is the time constant of a first-order mode, while ω_5 and ζ_5 are the natural frequency and damping ratio of a second-order mode (ϕ_4 and ϕ_5 due to lag terms add to ϕ_3, while lead terms subtract from ϕ_3). The approximations of Eqs. 17.22 and 17.23 are accurate to one degree or better for $1/\tau_4 > 3\omega_3$ and $\omega_5 > 3\omega_3$, respectively.

17.4   INNER FEEDBACK LOOPS

If the desired servo performance cannot be achieved using frequency compensation, as described in Section 17.3, the addition of inner feedback loops can often provide the needed improvement. This is particularly true when the open-loop damping ratio is poor and the forward loop has an inherent integration. Although inner feedback loops require additional transducers, they offer more flexibility in modifying the servo dynamics than does frequency compensation alone. The following sections discuss the merits of feeding back derivatives of the controlled variable, feeding back variables dynamically different than the controlled variable, and nonelectronic mechanizations of inner loops.

17.4.1   Derivatives of the Controlled Variable

Section 17.3 illustrates the benefits of lead compensation but also shows that its effectiveness is limited by the high-frequency gain amplification of the compensator itself. This problem can be alleviated by the use of transducers that directly measure derivatives of the controlled variable. To

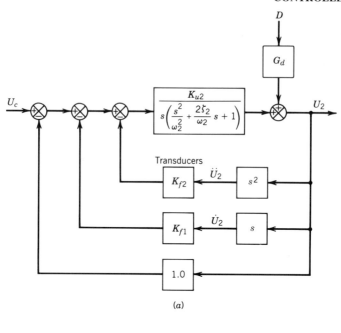

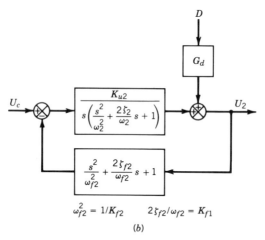

Fig. 17.22 Feedback of controlled-variable derivatives; (a) inner loops to improve damping; (b) combined feedback loops.

illustrate the potential benefits, consider again the example of Fig. 17.6b with additional feedback of the first and second derivatives of the controlled variable, as illustrated in Fig. 17.22a. As shown in Fig. 17.22b, the three feedbacks can be mathematically combined into a single loop having a pure second-order pair of zeros, without the added lag normally associated with lead compensators. The result is that much higher forward-loop gains can be achieved for a given level of closed-loop stability. Of course, each transducer will introduce higher-frequency dynamic effects that will eventually limit the maximum forward-loop gain, but these effects are usually less restrictive than those imposed by electronic lead compensators. Note that the natural frequency and damping ratio of the feedback zeros in Fig. 17.22 are determined by the magnitude of the derivative feedback gains relative to the primary feedback gain, which is 1.0 in this case.

There are a variety of uses for derivative feedback loops, including improved closed-loop accuracy, bandwidth, and stability, as well as reduced sensitivity to changes in system parameters. These uses can be illustrated with the aid of Figs. 17.23 and 17.24 which show the effects of

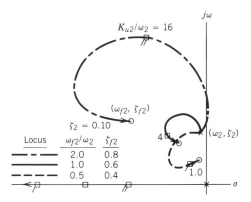

Fig. 17.23 Effects of derivative feedback on closed-loop roots.

various sets of feedback zeros on the closed-loop dynamics of Fig. 17.22b. Figure 17.23 gives root loci for the different zero locations. Because electrical noise and excitation of high-frequency dynamic modes usually limit the gain in the highest derivative loop, a forward-loop gain is selected for each locus that holds $(K_{u2}K_{f2})$ constant. Closed-loop frequency responses are then given in Fig. 17.24. It can be seen that closed-loop damping is improved in all cases. Placement of the feedback zeros at high frequency (low derivative feedback gains) yields high closed-loop bandwidth, but the second-order poles can easily be destabilized by a reduction in forward-loop gain or the presence of the inevitable higher-order modes described in Section 17.3.3. On the other hand, low-frequency feedback zeros (high derivative feedback gains) result in lower closed-loop bandwidth but offer greatly reduced gain sensitivity and are little affected by high-frequency modes. Also in the latter case, the closed-loop response is dominated by the low-frequency second-order poles, which are nearly the same as the feedback zeros. Therefore, the closed-loop characteristics do not vary significantly with large changes in the plant parameters $(K_{u2}, \omega_2, \zeta_2)$. A good compromise between bandwidth and parameter sensitivity is to set ω_{f2} approximately equal to ω_2.

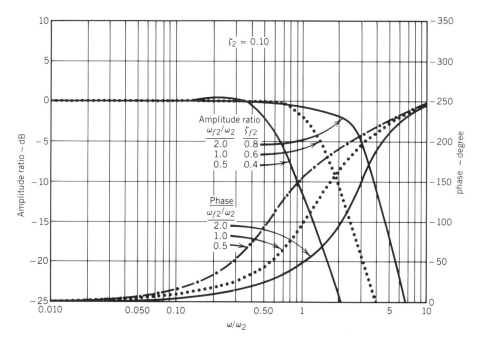

Fig. 17.24 Effects of derivative feedback on closed-loop frequency response of U_2/U_c.

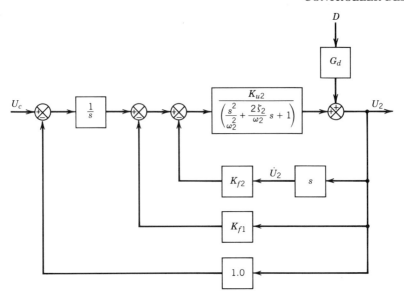

Fig. 17.25 Rearrangement of Fig. 17.22a inner loops (electronic integration).

Comparison of Figs. 17.23 and 17.24 with Figs. 17.18 and 17.19 show that derivative feedback offers better stability and more flexibility than lead compensation, particularly when the forward loop has an inherent integration. In addition, higher forward-loop gains can generally be used with derivative feedback, which improves the static accuracy of the system. However, forward-loop lead compensation can often produce higher closed-loop bandwidth if the forward-loop integration is electronic in nature, as illustrated in Fig. 17.18.

A useful way to optimize the system of Fig. 17.22a is to first establish rough estimates of the feedback and forward-loop gains using the combined feedback approach of Figs. 17.22b and 17.23. It is often helpful to include a first-order approximation of the phase lags caused by the higher-order modes (Eq. 17.20). Once the approximate gains are established, the closed-loop response and stability can be checked by analysis of a complete multiloop model, with the higher-order modes described more completely and placed in the appropriate loops.

If the forward-loop integrator is electronic in nature, the benefits of derivative feedback can be achieved with one less derivative than if the integrator is inherent. This is accomplished by feeding back the inner loops downstream of the integrator, as illustrated in Fig. 17.25. The static and dynamic characteristics of Fig. 17.25 are entirely equivalent to those of Fig. 17.22a and can also be mathematically reduced to the single-loop configuration of Fig. 17.22b. Note, however, that the practical implementation of Fig. 17.25 requires one less transducer than Fig. 17.22a.

Another way in which derivative feedback can be useful is to provide very smooth and repeatable dynamic response and high static accuracy when the primary control loop has an inherent integration. This is particularly useful when closed-loop bandwidth is not a major concern. The technique involves closing a tight integrating loop around the first derivative of the controlled variable, as illustrated in Fig. 17.26. This loop submerges the effects of forward-loop gain variations, static offsets, and external disturbances. The primary control loop gain can then be set at relatively low levels to ensure smooth, repeatable dynamic response without the usual concerns about reduced static accuracy. This technique is also useful if the mounting arrangement of the primary transducer results in gain-limiting higher-order dynamic characteristics in the outer feedback loop.

17.4.2 Alternative Inner-Loop Variables

Sometimes it is advantageous to close inner feedback loops around variables that are dynamically different than derivatives of the primary controlled variable. This might be done because of practical problems related to accurately transducing derivatives of the controlled variable or because it offers an inherent advantage related to feedback dynamic characteristics. Usually, there are advantages and disadvantages of using alternative feedback loops, as illustrated in the following example.

Consider an electrohydraulic position servo that has rather demanding requirements for dynamic response. Suppose that envelope or environmental constraints make it very difficult to mount velocity and acceleration transducers or that the mounting arrangement itself introduces undesirable higher-

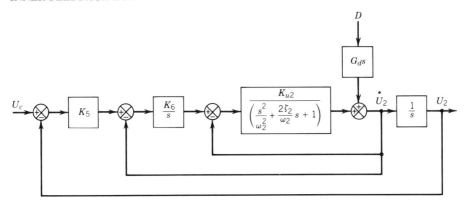

Fig. 17.26 Rearrangement of Fig. 17.6 to give smooth, repeatable, and accurate response (inherent integration).

order dynamics. In this case, consideration might be given to transducing cylinder differential pressure in place of the velocity or acceleration feedbacks. If the load can be represented as primarily a mass, cylinder pressure will be proportional to load acceleration. If cylinder or load friction is large, it may be necessary to use a load cell rather than a pressure transducer. However, the load is often more complex than a simple mass. For example, the load might also have substantial stiffness to ground and viscous damping. In this case, cylinder pressure would have components proportional to load acceleration, velocity, and position. If these components were of the proper relative size and not highly variable, a pressure transducer alone could replace the acceleration and velocity transducers. However, if the load dynamics were complex or highly variable, the use of cylinder pressure as an inner loop might do more harm than good.

Another potential problem associated with the use of alternative inner feedback loops is the influence of external disturbances. Consider again the example of the electrohydraulic position servo. Feedback of either velocity or acceleration will have no effect on closed-loop static stiffness because, by definition, all derivatives of position are zero in the steady state. However, an external force applied to the load will change the cylinder pressure, even in the steady state. Therefore, pressure feedback will reduce closed-loop stiffness unless the pressure feedback signal is high passed, which introduces its own set of dynamic characteristics. Of course, there are applications in which closed-loop stiffness is not critical in the first place.

It should also be mentioned that the mounting of the primary transducer can introduce its own dynamic peculiarities. For example, the controlled variable might be load position relative to ground, and the transducer might be integrally mounted within the servoactuator assembly. If there were substantial compliance in the structure to which the actuator is mounted, the position feedback loop would contain structural zeros. An integral velocity transducer would have the same problem, but a load-mounted accelerometer would not. In this case and in others where the derivative feedback loops are dynamically different from one another, the simplified techniques of Section 17.4.1 are of limited use, and the complete multiloop model must be analyzed directly. Again it should be noted that these more complex derivative feedback loops may result in better or poorer closed-loop performance than comparable inner loops that feed back pure derivatives of the controlled variable. Proper assessment of these trade-offs requires a good physical model of the system, showing the proper relationships of all the feedback variables being considered.

17.4.3 Nonelectronic Inner Loops

Occasionally, it is useful to implement inner-loop feedbacks by mechanical design rather than electronic means. For example, it may be possible to mount the servoactuator or primary transducer so that structural deflections under load produce favorable feedback zeros that improve closed-loop damping. In an electrohydraulic servo, improved damping can often be obtained by using hydraulic pressure feedback, implemented with a cross-port orifice or laminar leakage path. Both of the schemes will suffer loss of closed-loop stiffness. Sometimes it is possible to mount an electrohydraulic servoactuator so that its rod attaches to the mounting structure and its body attaches to the load. If a mass is then attached to the control valve spool and the spool is aligned with the actuator centerline, a form of acceleration feedback can be achieved (valve porting must be arranged to give proper feedback polarity). Mechanical feedback schemes offer the potential advantages of reduced costs and complexity, as well as alleviating the need for high open-loop bandwidth in the actuator and controller. However, the design of servos using these techniques often requires manipulation of rather complex physical models, which is beyond the scope of this chapter.

17.5 PREFILTERS AND FEED FORWARD

As can be seen from Sections 17.3 and 17.4, the business of obtaining good closed-loop performance can become rather complex. Sometimes feedback loops and frequency compensation are optimized to achieve the desired stability and accuracy, but the closed-loop response to commands is not particularly desirable. Rather than compromise stability or accuracy by altering the servoloop characteristics, it is often easier to shape the command signal before it enters the servoloop. The following sections discuss some commonly used techniques for accomplishing this, together with their limitations.

17.5.1 Lag Prefilters

High-gain servoloops are required to achieve static accuracy and rejection of load transient disturbances, but rapid response to commands is often unnecessary or undesirable. In this case, the addition of a simple lag prefilter will often provide the desired result:

$$G_{\mathrm{pf}} = \frac{1}{\tau_{\mathrm{pf}}s + 1} \tag{17.24}$$

If this prefilter is placed in the command path, as illustrated in Fig. 17.27, τ_{pf} can be made large enough that G_{pf} dominates the U/U_c response, with an appropriate rise time.

Another way that lag prefilters can be used is to provide a particular set of dynamics for U/U_c, which are easily settable and do not change with variations in the forward-loop parameters. Often called "model following," this technique requires the use of high-bandpass servoloops, so that the dynamics of the prefilter model (G_{pf}) dominate the U/U_c response. In concept, this makes it very easy to obtain any desired U/U_c transfer function by simply changing the electronic prefilter model. However, to have U/U_c faithfully reflect the model dynamics, it is often necessary for the bandpass of the servoloop to be an order of magnitude higher than the highest frequency singularity in the model transfer function. This is often impractical. If inner feedback loops are needed to achieve the desired servoloop bandpass, it may be more effective to tailor feedback loops to provide a combined feedback transfer function that is the inverse of the desired U/U_c transfer function. This can be seen with the aid of Fig. 17.27 by eliminating the prefilter. If the forward-loop gain is high enough, $U/U_c = 1/H_1$. This technique is further discussed in Section 17.4.1, and an example is given in Fig. 17.22.

17.5.2 Lead Prefilters

Since the closed-loop response characteristics of most servoloops are dominated by lag elements, lead prefilters are often used to improve the response to command inputs. Theoretically, this can be accomplished by simply making the prefilter transfer function equal to the reciprocal of the servoloop closed-loop transfer function. For the generalized system illustrated in Fig. 17.27, the ideal prefilter would be

$$G_{\mathrm{pf}} = \frac{1 + G_1 H_1}{G_1} \tag{17.25}$$

Unfortunately, the lead required to accomplish this will be limited by its associated poles, which must be selected to prevent excessive electrical noise (as discussed in Section 17.3.1). Also lead

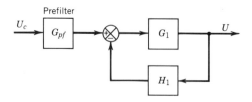

Fig. 17.27 Generalized use of prefilters.

prefilters can accentuate the oscillatory tendencies of a poorly damped servoloop. Even for a well-damped servoloop, overshooting response can occur if the servoloop parameters vary substantially over the operating envelope. In many cases, the lead network may be more effective if it is moved to the forward-loop so that higher servoloop gains can be used.

To illustrate the effects of lead prefilters, consider a servoloop of the configuration shown in Fig. 17.6a. Since the closed-loop transfer function will be a second-order lag, as shown by Eq. 17.8, a second-order prefilter lead would be appropriate. The use of such a prefilter is illustrated in Fig. 17.28. Assume that the forward-loop gain is selected to give a closed-loop damping ratio $\zeta_6 = 0.50$. This would require $K_{u1} = 1/\tau_1$, as determined from Fig. 17.12. The closed-loop natural frequency can then be determined from Eq. 17.8, as follows:

$$\omega_6 = \sqrt{\frac{K_{u1}}{\tau_1}} = \frac{1}{\tau_1} \qquad (17.26)$$

The prefilter zeros can now be set to cancel the closed-loop lag: $\omega_{pz} = 1/\tau_1$ and $\zeta_{pz} = 0.50$. As discussed in Section 17.3.1, a practical upper limit on the high-frequency gain amplification of a lead network is approximately a factor of 10. Therefore, the prefilter poles can be set to $\omega_{pp} = 3/\tau_1$ and $\zeta_{pp} = 0.50$. The net effect of the prefilter is an effective boost in closed-loop natural frequency by a factor of 3.

It should be noted that the use of lead prefilters can cause some peculiar effects when variations in the forward-loop characteristics are considered. For the example of Fig. 17.28, the effects of forward-loop gain variations are illustrated in Figs. 17.29 and 17.30. Because the closed-loop peaking at high

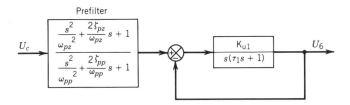

Fig. 17.28 Example of a lead prefilter.

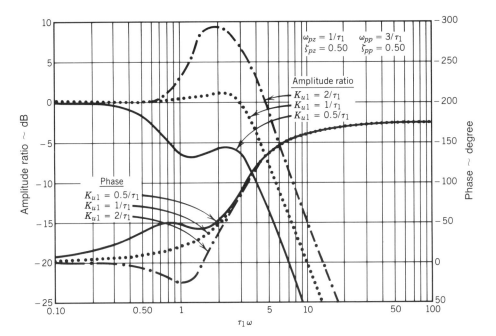

Fig. 17.29 Effects of loop gain variations on U_6/U_c frequency response.

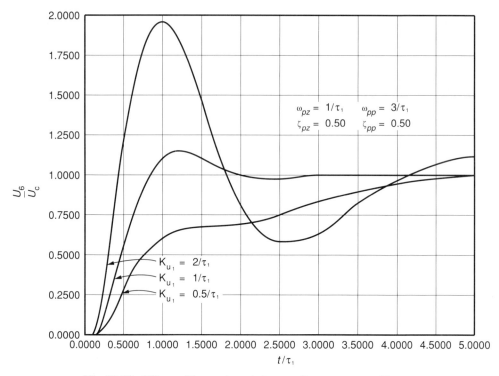

Fig. 17.30 Effects of loop gain variations on U_6 response to a U_c step.

gains is accentuated by the prefilter, it is usually best to optimize the prefilter for the maximum-gain situation and accept the degraded response at low gains. If the variations in forward-loop characteristics are large, it may be necessary to use inner feedback loops to minimize the variations in closed-loop response, as explained in Section 17.4.1.

17.5.3 Feed Forward

As explained in Section 17.5.2, the effectiveness of lead prefilters is limited by the fact that the command signal must be differentiated. In a system whose forward loop contains electronic lags, the number of command differentiations can often be reduced by the use of feed-forward techniques. Feed forward can also be used to reduce the following errors associated with a steady rate of change of the command signal. However, these techniques must be carefully applied to prevent adverse effects on system dynamic response.

The principles of feed forward are illustrated in Fig. 17.31a. If the electronic shaping in the feed-forward path approximates the reciprocal of the nonelectronic forward-loop elements, the command signal will nominally be reproduced at the output. Follow-up by the feedback loop will then try to minimize the effects of inaccuracies in the feed-forward signal. The use of a prefilter matched to the feedback path further improves the overall response to commands. The net effect of the feed-forward configuration can be seen by rearranging the block diagram into an equivalent one that has only a prefilter, as shown in Fig. 17.31b. From this figure, the system response can be written

$$\frac{U_7}{U_c} = \left[H_2 + (G_2 G_3)^{-1} \right] \frac{G_2 G_3}{1 + G_2 G_3 H_2}$$

$$= \frac{G_2 G_3 H_2 + 1}{G_2 G_3} \frac{G_2 G_3}{1 + G_2 G_3 H_2} = 1.0 \tag{17.27}$$

Of course, it is not possible to actually achieve the ideal result given by Eq. 17.27 because of the lags associated with the lead network in the feed-forward path. To illustrate the practical aspects of feed forward, it is useful to reexamine the example of Fig. 17.28. If the forward-loop integrator is electronic in nature, the use of feed forward offers some advantages over the straight prefilter. Figure 17.32a shows the appropriate form for the feed-forward model, and Fig. 17.32b shows it

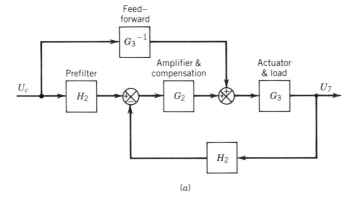

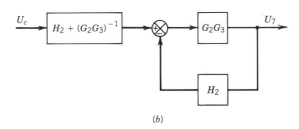

Fig. 17.31   Idealized use of feed forward; (a) feed-forward model; (b) equivalent prefilter model.

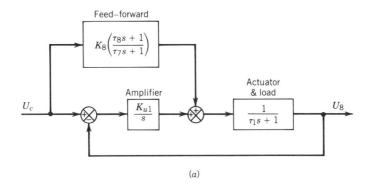

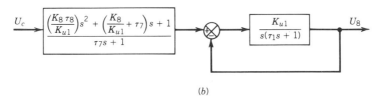

Fig. 17.32   Example of feed forward; (a) feed-forward model; (b) equivalent prefilter model.

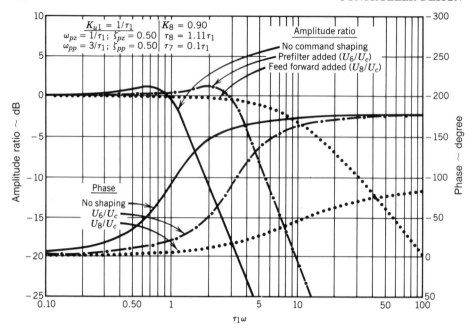

Fig. 17.33 Comparison of closed-loop frequency responses for the prefilter and feed-forward examples.

reduced to an equivalent prefilter model. Notice that Fig. 17.32a requires only one differentiation of the command signal, while Fig. 17.28 requires two. The result is that its equivalent prefilter (Fig. 17.32b) includes a first-order lag rather than a second-order lag (Fig. 17.28).

To allow direct comparison of the Figs. 17.28 and 17.32 examples, the same loop gain is used in each case ($K_{u1} = 1/\tau_1$). Noting that τ_8 will be approximately equal to τ_1 and that the high-frequency gain amplification of the feed-forward network should be limited to 10, it is sensible to set $\tau_7 = 0.1\tau_1$. Since the closed-loop poles are given by Eq. 17.8, the feed-forward parameters are selected as follows:

$$\frac{K_8 \tau_8}{K_{u1}} = \frac{\tau_1}{K_{u1}} \tag{17.28}$$

$$\frac{K_8}{K_{u1}} + \tau_7 = \frac{1}{K_{u1}} \tag{17.29}$$

Solving these equations after the appropriate substitutions, the feed-forward parameters are $K_8 = 0.90$, $\tau_8 = 1.11\tau_1$, $\tau_7 = 0.1\tau_1$. Using these parameters, U_8/U_c frequency and step responses are computed and presented in Figs. 17.33 and 17.34, along with the corresponding U_6/U_c responses. In this example, the use of feed-forward techniques offers a substantial improvement in system bandwidth, by comparison with a prefilter alone. Of course, the feed-forward scheme suffers from sensitivity to forward-loop gain variations similar to those of the prefilter scheme, as illustrated in Figs. 17.29 and 17.30.

As previously mentioned, feed-forward techniques can also reduce following errors. In general, the following error, U_e, can be calculated as follows:

$$\frac{U_e}{U_c} = \frac{1}{s}\left(1 - \frac{U}{U_c}\right) \tag{17.30}$$

For the example of Fig. 17.32, in which the closed-loop transfer function is determined by the feed-forward pole, the following error is determined from

$$\frac{U_{8e}}{U_c} = \frac{1}{s}\left(1 - \frac{1}{\tau_7 s + 1}\right) = \frac{\tau_7}{\tau_7 s + 1} \tag{17.31}$$

For steady command rates, the error is $\tau_7 \dot{U}_c$. Without feed forward, the following error can be determined from Eq. 17.8.

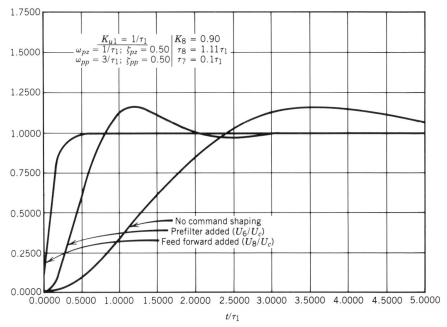

Fig. 17.34 Comparison of closed-loop step responses for the prefilter and feed-forward examples.

$$\frac{U_{1e}}{\dot{U}_c} = \frac{1}{s}\left\{1 - \left[\frac{1}{(\tau_1/K_{u1})s^2 + (1/K_{u1})s + 1}\right]\right\}$$

$$= \frac{(\tau_1/K_{u1})s + (1/K_{u1})}{(\tau_1/K_{u1})s^2 + (1/K_{u1})s + 1} \qquad (17.32)$$

For steady command rates, the error is $\dot{U}_c/K_{u1}$. For this example, $\tau_7 = 0.1/K_{u1}$. In this case, the proper use of feed forward has reduced the following errors by a factor of 10.

17.6 PID CONTROLLERS

A very popular form of controller is called PID (proportional integral differential). It is very simple in concept and is relatively easy to mechanize. Many essays have been written that describe rules of thumb for "tuning" the controller. Unfortunately these tuning procedures can be rather tedious and are usually applicable for only very simple actuator-load dynamics. The purpose of this section is to offer a unified rationale for applying PID controllers that is useful in synthesizing a control system. This rationale also provides insight for systematically adjusting PID parameters on actual hardware.

17.6.1 Equivalence to Frequency Compensation

The basis for the ensuing discussion is that a PID controller is simply a particular form of forward-loop frequency compensation. This can be seen from the generalized controller shown in Fig. 17.35. Note that the differential path is filtered to limit the high-frequency amplitude ratio. Combining the parallel paths of Fig. 17.35, a single transfer function for the controller can be written:

$$G_{\text{pid}} = \left(\frac{K_d s}{\tau_d s + 1}\right) + K_p + \left(\frac{K_i}{s}\right)$$

$$= \left(\frac{K_i}{s}\right)\left\{\frac{[K_d/K_i + (K_p/K_i)\tau_d]s^2 + [K_p/K_i + \tau_d]s + 1}{\tau_d s + 1}\right\}$$

$$= \left(\frac{K_i}{s}\right)\left[\frac{s^2/\omega_{\text{pid}}^2 + (2\zeta_{\text{pid}}/\omega_{\text{pid}})s + 1}{\tau_d s + 1}\right] \qquad (17.33)$$

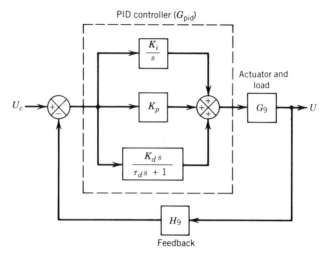

PID controller (G_{pid})

Actuator and load

Feedback

Fig. 17.35 Generalized PID controller.

Note that Eq. 17.33 is simply the transfer function of a lead compensator combined with an integrator, as given by Eq. 17.19. As discussed in Section 17.3.1, it is usually possible to place the lag break frequency a factor of 10 above the lead natural frequency. In this case, the τ_d terms in the numerator are usually small:

$$\omega_{pid} = \sqrt{\frac{K_i}{K_d}} \qquad \text{for } \tau_d \ll \frac{K_d}{K_p} \tag{17.34}$$

$$2\zeta_{pid}\omega_{pid} = \left(\frac{K_p}{K_d}\right) \qquad \text{for } \tau_d \ll \frac{K_p}{K_i} \tag{17.35}$$

In some applications, second-order lead compensation is not required. In such cases, a simplified version of the PID controller can often be useful. This so-called proportional-integral (PI) controller is formed by setting K_d to zero. The resulting transfer function can then be derived from Eq. 17.33:

$$G_{pi} = \left(K_p + \frac{K_i}{s}\right) = \left(\frac{K_i}{s}\right)\left(\frac{K_p}{K_i}s + 1\right)$$

$$= \left(\frac{K_i}{s}\right)(\tau_{pi}s + 1) \tag{17.36}$$

This result is similar to Eq. 17.18.

Another simplified version of the PID controller is obtained by setting $K_i = 0$. The transfer function of this so-called proportional-differential (PD) controller can also be derived from Eq. 17.33:

$$G_{pd} = \left(\frac{K_d s}{\tau_d s + 1}\right) + K_p$$

$$= K_p\left[\frac{(K_d/K_p + \tau_d)s + 1}{\tau_d s + 1}\right]$$

$$= K_p\left[\frac{\tau_{pd}s + 1}{\tau_d s + 1}\right] \tag{17.37}$$

This transfer function is the same as the first-order lead compensator of Eq. 17.16 and is normally used in systems having an inherent integration. Note that the lead break frequency can be a factor of 10 lower than the lag break frequency. In this case the τ_d term in the numerator of Eq. 17.37 is small:

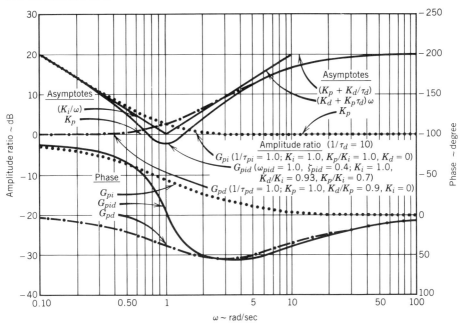

Fig. 17.36 Frequency response comparisons of PID, PI, and PD controllers.

$$\tau_{pd} = \left(\frac{K_d}{K_p}\right) \qquad \text{for } \tau_d \ll \frac{K_d}{K_p} \tag{17.38}$$

Frequency responses of representative PID, PI, and PD controllers are given in Fig. 17.36, which also shows the effects of the various controller parameters.

17.6.2 Systems Having No Inherent Integration

Electronic integrators are normally used in the forward loops of systems having no inherent integrations, as explained in Section 17.2.1. Section 17.3 explains the various ways in which lead compensation can be usefully applied. A PID controller offers a convenient method for combining the electronic integration with lead compensation. As discussed in Section 17.6.1, the PI scheme provides first-order lead, while the complete PID scheme offers second-order lead.

17.6.3 Systems Having an Inherent Integration

Section 17.2.1 explains that an electronic integrator in the forward loop minimizes static servo errors. However, the addition of an electronic integrator to a system that already has an inherent integration will usually cause dynamic instability. To prevent this type of instability, lead compensation must be combined with the electronic integrator. This can be accomplished with a PI controller.

As shown in Fig. 17.36, a PI controller contributes nearly 90° of phase lag at low frequencies. If this is added to the 90° of lag already contributed by the integrator inherent in the system, the total low-frequency lag approaches 180°. Since the other system dynamics will add even more phase lag at high frequencies, the PI break frequency must be set low enough to ensure an intermediate frequency range over which the phase lag is reduced. The open-loop frequency responses of Fig. 17.37a illustrates this effect for a system whose inherent characteristics consist of an integrator and a second-order lag. Generally, a PI break frequency greater than 10% of the system's lowest lag frequency will substantially reduce closed-loop stability. This is shown by the phase plots of Fig. 17.37a, by the root loci of Fig. 17.37b, and by the closed-loop frequency responses of 17.37c.

Using the 10% rule of thumb for the PI controller, it is interesting to examine some time histories of the closed-loop system. Figure 17.38 shows time responses to step and ramp commands for the

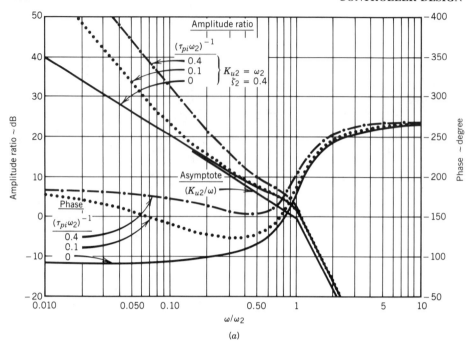

(a)

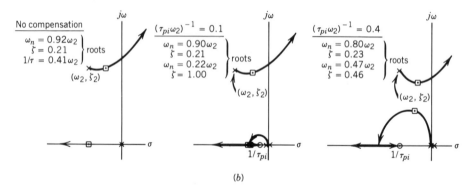

(b)

Fig. 17.37 PI controller added to the system of Fig. 17.6b—effects of lead break frequency; (a) open-loop frequency response; (b) root loci ($\zeta_2 = 0.4$, $K_{u2} = 0.35\omega_2$).

system of Fig. 17.37. As illustrated in Fig. 17.38a, the PI controller degrades the step response. Figure 17.38b shows that the PI's double integration at low frequencies eliminates the following error but causes larger overshoot of the steady state.

The 10% rule also applies to systems whose inherent characteristics consist of an integrator and a first-order lag. Frequency responses, root loci, and time histories for such a system are given in Fig. 17.39. It should also be noted that the effective lag break frequency of the system can be increased by using lead compensation techniques, as described in Section 17.3. For the system of Fig. 17.39, this can be accomplished by using a PID, rather than PI, controller. Reexamining the example of Fig. 17.39 using the PID characteristics of Fig. 17.36, a 10-fold increase in effective system bandwidth can be achieved by setting $\omega_{pid} = 1/\tau_1$ and $\zeta_{pid} = 1.0$. In this case, one of the two PID lead terms cancels the system lag at $1/\tau_1$. The resulting open-loop frequency response is shown in Fig. 17.40.

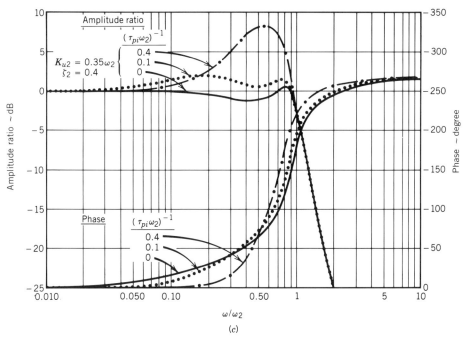

Fig. 17.37 (c) closed-loop frequency response.

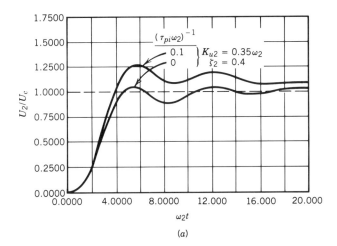

Fig. 17.38 PI controller added to the system of Fig. 17.6b—time histories; (a) step response.

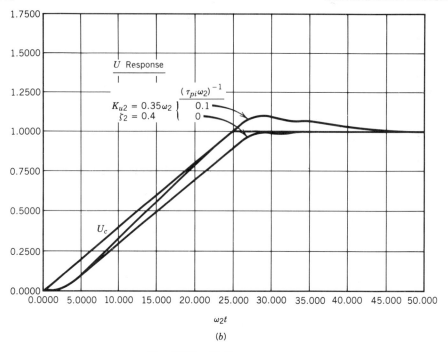

Fig. 17.38 (*b*) Ramp response.

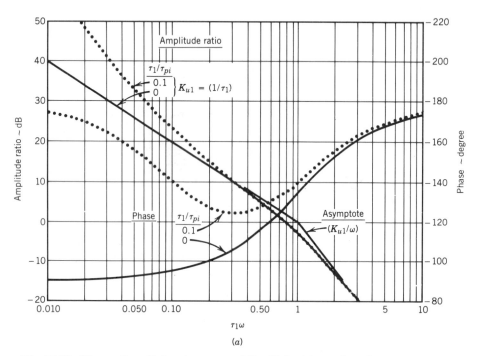

Fig. 17.39 PI controller added to the system of Fig. 17.6*a*; (*a*) open-loop frequency response.

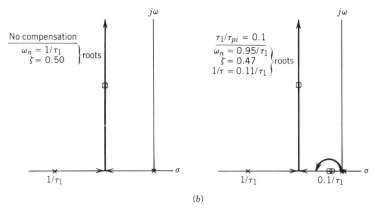

(b)

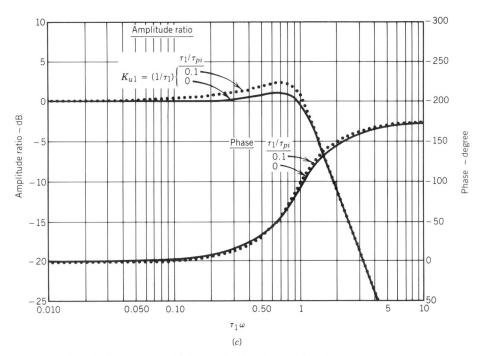

$\tau_1 \omega$

(c)

Fig. 17.39 (b) root loci ($K_{u1} = 1/\tau_1$). (c) closed-loop frequency response.

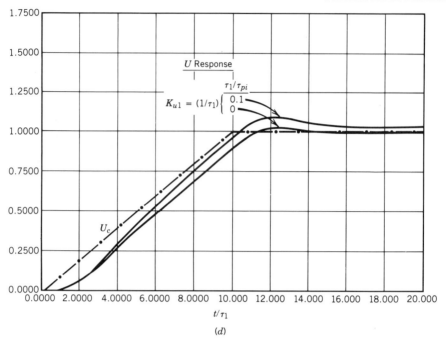

Fig. 17.39 (*d*) Ramp response.

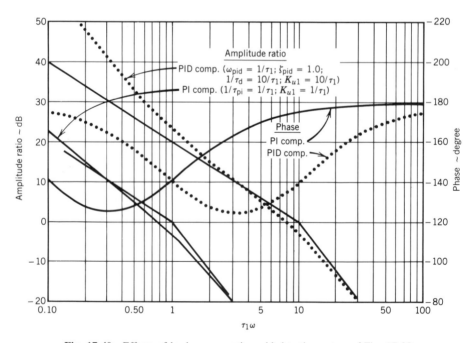

Fig. 17.40 Effects of lead compensation added to the system of Fig. 17.39.

17.7   EFFECTS OF NONLINEARITIES

The previous sections have concentrated on the design of controllers for linear systems. In practice, physical systems are never truly linear. If the nonlinearities are not large, design of the controller using linear techniques is very useful. However, it is important to understand the limitations of this approach. The following sections discuss these limitations and offer several approaches to dealing with nonlinearities.

Figure 17.41 illustrates idealized forms of several nonlinearities that are commonly encountered in systems controlling mechanical loads. Saturating nonlinearities can occur in transducers, electronics, and the servoactuator itself. Deadzone is the lack of output for small changes in input and is generally most significant in servoactuators and transducers. Resolution is the availability of a limited number of output values and is typical of digital electronics and many types of transducers, including encoders and wire-wound potentiometers. Most servoactuators provide output velocities that are force dependent or output forces that are velocity dependent. The load-velocity curves shown in Fig. 17.41*d*

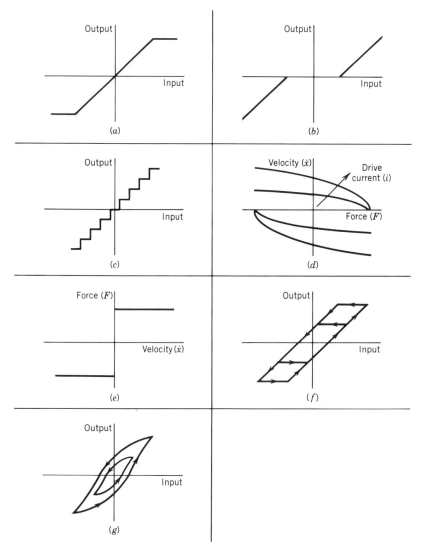

Fig. 17.41 Common nonlinearities; (*a*) saturation; (*b*) deadzone or threshold; (*c*) resolution; (*d*) load-velocity curves; (*e*) Coulomb friction; (*f*) mechanical backlash; (*g*) magnetic hysteresis.

are typical of an electrohydraulic servoactuator and are highly nonlinear. Coulomb friction is a constant force that always opposes motion. Static friction (stiction) is often larger than Coulomb friction but is very difficult to model.

Mechanical backlash is motion lost when the direction of motion is reversed, as in gear trains and bearings. Since most servoactuators make use of electromagnetic elements, magnetic hysteresis effects can cause some system performance anomalies. The width of the hysteresis band is dependent upon the amplitude of the input signal (the output is a function of the input's prior history as well as its present value).

17.7.1 Simple Nonlinearities

When the output of a nonlinear element depends only upon the present value of its input, the element can often be described by a simple relationship between input and output amplitude. If this function is single valued, it is often possible to assess its effect on the system by using linear approximations. One useful technique is to examine the small-perturbation behavior of the system at a series of operating points along the input-output curve, by performing a linear analysis using the local slope at each operating point.

Another technique is describing function analysis, which is useful in estimating the response of nonlinear systems to sinusoidal inputs. In general, a describing function is an amplitude-dependent, frequency-dependent transfer function of a nonlinear element, which allows the system to be analyzed by conventional frequency domain techniques. It is derived from a Fourier analysis of the output of the nonlinear element to a sinusoidal input.[1] For simple nonlinearities that can be described by a single-valued output amplitude versus input amplitude, the describing function is a simple gain that varies with input amplitude. In concept, this gain is the average slope of the input-output curve for the particular input amplitude being considered.

Saturation and deadzone are two of the most common nonlinearities encountered in control of mechanical systems. Referring to Fig. 17.41a, an operating-point analysis would idealize the nonlinearity as a simple gain when no saturation takes place and as zero gain when fully in the saturation region. Similar rational can be applied to Fig. 17.41b. Generally speaking, linear techniques can be used to ensure system stability by analyzing the system with a range of gains determined by the minimum and maximum slopes of the nonlinear amplitude curve. Small-perturbation step response and frequency response of the system around an operating point can also be determined by linear analysis.

Describing-function analysis can provide useful insight into the behavior of the combined dead-zone-saturation nonlinearity shown in Fig. 17.42. Its describing function is a gain that is zero in the deadzone region, increases to a maximum as the input amplitude approaches saturation, then decreases again as the input pushes well into the saturation region. If linear analysis predicts instability at the maximum value of the nonlinear gain, this type of describing function will result in a sustained oscillation at an amplitude corresponding to maximum gain. This behavior is called a stable limit cycle because any tendency of the oscillation to diverge will result in lower gain, which will reduce the tendency to oscillate.

Stable limit cycles can also result from deadzone in a system that is marginally stable at low gains. For example, Section 17.6.3 explains that a PI compensator used in a system having an inherent integration can exhibit 180° of phase lag at low gains, become stable at intermediate gains, then become unstable at high gains. In this case, a low-frequency oscillation can develop, whose amplitude will grow until the describing-function gain is high enough to produce a stable limit cycle.

The effects of saturation and deadzone on system stability are generally straightforward to analyze by operation-point analysis or describing functions, as long as the system consists of single control loops. However, when multiple feedback and feed-forward loops are present, the linearized analysis must be performed very carefully. For example, when an inner feedback loop is used to damp an open-loop resonant mode so that higher gains can be achieved in the outer feedback loop, hard

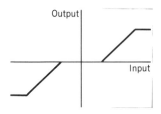

Fig. 17.42 Combined deadzone and saturation.

saturation or deadzone in the inner feedback path can cause the outer loop to become unstable. Similarly, saturation in a feed-forward path or in a lead network can greatly reduce the stabilizing effects they were designed to provide. Another type of saturation is acceleration limiting. Even if it does not create any stability problems, acceleration limiting can cause large overshoots when a position servo decelerates into final position following a large step command.

The resolution of digital systems and certain transducers creates its own set of problems. As shown in Fig. 17.41c, resolution nonlinearities can be described as alternating regions of zero gain and infinite gain. Therefore, resolution will cause most feedback systems to exhibit continuous stable limit cycles with an amplitude corresponding to the resolution increment (least significant bit in a digital system). In a high-resolution system, the magnitude of this limit cycle may be so small that it is not noticeable. Similarly, Coulomb friction (Fig. 17.41e) exhibits infinite gain around zero and a saturating behavior as amplitude increases. In this case, however, the nonlinearity is usually a feedback loop around a mechanical load. When the load is primarily inertial and has no backlash, friction may actually improve system stability rather than decrease it. Of course, friction will also decrease system accuracy.

17.7.2 Complex Nonlinearities

As nonlinear elements become more complex, linear analysis becomes more complicated and less realistic. However, linear techniques may still be of some use in estimating system stability. For example, a servoactuator's output velocity may be a nonlinear function of output force, as well as input drive current. For the example of Fig. 17.41d, system stability can be explored by using linearized characteristics at selected operating points:

$$\Delta \dot{X} = \left(\frac{\partial \dot{X}}{\partial i} \right) \Delta i + \left(\frac{\partial \dot{X}}{\partial F} \right) \Delta F \qquad (17.39)$$

The two derivatives in Eq. 17.39 can be used in a conventional linear model showing velocity as a function of current input and load force feedback (Fig. 17.43). Note that the derivative of velocity with respect to force is negative in this case.

Some nonlinear elements such as hysteresis and backlash, cannot be approximated by a simple relationship between input and output amplitude. Instead, the output depends upon the history of the input as well as its present value. The describing functions of such elements are typically frequency dependent, as well as amplitude dependent.[1] Describing-function analysis with such nonlinearities can become rather complicated and is beyond the scope of this chapter. Also, it can be argued that computer simulation yields more realistic results without much additional effort. This is particularly true if the control system has multiple nonlinearities that are significant.

17.7.3 Computer Simulation

If used properly, modern digital simulation techniques can provide realistic results for very complicated, highly nonlinear systems. Programming is typically accomplished quickly and easily. General-purpose simulation programs can have several drawbacks such as limited ability for on-line operator interaction and excessive time required to generate output data, especially frequency responses. Furthermore, most general-purpose programs are carefully designed to minimize limitations on what can be programmed. As a result, they will happily violate the laws of physics without complaint. As with all computer tools, it is good practice to check out all critical program functions prior to generating data, and to spot-check the early data with hand calculations wherever possible.

Analog computers offer unlimited opportunity for on-line operator interaction and real-time operation. Real-time capability means that there is no waiting for data and allows system development work to be accomplished using a combination of simulation and real hardware ("hardware in the

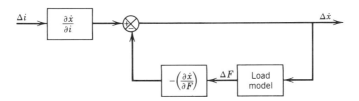

Fig. 17.43 Linearized model of servoactuator load-velocity characteristics (from Fig. 17.41d).

loop"). Unfortunately, considerable setup time is required because of the need for patching and scaling, and there are severe restrictions on the size of the simulation that can be handled. These limitations have caused analog computers to become virtually extinct. In their place, a variety of real-time digital and hybrid simulation tools have been developed. These typically offer convenient programming, fast run times, the ability to handle large simulations, and hardware-in-the-loop capability. Their disadvantages include the high cost and maintenance problems associated with specialized computer hardware, inability to time-share, inability to use for other kinds of engineering problems, and obsolescence.

In summary, modern computer simulation is a powerful tool to aid in the development of servo control systems, but it is not a panacea. Unless the control system is extremely complex or highly nonlinear, the use of a simplified linear model is often the best way to synthesize the function of the control system and to perform preliminary perfomance estimates. This approach is typically faster than simulation, is less prone to major errors, and promotes physical understanding of the system's behavior. It is true that simplifying assumptions must be made very carefully, but this process also promotes improved understanding of the system. With the basic system function defined, simulation can then be used to evaluate the simplifying assumptions of the linear analysis, "fine-tune" the system design, and generate detailed performance data over a wide range of operating conditions. A listing of various computer programs to perform linear analyses and nonlinear simulation is given in Ref. 6. Also see Section 14.8 for a discussion of simulation for control system analysis.

17.8 CONTROLLER IMPLEMENTATION

As mentioned in the introduction, it is not the intent of this chapter to address the electronic design of a servocontroller. Nevertheless, some basic understanding of controller implementation is required to properly specify and select a controller. The following discussion describes several basic implementation approaches, together with their relative advantages and disadvantages.

Since control of a mechanical load is inherently a continuous process, the use of a dc analog controller typically provides the highest servo bandwidth and smoothest operation. Furthermore, basic servoloops can be implemented with very simple circuits. Hard-wired digital controllers offer the potential for increased overall accuracy at the cost of degraded resolution and more complex electronic hardware. Microprocessor-based digital controllers offer increased flexibility, versatility, accuracy, and computer-interface capability but typically suffer reduced bandwidth as well as degraded resolution. The electronic hardware associated with processor-based controllers is typically more complex than a simple analog controller. However, in complex control applications, large amounts of analog circuitry can be replaced with software.

17.8.1 Analog Controllers

Analog controllers are typically used in servoloops for which high closed-loop bandwidth and smooth operation are required. Modern operational amplifiers have bandwidths of several hundred kilohertz and virtually infinite resolution. Because of this, they are free of the sampling delays, phase lags, and resolution problems associated with microprocessor-based controllers. Furthermore, they are relatively tolerant of electrical noise, and troubleshooting can be accomplished with simple equipment. On the other hand, analog implementation of complex controller functions such as nonlinearities, automatic gain changing, elaborate command processing, and complex failure detection can cause the electronic circuitry to become extremely complicated. In addition, analog controllers typically require periodic adjustment and calibration.

The basic functional elements of a typical analog controller are shown in Fig. 17.44. The functions of prefilters and compensation networks have been discussed previously in this chapter. Some form of signal conditioning is usually required for transducers. In the case of a simple dc transducer such as a potentiometer, the conditioning may consist of dc excitation together with an output buffer amplifier. Buffer amplifiers can be designed to protect against large voltages erroneously connected to the electronics, to provide consistent loading of the transducer output, to reject electrical cabling noise (electromagnetic interference), and to filter transducer ripple. Low-output transducers, such as those that employ strain gage elements, require high-gain, low-drift amplifiers. Linear variable differential transformers (LVDTs), resolvers, and other ac transducers require ac excitation, demodulation, and filtering to remove ripple. Greatly increased servo accuracy can be obtained with a combination of ac command generation and ac feedback, such as when synchros are utilized, but this approach has largely been replaced by the use of digital controllers and transducers. In any case, transducer specifications should be carefully studied to determine the proper signal conditioner characteristics. The dynamic characteristics of the signal conditioners, as well as the transducers themselves, can have significant impact on the stability and performance of the servoloops, as explained previously in this chapter.

There are many sources of information concerning the design of analog controllers. Reference 7 is an excellent source for the design of operational amplifier circuits for a wide variety of purposes,

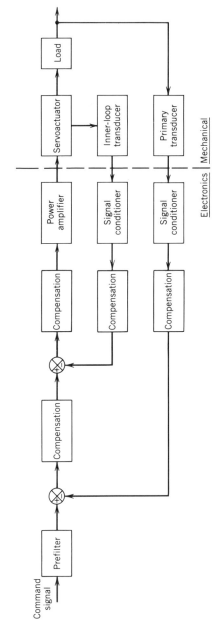

Fig. 17.44 Typical analog controller.

including compensation and signal conditioning. Furthermore, most manufacturers of operational amplifiers publish useful application handbooks. Also, application literature from transducer manufacturers often discusses signal conditioning techniques in some detail.

The design of power amplifiers varies widely with the type of servoactuator being driven. In the case of conventional electrohydraulic servoactuators and other types requiring low-power electrical inputs, the power amplifier can be a very simple linear (proportional) circuit. Typically, it utilizes an operational amplifier and a power boost stage consisting of a complimentary pair of transistors. Some operational amplifiers have enough output capability to provide the required electrical input directly. In the case of electromechanical devices that must provide a direct electrical-to-mechanical energy conversion, such as dc servomotors, the power amplifier can become very complex. In this case, its design should be left to an experienced electronics engineer. Linear amplifiers are still the most straightforward and offer the best servo performance but are severely limited in the size of the motor they can control because of the large amount of heat they must dissipate. Since switching transistors typically generate little heat in their full-on or full-off states, various time-modulated on/off power drivers have been developed. The most popular type for servo control applications seems to be pulsewidth modulation (PWM). In this approach, the power devices are switched on and off at a very high fixed frequency. The percent on-time during each cycle is proportional to the dc input voltage from the upstream analog controller circuitry. If the PWM frequency is high enough, only the average cycle voltage will affect the servoactuator output, thereby resulting in nearly proportional control. To accomplish this, the PWM frequency must usually be at least one order of magnitude higher than the bandwidth of the innermost feedback loop (typically a current feedback loop around the power amplifier) or 2 to 3 orders of magnitude higher than the bandwidth of the primary control loop. If PWM frequencies of this magnitude are impractical, it may be necessary to reduce bandwidths of the various control loops to prevent unacceptable servo output at the PWM frequency. Design of a PWM amplifier can be difficult, particularly with regard to polarity switching around zero.

It should be noted that servoactuators requiring high-power electrical inputs can create problems related to voltage saturation in the drive electronics. The reason for this is that the input inductive characteristics of many actuators tend to cause long L/R time constants. Current feedback is often used to improve the current response of the actuator, and this leads to large transient voltages for abrupt inputs to the power amplifier. Practical limits on amplifier voltage capability can result in voltage saturation during transients. It is often necessary to design an active voltage-limiting circuit to prevent stability problems during saturation. Careful attention must be devoted to tailoring the current feedback loop and voltage limiter circuits to the servoactuator's electrical dynamics if amplifier stability and performance problems are to be avoided.

Multiturn potentiometers are usually included in analog controller circuitry to allow proper calibration of transducer scale factors, compensate for electrical and mechanical offsets, and to adjust loop gains. The desire for system accuracy often drives the designer to providing many adjustments, but this practice can greatly complicate maintenance procedures. If adequate performance cannot be achieved with a limited number of well-placed adjustments, then serious consideration should be given to the use of a digital controller.

The use of integrating amplifiers in control loops requires some special consideration. First, pure integrators often cause low-frequency oscillations or "hunting" when backlash, deadzone, or friction exists in the system. This behavior can often be controlled by adding a large resistance across the integrator's feedback capacitor. This limits the amplifier's gain and makes it look proportional at low frequencies, while preserving an integrating characteristic in the crossover frequency range. If it is possible for an integrating amplifier to saturate during abrupt commands, it may "latch up" and exhibit a long recovery period, which can result in large servo overshoots. This behavior can be prevented by proper gain distribution in the servoloops or by providing the amplifier with a diode limiter. Of course, an integrating amplifier can drift into saturation if it is powered up before the servoactuator is allowed to move. For example, electronics are often powered up prior to releasing a mechanical brake or applying hydraulic power. In this case, integrator saturation can cause a large engagement transient. This can be prevented by shorting the integrator's feedback capacitor with a relay contact or electronic switch. The short is then opened when the actuator is mechanically or hydraulically engaged.

17.8.2 Hard-Wired Digital Controllers

The overall accuracy of a servomechanism can be greatly enhanced by the use of a digital transducer and a digital controller. Furthermore, the need for periodic calibration and adjustment can be virtually eliminated. Hard-wired digital electronics can provide this improved accuracy with bandwidths comparable to analog electronics. However, these digital circuits are considerably more complex than comparable analog circuits, are more susceptible to electrical noise, and have finite resolution. For these reasons, hard-wired digital electronics are usually used only in the primary control loop (accuracy is typically not critical in the inner loops). Furthermore, frequency compensation is difficult to implement in digital hardware and is usually left to analog circuitry.

Hard-wired digital electronics are commonly used with high-resolution incremental encoders, as illustrated in Fig. 17.45a. Two pulse trains are generated by the encoder, 90° out of phase with one another. The pulse-conditioning circuitry squares up the incoming pulses, determines the transducer's direction of motion, and often increases the resolution by a factor of four. The asynchronous counter is incremented up or down by each feedback pulse, depending upon the direction of motion. Similarly, command pulses also increment the counter up or down. The net count at any particular time represents the difference between the number of command and feedback pulses since the counter was initialized. After digital-to-analog conversion, this count becomes the error signal transmitted to the analog electronics.

The command pulse train can be generated by additional digital hardware or by a computer. However, if the transducer resolution is high and the desired maximum command rate is high, a computer may be hard-pressed to provide the required pulse rates. In this case, hardware comparators and rate multipliers may prove more satisfactory. It should be noted that an incremental system has no inherent knowledge of its absolute position. Therefore, power shutdowns and electrical noise can cause such a system to lose track of where it is. For this reason a "marker pulse" is often provided at some known position to reinitialize the counter periodically. Alternatively, the servo occasionally can be commanded to a mechanical "home" position. As with most transducers having finite resolution, encoders will usually cause limit cycling with an amplitude equal to the least significant bit (Section 17.7.1).

Absolute digital systems can also be implemented in electronic hardware, as shown in Fig. 17.45b. The encoding transducer outputs a digital word that represents its absolute position at all times. For an optical encoder, the resolution is typically between 12 and 24 bits. After buffering, the feedback word is digitally subtracted from a digital command, and the resulting error is converted to an analog signal that is transmitted to the analog electronics. The digital summing junction can be implemented in a number of ways, including the use of an arithmetic logic unit (ALU), which operates at very high speeds. The digital command and feedback information can be transmitted to the summing junction as parallel digital words or as serial data that must be multiplexed and then decoded. The command information can be generated from a computer or digital thumbwheels.

The absolute system is less susceptible to loss of position information than the incremental system, but the transducers are considerably more expensive and less reliable, and a wire is required for each bit. To improve reliability and reduce cost, resolvers or other sine/cosine output devices are often used with a resolver-to-digital (R/D) converter. The penalty is reduced overall accuracy, although this can be improved by using coarse/fine resolvers and appropriate additional hardware logic.

17.8.3   Computer-Based Digital Controllers

The hard-wired digital controllers of Fig. 17.45 have limited flexibility and functional capability. The use of a microprocessor may reduce electronic hardware complexity when elaborate system functions are required. There are many such functions that are well suited to microprocessor implementation:

- Command processing (nonlinear functions, limiting, switching, and communication with other computers)
- Redundancy management (fault detection, isolation, and reconfiguration)
- Adaptive control (self-adjustment of control loop parameters as operating conditions or environmental factors change)
- Built-in test (BIT) features

Once the need for a microprocessor is established, it may also become reasonable to implement the servoloops in software. Note that if the application requires only the closure of simple servoloops without the need for elaborate additional functions, the use of a microprocessor will usually result in more complex hardware than an all-analog system and may be more complex than a hard-wired digital system.

The loop closure architecture of a microprocessor-based controller is illustrated in Fig. 17.46. This block diagram implements virtually all the loop functions in software, including frequency compensation. Of course, it is not necessary to use a digital outer-loop transducer, but use of an analog transducer compromises the potential accuracy advantages of the digital controller. Several methods can be used to generate the software compensator designs. Perhaps the most straightforward technique is to generate Laplace transfer functions using the continuous frequency domain techniques outlined in Sections 17.1–17.6. These transfer functions can then be converted to equivalent z transforms, from which difference equations can be generated for implementation in software.[5]

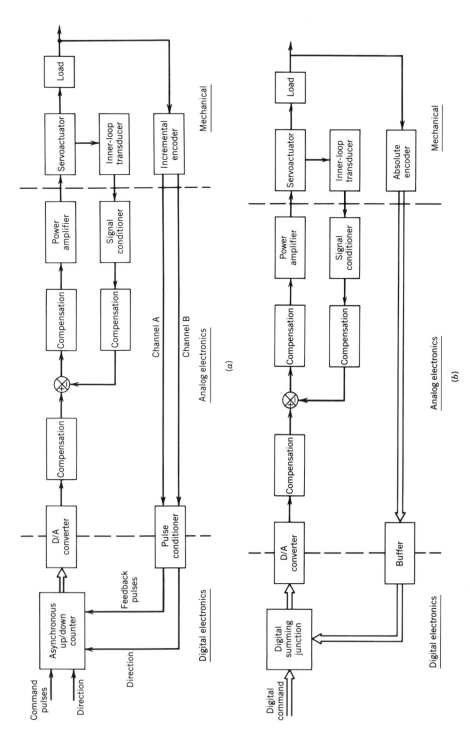

Fig. 17.45 Typical hard-wired digital controller; (*a*) incremental system; (*b*) absolute system.

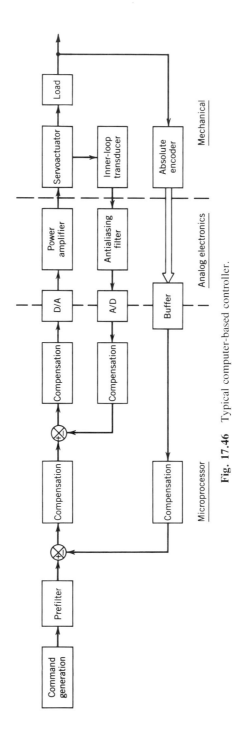

Fig. 17.46 Typical computer-based controller.

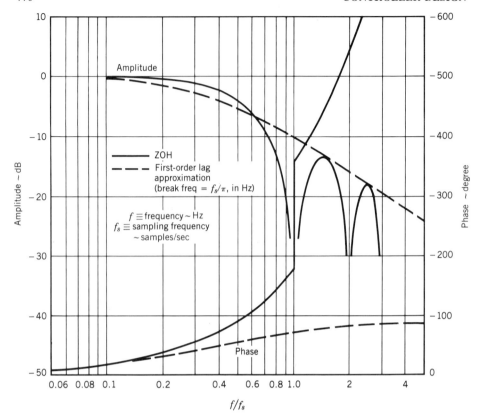

Fig. 17.47 Frequency response characteristics of a zero-order hold.

It should be noted that a microprocessor-based controller is a sampled-data system, and the digital-to-analog converter usually operates as a zero-order hold (ZOH). The sampling nature of the system together with computation times introduce time delay into the control loops, which can have a profound influence on system performance and stability. Figure 17.47 shows a frequency response of a ZOH operated in a sampled-data system. If the system has been designed using frequency domain techniques, this figure represents the additional phase lag and amplitude that will result from computer implementation of the design. The phase lag is linear with frequency, and an approximate transfer function is a first-order lag with a break frequency equal to (f_s/π) where f_s is the sampling frequency of the computer in hertz. This approximation is very accurate out to a frequency of $f_s/2\pi$. Note that $10°$ of phase lag exists at a frequency of $f_s/18$. This suggests that the sampling frequency should be at least 20 times the crossover frequency (in hertz) of the loop being implemented if the impact on phase margin is to be minimized. Furthermore, smooth operation of the servo may require heavy filtering at the output of the D/A converter to reduce sampling-induced ripple. This will add additional phase lag in the servoloops.

The need to minimize phase lags can place severe restrictions on the complexity of computations that the microprocessor can handle in one sampling interval. This problem can be partially overcome by using a separate processor to perform loop closure computations or by implementing compensators in the analog circuitry. Alternatively, the need for high sampling rates can be reduced by implementing high-gain inner feedback loops in analog circuitry, as shown in Fig. 17.45b. If the inner loops utilize analog transducers, the problem of aliasing[8] adds another reason for using analog electronics. To properly utilize the output of an analog transducer in the computer, an antialiasing filter is required at the input to the A/D converter. These filters are often first order with a break frequency equal to f_s/π. This doubles the effective phase lag of the computer.

REFERENCES

1. J. J. D'Azzo, and C. H. Houpis, *Feedback Control System Analysis and Synthesis*, McGraw-Hill, New York, 1966.
2. B. C. Kuo, *Automatic Control Systems*, Prentice-Hall, Englewood Cliffs, NJ, 1982.
3. E. O. Doebelin, *Dynamic Analysis and Feedback Control*, McGraw-Hill, New York, 1962.
4. B. C. Kuo, *Digital Control Systems*, Holt, Rinehart and Winston, New York, 1980.
5. J. A. Cadzow, and H. R. Martens, *Discrete-Time and Computer Control Systems*, Prentice-Hall, Englewood Cliffs, NJ, 1970.
6. Listing of software packages for linear analysis and simulation, *IEEE Control System Magazine*, December 1982.
7. J. G. Graeme, G. E. Tobey, and L. P. Huelsman, *Operational Amplifiers, Design and Applications*, McGraw-Hill, New York, 1971.
8. E. O. Doebelin, *System Modeling and Response*, Wiley, New York, 1980.

CHAPTER **18**
GENERAL PURPOSE CONTROL DEVICES

JAMES H. CHRISTENSEN
ODO J. STRUGER

Allen-Bradley Company, Inc.
Highland Heights, Ohio

SUJEET CHAND

Rockwell International Corporation
Thousand Oaks, California

MICHAEL D. MCEVOY

Nematron Corp.
Ann Arbor, Michigan

18.1 CHARACTERISTICS OF GENERAL PURPOSE CONTROL DEVICES

James H. Christensen and Odo J. Struger

18.1 Hierarchical Control

As shown in Fig. 18.1, **general purpose control devices** (GPCDs) occupy a place in the hierarchy of factory automation above the closed-loop control systems described in Chapters 14–17. The responsibility of the GPCD is the coordinated control of one or more machines or processes. Thus, a GPCD may operate at the "station" level, where it controls part or all of a single machine or process, or at the "cell" level, where it coordinates the operation of multiple stations. In fulfilling its responsibilities, the GPCD must be capable of performing the following functions, as shown in Fig. 18.2.

1. Issuing commands to and receiving status information from a set of closed-loop controllers that control individual machine and process variables such as velocity, position, temperature, et cetera. These closed-loop controllers may be separate devices or integral parts of the GPCD hardware and/or software architecture.

2. Issuing commands to and receiving status information from a set of actuators and sensors directly connected to the controlled operation. These actuators and sensors may include signal processing and transmission elements as described in Chapter 12. This capability may not be required if all interface to the controlled operation is through the closed-loop controllers described above.

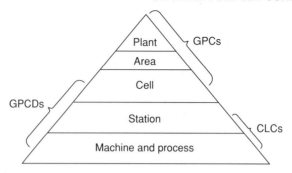

Fig. 18.1 A plant control hierarchy using general purpose control devices (GPCDs), general purpose computers (GPCs), and closed loop controllers (CLCs).

3. Receiving commands from and sending status information to a control panel or console for the operator of the machine or process.

4. Receiving commands from and sending status information to a manual or automated system with the responsibility of supervising the operation of a number of GPCDs within the boundaries of a "cell," for example, a number of coordinated machines or unit operations, or over a wider "area," such as a chemical process or production zone in a factory.

5. Interchanging status information with other GPCDs within the same cell or across cell boundaries.

It should be noted that not all of these capabilities are necessarily required for every application of a GPCD. For instance, special closed-loop controllers may not be necessary in an operation requiring only simple on/off or modulating control. Again, communication in an automation hierarchy may not be required for simple "stand-alone" applications such as an industrial trash compactor;

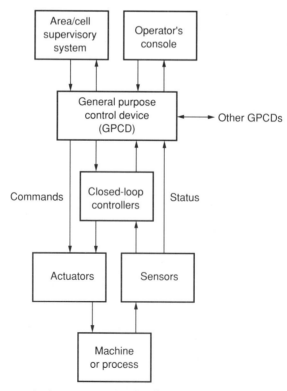

Fig. 18.2 Communication and control paths of a general purpose control device (GPCD).

however, the capability for expansion into a communicating hierarchy should be inherent in the GPCD architecture if retrofit of stand-alone systems into an integrated production system is considered a future possibility.

18.1.2 Programmability

In order to be truly general purpose, a GPCD must be programmable; that is, its operation is controlled by sequences of instructions and data stored in internal memory. The languages used for programming GPCDs are usually **problem oriented:** programs are expressed in terms directly related to the control to be performed, rather than in a general purpose programming language such as FORTRAN or BASIC. These languages will be described in appropriate sections for each type of GPCD.

Depending on the application, the responsibility for development and maintenance of GPCD programs may reside with

The **original equipment manufacturer** (OEM) of a machine that includes a GPCD as part of its control apparatus

The **system integrator** who designs and installs an integrated hierarchical control system, or

The **end user** who wishes to modify the operation of the installed system.

The degree to which the operation of the system can be modified by the end user is a function of

The complexity of the system

The degree to which the end user has been trained in the programming of the system, and

The extent to which the operation of the process must be modified over time.

For instance, in a high-volume chemical process, only minor modifications of set points may be required over the life of the plant. However, major modifications of the process may be required annually in an automotive assembly plant. In the latter case, complete user programmability of the system is required.

Depending on the complexity of the control program and the degree of reprogrammability required, GPCD programming may be supported by any of several means, including

An integral programming panel on the GPCD

Portable programming and debugging tools

Minicomputers, personal computers, or engineering workstations that may or may not be connected to the GPCD during control operation

An on-line computer system, for example, the area or cell controller shown in Fig. 18.2.

18.1.3 Device Architecture

Figure 18.3 illustrates a GPCD architecture capable of providing the required functional characteristics.

1. The **memory** provides storage for the programs and data entered into the system via the **communications processor.**
2. The **control processor** performs control actions under the direction of the stored program, as well as coordinating the operation of the other functions.
3. The **communications processor** provides the means of accepting commands from and providing status information to the supervisory system and operator's console, interchanging status information with other GPCDs, and interacting with program development and configuration tools.
4. The **I/O** (Input/Output) **processor** provides the means by which the control processor can issue commands to and receive status information from the closed-loop controllers, as well as interchanging information directly with the controlled operation via the **output** and **input interfaces.**
5. The **system bus** provides for communication among the functional blocks internal to the GPCD, while the **I/O bus** provides for communication between the internal functional blocks and the "outside world" via the closed-loop controllers and I/O interfaces. As an option, the I/O controller may extend the I/O bus functionality to remote locations, using data communications methods such as those described in Section 12.9.

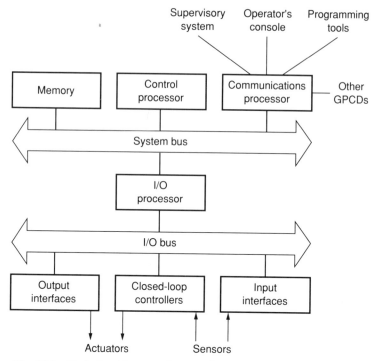

Fig. 18.3   Typical architecture of a general purpose control device (GPCD).

The architecture shown in Fig. 18.3 is not the only one possible for GPCDs, nor is it necessarily the most desirable for all applications. For instance

Large systems may require multiple control processors on the system bus.

In applications requiring only a few control loops, it may be more economical to perform the closed-loop control function directly in the software of the control processor.

A separate I/O processor may not be required if the number of separate I/O interface points is less than a few dozen.

In small systems, the programmer's console may be interfaced directly to the system or I/O bus.

However, if it is anticipated that control system requirements will grow substantially in the future, or if total control system requirements are only partially understood, the use of a flexible, extendable GPCD architecture such as that shown in Fig. 18.3 is recommended.

18.1.4   Sequential Control

It is obvious that GPCDs must perform complex sequences of control actions when they are applied to the coordination of material handling and machine operation in the fabrication and assembly of discrete parts, or in batch and semibatch processes such as blast furnace operation and pharmaceutical manufacture. However, sequential control is also increasing in importance in "continuous" process control, since no process is truly continuous. At the very least, the process must be started up and shut down for maintenance or emergencies by a predetermined sequence of control actions. In large, integrated processes, these sequences are too complicated to be carried out manually and must be performed automatically by GPCDs.

The increasing importance of sequential control, coupled with the increasing complexity of the controlled processes, have generated the need for graphical programming and documentation techniques for the representation of large, complex sequential control plans. These plans must provide a straightforward representation of the relationship between the operation of the control program in the GPCD and the operation of the controlled machine or process, as well as the interrelationships between multiple, simultaneous control sequences.

Recognizing this need, the International Electrotechnical Commission has undertaken the standardization of **sequential function charts** (SFCs) for the representation of sequential control plans.[1,2] This representation is an extension of the original French *GRAFCET* standard.[3]

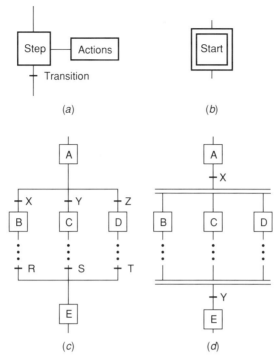

Fig. 18.4 Sequential function chart (SFC) constructs. (*a*) Step, actions, and transition; (*b*) Initial step; (*c*) Step selection and convergence; (*d*) Parallel sequence initiation and convergence. *Note:* Actions may be present but are omitted for simplicity in (*b*), (*c*), and (*d*).

As shown in Fig. 18.4*a*, an SFC is constructed from three basic elements

A STEP, representing the current state of the controller and controlled system within the sequential control plan

A set of associated **actions** at each step, and

A **transition condition** that determines when the state of the controller and controlled system is to evolve to another step or steps.

An SFC consists of a set of independently operating **sequences** of control actions built up out of these basic elements via two mechanisms.

selection of one of a number of alternate successors to a step based on mutually exclusive transition conditions as shown in Fig. 18.4*c*, or

initiation (or "spawning") of two or more independently executing sequences based on a transition condition, as shown in Fig. 18.4*d*.

The mechanisms for representing **convergence**, that is, resumption of a main sequence after step selection or parallel sequence initiation, are also shown in Figs. 18.4*c* and *d*, respectively.

Figure 18.4*b* illustrates the representation of the **initial step** of each sequence. The operation of control sequences can be visualized by placing a **token** in each initial step upon the initiation of system operation. A step is then said to be **active** while it possesses a token, and **reset** when it does not possess a token. The actions associated with the step are performed while the step is active, and are not performed when it is reset. The resetting of one or more steps, and the activation of one or more successor steps, can then be envisioned as the processes of token passing, consumption, and generation, as shown in Figs. 18.5 and 18.6. It should be noted that the selection and convergence of alternate paths within a sequence, as shown in Fig. 18.5, simply involve the passing of a single token. In contrast, the spawning of multiple sequences involves the consumption of a single token and the generation of multiple tokens as shown in Fig. 18.6*a*, with the converse operation for termination of multiple sequences shown in Fig. 18.6*b*.

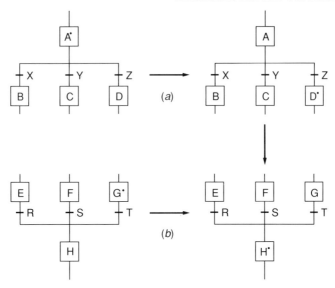

Fig. 18.5 Sequence selection and convergence. (*a*) Transition condition "Z" true; (*b*) Transition condition "T" true.

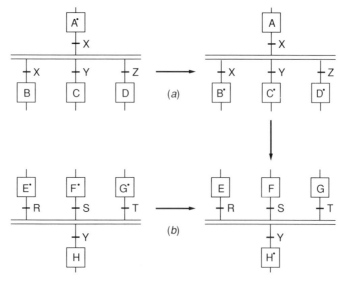

Fig. 18.6 Parallel sequence initiation and convergence. (*a*) Transition condition "X" true; (*b*) Transition condition "Y" true.

When an action associated with a step is a Boolean variable, its association with the step may be expressed in an **action block** as shown in Fig. 18.7. The qualifiers which can be used to specify the duration of the action are listed in Table 18.1. More complex actions can be specified via one of the programming languages described in Section 18.2.3; in this case, the action executes continuously while the associated "action control" shown in Table 18.1 has the Boolean value "1."

An example of the application of SFCs to the control and monitoring of a single motion, for example, in a robot control system, is given in Fig. 18.8. Here, the system waits until a motion command is received via the Boolean variable CMD_IN. It then initiates the appropriate motion by asserting the Boolean variable CMD_OUT. If a feedback signal DONE is not received within a

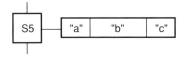

Fig. 18.7 Action block (with associated step): "a"–action qualifier; "b"–action name; "c"–feedback name.

time interval specified by CMD_TIME, the system enters an error step and issues the error message TIMEOUT_ERROR. This error condition is cleared when acknowledged by the signal ACK, for example, from an operator's console, combined with the feedback (DONE) that the motion has been accomplished. The system then reenters the initial step and waits for another command.

18.1.5 Path Control

General purpose control devices must often perform **path control**, that is, the coordinated control of several variables at once along a continuous path through time. Typical kinds of path control include

The path of a metal cutting tool or a robot manipulator

The trajectory of a set of continuous process variables such as temperature, pressure, and composition

The startup of a set of velocity and tension variables in a paper or steel processing line

The numerical control of metal cutting tools and robot manipulators is discussed in more detail in Sections 18.3 and 18.4, respectively.

A typical application to path control of the general purpose architecture shown in Fig. 18.3 has the control processor planning the motions to be accomplished and issuing commands to the closed-loop controllers to perform the required motions. Coordination between the closed-loop controllers may

TABLE 18.1 ACTION BLOCK QUALIFIERS

Qualifier/Meaning	Action Control
N/Nonstored	1 while step is active
S/Stored	1 until reset
R/Reset	0
L/Time Limited	1 until specified time delay after entering step or until exit from step
D/Time Delayed	0 until specified time delay after entering step, then 1 until exit from step
P/Pulse	1 (pulse) once upon entering step, then 0
SD/Stored and Time Delayed	0 until specified time delay after entering step, then 1 until reset (delay is canceled by reset)
DS/Time Delayed and Stored	0 until specified time delay after entering step, then 1 until reset (delay is canceled by exit from step)
SL/Stored and Time Limited	1 until specified time delay after entering step, then 0 (delay is canceled by reset)

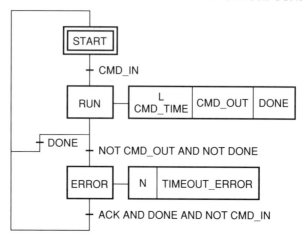

Fig. 18.8   Sequential function chart (SFC) example.

be performed by the control processor, or by direct interaction among the closed-loop controllers, using the I/O bus or special interconnections.

In addition to performing path planning, the control processor also performs sequencing of individual motions and coordination of the motions with other control actions, typically using programming mechanisms such as the Sequential Function Charts discussed in Section 18.1.4.

18.2   PROGRAMMABLE CONTROLLERS

James H. Christensen and Odo J. Struger

18.2.1   Principles of Operation

A **programmable controller** (PC) is defined by the International Electrotechnical Commission[1] as:

> "A digitally operating electronic system, designed for use in an industrial environment, which uses a programmable memory for the internal storage of user-oriented instructions for implementing specific functions such as logic, sequencing, timing, counting, and arithmetic, to control, through digital or analog inputs and outputs, various types of machines or processes. Both the PC and its associated peripherals are designed to be easily integrable into an industrial control system and easily used in all their intended functions."

The hardware architecture of almost all programmable controllers is the same as that for the general purpose control device (GPCD) shown in Fig. 18.3.

As illustrated in Fig. 18.9, the operation of most programmable controllers consists of a repeated cycle of four major steps.

1. All inputs from interfaces and closed-loop controllers on the I/O bus, and possibly from other GPCDs, are scanned to provide a consistent "image" of the inputs.
2. One "scan" of the user program is performed to derive a new "image" of the desired outputs, as well as internal program variables, from the image of the inputs and the internal and output variables computed during the previous program scan. Typically, the program scan consists of
   a. Determining the currently active steps of the sequential function chart (SFC) (see Section 18.1.4), if any, contained in the program
   b. Scanning the program elements or computing the outputs contained in the active actions of the SFC, if any (if the user program does not contain an SFC, then all program elements are scanned). Scanning of program elements in ladder diagrams or function block diagrams (see Section 18.2.3) typically proceeds from left to right, and from top to bottom. Programming elements are sometimes provided to enable skipping the evaluation of groups of program elements, or to force the outputs of a group of elements to zero.
   c. Evaluating transition conditions of the SFC (if any) at the end of the program scan, in preparation for Step 2.a in the next program scan.

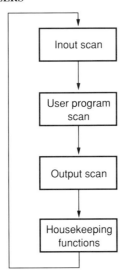

Fig. 18.9 Basic operation cycle of a programmable controller.

3. The data from the updated output image is then transferred to the interfaces and closed-loop controllers on the I/O bus, and possibly to other GPCDs as well.

4. Finally, "housekeeping" tasks are performed on a time-available basis. These typically include communication with the operator, a supervisory controller, a programming terminal, or other GPCDs.

After the performance of housekeeping tasks, the cyclic operation of the PC begins again with the input scan. This may follow immediately upon execution of the housekeeping tasks, or may be scheduled to repeat at a fixed execution interval.

Some programmable controller systems with separate I/O and/or communications processors provide for overlapping the scanning of the user program with the scanning of the inputs (Step 1) and outputs (Step 3) and communication functions (Step 4). In these cases, special programming mechanisms may be needed to achieve concurrency and synchronization between the program and I/O scans, and between the program and communications processing.

18.2.2 Interfaces

The International Electrotechnical Commission (IEC) has specified the standard voltage ratings shown in Table 18.2 for power supplies, digital inputs, and digital outputs of programmable controllers.[5] The IEC Standard also defines additional parameters for digital inputs and outputs; the parameters specified by the manufacturer should be checked against those defined in the IEC Standard in order to assure the suitability of a particular input or output module for its intended use in the control system.

The IEC-specified signal ranges for analog inputs and outputs for programmable controllers are shown in Tables 18.3 and 18.4, respectively. The IEC Standard lists a number of characteristics whose values are to be provided by the manufacturer, such as input impedance, maximum input error, conversion time and method, and which must be checked against the requirements of the particular control application.

In addition to simple digital and analog inputs and outputs, closed-loop controllers which can reside on the I/O bus of the programmable controller system may be provided, as illustrated for general purpose control devices in Fig. 18.2. In this case, the programming languages for the programmable controller typically provide language elements, in addition to those described in Section 18.2.3, to support the configuration and supervisory control of these "slave" closed-loop controllers.

Communication interfaces for programmable controllers provide many different combinations of connectors, signal levels, signaling rates, and communication services. The manufacturer's specifications of these characteristics should be checked against applicable standards, such as those discussed in Section 12.9 of this handbook, to assure the achievement of the required levels of system performance and compatibility of all general purpose control devices in the system.

TABLE 18.2 RATED VALUES AND OPERATING RANGES FOR INCOMING POWER SUPPLIES AND DIGITAL I/O INTERFACES OF PROGRAMMABLE CONTROLLERS[5]

Rated Voltage	Recommended for: Power Supply	I/O Signals	Notes
24 VDC	yes	yes	1
48 VDC	yes	yes	
24 VAC	no	no	2
48 VAC	no	no	
120 VAC	yes	yes	
230 VAC	yes	yes	
400 VAC	yes	no	

Notes: (1) Voltage tolerance for DC voltage ratings is -15 to $+20\%$.

(2) Voltage tolerance for AC voltage ratings is -15 to $+10\%$. AC voltage ratings are in Volts rms.

(3) See the IEC Programmable Controller standard[5] for additional notes and rating values.

TABLE 18.3 RATED VALUES FOR ANALOG INPUTS OF PROGRAMMABLE CONTROLLERS[5]

Signal Range	Input Impedance
-10 V to $+10$ V	≥ 10 K Ohms
0 V to $+10$ V	≥ 10 K Ohms
$+ 1$ V to $+ 5$ V	≤ 5 K Ohms
4 mA to 20 mA	≤ 300 Ohms

Note: See the IEC Programmable Controller standard[5] for additional notes and rating values.

TABLE 18.4 RATED VALUES FOR ANALOG OUTPUTS OF PROGRAMMABLE CONTROLLERS[5]

Signal Range	Load Impedance
-10 V to $+10$ V	≥ 1 K Ohm
0 V to $+10$ V	≥ 1 K Ohm
$+ 1$ V to $+ 5$ V	≥ 500 K Ohms
4 mA to 20 mA	≤ 600 Ohms

Note: See the IEC Programmable Controller standard[5] for additional notes and rating values.

18.2.3 Programming

Programmable controllers have historically been used as programmable replacements for relay and solid-state logic control systems. As a result, their programming languages have been oriented around the conventions used to describe the control systems they have replaced, that is, relay ladder logic and function block diagrams. Since these representations are fundamentally graphic in nature, programmable controllers provide one of the first examples of the practical application of graphic programming languages.

Much of the programming of programmable controllers is done in the factory environment while the controlled system is being installed or maintained. Hence, programming has been traditionally supported by special-purpose portable programming terminals. In recent years, much of the support for PC programming has migrated to software packages for personal computers. However, the need still exists for specialized terminals for programming and debugging of programmable controllers in the industrial environment, although most of this functionality can also be supplied by ruggedized, portable personal computers. The selection of support environments is thus an important consideration in the selection and implementation of programmable controller systems.

A working group of the International Electrotechnical Commission (IEC Subcommittee 65A/Working Group 6) is in the process of standardizing the programming languages for programmable controllers.[2] The working group has proposed a set of mutually compatible programming languages, taking into account the different courses of evolution of programmable controllers in North America, Europe, and Japan, and the wide variety of applications of programmable controllers in modern industry. These languages include:

The Sequential Function Chart (SFC) elements described in Section 18.1.4 for sequential control

Ladder Diagrams (LD) for relay replacement functions

Function Block Diagrams (FBD) for logic, mathematical, and signal-processing functions

Structured Text (ST) for data manipulation

Instruction List (IL) for assembly language level programming. This language will not be described in this handbook; for further details the IEC standard should be consulted.[2]

Figure 18.10 shows the application of the LD, FBD, and ST languages to implement a simple command execution and monitoring function. In general, a desired functionality can be programmed in any one of the IEC languages. Hence, languages can be chosen depending on their suitability for each particular application. An exception to this portability is the use of iteration and selection constructs (IF . . . THEN . . . ELSIF, CASE, FOR, WHILE, and REPEAT) in the ST language.

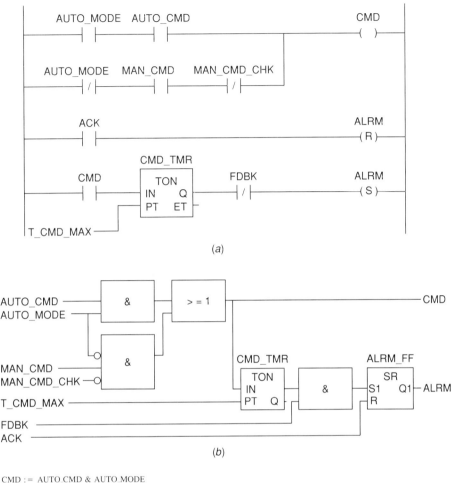

Fig. 18.10 Programmable controller programming example. (*a*) LD (Ladder Diagram) language; (*b*) FBD (Function Block Diagram) language; (*c*) ST (Structured Text) language.

The functionality shown in Fig. 18.10 can be "packaged" into a reusable "function block" by following the "declaration" process defined in the IEC language standard. An example of the graphical and textual declaration of this functionality is shown in Fig. 18.11.

In addition to providing mechanisms for the programming of mathematical functions and function blocks, the standard provides a large set of predefined standardized functions and function blocks, as listed in Tables 18.5 and 18.6, as "building blocks" for user programs.

In addition to being used directly for building functions, function blocks, and programs, the LD, FBD, ST, and IL languages can be used to program the "actions" to be performed under the control of Sequential Function Charts (SFCs) as described in Section 18.1.4. These SFCs can then be used to build programs and reusable function blocks, using the mechanisms defined in the IEC language standard.[2]

It will be noted in Fig. 18.11 that data types are defined for all variables. In accordance with modern programming practice, the IEC standard provides facilities for strong data typing, with a large set of predefined data types as listed in Table 18.7. In addition, facilities are provided for user-defined data types as listed in Table 18.8.

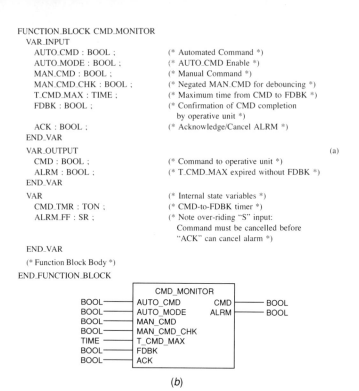

```
FUNCTION_BLOCK CMD_MONITOR
    VAR_INPUT
        AUTO_CMD : BOOL ;          (* Automated Command *)
        AUTO_MODE : BOOL ;         (* AUTO_CMD Enable *)
        MAN_CMD : BOOL ;           (* Manual Command *)
        MAN_CMD_CHK : BOOL ;       (* Negated MAN_CMD for debouncing *)
        T_CMD_MAX : TIME ;         (* Maximum time from CMD to FDBK *)
        FDBK : BOOL ;              (* Confirmation of CMD completion
                                      by operative unit *)
        ACK : BOOL ;               (* Acknowledge/Cancel ALRM *)
    END_VAR
    VAR_OUTPUT                                                     (a)
        CMD : BOOL ;               (* Command to operative unit *)
        ALRM : BOOL ;              (* T_CMD_MAX expired without FDBK *)
    END_VAR
    VAR                            (* Internal state variables *)
        CMD_TMR : TON ;            (* CMD-to-FDBK timer *)
        ALRM_FF : SR ;             (* Note over-riding "S" input:
                                      Command must be cancelled before
                                      "ACK" can cancel alarm *)
    END_VAR
    (* Function Block Body *)
END_FUNCTION_BLOCK
```

(b)

Fig. 18.11 Function block encapsulation of programming example in Fig. 18.10. (a) Textual representation; (b) Graphic representation.

TABLE 18.5   IEC STANDARD FUNCTIONS FOR PROGRAMMABLE CONTROLLERS[2]

Standard Name	Function		Standard Name	Function
Numeric Functions			Selection and Comparison Functions	
ABS	Absolute value		SEL	Binary (1 of 2) Selection
SQRT	Square root			
LN	Natural logarithm		MUX	Multiplexer (1 of N)
LOG	Logarithm Base 10		MIN	Minimum of N inputs
EXP	Natural exponential		MAX	Maximum of N inputs
SIN	Sine		LIM	Hard Upper/Lower
COS	Cosine			Limiter
TAN	Tangent		GT or $>$	Greater Than
ASIN	Arc sine		GE or $\geq$	Greater Than or Equal
ACOS	Arc cosine			To
ATAN	Arc tangent		EQ or $=$	Equal To
ADD or $+$	Addition		LE or $\leq$	Less Than or Equal To
SUB or $-$	Subtraction		LT or $<$	Less Than
MUL or $*$	Multiplication		NE or $<>$	Not Equal
DIV or $/$	Division		Character String Functions	
MOD	Modulo			
EXPT or $**$	Exponentiation		CONCAT	Concatenate N strings
Bit String Functions			INSERT	Insert one string into another
SHL	Shift left, zero-filled		DELETE	Delete a portion of a
SHR	Shift right, zero-filled			string
ROL	Rotate left circular		REPLACE	Replace a portion of
ROR	Rotate right circular			one string with another
AND or $\&$	Bitwise Boolean AND		FIND	Find the first occurence
OR or $> =1$	Bitwise Boolean OR			of one string in another
XOR or $=2k+1$	Bitwise Boolean Exclusive OR			
NOT	Bitwise Boolean Complement			

TABLE 18.6 IEC STANDARD FUNCTION BLOCKS FOR PROGRAMMABLE CONTROLLERS[2]

Standard Name	Function Block
Bistable Function Blocks	
SR	Flip-flop (Set Dominant)
RS	Flip-flop (Reset Dominant)
TRIGGER	Edge Detection
Counter Function Blocks	
CTU	Up-Counter
CTD	Down-Counter
Timer Function Blocks	
TP	One-shot (Pulse) Timer
TON	On-Delay Timer
TOF	Off-Delay Timer
Message Transfer and Synchronization	
SEND	Message Requester
RCV	Message Responder

TABLE 18.7 IEC STANDARD DATA TYPES FOR PROGRAMMABLE CONTROLLERS[2]

KEYWORD	DATA TYPE	BITS
BOOL	Boolean	1
EDGE	Edge-Triggered Boolean	—
SINT	Short Integer	8
INT	Integer	16
DINT	Double Integer	32
LINT	Long Integer	64
USINT	Unsigned Short Integer	8
UINT	Unsigned Integer	16
UDINT	Unsigned Double Integer	32
ULINT	Unsigned Long Integer	64
REAL	Real Numbers	32
LREAL	Long Reals	64
TIME	Duration	—
DATE	Date (only)	—
TIME_OF_DAY	Time of Day (only)	—
DATE_AND_TIME	Date and Time of Day	—
STRING	Variable-Length Character String	—
BYTE	Bit String of length 8	8
WORD	Bit String of length 16	16
DWORD	Bit String of length 32	32
LWORD	Bit String of length 64	64

TABLE 18.8   EXAMPLES OF USER-DEFINED DATA TYPES FOR PROGRAMMABLE CONTROLLERS[2]

DIRECT DERIVATION from elementary types, e.g.:
TYPE R : REAL ; END-TYPE
TYPE PI : REAL := 3.1415925 ; END-TYPE

ENUMERATED data types, e.g.:
TYPE
ANALOG-SIGNAL-TYPE : (SINGLE-ENDED, DIFFERENTIAL) ;
ANALOG-SIGNAL-RANGE :
  (BIPOLAR_10V,   (* −10 to +10 VDC   *)
  UNIPOLAR_10V,   (*   0 to +10 VDC   *)
  UNIPOLAR_1_5V,   (* + 1 to + 5 VDC   *)
  UNIPOLAR_0_5V,   (*   0 to + 5 VDC   *)
  UNIPOLAR_4_20_mA, (* + 4 to +20 mADC *)
  UNIPOLAR_0_20_mA   (*   0 to +20 mADC *)
) := UNIPOLAR_1_5V;
END-TYPE

SUBRANGE data types, e.g.:
TYPE
  ANALOG-DATA : INT (−4095 .. 4095) := 0 ;
END-TYPE

ARRAY data types, e.g.:
TYPE
  ANALOG_16_INPUT_DATA : ARRAY (1 ..16) OF ANALOG-DATA
                       := 8(−4095), 8(4095) ;
END-TYPE

STRUCTURED data types, e.g.:
TYPE
  ANALOG_CHANNEL_CONFIGURATION :
   STRUCTURE
    RANGE : ANALOG_SIGNAL_RANGE ;
    MIN_SCALE : ANALOG_DATA := −4095 ;
    MAX_SCALE : ANALOG_DATA := 4095 ;
   END-STRUCTURE ;
  ANALOG_16_INPUT_CONFIGURATION :
   STRUCTURE
    SIGNAL_TYPE : ANALOG_SIGNAL_TYPE ;
    FILTER_PARAMETER : SINT (0 ..99) ;
    CHANNEL : ARRAY (1 ..16) OF ANALOG_CHANNEL_CONFIGURATION ;
   END-STRUCTURE ;
END-TYPE

It is impossible, in an article of this length, to cover all the programming language features defined in the IEC language standard for programmable controllers.[2] Copies of the draft standard are currently available for review from the appropriate National Committees for the International Electrotechnical Commission. In the United States, the appropriate organization is:

U.S. National Committee for the IEC
American National Standards Institute
1430 Broadway
New York City, NY 10018

Final copies of the standard, when adopted by the IEC, will also be available from the appropriate National Committees. Since the standard allows manufacturers to specify the language features that they support, users should consult the standard to determine which language features are required by their application, and check their language requirements carefully against the manufacturers' specifications when making their choice of a programmable controller system.

18.3 NUMERICAL CONTROLLERS

Sujeet Chand

18.3.1 Introduction and Applications

The century from 1760 to 1860 saw the development of a large number of machine tools for shaping cylindrical and flat surfaces, threads, grooves, slots, and holes of many shapes and sizes in metals. Some of the machine tools developed were the lathe, the planer, the shaper, the milling machine, drilling machines, and power saws. With increasing applications for metal machining, the cost in terms of manpower and capital equipment grew rapidly. The attempt at automation of the metal removal process gave birth to numerical controllers.

The history of numerical controllers dates back to the late 1940s, when John T. Parsons proposed a method to automatically guide a milling cutter to generate a smooth curve. Parsons proposed that successive coordinates of the tool be punched on cards and fed into the machine. The idea was to move the machine in small incremental steps to achieve a desired path. In 1952, the U.S. Air Force provided funding for a project at the Massachusetts Institute of Technology (MIT) that developed the Whirlwind computer. In a subsequent project, the Servomechanisms Laboratory at MIT developed the concept of the first workable numerical control (NC) system. The NC architecture was designed to exploit the Whirlwind computer with emphasis on five-axis NC for machining complex aircraft parts.

The MIT NC architecture identified three levels of interaction with the numerical controller.[6,7,8] At the highest level is a machine-independent language, called APT (Automatically Programmed Tools). APT provides a symbolic description of the part geometry, tools, and cutting parameters. The next level, called the cutter location (CL) level, changes the symbolic specification of cutter path and tool control data to numeric data. The CL level is also machine-independent. The lowest level, called the G-code level, contains machine-specific commands for the tool and the NC axis motions.

The conversion from APT to CL data involves the computation of cutter offsets and resolution of symbolic constraints. The conversion from CL data to G-code is called *postprocessing*.[6,8] Postprocessing transforms the tool center line data to machine motion commands, taking into account the various constraints of the machine tool such as machine kinematics and limits on acceleration and speed. The APT-to-CL data conversion and the compilation of CL data to G-code are computationally intensive; these computationally intensive functions were envisioned to be performed by the Whirlwind computer. The numerical controller works with simple G-codes to keep computational requirements low. The G-codes, punched on perforated paper tape, would be the input medium to the numerical controller.

Since their inception in the fifties, numerical controllers have followed a similar pattern in the evolution of controller technology as the computer industry in the past 30 years. The first numerical controllers were designed with vacuum tube technology. The controllers were bulky and the logic inside the control was hardwired. The hardwired nature of the controller made it very difficult to change or modify its functionality. Vacuum tubes were replaced by semiconductors in the early sixties. In the early seventies, numerical controllers started using microprocessors for control. The first generation of NCs with microprocessor technology were mostly hybrid, with some hardwired logic and some control functions in software. Today, most of the NC functionality is in the software. The microprocessor-based numerical controllers are also called computer numerical controllers (CNC).

The concept of distributed numerical control (DNC) was introduced in the sixties to provide a single point of programming and interface to a large number of NCs. Most NC users agree that the paper tape reader on a numerical controller suffers the most in reliability. A DNC can transfer a program to an NC through a direct computer link, bypassing the paper tape reader. A DNC has two primary functions: (1) computer-assisted programming and storage of NC programs in computer memory, and, (2) transfer, storage, and display of status and control information from the NCs. A DNC can store and transfer programs to as many as 100 NCs. DNCs commonly connect to the NCs through a link called Behind the Tape Reader (BTR). The name BTR comes from the fact that the connection between the DNC and the CNC is made between the paper tape reader and the control unit.

Numerical controllers are widely used in industry today. The predominant application of NCs is still for metal cutting machine tools. Some of the basic operations performed by machine tools in metal machining are turning, boring, drilling, facing, forming, milling, shaping, and planing.[9] Turning is one of the most common operations in metal cutting. Turning is usually accomplished by lathe machines. The part or workpiece is secured in the chuck of a lathe machine and rotated. The tool, held rigidly in a tool post, is moved at a constant speed along the rotational axis of the workpiece, cutting away a layer of metal to form a cylinder or a surface of a more complex profile.

Applications other than metal machining for numerical controllers include flame cutting, water jet cutting, plasma arc cutting, laser beam cutting, spot and arc welding, and assembly machines.[10]

18.3.2   Principles of Operation

NC System Components

A block diagram of a numerical control system is shown in Fig. 18.12. The three basic components in a numerical control system are (1) a program input medium, (2) the controller hardware and software including the feedback transducers and the actuation hardware for moving the tool, and (3) the machine itself.

The controller hardware and software execute programmed commands, compute servo commands to move the tool along the programmed path, read machine feedback, close the servo control loops, and drive the actuation hardware for moving the tool. The actuation hardware consists of servomotors and gearing.

The feedback devices on an NC servo system provide information about the instantaneous position and velocity of the NC axes. The servo feedback devices can be linear transducers or rotary transducers. The two most common rotary transducers are *resolvers* and *encoders*.[11] Resolvers consist of an assembly that resembles a small electric motor with a stator-rotor configuration. As the rotor turns, the phase relationship between the stator and rotor voltages corresponds to the shaft angle such that one electrical degree of phase shift corresponds to one mechanical degree of rotation. An optical encoder produces pulsed output that is generated as a disk containing finely etched lines rotates between an exciter lamp and one or more photo diodes. The total number of pulses generated in a single revolution is a function of the number of lines etched on the disk. Typical disks contain two thousand to ten thousand lines. A resolver is an analog device with an analog output signal, whereas an encoder is a numerical device that produces a digital output signal.

The four input media for entering programs into an NC machine are (1) punched cards, (2) punched tape, (3) magnetic tape, and (4) direct entry of the program into the computer memory of the numerical controller.[12]

Originated by Herman Hollerith in 1887, the punched card as an input medium is almost obsolete. The standard "IBM" card's fixed dimensions are 3.25 in. wide, 7.375 in. long and 0.007 in. thick. Each card contains 12 rows of hole locations with 80 columns across the card. To edit a part program, cards in the deck are replaced with new cards. With a deck of cards, it is easy to lose sequence or have missing blocks due to the loss of a card. Also, punched cards are a low density storage medium with an input rate that is slower than most other media.

Punched tape is the most popular input medium for an NC controller. Although punched tape is slowly becoming obsolete, most NC controllers still provide a punched tape reader. The specifications of the punched tape are standardized by the EIA (Electronic Industries Association) and the AIA (Aerospace Industries Association). Tapes are made of paper, aluminum-plastic laminates, or other materials. Making editorial changes to the punched tape is difficult; only minor editing is possible by splicing new data into the tape. With the advent of on-line computer editing techniques, rapid editorial changes to a program can be made on a computer screen. At the end of an editing session, a tape can be automatically punched on command from the keyboard.

Magnetic tape is not used as much as punched tape because of its susceptibility to pollutants in the NC environment. Dust, metal filings and oil can cause read errors on the tape. Sealed magnetic tapes overcome some of these problems.

Direct entry of a part program into the controller memory is a common input medium for today's NCs. The programmer can either type in the NC program from a keyboard and a video display terminal or generate the NC program from an interactive graphics environment.[13] Part programming with the aid of interactive graphics is discussed in more detail in Section 18.3.5.

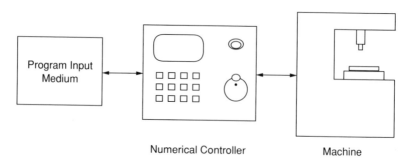

Program Input Medium   Numerical Controller   Machine

Fig. 18.12   Components of a numerical control system.

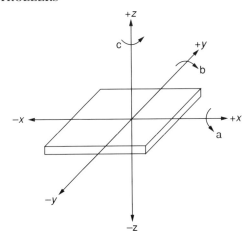

Fig. 18.13   NC coordinate system.

Operation of an NC: Machine Coordinate System

NCs require a point of reference and a coordinate system to express the coordinates of parts, tools, fixtures, and other components in the workspace of the machine tool. The commonly used coordinates are three orthogonal intersecting axes of a right-handed Cartesian coordinate frame, as shown in Fig. 18.13. The rotations a, b, and c about the x, y, and z axes are used for NCs with more than three axes.

In most of the older NCs, a coordinate frame is marked on the machine and all coordinates are with respect to this fixed frame. In the newer NCs, the machine-tool user can program or "teach" a location for the origin of the reference coordinate system. Since such a reference coordinate system is not permanently attached to the machine, it is sometimes called a floating coordinate system.

To illustrate the use of an NC coordinate system, let us consider a simple example of drilling a hole in a rectangular plate with a numerical controller. This example will illustrate the steps in the initial setup and operation of an NC machine. Figure 18.14 shows the drawing of a simple rectangular plate to be drilled by a drilling machine.

The first step in the programming of any NC operation is getting a drawing of the part. The drawing is usually a blueprint with the dimensions and geometrical attributes of the part. Figure 18.14 shows a rectangular part and the location of the center of the hole to be drilled in the part. Let the lower left-hand corner of the part be the origin of a two-dimensional Cartesian coordinate frame with the x-axis and the y-axis as shown in Fig. 18.14. The location of the center of the hole is specified by the coordinates $x = 1.500$ and $y = 2.500$.

The drilling machine must be told the location of the center of the hole. The first step is to establish the location of the NC coordinate system. With the part rigidly held in a fixture, the operator

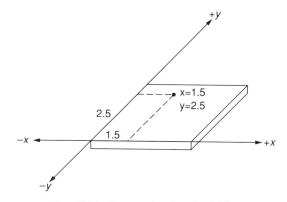

Fig. 18.14   Rectangular plate for drilling.

manually moves the machine tool to the lower left-hand corner of the part and presses a button on the machine control panel to teach this point as the origin of the Cartesian coordinate system. Now the machine can locate the center of the hole from the coordinates $x = 1.500$ and $y = 2.500$. In addition to specifying the location of the hole, a programmer can program the rotational speed of the drill, the direction of rotation of the drill (clockwise or counterclockwise), the feed rate, and the depth of cut. The *feed rate* is the distance moved by the tool in an axial direction for each revolution of the workpiece. The *depth of cut* is defined by the thickness of the metal removed from the workpiece, measured in a radial direction.

If the application were milling instead of drilling, the programmer also specifies a cutting speed and a rate of metal removal. The *cutting speed* in a turning or milling operation is the rate at which the uncut surface of the workpiece passes the cutting edge of the tool, usually expressed in millimeters per minute or inches per minutes. The *rate of metal removal* is given by the product of the cutting speed, the feed rate, and the depth of cut. The cutting speed and the feed rate are the two most important parameters that a machine operator can adjust to achieve optimum cutting conditions.

18.3.3 Point-to-Point and Contouring Numerical Controllers

In the example of the preceding section, the drill must be moved to the center of the hole before the drilling operation starts. The drill can start from the origin and traverse to the center of the hole. The machine tool may first move along the x-axis by 1.5 inches, followed by a movement along the y-axis by 2.5 inches. It may also simultaneously start moving along the x and y axes. An NC controller that can position the tool at specified locations without control over the path of the tool between locations is called a point-to-point or positioning controller.[6,8] A positioning controller may be able to move along a straight line at 45 degrees by simultaneously driving its axes.

A contouring controller provides control over the tool path between positions. For instance, the tool can trace the boundary of a complex part with linear and circular segments in one continuous motion without stopping. For this reason, contouring controllers are also called continuous path machines. The controller can direct the tool along straight lines, circular arcs, and several other geometric curves. The user specifies a desired contour which is typically the boundary or shape of a complex part, and the controller performs the appropriate calculations to continuously drive two or more axes at varying rates to follow this contour.

The specification of a contour in a point-to-point NC is tedious because it takes a large number of short connected lines to generate a contour such as a circular arc. The programmer must also provide the appropriate feed rates along these short straight line segments to control the speed along the contour.

NC Interpolators

In a multi-axis contoured NC, multiple independently driven axes are moved in a coordinated manner to direct the tool along a desired path. An interpolator generates the signals to drive the servo loops of multiple actuators along a desired tool path.[14] The interpolator generates a large number of intermediate points along the tool path. The spacing between the intermediate points determines the tool accuracy in tracking the desired path. The closer the points are, the better the accuracy. At each interpolated point along the tool path, the positions of the multiple axes are computed by a kinematic transformation. By successively commanding the NC axes to move from one intermediate point to the next, the tool traces a desired contour with respect to the part.

Older NC systems contained hardware interpolators such as integrators, exponential deceleration circuits, linear interpolators, and circular interpolators. Most of today's NCs use a software interpolator. The software interpolator takes as input the type of curve (linear, circular, parabola, spline, etc.), the start and end points, parameters of the curve such as the radius and the center of a circle, and the speed and acceleration along the curve. The interpolator feeds the input data into a computer program that generates the intermediate target points along the desired curve. At each intermediate target point, the interpolator generates the commands to drive the servo loops of the actuators. The servo loops are closed outside of the interpolator. Several excellent references provide details on closed-loop control of servoactuators.[15,16]

18.3.4 NC Programming

The two methods of programming numerical controllers are manual programming and the use of a high-level, computer-assisted part programming language.[17] In both cases, the part programmer uses an engineering drawing or a blueprint of the part to program the geometry of the part. Manual programming typically requires an extensive knowledge of the machining process and the capabilities of the machine tool. In manual programming, the part programmer determines the cutting parameters such as the spindle speed and feedrate from the characteristics of the workpiece, tool material, and the machine tool.

The most common programming medium for NC machines is perforated tape. EIA standards RS-273-A and RS-274-B define the formats and codes for punching NC programs on paper tape. The RS-273-A standard is called the interchangeable perforated tape variable block format for positioning and straight cut NC systems, while the RS-274-B is the interchangeable perforated tape variable block format for contouring and contouring/positioning NC systems. A typical line on a punched tape according to the RS-273-A standard is as follows:

$$n001 \quad g08 \quad x0.0 \quad y1.0 \quad f225 \quad S100 \quad t6322 \quad m03 \quad (EB)$$

where n is the sequential block number; g is a preparatory code used to prepare the NC for instructions to follow; x, y are dimensional words or coordinates; f is the feedrate; s is the code for the spindle rotation speed; t is the tool selection code; m denotes a miscellaneous function and (EB) is the end-of-block character, typically a carriage return. The dimension words, x and y, can be expanded for machines with greater than two axes. The order of the dimension words for machines with more than two axes is given as x, y, z, u, v, w, p, q, r, i, j, k, a, b, c, d, e.

Punched tape can be created manually or by means of a high-level computer programming system. Manual programming is more suited to simple point-to-point applications. Complex applications are usually programmed by a computer-assisted part programming language. The most common computer-assisted programming system for NCs is the Automatically Programmed Tools (APT) system.[17] APT programming is independent of the machine tool specific parameters such as tool dimensions. Some of the other programming systems are ADAPT, SPLIT, EXAPT, AUTOSPOT, and COMPACT II.

NC programming today is getting less specialized and less tedious with the help of interactive graphics programming techniques. The part programmer creates the NC program from a display of the part drawing on a high-resolution graphics monitor in a user-friendly, interactive environment. Graphical programming is discussed in Section 18.3.5.

Manual Programming

Manual programming of a machine tool is performed in units called blocks. Each block represents a machining operation, a machine function, or a combination of both. Each block is separated from the succeeding block by an end-of-block code.

The standard tape code characters are shown in Table 18.9. Preparatory functions or g-codes are used primarily to direct a machine tool through a machining operation. Examples are linear interpolation and circular interpolation. Miscellaneous functions or m-codes are generally on-off functions such as coolant-on, end-of-program, and program stop. Details on the g-codes and the m-codes and their corresponding functions are given in several excellent tutorials on NC programming.[12,17]

The feedrate function, f, is followed by a coded number representing the feedrate of the tool. A common coding scheme for the feedrate number is the *inverse-time code*, which is generated by multiplying by 10 the feedrate along the path divided by the length of the path. If FN denotes the feedrate number in inverse-time code, FN is given as follows.

$$FN = 10 \times \frac{\text{feedrate along the path}}{\text{length of the path}}$$

For example, if the desired feedrate is 900 mm/min and the length of the path is 180 mm, the feedrate number is 50 (f 0050). The inverse-time coded feedrate is expressed by a four-digit number ranging from 0001 through 9999. This range of feedrate numbers corresponds to a minimum interpolation time of 0.06 seconds and a maximum interpolation time of 10 minutes.

Spindle speeds are prefixed with the code "s" followed by a coded number denoting the speed. A common code for spindle speed is the *magic-three code*. To illustrate the steps in computing the magic-three code, let us convert a spindle speed of 2345 rpm to magic-three. The first step is to write the spindle speed as 0.2345×10^4. The next step is to round the decimal number to two decimal places and write the spindle speed as 0.23×10^4. The magic-three code can now be derived as 723, where the first digit 7 is given by adding three to the power of 10 (four plus three), and the second two digits are the rounded decimal numbers (23). Similarly, a spindle speed of 754 rpm can be written as 0.75×10^3, and the magic-three code is 675.

Manual programming of the machine tool requires the programmer to compute the dimensions of the part from a fixed reference point, called the origin. Since the numerical control unit controls the path of the tool center point, the programmer must take into account the dimensions of the machine tool before generating the program. For instance, a cylindrical cutting tool in a milling

TABLE 18.9   MANUAL PROGRAMMING CODES

CHARACTER DIGIT OR CODE	DESCRIPTION
0	Digit 0
1	Digit 1
2	Digit 2
3	Digit 3
4	Digit 4
5	Digit 5
6	Digit 6
7	Digit 7
8	Digit 8
9	Digit 9
a	Angular dimension around x-axis
b	Angular dimension around y-axis
c	Angular dimension around z-axis
d	Angular dimension around special axis or third feed function
e	Angular dimension around special axis or second feed function
f	Feed function
g	Preparatory function
h	Unassigned
i	Distance to arc center parallel to x
j	Distance to arc center parallel to y
k	Distance to arc center parallel to z
m	Miscellaneous function
n	Sequence number
p	Third rapid traverse direction parallel to x
q	Second rapid traverse direction parallel to y
r	First rapid traverse direction parallel to z
s	Spindle speed
t	Tool function
u	Secondary motion dimension parallel to x
v	Secondary motion dimension parallel to y
w	Secodary motion dimension parallel to z
x	Primary x motion dimension
y	Primary y motion dimension
z	Primary z motion dimension
.	Decimal point
,	Unassigned
Check	Unassigned
+	Positive sign
−	Negative sign
Space	Unassigned
Delete	Error delete
Carriage return	End of block
Tab	Tab
Stop Code	Rewind stop
/	Slash code

operation will traverse the periphery of the part at a distance equal to its radius. The programmer also takes into account the limits on acceleration and deceleration of the machine tool in generating a program. For example, the feedrate for straight-line milling can be much higher than the feedrate for milling an inside corner of a part. The tool must be slowed down as it approaches the corner to prevent overshoot. Formulas and graphs are often provided to assist the programmer in the above calculations.

To illustrate manual part programming, let us consider an example of milling a simple part shown in Fig. 18.15. The programmer initially identifies the origin for the Cartesian coordinate system and a starting point for the tool. We will assume that the tool starts from the origin (point PT. S). The programmer builds each block of the program, taking into account the tool diameter and the dynamic constraints of the machine tool. Let us assume that the milling tool in our example is cylindrical with a diameter of 12 mm.

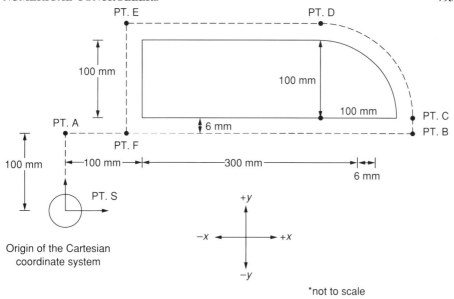

Fig. 18.15 Drawing of a part for machining.

In the first block in the program, the tool starts from rest at point PT. S and moves to point PT. A. Let the desired feedrate be 3000 mm/min. Since the distance from PT. S to PT. A is 100 mm, the inverse-time code for the feedrate is 300. Let the spindle speed be 2000 rpm. The magic three code for 2000 rpm is 720. We will use a preparatory function code, *g* 08, for exponential acceleration from rest to the desired feedrate. The first block of the program can now be written as follows.

Block #1 (from PT.S to PT.A)

n 0001	Block #1
g 08	Automatic acceleration
x 0.0	No displacement along the *x*-axis
y 100	A displacement of 100 mm along the *y*-axis
f 300	Feedrate in inverse-time code
s 720	Spindle speed in magic-three code
m 03	Clockwise spindle rotation
(EB)	End of block

Similarly, the remaining blocks for moving the tool center point along the dotted line of Fig. 18.15 can be derived as follows. Note that the feedrate and spindle speed need not be programmed for each block if there is no change.

Block #2 (from PT.A to PT.B)

n 0002	Block #2
x 406	Displacement of 406 mm along the *x*-axis
y 0.0	No displacement along the *y*-axis
m 08	Turn coolant on
(EB)	End of block

Block #3 (from PT.B to PT.C)

n 0003	Block #3
x 0.0	No displacement along the *x*-axis
y 6	Displacement of 6 mm along the *y*-axis
(EB)	End of block

Block #4 (from PT. C to PT. D)

n 0004	Block #4
g 03	Circular interpolation, counterclockwise
x −106	Displacement of −106 mm along the *x*-axis
y 106	Displacement of 106 mm along the *y*-axis
i 300	*x*-coordinate of the center of the circle
j 106	*y*-coordinate of the center of the circle
(EB)	End of block

Block #5 (from PT. D to PT. E)

n 0005	Block #5
x −206	Displacement of −206 mm along the *x*-axis
y 0.0	No displacement along the *y*-axis
(EB)	End of block

Block #6 (from PT. E to PT. F)

n 0006	Block #6
g 09	Automatic deceleration
x 0.0	No displacement along the *x*-axis
y −112	Displacement of −112 mm along the *y*-axis
m 30	Turn off spindle and coolant
(EB)	End of block

These six blocks are entered on a sheet called the process sheet, as shown in Fig. 18.16. The data from the process sheet is punched on a paper tape and input to the NC machine. Notice that each line on the process sheet represents one block of the program. Manual programming as illustrated above can be both tedious and error prone for an inexperienced programmer. In the next section, we illustrate the use of a computer to enter the part program.

Computer Assisted Programming: Programming in APT

APT was designed to be the common programming language standard for all NCs.[18] APT programs are machine-independent. An APT program is a series of English-like statements with a precise set of grammatical rules. Hundreds of keywords embody the huge expanse of numerical control knowledge into one language. An APT program typically contains process- or part-oriented information. The program does not contain control- and machine tool-oriented information. APT provides three-dimensional programming for up to five axes.

An APT program containing a description of the part geometry and tool motions is input to a computer program called the *postprocessor*. The postprocessor checks for errors, adds machine

PART NUMBER Sample					TAPE NUMBER 2003A						DATE 6/1/88
PART NAME Example					MACHINE Allen-Bradley 8200						PROGRAMMER S. Chand
N	G	X	Y	Z	I	J	K	F	S	T	M	COMMENTS
001	08	0.0	100					300	720		03	From PT. S to PT. A
002		406	0.0								08	From PT. A to PT. B
003		0.0	6									From PT. B to PT. C
004	0.3	-106	106		300	106						From PT. C to PT. D
005		-206	0.0									From PT. D to PT. E
006	0.9	0.0	-112								30	From PT. E to PT. F

Fig. 18.16 Process sheet for manual programming.

tool-specific control information, transforms the geometric description of the part into tool motion statements, and produces the proper codes for running the machine tool. Each special machine has its own postprocessor.

Most statements in APT are composed of two parts, separated by a slash. The word to the left of the slash is called the major word or the keyword, and the one to the right is called the minor word or the modifier. For example, COOLNT/OFF is an APT statement for turning the coolant off. Comments can be inserted anywhere in an APT program following the keyword REMARK.

An APT program is generated in two steps. The first step is to program the geometrical description of the part. The second step is to program the sequence of operations that defines the motion of the tool with respect to the part. The programmer defines the geometry of the part in terms of a few basic geometrical shapes such as straight lines and circles. The geometrical description comprises a sequence of connected shapes that defines the geometry of the part. No matter how complex a part is, it can always be described in terms of a few basic geometric shapes.

Step 1: Programming Part Geometry in APT

The programmer programs the geometry of a part in terms of points, lines, circles, planes, cylinders, ellipses, hyperbolas, cones, and spheres. APT provides 12 ways of defining a line; two of the 12 definitions are (1) a line defined by the intersection of two planes, and (2) a line defined by two points. There are 10 ways of defining a circle; for instance, by the coordinates of the center and the radius or by the center and a line to which the circle is tangent. The first part of an APT program typically contains the definitions of points, lines, circles, and other curves on the part. For the part shown in Fig. 18.17, an APT program may start as follows. Note that the line numbers in the leftmost column are for reference only, and are not a part of the APT program.

```
1     PARTNO      A345, Revision 2
2     REMARK      Part machined in 3/4 inch Aluminum
3     INTOL/0.00005
4     OUTTOL/0.00005
5     CUTTER/12
6     STPT = POINT/0,0,0
7     PT1 = POINT.0,106,0
8     LIN1 = LINE/(POINT/0, 106, 0), (POINT/400, 106, 0)
9     CIRC1 = CIRCLE/CENTER, (POINT/300, 106, 0), RADIUS, 100
10    LIN2 = LINE/(POINT/400, 206, 0), LEFT, TANTO, CIRC1
11    LIN3 = LINE/(POINT/400, 206, 0), (POINT/100, 206, 0)
12    LIN4 = LINE/(POINT/100, 206, 0), (POINT/100, 106, 0)
13    SPINDL/ON, CLW
14    FEDRAT/300
15    COOLNT/ON
```

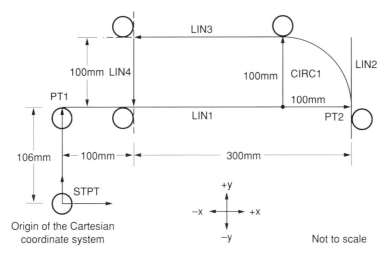

Fig. 18.17 APT programming of part for machining.

Line numbers 3 and 4 define the inside and outside tolerances in mm.[6,8] Line 5 defines the tool diameter as 12 mm. Lines 6 and 7 define the start point and point PT1. Lines 8 through 12 define the lines and the circular arc in the part geometry. Note that a temporary line, LIN2, is introduced to guide the tool to the corner at point PT2 (Fig. 18.17). Line 13 turns the spindle on with a clockwise rotation. Line 14 specifies the feedrate, and line 15 turns the coolant on.

Step 2: Programming the Tool Motion Statements in APT

In programming the motion of the tool, the programmer usually starts at a point on the part that is closest to the origin. The programmer then traverses the geometry of the part as viewed from the tool, indicating directions of turn such as GOLFT and GORGT. Sometimes the analogy of the programmer "riding" or "straddling" the tool is used to get the sense of direction around the periphery of the part.

An initial sense of direction for the tool is established by the statement INDIRP (in the direction of a point). For the part of Fig. 18.17, let us assume that the tool starts from the point STPT and moves to point PT1. A sense of direction is established by the following statements:

FROM/STPT

INDIRP/PT1

The statement GO is used in startup of the motion. The termination of the motion is controlled by three statements: TO, ON, and PAST. The difference between the statements GO/TO LIN1, GO/ON LIN1, and GO/PAST LIN1 is illustrated in Fig. 18.18.

For the example of Fig. 18.17, the first GO statement is as follows.

GO/TO, LIN1

The next statement moves the tool to PT2. The sense of direction is set by the previous motion from STPT to PT1. The tool must turn right at PT1 to traverse to PT2. Also, the tool must remain to the right of the line LIN1 as seen from the tool in the direction of motion between PT1 and PT2. This move statement is written as follows.

TLRGT, GORGT/LIN1, PAST, LIN2

The next step is to move the tool from PT2 to PT3 along the circle CIRC1.

TLRGT, GOLFT/CIRC1, PAST, LIN3

The next two motion statements to complete the motion around the part, are as follows.

GOFWD/LIN3, PAST, LIN4

TLRGT, GOLFT/LIN4, PAST, LIN1

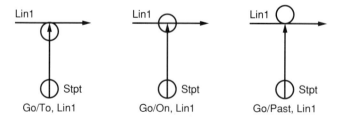

Fig. 18.18   GO statement in APT.

The complete program is shown below.

```
PARTNO      A345, Revision 2
REMARK        Part machined in 3/4 inch Aluminum
INTOL/0.00005
OUTTOL/0.00005
CUTTER/12
STPT = POINT/0,0,0
PT1 = POINT/0,106,0
LIN1 = LINE/(POINT/0,106,0), (POINT/400, 106, 0)
CIRC1 = CIRCLE/CENTER, (POINT/300, 106, 0), RADIUS, 100
LIN2 = LINE/(POINT/400, 206, 0), LEFT, TANTO, CIRC1
LIN3 = LINE/(POINT/400, 206, 0), (POINT/100, 206, 0)
LIN4 = LINE/(POINT/100, 206, 0), (POINT/100, 106, 0)
SPINDL/ON, CLW
FEDRAT/300
COOLNT/ON
FROM/STPT
INDIRP/PT1
GO/TO, LIN1
TLRGT, GORGT/LIN1, PAST, LIN2
TLRGT, GOLFT/CIRC1, PAST, LIN3
GOFWD/LIN3, PAST, LIN4
TLRGT, GOLFT/LIN4, PAST, LIN1
STOP
FINI
```

18.3.5 Numerical Controllers and CAD/CAM

A computer-aided design (CAD) system is used for the creation, modification, and analysis of designs. A CAD system typically comprises a graphics display terminal, a computer, and several software packages. The software packages aid the designer in the creation, editing, and analysis of the design. For instance, in the analysis of a part design, the designer can access a library of routines for finite element analysis, heat-transfer study, and dynamic simulation of mechanisms.

Computer-aided manufacturing (CAM) refers to the use of a computer to plan, manage, and control the operations in a factory. CAM typically has two functions: (1) monitoring and control of the data on the factory floor, and (2) process planning of operations on the factory floor.

Since the parts manufactured in a factory are usually designed on a CAD system, the design should be integrated with the programming of the machines on the factory floor. The process of integrating design and manufacturing is often referred to as CAD/CAM.

Today, CAD systems are commonly used to design the parts to be machined by an NC. The NC programmer uses a CAD drawing of the part in the generation of the part program. Several NCs provide an interactive graphic programming environment that displays the CAD drawing of the part on a graphics terminal. On the screen of the graphics terminal, a programmer can display the cross section of the part, rotate the part in three dimensions, and magnify the part. These operations help the programmer to better visualize the part in three dimensions; a similar visualization of the part from a two-dimensional blueprint is difficult.

In an interactive graphic programming environment, the programmer constructs the tool path from a CAD drawing. In many systems, the tool path is automatically generated by the system following an interactive session with the programmer. The output can be an APT program or a CLFILE, which can be postprocessed to generate the NC punched tape.[13]

There are two basic steps in an interactive graphics programming environment. The first step is the definition of the geometry of the part. The geometric definition of the part can be specified during the part design process in a CAD/CAM system. If the geometric definition does not exist, the part programmer must create it on the graphics terminal. This process is interactive, with the part programmer labeling the various edges and surfaces of the part on the graphics screen. After the labeling process is complete, the system automatically generates the APT geometry statements for the part. The definition of the part geometry is much more easily performed on the interactive graphic system than by the time-consuming, error-prone, step-by-step manual process.

The second step in the interactive graphic programming of NCs is generation of the tool path. The programmer starts by defining the starting position of the cutter. The cutter is graphically moved along the geometric surfaces of the part through an interactive environment. Most CAD/CAM systems have built-in software routines for many machining operations, such as surface contouring, profile milling around a part, and point-to-point motion. The programmer can enter feed rates and spindle speeds along each segment of the path. As the program is created, the programmer can visually verify the tool path with respect to the part surface on the graphics screen. The use of color graphics greatly aids in the visual identification of the part, tool, and program parameters.

The advantages of the CAD/CAM approach to NC programming are, (1) the geometric description of the part can be easily derived from the CAD drawing; (2) the programmer can better visualize the part geometry by manipulating the part drawing on the screen, such as rotating the part and viewing cross sections; (3) the interactive approach for the generation of part programs allows the user to visually verify a program on the graphics screen as it is being created; (4) the programmer has access to many machining routines on the CAD system to simplify tool path programming; and (5) the CAD environment allows the integration of part design and tool design with programming.[20]

18.4   ROBOT CONTROLLERS

Michael D. McEvoy

18.4.1   A Brief History of Industrial Robotics

The modern day industry of robotics and robot-based industrial automation finds its roots, like many areas of modern high technology, in the minds of the science fiction author. Today, book racks are filled with stories of robot soldiers, cyborgs, automated servants, and the like written by almost every science fiction author. Thanks to George Lucas and the Star Wars™ trilogy, two robots, R2-D2 and C-3PO, are more widely known by today's youth than George Washington.

There is almost universal agreement that the robot was brought to life by two individuals, Karel Capek and Isaac Asimov. In 1922, Karel Capek wrote a play called *Rossum's Universal Robots* (RUR) describing how robots would turn on mankind and eventually take over the world. Karel Capek also coined the word "robot." In the 1940s, Isaac Asimov freed us from a view of robots as malevolent beings and showed us a world where robots would be our helpmates, improving our lives and helping us to become the masters of our universe. At the same time, Asimov coined the term "robotics," giving, for the first time, a proper name for this future science and technology.

If Capek and Asimov are the prophets of modern robotics, then Joseph Engelberger and George Devol are most certainly the parents. Conception took place in 1954 when Devol applied for the first significant patent for a programmable manipulator. Birth occurred two years later when Devol and Engelberger met at a cocktail party and formed the basis for the world's first robot company, Unimation. In the following years, Unimation developed and released the first practical robots that took their place in 1961 on the production line doing simple, repetitive tasks. From that time until the mid-seventies, Unimation proceeded with pioneering the development of basic robot technology and applications. In the early seventies, a breakthrough came that would significantly change the course of the robot industry, the development of the microprocessor by Intel and other semiconductor companies. Up to that point, a robot's cost had been 70 percent control and 30 percent mechanical. The advent of the microprocessor reversed that ratio and created a whole new competitive arena. Robotics was heralded as the next superstar high technology market. During the seventies and early eighties, many large companies such as Cincinnati-Milacron, GE, Westinghouse, General Motors, and IBM entered the field, along with smaller startups such as Automatix, Cybotech, Advanced Robotics, and Adept. By the mid-eighties, the profits and sales for robot companies were failing to live up to their projected levels. This was followed in the mid- to late eighties by the sale, closure, and downsizing of many robot companies and divisions.

This pattern of breakthrough, rapid growth, saturation, and contraction followed by a reduced rate of growth is not a new pattern. Witness the same pattern in the machine tool, semiconductor, and video game marketplaces, to name a few. In the case of the robot industry, the root causes stem from several areas. First and foremost was a higher than expected failure rate and cost of system integration versus the perceived and promised low risk and cost of robotic technology by many companies. The second contributor was that the initial marketplace, automotive, has a fairly low number of early adapters, those who are willing and able to take on new technology. Because of this, the rate of technology penetration into new applications (beyond such proven applications as spot welding and spray painting) in this market failed to grow at anywhere near the expected rates. A final factor was the low gross margins that most robot companies actually achieved, typically ranging from 25 to 35 percent, caused by the oversupply of robot companies and the corresponding highly competitive nature of the business. With margins at this level, most robot companies operated at a significant loss.

At this time, a major shakeout of companies in the robotic field has occurred and most who remain have developed the products as well as the technical and market savvy needed to be successful. From this point on, we should see a steady, albeit slower, growth as this market matures.

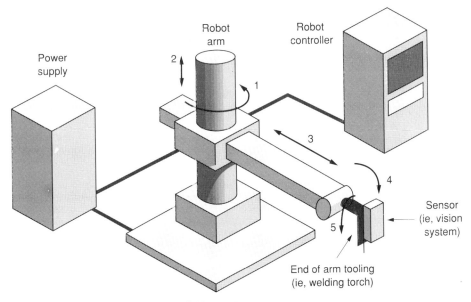

Fig. 18.19 Typical robot system.

18.4.2 Elements of a Robot System

A robot is a system that consists of a number of modules as shown in the diagram of a typical industrial robot in Fig. 18.19.

Arm

The most obvious component of a robot system is its one or more arms or manipulators. An arm consists of two or more linear or rotary joints and an associated actuator or motor for each joint. These actuator elements consist of electric, hydraulic, or pneumatic motors and/or linear devices along with, in most cases, a position sensor for each joint, and possibly limit switches for calibration and limit control.

Power Supply

Each of the actuators requires some source of specialized power, either of a switched or proportional nature, to each joint. In the case of DC or AC motor drives, each motor is supplied with carefully controlled power from a servoamplifier based on a command signal from the robot controller, typically a $\pm$ 10 VDC signal that is proportional to the desired motor speed or torque. In the case of stepping motors, a carefully controlled sequence of power is applied to the motor's windings. By changing the sequence and frequency of power applied to the motor, its speed and direction can be controlled. With hydraulic and pneumatic systems, either switched or proportional air or hydraulic power is applied to either a cylinder or motor. Pneumatic power is usually reserved for simple limited sequence, non-servo robots.

Robot Controller

A robot controller performs a variety of functions in the system. The priorities for a modern controller were defined quite well in 1940 by Isaac Asimov in what has become known as the "Three Laws of Robotics":

Asimov's Three Laws of Robotics

   I   A robot may not injure a human being, or, through inaction, allow a human being to come to harm.

   II A robot must obey the orders given it by human beings except where such orders would conflict with the First Law.

   III A robot must protect its own existence as long as such protection does not conflict with the First or Second Law.

At first glance, these "Three Laws of Robotics" may appear to be rules for a truly intelligent futuristic machine but, with additional examination, reordering, and rephrasing, they form an excellent set of priorities and guidelines for the designer of modern robot controls and systems.

Priorities for Modern Robot System Design

I   A robot must operate in a safe manner that cannot place at risk or harm a human being, either through the direct or unintentional actions of the individual.

II A robot must protect itself, its tooling, and the material being used as long as such protection does not conflict with the First Priority.

III A robot must accept and follow the valid commands and programming given it by its operator, faithfully executing these instructions up to the limits of the robot's capability, except where such orders would conflict with the First or Second Priority.

In essence, the safety of the individual should be given the highest priority in the design of robot controllers and systems. This is usually achieved through the use of physical barriers, software, and sensors for detecting hazardous conditions as well as fault detecting/tolerant computing techniques. Attention to this area is becoming even more critical as robots shift from high volume manufacturing applications, where the direct interaction with humans is minimal, to batch manufacturing, where the level of direct contact and interaction with the individual is much higher. The robot must next accept the instructions of the operator or higher level control, check these instructions for validity and feasibility, and execute them reliably. During the execution of these instructions, the robot should monitor its operation and the process it is performing to ensure that the robot, tooling, and parts are not being damaged and that quality parts are being produced.

End-of-Arm Tooling

No robot is complete without some form of end-of-arm tooling. This can range from a simple gripper for part handling to a complex tool-changing system equipped with multiple tools. Almost every tool that can be wielded by a person has found its way onto a robot arm.

Sensors

Sensors are the last piece of the robot system. A sensor can be quite simple, such as a switch that detects the position or presence of a part, to a 3D vision system that recognizes and measures the locations of a part and corrects the robot's programmed path. Sensors are used to make simple go/no-go decisions as a part of the robot program's execution. In this case, they are wired to the available controller I/O and interrogated by the robot's program as needed. In more sophisticated applications, they are also used to adapt the robot's path, speed, and process parameters to ensure that a quality part is being manufactured. Currently, when this higher level of integration occurs, the robot controller's internal software is modified to allow the sensor to become tightly coupled to the robot's real-time operation. This increased performance, therefore, comes at a higher cost and results in the transition of the robot from a general purpose device to a high performance, special purpose machine.

18.4.3   Types of Robots

Non-Servo Robots

Robots and their controllers come in a variety of shapes and sizes depending on the specific application for which they are intended. A simple robot controller can be built out of a programmable logic controller for simple, non-servo robots. In the sample three-axis system shown in Fig. 18.20, the PLC controls a series of on/off pneumatic valves to move the arm left, right, up, down, in, and out as well as to open or close the gripper. This represents the simplest type of robot and controller. Positions are not programmed, only on/off sequences for each valve and other devices. A simple manipulator like this is useful for a variety of material handling applications.

Tele-Operator

The next type of robot, the tele-operator, is not really a robot at all, but is a remote control device where an operator's action is directly reflected in the movement of the arm. It is used in a wide variety of applications from the handling of radioactive material, the collection of biological samples from a deep diving submarine at 2000 fathoms, to the repair of satellites 200 miles above earth on the space shuttle or anyplace else where it is hazardous or undesirable to place a human being. A simple tele-operator consists of two geometrically identical mechanisms, a master manipulated by the operator and a slave equipped with a set of actuators and sensors that allow it to precisely match and

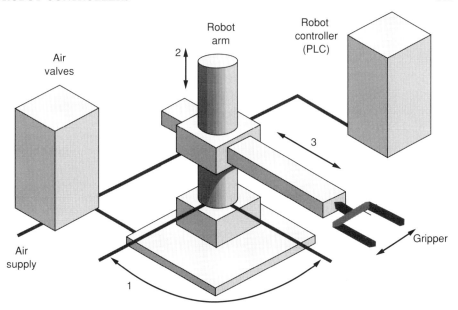

Fig. 18.20   Simple pneumatic robot system.

follow the position of the master. As tele-operators grow in sophistication, they gain features such as feedback, allowing the operator to control force as well as position. Advanced tele-operators, such as the space shuttle RMS arm, also allow for the arm to be operated from a hand control that instructs the arm motion at a higher, symbolic level such as in/out or left/right, rather than in terms of raw joint-by-joint motion. In this case, the controller provides the conversion between the kinematics of the hand control and that of the manipulator. Some tele-operators even allow for actual programmed operation, creating a hybrid of a tele-operator and robot.

Up to this point, we've been discussing devices that are related to the modern servo-controlled industrial robot that will be discussed in the balance of this section. These devices can be classified as continuous path, point-to-point, or kinematically controlled robots.

Continuous Path Robots

Continuous path robots are used for spray painting and other applications where a smooth, fluid motion is required to apply material. The continuous path robot is typically programmed by either pulling the manipulator through the desired path or by the use of a duplicate, lightweight version of the manipulator. As the manipulator or its duplicate is moved along the desired path, its joint positions are recorded by the controller. When the program is finished, the controller stores the recorded information for later recall.

During execution, the recorded joint positions are recalled by the controller and fed into the servo loop for each axis of the robot in the same order and timing as they were recorded, causing the robot to duplicate the previously taught motion. Due to the constant sampling and recording of joint positions, a large quantity of information must be stored. In order to reduce the amount of stored information, some manufacturers use a compression algorithm to reduce the overall storage requirements.

Point-to-Point Robots

A point-to-point robot is programmed in terms of a series of joint positions, which are typically taught using a teach pendant. These points are stored and are used as final targets where the control will move the robot's joints. The robot is also instructed how fast to move the joints, either in terms of time to complete the move or as a percentage of maximum speed. The controller interpolates a series of positions for the joints that results in smooth acceleration and speed as well as synchronization of the axes. Since only the target position is stored, the amount of memory required is considerably less than a continuous path, but the ability to program and control the robot's motion and speed is complicated. In addition to the control of the arm, the controller must also control and synchronize with the operation of other devices in the robot's workcell. This is done using digital and analog I/O channels that are integral to the robot controller. Due to its simplicity, this type of robot is typically found in spot welding or material handling systems.

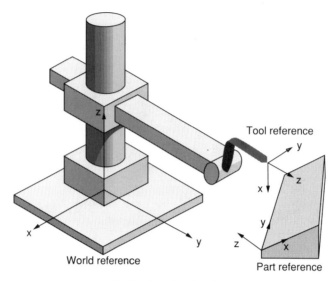

Fig. 18.21   Coordinate references.

Kinematic Robots

When discussing a modern industrial robot, the discussion is usually about robots that contain a kinematic model of the manipulator. This model, in addition to the functions provided by the point-to-point robot, allows the operator to interact with the robot in terms of Cartesian coordinate systems, providing the operator with the ability to both teach and operate the robot in terms of up/down, left/right, and in/out as well as roll, pitch, and yaw motions, instead of a series of joint positions. Frames of reference for these Cartesian coordinate systems can find their origin and orientation reference at the base of the robot, or World Coordinates, attached to the part, or Part Coordinates, or aligned with the end effector, or Tool Coordinates (see Fig. 18.21). When up/down (Z), left/right (X), and in/out (Y) motions are specified, the motion occurs in alignment to the specified frame of reference. This greatly simplifies the programming process, allowing the user to "back the torch away from the part," which in World Coordinates might require motion in X, Y, and Z, but if done in the Tool Coordinate reference, would only require one motion, a Z move.

Another characteristic typically found in this type of robot is the ability to program its operation in terms of a language. More advanced versions also allow for the direct input of sensor data to control the forces applied by the arm and to modify the path or trajectory of the tool tip.

18.4.4   The Structure of a Robot Controller

A basic robot controller is a fairly straightforward device as shown in Fig. 18.22. The highest level of operation is the program interpreter, which takes a previously written robot program and translates it into a series of commands. These commands provide the control with basic motion, sequence, and process control operations. Robot programs are typically in the form of a formal programming language or a data base with an intelligent, interactive editor. During execution, the program interpreter sends commands to the trajectory generator, or interpolator, which takes high level commands, such as

MOVE TO POSITION ($X = 10$, $Y = 20$, $Z = 10$, Roll $= 90$, Pitch $= 45$, Yaw $= 0$)

SPEED 20

MOVE TO POSITION ($X = 10$, $Y = 50$, $Z = 100$, Roll $= 90$, Pitch $= 45$, Yaw $= 45$)

and generates a series of points at a high and regular frequency (typically 10 ms or less) in the Cartesian coordinate system that defines a path and speed from the current position to another point in space.

These points, which are in Cartesian space, are converted to positions for each joint of the robot by a coordinate transformation (called the inverse kinematic solution) and sent to the servo control

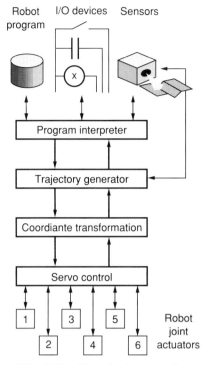

Fig. 18.22 Typical robot controller.

to close the position loop, moving the robot arm along the desired path. The paths may be in the form of a straight line and simple circular curves to complex paths defined by quadratic, cubic, general polynomials, or other functions. While a typical path may be represented by only a few points, the interpolated results may, just like in a machine tool, contain many times the number of defining points as shown in Fig. 18.23. If an articulated, or jointed arm is used, the kinematic solution may produce results that require speed from a given axis that is well out of proportion to the actual speed required to move along the path. In the previous path, if a planar, two-axis robot was used to move a tool along the path, the results would require a very wide speed range for axis 1 and a very low speed range for axis 2, as shown in Fig. 18.24.

If the path is translated or rotated with respect to the center of the robot (located at axis 1), the motion requirements of a given axis may change rather dramatically. In addition, when the tool tip is working near axis 1, the total inertia seen by the axis 1 motor is very small compared to what the axis 1 motor sees when the arm is near full extension (such as at point 1). Since varying inertia results in varying dynamics for the robot and its servo system, the controller may also need to provide

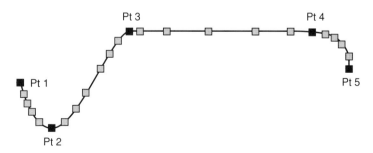

Fig. 18.23 Sample interpolated path.

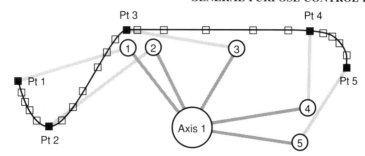

Fig. 18.24   Motion along path with two-axis arm.

dynamic modeling of the robot's arm to compensate for this change in dynamic performance, further complicating the design of the controller. Due to these types of problems, robot controllers can be built with a range of modeling, from a simple kinematic model of the arm to one that encompasses the dynamics of the entire system, in order to produce the desired level of performance.

In addition to the handling of basic path and motion generation, the robot controller is often configured with I/O for workcell control. This is also driven and interrogated directly by the Program Interpreter. In addition, sensors such as 3D vision, force, and others that provide external feedback on position and process may be available. These may be evaluated by programmed instructions or directly sent to the Trajectory Generator to adaptively control the path, speed, or force applied by the arm.

18.4.5   Programming a Modern Robot Controller

While a robot controller is responsible for controlling the motion of the arm, it is often required to perform many more advanced functions. This has resulted in the development and use by most manufacturers of a robust and powerful robot programming language such as Karel (GMF), VAL (Westinghouse/Adept), AML (IBM), RAIL (Automatix), and others. These languages have a wide range of capabilities in addition to motion control and often rival the capability of modern programming languages such as C and Pascal. Probably the most popular language is Karel, which was developed by GMF and has been recently adopted in modified form by Cybotech, Allen-Bradley, and Renault. Karel will be examined in more detail as a "typical" robot programming language.

As mentioned before, Karel, as well as other robot programming languages, incorporates structures and conventions used in high-level computer languages as well as features developed especially for robotics applications. These features show the typical capability of a modern robot controller's programming environment. They should be compared with machine tool programming as described in the previous section to illustrate one of the principle differences between a machine tool and robot controller, namely, that motion is not the only focus of a robot controller, but that process and sequence control also dominate.

Simple and Structured Data Types

The Karel programming language supports two types of data identifiers, constant and variable. Identifiers are declared at the beginning of the program or module. Constants, as the name implies, have an associated value. Variables, on the other hand, have an associated type, either integer, real, Boolean, or string, and can be defined as arrays. In more advanced versions of Karel, a Pascal-like record structure has been introduced, allowing complex data structures to be defined. Variables can also be defined as being "from" another program, allowing them to be shared with other programs. In addition, complex data is allowed to express three component vectors, six component positions, paths (an array of positions), and auxiliary positions, allowing the control and manipulation of nonrobot positions.

Arithmetic, Relational, and Boolean Operators

Expressions are allowed in a variety of forms. This includes assignment, where resulting evaluation of an expression is assigned to a variable; and relational, where two expressions can be compared resulting in a true or false evaluation. The Boolean operators AND, OR, and NOT are also provided. Vector and position operations are also provided to support more complex motion-related operations. Parentheses are allowed and expressions within these are evaluated first. The precedence of the operators follows in the order of most high-level languages.

Control Structures for Loops and Decisions

The Karel language provides a variety of control structures including:

```
IF condition THEN-ELSE . . .
SELECT n CASE . . .
FOR n = a TO b DO . . .
WHILE condition DO . . .
REPEAT . . . UNTIL condition
WAIT FOR condition
GOTO
ABORT
DELAY n
PAUSE
```

Input and Output Operations

Two types of I/O operations are provided, file and device. File I/O is used to read and write information to storage devices, while device I/O is used to read and write to logical devices such as keyboards, CRTs, and serial ports.

Input and Output to Ports

I/O operations can be performed on both digital and analog I/O devices, allowing the user to read inputs into the system and set outputs. In addition, internal or system defined devices can be accessed.

Motion Control

Motion control takes the form of MOVE statements in Karel and the defining of positions. Positions can either be taught by the teach pendant or created by the user's program. MOVE statements can specify motion TO a position, ALONG a path, NEAR, AWAY from, RELATIVE to, ABOUT, or BY a position. In addition, the speed and type of transition between segments of the path can be specified, as well as motion until a condition is satisfied. Motion can also be started and CANCELed, STOPped, or put on HOLD under program control. A RESUME can be issued to continue a motion that was stopped.

Procedures and Functions

Karel allows the definition of both routines and procedures which can be invoked directly by name. In addition, routines and procedures can be invoked from other programs using a FROM program name clause. A large library of system-defined routines are available to augment the language's capability.

Condition Handlers

Condition handlers allow the user to create a limited form of multitasking. Condition handlers have a test that is evaluated periodically and if found to be true, the programming code in the condition handler will be executed. Condition handlers can be defined, enabled, disabled, and purged from the system. In essence, they allow the user to define a simple task that is periodically evaluated and executed.

REFERENCES

1. International Electrotechnical Commission, *Preparation of Function Charts for Control Systems,* IEC Standard Publication 848, Geneva, 1987.

2. International Electrotechnical Commission Draft, *Programmable Controllers, Part 3: Programming Languages,* 65A(Secretariat)90, Geneva, December 1988.

3. UTE Standard PR C03-190, *Diagrams, Charts, Tables—Function Chart GRAFCET for the Description of Logic Control Systems,* Paris, 1981.

4. International Electrotechnical Commission, *Programmable Controllers, Part 1: General Information,* 65A(Central Office)21, Geneva, November 1988.

5. International Electrotechnical Commission, *Programmable Controllers, Part 2: Equipment Characteristics,* 65A(Central Office)22, Geneva, November 1988.

6. R. S. Pressman and J. E. Williams, *Numerical Control and Computer-Aided Manufacturing,* Wiley, New York, 1977.

7. W. Leslie, *Numerical Control Users Handbook*, McGraw-Hill, New York, 1970.

8. Y. Koren, *Computer Control of Manufacturing Systems*, McGraw-Hill, New York, 1983.

9. E. M. Trent, *Metal Cutting*, Butterworths, Woburn, MA, 1977.

10. G. Boothroyd, C. Poli, and L. E. Murch, *Automatic Assembly*, Marcel Dekker, New York, 1982.

11. A. Fitzgerald and C. Kingsley, *Electric Machinery*, 2nd ed., McGraw-Hill, New York, 1958.

12. J. Childs, *Numerical Control Part Programming*, Industrial Press, New York, 1973.

13. M. Groover, and E. Zimmers, *CAD/CAM Computer-Aided Design and Manufacturing*, Prentice-Hall, Englewood Cliffs, NJ, 1984.

14. Y. Koren, A. Shani, and J. Ben-Uri, "Interpolator for a CNC System," *IEEE Trans. Comp.*, Vol. C-25, No. 1, January 1976, pp. 32–37.

15. General Electric, *Pulse Width Modulated Servo Drive*, GEK-36203, March, 1973.

16. J. Beckett and G. Mergler, "Analysis of an Incremental Digital Positioning Servosystem with Digital Rate Feedback," *Trans. ASME J. Dyn. Sys. Meas. Contr.*, Vol. 87, March, 1965.

17. A. Roberts and R. Prentice, *Programming for Numerical Control Machines*, 2nd ed., McGraw-Hill, New York, 1978.

18. Illinois Institute of Technology Research Institute (IITRI), *APT Part Programming*, McGraw-Hill, New York, 1967.

19. R. Neil, "CAD/CAM Use in Numerical Control," *Proceedings, Eighteenth Annual Meeting and Technical Conference*, Numerical Control Society, Dallas, May 1981, pp. 56–82.

20. D. Grossman, "Opportunities for Research on Numerical Control Machining," *Communications of the ACM*, Vol. 29, No. 6, June 1986.

21. R. C. Paul, *Robot Manipulators*, MIT Press, 1982.

22. S. Y. Nof, *Handbook of Industrial Robotics*, Wiley, 1985.

23. M. Brady and R. C. Paul, *Robotics Research*, MIT Press, 1984.

24. GMF Corporation, *Karel Language Manual*, Troy, MI, 1987.

BIBLIOGRAPHY

Jones, C. T., and L. A. Bryan, *Programmable Controllers: Concepts and Applications*, IPC/ASTEC, Atlanta, 1983.

Wilhelm, Robert E., Jr., *Programmable Controller Handbook*, Hayden, Hasbrouck Heights, NJ, 1985.

Control Engineering, Des Plaines, IL.

I & CS, Chilton, Radnor, PA.

InTech, Instrument Society of America, Research Triangle Park, NC.

Programmable Controls, Instrument Society of America, Research Triangle Park, NC.

CHAPTER **19**

STATE-SPACE METHODS FOR DYNAMIC SYSTEMS ANALYSIS

KRISHNASWAMY SRINIVASAN

Department of Mechanical Engineering
The Ohio State University
Columbus, Ohio

19.1 INTRODUCTION

The use of the state-space approach for the dynamic analysis and control of systems results in analysis and design techniques based in the time domain, as opposed to frequency-domain-based transform techniques. The state-space approach has the following characteristics:

1. It employs a more complete internal representation of dynamic systems as compared to transform methods that use input-output representations. The state of a system represents complete information about the current dynamic condition of the system. It incorporates the effect of all past inputs on the system. When combined with a complete description of the system dynamics in the form of state-space equations and knowledge of all future inputs, the future behavior of the system can be determined. More precise definitions of the notion of state are given in standard textbooks.[1−4]

2. It offers a unified approach to the analysis and synthesis of linear and nonlinear, time-invariant and time-varying, continuous-time and discrete-time, single-input and single-output, and multiple-input and multiple-output systems. Available techniques, however, are more plentiful for some categories of systems.

3. State-space-based methods rely more heavily on digital computers than classical transform-based techniques for dynamic systems analysis and control. In fact, the availability of digital computers both for analysis and control synthesis and for implementation of the controllers has been an important factor underlying the growing use of state-space-based methods.

4. State-space-based methods have the potential to improve the performance of controlled systems if such systems can be modeled accurately. They have been less successful in cases where system models are characterized by significant uncertainty. Classical transform-based techniques have been and continue to be widely used in such cases. In fact, one of the more encouraging trends in control systems development has been the establishment of links between state-space-based methods and transform-based methods.[5]

Though the concept of state has been invoked by a number of methods of classical mechanics and is implicit in the phase-plane concept used for nonlinear system stability analysis, the effective application of state-space-based methods for analysis and control of dynamic system behavior has occurred only over the last three decades. Pioneering theoretical work by Kalman[8-11] and others and the availability of digital computers for performing analysis and design computations have been important underlying factors. State-space methods have been most successful in aerospace control applications and less so in a variety of industrial control applications. Among the factors favoring increased emphasis in the future on state-space methods are:

1. The emphasis on controlled system performance improvement resulting from imperatives such as improved efficiency of energy utilization and improved productivity
2. The increasing availability of inexpensive but powerful digital computers for off-line analysis and design computations and on-line control computations

In Sections 19.2–19.7, methods for analysis of dynamic systems using state-space methods are described. Even though most of the results presented in the literature use the continuous-time formulation, the fact that digital computers will be increasingly used for controller implementation implies that discrete-time formulations have significant practical importance. Hence, both continuous-time and discrete-time formulations are presented to the fullest extent possible.

19.2 STATE-SPACE EQUATIONS FOR CONTINUOUS-TIME AND DISCRETE-TIME SYSTEMS

The differential equations describing the input-output behavior of an nth-order, continuous-time, nonlinear, time-varying, lumped-parameter system can be written in the form of a first-order vector ordinary differential equation and a vector output equation:

$$\mathbf{x}(t) = \mathbf{f}[\mathbf{x}(t), \mathbf{u}(t), t] \qquad t \geq t_0 \tag{19.1}$$

$$\mathbf{y}(t) = \mathbf{g}[\mathbf{x}(t), \mathbf{u}(t), t] \qquad t \geq t_0 \tag{19.2}$$

where $\mathbf{x}(t) = n$-dimensional state vector
$\mathbf{y}(t) = p$-dimensional output vector
$\mathbf{u}(t) = r$-dimensional input vector
$\mathbf{f}, \mathbf{g} = $ vectors of appropriate dimension whose elements are single-valued nonlinear functions of the arguments noted.

Equation 19.1 is the state equation and Eq. 19.2 is the output equation. The state, output, and input vectors are

$$\mathbf{x}(t) = \begin{bmatrix} x_1(t) \\ x_2(t) \\ \vdots \\ x_n(t) \end{bmatrix} \quad \mathbf{y}(t) = \begin{bmatrix} y_1(t) \\ y_2(t) \\ \vdots \\ y_p(t) \end{bmatrix} \quad \mathbf{u}(t) = \begin{bmatrix} u_1(t) \\ u_2(t) \\ \vdots \\ u_r(t) \end{bmatrix} \tag{19.3}$$

The elements $x_1(t), x_2(t), \ldots, x_n(t)$ of the state vector are the state variables of the system.

Formulation of the higher-order system differential equations as a set of first-order differential equations has the advantage that the latter are easier to solve by numerical methods than the former. If the functions $\mathbf{f}$ and $\mathbf{g}$ are linear functions of $\mathbf{x}(t)$ and $\mathbf{u}(t)$, the system can be described by linear ordinary differential equations. Matrix notation can then be employed to simplify their representation:

$$\dot{\mathbf{x}}(t) = \mathbf{A}(t)\mathbf{x}(t) + \mathbf{B}(t)\mathbf{u}(t) \qquad t \geq t_0 \tag{19.4}$$

$$\mathbf{y}(t) = \mathbf{C}(t)\mathbf{x}(t) + \mathbf{D}(t)\mathbf{u}(t) \qquad t \geq t_0 \tag{19.5}$$

where $\mathbf{A}(t) = n \times n$ system matrix
$\mathbf{B}(t) = n \times r$ input-state coupling matrix
$\mathbf{C}(t) = p \times n$ state-output coupling matrix
$\mathbf{D}(t) = p \times r$ input-output transmission matrix

A block diagram representation of Eqs. 19.4 and 19.5 is given in Fig. 19.1, using standard symbols appropriate for simulation diagrams. If the system is linear and time invariant (LTI), the matrices noted become constant matrices, as indicated by the following equations:

$$\dot{\mathbf{x}}(t) = \mathbf{A}\mathbf{x}(t) + \mathbf{B}\mathbf{u}(t) \qquad t \geq t_0 \tag{19.6}$$

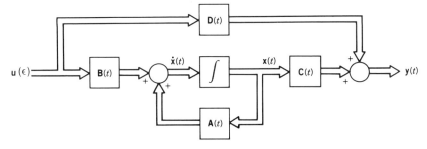

Fig. 19.1   Linear continuous-time system.

$$\mathbf{y}(t) = \mathbf{C}\mathbf{x}(t) + \mathbf{D}\mathbf{u}(t) \qquad t \geq t_0 \tag{19.7}$$

If only values of the input and output variables at discrete instants in time are of interest, difference equations are appropriate for describing their relationship. The difference equations describing the input-output behavior of an nth-order, discrete-time, nonlinear, time-varying, lumped-parameter system can be written in the form of a first-order vector difference equation and a vector output equation:

$$\mathbf{x}(t_{k+1}) = \mathbf{f}[\mathbf{x}(t_k), \mathbf{u}(t_k), t_k] \qquad t_k \geq t_0 \tag{19.8}$$

$$\mathbf{y}(t_k) = \mathbf{g}[\mathbf{x}(t_k), \mathbf{u}(t_k), t_k] \qquad t_k \geq t_0 \tag{19.9}$$

where $\mathbf{x}$, $\mathbf{y}$, $\mathbf{u}$ = state, output, and input vectors of the same dimensions as noted in connection with Eqs. 19.1 and 19.2

$\mathbf{f}$, $\mathbf{g}$ = same functions as in the equations already mentioned

t_k, t_{k+1} = the kth and $(k + 1)$th discrete-time instants, respectively

If the interval between consecutive discrete-time instants is constant and equal to T, and if the functions $\mathbf{f}$ and $\mathbf{g}$ are linear, the state-space equations become

$$\mathbf{x}(k + 1) = \mathbf{F}(k)\mathbf{x}(k) + \mathbf{G}(k)\mathbf{u}(k) \qquad k \geq k_0 \tag{19.10}$$

$$\mathbf{y}(k) = \mathbf{C}(k)\mathbf{x}(k) + \mathbf{D}(k)\mathbf{u}(k) \qquad k \geq k_0 \tag{19.11}$$

where the time instants kT and $(k + 1)T$ are represented by the corresponding sequence numbers k and $k + 1$, for notational convenience. In the preceding equations, $\mathbf{F}$, $\mathbf{G}$, $\mathbf{C}$, and $\mathbf{D}$ take the place of the matrices $\mathbf{A}$, $\mathbf{B}$, $\mathbf{C}$, and $\mathbf{D}$ in Eqs. 19.4 and 19.5. A block diagram representation of the system equations is given in Fig. 19.2. The matrices $\mathbf{F}$, $\mathbf{G}$, $\mathbf{C}$, and $\mathbf{D}$ become constant matrices for time-invariant systems:

$$\mathbf{x}(k + 1) = \mathbf{F}\mathbf{x}(k) + \mathbf{G}\mathbf{u}(k) \qquad k \geq k_0 \tag{19.12}$$

$$\mathbf{y}(k) = \mathbf{C}\mathbf{x}(k) + \mathbf{D}\mathbf{u}(k) \qquad k \geq k_0 \tag{19.13}$$

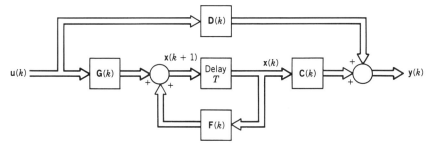

Fig. 19.2   Linear discrete-time system.

19.3   STATE-VARIABLE SELECTION AND CANONICAL FORMS

The state vector of a system is comprised of the minimum set of variables necessary to describe the system behavior in the form of the state-space equations already given. It can be shown that the selection of the state vector for a system is not unique.

For linear continuous-time systems, the following development shows that any vector $\mathbf{q}(t)$ related to a valid state vector selection $\mathbf{x}(t)$ by a constant nonsingular transformation matrix $\mathbf{T}$ is also a valid state vector:

$$\mathbf{q}(t) = \mathbf{Tx}(t) \tag{19.14}$$

where $\mathbf{T}$ is a nonsingular $(n \times n)$ matrix. Equations 19.4 and 19.5 may be rewritten in terms of the vector $\mathbf{q}(t)$ as

$$\dot{\mathbf{q}}(t) = [\mathbf{TA}(t)\mathbf{T}^{-1}]\mathbf{q}(t) + [\mathbf{TB}(t)]\mathbf{u}(t) \qquad t \geq t_0 \tag{19.15}$$

$$\mathbf{y}(t) = [\mathbf{C}(t)\mathbf{T}^{-1}]\mathbf{q}(t) + [\mathbf{D}(t)]\mathbf{u}(t) \qquad t \geq t_0 \tag{19.16}$$

$\mathbf{q}(t)$ satisfies the definition of a state vector since it has the same dimension as $\mathbf{x}(t)$. Equations 19.15 and 19.16 are the state-space equations in terms of $\mathbf{q}(t)$. The matrices within parentheses in these equations are the modified system and coupling matrices.

As for continuous-time systems, the state vector for a given linear, discrete-time system is not unique. Any vector $\mathbf{q}(k)$ related to a valid state vector $\mathbf{x}(k)$ by a constant, nonsingular matrix $\mathbf{T}$ is also a valid state vector:

$$\mathbf{q}(k) = \mathbf{Tx}(k) \tag{19.17}$$

The corresponding state-space equations are

$$\mathbf{q}(k + 1) = [\mathbf{TF}(k)\mathbf{T}^{-1}]\mathbf{q}(k) + [\mathbf{TG}(k)]\mathbf{u}(k) \qquad k \geq k_0 \tag{19.18}$$

$$\mathbf{y}(k) = [\mathbf{C}(k)\mathbf{T}^{-1}]\mathbf{q}(k) + [\mathbf{D}(k)]\mathbf{u}(k) \qquad k \geq k_0 \tag{19.19}$$

Since the state vector of a system is not unique, the selection of state variables for a given application is governed by considerations such as ease of measurement of state variables or simplification of the resulting state-space equations. If the independent energy storage elements in the system of interest are readily identified, selection of state variables directly related to energy storage in the system is appropriate. An nth-order system has n independent energy storage elements that would enable the selection of n state variables. Examples of energy storage elements are springs and masses in mechanical systems, capacitors and inductors in electrical systems, and capacitance and inertance elements in fluid (hydraulic and pneumatic) systems.

Consider the RLC circuit shown in Fig. 19.3. Let $e_{in}(t)$ be the input and $e_{out}(t)$ be the output. The current i_{out} is assumed to be negligible. Kirchhoff's voltage law for the loop yields

$$e_{in}(t) - Ri_{in}(t) - L\frac{di_{in}(t)}{dt} - \frac{1}{C_1}\int i_{in}(t)\, dt = 0 \tag{19.20}$$

The current $i_{in}(t)$ through the inductor and the voltage $e_{out}(t)$ across the capacitor are directly related to energy storage in the system and are chosen as state variables:

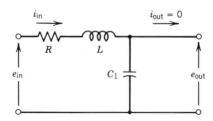

Fig. 19.3 RLC circuit.

$$x_1(t) = i_{in}(t) \tag{19.21}$$

$$x_2(t) = e_{out}(t) = \frac{1}{C_1} \int i_{in}(t) \, dt \tag{19.22}$$

The state equations can be determined from Eqs. 19.20–19.22 as

$$\dot{x}_1(t) = -\frac{R}{L} x_1(t) - \frac{1}{L} x_2(t) + \frac{e_{in}(t)}{L} \tag{19.23}$$

$$\dot{x}_2(t) = \frac{x_1(t)}{C_1} \tag{19.24}$$

The output equation is

$$y(t) = e_{out}(t) = x_2(t) \tag{19.25}$$

The system and coupling matrices for the electrical circuit are

$$\mathbf{A} = \begin{bmatrix} -\dfrac{R}{L} & -\dfrac{1}{L} \\ \dfrac{1}{C_1} & 0 \end{bmatrix} \qquad \mathbf{B} = \begin{bmatrix} \dfrac{1}{L} \\ 0 \end{bmatrix} \tag{19.26}$$

$$\mathbf{C} = [0 \quad 1] \qquad \mathbf{D} = 0$$

For sampled-data control applications involving digital control of continuous-time systems, state-variable selection is often based on continuous-time system equations and needs to be retained in the discrete-time formulation. Let the time-invariant, continuous-time system equations be of the form given by Eqs. 19.6 and 19.7. The solution for the vector equation of state $\mathbf{x}(t)$ is given by Eqs. 19.32 and 19.33 later in this chapter. Applying this solution form over the time interval from kT to $(k + 1)T$, we get

$$\mathbf{x}[(k + 1)T] = e^{\mathbf{A}T}\mathbf{x}(kT) + \int_{kT}^{(k+1)T} e^{\mathbf{A}[(k+1)T-t]}\mathbf{B}\mathbf{u}(t)dt \tag{19.27}$$

In sampled-data control applications, the input vector $\mathbf{u}(t)$ is the control input computed by the digital computer and applied to the continuous-time system by a digital-analog converter. Usually, the digital-analog converter has a latch that maintains the output constant between the time instants kT and $(k + 1)T$:

$$\mathbf{u}(t) = \mathbf{u}(kT) \qquad kT < t \leq (k + 1)T \tag{19.28}$$

Equation 19.27 then simplifies to

$$\mathbf{x}(k + 1) = (e^{\mathbf{A}T})\mathbf{x}(k) + \left(\int_0^T e^{\mathbf{A}(T-t)}\mathbf{B}dt \right)\mathbf{u}(k)$$

$$= \mathbf{F}\mathbf{x}(k) + \mathbf{G}\mathbf{u}(k) \tag{19.29}$$

Equation 19.29 establishes the relationship between the system matrices in the discrete-time and continuous-time formulations of the system equations.

19.3.1 Canonical Forms for Continuous-Time Systems

For high-order, single-input–single-output (SISO) systems or multiple-input–multiple-output (MIMO) systems, the number of elements in the system matrices is large. Selection of state variables that simplify the state-space representation is thus desirable. Such representations of the state-space equations also exhibit significant properties of the system more clearly and are referred to as canonical forms. However, the names and corresponding structures of the canonical forms are not completely standardized.

The controllable canonical form is useful in control system design applications. The controllable canonical form for an nth-order, SISO, LTI, continuous-time system is described in Table 19.1.

TABLE 19.1 STATE-SPACE CANONICAL FORMS FOR SISO CONTINUOUS-TIME SYSTEMS

$$H(s) = \frac{Y(s)}{U(s)} = \frac{\beta_n s^n + \beta_{n-1} s^{n-1} + \cdots + \beta_1 s + \beta_0}{s^n + \alpha_{n-1} s^{n-1} + \cdots + \alpha_1 s + \alpha_0}$$

I. Controllable Canonical Form
Simulation diagram

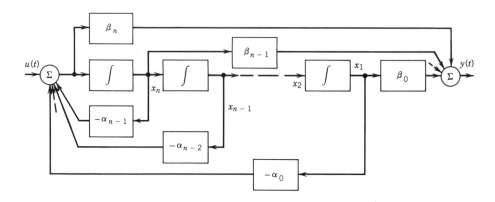

System matrices

$$A = \begin{bmatrix} 0 & & & \\ 0 & & & \\ \vdots & & \mathbf{I}_{n-1} & \\ 0 & & & \\ \hdashline -\alpha_0 & -\alpha_1 & \cdots & -\alpha_{n-1} \end{bmatrix} \qquad B = \begin{bmatrix} 0 \\ 0 \\ \vdots \\ 0 \\ 1 \end{bmatrix} \qquad D = \beta_n$$

$$C = [\beta_0 - \beta_n \alpha_0 \quad \beta_1 - \beta_n \alpha_1 \quad \cdots \quad \beta_{n-1} - \beta_n \alpha_{n-1}]$$

II. Observable Canonical Form
Simulation diagram

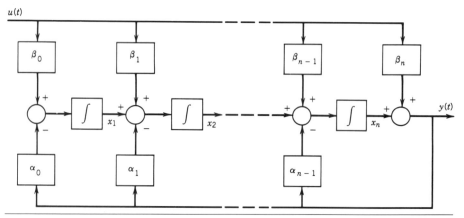

(Continued)

System matrices

$$
A = \begin{bmatrix} 0 & 0 & \cdots & 0 & \vdots & -\alpha_0 \\ & & & & \vdots & -\alpha_1 \\ & I_{n-1} & & & \vdots & \vdots \\ & & & & \vdots & -\alpha_{n-1} \end{bmatrix}
\qquad
B = \begin{bmatrix} \beta_0 - \beta_n \alpha_0 \\ \beta_1 - \beta_n \alpha_1 \\ \vdots \\ \beta_{n-1} - \beta_n \alpha_{n-1} \end{bmatrix}
$$

$$C = [0 \ 0 \ \ldots \ 0 \ 1] \qquad\qquad D = \beta_n$$

III.   Normal or Diagonal Jordan Canonical Form

Related conditions

   (i)   Characteristic equation roots s_i, $i = 1, \ldots, n$ are real and distinct

Simulation diagram

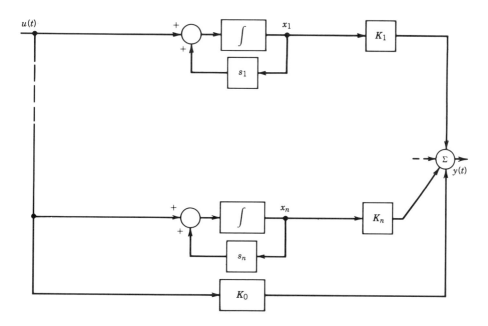

System matrices

$$
A = \begin{bmatrix} s_1 & & & \\ & s_2 & & \mathbf{0} \\ & & \ddots & \\ \mathbf{0} & & & s_n \end{bmatrix}
\qquad
B = \begin{bmatrix} 1 \\ 1 \\ \vdots \\ 1 \end{bmatrix}
$$

$$C = [K_1 \quad K_2 \quad \ldots \quad K_n] \qquad\qquad D = K_0$$

where $K_0 = \lim_{s \to \infty} [H(s)]$ and $K_i = \lim_{s \to s_i} [(s - s_i)H(s)]$ $i = 1, \ldots, n$.

(Continued)

813

IV.   Near-Normal Canonical Form
Related conditions
(i)   One pair of complex conjugate characteristic equation roots

$$s_k = s_{kr} + js_{ki}$$

$$s_{k+1} = s_{kr} - js_{ki}$$

(ii)   All other roots are real and distinct.
Simulation diagram

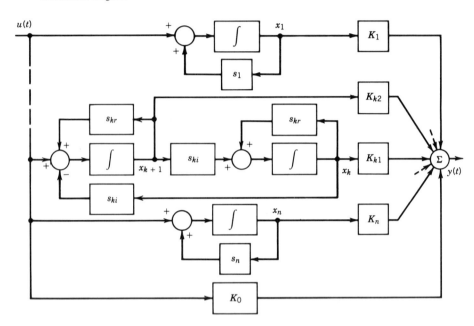

System Matrices

$$A = \begin{bmatrix} s_1 & & & & & & \\ & \ddots & & 0 & & 0 & \\ 0 & & s_{k-1} & & & & \\ \hline & & & s_{kr} & s_{ki} & & \\ & 0 & & -s_{ki} & s_{kr} & & 0 \\ \hline & & & & & s_{k-2} & \\ & 0 & & 0 & & & \ddots & 0 \\ & & & & & 0 & & s_n \end{bmatrix} \qquad B = \begin{bmatrix} 1 \\ \vdots \\ 1 \\ \hline 0 \\ 1 \\ \hline 1 \\ \vdots \\ 1 \end{bmatrix}$$

$$C = \begin{bmatrix} K_1 & \cdots & K_{k-1} & K_{k1} & K_{k2} & K_{k+2} \cdots K_n \end{bmatrix} \qquad D = K_0$$

where $K_0 = \lim_{s \to \infty} [H(s)]$ $K_i = \lim_{s \to s_i} [(s - s_i)H(s)]$

$$i = 1, k - 1 \text{ and } i = k + 2, \ldots, n$$

$$K_{k1} = -\frac{1}{2}\text{Im}[(s - s_k)H(s)]_{s = s_k} \qquad K_{k2} = \frac{1}{2}\text{Re}[(s - s_k)H(s)]_{s = s_k}$$

(*Continued*)

V.   Nondiagonal Jordan Canonical Form
   Related conditions
   (i) One real characteristic equation root s_k repeated m times
      (i.e., $s_k = s_{k+1} = \cdots = s_{k+m-1}$).
   (ii) All other roots real and distinct.
   Simulation diagram

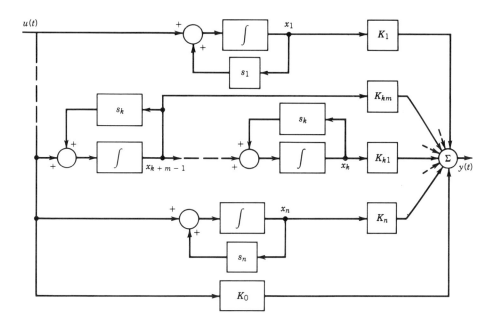

System matrices

$$
\mathbf{A} = \begin{bmatrix}
\begin{matrix} s_1 \\ & \ddots & \mathbf{0} \\ \mathbf{0} & & s_{k-1} \end{matrix} & \mathbf{0} & \mathbf{0} \\
\mathbf{0} & \begin{matrix} s_k & 1 & \mathbf{0} \\ & \ddots & \ddots & 1 \\ \mathbf{0} & & s_k \end{matrix} & \mathbf{0} \\
\mathbf{0} & \mathbf{0} & \begin{matrix} s_{k+m} \\ & \ddots & \mathbf{0} \\ \mathbf{0} & & s_n \end{matrix}
\end{bmatrix}
\qquad
\mathbf{B} = \begin{bmatrix} 1 \\ \vdots \\ 1 \\ \hline 0 \\ 1 \\ \hline 1 \\ \vdots \\ 1 \end{bmatrix}
$$

$$\mathbf{C} = \begin{bmatrix} K_1 & \cdots & K_{k-1} & K_{k1} \cdots K_{km} & K_{k+m} & \cdots & K_n \end{bmatrix} \qquad \mathbf{D} = K_0$$

where $K_0 = \lim_{s \to \infty} [H(s)]$ $K_i = \lim_{s \to s_i} [(s - s_i)H(s)]$

$$i = 1, k - 1 \text{ and } i = k + m, \ldots, n$$

and $K_{kj} = \dfrac{1}{(j-1)!} \left\{ \dfrac{d^{j-1}}{ds^{j-1}} [(s - s_k)^m H(s)] \right\}_{s = s_k}$ $j = 1, \ldots, m$

The selection of state variables $x_1, x_2, \ldots, x_n$ that results in the controllable canonical form of the state equations is indicated on the simulation diagram in the table. For the case where all the $\beta_i, i = 1, n$ are zero and $\beta_0 \neq 0$, the transfer function $Y(s)/U(s)$ has no numerator dynamics. The n state variables are then simply the output and $n - 1$ successive derivatives of the output. These are referred to as the phase variables. The phase-variable canonical form is thus a special case of the controllable canonical form. The system and coupling matrices for the controllable canonical form are listed in Table 19.1. The special form of the $\mathbf{A}$ matrix is referred to as the companion form. $\mathbf{I}_{n-1}$ is the $(n - 1) \times (n - 1)$ identity matrix.

For a SISO, LTI system described by Eqs. 19.6 and 19.7, the state transformation matrix $\mathbf{T}$ in Eq. 19.14, which transforms the state-space equations into the controllable canonical form exists if the controllability matrix $\mathbf{P}_c$ in Eq. 19.30 is nonsingular:[1]

$$\mathbf{P}_c = [\mathbf{B} \ \mathbf{AB} \ \ldots \ \mathbf{A}^{n-1}\mathbf{B}] \tag{19.30}$$

The transformation matrix $\mathbf{T}$ is defined in Table 19.2.

The observable canonical form is useful in state-estimator or observer design applications. The observable canonical form for an nth-order, SISO, LTI system is described in Table 19.1 in a manner similar to the controllable canonical form. The corresponding $\mathbf{A}$ matrix is the transpose of the $\mathbf{A}$ matrix for the controllable canonical form and is also referred to as a companion matrix. The state transformation matrix $\mathbf{T}$ in Eq. 19.14, which transforms given state-space Eqs. 19.6 and 19.7 into the observable canonical form, exists if the observability matrix $\mathbf{P}_0$ is nonsingular:[1]

$$\mathbf{P}_0 = \begin{bmatrix} \mathbf{C} \\ \mathbf{CA} \\ \vdots \\ \mathbf{CA}^{n-1} \end{bmatrix} \tag{19.31}$$

The transformation matrix $\mathbf{T}$ is defined in Table 19.2.

State variables can also be chosen to diagonalize or nearly diagonalize the state matrix $\mathbf{A}$. The resulting state-space equations are completely or almost completely decoupled from one another and hence show very clearly the effect of initial conditions or forcing inputs on the different characteristic modes of the system response. The resulting physical insight into the system behavior makes the corresponding form of the system equations, called the normal form or diagonal Jordan form, valuable in vibration analysis applications and in control applications involving modal control. The normal form of the state-space equations for a SISO, LTI system with real, distinct characteristic roots is given in Table 19.1. The diagonal elements of the $\mathbf{A}$ matrix in the table are the system characteristic roots or eigenvalues. The state variables $x_1, x_2, \ldots, x_n$ lie along the eigenvectors of the $\mathbf{A}$ matrix, in state space. As the corresponding simulation diagram indicates, the behavior of each state variable is governed solely by one eigenvalue, the initial condition on that state variable, and the forcing input.

If some of the distinct characteristic roots of a SISO, LTI system are complex, the matrices $\mathbf{A}$ and $\mathbf{C}$ in Eqs. 19.6 and 19.7 have complex elements when represented in the normal form just described. Since this could be inconvenient in subsequent matrix manipulations, a nearly diagonal $\mathbf{A}$ matrix can be obtained for cases where the complex characteristic roots occur in complex conjugate pairs. This would be the case for system differential equations with only real coefficients. The near-normal form of the system equations for a system with one pair of complex conjugate characteristic roots is given in Table 19.1. Extension to the case of multiple complex root pairs is straightforward. The complex characteristic roots result in a few nonzero off-diagonal elements in the $\mathbf{A}$ matrix; otherwise, the decoupled nature of the system equations is retained.

If one characteristic root of a SISO, LTI system is real and repeated m times, the state equations can only be partially decoupled by appropriate state-variable selection, as shown in Table 19.1. The resulting state-space equations are said to be in the Jordan canonical form. The corresponding $\mathbf{A}$ matrix has one submatrix with the repeated eigenvalue at the diagonal positions, ones immediately to the right of the repeated diagonal elements within the submatrix and zero elements at all other nondiagonal positions.[21] The $\mathbf{A}$ matrix is then said to have one Jordan block. The extension of the result in Table 19.1 to the case of many different repeated characteristic roots is straightforward.

For SISO or MIMO, LTI systems described by state-space Eqs. 19.6 and 19.7, the transformation matrix $\mathbf{T}$ in Eq. 19.14, which transforms the state-space equations into the diagonal or nondiagonal Jordan form, can be determined. As in Table 19.1, there are a number of different cases to be considered.[2] If the $\mathbf{A}$ matrix has real, distinct eigenvalues $s_i, i = 1, \ldots, n$, the eigenvectors are linearly independent and can be used to form the modal matrix $\mathbf{M}$ as indicated in Table 19.2. The transformation matrix $\mathbf{T}$ is then taken to be $\mathbf{M}^{-1}$. If the $\mathbf{A}$ matrix has one pair of complex, conjugate eigenvalues and if system matrices with real elements only are desired, the transformation matrix $\mathbf{T}$ is defined in a slightly different form as indicated in Table 19.2. The resulting transformed state matrix will have two nonzero off-diagonal elements as indicated.

State-space equations (SISO, LTI system)	Characteristic equation
$\dot{x}(t) = Ax(t) + Bu(t)$	$\det(sI - A) =$
$y(t) = Cx(t) + Du(t)$	$s^n + \alpha_{n-1}s^{n-1} + \cdots + \alpha_1 s + \alpha_0 = 0$

I. Controllable Canonical Form
　　Transformation conditions
　　　(i) $P_c = [B \quad AB \ldots A^{n-1}B]$ must be nonsingular

　　Transformation matrices
　　　(i) $q = Tx, \ T = R^{-1}P_c^{-1}$　　　　　　(ii) New state matrix $= TAT^{-1}$
　　　　where

$$R = \begin{bmatrix} \alpha_1 & \alpha_2 & \cdots & \alpha_{n-1} & 1 \\ \alpha_2 & \alpha_3 & \cdots & 1 & 0 \\ \vdots & \vdots & & \vdots & \vdots \\ \alpha_{n-1} & 1 & \cdots & 0 & 0 \\ 1 & 0 & \cdots & 0 & 0 \end{bmatrix} \qquad = \begin{bmatrix} 0 & & & \\ 0 & & & \\ \vdots & & I_{n-1} & \\ 0 & & & \\ \hline -\alpha_0 & -\alpha_1 & \cdots & -\alpha_{n-1} \end{bmatrix}$$

II. Observable Canonical Form
　　Transformation conditions

　　　(i) $P_0 = \begin{bmatrix} C \\ CA \\ \vdots \\ CA^{n-1} \end{bmatrix}$ must be nonsingular

　　Transformation matrices
　　　(i) $q = Tx, \ T = RP_0$　　　　　　(ii) New state matrix $= TAT^{-1}$
　　　　where

$$R = \begin{bmatrix} \alpha_1 & \alpha_2 & \cdots & \alpha_{n-1} & 1 \\ \alpha_2 & \alpha_3 & \cdots & 1 & 0 \\ \vdots & \vdots & & \vdots & \vdots \\ \alpha_{n-1} & 1 & \cdots & 0 & 0 \\ 1 & 0 & \cdots & 0 & 0 \end{bmatrix} \qquad = \begin{bmatrix} 0 & 0 & \cdots & 0 & -\alpha_0 \\ \hline & & & & -\alpha_1 \\ & I_{n-1} & & & \vdots \\ & & & & -\alpha_{n-1} \end{bmatrix}$$

III. Normal or Diagonal Jordan Canonical Form
　　Transformation conditions
　　　(i) A matrix has only distinct, real eigenvalues $s_i, \ i = 1, \ldots, n$.

　　Transformation matrices
　　　(i) $q = Tx, \ T^{-1} = M = [v_1 \ v_2 \ \ldots \ v_n]$
　　　　　where (a) v_i are the n linearly independent eigenvectors corresponding to s_i and
　　　　　　　　(b) v_i are taken to be equal or proportional to any nonzero column of
　　　　　　　　　$\text{Adj}(s_i I - A)$
　　　(ii) New state matrix $= TAT^{-1}$

$$= \begin{bmatrix} s_1 & & & \\ & s_2 & & 0 \\ & & \ddots & \\ 0 & & & s_n \end{bmatrix}$$

(Continued)

IV.   Normal or Diagonal Jordan Canonical Form
Transformation conditions
    (i) **A** matrix has one repeated, real eigenvalue s_k of multiplicity m (i.e., $s_k = s_{k+1} = \ldots = s_{k+m-1}$). All other eigenvalues are real and distinct.
    (ii) Degeneracy $d = n - \text{rank}(s_k\mathbf{I} - \mathbf{A}) = m$. Full degeneracy.

Transformation matrices
    (i) $\mathbf{q} = \mathbf{Tx}$, $\mathbf{T}^{-1} = \mathbf{M} = [\mathbf{v}_1 \ \mathbf{v}_2 \ \ldots \ \mathbf{v}_n]$
        where (a) $\mathbf{v}_i$, $i = 1, k-1$ and $i = k + m, n$ are the linearly independent eigenvectors corresponding to the real, distinct eigenvalues,
            (b) $\mathbf{v}_i$, $i = 1, k-1$ and $i = k + m, n$ are taken to be equal or proportional to any nonzero column of $\text{Adj}(s_i\mathbf{I} - \mathbf{A})$, and
            (c) $\mathbf{v}_i$, $i = k, k + m - 1$ are the m linearly independent eigenvectors corresponding to the repeated eigenvalue. They are equal or proportional to the nonzero linearly independent columns of

$$\left\{ \frac{d^{m-1}}{ds^{m-1}}[\text{Adj}(s\mathbf{I} - \mathbf{A})] \right\}_{s=s_k}$$

    (ii) New state matrix $= \mathbf{TAT}^{-1}$

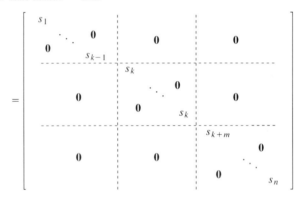

V.   Near-Normal Canonical Form
Transformation conditions
    (i) **A** matrix has one pair of complex, conjugate eigenvalues, s_k, s_{k+1}

$$s_k = s_{kr} + j s_{ki}$$

$$s_{k+1} = s_{kr} - j s_{ki}$$

    (ii) All other eigenvalues are real and distinct.

Transformation matrices
    (i) $\mathbf{q} = \mathbf{Tx}$, $\mathbf{T}^{-1} = [\mathbf{v}_1 \ \ldots \ \mathbf{v}_{k-1} \ \mathbf{v}_{kr} \ \mathbf{v}_{ki} \ \mathbf{v}_{k+2} \ \ldots \ \mathbf{v}_n]$
        where (a) $\mathbf{v}_i$, $i = 1, k-1$ and $i = k + 2, n$ are the linearly independent eigenvectors corresponding to the real, distinct eigenvalues.
            (b) $\mathbf{v}_i$, for $i = 1, \ldots, n$ are taken to be equal or proportional to any nonzero column of $\text{Adj}(s_i\mathbf{I} - \mathbf{A})$, and
            (c) $\mathbf{v}_k = \mathbf{v}_{kr} + j\mathbf{v}_{ki}$
               $\mathbf{v}_{k+1} = \mathbf{v}_{kr} - j\mathbf{v}_{ki}$
            are the complex conjugate eigenvectors corresponding to s_k and s_{k+1}, respectively.
    (ii) New state matrix $= \mathbf{TAT}^{-1}$

(Continued)

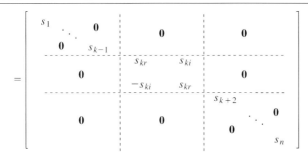

VI. Nondiagonal Jordan Canonical Form

Transformation conditions

    (i) **A** matrix has one repeated, real eigenvalue s_k of multiplicity m (i.e., $s_k = s_{k+1} = \cdots = s_{k+m-1}$). All other eigenvalues are real and distinct.

    (ii) Degeneracy $d = n - \text{rank}(s_k\mathbf{I} - \mathbf{A}) = 1$. Simple degeneracy.

Transformation matrices

    (i) $\mathbf{q} = \mathbf{Tx}$, $\mathbf{T}^{-1} = [\mathbf{t}_1 \quad \mathbf{t}_2 \quad \cdots \quad \mathbf{t}_n]$

        where (a) $\mathbf{t}_i$, $i = 1, k - 1$ and $i = k + m, n$ are the linearly independent eigenvectors corresponding to the real, distinct eigenvalues,

            (b) $\mathbf{t}_i$, $i = 1, k - 1$ and $i = k + m, n$ are taken to be equal or proportional to any nonzero column of $\text{Adj}(s_i\mathbf{I} - \mathbf{A})$, and

            (c) $\mathbf{t}_i$, $i = k, k + m - 1$ are obtained by solution of the equation $\mathbf{AT}^{-1} = \mathbf{T}^{-1}\mathbf{J}$, where $\mathbf{J}$ is the Jordan canonical matrix given. Each $\mathbf{t}_i$ is determined to within a constant of proportionality.

    (ii) New state matrix $= \mathbf{J} = \mathbf{TAT}^{-1}$

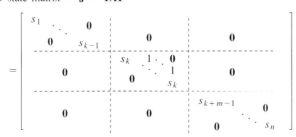

VII. Nondiagonal Jordan Canonical Form

Transformation conditions

    (i) **A** matrix has one repeated, real eigenvalue of multiplicity m (i.e., $s_k = s_{k+1} = \cdots = s_{k+m-1}$). All other eigenvalues are real and distinct.

    (ii) Degeneracy $d = n - \text{rank}(s_k\mathbf{I} - \mathbf{A})$ is between 1 and m. General degeneracy.

Transformation matrices

    (i) $\mathbf{q} = \mathbf{Tx}$, $\mathbf{T}^{-1} = [\mathbf{t}_1 \quad \mathbf{t}_2 \quad \cdots \quad \mathbf{t}_n]$

        where (a) $\mathbf{t}_i$, $i = 1, k - 1$ and $i = k + m, n$ are the linearly independent eigenvectors corresponding to the real, distinct eigenvalues,

            (b) $\mathbf{t}_i$, $i = 1, k - 1$ and $i = k + m, n$ are taken to be equal or proportional to any nonzero column of $\text{Adj}(s_i\mathbf{I}-\mathbf{A})$,

            (c) $\mathbf{t}_i$, $i = k, k + m - 1$ are m linearly independent vectors, corresponding to the repeated eigenvalue of multiplicity m. Only d of these vectors are eigenvectors, and

            (d) $\mathbf{t}_i$, $i = k, k + m - 1$ obtained by solution of the equation $\mathbf{AT}^{-1} = \mathbf{T}^{-1}\mathbf{J}$, where $\mathbf{J}$ is a Jordan canonical matrix for the problem, with d Jordan blocks. There are d possible choices for $\mathbf{J}$. Each of these choices needs to be tried and the $\mathbf{t}_i$ vectors solved for. Only the correct $\mathbf{J}$ will give the m linearly independent vectors $\mathbf{t}_i$, $i = k, k + m - 1$. Each $\mathbf{t}_i$ is determined to within a constant of proportionality.

    (ii) New state matrix $\mathbf{J} = \mathbf{TAT}^{-1}$. Correct $\mathbf{J}$ determined by trial-and-error, as previously described.

The transformed state-space equations may be in the nondiagonal Jordan canonical form if the $\mathbf{A}$ matrix has repeated eigenvalues. The procedure for determining the transformation matrix depends on the degeneracy of the matrix $s_k\mathbf{I} - \mathbf{A}$ corresponding to the repeated eigenvalue s_k. If the degeneracy d of $s_k\mathbf{I} - \mathbf{A}$, defined in Table 19.2, is equal to m where m is the multiplicity of the repeated eigenvalue s_k, m linearly independent eigenvectors can be found for the repeated eigenvalue. The procedure for doing so is indicated in Table 19.2. The transformed state matrix is then diagonal. If the degeneracy of $s_k\mathbf{I} - \mathbf{A}$ is one, only one eigenvector can be determined. Since it can be shown that the degeneracy is equal to the number of Jordan blocks associated with the eigenvector, the transformed state matrix $\mathbf{J}$ has only one Jordan block and is uniquely defined.[2] The nonsingular transformation matrix $\mathbf{T}$ is then determined as indicated in Table 19.2. If the degeneracy d of $s_k\mathbf{I} - \mathbf{A}$ is greater than one but less than m, there are d linearly independent eigenvectors and d Jordan blocks associated with the eigenvalue s_k. In this case, the transformed state matrix $\mathbf{J}$ cannot be uniquely defined but can be one of a finite number of possibilities. A trial-and-error formulation of $\mathbf{J}$ and solution for $\mathbf{T}$, as indicated in Table 19.2, is necessary until a nonsingular transformation matrix $\mathbf{T}$ is obtained.[2]

19.3.2 Canonical Forms for Discrete-Time Systems

Canonical forms of discrete-time state-space equations have the same uses that such forms have for continuous-time systems. The development of these canonical forms closely parallels that for continuous-time systems and is therefore summarized here. Table 19.3 indicates the state-variable selection for SISO, LTI systems described by pulse transfer functions that yield the controllable, observable, and Jordan canonical forms of the state-space equations. For systems already described by discrete-time state-space Eqs. 19.12 and 19.13, Table 19.4 indicates the transformation matrix $\mathbf{T}$ in Eq. 19.17, which transforms the state-space equations into the canonical forms named previously.

With the exception of the procedures for selecting transformation matrices to convert state-space equations for MIMO, LTI systems to the Jordan canonical form, the procedures and results presented in Tables 19.1–19.4 are restricted to SISO, LTI systems. A procedure for representing a SISO, linear time-varying (LTV) system, described by a differential equation, in controllable canonical form has been described by DeRusso, Roy, and Close.[2]

Canonical forms for MIMO, LTI systems cannot, in general, be specified uniquely as is the case for SISO systems. Kailath,[3] Fortmann and Hitz,[1] and Kalman[9] have specified some canonical forms for MIMO systems and have described procedures for representing such systems in these forms, given their transfer function matrix descriptions or state-space equations. The problem of selection of state variables given the transfer function matrix description of MIMO systems is the problem of realization and is considered in Section 19.7.

19.4 SOLUTION OF SYSTEM EQUATIONS

19.4.1 Continuous-Time Systems

The state-space Eqs. 19.1 and 19.2 for time-varying, nonlinear systems described by ordinary differential equations can be solved by numerical integration techniques. Such numerical integration would, however, have to be repeated if the initial conditions $\mathbf{x}(t_0)$ or the forcing function $\mathbf{u}(t)$ were to be changed. The computational burden can be reduced for linear systems by using the concept of the state transition matrix.

For LTI systems described by the state Eqs. 19.6 and 19.7, the solution is given by

$$\mathbf{x}(t) = \boldsymbol{\varphi}(t - t_0)\mathbf{x}(t_0) + \int_{t_0}^{t} \boldsymbol{\varphi}(t - \tau)\mathbf{B}\mathbf{u}(\tau)\,d\tau \tag{19.32}$$

$$\mathbf{y}(t) = \mathbf{C}\boldsymbol{\varphi}(t - t_0)\mathbf{x}(t_0) + \int_{t_0}^{t} \mathbf{C}\boldsymbol{\varphi}(t - \tau)\mathbf{B}\mathbf{u}(\tau)\,d\tau + \mathbf{D}\mathbf{u}(t) \tag{19.33}$$

where the $n \times n$ matrix $\boldsymbol{\varphi}(t)$ is defined as the state transition matrix of the system. Derivation of this result is available in many standard textbooks on state-space methods[1,4] The first terms on the right-hand sides of the preceding equations represent the response of the homogeneous system to the initial condition $\mathbf{x}(t_0)$ whereas the second terms represent the forced response of the system. Comparison

TABLE 19.3 STATE-SPACE CANONICAL FORMS FOR SISO DISCRETE-TIME SYSTEMS

$$H(z) = \frac{Y(z)}{U(z)} = \frac{\beta_n z^n + \beta_{n-1} z^{n-1} + \cdots + \beta_1 z + \beta_0}{z^n + \alpha_{n-1} z^{n-1} + \cdots + \alpha_1 z + \alpha_0}$$

I. Controllable Canonical Form
Simulation diagram

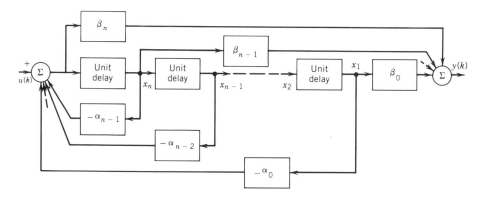

System matrices

$$\mathbf{F} = \begin{bmatrix} 0 & & & & \\ 0 & & & & \\ \vdots & & \mathbf{I}_{n-1} & & \\ 0 & & & & \\ \hline -\alpha_0 & -\alpha_1 & \cdots & -\alpha_{n-1} \end{bmatrix} \qquad \mathbf{G} = \begin{bmatrix} 0 \\ 0 \\ \vdots \\ 0 \\ 1 \end{bmatrix} \qquad \mathbf{D} = \beta_n$$

$$\mathbf{C} = [\beta_0 - \beta_n \alpha_0 \quad \beta_1 - \beta_n \alpha_1 \quad \cdots \quad \beta_{n-1} - \beta_n \alpha_{n-1}]$$

II. Observable Canonical Form
Simulation diagram

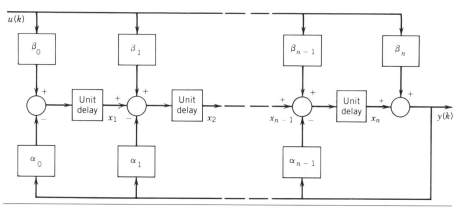

(Continued)

System matrices

$$F = \begin{bmatrix} 0 & 0 & \cdots & 0 & \vdots & -\alpha_0 \\ & & & & \vdots & -\alpha_1 \\ & & I_{n-1} & & \vdots & \vdots \\ & & & & \vdots & -\alpha_{n-1} \end{bmatrix} \qquad G = \begin{bmatrix} \beta_0 - \beta_n\alpha_0 \\ \beta_1 - \beta_n\alpha_1 \\ \vdots \\ \beta_{n-1} - \beta_n\alpha_{n-1} \end{bmatrix}$$

$$C = [0 \ 0 \ \ldots \ 0 \ 1] \qquad\qquad D = \beta_n$$

III. Normal or Diagonal Jordan Canonical Form
Related conditions
   (i) Characteristic equation roots z_i, $i = 1, \ldots, n$ are real and distinct
Simulation diagram

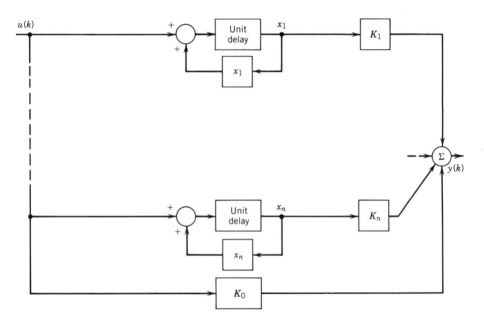

System matrices

$$F = \begin{bmatrix} z_1 & & & \\ & z_2 & & \mathbf{0} \\ & & \ddots & \\ \mathbf{0} & & & z_n \end{bmatrix} \qquad\qquad G = \begin{bmatrix} 1 \\ 1 \\ \vdots \\ 1 \end{bmatrix}$$

$$C = [K_1 \quad K_2 \quad \ldots \quad K_n] \qquad\qquad D = K_0$$

where $K_0 = \lim_{z \to \infty} [H(z)]$ and $K_i = \lim_{z \to z_i} [(z - z_i)H(z)]$ $\qquad i = 1, \ldots, n$.

(Continued)

IV. Near-Normal Canonical Form
   Related conditions
   (i) One pair of complex conjugate characteristic equation roots

$$z_k = z_{kr} + jz_{ki}$$

$$z_{k+1} = z_{kr} - jz_{ki}$$

   (ii) All other roots are real and distinct.
   Simulation diagram

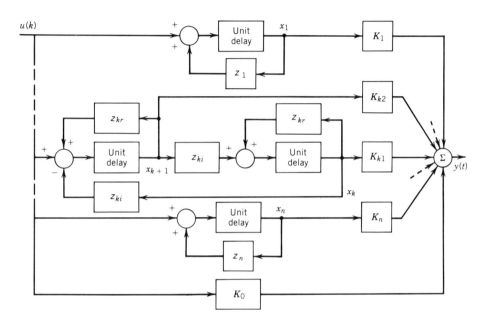

System matrices

$$
\mathbf{F} = \begin{bmatrix}
z_1 & & & & & & \\
& \ddots & \mathbf{0} & & \mathbf{0} & & \mathbf{0} \\
& \mathbf{0} & z_{k-1} & & & & \\
\hline
& & & z_{kr} & z_{ki} & & \\
& \mathbf{0} & & & & & \mathbf{0} \\
& & & -z_{ki} & z_{kr} & & \\
\hline
& & & & & z_{k-2} & \\
& \mathbf{0} & & \mathbf{0} & & & \ddots & \mathbf{0} \\
& & & & & & z_n
\end{bmatrix}
\qquad
\mathbf{G} = \begin{bmatrix}
1 \\ \vdots \\ 1 \\ \hline 0 \\ 1 \\ \hline 1 \\ \vdots \\ 1
\end{bmatrix}
$$

$$\mathbf{C} = \begin{bmatrix} K_1 & \cdots & K_{k-1} & K_{k1} & K_{k2} & K_{k+2} \cdots & K_n \end{bmatrix} \qquad \mathbf{D} = K_0$$

where $K_0 = \lim\limits_{z \to \infty} [H(z)]$ $\qquad\qquad$ $K_i = \lim\limits_{z \to z_i} [(z - z_i)H(z)]$

$\qquad i = 1, k-1$ and $i = k+2, \ldots, n$

$$K_{k1} = -\frac{1}{2}\text{Im}[(z - z_k)H(z)]_{z=z_k} \qquad K_{k2} = \frac{1}{2}\text{Re}[(z - z_k)H(z)]_{z=z_k}$$

(Continued)

V.   Nondiagonal Jordan Canonical Form
   Related conditions
   (i) One real characteristic equation root z_k repeated m times
   (i.e., $z_k = z_{k+1} = \cdots = z_{k+m-1}$).
   (ii) All other roots real and distinct.
   Simulation diagram

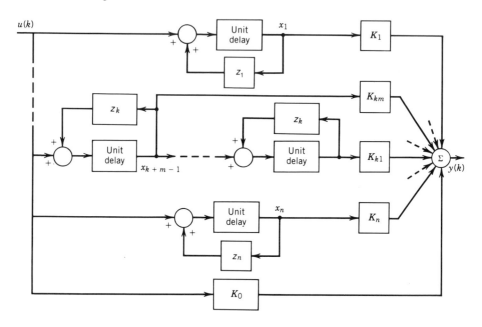

System matrices

$$\mathbf{F} = \begin{bmatrix} z_1 & & & & & & \\ & \ddots & \mathbf{0} & & \mathbf{0} & & \mathbf{0} \\ \mathbf{0} & & z_{k-1} & & & & \\ \hline & & & z_k & 1 & \mathbf{0} & \\ \mathbf{0} & & & & \ddots & 1 & \mathbf{0} \\ & & & \mathbf{0} & & z_k & \\ \hline & & & & & z_{k+m} & \mathbf{0} \\ \mathbf{0} & & \mathbf{0} & & & & \ddots \\ & & & & & \mathbf{0} & z_n \end{bmatrix} , \quad \mathbf{G} = \begin{bmatrix} 1 \\ \vdots \\ 1 \\ \hline 0 \\ 1 \\ \vdots \\ 1 \\ \hline 1 \\ \vdots \\ 1 \end{bmatrix}$$

$$\mathbf{C} = \begin{bmatrix} K_1 & \cdots & K_{k-1} & K_{k1} & \cdots & K_{km} & K_{k+m} & \cdots & K_n \end{bmatrix} \qquad \mathbf{D} = K_0$$

where $K_0 = \lim_{z \to \infty} [H(z)]$ $K_i = \lim_{z \to z_i} [(z - z_i)H(z)]$

$$i = 1, k-1 \text{ and } i = k+m, \ldots, n$$

and $K_{kj} = \dfrac{1}{(j-1)!} \left\{ \dfrac{d^{j-1}}{dz^{j-1}} [(z - z_k)^m H(z)] \right\}_{z=z_k}$ $j = 1, \ldots, m$

TABLE 19.4 TRANSFORMATION MATRICES FOR DISCRETE-TIME STATE-SPACE CANONICAL FORMS

State-space equations (SISO system)	Characteristic equation
$\mathbf{x}(k+1) = \mathbf{Fx}(k) + \mathbf{G}u(k)$	
$y(k) = \mathbf{Cx}(k) + \mathbf{D}u(k)$	$\det(z\mathbf{I} - \mathbf{F})$
	$= z^n + \alpha_{n-1}z^{n-1} + \cdots + \alpha_1 z + \alpha_0 = 0$

I. Controllable Canonical Form

Transformation conditions

    (i) $\mathbf{P}_c = [\mathbf{G} \quad \mathbf{FG} \ldots \mathbf{F}^{n-1}\mathbf{G}]$ must be nonsingular

Transformation matrices

    (i) $\mathbf{q} = \mathbf{Tx}, \mathbf{T} = \mathbf{R}^{-1}\mathbf{P}_c^{-1}$ (ii) New state matrix $= \mathbf{TFT}^{-1}$

where

$$\mathbf{R} = \begin{bmatrix} \alpha_1 & \alpha_2 & \cdots & \alpha_{n-1} & 1 \\ \alpha_2 & \alpha_3 & \cdots & 1 & 0 \\ \vdots & \vdots & & \vdots & \vdots \\ \alpha_{n-1} & 1 & \cdots & 0 & 0 \\ 1 & 0 & \cdots & 0 & 0 \end{bmatrix} = \begin{bmatrix} 0 & & & \\ 0 & & & \\ \vdots & & \mathbf{I}_{n-1} & \\ 0 & & & \\ \hline -\alpha_0 & -\alpha_1 & \cdots & -\alpha_{n-1} \end{bmatrix}$$

II. Observable Canonical Form

Transformation conditions

    (i) $\mathbf{P}_0 = \begin{bmatrix} \mathbf{C} \\ \mathbf{CF} \\ \vdots \\ \mathbf{CF}^{n-1} \end{bmatrix}$ must be nonsingular

Transformation matrices

    (i) $\mathbf{q} = \mathbf{Tx}, \mathbf{T} = \mathbf{RP}_0$ (ii) New state matrix $= \mathbf{TFT}^{-1}$

where

$$\mathbf{R} = \begin{bmatrix} \alpha_1 & \alpha_2 & \cdots & \alpha_{n-1} & 1 \\ \alpha_2 & \alpha_3 & \cdots & 1 & 0 \\ \vdots & \vdots & & \vdots & \vdots \\ \alpha_{n-1} & 1 & \cdots & 0 & 0 \\ 1 & 0 & \cdots & 0 & 0 \end{bmatrix} = \begin{bmatrix} 0 & 0 & \cdots & 0 & -\alpha_0 \\ \hline & & & & -\alpha_1 \\ & \mathbf{I}_{n-1} & & & \vdots \\ & & & & -\alpha_{n-1} \end{bmatrix}$$

III. Normal or Diagonal Jordan Canonical Form

Transformation conditions

    (i) $\mathbf{F}$ matrix has only distinct, real eigenvalues z_i, $i = 1, \ldots, n$

Transformation matrices

    (i) $\mathbf{q} = \mathbf{Tx}, \mathbf{T}^{-1} = \mathbf{M} = [\mathbf{v}_1 \ \mathbf{v}_2 \ldots \mathbf{v}_n]$

         where (a) $\mathbf{v}_i$ are the n linearly independent eigenvectors corresponding to z_i and

               (b) $\mathbf{v}_i$ are taken to be equal or proportional to any nonzero column of

                 $\text{Adj}(z_i\mathbf{I} - \mathbf{F})$

    (ii) New state matrix $= \mathbf{TFT}^{-1}$

$$= \begin{bmatrix} z_1 & & & \mathbf{0} \\ & z_2 & & \\ & & \ddots & \\ \mathbf{0} & & & z_n \end{bmatrix}$$

IV. Normal or Diagonal Jordan Canonical Form

Transformation conditions

    (i) $\mathbf{F}$ matrix has one repeated, real eigenvalue z_k of multiplicity m (i.e., $z_k = z_{k+1} = \cdots = z_{k+m-1}$). All other eigenvalues are real and distinct.

    (ii) Degeneracy $d = n - \text{rank}(z_k\mathbf{I} - \mathbf{F}) = m$. Full degeneracy.

(Continued)

Transformation matrices

(i) $\mathbf{q} = \mathbf{Tx}$, $\mathbf{T}^{-1} = \mathbf{M} = [\mathbf{v}_1 \ \mathbf{v}_2 \ \cdots \ \mathbf{v}_n]$

where (a) $\mathbf{v}_i$, $i = 1, k-1$ and $i = k+m, n$ are the linearly independent eigenvectors corresponding to the real, distinct eigenvalues,

(b) $\mathbf{v}_i$, $i = 1, k-1$ and $i = k+m, n$ are taken to be equal or proportional to any nonzero column of $\text{Adj}(z_i \mathbf{I} - \mathbf{F})$, and

(c) $\mathbf{v}_i$, $i = k, k+m-1$ are the m linearly independent eigenvectors corresponding to the repeated eigenvalue. They are equal or proportional to the nonzero linearly independent columns of

$$\left\{ \frac{d^{m-1}}{dz^{m-1}} [\text{Adj}(z\mathbf{I} - \mathbf{F})] \right\}_{z=z_k}$$

(ii) New state matrix $= \mathbf{TFT}^{-1}$

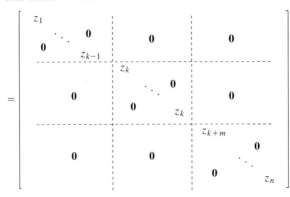

V.   Near-Normal Canonical Form

Transformation conditions

(i) $\mathbf{F}$ matrix has one pair of complex, conjugate eigenvalues, z_k, z_{k+1}

$$z_k = z_{kr} + jz_{ki}$$

$$z_{k+1} = z_{kr} - jz_{ki}$$

(ii) All other values are real and distinct.

Transformation matrices

(i) $\mathbf{q} = \mathbf{Tx}$, $\mathbf{T}^{-1} = [\mathbf{v}_1 \ \cdots \ \mathbf{v}_{k-1} \ \mathbf{v}_{kr} \ \mathbf{v}_{ki} \ \mathbf{v}_{k+2} \ \cdots \ \mathbf{v}_n]$

where (a) $\mathbf{v}_i, i = 1, k-1$ and $i = k+2, \ldots, n$ are the linearly independent eigenvectors corresponding to the real, distinct eigenvalues.

(b) $\mathbf{v}_i$, for $i = 1, \ldots, n$ are taken to be equal or proportional to any nonzero column of $\text{Adj}(z_i \mathbf{I} - \mathbf{F})$, and

(c) $\mathbf{v}_k = \mathbf{v}_{kr} + j\mathbf{v}_{ki}$
$\mathbf{v}_{k+1} = \mathbf{v}_{kr} - j\mathbf{v}_{ki}$
are the complex conjugate eigenvectors corresponding to z_k and z_{k+1}, respectively.

(ii) New state matrix $= \mathbf{TFT}^{-1}$

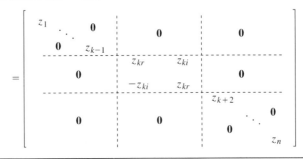

(Continued)

VI. Nondiagonal Jordan Canonical Form
  Transformation conditions
  (i) **F** matrix has one repeated, real eigenvalue z_k of multiplicity m (i.e., $z_k = z_{k+1} = \cdots = z_{k+m-1}$). All other eigenvalues are real and distinct.
  (ii) Degeneracy $d = n - \text{rank}(z_k\mathbf{I} - \mathbf{F}) = 1$. Simple degeneracy.

  Transformation matrices
  (i) $\mathbf{q} = \mathbf{Tx}$, $\mathbf{T}^{-1} = [\mathbf{t}_1 \quad \mathbf{t}_2 \quad \ldots \quad \mathbf{t}_n]$
    where (a) $\mathbf{t}_i$, $i = 1, k - 1$ and $i = k + m, n$ are the linearly independent eigenvectors corresponding to the real, distinct eigenvalues,
      (b) $\mathbf{t}_i$, $i = 1, k - 1$ and $i = k + m, n$ are taken to be equal or proportional to any nonzero column of $\text{Adj}(z_i\mathbf{I} - \mathbf{F})$, and
      (c) $\mathbf{t}_i$, $i = k, k + m - 1$ are obtained by solution of the equation $\mathbf{FT}^{-1} = \mathbf{T}^{-1}\mathbf{J}$, where $\mathbf{J}$ is the Jordan canonical matrix given. Each $\mathbf{t}_i$ is determined to within a constant of proportionality.
  (ii) New state matrix $= \mathbf{J} = \mathbf{TFT}^{-1}$

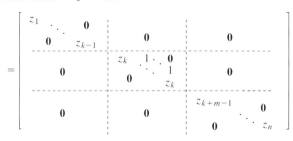

VII. Nondiagonal Jordan Canonical Form
  Transformation conditions
  (i) **F** matrix has one repeated, real eigenvalue of multiplicity m (i.e., $z_k = z_{k+1} = \cdots = z_{k+m-1}$). All other eigenvalues are real and distinct.
  (ii) Degeneracy $d = n - \text{rank}(z_k\mathbf{I} - \mathbf{F})$ is between 1 and m. General degeneracy.

  Transformation matrices
  (i) $\mathbf{q} = \mathbf{Tx}$, $\mathbf{T}^{-1} = [\mathbf{t}_1 \quad \mathbf{t}_2 \quad \cdots \quad \mathbf{t}_n]$
    where (a) $\mathbf{t}_i$, $i = 1, k - 1$ and $i = k + m, n$ are the linearly independent eigenvectors corresponding to the real, distinct eigenvalues,
      (b) $\mathbf{t}_i$, $i = 1, k - 1$ and $i = k + m, n$ are taken to be equal or proportional to any nonzero column of $\text{Adj}(z_i\mathbf{I}-\mathbf{F})$,
      (c) $\mathbf{t}_i$, $i = k, k + m - 1$ are m linearly independent vectors, corresponding to the repeated eigenvalue of multiplicity m. Only d of these vectors are eigenvectors, and
      (d) $\mathbf{t}_i$, $i = k, k + m - 1$ obtained by solution of the equation $\mathbf{FT}^{-1} = \mathbf{T}^{-1}\mathbf{J}$, where $\mathbf{J}$ is a Jordan canonical matrix for the problem, with d Jordan blocks. There are d possible choices for $\mathbf{J}$. Each of these choices needs to be tried; and the $\mathbf{t}_i$ vectors solved for. Only the correct $\mathbf{J}$ will give the m linearly independent vectors $\mathbf{t}_i$, $i = k, k + m - 1$. Each $\mathbf{t}_i$ is determined to within a constant of proportionality.
  (ii) New state matrix $\mathbf{J} = \mathbf{TFT}^{-1}$. Correct $\mathbf{J}$ determined by trial-and-error, as described.

of the second term in Eq. 19.33, for the case $\mathbf{D}=0$, with Eq. 19.34 for the forced response of a SISO, LTI system indicates that the matrix $\mathbf{C\phi}(t)\mathbf{B}$ is a matrix of impulse responses:

$$y(t_0) = y(t_0) + \int_{t_0}^{t} h(t - \tau)u(\tau)\,d\tau \tag{19.34}$$

The variable $h(t)$, in Eq. 19.34, is the impulse response of the system. The interpretation of $\mathbf{C\phi}(t)\mathbf{B}$ as a matrix of impulse responses forms the basis of one of the methods for determining the elements of the transition matrix.[2,4]

The state transition matrix $\mathbf{\phi}(t - t_0)$ is the solution of the matrix differential equation:

$$\dot{\mathbf{\phi}}(t - t_0) = \mathbf{A\phi}(t - t_0) \qquad t \geq t_0 \tag{19.35}$$

with the initial condition

$$\mathbf{\phi}(t_0 - t_0) = \mathbf{\phi}(0) = \mathbf{I} \tag{19.36}$$

It has the following properties:

$$\mathbf{\phi}(t + \tau) = \mathbf{\phi}(t)\mathbf{\phi}(\tau) = \mathbf{\phi}(\tau)\mathbf{\phi}(t) \tag{19.37}$$

$$\mathbf{\phi}^{-1}(t) = \mathbf{\phi}(-t) \tag{19.38}$$

The following expressions for $\mathbf{\phi}(t)$ can be verified and are useful in its evaluation:

$$\mathbf{\phi}(t) = e^{\mathbf{A}t} = \mathbf{I} + \mathbf{A}t + \frac{\mathbf{A}^2 t^2}{2!} + \frac{\mathbf{A}^3 t^3}{3!} + \cdots \tag{19.39}$$

and

$$\mathbf{\phi}(t) = \mathcal{L}^{-1}\left[(s\mathbf{I} - \mathbf{A})^{-1}\right] \tag{19.40}$$

where $\mathcal{L}^{-1}$ denotes the inverse Laplace transform. Details related to Eqs. 19.35–19.40 have been described by DeRusso et al.[2] and Brogan.[4]

Knowledge of the state transition matrix for a given system simplifies the task of determining the response of the system to a variety of initial conditions $\mathbf{x}(t_0)$ and forcing functions $\mathbf{u}(t)$. A number of analytical and numerical techniques for its evaluation are available.

Equation 19.39 forms the basis for a numerical method of determining $\mathbf{\phi}(t)$. Closed-form evaluation of $e^{\mathbf{A}t}$ is possible only for special forms of the $\mathbf{A}$ matrix. For example, if $\mathbf{A}$ is a diagonal matrix with diagonal elements equal to the eigenvalues s_i, it can be shown that $\mathbf{\phi}(t)$ is also diagonal[4] and is given by Eq. 19.41:

$$\mathbf{\phi}(t) = e^{\mathbf{A}t} = \begin{bmatrix} e^{s_1 t} & & & \\ & e^{s_2 t} & & \mathbf{0} \\ & & \ddots & \\ \mathbf{0} & & & e^{s_n t} \end{bmatrix} \tag{19.41}$$

Closed-form evaluation of $e^{\mathbf{A}t}$ is only slightly more complex if $\mathbf{A}$ is in the nondiagonal Jordan canonical form.[4] If the transformation matrix $\mathbf{T}$ (Eq. 19.14) was used to obtain the diagonal or nondiagonal Jordan matrix $\mathbf{A}$, the transition matrix, for the original state vector $\mathbf{T}^{-1}\mathbf{x}$, is $\mathbf{T}^{-1}e^{\mathbf{A}t}\mathbf{T}$.

Equation 19.40 provides the basis for an analytical evaluation of $\mathbf{\phi}(t)$ that is suitable for low-order dynamic systems. This method requires the inversion of the $n \times n$ matrix $s\mathbf{I} - \mathbf{A}$, followed by the inverse Laplace transformation of the n^2 elements. The matrix inversion is especially cumbersome since the elements of the matrix are functions of s. The matrix inversion can be avoided altogether by using simulation diagrams of the system, in conjunction with block diagram reduction techniques,[2] to determine elements of the matrix $(s\mathbf{I} - \mathbf{A})^{-1}$. Alternative analytical techniques for the evaluation of $\mathbf{\phi}(t)$ based on Sylvester's theorem and the Cayley–Hamilton theorem have been described by DeRusso et al.[2] and Brogan.[4]

Numerical evaluation of $\mathbf{\phi}(t)$, for a specified value of t, can be performed using Eq. 19.39 and retaining a finite number of terms from the series expansion. The number of terms retained increases with the desired degree of accuracy. An iterative procedure for determining the number of terms to be retained for a specified degree of accuracy has been described by Shinners.[5]

For linear, time-varying systems described by state-space Eqs. 19.4 and 19.5, the solution is given by[2,4]

$$\mathbf{x}(t) = \boldsymbol{\phi}(t, t_0)\mathbf{x}(t_0) + \int_{t_0}^{t} \boldsymbol{\phi}(t, \tau)\mathbf{B}(\tau)\mathbf{u}(\tau)\, d\tau \qquad t \geq t_0 \tag{19.42}$$

$$\mathbf{y}(t) = \mathbf{C}(t)\boldsymbol{\phi}(t, t_0)\mathbf{x}(t_0) + \int_{t_0}^{t} \mathbf{C}(t)\boldsymbol{\phi}(t, \tau)\mathbf{B}(\tau)\mathbf{u}(\tau)\, d\tau + \mathbf{D}(t)\mathbf{u}(t) \tag{19.43}$$

where the $n \times n$ state transition matrix $\boldsymbol{\phi}(t, t_0)$ for the time-varying system depends on both arguments t and t_0 and not merely on the difference between these two time instants as in the time-invariant system.

The state transition matrix $\boldsymbol{\phi}(t, t_0)$ is the solution of the partial differential equation[2,4]

$$\frac{\partial \boldsymbol{\phi}(t, t_0)}{\partial t} = \mathbf{A}(t)\boldsymbol{\phi}(t, t_0) \tag{19.44}$$

with the initial condition

$$\boldsymbol{\phi}(t_0, t_0) = \mathbf{I} \tag{19.45}$$

It has the following properties:

$$\boldsymbol{\phi}(t_2, t_0) = \boldsymbol{\phi}(t_2, t_1)\boldsymbol{\phi}(t_1, t_0) \tag{19.46}$$

$$\boldsymbol{\phi}(t_1, t_0) = \boldsymbol{\phi}^{-1}(t_0, t_1) \tag{19.47}$$

Techniques for evaluating the state transition matrix for time-varying systems are considerably more involved than for time-invariant systems and are less widely applicable. A number of analytical methods for determining $\boldsymbol{\phi}(t, t_0)$ for special cases of linear time-varying systems have been described by DeRusso et al.[2]

A simple numerical procedure has been suggested by Palm[6] for computing the transition matrix when analytical determination is not possible. Let the ith column of $\boldsymbol{\phi}(t, t_0)$ be denoted by $\boldsymbol{\psi}_i(t)$, for a specified value of t_0. The matrix partial differential Eq. 19.44 becomes n vector ordinary differential equations:

$$\dot{\boldsymbol{\psi}}_i(t) = \mathbf{A}(t)\boldsymbol{\psi}_i(t) \qquad i = 1, \ldots, n \qquad t \geq t_0 \tag{19.48}$$

with the initial conditions

$$[\boldsymbol{\psi}_1(t_0)\ \boldsymbol{\psi}_2(t_0)\ \ldots\ \boldsymbol{\psi}_n(t_0)] = \mathbf{I} \tag{19.49}$$

Numerical solution of the ordinary differential equations gives $\boldsymbol{\psi}_i(t)$ and hence $\boldsymbol{\phi}(t, t_0)$. Note that the computed $\boldsymbol{\phi}(t, t_0)$ would be different for different values of t_0 for time-varying systems.

19.4.2 Discrete-Time Systems

The solutions of the system equations for linear, discrete-time systems described either by Eqs. 19.10 and 19.11 or by Eqs. 19.12 and 19.13 are given in Table 19.5. Expressions for the state transition matrix $\boldsymbol{\phi}(k - k_0)$ for time-invariant systems and $\boldsymbol{\phi}(k, k_0)$ for time-varying systems and the properties of the state transition matrix are also included in the table. These results can be found in standard textbooks on state-space methods.[2,4]

The state transition matrix $\boldsymbol{\phi}(k - k_0)$ for a time-invariant system depends only on the difference in the sequence number $(k - k_0)$. The transition matrix is given by

$$\boldsymbol{\phi}(k - k_0) = \mathbf{F}^{k-k_0} \tag{19.50}$$

Numerical computation of $\boldsymbol{\phi}(k - k_0)$ is thus straightforward. Analytical evaluation of $\boldsymbol{\phi}(k - k_0)$ using Eq. 19.50 is feasible if $\mathbf{F}$ is in the diagonal or nondiagonal Jordan canonical form. If $\mathbf{F}$ is a diagonal matrix with diagonal elements equal to its eigenvalues z_i, $i = 1, \ldots, n$, then

TABLE 19.5 SOLUTION OF THE STATE-SPACE EQUATIONS FOR LINEAR, DISCRETE-TIME SYSTEMS

	Time-Invariant System (Eqs. 19.12 and 19.13)	Time-Varying System (Eqs. 19.10 and 19.11)
Solution $k > k_0$	$\mathbf{x}(k) = \boldsymbol{\phi}(k - k_0)\mathbf{x}(k_0)$ $+ \sum_{m=k_0}^{k-1} \boldsymbol{\phi}(k - m - 1)\mathbf{G}\mathbf{u}(m)$ $\mathbf{y}(k) = \mathbf{C}\boldsymbol{\phi}(k - k_0)\mathbf{x}(k_0)$ $+ \sum_{m=k_0}^{k-1} [\mathbf{C}\boldsymbol{\phi}(k - m - 1)\,\mathbf{G}\mathbf{u}(m)]$ $+ \mathbf{D}\mathbf{u}(k)$	$\mathbf{x}(k) = \boldsymbol{\phi}(k, k_0)\mathbf{x}(k_0)$ $+ \sum_{m=k_0}^{k-1} \boldsymbol{\phi}(k, m + 1)\mathbf{G}(m)\mathbf{u}(m)$ $\mathbf{y}(k) = \mathbf{C}(k)\boldsymbol{\phi}(k, k_0)\mathbf{x}(k_0)$ $+ \sum_{m=k_0}^{k-1} [\mathbf{C}(k)\boldsymbol{\phi}(k, m + 1)\mathbf{G}(m)\mathbf{u}(m)]$ $+ \mathbf{D}(k)\mathbf{u}(k)$
State transition matrix	$\boldsymbol{\phi}(k - k_0) = \mathbf{F}^{k-k_0} \quad k > k_0$ or $\boldsymbol{\phi}(k) = \mathscr{L}^{-1}[z(z\mathbf{I} - \mathbf{F})^{-1}]$	$\boldsymbol{\phi}(k, k_0) = \begin{cases} \prod_{l=k}^{k-1} \mathbf{F}(l) & k > k_0 \\ \mathbf{I} & k = k_0 \end{cases}$
Properties of the state transition matrix	$\boldsymbol{\phi}(0) = \mathbf{I}$ $\boldsymbol{\phi}(k_1 + k_2) = \boldsymbol{\phi}(k_1)\boldsymbol{\phi}(k_2)$ $\boldsymbol{\phi}(k) = \boldsymbol{\phi}^{-1}(-k)$ when the inverse exists	$\boldsymbol{\phi}(k_0, k_0) = \mathbf{I}$ $\boldsymbol{\phi}(k_2, k_1)\boldsymbol{\phi}(k_1, k_0) = \boldsymbol{\phi}(k_2, k_0)$ $\boldsymbol{\phi}(k_1, k_2) = \boldsymbol{\phi}^{-1}(k_2, k_1)$ when the inverse exists

$$\boldsymbol{\phi}(k) = \mathbf{F}^k = \begin{bmatrix} z_1^k & & & \\ & z_2^k & & \mathbf{0} \\ & & \ddots & \\ \mathbf{0} & & & z_n^k \end{bmatrix} \tag{19.51a}$$

If $\mathbf{F}$ is in the nondiagonal Jordan canonical form, analytical evaluation of $\boldsymbol{\phi}(k)$ is only slightly more complex. If the transformation matrix $\mathbf{T}$ (Eq. 19.17) was used to obtain the diagonal or nondiagonal Jordan matrix $\mathbf{F}$, the transition matrix, for the original state vector $\mathbf{T}^{-1}\mathbf{x}$, is $\mathbf{T}^{-1}\mathbf{F}^k\mathbf{T}$.

An alternative analytical evaluation of the transition matrix uses the relationship

$$\boldsymbol{\phi}(k) = \mathscr{L}^{-1}[z(z\mathbf{I} - \mathbf{F})^{-1}] \tag{19.51b}$$

where $\mathscr{L}^{-1}$ denotes the inverse z transform. This method is useful only for low-order systems because of the need for inverting the matrix $z\mathbf{I} - \mathbf{F}$, which has symbolic elements. The matrix inversion is particularly simple if $\mathbf{F}$ is in the diagonal or Jordan canonical form. As for continuous-time systems, the matrix inversion can be avoided altogether by using simulation diagrams of the discrete-time system and block diagram reduction,[2] to directly determine elements of the matrix $(z\mathbf{I} - \mathbf{F})^{-1}$.

The computation of the state transition matrix $\boldsymbol{\phi}(k, k_0)$ for time-varying, discrete-time systems is difficult, as it is for continuous-time systems. For small or moderate values of the order n of the system, numerical evaluation of Eq. 19.52 is appropriate:

$$\boldsymbol{\phi}(k, k_0) = \begin{cases} \prod_{l=k_0}^{k-1} \mathbf{F}(l) & k > k_0 \\ \mathbf{I} & k = k_0 \end{cases} \tag{19.52}$$

For larger values of n, analytical and numerical methods for determining $\boldsymbol{\phi}(k, k_0)$ in special cases are available.[2]

19.5 STABILITY

Since state-space formulation is applicable to a large class of dynamic systems, the question of stability for systems represented in state space is quite a complex one. A more general consideration of stability than that used for SISO, LTI systems would indicate that stability of dynamic systems is not really a property of the systems but is more properly associated with isolated equilibrium points of dynamic systems.[4] A particular point $\mathbf{x}_e$ in state space is an equilibrium point of a dynamic system if, in the absence of inputs, the system state $\mathbf{x}$ is equal to $\mathbf{x}_e$ for time $t \geq t_0$ for continuous-time systems or for $k \geq k_0$, for discrete-time systems. For linear systems described by the state-space equations given in Section 19.2, the only isolated equilibrium point is at the origin in state space. For nonlinear systems, there may be a number of isolated equilibrium points. Any isolated equilibrium point can be shifted to the origin in state space by a simple change of state variables.[4] The stability definitions to be given assume therefore that the equilibrium point is at the origin in state space and that the system is unforced. Only the more commonly used types of stability will be defined.

The origin is a stable equilibrium point if, for any given value $\epsilon > 0$, there exists a number $\kappa(\epsilon, t_0) > 0$ such that if the norm $\|\mathbf{x}(t_0)\| < \kappa$, then the norm $\|\mathbf{x}(t)\| < \epsilon$ for all $t > t_0$. The norm of a vector $\mathbf{x}$ maybe defined as the Euclidean norm.

$$\|\mathbf{x}(t)\| = \sqrt{\sum_{i=1}^{n} x_i^2(t)} \tag{19.53}$$

The origin is asymptotically stable if, in addition to being stable, there exists a number $\gamma(t_0) > 0$ such that whenever $\|\mathbf{x}(t_0)\| < \gamma(t_0)$, the following condition is satisfied:

$$\lim_{t \to \infty} \|\mathbf{x}(t)\| = 0 \tag{19.54}$$

If κ and γ are not functions of t_0 in the previous definitions, the origin is said to be uniformly stable or uniformly asymptotically stable, respectively. If $\gamma(t_0)$ can be arbitrarily large, the origin is said to be globally asymptotically stable. Extension of these stability definitions to discrete-time systems is straightforward and merely requires that the sequence numbers k, k_0 be used instead of the time instants t, t_0, respectively, in the definitions already given. Additional types of stability that depend on the inputs to the system have been defined by Brogan[4] and Kuo.[7]

For LTI systems, the conditions for stability reduce to conditions on the eigenvalues of the system matrix $\mathbf{A}$ or $\mathbf{F}$ and are summarized in Table 19.6. These eigenvalues are the roots of the system characteristic equation as well, as shown in Section 19.7. They may be computed explicitly by numerical methods. Alternatively, stability criteria such as the Routh–Hurwitz criterion[5] for continuous-time systems or the Jury test[7] for discrete-time systems may be applied. The conditions for asymptotic stability of such systems can also be shown to be sufficient for other types of stability depending on the input, such as bounded-input, bounded-output stability.[4]

For continuous-time, LTV systems, the necessary and sufficient condition for the origin to be a stable equilibrium point is that there exists a number $N(t_0)$ such that the norm of the transition matrix satisfies the following condition:

$$\|\boldsymbol{\phi}(t, t_0)\| \leq N(t_0) \quad \text{for } t \geq t_0 \tag{19.55}$$

If, in addition, $\|\boldsymbol{\phi}(t, t_0)\| \to 0$ as $t \to \infty$, the system is globally asymptotically stable.[4] The norm of the matrix $\boldsymbol{\phi}$ may be defined as the spectral norm.

$$\|\boldsymbol{\phi}(t, t_0)\| = \sqrt{\max_{\|x\|=1}(\mathbf{x}^T \boldsymbol{\phi}^T \boldsymbol{\phi} \mathbf{x})} \tag{19.56}$$

The corresponding stability conditions for linear, discrete-time systems are obtained simpy by substituting the sequence numbers k and k_0 for time instants t and t_0, respectively, in the development. Time-varying systems that satisfy the property that the state converges exponentially with time to the zero state are said to be exponentially stable.[12] For LTI systems, of course, asymptotic stability is the same as exponential stability.

Stability considerations for nonlinear systems are more complex. For unforced second-order nonlinear systems, the phase-plane method is useful for examining the stability of equilibrium points of the system. The phase plane has the state variables as the coordinates. The state-space equations are used to derive analytical expressions for the trajectories or to draw the trajectories by graphical means. The phase portraits can then be examined to determine the equilibrium points and their stability. Application of the phase-plane method is described by DeRusso et al.[2] for continuous-time systems and by Kuo[7] for discrete-time systems.

Stability analysis of high-order nonlinear systems represented in state space can be done using the

TABLE 19.6   STABILITY CRITERIA FOR LINEAR, TIME-INVARIANT SYSTEMS

	Continuous-Time System $\dot{\mathbf{x}}(t) = \mathbf{A}\mathbf{x}(t) + \mathbf{B}\mathbf{u}$ Eigenvalues of $\mathbf{A}$ are $s_i = \alpha_{ic} \pm j\omega_{ic}$	Discrete-Time System $\mathbf{x}(k+1) = \mathbf{F}\mathbf{x}(k) + \mathbf{G}\mathbf{u}(k)$ Eigenvalues of $\mathbf{F}$ are $z_i = \beta_{ic} \pm j\omega_{ic}$
Asymptotically stable	$\alpha_{ic} < 0$ for all roots	$|z_i| < 1$ for all roots
Stable	$\alpha_{ic} < 0$ for all repeated roots and $\alpha_{ic} \leq 0$ for all simple roots	$|z_i| < 1$ for all repeated roots and $|z_i| \leq 1$ for all simple roots
Unstable	$\alpha_{ic} > 0$ for any simple root or $\alpha_{ic} \geq 0$ for any repeated root	$|z_i| > 1$ for any simple root or $|z_i| \geq 1$ for any repeated root

second method of Lyapunov. This is a technique requiring considerable ingenuity for effective use and provides sufficient conditions for stability rather than necessary and sufficient conditions.[8]

Lyapunov's method for nonlinear, unforced, time-invariant systems requires the definition of a scalar function of state $V(\mathbf{x})$ called the Lyapunov function. The latter may be thought of as a generalized energy function. The requirement on the Lyapunov function is that it be positive definite in some region about the origin in state space, the origin having been assumed to be an isolated equilibrium point here. A function $V(\mathbf{x})$, which is continuous and has continuous partial derivatives, is said to be positive (negative) definite in some region about the origin if it is zero at the origin and greater than (less than) zero everywhere else in the specified region. If the function is greater than (less than) or equal to zero everywhere in the specified region, it is said to be positive (negative) semidefinite.[4]

Consider the unforced continuous-time system represented by the state equation

$$\dot{\mathbf{x}}(t) = \mathbf{f}[\mathbf{x}(t)] \tag{19.57}$$

where

$$\mathbf{f}(\mathbf{0}) = \mathbf{0} \tag{19.58}$$

If a positive definite function $V(\mathbf{x})$ can be determined in some region Γ about the origin such that its derivative with respect to time is negative semidefinite in Γ, then the origin is a stable equilibrium point. If dV/dt is negative definite, the origin is asymptotically stable. If the region Γ can be arbitrarily large and the conditions for asymptotic stability hold and if, in addition, $V(\mathbf{x}) \to \infty$ as $\|\mathbf{x}\| \to \infty$, the origin is a globally asymptotically stable equilibrium point. Table 19.7 gives the corresponding stability conditions for nonlinear, time-invariant, discrete-time systems. Extensions of the stability conditions for time-varying systems have been described by Kalman and Bertram[8] and DeRusso et al.[2]

As an example of the application of the second method of Lyapunov, consider the following nonlinear system:

$$\dot{x}_1 = x_2$$
$$\dot{x}_2 = -a_0 x_2 - b_0 x_2^3 - x_1 \tag{19.59}$$

TABLE 19.7   APPLICATION OF THE SECOND METHOD OF LYAPUNOV TO NONLINEAR, TIME-INVARIANT, DISCRETE-TIME SYSTEMS

State equation	$\mathbf{x}(k+1) = \mathbf{f}[\mathbf{x}(k)]$ $\mathbf{f}(\mathbf{0}) = \mathbf{0}$
Lyapunov function	Scalar function $V[\mathbf{x}(k)]$ positive definite in some region about the origin
Condition for stability in Γ	$\Delta V = V[\mathbf{x}(k+1)] - V[\mathbf{x}(k)]$ is negative semidefinite in Γ
Condition for asymptotic stability in Γ	$\Delta V = V[\mathbf{x}(k+1)] - V[\mathbf{x}(k)]$ is negative definite in Γ
Condition for global asymptotic stability	(i) Γ can be arbitrarily large (ii) $V[\mathbf{x}(k)] \to \infty$ as $\|\mathbf{x}(k)\| \to \infty$

where $a_0, b_0 \geq 0$ and both are not zero. The origin is an equilibrium point for this system since, if both x_1 and x_2 are zero,

$$\dot{x}_1 = \dot{x}_2 = 0 \tag{19.60}$$

Consider the following Lyapunov function:

$$V(x_1, x_2) = x_1^2 + x_2^2 \tag{19.61}$$

It satisfies the conditions for positive definiteness in an arbitrarily large region about the origin:

$$\frac{dV(x_1, x_2)}{dt} = 2x_1\dot{x}_1 + 2x_2\dot{x}_2 \tag{19.62}$$

$$= -2(a_0 x_2^2 + b_0 x_2^4)$$

after using the state equations to substitute for $\dot{x}_1$ and $\dot{x}_2$. If a_0, b_0 satisfy the inequalities stated, dV/dt is negative semidefinite in an arbitrarily large region about the origin. The origin is thus a stable equilibrium point. In fact, using a corollary to the main stability theorem provided by Kalman and Bertram[8], it can be shown that the origin is a global asymptotically stable equilibrium point.

The limitations of Lyapunov's method are that the Lyapunov function is not unique for a system and there are no systematic procedures for finding a suitable Lyapunov function. Since only sufficient conditions for stability are determined, some choices of Lyapunov functions are better in that they provide more information about system stability than others. Also, appropriate choice of the Lyapunov function can lead to an estimate of the system speed of response.[8] In practice, therefore, the second method of Lyapunov is used primarily to analyze the stability of systems such as high-order, nonlinear systems for which other methods of stability analysis are not available.

19.6 CONTROLLABILITY AND OBSERVABILITY

The controllability of a linear system is a measure of the coupling between the inputs to the system and the system state. The concept of state controllability was introduced by Kalman[11] in order to clarify conditions for the existence of solutions to specific control problems.

A linear, continuous-time system is said to be state controllable at time t_0 if there exists a finite time $t_1 > t_0$ and a control function $\mathbf{u}(t)$, $t_0 < t < t_1$ that can drive the system state from any initial value to any final value at $t = t_1$. If the system is controllable for all times t_0, the system is completely state controllable.[4] A linear, discrete-time system is said to be state controllable and completely state controllable, respectively, if the sequence numbers k, k_0, k_1 are substituted for the times t, t_0, t_1, respectively, in the two previously given definitions. An additional form of controllability for continuous-time and discrete-time LTV systems is that of uniformly complete state controllability. The mathematical definition of this form of controllability may be found in Kalman.[11] This property implies that the control effort and time interval required to drive the system state to the final value is relatively independent of the initial time. For LTI systems, of course, complete state controllability is the same as uniformly complete state controllability.

Though the control problems formulated above are open-loop control problems, the property of controllability has very significant implications for closed-loop control problems. Section 20.2 indicates that the closed-loop poles of a completely state-controllable time-invariant system can be specified and placed arbitrarily in the complex s plane (or z plane for discrete-time systems) by proportional state-variable feedback. Moreover, satisfaction of the controllability conditions to be defined in this section for time-invariant systems ensures that the optimal control law for a quadratic performance index is a proportional state-variable feedback law and yields an asymptotically stable closed-loop system.[10]

Direct application of the definition of state controllability to LTI systems yields controllability conditions involving the transition matrices. Simple algebraic conditions are usually available for such systems and are used more often in practice to evaluate controllability.

The controllability condition for LTI systems with distinct eigenvalues may be stated very simply if the state equations are transformed to the diagonal Jordan canonical form. Such systems are completely controllable if there are no zero rows in the transformed $\mathbf{B}$ matrix for continuous-time systems or in the transformed $\mathbf{G}$ matrix for discrete-time systems.[4] The presence of a zero row in either of these matrices would indicate that the inputs are not coupled to and cannot control the corresponding mode. Algebraic controllability conditions for systems with repeated eigenvalues are given by Palm.[6]

TABLE 19.8 CONTROLLABILITY CONDITIONS FOR LINEAR DYNAMIC SYSTEMS

	Continuous Time	Discrete Time
	Time-Invariant System	
Necessary and sufficient condition for state controllability	(i) $\text{rank}(\mathbf{B}\ \mathbf{AB}\ \ldots\ \mathbf{A}^{n-1}\mathbf{B})$ $= \text{rank}\ (\mathbf{B}\ \mathbf{AB}\ \ldots\ \mathbf{A}^{n-r}\mathbf{B})$ $= n$ or (ii) $\det(\mathbf{P}_c\mathbf{P}_c^{\mathrm{T}}) \neq 0$	(i) $\text{rank}(\mathbf{G}\ \mathbf{FG}\ \ldots\ \mathbf{F}^{n-1}\mathbf{G})$ $= \text{rank}\ (\mathbf{G}\ \mathbf{FG}\ \ldots\ \mathbf{F}^{n-r}\mathbf{G})$ $= n$ or (ii) $\det(\mathbf{P}_c\mathbf{P}_c^{\mathrm{T}}) \neq 0$
Necessary condition for state controllability	$\text{rank}(\mathbf{B}\ \mathbf{A}) = n$	$\text{rank}(\mathbf{G}\ \mathbf{F}) = n$
Necessary and sufficient condition for output controllability	(i) $\text{rank}(\mathbf{CP}_c) = p$ or (ii) $\det(\mathbf{CP}_c\mathbf{P}_c^{\mathrm{T}}\mathbf{C}^{\mathrm{T}}) \neq 0$	(i) $\text{rank}(\mathbf{CP}_c) = p$ or (ii) $\det(\mathbf{CP}_c\mathbf{P}_c^{\mathrm{T}}\mathbf{C}^{\mathrm{T}} \neq 0$
	Time-Varying System: *Time Interval of Interest $[t_0, t_1]$ or $[k_0, k_1]$*	
Necessary and sufficient condition for state controllability	(i) $\mathbf{W}_c(t_1, t_0)$ is positive definite or (ii) Zero is not an eigenvalue of $\mathbf{W}_c(t_1, t_0)$ or (iii) $\|\mathbf{W}_c(t_1, t_0)\| \neq 0$ where $\mathbf{W}_c(t_1, t_0)$ $$\triangleq \int_{t_0}^{t_1} \boldsymbol{\phi}(t_1, \tau)\mathbf{B}(\tau)\mathbf{B}^T(\tau)$$ $$\times \boldsymbol{\phi}^{\mathrm{T}}(t_1, \tau)d\tau$$	(i) $\mathbf{W}_c(k_1, k_0)$ is positive definite or (ii) Zero is not an eigenvalue of $\mathbf{W}_c(k_1, k_0)$ or (iii) $\|\mathbf{W}_c(k_1, k_0)\| \neq 0$ where $\mathbf{W}_c(k_1, k_0)$ $$\triangleq \sum_{k=k_0}^{k_1} \boldsymbol{\phi}(k_1, k)\mathbf{G}(k)\mathbf{G}^T(k)$$ $$\times \boldsymbol{\phi}^{\mathrm{T}}(k_1, k)$$
Necessary and sufficient condition for output controllability	$\det \mathbf{W}_y(t_1, t_0) \neq 0$ where $\mathbf{W}_y(t_1, t_0)$ $$\triangleq \int_{t_0}^{t_1} \mathbf{C}(\tau)\boldsymbol{\phi}(t_1, \tau)\mathbf{B}(\tau)\mathbf{B}^T(\tau)$$ $$\times \boldsymbol{\phi}^{\mathrm{T}}(t_1, \tau)\mathbf{C}^{\mathrm{T}}(\tau)\, d\tau$$	$\det \mathbf{W}_y(k_1, k_0) \neq 0$ where $\mathbf{W}_y(k_1, k_0)$ $$\triangleq \sum_{k=k_0}^{k_1} \mathbf{C}(k)\boldsymbol{\phi}(k_1, k)\mathbf{B}(k)$$ $$\times \mathbf{B}^{\mathrm{T}}(k)\boldsymbol{\phi}^{\mathrm{T}}(k_1, k)\,\mathbf{C}^{\mathrm{T}}(k)$$

The controllability conditions for LTI systems in general are stated in terms of the matrix $\mathbf{P}_c$, referred to as the controllability matrix in Section 19.3, and are summarized in Table 19.8. The $n \times nr$ controllability matrix for a MIMO system is defined by

$$\mathbf{P}_c = [\mathbf{B}\ \mathbf{AB}\ \ldots\ \mathbf{A}^{n-1}\ \mathbf{B}] \tag{19.63}$$

for continuous-time systems and by

$$\mathbf{P}_c = [\mathbf{G}\ \mathbf{FG}\ \ldots\ \mathbf{F}^{n-1}\mathbf{G}| \tag{19.64}$$

for discrete-time systems. The condition for complete state controllability is simply that the matrix $\mathbf{P}_c$ has rank n. The controllable canonical form for SISO systems, described in Section 19.3, derives its name from the fact that transformation to that form is possible if and only if the system is completely state controllable. The transformation matrix $\mathbf{T}$ in Tables 19.2 and 19.4 required to transform the state-space equations for a SISO system to the controllable canonical form exists if and only if the $n \times n$ matrix $\mathbf{P}_c$ is nonsingular; that is, the system is controllable. Equivalent controllability conditions that are simpler to evaluate than the one previously stated are also listed in Table 19.8 along with a necessary (but not sufficient) condition for complete controllability.

The controllability conditions for LTV systems over a specified time interval are more cumbersome to evaluate in practice as they involve the system transition matrix.[4] These conditions are listed in Table 19.8. In contrast to time-invariant systems, the controllability of time-varying systems depends on the time interval under consideration.

The concept of output controllability, as opposed to state controllability described earlier, was introduced by Kreindler and Sarachik.[13] A linear, continuous-time system is said to be output controllable at time t_0 if there exists a finite time $t_1 > t_0$ and a control function $\mathbf{u}(t)$, $t_0 < t < t_1$, that drives the system output from any initial value $\mathbf{y}(t_0)$ to any final value $y(t_1)$. If this condition holds true for all times t_0, the system is completely output controllable. Extension of the concept to linear, discrete-time systems is straightforward as before.

Output controllability conditions[4] for linear systems that are purely dynamic (i.e., the matrix $\mathbf{D} = \mathbf{0}$ in Eqs. 19.5, 19.7, 19.11, and 19.13) are summarized in Table 19.8. These conditions are weaker than the corresponding conditions for state controllability if the number of outputs p is less than the number of state variables n. Since this is true in practice, state controllability implies output controllability. On the other hand, output controllability does not imply state controllability in general. It can be shown, however, that for time-invariant systems if the matrix $(\mathbf{CC}^T)$ is nonsingular, output controllability is equivalent to state controllability.

The observability of a linear system is a measure of the coupling between the system state and its outputs. The concept of observability was introduced by Kalman[11] and is relevant to the problem of estimation of system state based on the output vector. The output vector is usually chosen to correspond to measurable variables.

A linear, continuous-time system is said to be observable[4] at time t_0 if there exists a finite time $t_1 > t_0$ such that $\mathbf{x}(t_0)$ can be determined from the history of inputs $\mathbf{u}(t)$ and outputs $\mathbf{y}(t)$ over the time interval $t_0 \le t \le t_1$. If the system is observable for all times t_0 and all initial states $\mathbf{x}(t_0)$, the system is completely observable. Extension of the observability concept to discrete-time systems simply requires that the sequence numbers k, k_0, k_1 be substituted for the time t, t_0, t_1, respectively, in the previous definitions. A stronger form of observability for LTV systems is that of uniformly complete observability. The mathematical definition of this form of observability is given by Kalman.[11] This property guarantees that the time interval required to estimate the state is relatively independent of the initial time. For LTI systems, of course, complete observability is the same as uniformly complete observability. A property complementary to observability for LTV systems, is that of reconstructibility,[14,12] which concerns the estimation of the state of the system from past measurements of the state. In contrast to this, observability concerns the estimation of the state from future measurements of the output. For time-invariant systems, the two properties of reconstructibility and observability are identical to one another.

As was the case for controllability, direct application of the definition of observability already stated yields conditions involving the transition matrix.[4] Simpler algebraic conditions are available for time-invariant systems. The observability condition for LTI systems with distinct eigenvalues can be stated very simply if the state equations are transformed to the Jordan canonical form. Such systems are completely observable if each column in the transformed $\mathbf{C}$ matrix has at least one nonzero element.[4] The presence of a column of zeros in this matrix would indicate that the corresponding state variable cannot be estimated from the measured output and input vectors.

More general observability conditions for LTI systems are stated in terms of a matrix $\mathbf{P}_0$ referred to as the observability matrix and are summarized in Table 19.9. The $np \times n$ observability matrix is defined by

$$\mathbf{P}_0 = \begin{bmatrix} \mathbf{C} \\ \mathbf{CA} \\ \vdots \\ \mathbf{CA}^{n-1} \end{bmatrix} \tag{19.65}$$

for continuous-time systems and by

$$\mathbf{P}_0 = \begin{bmatrix} \mathbf{C} \\ \mathbf{CF} \\ \vdots \\ \mathbf{CF}^{n-1} \end{bmatrix} \tag{19.66}$$

for discrete-time systems. The condition for complete observability is simply that the matrix $\mathbf{P}_0$ have rank n. The observable canonical form for SISO systems, described in Section 19.3, derives its name from the fact that transformation to that form is possible if and only if the system is observable. The transformation matrix $\mathbf{T}$ in Tables 19.2 and 19.4, required to transform the state-space equations for a SISO system to the observable canonical form, exists if and only if the $n \times n$ matrix $\mathbf{P}_0$ is nonsingular, that is, the system is observable. Equivalent observability conditions, which are simpler to evaluate than the one stated previously are also listed in Table 19.9 along with a necessary (but not sufficient) condition for complete observability. It should be noted that the observability conditions are independent of time for time-invariant systems. In contrast, the observability conditions for time-varying systems over a specified time interval involve the system transition matrix and hence depend on the time interval.[4] They are also listed in Table 19.9.

Though the definition of observability above involves an open-loop state estimation problem, the property of observability has important implications for closed-loop realizations of the state estimation problem. It will be shown in Section 20.5 that, if a time-invariant system is completely

TABLE 19.9   OBSERVABILITY CONDITIONS FOR LINEAR DYNAMIC SYSTEMS

	Continuous Time	Discrete Time
	Time-Invariant System	
Necessary and sufficient condition for observability	(i) rank$[\mathbf{C}^T \ \mathbf{A}^T\mathbf{C}^T \ \ldots \ (\mathbf{A}^T)^{n-1}\mathbf{C}^T]$ $= \text{rank}[\mathbf{C}^T \ \mathbf{A}^T\mathbf{C}^T \ \ldots \ (\mathbf{A}^T)^{n-p}\mathbf{C}^T]$ $= n$ or (ii) $\det(\mathbf{P}_0^T\mathbf{P}_0) \neq 0$	(i) rank$[\mathbf{C}^T \ \mathbf{F}^T\mathbf{C}^T \ \ldots \ (\mathbf{F}^T)^{n-1}\mathbf{C}^T]$ $= \text{rank}[\mathbf{C}^T \ \mathbf{F}^T\mathbf{C}^T \ \ldots \ (\mathbf{F}^T)^{n-p}\mathbf{C}^T]$ $= n$ or (ii) $\det(\mathbf{P}_0^T\mathbf{P}_0) \neq 0$
Necessary condition	$\text{rank}(\mathbf{C}^T \ \mathbf{A}^T) = n$	$\text{rank}(\mathbf{C}^T \ \mathbf{F}^T) = n$
	Time-Varying Systems:	
Necessary and sufficient condition for observability	Observable at t_0 if and only if there exists a finite time $t_1, t_1 > t_0$ such that (i) $\mathbf{W}_0(t_1, t_0)$ is positive definite or (ii) Zero is not an eigenvalue of $\mathbf{W}_0(t_1, t_0)$ or (iii) $\|\mathbf{W}_0(t_1, t_0)\| \neq 0$ where $\mathbf{W}_0(t_1, t_0)$ $$\triangleq \int_{t_0}^{t_1} \boldsymbol{\phi}^T(\tau, t_0)\mathbf{C}^T(\tau)\mathbf{C}(\tau)$$ $$\times \boldsymbol{\phi}(\tau, t_0)\, d\tau$$	Observable at k_0 if and only if there exists a finite time $k_1, k_1 > k_0$ such that (i) $\mathbf{W}_0(k_1, k_0)$ is positive definite or (ii) Zero is not an eigenvalue of $\mathbf{W}_0(k_1, k_0)$ or (iii) $\|\mathbf{W}_0(k_1, k_0)\| \neq 0$ where $\mathbf{W}_0(k_1, k_0)$ $$\triangleq \sum_{k=k_0}^{k_1} \boldsymbol{\phi}^T(k_1, k_0)\mathbf{C}^T(k)\mathbf{C}(k)\boldsymbol{\phi}(k_1, k)$$

observable, a closed-loop state estimator can be constructed such that the estimation error transients can be made to decay to zero as rapidly as possible.

The conditions for controllability and observability noted in Tables 19.8 and 19.9 have obvious similarities. The two properties can be shown to be duals of each other by formulating the concept of the dual of a dynamic system. Interested readers are referred to Kalman et al.[11,14]

A linear system can, in general, be divided into four subsystems as indicated by Fig. 19.4. The state vector $\mathbf{x}$ can be written as

$$x^T = x_C^T + x_{CO}^T + x_N^T + x_O^T \tag{19.67}$$

where the subscripts have the meaning assigned in the figure. The corresponding state-space equations for a time-invariant, continuous-time system are

$$\begin{bmatrix} \dot{\mathbf{x}}_C \\ \dot{\mathbf{x}}_{CO} \\ \dot{\mathbf{x}}_N \\ \dot{\mathbf{x}}_O \end{bmatrix} = \begin{bmatrix} \mathbf{A}_{11} & \mathbf{A}_{12} & \mathbf{A}_{13} & \mathbf{A}_{14} \\ 0 & \mathbf{A}_{22} & 0 & \mathbf{A}_{24} \\ 0 & 0 & \mathbf{A}_{33} & \mathbf{A}_{34} \\ 0 & 0 & 0 & \mathbf{A}_{44} \end{bmatrix} \begin{bmatrix} \mathbf{x}_C \\ \mathbf{x}_{CO} \\ \mathbf{x}_N \\ \mathbf{x}_O \end{bmatrix} + \begin{bmatrix} \mathbf{B}_{11} \\ \mathbf{B}_{21} \\ 0 \\ 0 \end{bmatrix} \mathbf{u} \tag{19.68}$$

$$\mathbf{y} = [0 \ \mathbf{C}_{12} \ 0 \ \mathbf{C}_{14}] \begin{bmatrix} \mathbf{x}_C \\ \mathbf{x}_{CO} \\ \mathbf{x}_N \\ \mathbf{x}_O \end{bmatrix} + \mathbf{Du}$$

The zero matrices in the $\mathbf{B}$ matrix correspond to the fact that $\mathbf{x}_N$ and $\mathbf{x}_O$ are not controllable. The zero matrices in the $\mathbf{C}$ matrix correspond to the fact that $\mathbf{x}_C$ and $\mathbf{x}_N$ are not observable. If the eigenvalues of the $\mathbf{A}$ matrix are distinct, all off-diagonal elements in the $\mathbf{A}$ matrix would be zero. The procedure for determining the transformation matrix to convert the state-space equations into the canonical form (Eq. 19.68) has been described by Kalman.[9]

The extension of this discussion to time-invariant, discrete-time systems is straightforward. For time-varying systems, the state-space decomposition is a function of time but is similar in structure to that already described.

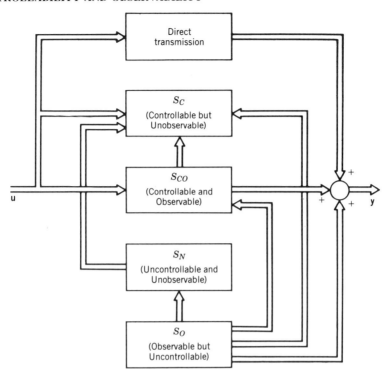

Fig. 19.4 Decomposition of linear system based on controllability and observability.

The significance of the system decomposition as shown in Fig. 19.4 is that it helps relate the state-space description of linear dynamic systems to transfer function or transfer function matrix descriptions of such systems. The transfer function matrix relating **y** to **u** is a description only of the controllable and observable part of the system and masks other modes that are either not observable or not controllable or neither controllable nor observable. The relationship of the state-space description of dynamic systems to the transfer function matrix description of such systems is discussed in greater detail in Section 19.7.

Loss of controllability or observability could occur when controllable and observable subsystems are connected together to form composite systems. Gilbert[15] has formulated rules relating the composite system properties to those of the individual open-loop systems. These rules provide greater insight into the conditions leading to loss of controllability or observability than the simple application of the conditions noted in Tables 19.8 and 19.9.

The concepts of controllability and observability are obviously very important for MIMO systems since the complexity of such systems frequently masks the nature of the coupling of the system state to the inputs and outputs. For SISO systems, lack of complete controllability or observability is a less common occurrence. Conclusions concerning controllability and observability are obvious in many of these cases but less so in others.

Consider the electrical circuit in Fig. 19.5. The state-space equation for the system is given by

$$
\begin{bmatrix} \dot{x}_1 \\ \dot{x}_2 \end{bmatrix} = \begin{bmatrix} -\dfrac{R_1 + R_3}{L_1} & \dfrac{R_3}{L_1} \\ \dfrac{R_3}{L_2} & -\dfrac{R_3 + R_2}{L_2} \end{bmatrix} \begin{bmatrix} x_1 \\ x_2 \end{bmatrix} + \begin{bmatrix} \dfrac{1}{L_1} \\ 0 \end{bmatrix} u(t) \tag{19.69}
$$

The system is completely controllable except for the trivial case where R_3 is zero. Similarly, if x_1 or x_2 is chosen as the only output, the system is completely observable as long as R_3 is nonzero. If the voltage across the resistor R_3 is chosen as the output, the corresponding output equation is

$$
y(t) = [R_3 \quad -R_3] \begin{bmatrix} x_1 \\ x_2 \end{bmatrix} \tag{19.70}
$$

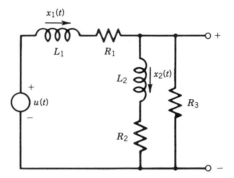

Fig. 19.5 State variables for RL circuit.

The observability matrix $\mathbf{P}_0$ in Eq. 19.65 is then nonsingular and the system is observable if and only if

$$\frac{R_1}{L_1} \neq \frac{R_2}{L_2} \tag{19.71}$$

This observability condition is not an obvious one and is equivalent to the requirement that the time constants associated with the two $R - L$ pairs not be identical. However, it should be noted that, given component tolerances in practice, the inequality (19.71) will be satisfied almost always and the corresponding system will be observable. A discussion on conditions leading to loss of controllability or observability is given by Friedland.[16]

Despite the fact that lack of complete controllability or observability is infrequent for SISO systems, these concepts have practical significance for SISO as well as MIMO systems because of the relationship of these concepts to closed-loop control and state estimation problems. Measures of the degree of controllability and observability can be defined for time-invariant and time-varying systems.

The controllability and observability conditions in Tables 19.8 and 19.9 relate these properties to the nonsingularity of square matrices for SISO systems. Measures of the degree of controllability and observability are related to the closeness of these matrices to the singularity condition. Such measures have been defined for time-invariant systems by Johnson[17] and Friedland[18] and are significant for SISO as well as MIMO systems. A system with a better degree of controllability can in general be controlled more effectively. Similarly, a better degree of observability implies that state estimation can be performed more accurately. The proposed measures of the degree of controllability and observability are not in common use but have the potential to quantitatively evaluate proposed control strategies and measurement schemes.[19]

Additional concepts of degrees of controllability and observability for time-varying systems have been described by Silverman and Meadows[20] and for MIMO systems by Kreindler and Sarachik.[13] Properties weaker than state controllability and observability have also been defined[12] and are useful in ensuring that closed-loop control and state estimation problems are well-posed. A linear system is said to be stabilizable if the uncontrollable subsystems S_N and S_O in the decomposition of Fig. 19.4 are stable. Similarly, if the unobservable subsystems S_C and S_N are stable, the system is said to be detectable.[12]

19.7 RELATIONSHIP BETWEEN STATE-SPACE AND TRANSFER FUNCTION DESCRIPTIONS

The state-space representation of dynamic systems is an accurate representation of the internal structure of a system and its coupling to the system inputs and outputs. For LTI systems, transfer functions (for SISO systems) or transfer function matrices (for MIMO systems) are useful in practice since the dimensions of these matrices are invariably smaller than the dimensions of the corresponding system matrices $\mathbf{A}$ or $\mathbf{F}$ in Eqs. 19.6 and 19.12. Analysis and design procedures based on the transfer function matrix descriptions are therefore simpler. The relationship between these two alternative descriptions of LTI systems is described in this section.

Determination of the transfer function matrix from the state-space equations is straightforward. For continuous-time systems, Laplace transformation of Eqs. 19.6 and 19.7 with zero initial conditions $\mathbf{x}(0)$ and elimination of $\mathbf{X}(s)$ yields

$$\mathbf{H}(s) = \frac{\mathbf{Y}(s)}{\mathbf{U}(s)} = \mathbf{C}(s\mathbf{I}_n - \mathbf{A})^{-1}\mathbf{B} + \mathbf{D} \qquad (19.72)$$

where $\mathbf{I}_n$ is the $n \times n$ identity matrix. For discrete-time systems, a similar procedure using z transforms and applied to Eqs. 19.12 and 19.13 yields the pulse transfer function matrix

$$\mathbf{H}(z) = \frac{\mathbf{Y}(z)}{\mathbf{U}(z)} = \mathbf{C}(z\mathbf{I}_n - \mathbf{F})^{-1}\mathbf{G} + \mathbf{D} \qquad (19.73)$$

The transfer function matrix corresponding to a given state-space description is therefore unique. However, as indicated in the previous section, the former represents only the controllable and observable part of a system. Unless the entire system is completely controllable and observable, a transfer function matrix description is not a complete characterization of the system dynamic behavior. It can be shown that, for SISO systems, a necessary and sufficient condition for controllability and observability of the system is that there are no pole-zero cancellations between the numerator and denominator of the transfer function matricies in Eqs. 19.72 and 19.73. For MIMO systems, this is only a sufficient condition and not a necessary one.[4]

The determination of the state-space description corresponding to a given transfer function matrix description is more complex and is referred to as the problem of realization. Since the transfer function matrix represents only the controllable and observable part of a system, the problem of realization does not have a unique solution. In fact, the transfer function matrix description does not even determine the dimension of the corresponding state vector uniquely. The minimal dimension of the state vector corresponding to a given transfer function matrix is, however, uniquely determined. The associated state-space equations are said to constitute the minimal or irreducible realization of the transfer function matrix. It can be shown that a realization is minimal if and only if it is both controllable and observable.[4] Minimal realizations are not unique. However, any two different minimal realizations of a given transfer function matrix are equivalent in that the corresponding state vectors are related by a nonsingular transformation matrix.

The canonical forms of the state-space equations for SISO systems in Tables 19.1 and 19.3 represent minimal realizations if there is no pole-zero cancellation. Techniques for obtaining minimal realizations for MIMO systems are more involved. Brogan[4] has described a procedure for obtaining a Jordan form realization for a given transfer function matrix. When applied to cases where elements of the transfer function matrix have only simple poles, the resulting realization is controllable and observable as shown in the following example. If one or more elements of the transfer function matrix have repeated poles, the realization that results is controllable but may or may not be observable.

Consider a continuous-time system with two inputs and two outputs and the following transfer function matrix:

$$\mathbf{H}(s) = \begin{bmatrix} \dfrac{1}{s+1} & \dfrac{s}{(s+1)(s+3)} \\ \dfrac{1}{s+3} & \dfrac{1}{s+1} \end{bmatrix} \qquad (19.74)$$

Expand $\mathbf{H}(s)$ using a matrix version of partial fraction expansion as

$$\mathbf{H}(s) = \frac{\begin{bmatrix} 1 & -\dfrac{1}{2} \\ 0 & 1 \end{bmatrix}}{s+1} + \frac{\begin{bmatrix} 0 & \dfrac{3}{2} \\ 1 & 0 \end{bmatrix}}{s+3}$$

$$= \frac{\begin{bmatrix} 1 \\ 0 \end{bmatrix}\begin{bmatrix} 1 & -\dfrac{1}{2} \end{bmatrix} + \begin{bmatrix} 0 \\ 1 \end{bmatrix}\begin{bmatrix} 0 & 1 \end{bmatrix}}{s+1} + \frac{\begin{bmatrix} 1 \\ 0 \end{bmatrix}\begin{bmatrix} 0 & \dfrac{3}{2} \end{bmatrix} + \begin{bmatrix} 0 \\ 1 \end{bmatrix}\begin{bmatrix} 1 & 0 \end{bmatrix}}{s+3} \qquad (19.75)$$

It should be noted that the number of vector products each coefficient matrix is factored into is equal to the rank of the matrix. $\mathbf{H}(s)$ is then written in a form that indicates the matrices $\mathbf{A}$, $\mathbf{B}$, $\mathbf{C}$ clearly, by comparison with $\mathbf{C}(s\mathbf{I} - \mathbf{A})^{-1}\mathbf{B}$:

$$\mathbf{H}(s) = \begin{bmatrix} 1 & 0 & 1 & 0 \\ 0 & 1 & 0 & 1 \end{bmatrix} \begin{bmatrix} \dfrac{1}{s+1} & & & 0 \\ & \dfrac{1}{s+1} & & \\ & 0 & \dfrac{1}{s+3} & \\ & & & \dfrac{1}{s+3} \end{bmatrix} \begin{bmatrix} 1 & -\frac{1}{2} \\ 0 & 1 \\ 0 & \frac{3}{2} \\ 1 & 0 \end{bmatrix}$$

(19.76)

$$= \begin{bmatrix} 1 & 0 & 1 & 0 \\ 0 & 1 & 0 & 1 \end{bmatrix} \begin{bmatrix} s+1 & & & \\ & s+1 & 0 & \\ 0 & & s+3 & \\ & & & s+3 \end{bmatrix}^{-1} \begin{bmatrix} 1 & -\frac{1}{2} \\ 0 & 1 \\ 0 & \frac{3}{2} \\ 1 & 0 \end{bmatrix}$$

Thus, the corresponding realization is

$$\mathbf{A} = \begin{bmatrix} -1 & & & \\ & -1 & 0 & \\ & 0 & -3 & \\ & & & -3 \end{bmatrix} \qquad \mathbf{B} = \begin{bmatrix} 1 & -\frac{1}{2} \\ 0 & 1 \\ 0 & \frac{3}{2} \\ 1 & 0 \end{bmatrix}$$

(19.77)

$$\mathbf{C} = \begin{bmatrix} 1 & 0 & 1 & 0 \\ 0 & 1 & 0 & 1 \end{bmatrix}$$

The realization is controllable and observable and hence is minimal. Modifications of this procedure for cases where $\mathbf{H}(s)$ has elements with repeated poles are described by Brogan.[4] Extensions to discrete-time systems are straightforward.

An alternative two-step procedure for determining a minimal realization for a transfer function matrix involves obtaining a nonminimal realization by any one method, as the first step. For example, one of the many realizations in Table 19.1 (Table 19.3 for discrete-time systems) can be chosen to represent each of the elements of the transfer function matrix. The state-space descriptions of the elements can then be combined to get the state-space equations for the MIMO system. The resulting realization would, in general, be nonminimal. The second step requires transformation of the state-space equations to the form given by Eq. 19.68 or an equivalent one for discrete-time systems. Techniques for selecting the transformation matrix are described by Kalman[9] and Fortmann and Hitz.[1] The minimal realization is then given by the controllable and observable subsystem in Fig. 19.4. The resulting equations for a continuous-time system are

$$\dot{\mathbf{x}}_m = \mathbf{A}_{22}\mathbf{x}_m + \mathbf{B}_{21}\mathbf{u} \qquad (19.78)$$

$$\mathbf{y}_m = \mathbf{C}_{12}\mathbf{x}_m + \mathbf{D}\mathbf{u} \qquad (19.79)$$

where the subscript m indicates a minimal realization. Similar results for discrete-time systems are given by Brogan,[4] Kuo,[7] and Kalman.[9]

19.8 CONCLUSION

The state-space methods presented in this chapter offer a unifying framework for the dynamic analysis and control of a variety of systems. The primary emphasis, in these methods, on linear time-invariant systems is a reflection of the state of the literature on the subject and the practice of the art. Results for linear time-varying systems have been given in some of the standard texts[2-4] referred to. The application of state-space methods to nonlinear system analysis and control is treated at some length by Hedrick and Paynter.[21]

Distributed parameter systems are examples of systems with infinite-dimensional states. Application of state-space methods to these systems has been described by Tzafestas et al.[22] Time-delayed systems are also examples of systems with infinite-dimensional states. The analysis and control of such systems and of many of the other types of systems referred to in this section remains a subject of current research. For current research results in these areas, the reader is referred to journals such as the *ASME Journal of Dynamic Systems, Measurements and Controls*; *IEEE Transactions on Automatic Control*; *AIAA Journal of Guidance, Control and Dynamics*; *SIAM Journal on Control*; and *Automatica, the Journal of the International Federation of Automatic Control*.

REFERENCES

1. T. E. Fortmann and K. L. Hitz, *An Introduction to Linear Control Systems*, Marcel Dekker, New York, 1977.

2. P. M. DeRusso, R. J. Roy and C. M. Close, *State Variables for Engineers*, Wiley, New York, 1965.

3. T. Kailath, *Linear Systems*, Prentice Hall, Englewood Cliffs, NJ, 1980.

4. W. L. Brogan, *Modern Control Theory*, Prentice Hall, Englewood Cliffs, NJ, 1982.

5. J. C. Doyle and G. Stein, "Multivariable Feedback Design: Concepts for a Classical/Modern Synthesis," IEEE Transactions on Automatic Control, Vol. AC-26, No. 1, pp. 4-16, Feb. 1981.

6. William J. Palm III, *Modeling, Analysis and Control of Dynamic Systems*, Wiley, New York, 1983.

7. B. C. Kuo, *Digital Control Systems*, SRL Publishing, Champaign, IL, 1977.

8. R. E. Kalman and J. E. Bertram, "Control-System Analysis and Design Via the Second Method of Lyapunov. I—Continuous Time Systems. II—Discrete-Time Systems," *Transactions of the ASME, Journal of Basic Engineering*, Vol. 82D, pp. 371–400, June 1960.

9. R. E. Kalman, "Mathematical Description of Linear Dynamical Systems," *SIAM Journal on Control*, Series A, Vol. 1, No. 2, pp. 153–192, 1963.

10. R. E. Kalman, "When Is a Linear Control System Optimal?" *Transactions of the ASME, Journal of Basic Engineering*, Vol. 86D, pp. 51–60, 1964.

11. R. E. Kalman, "On the General Theory of Control Systems," *Proceedings of the First International Congress on Automatic Control*, pp. 481–493, Butterworth's, London, 1960.

12. H. Kwakernaak and R. Sivan, *Linear Optimal Control Systems*, Wiley, New York, 1972.

13. E. Kreindler and P. Sarachik, "On the Concepts of Controllability and Observability of Linear Systems," *IEEE Transactions on Automatic Control*, Vol. AC-9, No. 1, pp. 129–136, Feb. 1964.

14. R. E. Kalman, P. L. Falb, and M. Arbib, *Topics in Mathematical System Theory*, McGraw-Hill, New York, 1969.

15. E. G. Gilbert, "Controllability and Observability in Multi-variable Control Systems," *SIAM Journal on Control*, Series A, Vol. 2, No. 1, pp. 128–151, 1963.

16. B. Friedland, *Control System Design, An Introduction to State-Space Methods*, McGraw-Hill, New York, 1986.

17. C. D. Johnson, "Optimization of a Certain Quality of Complete Controllability and Observability for Linear Dynamical Systems," *Transactions of the ASME, Journal of Basic Engineering*, Vol. 91D, pp. 228–238, 1969.

18. B. Friedland, "Controllability Index Based on Conditioning Number," *ASME Transactions, Journal of Dynamic Systems, Measurement and Control*, Vol. 97, pp. 444–445, December 1975.

19. P. C. Muller and H. I. Weber, "Analysis and Optimization of Certain Qualities of Controllability and Observability for Linear Dynamical Systems," *Automatica*, Vol. 8, pp. 237–246, 1972.

20. L. M. Silverman and H. E. Meadows, "Controllability and Observability in Time-Variable Linear Systems," *SIAM Journal on Control*, Vol. 5, No. 1, pp. 64–73, 1967.

21. J. K. Hedrick and H. M. Paynter, Eds., *Nonlinear System Analysis and Synthesis: Volume 1— Fundamental Principles*, Workshop/Tutorial Session at the Winter Annual Meeting of ASME, New York, December, 1976.

22. S. G. Tzafestas, Ed., *Distributed Parameter Control Systems, Theory and Application*, Vol. 6, International Series on Systems and Control, Pergamon Press, Oxford, England, 1982.

CHAPTER 20

CONTROL SYSTEM DESIGN USING STATE-SPACE METHODS

KRISHNASWAMY SRINIVASAN

Department of Mechanical Engineering
The Ohio State University
Columbus, Ohio

20.1 INTRODUCTION

The advantages of feedback control in achieving desired input/output relationships are well known. Control system theory based on a frequency domain approach[1] illustrates clearly that the following aspects of SISO system performance can be improved by feedback: (1) the ability to follow reference inputs accurately in the steady state or under transient conditions and (2) the ability to reject disturbance inputs and reduce sensitivity of the overall controlled system behavior to plant parameter variations and modeling errors. For MIMO systems, the coupling between individual inputs and outputs can be modified in a desired manner, in addition to the performance features already mentioned, by appropriate control system design.[2]

State-space methods for control system design result in solutions that utilize the state of the system most effectively for feedback. The resulting state-variable feedback control systems improve the same aspects of system performance as previously mentioned. However, the available state-space design procedures accommodate some performance specifications more readily than others. For instance, performance specifications in the form of desired closed-loop pole locations are readily accommodated. Similarly, performance specifications in the form of an index of performance to be optimized can be accommodated by optimal control theory if the index of performance belongs to a restricted class of performance measures. In fact, recent efforts in control system design using state-space methods have been directed at enhancing the problem formulation to accommodate a greater variety of performance specifications. In spite of these enhancements, performance specifications such as sensitivity of the controlled system performance to plant parameter variations and modeling errors are accommodated more readily by frequency-domain-based design procedures than by state-space or time-domain-based design procedures. Thus, control system design techniques based on frequency domain and time domain approaches should be viewed as being complementary to each other in some ways.

20.2 THE POLE PLACEMENT DESIGN METHOD

20.2.1 Regulation Problem

It can be shown that, if a linear time-invariant system is completely state controllable and if linear instantaneous state-variable feedback is used, the associated feedback gains can be chosen to place the closed-loop poles of the controlled system at any arbitrarily specified locations in the s or z

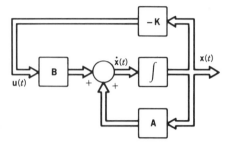

Fig. 20.1 Linear state-variable feedback for continuous-time system.

plane,[3] depending on whether the system is continuous time or discrete time. Thus, if the continuous-time and discrete-time systems described by Eqs. 19.6 and 19.12, respectively, are completely state controllable and the control law is given by (Figs. 20.1 and 20.2)

$$\mathbf{u} = -\mathbf{Kx} \tag{20.1}$$

then the eigenvalues of the matrices $\mathbf{A} - \mathbf{BK}$ and $\mathbf{F} - \mathbf{GK}$ are the closed-loop pole locations and can be assigned any specified locations in the complex plane by appropriate selection of the gain matrix $\mathbf{K}$. If $\mathbf{K}$ is constrained to be a real matrix, the desired eigenvalues should be specified either as real or as complex conjugate pairs. The resulting design procedure is referred to as the pole placement method and is useful for regulation problems where the objective of the controller is to return the system to equilibrium conditions following an initial disturbance. Specification of the closed-loop poles is equivalent to specification of the damping and speed of response of the closed-loop system transients as the system returns to equilibrium.

For single-input systems, specification of the desired closed-loop pole locations uniquely specifies the gain vector $\mathbf{K}$. A formula for the gain vector $\mathbf{K}$, convenient to evaluate and applicable to both continuous-time and discrete-time systems, is

$$\mathbf{K} = (0 \ldots 0\ 1)(\mathbf{B}\ \mathbf{AB} \ldots \mathbf{A}^{n-1}\mathbf{B})^{-1}\alpha_c(\mathbf{A}) \tag{20.2}$$

for continuous-time systems and

$$\mathbf{K} = (0 \ldots 0\ 1)(\mathbf{G}\ \mathbf{FG} \ldots \mathbf{F}^{n-1}\mathbf{G})^{-1}\alpha_c(\mathbf{F}) \tag{20.3}$$

for discrete-time systems. In these equations,

$$\alpha_c(\mathbf{A}) = \mathbf{A}^n + \sum_{i=0}^{n-1} \alpha_i \mathbf{A}^i \tag{20.4}$$

for continuous-time systems and $\alpha_c(\mathbf{F})$ is a similar function of $\mathbf{F}$ for discrete-time systems. The α_i's are the coefficients of the desired characteristic equations of the closed-loop systems. For continuous-time systems we have

$$\alpha_c(s) = \det(s\mathbf{I} - \mathbf{A} + \mathbf{BK})$$

$$= s^n + \sum_{i=0}^{n-1} \alpha_i s^i = 0 \tag{20.5}$$

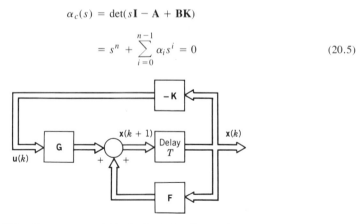

Fig. 20.2 Linear state-variable feedback for discrete-time system.

A similar equation describes the discrete-time system characteristic equation. Computer-aided control system design (CACSD) packages supporting state-space methods usually support pole placement designs[4−6] and require only that the designer input information about the system matrices and the desired closed-loop pole locations. The gain vector $\mathbf{K}$ is then computed and output to the designer.

For multi-input systems, specification of the closed-loop poles does not specify the gain matrix $\mathbf{K}$ uniquely. The additional freedom in the gain matrix selection can be used to assign eigenvectors (or generalized eigenvectors) or individual transfer function zeros to improve the transient response to nonzero reference inputs.[7] Alternative criteria for gain matrix selection are optimization of feedback gain magnitudes and stability of the closed-loop system in the absence or failure of some of the inputs. Brogan[8] has outlined a procedure for gain matrix selection for multi-input systems, based on closed-loop eigenvector specification in addition to eigenvalue specification. For continuous-time systems described by Eqs. 19.6 and 20.1, the feedback gain matrix is given by

$$\mathbf{K} = -(\mathbf{e}_{j_1} \ \mathbf{e}_{j_2} \ldots \ \mathbf{e}_{j_n})[\ \boldsymbol{\psi}_{j_1}(s_1) \ \boldsymbol{\psi}_{j_2}(s_2) \ldots \ \boldsymbol{\psi}_{j_n}(s_n)]^{-1} \qquad (20.6)$$

where the desired closed-loop eigenvalues and the corresponding eigenvectors are s_i and $\boldsymbol{\psi}_{j_i}(s_i)$, $i = 1, \ldots, n$, respectively. The eigenvectors are chosen to be n linearly independent columns from the $n \times nr$ matrix $[\boldsymbol{\psi}(s_1) \ \boldsymbol{\psi}(s_2) \ \ldots \ \boldsymbol{\psi}(s_n)]$ where

$$\boldsymbol{\psi}(s) = (s\mathbf{I}_n - \mathbf{A})^{-1}\mathbf{B} \qquad (20.7)$$

If the desired s_i are distinct, it will always be possible to find n linearly independent columns as already described. Here $\mathbf{e}_{j_i}$ is defined as the j_ith column of the $r \times r$ identity matrix $\mathbf{I}_r$ and is uniquely determined once j_i is determined. When repeated eigenvalues are desired, the procedure for specifying n linearly independent generalized eigenvectors is different and has been described by Brogan.[8] The results given here are readily applicable to multi-input discrete-time systems described by Eqs. 19.12 and 20.1. The specified eigenvalues are z_i instead of s_i and the $\mathbf{A}$ and $\mathbf{B}$ matrices are replaced by $\mathbf{F}$ and $\mathbf{G}$, respectively. Also, s is replaced by z in Eq. 20.7. Alternative methods for gain matrix selection for multi-input systems have been described by Kailath.[7]

If a single-input, LTV system is completely state controllable, linear state-variable feedback can be used to ensure that the closed-loop transition matrix corresponds to that of any desired nth-order linear differential equation with time-varying coefficients. The state-variable feedback gains are time varying in general and can be computed using a procedure described by Wiberg.[9]

If the complete state is not available for feedback, linear instantaneous feedback of the measured output can be used to place some of the closed-loop poles at specified locations in the complex plane. If the continuous-time and discrete-time systems described by Eqs. 19.6, 19.7, 19.12, and 19.13, respectively, satisfy the output controllability conditions listed in Table 19.8, then p of the n eigenvalues of the closed-loop system can approach arbitrarily specified values to within any degree of accuracy but not always exactly. The control law is

$$\mathbf{u} = -\mathbf{K}\mathbf{y} \qquad (20.8)$$

where $\mathbf{K}$ is a $r \times p$ gain matrix. Brogan[8] has described an algorithm for computing $\mathbf{K}$, given the desired values of p closed-loop eigenvalues. The corresponding characteristic equation is

$$\det[s\mathbf{I}_n - \mathbf{A} + \mathbf{BK}(\mathbf{I}_p + \mathbf{DK})^{-1}\mathbf{C}] = 0 \qquad (20.9)$$

for continuous-time systems and

$$\det[z\mathbf{I}_n - \mathbf{F} + \mathbf{GK}(\mathbf{I}_p + \mathbf{DK})^{-1}\mathbf{C}] = 0 \qquad (20.10)$$

for discrete-time systems.

An alternative approach to control system design in the case of incomplete state measurement is to use an observer or a Kalman filter for state estimation. The estimated state is then used for feedback. This procedure is discussed in Section 20.5.

The advantages of the pole placement design method already described are that the controller achieves desired closed-loop pole locations without using pole-zero cancellation and without increasing the order of the system. The desired pole locations can be chosen to ensure a desired degree of stability or damping and speed of response of the closed-loop system. However, there is no convenient way to ensure a priori that the closed-loop system satisfies other important performance specifications such as a desired level of insensitivity to plant parameter variations, acceptable disturbance rejection, and compatibility of control effort with actuator limitations. In addition, for single-input LTI systems, instantaneous state feedback of the form given by Eq. 20.1 does not affect the locations of zeros of the transfer functions between the system input and system outputs.[3] Thus, the pole placement design method does not afford complete control over the system response to the reference input or disturbance inputs. For multi-input systems, the available freedom in the gain matrix selection can be used to assign individual transfer function zeros, in addition to achieving desired closed-loop pole locations. However, systematic procedures to do this are not available. The consequence of these limitations of the pole placement method is that the design process involves considerable trial and error.

20.2.2 Modification for Constant Reference and Disturbance Inputs

The pole placement method described is appropriate for regulation problems. For the case of nonzero reference inputs that may be constant or varying with time, the system outputs are required to follow the reference inputs. The control law, Eq. 20.1, needs to be modified for such problems. If the output vector $\mathbf{y}$ and the input vector $\mathbf{u}$ have the same dimension and if the $\mathbf{D}$ matrix is zero in Eqs. 19.7 and 19.13, the modified control law has the form

$$\mathbf{u} = -\mathbf{Kx} + \mathbf{Ny}_d \qquad (20.11)$$

where $\mathbf{y}_d$ is a vector of reference inputs. For constant reference inputs $\mathbf{y}_d$, the error $\mathbf{y}_d - \mathbf{y}$ can be reduced to zero under steady-state conditions by selecting

$$\mathbf{N} = [\mathbf{C}(-\mathbf{A} + \mathbf{BK})^{-1}\mathbf{B}]^{-1} \qquad (20.12)$$

for continuous-time systems and

$$\mathbf{N} = [\mathbf{C}(\mathbf{I}_n - \mathbf{F} + \mathbf{GK})^{-1}\mathbf{G}]^{-1} \qquad (20.13)$$

for discrete-time systems. The matrices to be inverted on the right-hand sides of the preceding equations exist if and only if the corresponding open-loop transfer matrices $[\mathbf{C}(s\mathbf{I}_n - \mathbf{A})^{-1}\mathbf{B}]$ and $[\mathbf{C}(z\mathbf{I}_n - \mathbf{F})^{-1}\mathbf{G}]$ have no zeros at the origin and at $z = 1$, respectively.[3] The $\mathbf{K}$ matrix in Eq. 20.11 is chosen to give the desired closed-loop poles as before. It should be noted that the gain matrix $\mathbf{N}$ is outside the feedback loop. Hence, the controlled system performance, particularly the steady-state error, would be sensitive to modeling error or error in the elements of the system matrices.

It is well known from classical control theory that integral controller action on the error has the effect of reducing the steady-state error to reference and disturbance inputs. In particular, the steady-state error is reduced to zero for constant reference and disturbance inputs. A similar result can be obtained within the framework of state-variable feedback[3] and will be described for the case where the $\mathbf{y}$ and $\mathbf{u}$ vectors have the same dimension and the $\mathbf{D}$ matrix is zero in Eqs. 19.7 and 19.13.

Consider the case of constant but unknown disturbance inputs:

$$\dot{\mathbf{x}}(t) = \mathbf{Ax}(t) + \mathbf{Bu}(t) + \mathbf{w}(t) \qquad (20.14)$$

$$\mathbf{y}(t) = \mathbf{Cx}(t) \qquad (20.15)$$

for continuous-time systems and

$$\mathbf{x}(k + 1) = \mathbf{Fx}(k) + \mathbf{Gu}(k) + \mathbf{w}(k) \qquad (20.16)$$

$$\mathbf{y}(k) = \mathbf{Cx}(k) \qquad (20.17)$$

for discrete-time systems. The state-space equations are augmented by

$$\dot{\mathbf{q}}_e(t) = \mathbf{y}(t) = \mathbf{Cx}(t) \qquad (20.18)$$

for continuous-time systems and

$$\mathbf{q}_e(k + 1) = \mathbf{q}_e(k) + \mathbf{y}(k)$$
$$= \mathbf{q}_e(k) + \mathbf{Cx}(k) \qquad (20.19)$$

for discrete-time systems. The control law is modified to include feedback of the additional states (Figs. 20.3 and 20.4).

$$\mathbf{u} = -\mathbf{Kx} - \mathbf{K}_q\mathbf{q}_e \qquad (20.20)$$

If the feedback gains $\mathbf{K}$, $\mathbf{K}_q$ are chosen to ensure asymptotic stability of the resulting closed-loop systems, then

$$\lim_{t \to \infty} \dot{\mathbf{q}}_e(t) = 0 \qquad (20.21)$$

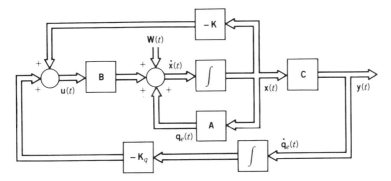

Fig. 20.3 Proportional plus integral control, continuous-time system.

for continuous-time systems and

$$\lim_{k \to \infty} \mathbf{q}_e(k) = \text{const.} \tag{20.22}$$

for discrete-time systems, regardless of the value of the disturbance input. When combined with Eqs. 20.19 and 20.20, the preceding equations indicate that the output $\mathbf{y}$ and hence the error goes to zero in the steady state for continuous-time and discrete-time systems, respectively. The necessary and sufficient conditions for the existence of an aymptotically stable control law of the form of Eq. 20.20 are that the continuous-time systems, Eq. 20.14, and discrete-time systems, Eq. 20.16, be stabilizable and that the corresponding open-loop transfer matrices $[\mathbf{C}(s\mathbf{I}_n - \mathbf{A})^{-1}\mathbf{B}]$ and $[\mathbf{C}(z\mathbf{I}_n - \mathbf{F})^{-1}\mathbf{G}]$ have no zeros at the origin and at $z = 1$, respectively. It should also be clear that if a constant reference input $\mathbf{y}_d$ is to be included in the problem, the error is $\mathbf{y} - \mathbf{y}_d$ and should be used instead of $\mathbf{y}$ in Eqs. 20.18 and 20.19. In this case, the error will go to zero in the steady state while the ouput $\mathbf{y}$ reaches $\mathbf{y}_d$. The advantage of the integral control action is that it reduces the steady-state error to zero without requiring knowledge of the constant disturbance input or accurate values of the system parameters as in Eqs. 20.11–20.13. The disadvantage is that it increases the order of the system and, in practice, would degrade system stability or speed of response.

20.3 THE STANDARD LINEAR QUADRATIC REGULATOR PROBLEM

Controller design in regulation applications using pole placement specifications emphasizes only the transient behavior of the state variables as the system returns to equilibrium. There is no explicit consideration of the required control effort. Control effort can be considered if the controller design problem is formulated as an optimal control problem with weighting of both control effort and state-variable transients. For regulation applications, the index of performance to be optimized, J, is

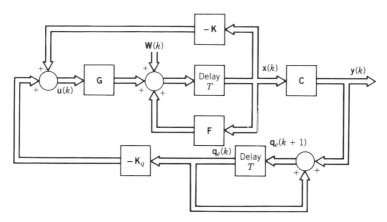

Fig. 20.4 Proportional plus integral control, discrete-time system.

usually chosen to be a quadratic function of the control inputs and the state variables. The resulting optimal control law for the control input, when expressed in feedback form, is a linear function of the system state. Hence, this approach to control system design is referred to as the linear quadratic regulator (LQR) problem.

20.3.1 The Continuous-Time LQR Problem

Consider the continuous-time, LTV system described by Eq. 19.4 and the initial condition, for regulation problems, of

$$\mathbf{x}(t_0) = \mathbf{x}_0 \tag{20.23}$$

The controller design problem is to choose the control input $\mathbf{u}(t)$ to minimize the quadratic index of performance:[3]

$$J = \int_{t_0}^{t_1} \left[\mathbf{x}^T(t)\mathbf{R}_1(t)\mathbf{x}(t) + \mathbf{u}^T(t)\mathbf{R}_2(t)\mathbf{u}(t) \right] dt + \mathbf{x}^T(t_1)\mathbf{P}_f\mathbf{x}(t_1) \tag{20.24}$$

$\mathbf{R}_1(t)$ is a positive semidefinite symmetric weighting matrix on the state variables, $\mathbf{R}_2(t)$ is a positive definite symmetric weighting matrix on the control inputs, and $\mathbf{P}_f$ is a positive semidefinite symmetric weighting matrix on the terminal state. Times t_0, t_1, are initial and terminal time instants. The problem, as previously formulated, is a finite time, deterministic, LQR problem. Kwakernaak and Sivan[3] have also considered a more general J that includes an additional term of the form $\mathbf{x}^T(t)\mathbf{R}_{12}(t)\mathbf{u}(t)$ within the integral. They have shown that this J can be reduced to the form of Eq. 20.24 by appropriate redefinition of the weighting matrices and control vector.

The solution of this control problem, using methods from calculus of variations, can be obtained from standard textbooks on optimal control,[3,12] along with conditions for its existence and uniqueness. The optimal control law, given in feedback form, is a linear, time-varying function of the system state:

$$\mathbf{u}(t) = -\mathbf{R}_2^{-1}(t)\mathbf{B}^T(t)\mathbf{P}(t)\mathbf{x}(t) \tag{20.25}$$

where $\mathbf{P}(t)$ is a $n \times n$ symmetric positive semidefinite matrix satisfying the matrix Riccati equation:

$$-\dot{\mathbf{P}}(t) = \mathbf{R}_1(t) - \mathbf{P}(t)\mathbf{B}(t)\mathbf{R}_2^{-1}(t)\mathbf{B}^T(t)\mathbf{P}(t) + \mathbf{P}(t)\mathbf{A}(t) + \mathbf{A}^T(t)\mathbf{P}(t) \tag{20.26}$$

and the terminal condition

$$\mathbf{P}(t_1) = \mathbf{P}_f \tag{20.27}$$

Numerical solution of the matrix Riccati equation is a subject of great importance and of considerable research. Some useful techniques have been briefly described by Kwakernaak and Sivan.[3]

Solution of the LQR problem simplifies as the terminal time t_1 approaches infinity. It can be shown then that the solution of the matrix Riccati equation approaches a steady-state solution $\mathbf{P}_s(t)$ that is independent of $\mathbf{P}_f$. The resulting steady-state control law

$$\mathbf{u}(t) = -\mathbf{R}_2^{-1}\mathbf{B}^T(t)\mathbf{P}_s(t)\mathbf{x}(t) \tag{20.28}$$

results in an exponentially stable closed-loop system if:

1. The linear system of Eq. 19.4 is uniformly completely state controllable.
2. The pair $\mathbf{A}(t)$, $\mathbf{H}_r^T(t)$ is uniformly completely reconstructible where $\mathbf{H}_r(t)$ is any matrix such that $\mathbf{H}_r(t)\mathbf{H}_r^T(t)$ equals $\mathbf{R}_1(t)$.

The matrix Riccati equation for the steady-state LQR problem simplifies to an algebraic equation and $\mathbf{P}_s$ is a constant if the system matrices and the weighting matrices in the index of performance are constant. The resulting algebraic Riccati equation is

$$\mathbf{R}_1 - \mathbf{P}_s\mathbf{B}\mathbf{R}_2^{-1}\mathbf{B}^T\mathbf{P}_s + \mathbf{A}^T\mathbf{P}_s + \mathbf{P}_s\mathbf{A} = 0 \tag{20.29}$$

where $\mathbf{P}_s$ is a unique positive definite solution of Eq. 20.29 and the resulting time-invariant closed-loop system is asymptotically stable if:

1. The linear system of Eq. 19.6 is completely state controllable.
2. The pair $\mathbf{A}$, $\mathbf{H}_r^T$ is completely observable (reconstructible), where $\mathbf{H}_r$ is any matrix such that $\mathbf{H}_r \mathbf{H}_r^T$ equals $\mathbf{R}_1$.

Another version of the LQR problem involves minimization of the quadratic index of performance for an LTI system over a finite time interval. If the weighting matrices are also time invariant, in many cases the optimal feedback gains are constant over most of the time interval of interest and vary with time only near the terminal time. Since constant feedback gains are easier to implement in practice, implementation of constant gains over the entire time interval would represent a nearly optimal solution that is practically more convenient.[3]

20.3.2 The Discrete-Time LQR Problem

The results of the LQR problem for discrete-time systems parallel those for continuous-time systems already stated. They are summarized here and described in greater length by Kwakernaak and Sivan.[3] The time-varying, discrete-time system is described by Eq. 19.10 and the initial condition

$$\mathbf{x}(k_0) = \mathbf{x}_0 \qquad (20.30)$$

The index of performance to be minimized by controller design, for the finite-time LQR problem, is

$$J = \sum_{k=k_0}^{k_1-1} [\mathbf{x}^{T-1}(k+1)\mathbf{R}_1(k+1)\mathbf{x}(k+1) + \mathbf{u}^T(k)\mathbf{R}_2(k)\mathbf{u}(k)] + \mathbf{x}^T(k_1)\mathbf{P}_f\mathbf{x}(k_1) \qquad (20.31)$$

where the weighting matrices $\mathbf{R}_1(k)$, $\mathbf{R}_2(k)$, and $\mathbf{P}_f$ serve the same functions as $\mathbf{R}_1(t)$, $\mathbf{R}_2(t)$, and $\mathbf{P}_f$ did for continuous-time systems and satisfy the same conditions. The values k_0, k_1 are the initial and final time instants. A more general version of the J, including the term $\mathbf{x}^T(k)\mathbf{R}_{12}(k)\mathbf{u}(k)$ within the summation sign, can be reduced to the form of Eq. 20.31 by appropriate redefinition of the weighting matrices and control vector.[3]

The solution of this control problem can be obtained using dynamic programming methods and is given by

$$\mathbf{u}(k) = -\mathbf{K}(k)\mathbf{x}(k) \qquad (20.32)$$

where

$$\mathbf{K}(k) = \{\mathbf{R}_2(k) + \mathbf{G}^T(k)[\mathbf{R}_1(k+1) + \mathbf{P}(k+1)]\mathbf{G}(k)\}^{-1} \times \mathbf{G}^T(k)[\mathbf{R}_1(k+1) + \mathbf{P}(k+1)]\mathbf{F}(k)$$

$$(20.33)$$

$\mathbf{P}(k)$ is a $n \times n$ symmetric, positive semidefinite matrix satisfying the matrix difference equation

$$\mathbf{P}(k) = \mathbf{F}^T(k)[\mathbf{R}_1(k+1) + \mathbf{P}(k+1)][\mathbf{F}(k) - \mathbf{G}(k)\mathbf{K}(k)] \quad k = k_0, k_1 - 1 \qquad (20.34)$$

with the terminal condition

$$\mathbf{P}(k_1) = \mathbf{P}_f \qquad (20.35)$$

Unlike the matrix Riccati Eq. 20.26 for continuous-time systems, numerical solution of the preceding matrix difference equations is straightforward for finite-time LQR problems. The procedure involves solution of the difference equations backward in time:

1. Let $k = k_1 - 1$. Then $\mathbf{P}(k+1)$ is equal to $\mathbf{P}_f$ and hence is known.
2. Compute $\mathbf{K}(k)$ using Eq. 20.33 and the known value of $\mathbf{P}(k+1)$.
3. Compute $\mathbf{P}(k)$ using Eq. 20.34 and the known values of $\mathbf{K}(k)$ and $\mathbf{P}(k+1)$.
4. Reduce k by one and repeat (2) and (3) until $k = k_0$.

The solution to the discrete-time LQR problem also simplifies as the terminal time k_1 approaches infinity. The solutions of the matrix difference Eqs. 20.33 and 20.34 converge to steady-state solutions $\mathbf{K}_s(k)$, $\mathbf{P}_s(k)$, which are independent of $\mathbf{P}_f$. The resulting steady-state control law

$$\mathbf{u}(k) = -\mathbf{K}_s(k)\mathbf{x}(k) \qquad (20.36)$$

results in an exponentially stable closed-loop system if:

1. The linear system of Eq. 19.10 is uniformly completely state controllable.
2. The pair $\mathbf{F}(k), \mathbf{H}_r^T(k)$ is uniformly completely reconstructible where $\mathbf{H}_r(k)$ is any matrix such that $\mathbf{H}_r(k)\mathbf{H}_r^T(k)$ equals $\mathbf{R}_1(k)$.

Also, the matrices $\mathbf{K}_s$ and $\mathbf{P}_s$ are constants if the system matrices and the weighting matrices in the index of performance of Eq. 20.31 are constants. They are given by solution of the following algebraic equations:

$$\mathbf{K}_s = [\mathbf{R}_2 + \mathbf{G}^T(\mathbf{R}_1 + \mathbf{P}_s)\mathbf{G}]^{-1}\mathbf{G}^T(\mathbf{R}_1 + \mathbf{P}_s)\mathbf{F} \tag{20.37}$$

$$\mathbf{P}_s = \mathbf{F}^T(\mathbf{R}_1 + \mathbf{P}_s)(\mathbf{F} - \mathbf{G}\mathbf{K}_s) \tag{20.38}$$

The optimal control law for the infinite-time LQR problem is

$$\mathbf{u}(k) = -\mathbf{K}_s\mathbf{x}(k) \tag{20.39}$$

and requires only constant state feedback gains. The solution $\mathbf{P}_s$ of Eqs. 20.37 and 20.38 is positive definite, and the optimal control law results in an asymptotically stable closed-loop system if:

1. The linear system of Eq. 19.12 is completely state controllable.
2. The pair $\mathbf{F}, \mathbf{H}_r^T$ is completely observable (reconstructible), where $\mathbf{H}_r$ is any matrix such that $\mathbf{H}_r\mathbf{H}_r^T$ equals $\mathbf{R}_1$.

Also, as in the case of continuous-time systems, the optimal feedback gains are nearly constant even for finite-time LQR problems, if the system matrices and weighting matrices in J are constant.[10] Finally, a number of techniques for solving the matrix algebraic Eqs. 20.37 and 20.38 and the matrix difference Eqs. 20.33–20.35 are described by Kuo.[11]

20.3.3 Stability and Robustness of the Optimal Control Law

An important consideration in the practical usefulness of the optimal control laws for the LQR problems described is the implication of these laws for performance features of the controlled systems not included in J, such as relative stability and sensitivity of the controlled system to unmodeled dynamics or plant parameter variations. Reference has already been made to the fact that the optimal control laws for the continuous-time and discrete-time infinite-time LQR problems described result in asymptotically stable closed-loop systems provided that specified controllability and reconstructibility or observability conditions are satisfied. Closed-loop systems with a prescribed degree of stability can be obtained by modifying the performance index J for linear, time-invariant, continuous-time systems:[12]

$$J = \int_0^\infty e^{2\alpha t}(\mathbf{x}^T\mathbf{R}_1\mathbf{x} + \mathbf{u}^T\mathbf{R}_2\mathbf{u})dt \tag{20.40}$$

where α is a positive scalar constant. If the pair $\mathbf{A}, \mathbf{B}$ is completely state controllable and the pair $\mathbf{A}, \mathbf{H}_r^T$ is completely observable where $\mathbf{H}_r\mathbf{H}_r^T$ is equal to $\mathbf{R}_1$, the solution to this LQR problem results in a finite value of J. Hence, the transients decay at least as rapidly as $e^{-\alpha t}$. Larger values of α would therefore ensure a more rapid return of the system to equilibrium. The corresponding algebraic Riccati equation is

$$\mathbf{R}_1 - \mathbf{P}_s\mathbf{B}\mathbf{R}_2^{-1}\mathbf{B}^T\mathbf{P}_s + \mathbf{A}^T\mathbf{P}_s + \mathbf{P}_s\mathbf{A} + 2\alpha\mathbf{P}_s = 0 \tag{20.41}$$

and the optimal feedback control law is given by Eq. 20.28. A similar procedure for discrete-time LTI systems is described by Franklin and Powell.[10]

Additional results concerning the stability properties of the optimal control law for continuous-time, LTI systems described by Eq. 19.6 and employing only constant weighting matrices in the index of performance, Eq. 20.24, are available and will be summarized. Anderson and Moore[12] have shown that for single-input systems the optimal control law for the infinite-time LQR problem has $\pm 60°$ phase margin, an infinite gain margin, and 50% gain reduction tolerance before the closed-loop system becomes unstable. These results are best explained with the aid of Fig. 20.5 a, where $G_p(s)$ is normalized to be unity at $s = 0$. The transfer function $K_pG_p(s)$ characterizes the modeling accuracy and is unity for an exact model. The result stated previously indicates that modeling errors that result either in phase shifts $\underline{/G_p(j\omega)}$ of less than $60°$ in magnitude for all frequencies or in

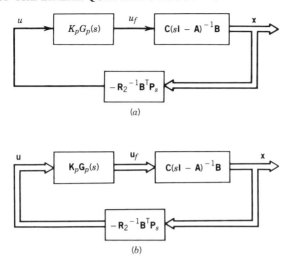

Fig. 20.5 Robustness of optimal LQR control (*a*) single input system and (*b*) multi-input system.

values of the magnitude ratio K_p greater than one-half would not destabilize the closed-loop system. The result on gain margins extends also to static nonlinear gain relationships between u_f and u.[12]

Safonov[13] has extended these results on gain and phase margins to multi-input infinite-time LQR problems. As shown in Fig. 20.5*b*, the quantities $\mathbf{u}_f(s)$ and $\mathbf{u}(s)$ are vectors and the modeling accuracy, if linear and time invariant, is represented by a transfer matrix $\mathbf{K}_p \mathbf{G}_p^{(s)}$. For an exact model, $\mathbf{K}_p \mathbf{G}_p(s)$ is the identity matrix. The results stated here are special cases of the results derived by Safonov[13] and are valid for the case where $\mathbf{K}_p$ and $\mathbf{G}_p(s)$ are diagonal matrices and $\mathbf{G}_p(s)$ consists of normalized transfer functions $\mathbf{G}_{pj}(s)$ [i.e., $\mathbf{G}_p(0)$ is the identity matrix]. For such a case, as long as all of the phase shifts $\underline{/G_{pj}(j\omega)}$ are less than 60° in magnitude for all frequencies or as long as all of the elements of the $\mathbf{K}_p$ matrix are greater than one-half, the closed-loop system using the optimal control law is asymptotically stable. More general robustness results, which are more difficult to apply, are also available.[13,36]

Perkins and Cruz[14] have examined frequency domain characterizations of the infinite-time LQR problem for continuous-time LTI systems and have shown that feedback realization of the optimal control law results in lower sensitivity to plant parameter variations than an equivalent open-loop realization. Kwakernaak and Sivan[3] have provided other results that are somewhat more useful in relating the sensitivity properties of optimal control laws to the weighting matrices in J. As the elements of the control effort weighting matrix $\mathbf{R}_2$ are decreased, the control law sensitivity decreases or improves since the optimal feedback gains are higher. However, higher feedback gains naturally imply greater likelihood of actuator saturation. The relative sensitivities of the different state variables depend on the elements of the state weighting matrix $\mathbf{R}_1$. State variables that are weighted more heavily would have lower sensitivity. Also, the sensitivity characteristics of the optimal control law for nonminimum-phase systems are shown to be inferior to that of minimum-phase systems. Finally, Kwakernaak and Sivan[3] have illustrated that the sensitivity results described do not necessarily extend to discrete-time systems.

20.4 EXTENSIONS OF THE LINEAR QUADRATIC REGULATOR PROBLEM

The optimal control law for the LQR problem and the pole placement design method described in the preceding sections have a number of limitations. First, as Horowitz[15] has pointed out, control system design by pole placement or quadratic performance index minimization obscures some practically important aspects and objectives. Among these are sensor noise, loop bandwidths, and sensitivity of system performance to significant plant parameter variations. Rosenbrock and McMorran[16] have pointed out that unconditional stability (i.e., stability for all values of the control gains between zero and design values) is essential for industrial control systems but is not guaranteed by the optimal control law. Hence, for multivariable optimal control systems, the failure of a single feedback measuring instrument could destabilize the closed-loop system. Moreover, the achievement of more modest sensitivity requirements is complicated by the lack of clear guidelines for weighting matrix selection. Available procedures for weighting matrix selection enable the achievement of desired transient response characteristics either by specification of a few dominant closed-loop system poles[12] or by implicit model reference following methods.[17] In the latter case, the reference model is chosen to have desired transient response characteristics. However, there is no available method to ensure a priori that the optimal control law has other desirable performance characteristics such as low sensitivity. The consequence of the lack of satisfactory guidelines for weighting matrix selection is that practical control system design using the LQR formulation involves considerable trial and error.

Second, the formulation of the standard LQR problem needs to be extended to be able to effectively handle control problems other than regulation. Examples of such problems include regulation in the presence of persistent disturbances, tracking problems, and vibration control problems. Even though some of these problems can be handled by simple extensions of the LQR problem, effective solutions to these problems require significant extensions of the LQR problem formulation.

Extensions of the standard LQR problem formulation addressing some of its limitations will be described here. The extensions involve alternative formulations of the quadratic index of performance to be minimized such that the resulting solutions have desired features. Additionally, the problem formulation utilizes more completely the available information on the systems to be controlled and their environments. One of the measures for evaluating the effectiveness of the resulting problem formulations and solutions is their ability to accommodate a greater variety of problems and performance specifications. Another such measure is their ability to incorporate in the proposed solutions features that are known to be effective in practice. The resulting variety of problem formulations and solutions runs somewhat counter to the unifying nature of the standard LQR problem formulation and constitutes a recognition of its limitations in practice.

20.4.1 Disturbance Accommodation

Extensions of the standard LQR problem to accommodate unknown disturbance inputs have been proposed by Anderson and Moore,[12] Johnson,[18,19] and Davison and Ferguson.[20] Anderson and Moore[12] consider LTI systems of the following form:

$$\dot{\mathbf{x}}(t) = \mathbf{A}\mathbf{x}(t) + \mathbf{B}[\mathbf{u}(t) + \mathbf{M}_w\mathbf{w}] \tag{20.42}$$

where $\mathbf{w}$ is a constant, unknown disturbance vector. The restriction that $\mathbf{u}(t)$ and $\mathbf{M}_w\mathbf{w}$ occur additively ensures that the equilibrium $\mathbf{x} = 0$ can be achieved and maintained even if $\mathbf{w}$ is nonzero. The following index of performance with input derivative constraints is to be minimized by the control law:

$$J = \int_0^\infty [\mathbf{x}^T\mathbf{R}_1\mathbf{x} + (\mathbf{u} + \mathbf{M}_w\mathbf{w})^T\mathbf{R}_2(\mathbf{u} + \mathbf{M}_w\mathbf{w}) + \dot{\mathbf{u}}^T\mathbf{R}_3\dot{\mathbf{u}}] \, dt \tag{20.43}$$

where $\mathbf{R}_1$, $\mathbf{R}_2$ are symmetric positive semidefinite matrices and $\mathbf{R}_3$ is a symmetric positive definite matrix. When the system state is augmented to include the vector $\mathbf{u} + \mathbf{M}_w\mathbf{w}$ and the input derivative $\dot{\mathbf{u}}$ is defined to be the new input, the problem reduces to the standard infinite-time LQR problem. The optimal control law is a proportional-plus-integral state feedback law:

$$\mathbf{u}(t) = -\mathbf{K}\mathbf{x}(t) - \mathbf{K}_I \int \mathbf{x} \, dt \tag{20.44}$$

where the constant gain matrices $\mathbf{K}, \mathbf{K}_I$ are known linear functions of matrices satisfying algebraic Riccati equations. In cases where the complete state $\mathbf{x}$ is not measurable, the state would be estimated from the measured outputs using the methods described in Section 20.5. The closed-loop system is asymptotically stable provided that certain controllability and observability conditions on the matrices $\mathbf{A}$, $\mathbf{B}$, $\mathbf{R}_1$, and $\mathbf{R}_2$ are satisfied. The gain and phase margin results noted earlier for the standard infinite-time LQR problem are valid here as well.

Johnson[18] considers a more general class of disturbances and state-space equations:

$$\dot{\mathbf{x}}(t) = \mathbf{A}(t)\mathbf{x}(t) + \mathbf{B}(t)\mathbf{u}(t) + \mathbf{B}_w(t)\mathbf{w}(t) \tag{20.45}$$

$$\mathbf{y}(t) = \mathbf{C}(t)\mathbf{x}(t) + \mathbf{D}(t)\mathbf{u}(t) + \mathbf{D}_w(t)\mathbf{w}(t) \tag{20.46}$$

where $\mathbf{w}(t)$ is a vector of disturbance inputs. The disturbance inputs are assumed to be described by linear time-varying differential equations that constitute a state-space model for the disturbances:

$$\dot{\mathbf{z}}_\sigma(t) = \mathbf{A}_\sigma(t)\mathbf{z}_\sigma(t) + \mathbf{B}_\sigma(t)\mathbf{x}(t) + \sigma(t) \tag{20.47}$$

$$\mathbf{w}(t) = \mathbf{C}_\sigma(t)\mathbf{z}_\sigma(t) + \mathbf{D}_\sigma(t)\mathbf{x}(t) \tag{20.48}$$

$\mathbf{z}_\sigma(t)$ represents the state of the disturbance and $\boldsymbol{\sigma}(t)$ is a vector of Dirac delta impulses occurring at unknown times. The terms including $\mathbf{x}(t)$ in the preceding equations enable cases of state-dependent disturbances to be considered within this framework. The coefficient matrices are determined experimentally by examination of the records of the disturbances. This type of description of disturbances constitutes a waveform mode description and is applicable to a broad class of disturbances of practical interest that are not described well either by deterministic process models or by stochastic process models. Examples of such disturbances are piecewise linear, piecewise polynomial, or piecewise periodic signals.

The waveform mode description of disturbances is combined with the system state-space equations to provide a rather complete description of the system to be controlled and the inputs affecting its behavior. The exact design of the controller depends on the specific objectives used to govern the design. If one of the objectives of the control system design is to counteract as completely as possible the effects of the disturbance inputs, then the control input is considered to be composed of two parts:

$$\mathbf{u}(t) = \mathbf{u}_m(t) + \mathbf{u}_d(t) \tag{20.49}$$

where $\mathbf{u}_d(t)$ is the component used to counteract disturbance effects either completely or partially and $\mathbf{u}_m(t)$ is the component used to accomplish other objectives such as closed-loop pole placement. Alternatively, the objective of control system design may be the minimization of a quadratic index of performance. In either of these cases, the control law would require the feedback of the system state $\mathbf{x}$ as well as the disturbance state $\mathbf{z}_\sigma$ (Fig. 20.6). The state estimation methods described in Section 20.5 can be used to generate these state estimates from available measurements. The extension of this disturbance accommodation approach to discrete-time systems has also been considered by Johnson.[19]

The formulation of disturbance state models and their incorporation in controller design results in controller features familiar from more classical approaches to disturbance suppression. Examples are integral control for constant disturbances, notch filter control for sinusoidal disturbances and disturbance feedforward if some components of the disturbance inputs are measurable.[18] Hence, the disturbance accommodation controllers described here may be viewed as generalizations of classical solutions to disturbance suppression. A similar comment may be made concerning the mechanism

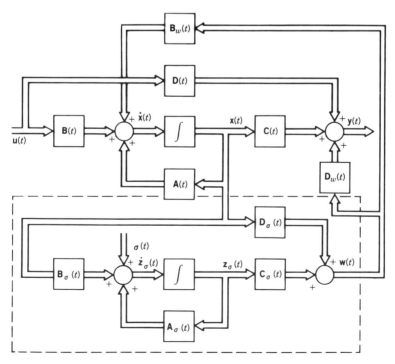

Fig. 20.6 Disturbance state model, continuous-time system.

for disturbance suppression inherent in the robust servomechanism structure described by Davison and Ferguson.[20] The robust controller structure is described later in this section.

20.4.2 Tracking Applications

Anderson and Moore,[12] Davison and Ferguson,[20] Trankle and Bryson,[21] and Tomizuka et al.[22-25] have considered extensions of the LQR formulation to accommodate tracking applications. Anderson and Moore[12] have considered the servomechanism problem where the linear system state equations are given by Eqs. 19.4 and 19.5 with $\mathbf{D}(t) = \mathbf{0}$ and the class of desired trajectories $\mathbf{y}_r$ is given by

$$\dot{\mathbf{x}}_r(t) = \mathbf{A}_r(t)\mathbf{x}_r(t) \tag{20.50}$$

$$\mathbf{y}_r(t) = \mathbf{C}_r(t)\mathbf{x}_r(t) \tag{20.51}$$

and a specified initial condition $\mathbf{x}_r(t_0)$. The index of performance to be optimized for the finite-time problem is

$$J = \int_{t_0}^{t_1} \left\{ \mathbf{x}^T \left[\mathbf{I} - \mathbf{C}^T(\mathbf{CC}^T)^{-1}\mathbf{C} \right]^T \mathbf{Q}_1 \left[\mathbf{I} - \mathbf{C}^T(\mathbf{CC}^T)^{-1}\mathbf{C} \right] \mathbf{x} + (\mathbf{y} - \mathbf{y}_r)^T\mathbf{Q}_2(\mathbf{y} - \mathbf{y}_r) + \mathbf{u}^T\mathbf{R}_2\mathbf{u} \right\} dt \tag{20.52}$$

where the time dependencies of the vectors and matrices have been omitted for convenience. The matrices $\mathbf{Q}_1$ and $\mathbf{Q}_2$ are positive semidefinite matrices and $\mathbf{R}_2$ is a positive definite matrix. The weighting on the tracking error $\mathbf{y} - \mathbf{y}_r$ helps reduce it, whereas the weighting on the state $\mathbf{x}$ achieves a smooth response. The optimal control law involves linear feedback of the system state as well as feedforward of the state of the trajectory model:

$$\mathbf{u} = -\mathbf{K}(t)\mathbf{x}(t) - \mathbf{K}_r(t)\mathbf{x}_r(t) \tag{20.53}$$

where the gain matrices $\mathbf{K}(t)$ and $\mathbf{K}_r$ are linearly related to solutions of the matrix Riccati differential equations. Conditions for the time-invariant version of this servo problem to reduce to the standard infinite-time LQR problem have also been noted.[12]

Trankle and Bryson[21] have considered the time-invariant servomechanism problem for the case where $\mathbf{y}$, $\mathbf{y}_r$, $\mathbf{u}$ have the same dimension and have proposed the following index of performance:

$$J = \int_0^\infty [(\mathbf{y} - \mathbf{y}_r)^T\mathbf{Q}_y(\mathbf{y} - \mathbf{y}_r) + (\mathbf{u} - \mathbf{U}_1\mathbf{x}_r)^T\mathbf{R}_u(\mathbf{u} - \mathbf{U}_1\mathbf{x}_r)] \, dt \tag{20.54}$$

where $\mathbf{Q}_y$ is positive semidefinite and $\mathbf{R}_u$ is positive definite. A modification of the index of performance to add integral error feedback can also be devised. A matrix $\mathbf{U}_1$ and another matrix $\mathbf{X}$ to occur later in the development are defined by

$$\mathbf{CX} = \mathbf{C}_r \tag{20.55}$$

$$\mathbf{AX} + \mathbf{BU}_1 = \mathbf{XA}_r \tag{20.56}$$

The optimal control law is asymptotically stable if the pair $\mathbf{A}$, $\mathbf{B}$ is completely state controllable and the pair $\mathbf{A}$, $\mathbf{C}$ is completely observable. The control law is given by

$$\mathbf{u} = (\mathbf{U}_1 + \mathbf{KX})\mathbf{x}_r(t) - \mathbf{Kx}(t) \tag{20.57}$$

where $\mathbf{K}$ is related, in the usual manner, to the solution of an algebraic Riccati equation. The first term on the right-hand side represents feedforward control action, and the second term represents feedback control action (Fig. 20.7). The feedforward action yields faster and more accurate tracking of the desired trajectory than other control schemes that depend more on integral error feedback. Finally, model and system state feedback is required by both Eqs. 20.53 and 20.57. If these states are not available for measurement, state estimators such as those described in Section 20.5 would be needed.

A variation on the servomechanism problem already described is that of tracking where the desired trajectory $\mathbf{y}_r$ is known a priori rather than being defined by a model as in Eqs. 20.50 and 20.51. Anderson and Moore[12] have determined the optimal control law for the index of performance Eq. 20.52:

$$u = -\mathbf{K}(t)\mathbf{x}(t) - \mathbf{R}_2(t)^{-1}\mathbf{B}^T(t)\mathbf{b}(t) \tag{20.58}$$

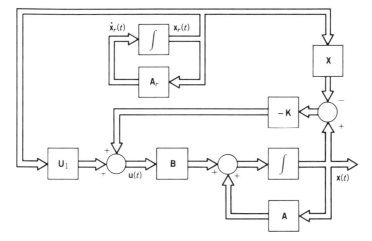

Fig. 20.7 Extension of LQR for time-invariant servomechanism problem.

where $\mathbf{b}(t)$ is the solution of a linear ordinary differential equation with $\mathbf{y}_r(t)$ as the forcing function, and $\mathbf{K}(t)$ is related, in the usual manner, to the solution of a matrix Riccati differential equation.

The control law, Eq. 20.58, incorporates information about future inputs over the entire interval of interest (t_0, t_1) and is said to have infinite preview control in addition to feedback control. A related problem is one where only finite preview of the desired trajectory is available; that is, at any time τ, $\mathbf{y}_r(t)$ is known for $\tau \leq t \leq \Delta T + \tau$, where ΔT is called the preview length. Tomizuka[22] has examined the continuous-time finite preview problem and determined the optimal control law for a quadratic index of performance over the entire interval of interest (t_0, t_1) where t_1 is greater than $t_0 + \Delta T$. The desired trajectory, not known from preview at any time t, is assumed to be modeled by a stochastic process. A discrete-time version of the problem is given by Tomizuka and Whitney.[23] Discrete-time finite preview of disturbance inputs in addition to the desired trajectory has also been considered by Tomizuka et al.[24,25] The results indicate that preview control improves the control system performance, especially in the low-frequency range and that there exists a critical preview length beyond which preview information is less important.

20.4.3 Frequency Shaping of Cost Functionals

Extensions of the standard LQR problem and the resulting control laws have certain limitations related to the nature of the index of performance used. These limitations become clear when the index of performance is viewed in the frequency domain.[26] The index of performance for the infinite-time, LQR problem for an LTI system

$$J = \int_0^\infty (\mathbf{x}^T \mathbf{R}_1 \mathbf{x} + \mathbf{u}^T \mathbf{R}_2 \mathbf{u}) \ dt \tag{20.59}$$

can be transformed to the frequency domain using Parseval's theorem:

$$J = \frac{1}{2} \int_{-\infty}^{\infty} [\mathbf{x}^*(j\omega) \mathbf{R}_1 \mathbf{x}(j\omega) + \mathbf{u}^*(j\omega) \mathbf{R}_2 \mathbf{u}(j\omega)] \ d\omega \tag{20.60}$$

where the superscript * implies a complex conjugate transpose of a complex matrix. The index of performance thus weights control and state transients at all frequencies equally despite the fact that, in practice, model accuracy is poorer at high frequencies. Model inaccuracy, sensor noise, and actuator bandwidth limitations are better accommodated by an index of performance that penalizes high-frequency control activity more heavily than low-frequency control activity. In fact, it is recognized good practice in classical control system design to have a steep enough rolloff or attenuation rate of the open-loop transmission functions at high frequencies for adequate noise suppression and robustness to model errors. In contrast, the closed-loop amplitude ratio frequency response curve corresponding to the optimal control law for an LQR problem may drop off only as slowly as 20 db/decade.[26] Another limitation of the standard LQR problem formulation is that, in many applications such as vibration control, the objectives of control are stated better in the frequency domain than in the time domain.

Specification of the index of performance in the frequency domain with the weighting matrices being functions of the frequency ω enables many of these limitations to be removed. The frequency-shaped cost functional method described by Gupta[26] is one such method and allows the generalized index of performance:

$$J = \frac{1}{2}\int_{-\infty}^{\infty} [\mathbf{x}^*(j\omega)\mathbf{R}_1(j\omega)\mathbf{x}(j\omega) + \mathbf{u}^*(j\omega)\mathbf{R}_2(j\omega)\mathbf{u}(j\omega)]\, d\omega \qquad (20.61)$$

The restrictions on the weighting matrices are:

1. $\mathbf{R}_1(j\omega)$ and $\mathbf{R}_2(j\omega)$ are positive semidefinite and positive definite respectively at all frequencies.
2. $\mathbf{R}_1(j\omega)$ and $\mathbf{R}_2(j\omega)$ are Hermitian matrices at all frequencies and, in fact, are rational functions of ω^2.
3. $\mathbf{R}_2(j\omega)$ has rank r, where r is the dimension of control vector $\mathbf{u}$.

Two new matrices $\mathbf{P}_1(j\omega)$ and $\mathbf{P}_2(j\omega)$ are defined based on $\mathbf{R}_1(j\omega)$ and $\mathbf{R}_2(j\omega)$:

$$\mathbf{R}_1(j\omega) = \mathbf{P}_1^*(j\omega)\mathbf{P}_1(j\omega) \qquad (20.62)$$

$$\mathbf{R}_2(j\omega) = \mathbf{P}_2^*(j\omega)\mathbf{P}_2(j\omega) \qquad (20.63)$$

where $\mathbf{P}_1$ is $m \times n$, m being the rank of $\mathbf{R}_1$, and $\mathbf{P}_2$ is $r \times r$, r being the rank of $\mathbf{R}_2$ and the dimension of the control vector $\mathbf{u}$. New vectors $\mathbf{x}_p$ and $\mathbf{u}_p$, dynamically related to $\mathbf{x}$ and $\mathbf{u}$ respectively, are then defined

$$\mathbf{x}_p(s) = \mathbf{P}_1(s)\mathbf{x}(s) \qquad (20.64)$$

$$\mathbf{u}_p(s) = \mathbf{P}_2(s)\mathbf{u}(s) \qquad (20.65)$$

Minimal realizations of $\mathbf{P}_1(s)$ and $\mathbf{P}_2(s)$ are then determined as described in Section 19.7 and new states $\mathbf{z}_x$ and $\mathbf{z}_u$ defined:

$$\dot{\mathbf{z}}_x = \mathbf{A}_x\mathbf{z}_x + \mathbf{B}_x\mathbf{x} \qquad (20.66)$$

$$\mathbf{x}_p = \mathbf{C}_x\mathbf{z}_x + \mathbf{D}_x\mathbf{x} \qquad (20.67)$$

$$\dot{\mathbf{z}}_u = \mathbf{A}_u\mathbf{z}_u + \mathbf{B}_u\mathbf{u} \qquad (20.68)$$

$$\mathbf{u}_p = \mathbf{C}_u\mathbf{z}_u + \mathbf{D}_u\mathbf{u} \qquad (20.69)$$

An augmented state $\mathbf{x}_a$ is then defined:

$$\mathbf{x}_a = \begin{bmatrix} \mathbf{x} \\ \mathbf{z}_x \\ \mathbf{z}_u \end{bmatrix} \qquad (20.70)$$

Using Parseval's theorem, the index of performance, Eq. 20.61, is transformed to the time domain to yield a standard LQR problem formulation with constant weighting matrices:

$$J = \int_0^\infty (\mathbf{x}_a^T\mathbf{R}_{1a}\mathbf{x}_a + 2\mathbf{x}_a^T\mathbf{R}_{12a}\mathbf{u} + \mathbf{u}^T\mathbf{R}_{2a}\mathbf{u})dt \qquad (20.71)$$

The optimal control law is obtained by solving the corresponding algebraic matrix Riccati equation:

$$\mathbf{u} = -(\mathbf{K}\mathbf{x} + \mathbf{K}_x\mathbf{z}_x + \mathbf{K}_u\mathbf{z}_u) \qquad (20.72)$$

The states $\mathbf{z}_x$ and $\mathbf{z}_u$ are dynamically related to $\mathbf{x}$ and $\mathbf{u}$, respectively. Hence, the optimal control law has dynamic compensators in addition to linear instantaneous state-variable feedback (Fig. 20.8). If the state $\mathbf{x}$ is not completely measurable, state estimation is required as described in Section 20.5.

The utility of this design method is that it establishes a clear link between features of the weighting matrices $\mathbf{R}_1(j\omega)$ and $\mathbf{R}_2(j\omega)$ and the resulting controllers. Gupta[26] has shown that the compensator poles and zeros are the same as poles and zeros of the transfer functions $\mathbf{P}_1(s)$ and $\mathbf{P}_2(s)$. For

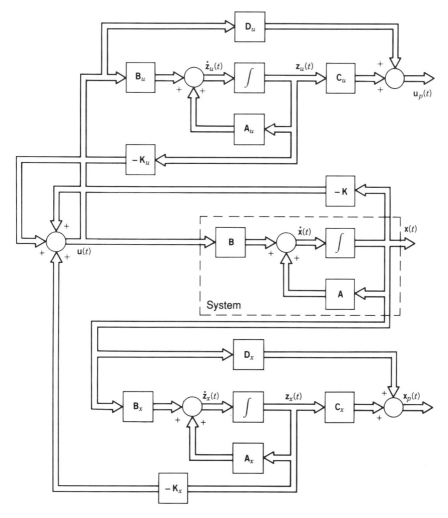

Fig. 20.8 Extension of LQR using frequency-dependent cost functionals.

example, if $\mathbf{R}_1(j\omega)$ is singular at $\omega = 0$, integral control results. If $\mathbf{R}_1(j\omega)$ is singular at any other frequency ω_1, the controller has a notch filter at that frequency. This would be desirable if the controlled system has a known resonant frequency at ω_1 and we wish to minimize the excitation of the resonance by disturbances. Finally, if $\mathbf{R}_2(j\omega)$ is chosen to increase with frequency ω, the optimal control law would have reduced control action at high frequencies. As indicated previously, this is a desirable feature if the controller is to have good noise suppression and robustness to model errors.

20.4.4 Robust Servomechanism Control

Davison and Ferguson[20] have proposed a controller structure and formulated a design procedure for the robust control of servomechanisms. Robustness is defined here to imply asymptotic stability of the closed-loop system and asymptotic tracking of the desired trajectory for all initial conditions of the controller used and for all variations in the system model parameters that do not cause the controlled system to become unstable. The system equations are the time-invariant versions of Eqs. 20.45 and 20.46. The disturbance inputs $\mathbf{w}(t)$ are modeled by time-invariant versions of Eqs. 20.47 and 20.48 with no provision for either state-dependent disturbances $[\mathbf{B}_\sigma(t) = \mathbf{0} = \mathbf{D}_\sigma(t)]$ or impulsive inputs $[\sigma(t) = 0]$. The desired trajectory $\mathbf{y}_r(t)$ is described by time-invariant versions of Eqs. 20.50 and 20.51.

Under certain specified conditions,[20] the robust servomechanism problem is assured of a solution. The resulting controller structure consists of a servocompensator and stabilizing compensator (Fig. 20.9), and the robust control input is given by

$$\mathbf{u} = -\mathbf{K}_\eta \eta - \mathbf{K}_\theta \theta \tag{20.73}$$

where η and θ are the outputs of the servocompensator and stabilizing compensator, respectively, and $\mathbf{K}_\eta$ and $\mathbf{K}_\theta$ are constant-gain matrices. The servocompensator is a dynamic controller with the trajectory error as input and its form and parameters are determined from the state-space models of the system and the disturbance and trajectory inputs. The servocompensator is a generalization of the integral controller from classical control theory. The stabilizing compensator has the function of stabilizing the augmented system consisting of the servocompensator and the system to be controlled. In general, the stabilizing compensator has a number of inputs as shown in Fig. 20.9. It is not uniquely defined, however, and is usually chosen to have as simple a form as is possible given the performance requirements on the controlled system. Complete state feedback, if measurable, and observer-based controllers of the type described in Section 20.6 are among the more elaborate forms of the stabilizing compensator.

Once the structure of the stabilizing compensator is determined, the unknown controller parameters are determined by minimization of a quadratic index of performance. The index of performance is specified such that minimizing it gives a system with fast response and low interaction for MIMO systems. The optimum value of the index of performance is given in terms of the controller parameters. Since controller parameters are not known a priori, the optimal controller design reduces to a multiparameter optimization problem where the quantity being optimized is the quadratic index of performance. The parameter optimization can be constrained to allow the designer to handle closed-loop system damping constraints, controller gain constraints to avoid saturation effects, controller integrity constraints for sensor and actuator failures, and tolerance constraints to system parameter variations. When applied to example systems,[20] the robust control approach yields controller features commonly obtained from other frequency-domain-based design procedures, such as error integral control, phase-lead compensation, pole-zero cancellation, and low interaction for MIMO systems. Its ability to accommodate a variety of constraints on the controllers to ensure their practical utility makes the robust controller design approach a practically useful approach. Care is needed, however, to keep the resulting controller as simple as possible.

The recent extensions of the LQR problem described have addressed many of the limitations of state-space-based approaches to control system design noted by Horowitz[15] and others. As a result, the state-space approach is expected to be more useful in a greater variety of practical applications. It should be noted, however, that there are still no specific guidelines for the selection of weighting matrices in the quadratic indices of performance used by the extended versions of the LQR problem. Consequently, control system design using these methods would still involve considerable trial and error.

20.5 DESIGN OF LINEAR STATE ESTIMATORS

The optimal control laws for the standard LQR problem and the extensions described earlier require feedback of the entire system state. Pole placement algorithms require state feedback as well. Complete state measurement is not possible or practical in many instances. Therefore, the state variables often must be estimated from the measured output variables. Closed-loop dynamic realizations of such state estimators are described here. The property of reconstructibility (or equivalently, observability for LTI systems) is critical for the design of such estimators. It plays a role relative to state estimators that is very similar to the role played by state controllability relative to state feedback controller design. This similarity is the consequence of the duality of the two properties, referred to earlier in Section 19.6. Therefore, many of the results presented in this section parallel those of the preceding section on controller design.

20.5.1 The Observer

The structure of a closed-loop dynamic system, called an observer, to estimate the state $\mathbf{x}(t)$ of an LTV system described by Eqs. 19.4 and 19.5, from output measurements $\mathbf{y}(t)$ and input measurements $\mathbf{u}(t)$, was proposed by Luenberger[27] (Fig. 20.10):

$$\dot{\hat{\mathbf{x}}}(t) = \mathbf{A}(t)\,\hat{\mathbf{x}}(t) + \mathbf{B}(t)\mathbf{u}(t) + \mathbf{L}(t)[\mathbf{y}(t) - \mathbf{C}(t)\hat{\mathbf{x}}(t) - \mathbf{D}(t)\mathbf{u}(t)] \tag{20.74}$$

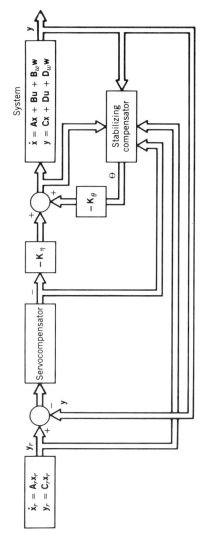

Fig. 20.9 Extension of LQR for the robust LTI servomechanism problem.

System

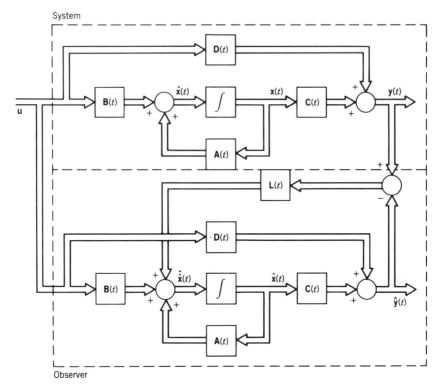

Observer

Fig. 20.10   Observer for LTV continuous-time system.

where $\hat{\mathbf{x}}(t)$ is the estimated state and $\mathbf{L}(t)$ is a time-varying matrix of observer gains. Combination of Eqs. 20.74, 19.4, and 19.5 yields a dynamic equation for the state estimation error $\mathbf{e}(t)$:

$$\dot{\mathbf{e}}(t) \overset{\triangle}{=} \dot{\mathbf{x}}(t) - \dot{\hat{\mathbf{x}}}(t) = [\mathbf{A}(t) - \mathbf{L}(t)\mathbf{C}(t)]\mathbf{e}(t) \tag{20.75}$$

with the initial condition

$$\mathbf{e}(t_0) = \mathbf{x}(t_0) - \hat{\mathbf{x}}(t_0) \tag{20.76}$$

Thus, if the observer is asymptotically stable

$$\lim_{t \to \infty} \mathbf{e}(t) = 0 \tag{20.77}$$

for all $\mathbf{e}(t_0)$. If the system under consideration is time invariant, the eigenvalues of the matrix $\mathbf{A} - \mathbf{LC}$ govern the transient behavior of the estimation error. These eigenvalues, referred to as observer poles, can be arbitrarily located in the complex plane by appropriate choice of the constant observer gain matrix $\mathbf{L}$, if and only if the system given by Eqs. 19.4 and 19.5 is completely reconstructible.[27] For LTI systems, reconstructibility is exactly equivalent to observability. Also, if the $\mathbf{L}$ matrix is to be real, the complex eigenvalues of $\mathbf{A} - \mathbf{LC}$ should be specified as conjugate pairs.

The discrete-time version of the observer is given below.[3] The structure of the observer to estimate the state $\mathbf{x}(k)$ of a linear, time-varying system described by Eqs. 19.10 and 19.11, is given by (Fig. 20.11)

$$\hat{\mathbf{x}}(k + 1) = \mathbf{F}(k)\,\hat{\mathbf{x}}(k) + \mathbf{G}(k)\mathbf{u}(k) + \mathbf{L}(k)[\mathbf{y}(k) - \mathbf{C}(k)\,\hat{\mathbf{x}}(k) - \mathbf{D}(k)\mathbf{u}(k)] \tag{20.78}$$

The resulting equation for the state estimation error $\mathbf{e}(k)$ is

$$\mathbf{e}(k + 1) \overset{\triangle}{=} \mathbf{x}(k + 1) - \hat{\mathbf{x}}(k + 1) = [\mathbf{F}(k) - \mathbf{L}(k)\mathbf{C}(k)]\mathbf{e}(k) \tag{20.79}$$

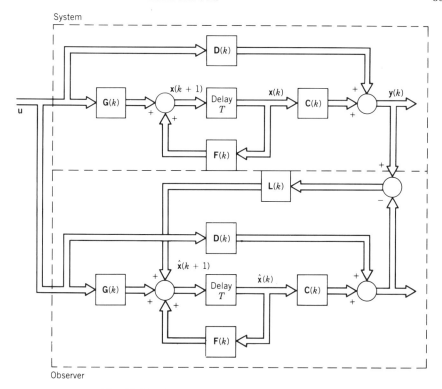

Fig. 20.11 Observer for LTV discrete-time system.

with the initial condition

$$\mathbf{e}(k_0) = \mathbf{x}(k_0) - \hat{\mathbf{x}}(k_0) \tag{20.80}$$

If the observer is asymptotically stable

$$\lim_{k \to \infty} \mathbf{e}(k) = 0 \tag{20.81}$$

for all $\mathbf{e}(k_0)$. If the system under consideration is time invariant also, the eigenvalues of the matrix $\mathbf{F} - \mathbf{LC}$ govern the estimation error transients. Observability of the system given by Eqs. 19.12 and 19.13 is equivalent to arbitrary pole assignability for the observer.

For single-output systems, specification of the desired observer pole locations uniquely specifies the gain vector $\mathbf{L}$. Formulas for the gain vector $\mathbf{L}$, applicable to both continuous-time and discrete-time systems, will be given here. They are obtained by analogy with Eqs. 20.2–20.5. The object is to get the observer gain vector $\mathbf{L}$ so that the characteristic equation governing the estimation error transients has specified coefficients:

$$\beta_e(s) = \det(s\mathbf{I} - \mathbf{A} + \mathbf{LC})$$

$$= \det(s\mathbf{I} - \mathbf{A}^{\mathrm{T}} + \mathbf{C}^{\mathrm{T}}\mathbf{L}^{\mathrm{T}})$$

$$= s^n + \sum_{i=0}^{n-1} \beta_{ie}s^i = 0 \tag{20.82}$$

where the matrix $\mathbf{A} - \mathbf{LC}$ is transposed so that the observer gain selection problem can be reduced to a form similar to the state feedback controller gain selection problem. Comparison of Eqs. 20.82 and 20.5 suggests that the following substitutions are necessary to convert algorithms for state feedback controller gain selection to observer gain selection.

$$
\begin{array}{ccc}
\mathbf{A} & \longrightarrow & \mathbf{A}^{\mathrm{T}} \\
\mathbf{B} & \longrightarrow & \mathbf{C}^{\mathrm{T}} \\
\mathbf{K} & \longrightarrow & \mathbf{L}^{\mathrm{T}}
\end{array}
\tag{20.83}
$$

Similarly, for discrete-time systems, we get the following transformations:

CONTROLLER GAIN SELECTION   OBSERVER GAIN SELECTION

$$
\begin{array}{ccc}
\mathbf{F} & \longrightarrow & \mathbf{F}^{\mathrm{T}} \\
\mathbf{G} & \longrightarrow & \mathbf{C}^{\mathrm{T}} \\
\mathbf{K} & \longrightarrow & \mathbf{L}^{\mathrm{T}}
\end{array}
\tag{20.84}
$$

Using the preceding transformations, we get

$$
\mathbf{L} = \beta_e(\mathbf{A}) \begin{bmatrix} \mathbf{C} \\ \mathbf{CA} \\ \vdots \\ \mathbf{CA}^{n-1} \end{bmatrix}^{-1} \begin{bmatrix} 0 \\ 0 \\ \vdots \\ 1 \end{bmatrix}
\tag{20.85}
$$

for continuous-time systems and

$$
\mathbf{L} = \beta_e(\mathbf{F}) \begin{bmatrix} \mathbf{C} \\ \mathbf{CF} \\ \vdots \\ \mathbf{CF}^{n-1} \end{bmatrix}^{-1} \begin{bmatrix} 0 \\ 0 \\ \vdots \\ 1 \end{bmatrix}
\tag{20.86}
$$

for discrete-time systems. In the preceding equations,

$$
\beta_e(\mathbf{A}) = \mathbf{A}^n + \sum_{i=0}^{n-1} \beta_{ie}\mathbf{A}^i
\tag{20.87}
$$

for continuous-time systems and $\beta_e(\mathbf{F})$ is a similar function of $\mathbf{F}$ for discrete-time systems. The β_{ie}'s are the coefficients of the desired characteristic Eq. 20.82 for the observer. A similar equation describes the discrete-time observer characteristic equation. If CACSD packages[4-6] supporting only controller pole placement algorithms are available, the transformations of Eqs. 20.83 and 20.84 are needed to select observer gains using these algorithms.

The observers of Eqs. 20.74 and 20.78 are called full-order observers since their dimensions are the same as those of the systems whose states are being estimated. Reduction of the observer dimension can be achieved by using the fact that the output equation provides us with p linear equations in the unknown state, where p is the number of output variables. Therefore, the observer need only provide $n - p$ additional linear equations and thus need only be of dimension $n - p$. For time-invariant systems, the corresponding observer gain matrix can be chosen to place the $n - p$ observer poles at any desired locations in the complex plane, if the original systems of Eqs. 19.6 and 19.7 or 19.12 and 19.13 are completely observable. Equations for reduced-order observers for continuous-time systems are given by Kwakernaak and Sivan[3] and for discrete-time systems by Franklin and Powell.[10] In the presence of measurement noise, the state estimates $\mathbf{x}(t)$ or $\mathbf{x}(k)$ obtained using reduced-order observers are more sensitive than those obtained using full-order observers.

The discrete-time observer of Eq. 20.78 is also referred to as a prediction estimator since $\mathbf{x}(k + 1)$ is ahead of the last measurements used, $\mathbf{y}(k)$ and $\mathbf{u}(k)$. A variation of this observer, called the current estimator, is useful if the computation time associated with the observer is very short compared to the sampling interval for the sampled-data system. The corresponding observer equation for an LTI system is[10]

$$
\hat{\mathbf{x}}(k + 1) = \mathbf{F}\hat{\mathbf{x}}(k) + \mathbf{Gu}(k) + \mathbf{L}\big[y(k + 1) - \mathbf{CF}\,\hat{\mathbf{x}}(k) - \mathbf{CBu}(k) - \mathbf{Du}(k + 1)\big]
\tag{20.88}
$$

The state estimate $\hat{\mathbf{x}}(k + 1)$ therefore depends on the current measurements $y(k + 1)$ and $\mathbf{u}(k + 1)$. In practice, however, the measurements precede the estimate by a very small computation time. The observer gain formula, for a single-output system, is then given by

$$
\mathbf{L} = \beta_e(\mathbf{F}) \begin{bmatrix} \mathbf{CF} \\ \mathbf{CF}^2 \\ \vdots \\ \mathbf{CF}^n \end{bmatrix}^{-1} \begin{bmatrix} 0 \\ 0 \\ \vdots \\ 1 \end{bmatrix}
\tag{20.89}
$$

instead of Eq. 20.86. The term $\beta_e(\mathbf{F})$ has the same meaning as before.

For multioutput LTI systems, specification of the desired observer poles does not specify the gain matrix $\mathbf{L}$ uniquely. The additional freedom in the gain matrix selection can be used to assign the eigenvectors (or generalized eigenvectors) of the matrix $\mathbf{A} - \mathbf{LC}$ or $\mathbf{F} - \mathbf{LC}$ in addition to the eigenvalues. The corresponding procedure would be very similar to that used for gain matrix selection for multi-input systems and described in Section 20.2. The transformations of Eqs. 20.83 and 20.84 can be used to adapt the controller gain selection procedure to observer gain selection. Alternatively, the observer gains can be chosen to design observers whose state estimates have low sensitivity to unmeasured disturbance inputs.[28]

The observer designs described work well in the absence of significant levels of measurement noise or unknown disturbance signals. The sensitivity of the state estimates to measurement noise and unknown disturbance signals depends on the specified location of the observer poles. If these pole locations are too far into the left half of the complex s plane or too close to the origin in the complex z plane, the state estimates would be unduly sensitive to measurement noise and disturbance signals. In the limiting case, the observers can be shown to reduce to ideal differentiators or differencing devices. The observer pole locations should therefore be chosen to avoid such high sensitivities of the state estimates, but at the same time ensure that the estimation error transients decay more rapidly than the state variables being estimated. Specification of the observer poles in practice involves considerable trial and error in much the same manner as specification of closed-loop poles does for state feedback controller design. If some information is available concerning the disturbance signals affecting the system and the measurement noise, the observer gain matrix selection problem can be formulated as an optimal estimation problem.

20.5.2   The Optimal Observer

Optimization of observer design has been primarily performed assuming stochastic models for the disturbance inputs and measurement noise, though the effect of deterministic model errors on observer design is also important.[29] Consider the continuous-time system[3]

$$\dot{\mathbf{x}}(t) = \mathbf{A}(t)\mathbf{x}(t) + \mathbf{B}(t)\mathbf{u}(t) + \mathbf{S}(t)\mathbf{w}_1(t) \qquad t \geq t_0 \tag{20.90}$$

$$\mathbf{y}(t) = \mathbf{C}(t)\mathbf{x}(t) + \mathbf{w}_2(t) \qquad t \geq t_0 \tag{20.91}$$

where $\mathbf{w}_1(t)$ is the random disturbance input, $\mathbf{w}_2(t)$ is the random measurement noise, and their joint probabilities are assumed to be known. The column vector $[\mathbf{w}_1^T(t) \; \mathbf{w}_2^T(t)]^T$ is assumed to be a white-noise process with intensity

$$\mathbf{V}(t) = \begin{bmatrix} \mathbf{V}_1(t) & \mathbf{V}_{12}(t) \\ \mathbf{V}_{12}^T(t) & \mathbf{V}_2(t) \end{bmatrix} \tag{20.92}$$

that is, the expected value

$$E\left\{ \begin{bmatrix} \mathbf{w}_1(t) \\ \mathbf{w}_2(t_1) \end{bmatrix} [\mathbf{w}_1^T(t_2) \; \mathbf{w}_2^T(t_2)] \right\} = \mathbf{V}(t_1)\delta(t_1 - t_2) \tag{20.93}$$

where $\delta(t_1 - t_2)$ is the Dirac delta function.

The initial state $\mathbf{x}(t_0)$ is assumed to be a random variable uncorrelated with $\mathbf{w}_1$ and $\mathbf{w}_2$ and its probability given by

$$E[\mathbf{x}(t_0)] = \mathbf{x}_0 \quad \text{and} \quad E\{[\mathbf{x}(t_0) - \mathbf{x}_0][\mathbf{x}(t_0) - \mathbf{x}_0]^T\} = \mathbf{Q}_0 \tag{20.94}$$

The observer form is given by Eq. 20.74 and Fig. 20.10. The optimal observer problem consists of determining $\mathbf{L}(\tau)$, $t_0 \leq \tau \leq t$ and the initial condition on the observer $\hat{\mathbf{x}}(t_0)$ so as to minimize the expected value $E\{[\mathbf{x}(t) - \hat{\mathbf{x}}(t)]^T\mathbf{W}(t)[\mathbf{x}(t) - \hat{\mathbf{x}}(t)]\}$, where $\mathbf{W}(t)$ is a symmetric positive definite weighting matrix.

If the problem as stated is nonsingular

$$\det[\mathbf{V}_2(t)] > 0 \qquad t \geq t_0 \tag{20.95}$$

and if the disturbance and measurement noise are uncorrelated

$$\mathbf{V}_{12}(t) = \mathbf{0} \tag{20.96}$$

the optimal observer gain matrix is given by Kalman and Bucy[30] as

$$\mathbf{L}(t) = \mathbf{Q}(t)\mathbf{C}^T(t)\mathbf{V}_2^{-1}(t) \qquad t \geq t_0 \tag{20.97}$$

where $\mathbf{Q}$ is a solution of the matrix Riccati equation:

$$\dot{\mathbf{Q}}(t) = \mathbf{A}(t)\mathbf{Q}(t) + \mathbf{Q}(t)\mathbf{A}^T(t) + \mathbf{S}(t)\mathbf{V}_1(t)\mathbf{S}^T(t)$$

$$-\mathbf{Q}(t)\mathbf{C}^T(t)\mathbf{V}_2^{-1}(t)\mathbf{C}(t)\mathbf{Q}(t) \qquad t \geq t_0 \tag{20.98}$$

with the initial condition

$$\mathbf{Q}(t_0) = \mathbf{Q}_0 \tag{20.99}$$

and the observer initial condition

$$\hat{\mathbf{x}}(t_0) = \mathbf{x}_0 \tag{20.100}$$

The resulting state estimator is called the Kalman–Bucy filter. The similarity of Eqs. 20.97 and 20.98 to the corresponding Eqs. 20.25 and 20.26, respectively, for the LQR problem is a result of the duality of the state estimation and state feedback control problems noted earlier. One difference, however, is that the Riccati equation for the optimal observer can be implemented in real time since Eq. 20.99 is an initial condition for Eq. 20.98. In contrast, for the finite-time LQR problem, Eq. 20.27 gives the terminal condition for the Riccati Eq. 20.26.

The steady-state properties of the optimal observer for linear time-varying and time-invariant systems parallel those of the optimal controller for the LQR problem and are described by Kwakernaak and Sivan.[3] If the time-varying system

$$\dot{\mathbf{x}}(t) = \mathbf{A}(t)\mathbf{x}(t) + \mathbf{S}(t)\mathbf{w}_1(t) \tag{20.101}$$

$$\mathbf{y}(t) = \mathbf{C}(t)\mathbf{x}(t) + \mathbf{w}_2(t) \tag{20.102}$$

is uniformly completely controllable by $\mathbf{w}_1(t)$ and uniformly completely reconstructible, the solution $\mathbf{Q}(t)$ of Eq. 20.98 converges to a steady-state solution $\mathbf{Q}_s(t)$ as $t_0 \to -\infty$ for any positive semidefinite $\mathbf{Q}_0$. The corresponding steady-state optimal observer

$$\dot{\hat{\mathbf{x}}}(t) = \mathbf{A}(t)\,\hat{\mathbf{x}}(t) + \mathbf{L}_s(t)[\mathbf{y}(t) - \mathbf{C}(t)\,\hat{\mathbf{x}}(t)] \tag{20.103}$$

where

$$\mathbf{L}_s(t) = \mathbf{Q}_s(t)\mathbf{C}^T(t)\mathbf{V}_2^{-1}(t) \tag{20.104}$$

is exponentially stable. Also, if the system and the noise statistics are invariant, the matrix Riccati differential Eq. 20.98 becomes an algebraic equation as $t_0 \to -\infty$:

$$\mathbf{A}\mathbf{Q}_s + \mathbf{Q}_s\mathbf{A}^T + \mathbf{S}\mathbf{V}_1\mathbf{S}^T - \mathbf{Q}_s\mathbf{C}^T\mathbf{V}_2^{-1}\mathbf{C}\mathbf{Q}_s = 0 \tag{20.105}$$

If the corresponding time-invariant system is completely controllable by the input $\mathbf{w}_1(t)$ and completely observable, Eq. 20.105 has a unique positive definite solution $\mathbf{Q}_s$ and the corresponding steady-state optimal observer of Eqs. 20.103 and 20.104 is asymptotically stable. Note that the measurable input $\mathbf{u}(t)$ has been omitted from Eqs. 20.101 and 20.102 for simplicity but does not change the substance of the results.

The discrete-time version of the optimal linear observer follows.[3] Consider the discrete-time system

$$\mathbf{x}(k + 1) = \mathbf{F}(k)\mathbf{x}(k) + \mathbf{G}(k)\mathbf{u}(k) + \mathbf{S}(k)\mathbf{w}_1(k) \tag{20.106}$$

$$\mathbf{y}(k) = \mathbf{C}(k)\mathbf{x}(k) + \mathbf{D}(k)\mathbf{u}(k) + \mathbf{w}_2(k) \tag{20.107}$$

where $\mathbf{w}_1(k)$, $\mathbf{w}_2(k)$ are zero-mean, uncorrelated vector random variables representing disturbance and measurement noise, respectively. Their joint probabilities are assumed to be known. The column vector $[\mathbf{w}_1^T(k)\ \mathbf{w}_2^T(k)]^T$ has the variance matrix

$$E\left\{\begin{bmatrix}\mathbf{w}_1(k)\\\mathbf{w}_2(k)\end{bmatrix}[\mathbf{w}_1^T(k)\ \ \mathbf{w}_2^T(k)]\right\} = \begin{bmatrix}\mathbf{V}_1(k) & \mathbf{V}_{12}(k)\\\mathbf{V}_{12}^T(k) & \mathbf{V}_2(k)\end{bmatrix} \tag{20.108}$$

The initial state $\mathbf{x}(k_0)$ is considered to be a random variable, uncorrelated with $\mathbf{w}_1$ and $\mathbf{w}_2$ with

$$E|\mathbf{x}(k_0)| = \mathbf{x}_0 \quad \text{and} \quad E\left\{[\mathbf{x}(k_0) - \mathbf{x}_0][\mathbf{x}(k_0) - \mathbf{x}_0]^T\right\} = \mathbf{Q}_0 \tag{20.109}$$

The observer form is given by Eq. 20.78 and Fig. 20.11. The optimal observer problem consists of determining $\mathbf{L}(k_7)$, $k_0 \leq k_7 \leq k$, and initial condition on the observer $\hat{\mathbf{x}}(k_0)$ so as to minimize the expected value $E\{[\mathbf{x}(k) - \hat{\mathbf{x}}(k)]\mathbf{W}(k)[\mathbf{x}(k) - \hat{\mathbf{x}}(k)^{\mathrm{T}}]\}$, where $\mathbf{W}(k)$ is a symmetric positive definite weighting matrix.

If the problem as stated is nonsingular

$$\det[\mathbf{V}_2(k)] > 0 \qquad k \geq k_0 \tag{20.110}$$

the optimal observer gain matrix is given by the recurrence relations

$$\mathbf{L}(k) = [\mathbf{F}(k)\mathbf{Q}(k)\mathbf{C}^{\mathrm{T}}(k) + \mathbf{V}_{12}(k)][\mathbf{V}_2(k) + \mathbf{C}(k)\mathbf{Q}(k)\mathbf{C}^{\mathrm{T}}(k)] \tag{20.111}$$

$$\mathbf{Q}(k + 1) = [\mathbf{F}(k) - \mathbf{L}(k)\mathbf{C}(k)]\mathbf{Q}(k)\mathbf{F}^{\mathrm{T}}(k) + \mathbf{V}_1(k) - \mathbf{L}(k)\mathbf{V}_{12}^{\mathrm{T}}(k) \tag{20.112}$$

with the initial condition

$$\mathbf{Q}(k_0) = \mathbf{Q}_0 \tag{20.113}$$

and $k \geq k_0$. The initial condition on the observer state should be

$$\hat{\mathbf{x}}(k_0) = \mathbf{x}_0 \tag{20.114}$$

Again, the similarity of the optimal observer Eqs. 20.111–20.113 to the optimal controller Eqs. 20.33–20.35 for the LQR problem results from the duality of state estimation and state feedback control problems.

The similarity extends to the steady-state behavior of the optimal observer and the optimal controller.[3] If the time-varying system

$$\mathbf{x}(k + 1) = \mathbf{F}(k)\mathbf{x}(k) + \mathbf{S}(k)\mathbf{w}_1(k) \tag{20.115}$$

$$\mathbf{y}(k) = \mathbf{C}(k)\mathbf{x}(k) + \mathbf{w}_2(k) \tag{20.116}$$

is uniformly completely controllable by $\mathbf{w}_1(k)$ and uniformly completely reconstructible, the solution $\mathbf{Q}(k)$ of Eqs. 20.111 and 20.112 converges to a steady-state solution $\mathbf{Q}_s(k)$ as $k_0 \to -\infty$ for any positive semidefinite $\mathbf{Q}_0$. The corresponding steady-state observer

$$\hat{\mathbf{x}}(k + 1) = \mathbf{F}(k)\,\hat{\mathbf{x}}(k) + \mathbf{L}_s(k)[\mathbf{y}(k) - \mathbf{C}(k)\,\hat{\mathbf{x}}(k)] \tag{20.117}$$

where $\mathbf{L}_s(k)$ is obtained using $\mathbf{Q}_s(k)$ for $\mathbf{Q}(k)$ in Eq. 20.111 is exponentially stable. Also, if the system and the noise statistics are time invariant, the matrix difference Eqs. 20.111 and 20.112 become algebraic equations as $k_0 \to -\infty$. If the corresponding time-invariant system is completely controllable by the input $\mathbf{w}_1(k)$ and completely observable, the resulting steady-state optimal observer is asymptotically stable. Again, note that the measurable input $\mathbf{u}(k)$ has been omitted from Eqs. 20.115 and 20.116 for simplicity but does not change the substance of the results.

Interested readers are referrred to Kwakernaak and Sivan[38] for a more complete consideration of the optimal observer. There is also extensive literature available on Kalman filters.[30–32]

20.6 OBSERVER-BASED CONTROLLERS

The observers described in the preceding section can be used to provide estimates of system state that, in turn, can be used to provide state feedback as described in Sections 20.2–20.4. The resulting controllers are dynamic compensators and are referred to as observer-based controllers.

The design of such observer-based controllers for LTI systems is simplified somewhat by the fact that their modes or eigenvalues satisfy the separation property, that is, the eigenvalues of the observer-controller are the same as the eigenvalues of the observer and the eigenvalues of the controller, the latter evaluated assuming perfect state measurement. For continuous-time LTI systems described by Eqs. 19.6 and 19.7, the control law given by Eq. 20.1, and the observer given by Eq. 20.74 with constant coefficient matrices, the observer-controller is given by Fig. 20.12 and the characteristic equation of the corresponding closed-loop system is[7]

$$\det(s\mathbf{I} - \mathbf{A} + \mathbf{B}\mathbf{K})\det(s\mathbf{I} - \mathbf{A} + \mathbf{L}\mathbf{C}) = 0 \tag{20.118}$$

A similar result can be shown to be true for discrete-time LTI systems.[10] The corresponding closed-loop system is shown in Fig. 20.13 and has the characteristic equation

$$\det(z\mathbf{I} - \mathbf{F} + \mathbf{G}\mathbf{K})\det(z\mathbf{I} - \mathbf{F} + \mathbf{L}\mathbf{C}) = 0 \tag{20.119}$$

System

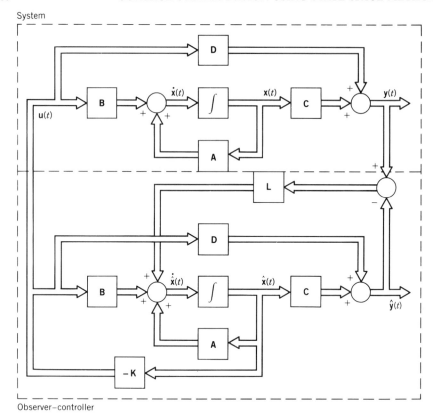

Observer–controller

Fig. 20.12   Observer-based controller for LTI continuous-time system.

For LTI systems subjected to unmeasured randomly varying disturbance inputs and measurement errors, if the statistics of these signals are known, state estimators of the Kalman–Bucy type will be used. The resulting estimator-based controllers have eigenvalues that also satisfy a separation property. As a result of the separation property for controllers based on observers or Kalman–Bucy filters, the design of the controllers can be treated independently of the observer.

The use of observers or Kalman filters to provide state estimates for state feedback controllers does, however, impair overall controller performance. For instance, the transient response of such controllers is poorer than that of controllers using complete state feedback. Moreover, the properties relating to the gain and phase margins of optimal controllers for the LQR problem are not applicable if the controllers use estimated states for feedback rather than measured states.[13] There is a definite loss of controller robustness associated with the use of state estimators. The robustness of such controllers is more properly evaluated by considering them as dynamic compensators and using methods common to the frequency domain approach.[33]

Observer-based controllers for LTI systems can naturally be examined using transfer function or frequency-domain-based methods. Such a linking of time-domain- and frequency-domain-based controller designs offers a number of advantages. The state-space-based design approach leads to consideration of controller structures that are not obvious from a transfer function approach. In addition, the state-space approach alerts the designer to the problem of loss of controllability or observability via pole-zero cancellation. On the other hand, the transfer function approach can result in controllers that cannot be obtained by using observer-based controllers.[7] In addition, the consideration of observer-based controllers from a transfer function perspective helps evaluate the controllers to see whether proven and practical design guidelines are violated. Such guidelines invariably use transfer function terminology since they have evolved from years of experience using classical control techniques. If such guidelines are violated, the state-space-based controller design procedure can be modified appropriately.[34]

A recent example of a MIMO controller design method that has evolved from a combination of frequency domain methods and state-space methods is the linear-quadratic-Gaussian method with loop-transfer-recovery (LQG/LTR), developed by Athans.[35] The procedure relies on the fact that

System

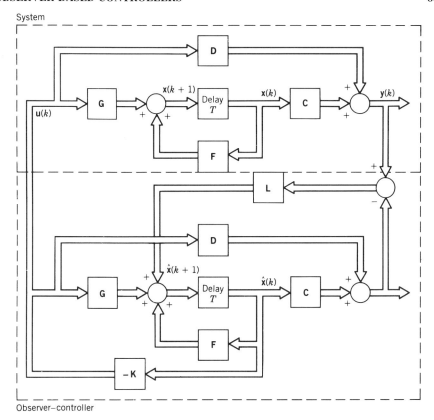

Observer–controller

Fig. 20.13 Observer-based controller for LTI discrete-time system.

results and requirements relating to control system robustness to modeling errors are best presented in the frequency domain. The powerful controller and estimator structures resulting from LQR formulations of the control and state estimation problem are used. These structures are useful in this design method because their robustness and performance have been well studied, using frequency domain measures.[36] The resulting method relies upon designer expertise in formulating good performance specifications at the outset. The design method therefore avoids the main weakness of state-space methods—namely, the weak connection between performance measures used by these methods and performance measures of engineering significance. The computation of the controller is, however, straightforward in this method, since the controller structures are derived from well-established state-space methods and are well supported by commercial CACSD packages.[4-6]

The first step in the design method is the definition of the design plant model. This model includes not only the nominal model of the system to be controlled but also includes scaling of the variables and augmentation of the dynamics, such as the inclusion of integrators dictated by control objectives. The number of control inputs r and the number of outputs p are assumed to be equal in the following development. The model is also linear and time invariant and strictly proper, that is, the transmission matrix $\mathbf{D}$ in the system Eqs. 19.6 and 19.7 is zero. The transfer function matrix of the system is

$$\mathbf{H}(s) = \mathbf{C}(s\mathbf{I}_n - \mathbf{A})^{-1}\mathbf{B} \tag{20.120}$$

Modeling inaccuracy is treated as follows. The actual transfer function matrix is given by

$$\mathbf{H}_A(s) = [\mathbf{I}_n + \mathbf{E}(s)]\mathbf{H}(s) \tag{20.121}$$

where $\mathbf{E}(s)$ characterizes the modeling error. The maximum singular value $\sigma_{\max}$ of the matrix $\mathbf{E}(j\omega)$ is assumed to be bounded by a known bound $e_m(\omega)$.

$$\sigma_{\max}[\mathbf{E}(i\omega)] < e_m(\omega) \tag{20.122}$$

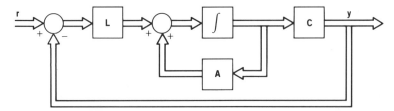

Fig. 20.14   Target feedback-loop block diagram.

The second step in the design procedure is the specification of a target feedback loop that has satisfactory robustness, stability, and performance specifications. The target feedback loop is shown in Fig. 20.14. It is obviously not directly implementable since the control inputs **u** do not appear in the system. The matrix **L** is a constant matrix and is chosen as described later. It is the designer's task to experiment with different choices of **L** and evaluate whether the resulting system has satisfactory performance. This stage of the design therefore requires considerable trial and error.

Athans[34] has suggested using the steady-state Kalman–Bucy filter formulation with stationary system and noise characteristics. It should be noted, however, that the formulation is not used here to perform an estimation task. It is being used here to help calculate the matrix **L** because the resulting target feedback loop has well-known performance and robustness characteristics. At the minimum, selection of **L** as described by Athans[34] guarantees that the target feedback loop will not amplify disturbances entering the system at the output. In addition, the target feedback loop will not go unstable as long as the modeling uncertainty $e_m(\omega)$ in Eq. 20.122 is below 0.5.

Once the target feedback loop is chosen, the final step in the design procedure is the design of a compensator to enable the controlled system to approximate the behavior of the target feedback loop closely. Athans[34] has proposed the compensator structure shown in Fig. 20.15. The similarity of the controller to the observer-based controller in Fig. 20.12 is clear. The gain matrix **K** is computed via the solution of a version of the LQR problem. The loop transfer recovery (LTR) result, credited by Athans to other researchers, guarantees that for minimum-phase systems, the procedure described yields a controlled system behavior that approximates the target feedback loop behavior as closely as desired.

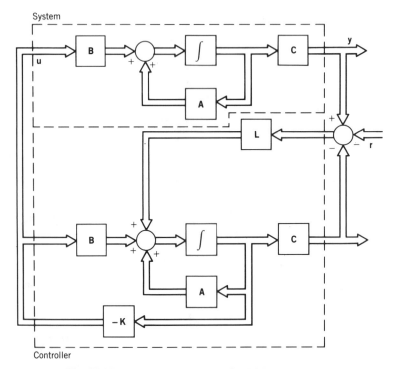

Fig. 20.15   Compensator structure for LQG/LTR method.

For nonminimum-phase systems, the design procedure remains the same. The only difference is that the final design may not approximate the behavior of the target feedback loop closely. In effect, this would result in additional design iterations to arrive at a satisfactory final design.

The LQG/LTR design procedure has been applied successfully to evaluate the feasibility of MIMO control for aircraft and helicopter flight control, jet engine control, and submersible control.

20.7   CONCLUSION

The primary emphasis, in this chapter, on linear time-invariant finite dimensional systems is a reflection of the state of the literature on the subject and the practice of the art. The reader is referred to the following works for more exhaustive treatment of some of the topics not covered at great length here. The subject of multivariable control using state-space methods has been addressed at much greater length by Kailath[7] and others.[37] The application of state-space methods to nonlinear system analysis and control is treated at some length by Ramnath et al.[38] Optimal control problems other than the LQR formulation have been described in detail in a number of textbooks.[39,40] The subject of adaptive control refers to control situations where the controller parameters are adapted or adjusted as the behavior of the system being controlled changes. One approach to adaptive control, termed the Model Reference Approach and employing state-space description, has been described at length by Landau.[41] Distributed-parameter systems are examples of systems with infinite-dimensional states. Application of state-space methods to these systems has been described by Tzafestas et al.[42] Time-delayed systems are also examples of systems with infinite-dimensional states. The analysis and control of such systems and of many of the other types of systems referred to in this section remains a subject of current research. For current research results in these areas, the reader is referred to journals such as the ASME Journal of Dynamic Systems, Measurement and Control, IEEE Transactions on Automatic Control, AIAA Journal of Guidance, Control and Dynamics, SIAM Journal on Control, and Automatica, The Journal of the International Federation of Automatic Control.

REFERENCES

1. I. M. Horowitz, *Synthesis of Feedback Control Systems*, Academic Press, New York, 1963.
2. A. G. J. MacFarlane, *Frequency Response Methods in Control Systems*, IEEE Reprint Series, IEEE, New York, 1979.
3. H. Kwakernaak and R. Sivan, *Linear Optimal Control Systems*, Wiley-Interscience, New York, 1972.
4. Anonymous, *CTRL-C, A Language for the Computer-Aided Design of Multivariable Control Systems*, Palo Alto, CA, Systems Control Technology, 1983.
5. R. Walker, C. Gregory, Jr., and S. Shah, "MATRIX$_x$: A Data Analysis, System Identification, Control Design and Simulation Package," *IEEE Control Systems Magazine*, pp. 30–36, December 1982.
6. K. J. Astrom, "Computer Aided Modeling, Analysis and Design of Control Systems—A Perspective," *IEEE Control Systems Magazine*, pp. 4–16, May 1983.
7. T. Kailath, *Linear Systems*, Prentice-Hall, Englewood Cliffs, NJ, 1980.
8. W. L. Brogan, *Modern Control Theory*, Prentice-Hall, Englewood Cliffs, NJ, 1982.
9. D. M. Wiberg, *State Space and Linear Systems*, Schaum's Outline Series, McGraw-Hill, New York, 1971.
10. G. F. Franklin and J. D. Powell, *Digital Control of Dynamic Systems*, Addison-Wesley, Reading, MA., 1980.
11. B. C. Kuo, *Digital Control Systems*, SRL Publishing, Champaign, IL, 1977.
12. B. D. O. Anderson and J. B. Moore, *Linear Optimal Control*, Prentice-Hall, Englewood Cliffs, NJ, 1971.
13. M. G. Safonov, *Stability and Robustness of Multivariable Feedback Systems*, MIT Press, Cambridge, MA, 1980.
14. W. R. Perkins and J. B. Cruz, Jr., "Feedback Properties of Linear Regulators," *IEEE Transactions on Automatic Control*, Vol. AC-16, No. 6, pp. 659–664, December 1971.
15. I. M. Horowitz and U. Shaked, "Superiority of Transfer Function over State-Variable Methods in Linear Time-Invariant Feedback System Design," *IEEE Transactions on Automatic Control*, Vol. AC-20, No. 1, pp. 84–97, February 1975.
16. H. H. Rosenbrock and P. D. McMorran, "Good, Bad or Optimal?" *IEEE Transactions on Automatic Control*, Vol. AC-16, No. 6, pp. 552–554, December 1971.
17. J. S. Tyler, Jr., "The Characteristics of Model Following Systems as Synthesized by Optimal Control," *IEEE Transactions on Automatic Control*, Vol. AC-9, No. 5, pp. 485–498, October 1964.

18. C. D. Johnson, "Theory of Disturbance-Accommodating Controllers," in *Control and Dynamic Systems, Advances in Theory and Applications*, C. T. Leondes Ed., Vol. 12, Academic Press, New York, pp. 387–489, 1976.

19. C. D. Johnson, "A Discrete-Time Disturbance-Accommodating Control Theory for Digital Control of Dynamical Systems," in *Control and Dynamic Systems, Advances in Theory and Applications*, C. T. Leondes, Ed., Vol. 18, Academic Press, New York, pp. 223–315, 1982.

20. E. J. Davison and I. J. Ferguson, "The Design of Controllers for the Multivariable Robust Servomechanism Problem Using Parameter Optimization Methods," *IEEE Transactions on Automatic Control*, Vol. AC-26, No. 1, pp. 93–110, February 1981.

21. T. L. Trankle and A. E. Bryson, Jr., "Control Logic to Track Outputs of a Command Generator," *AIAA Journal of Guidance and Control*, Vol. 1, No. 2, pp. 130–135, March–April 1978.

22. M. Tomizuka, "Optimal Continuous Finite Preview Problem," *IEEE Transactions on Automatic Control*, Vol. AC-20, No. 3, pp. 362–365, June 1975.

23. M. Tomizuka and D. E. Whitney, "Optimal Finite Preview Problems (Why and How Is Future Information Important?)" *ASME Transactions, Journal of Dynamic Systems, Measurement and Control*, Vol. 97, No. 4, pp. 319–325, December 1975.

24. M. Tomizuka and D. E. Rosenthal, "On the Optimal Digital State Vector Feedback Controller with Integral and Preview Actions," *ASME Transactions, Journal of Dynamic Systems, Measurement and Control*, Vol. 101, No. 2, pp. 172–178, June 1979.

25. M. Tomizuka and D. H. Fung, "Design of Digital Feedforward/Preview Controllers for Processes with Predetermined Feedback Controllers," *ASME Transactions, Journal of Dynamic Systems, Measurement and Control*, Vol. 102, No. 4, pp. 218–225, December 1980.

26. N. K. Gupta, "Frequency Shaped Cost Functionals: Extension of Linear-Quadratic Gaussian Design Methods," *AIAA Journal of Guidance and Control*, Vol. 3, No. 6, pp. 529–535, December 1980.

27. D. G. Luenberger, "Observing the State of a Linear System," *IEEE Transactions on Military Electronics*, Vol. 8, pp. 74–80, April 1964.

28. S. L. Shah, D. E. Seborg, and D. G. Fisher, "Design and Application of Controllers and Observers for Disturbance Minimization and Pole Assignment," *ASME Transactions, Journal of Dynamic Systems, Measurement and Control*, Vol. 102, No. 1, pp. 21–27, March 1980.

29. F. E. Thau and A. Kestenbaum, "The Effect of Modeling Errors on Linear State Reconstructors and Regulators," *ASME Transactions, Journal of Dynamic Systems, Measurement and Control*, Vol. 46, No. 4, pp. 454–459, December 1974.

30. R. E. Kalman and R. J. Bucy, "New Results in Linear Filtering and Prediction Theory," *ASME Transactions, Journal of Basic Engineering*, Series D, Vol. 83, pp. 45–108, March 1961.

31. J. M. Mendel and D. L. Gieseking, "Bibliography on the Linear-Quadratic-Gaussian Problem," *IEEE Transactions on Automatic Control*, Vol. AC-16, No. 6, pp. 847–869, December 1971.

32. B. O. Anderson and J. B. Moore, *Optimal Filtering*, Prentice-Hall, Englewood-Cliffs, NJ, 1979.

33. J. C. Doyle and G. Stein, "Robustness with Observers," *IEEE Transactions on Automatic Control*, Vol. AC-24, No. 4, pp. 607–611, August 1979.

34. A. E. Bryson, Jr., "Some Connections Between Modern and Classical Control Concepts," *ASME Transactions, Journal of Dynamic Systems, Measurement and Control*, Vol. 101, No. 3, pp. 91–98, June 1979.

35. M. Athans, "A Tutorial on the LQG/LTR Method," *Proceedings of the 1986 American Control Conference*, Seattle, WA, pp. 1289–1296, June 1986.

36. N. A. Lehtomaki, N. R. Sandell Jr., and M. Althaus, "Robustness Results in Linear-Quadratic-Gaussion Based Multivariable Control Designs," *IEEE Transactions on Automatic Control*, Vol. AC-26, No. 1, pp. 75–93, February 1981

37. M. Sain, Ed. "Special Issue on Linear Multivariable Control," *IEEE Transactions on Automatic Control*, Vol. AC-26, No. 6, pp. 1–295, February 1981.

38. R. V. Ramnath and H. M. Paynter, Eds., *Nonlinear System Analysis and Synthesis: Volume 2—Techniques and Applications*, Workshop/Tutorial Session at the Winter Annual Meeting of ASME, New York, December 1980.

39. M. Athans and P. Falb, *Optimal Control*, McGraw-Hill, New York, 1966.

40. A. P. Sage, *Optimum Systems Control*, Prentice-Hall, Englewood Cliffs, NJ, 1968.

41. Y. D. Landau, *Adaptive Control. The Model Reference Approach*, Marcel-Dekker, New York, 1979.

42. S. G. Tzafestas, Ed., *Distributed Parameter Control Systems, Theory and Application*, Vol. 6, International Series on Systems and Control, Pergamon Press, Oxford, England, 1982.

INDEX